Lecture Notes in Civil Engineering

Volume 386

Lecture Notes in Civil Engineering (LNCE) publishes the latest developments in Civil Engineering—quickly, informally and in top quality. Though original research reported in proceedings and post-proceedings represents the core of LNCE, edited volumes of exceptionally high quality and interest may also be considered for publication. Volumes published in LNCE embrace all aspects and subfields of, as well as new challenges in, Civil Engineering. Topics in the series include:

- Construction and Structural Mechanics
- Building Materials
- Concrete, Steel and Timber Structures
- Geotechnical Engineering
- Earthquake Engineering
- Coastal Engineering
- Ocean and Offshore Engineering; Ships and Floating Structures
- Hydraulics, Hydrology and Water Resources Engineering
- Environmental Engineering and Sustainability
- Structural Health and Monitoring
- Surveying and Geographical Information Systems
- Indoor Environments
- Transportation and Traffic
- Risk Analysis
- Safety and Security

To submit a proposal or request further information, please contact the appropriate Springer Editor:

- Pierpaolo Riva at pierpaolo.riva@springer.com (Europe and Americas);
- Swati Meherishi at swati.meherishi@springer.com (Asia—except China, Australia, and New Zealand);
- Wayne Hu at wayne.hu@springer.com (China).

All books in the series now indexed by Scopus and EI Compendex database!

Nuridah Sabtu
Editor

Proceedings of AWAM International Conference on Civil Engineering 2022—Volume 3

AICCE, Sustainability and Resiliency

Springer

Editor
Nuridah Sabtu
School of Civil Engineering
Universiti Sains Malaysia
Nibong Tebal, Pulau Pinang, Malaysia

ISSN 2366-2557 ISSN 2366-2565 (electronic)
Lecture Notes in Civil Engineering
ISBN 978-981-99-6025-5 ISBN 978-981-99-6026-2 (eBook)
https://doi.org/10.1007/978-981-99-6026-2

This Springer imprint is published by the registered company Springer Nature Singapore Pte Ltd.
The registered company address is: 152 Beach Road, #21-01/04 Gateway East, Singapore 189721, Singapore

Collaborators

REDAC
RIVER ENGINEERING AND URBAN DRAINAGE RESEARCH CENTRE

UNESCO
United Nations Educational, Scientific and Cultural Organization

uniTwin
UNESCO Chair on Ecohydraulics for Sustainable Water Infrastructures for SDG 6 in the Asia and the Pacific Region
Universiti Sains Malaysia
Penang, Malaysia

Kyutech
Kyushu Institute of Technology

Keynote Speakers

Dr. Tetsushi Kanda

Title: "Sustainable Supply Chain in Construction Industry and Approach for Carbon Neutralization in Construction Materials"

Biography

Dr. Tetsushi Kanda is a Deputy Director at Kajima Technical Research Institute. Dr. Kanda received his Ph.D. in Civil Engineering from the University of Michigan, USA, 1998. KaTRI is well known to initiate industrializing emerging advanced structural materials in construction such as ultra-high strength concrete and high-performance fiber reinforced cementitious composites in its history, with which Dr. Kanda took a leading role in large scale application. Dr. Kanda's research interests are extended to shrinkage cracking controlling technologies and carbon neutral technologies for structures. In these research areas, Dr. Kanda maintains an approach in which advanced fundamental technologies are transferred to construction projects and optimized there.

Prof. Dr. Nor Azazi Zakaria

Title: "Integrated Water Resources for Sustainability Management of Stormwater Towards SDG 6"

Biography

Professor Dr. Nor Azazi Zakaria has served in Universiti Sains Malaysia since 1994 and received his MSc. in 1990 and Ph.D. in December 1994. He then established the River Engineering and Urban Drainage Research Centre (REDAC) in 2001 and has since remained as the Director. Under the leadership of Prof. Dr. Nor Azazi, REDAC acts as a Self-Sustaining Centre as expected by USM in this corporatization era. REDAC is the first research centre at the USM Engineering Campus and was awarded the 1st Higher Institution Centre of Excellence (HICoE) in Service Thrust in USM as well as in Malaysia in October 2014 and received a special grant from the Ministry of Education (MOE) with a niche area on Stormwater Management.

Professor Dr. Nor Azazi is currently the Vice Chairman of the Executive Committee for the Malaysian Stormwater Organisation (MSO), the Malaysian National on Irrigation and Drainage (MANCID), and also UNESCO chair holder on Ecohydraulics for Sustainable Water Infrastructures for SDG 6 in the Asia and the Pacific Region. Furthermore, he has been actively organizing short courses, seminars, briefings, exhibitions, acting as guest and invited speaker and actively delivering keynotes at International Conferences in the area related to Sustainable Urban Drainage Management and Green Technology.

Editorial

Chief Editor

Dr. Nuridah Sabtu

Editors

Dr. Herni Halim
Dr. Noorhazlinda Abd Rahman
Dr. Nik Azimatolakma Awang
Dr. Mohamad Fared Murshed
Dr. Nurul Hana Mokhtar Kamal

Committee Members

Patron

Prof. Dr. Mohd Suffian Yusoff

Advisory

Ts. Dr. Mastura Azmi
Assoc. Prof. Ir. Ts. Dr. Leong Lee Vien

Event

Assoc. Prof. Dr. Puganeshwary Palaniandy
Ts. Dr. Mastura Azmi

Wellbeing

Assoc. Prof. Dr. Fadzli Mohamed Nazri
Ms. Zalika Ramly

International Advisory Board

Prof. Dr. Eng. Pei-Shan Chen—Graduate School of Engineering, Kyushu Institute of Technology, Japan
Prof. Dr. Hainian Wang—Chang'an University, China
Assoc. Prof. Dr. Haider M. Zwain—College of Water Resources Engineering, Al-Qasim Green University, Iraq
Assoc. Prof. Aboelkasim Diab—Aswan University, Egypt
Assoc. Prof. Dr. Baki Ozturk—Hacettepe University, Ankara, Turkey
Dr. Wesam Al Madhoun—Director for Center of Sustainable Development, Gaza University, Palestine

Local Advisory Board

Assoc. Prof. Ir. Dr. Norazura Muhamad Bunnori—University Malaya (UM)
Assoc. Prof. Dr. Azhar Abdul Halim—Universiti Kebangsaan Malaysia (UKM)
Assoc. Prof. Ts. Dr. Aeslina Abdul Kadir—Universiti Tun Hussein Onn Malaysia (UTHM)
Dr. Ir. Safari Hj. Mat Desa—National Hydraulic Research Institute of Malaysia (NAHRIM)
Dr. Zulkiflee Ibrahim—Universiti Teknologi Malaysia (UTM)
Assoc. Prof.Ramadhansyah Putra Jaya—Universiti Malaysia Pahang (UMP)
Ir. Ts. Dr. Hjh. Siti Fairuz Haji Zakaria—Department of Irrigation and Drainage
Ir. Dr. Mohd Azhar Abd Hamid—Bahagian Penyelidikan dan Teknologi, SWCORP

Nuridah Sabtu
Editor

Proceedings of AWAM International Conference on Civil Engineering 2022—Volume 3

AICCE, Sustainability and Resiliency

Springer

Editor
Nuridah Sabtu
School of Civil Engineering
Universiti Sains Malaysia
Nibong Tebal, Pulau Pinang, Malaysia

ISSN 2366-2557 ISSN 2366-2565 (electronic)
Lecture Notes in Civil Engineering
ISBN 978-981-99-6025-5 ISBN 978-981-99-6026-2 (eBook)
https://doi.org/10.1007/978-981-99-6026-2

This Springer imprint is published by the registered company Springer Nature Singapore Pte Ltd.
The registered company address is: 152 Beach Road, #21-01/04 Gateway East, Singapore 189721, Singapore

Collaborators

REDAC
RIVER ENGINEERING AND URBAN DRAINAGE RESEARCH CENTRE

United Nations
Educational, Scientific and
Cultural Organization

uniTwin

UNESCO Chair on Ecohydraulics for
Sustainable Water Infrastructures for
SDG 6 in the Asia and the Pacific Region
Universiti Sains Malaysia
Penang, Malaysia

Kyutech
Kyushu Institute of Technology

UMY
UNIVERSITAS MUHAMMADIYAH YOGYAKARTA

INTERNATIONAL ISLAMIC UNIVERSITY MALAYSIA
Garden of Knowledge and Virtue

UNIVERSITI TENAGA NASIONAL
The Energy University

Keynote Speakers

Dr. Tetsushi Kanda

Title: "Sustainable Supply Chain in Construction Industry and Approach for Carbon Neutralization in Construction Materials"

Biography
Dr. Tetsushi Kanda is a Deputy Director at Kajima Technical Research Institute. Dr. Kanda received his Ph.D. in Civil Engineering from the University of Michigan, USA, 1998. KaTRI is well known to initiate industrializing emerging advanced structural materials in construction such as ultra-high strength concrete and high-performance fiber reinforced cementitious composites in its history, with which Dr. Kanda took a leading role in large scale application. Dr. Kanda's research interests are extended to shrinkage cracking controlling technologies and carbon neutral technologies for structures. In these research areas, Dr. Kanda maintains an approach in which advanced fundamental technologies are transferred to construction projects and optimized there.

Prof. Dr. Nor Azazi Zakaria

Title: "Integrated Water Resources for Sustainability Management of Stormwater Towards SDG 6"

Biography

Professor Dr. Nor Azazi Zakaria has served in Universiti Sains Malaysia since 1994 and received his MSc. in 1990 and Ph.D. in December 1994. He then established the River Engineering and Urban Drainage Research Centre (REDAC) in 2001 and has since remained as the Director. Under the leadership of Prof. Dr. Nor Azazi, REDAC acts as a Self-Sustaining Centre as expected by USM in this corporatization era. REDAC is the first research centre at the USM Engineering Campus and was awarded the 1st Higher Institution Centre of Excellence (HICoE) in Service Thrust in USM as well as in Malaysia in October 2014 and received a special grant from the Ministry of Education (MOE) with a niche area on Stormwater Management.

Professor Dr. Nor Azazi is currently the Vice Chairman of the Executive Committee for the Malaysian Stormwater Organisation (MSO), the Malaysian National on Irrigation and Drainage (MANCID), and also UNESCO chair holder on Ecohydraulics for Sustainable Water Infrastructures for SDG 6 in the Asia and the Pacific Region. Furthermore, he has been actively organizing short courses, seminars, briefings, exhibitions, acting as guest and invited speaker and actively delivering keynotes at International Conferences in the area related to Sustainable Urban Drainage Management and Green Technology.

Prof. Dr. Hemanta Hazarika

Title: "Sustainable Early Warning System for Mitigating Rainfall Induced Landslide Disasters"

Biography

Hemanta Hazarika is currently a professor in the Graduate School of Engineering and Department of Interdisciplinary Science and Innovation, Kyushu University, Fukuoka, Japan. He obtained his Bachelor of Technology (B.Tech.) degree in Civil Engineering from the Indian Institute of Technology (IIT), Madras, India in 1990. He obtained his master's degree in Geotechnical Engineering in 1993 from Nagoya University, Japan and Ph.D. from the same university in 1996. Currently he is also adjunct professor of IIT Madras, India.

Professor Hazarika's research activities include disaster prevention and mitigation, soil-structure interaction, stability of soil-structures during earthquakes and tsunami, ground improvement, application of recycled waste and lightweight materials in constructions, stability of cut slopes, and landslides and their protection.

He has more than 350 technical papers in various international journals, international conferences, workshops and symposia to date. He has also authored two books on Soil Mechanics: "Soil Mechanics Fundamentals and Applications" (Published by CRC Press, USA) and its Japanese version (Published by Kyoritsu Publisher, Tokyo). In addition, Prof. Hazarika also served as the editor of four books: (1) "Advances in Sustainable Construction and Resource Management", (2) "Geotechnical Hazards from Large Earthquakes and Heavy Rainfall", (3) "Earthquake Hazards and Mitigations" and (4) "Scrap tire derived geomaterials—opportunities and challenges".

Professor Hazarika has several years of teaching, research as well as practical and consulting experiences both in Japan and outside Japan. He is currently a foreign expert of the world's first research centre on liquefaction research called "The National Research Centre for Liquefaction", which was established in Palu, Indonesia aftermath of the devastating damage by liquefaction during the 2018 Palu Earthquake. He also successfully guided a world bank project on "Improving the Resilience of Nepal's Strategic Roads Network" as the leader of the project.

He also successfully organized the first International Symposium on Construction Resources for Environmentally Sustainable Technologies (CREST 2020) in March this year. Currently, Prof. Hazarika is the chairman of Asian Technical Committee on "Geotechnical Mitigation and Adaptation to Climate Change-induced Geo-disasters in Asia-Pacific Regions" of International Society of Soil Mechanics and Geotechnical Engineering (ISSMGE). He is also the founding president of a general incorporated corporation called GLOSS (Global Society for Smart Geo-Sustainnovation) which was established in August 2021.

About the Conference

AWAM International Conference on Civil Engineering (AICCE) is the flagship conference of the School of Civil Engineering, Universiti Sains Malaysia (SoCE). SoCE was established in 1989, and since then, SoCE has organized numerous seminars, conferences and short courses. These activities have encouraged and contributed towards the creation of knowledge, innovativeness and exchange of ideas. To date, SoCE has successfully organized five national-level conferences (AWAM'99, AWAM'01, AWAM'04, AWAM'07 and AWAM'09) and three AICCE in collaboration with other collaborating partners which were Geohazard Information Zonation (GIZ) for the AICCE-GIZ'12, Construction Industry Development Board (CIDB) for eco-AICCE'15, Toyohashi University of Technology (TUT) for IGNITE-AICCE'17 and another collaboration with CIDB for AICCE'19.

In 2022, SoCE hosted the AICCE 2022 in Penang, Malaysia, during 15–17 February 2022. Following the success of the previous AICCE in 2019, this AICCE'22 provided another platform for academics and industries internationally for future collaboration opportunities. The scope of the conference was SUSTAINABILITY AND RESILIENCY: RE-ENGINEERING THE FUTURE.

Sustainability and Resiliency: Re-engineering the Future

Sustainability and resiliency are key concepts to new approaches in urban development. In recent years, the integrated use of sustainability and resiliency concepts in engineering are becoming more pronounced as a response to the increasing environmental threats due to climate changes and natural disasters. The application of sustainability and integration of resiliency are important to support and develop natural and built environments in a robust and enduring way.

Therefore, the AWAM International Conference in Civil Engineering (AICCE'22) served as the platform to bring together academics, scientists, engineers, research scholars, government officials, practitioners and learned individuals from multi-disciplines to present and discuss the most recent research, innovations, trends and concerns, practical challenges encountered as well as engineering solutions adopted in built environment. It aspires to advocate the integrated use of these key concepts in natural and built environment and endeavours towards re-engineering a better future.

Contents

Geotechnical Engineering

Effect of Soil Restraint to Long Slender Pile

Ismacahyadi Bagus Mohamed Jais, Diana Che Lat, Abubakar Ibrahim Alasan, and Fairul Zahri Mohamad Abas

Abstract Soil restraint has been identified as a possible mechanism that can pose a positive effect to long slender pile in different soil deposits, and this effect has not been explicitly mentioned in most design codes. To carry out routine design and checking, it is necessary to reliably estimate the soil restraint effect on the pile for a given site. However, the existing design methods of piles do not consider the influence of soil restraint to piles with different slenderness ratio as the piles are designed as a beam-column, thus limiting the pile to take up more load than it should. In this study, a finite element software PLAXIS 2D is used to carry out a parametric analysis on the effect of soil restraint to long slender pile in different soil conditions and slenderness ratio. A simplified estimation method based on Euler's buckling theory is provided for predicting critical buckling load of pile. Results are discussed in this paper.

Keywords Soil restraint · Pile · Buckling · Soil foundation · Settlement

1 Introduction

Pile foundations are deep foundations, formed by long and slender columnar elements typically made from steel or reinforced concrete, or sometimes timber. A foundation is described as "pile" when its depth is more than three times its breadth [1]. Piles

I. B. M. Jais (✉) · A. I. Alasan
School of Civil Engineering, College of Engineering, Universiti Teknologi MARA, 40450 Shah Alam, Selangor, Malaysia
e-mail: ismac821@uitm.edu.my

D. C. Lat (✉)
School of Civil Engineering, Universiti Teknologi MARA Pasir Gudang, 81750 Masai, Johor, Malaysia
e-mail: dianacl@uitm.edu.my

F. Z. M. Abas
Road Branch, Public Works Department, Menara PJD, Kuala Lumpur Jabatan Kerja Raya, Jalan Sultan Salahuddin, 50582 Kuala Lumpur, Malaysia

N. Sabtu (ed.), *Proceedings of AWAM International Conference on Civil Engineering 2022 - Volume 3*, Lecture Notes in Civil Engineering 386,
https://doi.org/10.1007/978-981-99-6026-2_1

are often required to resist lateral loads and moments in addition to their primary use as axially loaded members. The goals of designers are to determine deflections and stresses in the selected soil-pile system so that they are controlled within tolerable limits [2]. The types of soil that most often require piling are those with high clay or silt content since these are fine-grained soils. They tend to collapse more easily or else cause more uplift when they freeze. Sandy soils are stronger but can still cause problems, while soils with high organic content can also be too weak to support traditional foundations. Clay and silt are especially vulnerable if they are saturated since they liquefy easily. However, anywhere that has a high-water table or shows signs of being marshy are potentially vulnerable, and piling should be considered. Bearing capacity is the capacity of soil to support the loads applied to the ground. The bearing capacity of soil is the maximum average contact pressure between the foundation and the soil which should not produce shear failure in the soil. Aird [3] stated in his study that the axial capacity is largely dependent on the shaft friction developed between the piles and the soil. Certain loads on slender pile foundations can cause the piles to buckle. A study by Michele et al. [4] shows that there are slenderness thresholds under which the ultimate behaviour is governed by the critical load and not the bearing capacity. The critical load may be the main thing responsible for the failure mode, and it should not be neglected in the long pile design. They also found out that the slenderness thresholds vary nonlinearly with diameters, due to the specific volume involved in the contact, which depends exclusively on diameter.

To understand a slender pile, slenderness ratio must be understood first. In structural engineering, slenderness is a measure of the propensity of a column to buckle. It is defined as l/k, where l is the effective length of the column, and k is the least radius of gyration, the latter defined by $K_2 = I/A$, where A is the area of the cross-section of the column, and I is the second moment of area of the cross-section. The effective length is calculated from the actual length of the member considering the rotational and relative translational boundary conditions at the ends. Slenderness captures the influence on buckling of all the geometric aspects of the column, namely its length, area, and second moment of area. The influence of the material is represented separately by the material's modulus of elasticity, E [5]. In a comparative study, Kavitha et al. [6] found out that pile stiffness contributes towards the lateral load carrying capacity of pile. Stiffness of pile is mainly governed by the material property (E) and the moment of inertia (I) of the pile. Hence, pile characteristics such as cross-sectional shape, cross-sectional area, and material properties are also significant in the structural behaviour of laterally loaded piles, hence influencing soil restraint.

Broms [7] presented a method to determine the ultimate lateral load in cohesive and cohesionless soils. Hazzar et al. [8] in their study investigated the influence of pile diameter to lateral resistance by adopting the failure criteria suggested by Broms [7] in cohesive soil. The variation of pile diameter, especially when D/b is less than 14, does not significantly affect the normalized ultimate lateral bearing capacity of pile. The length D and the diameter b of pile are variable in order to investigate their influences on the lateral bearing capacity of pile. The ultimate lateral load of pile, Qu, is represented by the dimensionless factor defined by "Qu/Cu b^2" for which the influence of several parameters was studied. Michele et al. [4] investigated whether the critical load, i.e. the buckling load of a pile embedded in the soil, can be less than the bearing capacity for certain slenderness and they found out that there are slenderness thresholds under which the ultimate behaviour is governed by the critical load and not the bearing capacity and also the slenderness thresholds vary nonlinearly with diameters, due to the specific volume involved in the contact, which depends exclusively on diameter. High slenderness piles with a very small diameter therefore need particular attention in the evaluation of the ultimate load, and the employment of the bearing capacity for design purposes should be restricted by observations on the buckling.

2 Research Methodology

2.1 Problem Identification

The types of soil that most often require piling are those with high clay or silt content since these fine-grained soils tend to collapse more easily or else cause more uplift when they freeze. Soft soil in Malaysia is considered as quaternary sediments consist of alluvial deposits and organic or peat soils [9]. According to Geological Map of Malaysia, soft ground usually found in the coastal plain of the country covers large area of west coast and east coast of Peninsular and East Malaysia [10]. Due to significant proportion of soft ground to the country total land area and major economic activities and social developments are concentrating along the coastal area, construction projects on these problematic deposits are unavoidable. Construction on soft ground area is a great challenge in the field of geotechnical engineering. Many engineering problems in the form of slope instability, bearing capacity failure, or excessive settlement could occur either during or after the construction phase due to low shear strength and high compressibility of this soil [11]. At the moment, pile foundations in Malaysia are designed as a beam-column. This study will simulate the situation of long pile founded on a weak soil where the long pile foundations are embedded in Malaysian soft soil and compared with stiffer soil. One of the piles will be modelled as a free-standing pile, while the other 3 piles will be modelled in a soft, stiff, and very stiff soil layers, respectively. The foundations will be analysed and subjected to Euler's buckling equation to check if they are conservative or not. This study will provide an insight behaviour of certainty to long and slender piles.

2.2 Properties of Soil and Pile

The Client has appointed the Consultant to undertake the design for civil works, foundation, and overall site supervision for the construction of Klinik Kesihatan Pekan Simpang. The proposed site is situated approximately at Federal Road km

Table 1 Young's modulus used in the numerical analyses

Material	E (kN/m^2)
Soft soil	
Layer 1	9000
Layer 2	12,600
Layer 3	41,400
Layer 4	50,400
Stiff soil	
Layer 1	12,000
Layer 2	35,000
Layer 3	66,000
Very stiff soil	
Layer 1	28,000
Layer 2	30,000
Layer 3	32,400
Layer 4	106,800
Pile	40×10^6

88, from Ipoh-Butterworth. From the borelog results, the soil profile can be said to be quite inconsistent along the six boreholes. Since the laboratory data is not available, it is a common practice to estimate the shear strength parameters from the SPT results. There are many charts, tables, and empirical relationships that are available in the literature between the SPT-N value and the angle of internal friction (Ø) and undrained cohesion (Cu) by different researchers [12, 13]. Generally, the subsoil consists of interlays between sand, clay, and silt. The soil properties of the different types soil layer are based on the information gathered from SPT, manual calculations, and typical values. Table 1 shows typical values of Young's modulus for different type of soil used in the PLAXIS simulation. Pile load test is performed on the piles embedded in soft, stiff, and very stiff soil with first cycle of up to 100% pile working load and second cycle of up to 200% pile working load with the pile load test loading/unloading time as shown in Table 2. The spun pile estimated length and capacity are calculated based on SPT-N value obtained from the borehole test.

2.3 *Numerical Analysis*

Some of the essential features of PLAXIS finite element (PLAXIS FE) required for this study include graphical input of geometry models, automatic mesh generation, high-order elements, plates, interfaces, staged construction, consolidation analysis, and so on. In geometry modelling, the input of soil layers, structures, construction stages, loads, and boundary conditions are established on handy CAD drawing

Table 2 Pile load test loading/unloading time

Time (min)	Pile working load (%)	Pile working load (kN)
1st cycle	0	0
30	25	162.5
30	50	325
30	75	487.5
360	100	650
10	50	325
60	0	0
2nd cycle	0	0
60	100	650
60	125	812.5
60	150	975
60	175	1137.5
360	200	1300
10	150	975
60	0	0

methods, which allow for an intricate modelling of the geometry cross-section. From this geometry model, a 2D finite element mesh is easily generated. The automatic mesh generation allocates for automatic generation of unstructured 2D finite element meshes with choices for global and local mesh refinement [14]. Plates which are recognized as special beam elements are used to model the bending of pile linings in this study. The behaviour of these elements is characterized using a flexural rigidity, a normal stiffness, and an ultimate bending moment. Joint elements in interfaces are available to model soil–structure interaction such as simulating the thin zone of shearing material at the interaction between a pile lining and the surrounding soil as the values of interface friction angle and adhesion are usually different compared to the friction angle and cohesion of the surrounding soil. Staged construction feature facilitates a realistic simulation of construction and excavation procedures by activating and deactivating clusters of elements, application of loads, changing of water tables, etc. This process allows for a realistic evaluation of stresses and displacements. Automatic time stepping procedures make the analysis robust and easy-to-use.

PLAXIS 2D calculation is used in this research to model the simulation of interaction between soil and pile foundation. In order to investigate the displacement or settlement problem, the analysis must be done in a number of stages of construction. Figure 1 shows the general tab sheet of calculation window for initial condition.

There are two conditions of pile those are free-standing pile (unrestrained pile) as shown in Fig. 2 and restrained pile. The restrained piles in this study are the spun pile with 800 mm diameter and 24 m in length fully embedded in soft soil, stiff soil, and very stiff layer as shown in Figs. 3, 4 and 5, respectively. The piles are modelled in PLAXIS 2D software using elastic–perfectly plastic hardening soil constitutive

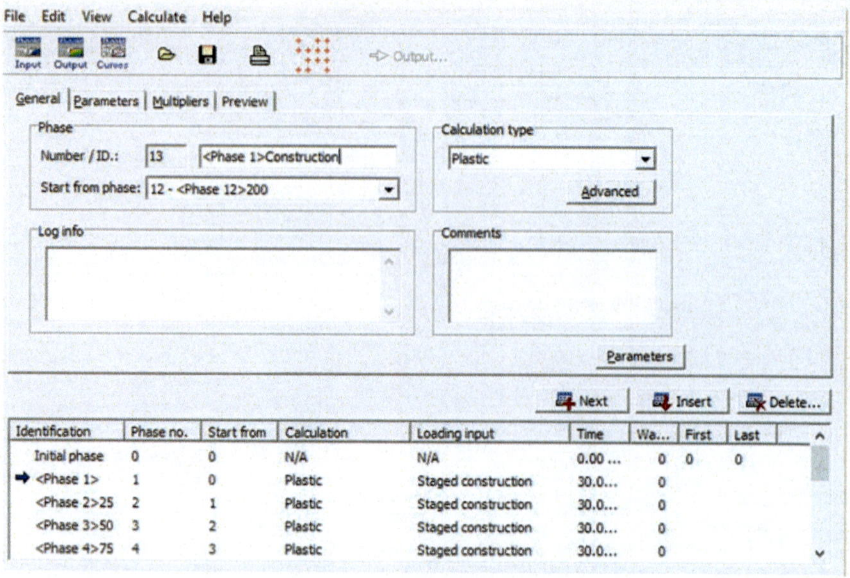

Fig. 1 General tab sheet of calculation window—initial condition (Plaxis 2D V8)

model. For restrained pile foundations, the geometry of soil–structure interaction was captured from the subsoil information gathered from boreholes in this study. For the soft soil, the layers the depth of each layer are; first layer 6.5 m, second layer 10 m, third layer 7.5 m and fourth layer 6 m. For stiff soil model, the first layer is 10 m deep, second layer 8 m deep and third layer 7 m deep. For very stiff soil, the depth of each layer is as follows: first layer 1 m, second layer 4 m, third layer 7 m, and fourth layer 3 m. A point load of 10 kPa is applied for all the piles.

2.4 Buckling Evaluation Using Euler's Theory

The Euler's buckling formula can be used to predict buckling load of column or pile pinned ends, fixed ends, pinned–fixed, or fixed-free. Pile foundations embedded in soil are regarded as pinned–fixed as shown in Fig. 6.

As shown in Fig. 6, the effective length is taken as Le = 0.7L, where 0.7 is a constant for pinned–fixed ends effective length factors.

$$Pcr = \frac{\pi^2 EI}{(KL)^2} \quad \text{(Units: N or KN)} \tag{1}$$

where

E represents Young's modulus of spun pile (in GPa)

Fig. 2 Geometry model for free-standing pile (PLAXIS 2D V8)

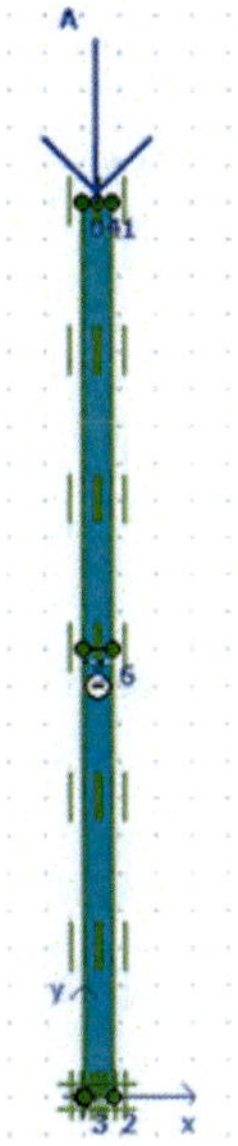

Fig. 3 Geometry model for pile embedded in soft soil (PLAXIS 2D V8)

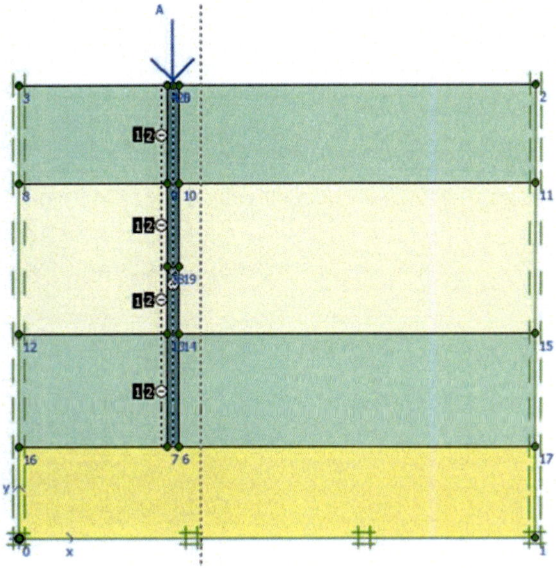

I represents moment of inertia that resists direction of buckling
L is a length of slender member
K is an effective length factor.

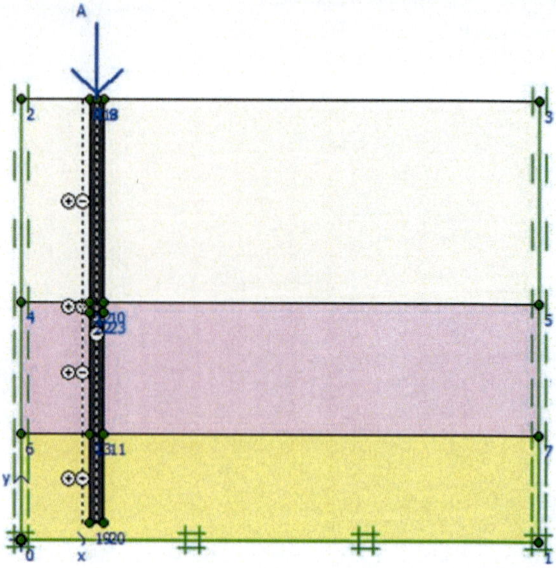

Fig. 4 Geometry model for pile embedded in stiff soil (PLAXIS 2D V8)

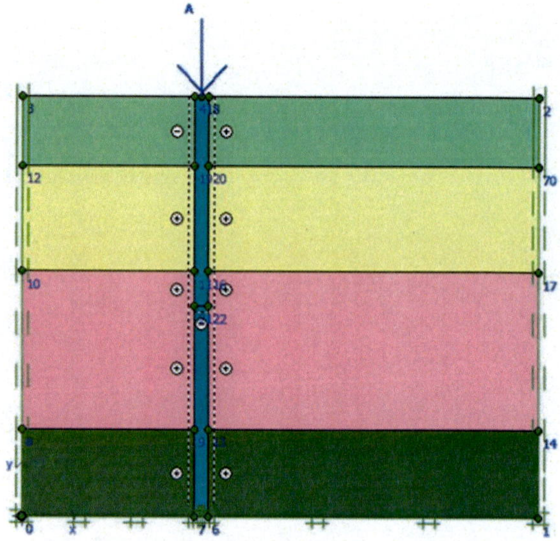

Fig. 5 Geometry model for pile embedded in very stiff soil (PLAXIS 2D V8)

3 Results and Discussion

The results of PLAXIS FE show that the free-standing pile is completely collapsed as the pile could not take up any load without being surrounded by some soil as shown in Fig. 7.

Fig. 6 Pinned–fixed buckling scenario

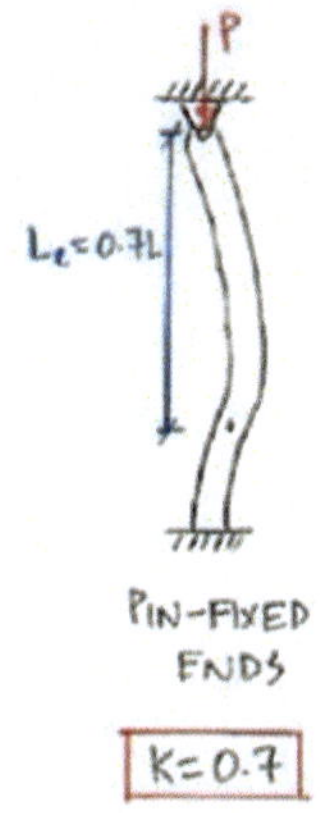

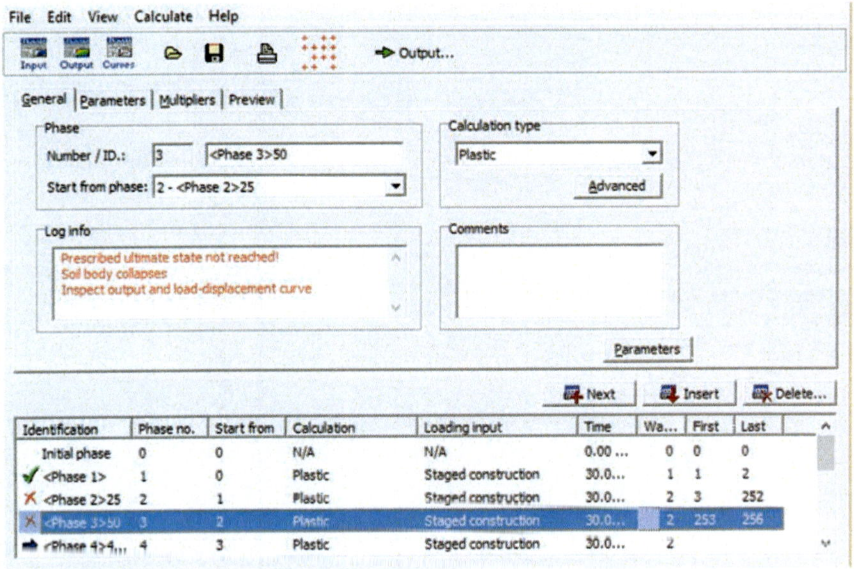

Fig. 7 Collapsed free-standing pile window (PLAXIS 2D V8)

The axial force results of the PLAXIS 2D simulation are shown in Figs. 8, 9 and 10. It can be seen that that the maximum axial forces which the pile can take in a soft soil are 773 kN per metre. In a stiff soil, the pile can take up to 786 kN per metre, and lastly in a very stiff soil, the pile can take a load of up to 1240 kN per metre. One of the reasons that influenced the pile settlement is the soil condition it is embedded in. The total displacement, horizontal displacement, and the pile axial force at the pile working load are also obtained from PLAXIS, and the summary of the results are shown in Table 3.

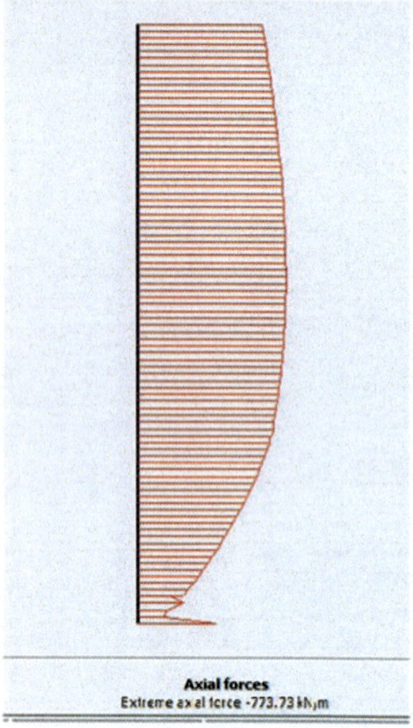

Fig. 8 Axial forces cross-section—pile embedded in soft soil at 100% working load

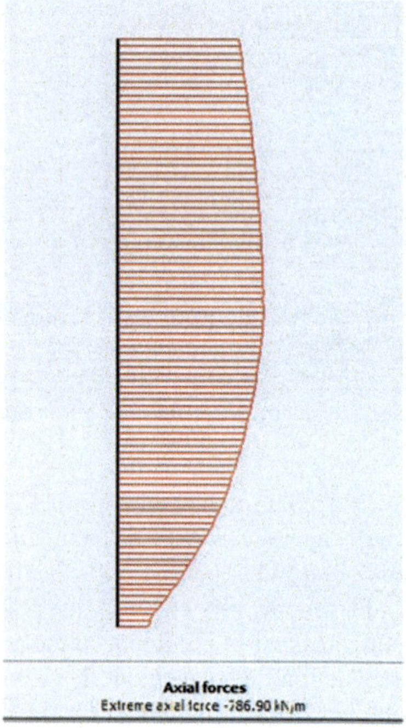

Fig. 9 Axial forces cross-section—pile embedded in stiff soil at 100% working load

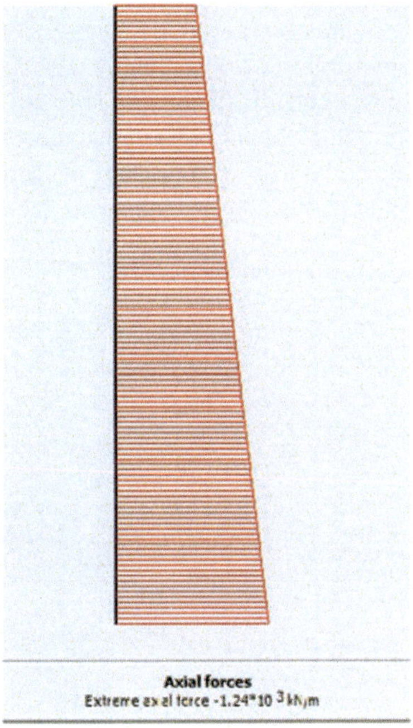

Fig. 10 Axial forces cross-section—pile embedded in very stiff soil at 100% working load

Table 3 PLAXIS results for pile embedded in different conditions of soil at 100% pile working load

100% working load	Soft soil	Stiff soil	Very stiff soil
Total displacement (mm)	12.07	5.41	0.748
Horizontal displacement (mm)	4.55	2.96	0.0107
Axial forces of pile (kN/m)	773.73	786.9	1240

The results of pile load test performed for the piles in different soil conditions are shown in Figs. 11, 12, and 13. From the figures, the first cycle represents the settlement at a pile working load, and the second cycle represents the twice working load. Tables 3 and 4 show total displacement at pile working load as well as at twice pile working load for different soil conditions. Figure 11 presents the settlement graph of pile in soft soil, where the pile experienced a settlement of 12.07 mm at pile working load and 26.29 mm at twice the pile working load. Figure 12 shows the settlement graph of pile in stiff soil, where the pile experienced a settlement of 5.41 mm at pile working load and 14.81 mm at twice the pile working load. Figure 13 shows the settlement graph of pile in a very stiff soil, where the pile experienced a settlement of 0.75 mm at pile working load and 1.42mmm at twice the pile working load. The pile settles the most in soft soil due to the effect of negative skin friction acting around the pile. According to JKR settlement criteria, a pile tested shall deem to have failed if the residual settlement after the test load has been removed exceeds

6.50 mm, total settlement under working load exceeds 12.50 mm or total settlement under twice working load exceeds 38 mm. Based on JKR settlement criteria, the piles in soft, stiff, and very stiff soil do not fail. It is worth noting that for this study, the water table is simulated at the ground surface and shows the worst-case scenario that could happen. However, if the position of the water table is changed to deeper below ground surface, the settlement of the piles will significantly reduce.

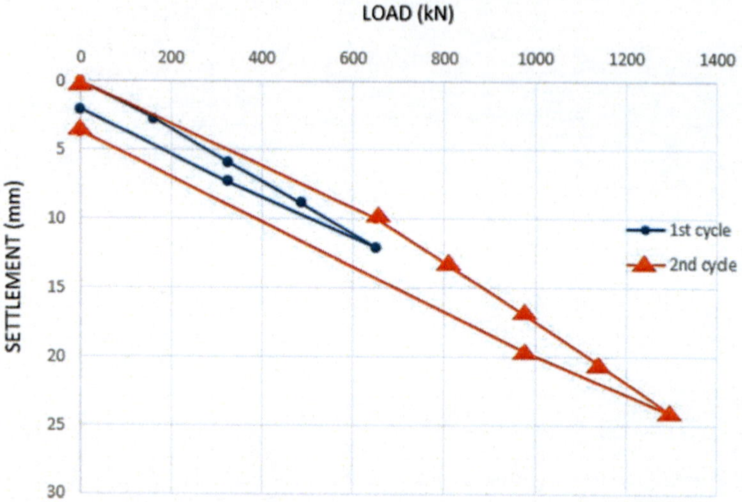

Fig. 11 Settlement graph of pile in soft soil

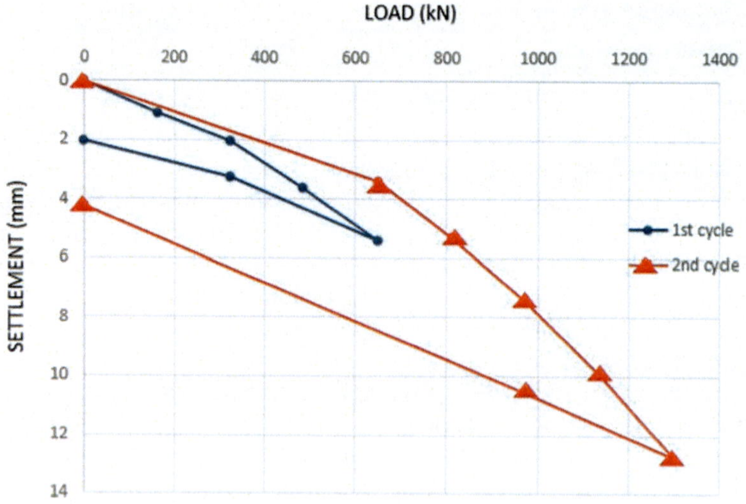

Fig. 12 Settlement graph of pile in stiff soil

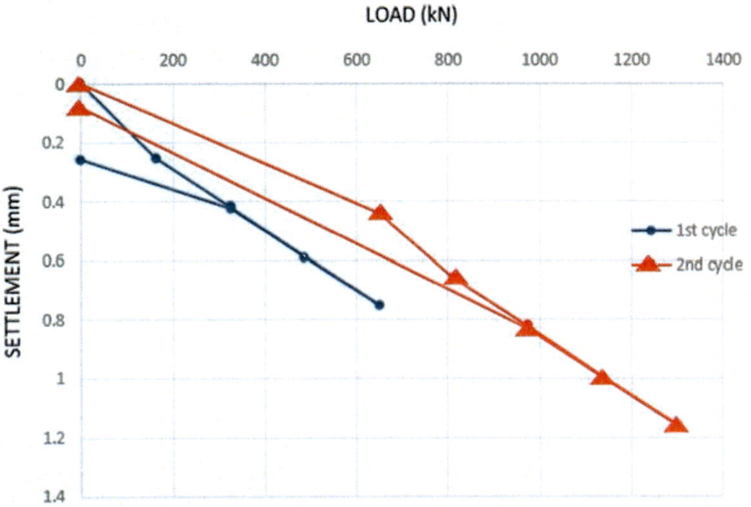

Fig. 13 Settlement graph of pile in very stiff soil

Table 4 PLAXIS total displacement for pile embedded in different conditions of soil at 200% pile working load

200% working load	Soft soil	Stiff soil	Very stiff soil
Total displacement (mm)	26.29	14.81	1.42

It was noted that the soil restraining effects of the soft soil layer improved the carrying capacity of the pile. However, the piles in stiff soil and very stiff soil can take more loads compared to soft soil. Although pile is embedded in a soft soil and tends to displace more, the soil restraining effects improve the soil carrying capacity. In deep soil layer, the soil restraint has a positive impact to the pile carrying capacity. From the horizontal displacement of the piles, it is noted that pile embedded in soft soil layer buckle more followed by in stiff soil and in very stiff soil. The reason for this is concluded to be due to interactive movement between pile and soil which produces a shear stress on the pile–soil interface. The downward movement of the consolidating soft soil causes a downward force along the pile–soil interface, hence making pile in soft soil buckle more and less in stiff soil and very stiff soil layers.

A comparison of the pile analysis of free-standing pile and pile in different soil conditions shows that the free-standing pile in the PLAXIS FE software completely failed. Free-standing pile could not take up any load without being surrounded by some soil even some part of the pile being restrained. It is noted that the pile in very stiff soil performed better and tends to take up more load than pile fully embedded into soft soil and stiff layer. However, it is worth to mention that all of the piles do not fail. According to the results, the pile in soft soil, stiff soil, and very stiff soil can take up to 773 kN/m, 786 kN/m, and 1240 kN/m, respectively. Also, comparison with PLAXIS FE buckling of the piles in three different conditions and free-standing pile are done with Euler's buckling equation. According to Euler's formula, it is proved that the free-standing pile can take up the required working load of up to 753 kN before it fails, while the piles embedded piles can take up to 1538 kN before it fails. It can be noted that according to the Euler's buckling formula in this study, free-standing pile can take the required working load, but the pile surrounded by either soft, stiff, or very stiff soil can take up more loads.

4 Conclusion

Soil restraint effect should always be considered when designing a pile foundation. Analysis of the pile settlement and buckling carried out in the PLAXIS FE software shows that the soil restraining effects of the soil layer improved the carrying capacity, reduce settlement and buckling of the pile compared to a free-standing pile. For further study, the next researcher should use PLAXIS 3D software to perform more innovative finite element analysis.

References

1. Atkinson, J.: The Mechanics of Soils and Foundations, 2nd edn (2007)
2. Davisson, M.T.: Lateral Load Capacity of Piles. University of Illinois, Urbana, Department of Civil Engineering (2007)
3. Aird, P.: Deepwater Drilling: Well Planning, Design, Engineering, Operations, and Technology Application (2019)
4. Michele, P., Antonio, G., Lorella, M.: Analysis of the Behaviour of Very Slender Piles: Focus on the Ultimate Load (2020)
5. Willis, C.: The Skyscraper Museum: Sky High & The Logic of Luxury Walkthrough (2016)
6. Kavitha, P.E., Beena, K.S., Narayanan, K.P.: A review on soil–structure interaction analysis of laterally loaded piles (2016)
7. Broms, B.B.: Lateral resistance of piles in cohesive soils. J. Soil Mech. Found. Div. **90**(2), 27–64 (1964)
8. Hazzar et al.: Ultimate lateral resistance of piles in cohesive soil. International Conference Soft Soil Engineering, Vancouver, Canada (2013)
9. Kaniraj, S.R., Josept, R.R.: Geotechnical Behavior of Organic Soils of Sarawak (2006)
10. Mineral, J., Malaysia, G.: Garispanduan Pemetaan Geologi Kejuruteraan Kawasan Tanah Gambut dan Tanih Lembut, Kementerian Sumber Asli dan Alam Sekitar (2010)
11. Mohamad, N.O., Razali, C.E., Hadi, A.A., Som, P.P., Rusli, M.B., Mohamad, F.R.: Challenges in Construction Over Soft Soil—Case Studies in Malaysia (2016)
12. Terzaghi, K., Peck, R.B.: Soil Mechanics in Engineering Practice, 2nd edn. Wiley, New York (1967)
13. Hatanaka, M., Uchida, A.: Empirical correlation between penetration resistance and internal friction angle of sandy soils. J. Jpn. Geotechn. Soc. **36**(4), 12–15 (1996)
14. Brinkgreve, R.B.J., Broere, W., Waterman, D.: PLAXIS V8 Reference Manual. Delft University of Technology and PLAXIS bv, The Netherlands (2002)

Physicochemical and Engineering Properties of Coal Bottom and Fly Ash Mixtures Modified with Poly-Vinyl Alcohol and Sodium Lauryl Sulfate

Afizah Ayob, Crystal Gayle's Robert, Hamizah Mokhtar, Mustaqqim Abdul Rahim, Senja Rum Harnaeni, Munif Bahatin, Farahiyah Abdul Rahman, Shahrul Azwan Shakrani, and Dayang Siti Hazimah Ali

Abstract Re-using bottom ash (BA) and fly ash (FA) can decrease disposal volumes and cost while also reducing dependence on non-renewable resources for sustainable aims in geotechnical applications. This study investigated the modification of the physicochemical and engineering properties of coal bottom and fly ash (BFA) with poly-vinyl alcohol (PVA) and sodium lauryl sulfate (SLS) at concentrations

---

A. Ayob (✉) · C. G. Robert · M. A. Rahim · M. Bahatin · F. A. Rahman · S. A. Shakrani
Faculty of Civil Engineering Technology, University Malaysia Perlis, 02600 Arau, Perlis, Malaysia
e-mail: afizah@unimap.edu.my

M. A. Rahim
e-mail: mustaqqim@unimap.edu.my

M. Bahatin
e-mail: munif@unimap.edu.my

F. A. Rahman
e-mail: farahiyahrahman@unimap.edu.my

S. A. Shakrani
e-mail: shahrulazwan@unimap.edu.my

H. Mokhtar
School of Civil Engineering, College of Engineering, Universiti Teknologi MARA Pahang, 26400 Bandar Tun Razak, Pahang, Malaysia
e-mail: hamizah1161@uitm.edu.my

S. R. Harnaeni
Civil Engineering Department, Faculty of Engineering, Universitas Muhammadiyah Surakata, 57102 Surakata, Indonesia
e-mail: srh289@ums.ac.id

D. S. H. Ali
School of Engineering and Technology, University of Technology Sarawak, 868 Persiaran Brooke, 96000 Sibu, Malaysia
e-mail: siti.hazimmah@ucts.edu.my

N. Sabtu (ed.), *Proceedings of AWAM International Conference on Civil Engineering 2022 - Volume 3*, Lecture Notes in Civil Engineering 386,
https://doi.org/10.1007/978-981-99-6026-2_2

ranging from 1.5 to 2.5%. The different mixture ratios were denoted as BA100, BA60 FA40, BA40 FA60, and FA100. The proportion of BA40 FA60 modified with SLS at a concentration of 2.5% yielded a mildly alkaline pH with a distinct surface morphology consisting of well-distributed, smooth, and fine particles. The modification by using PVA and SLS generated high maximum dry density, high optimum moisture content, and improved unconfined compressive strength values ranging from 8.36 to 22.14% as compared to unmodified specimens after a 28-day curing period. These results were attributed to PVA/SLS–coal ash electrostatic physical bonding. Lower permeabilities over time of similar mixtures were recorded, ranging from 5.22×10^{-4} to 5.29×10^{-4} cm s^{-1} in SLS solution concentrations ranging from 1.5 to 2.5%. These coal ash wastes showed no unique changes in terms of the variation of main oxide content and crystalline phase for alteration concentrations for both modifiers' solutions. These findings indicate that the physicochemical and engineering properties can be strengthened via a modification process achieved by inducing the novel stability of electrostatic suspension. In turn, this can actively interact with other matrices, such as those intended for soil as well as polymer or composite materials.

Keywords Modified mixtures · Fly ash · Bottom ash · Engineering properties

1 Introduction

Driven by robust economic development, urbanization, and higher energy demand, global energy-associated carbon dioxide (CO_2) emissions increased to a notably high level of 33.1 Gt CO_2 in 2018, 70% higher than the average increase recorded since 2010. Greater economic development was not met by higher energy productivity, and low-carbon technology deployment did not scale rapidly enough to adequately meet the increase in demand. Meanwhile, coal, fossil gas, and biomass are considered the three important energy sources for electricity generation. In 2019, the International Energy Agency stated that coal-fired electricity generated 30% of global CO_2 emissions. Coal consumption in the power plant industry accounted for nearly two-thirds of emissions [1]. It is used in power plants that exceed 10 Gt and are mostly located in Asian countries, including Vietnam, Indonesia, and Malaysia [2]. These plants have been operating for an average of 12 years, making them decades younger than their standard economic lifespan of an estimated 40 years [1, 3].

In Malaysia, the nation's CO_2 emission from its seven coal-fired electricity power plants operating since 1988 accounted for 43.4% of its total emission [4]. Total raw

A. Ayob
Centre of Excellence, Water Research and Environmental Sustainability Growth (WAREG), University Malaysia Perlis, 02600 Arau, Perlis, Malaysia

Cluster of Construction Management Technology, Uninersiti Malaysia Perlis, Arau, Perlis, Malaysia coal use was anticipated to rise from 12.4 million tons in 2005 to 36 million tons in 2020 [5]. In the power plant, coal is pulverized and blown into the combustion chamber by a spurt of hot rising air. Following the combustion process of electricity production, about 10% of the coal will change into coal combustion residuals (CCRs), including fly ash (FA), bottom ash (BA), boiler slag, flue gas desulfurization materials, fluidized bed combustion ash, and other types of by-products. About 75–85% of these CCRs are FA that passed through the tube gases and were then removed from the gas by electrostatic precipitators. Meanwhile, about 15–25% of BA would pile up from the combustion compartment as coarser materials [6, 7]. Even though FA is not classified as toxic waste in terms of metal content, some of its metallic constituents, including mercury, cadmium, and arsenic, would have an enormous impact on the global ecological environment if they percolate into the environment in excessive concentrations [8].

FA mainly comprises aluminosilicates, magnetites, unburned carbon, and cenospheres. It is commonly used in the construction industry due to its advantages of low density, high compressive strength, strong filling ability, smooth spherical surface, lower water absorption, chemical inertness, well-dispersed internal stress in the products, good thermal resistance, and good processability as supplementary cementitious constituents in concrete [9–13]. According to ASTM C 618 [14] the higher content of CaO in the FA of Class C ash results in higher permeability compared to Class F with similar grain size distribution due to its self-setting properties [15]. Many other potential use options for FA particles involve geotechnical applications [11, 16, 17], such as soil stabilization for roads, backfill excavations, mine fill, trenches, retaining walls, and landfill cover as a geopolymer material. Furthermore, along with the sphericity of FA particles, which makes it a suitable filler for polymers as it can facilitate dispersion and fluidity with the matrices, its cost also adds to the list of advantages compared to conventional fillers. However, utilizing FA as a filler in different polymeric materials may not create a solid bond between the polymer and fine FA particles; thus, the standard of the products may be inadequate. This can be attributed to the delicate interfacial bonding between FA and polymer matrix and to the weak friction of the FA surface [12].

As for BA, even though it is not customarily adopted in any form due to the relatively higher content of unburned carbon and varying structural quality, it has potential as a sustainable construction material. For example, one study reported that the non-ground BA substitution of up to 100% in concrete can alter the ratio of shrinkage cracking due to the porous structure morphology and thus affect the concrete deformation via plastic shrinkage [13]. However, a found that the replacement of BA in concrete affected the setting time and led to water loss by bleeding at a fresh phase [18].

Another impediment of using these materials, particularly FA as the primary cementitious material, is that they cannot be used alone due to the lack of strength and durability. The higher permeability of these materials allows water to seep via the interconnected voids in soil or concrete, thus reducing the durability of reinforced concrete structures [19]. Additionally, flocculation and agglomeration between FA particles have an unfavorable effect on the material's strength, because the presence of water can reduce dry densities, which in turn, correspond to an increase of void [11].

Meanwhile, grain size distribution, particle size and morphology, void ratio, mineralogical degree, and compaction are commonly appraised as the main influencing factors. Therefore, if the surface characteristics of the coal FA and BA mixtures can be improved in such a method that these problems are solved, a new innovation for the bulk application of these industrial wastes can be materialized. To increase durability, surface morphology, and engineering properties, various additives [20, 21], chemicals [22], nanoparticles [23], organic polymers [24–26], and micro-organisms [27] have been used in different laboratories to improve the properties. Several standard surface modification techniques, which involve the addition of either aqueous polymer modification on polymer–soil interaction or surfactant mixing, have been extensively used to modify the surface properties of FA [26]. To a certain extent, these chemical modification methods can enhance the interfacial properties between FA and a modifier, such as a polymer or surfactant.

Poly-vinyl alcohol (PVA) was initially synthetically formulated by Hermann and Haelnel in 1924 [28]. Due to its brilliant physical and chemical properties, approximately 650,000 tons of PVA are manufactured annually for various industrial applications, such as polymerization, textile processing, and manufacture of wood preservatives. PVA is also used in the building sector as a building binding agent, for the pre-treatment of aggregate surfaces, and as a fiber reinforcement agent in cementitious composite materials [3]. Meanwhile, using surfactants as stimulators can affect configurations and the microstructure of a matrix as well as enhance the inert particles in the deposit, thereby improving the stability of suspension via the hydrophobicity of the surface charge of suspended particles and increasing the electrostatic adsorption between the polar inorganic and non-polar organic [12].

Meanwhile, sodium lauryl sulfate (SLS) is an anionic surfactant with the formula $CH_3(CH_2)11OSO_3Na$. It is an organ sulfate consisting of a 12-carbon tail affixed to the sulfate group, thus giving the material its amphiphilic properties [29]. Nath et al. [24] investigated the effect of SLS-modified FA on the properties of composite films fabricated with PVA. They reported an improvement in the tensile strength properties of the polymer, which is due to the elimination of particle–particle interaction and the reduced size of FA. These also facilitated the uniform distribution of the materials within the polymer.

Therefore, the current study investigated the characterization of bottom and fly ash (BFA) from a coal-fired power plant in Janamanjung, Perak, which first fully operated in 2003. This plant was designed for a baseload operation system in compliance with the Malaysian Grid Code standards of operating while minimizing environmental impacts. The mineralogical content, surface morphology, and physicochemical and engineering properties of modified BFA at different mix ratios using PVA and SLS are investigated through X-ray fluorescence (XRF), X-ray diffraction (XRD), scanning electron microscopy (SEM), the Proctor method, and unconfined compressive strength (UCS) test for a curing period of 0–28 days.

2 Materials and Methods

2.1 *Specimens Preparation, Grain Size Distribution, Specific Gravity, and pH*

The BA and FA specimens were oven-dried for 48 h before use. The mixture proportions in different percentages were denoted as BA100, BA60 FA40, BA40 FA60, and FA100 and were measured based on the weight of the total BFA. Sieve analysis and grain size distribution were conducted to achieve the gradation curve of the BFA mix according to ASTM D422 [30]. The specific gravity (Gs) of the BFA mixtures was measured using a small pycnometer method, as stated in ASTM D854 [31]. PVA and SLS of $M_W \approx 120$ K and 200 K, respectively, were fully hydrolyzed. These were purchased from MERCK and used as received. Each set of BFA mixture was modified with PVA and SLS at concentrations of 1.5%, 2.0%, and 2.5%. SLS was liquefied in distilled water at 80 °C to get solutions of approximately 1.5 wt%.

Under a temperature of 50 °C, the BFA aqueous solution comprising SLS solution was agitated for about 6 h by a water bath shaker at 130 rpm. Then, pH values were determined after agitating the specimens of both unmodified and modified BFA of SLS for about 15 min to enable the pH of the solution to stabilize. Thereafter, the specimens were filtered with filter funnel, washed thrice with distilled water, and transferred into cylinder molds for 48 h. Then, they were wrapped and packed in cling film and left to cure for 7, 14, and 28 days. A similar process was followed in the other BFA mixtures modified with PVA.

2.2 *Morphological, Mineralogical, and Engineering Properties Tests*

The particle morphology and external surface structure of unmodified BFA and BFA modified with PVA and SLS at 2.5% concentration for a 28-day curing period was studied through SEM (JEOL JSM 6440, LA). The samples that passed through the 75 µm sieve tray were analyzed through XRF (MiniPAL 4, Analytical) for major and minor mineralogical contents. The crystalline phase was identified via XRD (Bruker, D2 Phaser) at 2θ values at a range of 20°–80°, scan rate of 2° per minute, and scan steps of 0.02°. The Standard Proctor Compaction Test [32] was conducted to assess the maximum dry density (MDD) together with optimum moisture content (OMC) of each BFA mixed specimen. For this test, a mold with a 152 mm diameter and a height of 127 mm was used for each specimen. Every specimen was compacted into three layers by applying 25 blows per layer using a 2.5 kg metal hammer. Two specimens, one from the top and the other from the bottom, were taken to examine the moisture content. The permeability test was measured by using a falling head test based on standard reference of ASTM D5084-16a [33], wherein each specimen was compacted in the mold (length: 17 cm, diameter: 6.4 cm) at MDD of 95% and OMC value. Next, the BFA specimens were subjected to the UCS test with reference to ASTM D2166 [34]. The axial strain load of 0.5–2.0% per minute was applied to the specimens until they reached the failure stage.

3 Results and Discussion

3.1 Particle Size Distribution, Specific Gravity, and pH Values

Particle Size Distribution and Specific Gravity. The fine grain composition increases when mixtures contain more FA particles. The grain size distributions of BFA mixtures cover the range between fine sand to gravel [35] with BA40 FA60 composition giving better grain size distributions ranging from 0.08 to 5.0 mm (Fig. 1). The fine size of round and spherical FA particles donates to excellence packing of the aggregate materials thus decreasing porosity and significantly enhancing the applicability of BFA [7] in civil engineering applications and beyond.

The Gs values for BA100, BA60 FA40, BA40 FA60, and FA100 are 2.40, 2.68, 2.45, and 2.27, respectively. The results show that the Gs value reduces as the FA composition in the mixture increases. The compositions of BFA at different percentages demonstrate that the Gs value depends on the grain size in mixture. In this case, the Gs value of FA is lower than that of BA, which due to the higher compositions of iron oxide together with greater number of BA particles. Given that higher iron

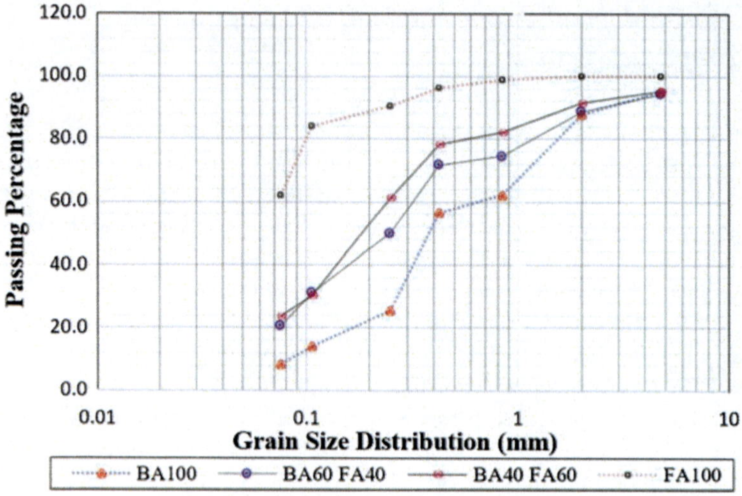

Fig. 1 Particle size distribution of BFA mixes

Table 1 pH values for BFA mixes modified with PVA and SLS

BFA mixes	pH			
	BA100	BA60 FA40	BA40 FA60	FA100
Unmodified	8.6	8.9	8.8	10.0
Modified with PVA (%)				
1.5	8.9	9.5	9.9	10.3
2.0	9.1	9.6	9.9	10.1
2.5	9.4	9.6	9.9	10.1
Modified with SLS (%)				
1.5	9.5	9.6	9.7	9.8
2.0	9.6	10.1	9.9	10.0
2.5	9.8	10.7	10.0	10.1

oxide contents result in higher Gs of BA, the amount of iron oxide contents of coal ash specimens depends on the source of coal used in the burning process [35].

pH Values for BFA Mixtures Modified with PVA and SLS. Both the unmodified BFA and BFA mixtures modified with PVA and SLS at concentrations of 1.5, 2.0, and 2.5% are mildly alkaline ash, with pH values ranging from 8.6 to 10.70 (Table 1). This is due to the FA containing higher amounts of free lime and alkaline oxides. The pH value of the coal ash-modifiers approach depends greatly on the calcium/sulfur molar ratio in coal ash, even though other insignificant alkaline earth cations could increase the balance.

3.2 *Mineralogical Composition of BFA Mixtures Modified with PVA and SLS*

Main Oxide Compositions. Almost all known natural inorganic compositions are detected in unmodified BFA and BFA mixtures modified with PVA and SLS at concentrations of 1.5, 2.0, and 2.5% via XRF (Table 2). The main oxide composition ($MOC = SiO_2 + Fe_2O_3 + CaO + Al_2O_3$) ranges from 92.22 to 94.58% for unmodified BFA. The amounts of other oxide components, such as SO_3, KO_2, TiO_2, and MnO, which are identified in unmodified BFA, range from 3.13 to 4.37% (data not shown). The recorded values of MOC for PVA-modified BFA at concentrations of 1.5%, 2.0%, and 2.5% range from 90.67% to 91.62%, 91.37% to 93.12%, and 92.06% to 94.57%, respectively. The modification by the counterpart SLS recorded MOC values ranging from 90.53% to 93.77%, 91.29% to 92.31%, and 90.62% to 92.08% for concentrations of 1.5%, 2.0%, and 2.5%, respectively. No change in the MOC percentage is noticed as a result of differences in the modified BFA samples and curing periods [36].

Table 2 Main oxide composition analyzed by XRF

BFA mixes	Main oxide compositions (%)			
	Al_2O_3	SiO_2	CaO	Fe_2O_3
Unmodified				
BA100	11.10	33.80	19.40	30.77
BA60 FA40	13.20	34.50	17.30	27.22
BA40 FA60	14.20	37.10	17.00	26.49
FA100	17.40	39.50	14.60	23.08
Modified with PVA (%)				
1.5				
BA100	7.93	20.40	22.70	40.52
BA60 FA40	11.50	24.90	19.80	35.42
BA40 FA60	12.40	26.40	18.90	33.10
FA100	12.70	26.70	18.10	33.17
2.0				
BA100	7.94	21.00	23.15	41.03
BA60 FA40	11.60	25.75	20.40	34.77
BA40 FA60	12.30	26.55	19.25	33.52
FA100	12.95	27.45	18.25	32.72
2.5				
FA100	13.20	28.20	18.40	32.26
FA60 BA40	12.20	26.70	19.60	33.94
FA40 BA60	11.70	26.60	21.00	34.12
BA100	7.94	21.50	23.60	41.53
Modified with SLS (%)				
1.5				
BA100	8.22	23.50	23.10	38.78
BA60 FA40	10.60	24.50	21.10	35.36
FA40 BA60	12.10	27.70	20.40	33.57
FA100	12.80	27.70	18.70	31.33
2.0				
BA100	8.22	22.25	22.70	39.14
BA60 FA40	11.80	26.10	19.80	34.12
BA40 FA60	12.20	27.30	19.85	32.85
FA100	13.00	28.00	18.55	31.74
2.5				
BA100	8.21	21.00	22.30	39.50
BA60 F40	13.00	27.70	18.50	32.88
BA40 FA60	12.30	26.90	19.30	32.12
FA100	13.20	28.30	18.40	32.14

The chemical characteristics of coal ash are greatly dependent on the initial chemical content of the burned coal. It is determined by the type and category of coal used, combustion procedure, and the collector system. Owing to the higher self-cementing properties, and in the presence of water, the burned coal ash could be anthracite, bituminous, sub-bituminous coal, or lignite, which are generated from the burning of younger lignite or sub-bituminous coal [14] bituminous coal commonly produces Class F ash [19].

Crystalline Compositions. Table 3 shows the crystalline compositions of unmodified BFA and BFA mixtures modified with PVA and SLS at concentrations of 1.5, 2.0, and 2.5%. The unmodified Jana Manjung coal ash mixtures consist predominantly of crystalline quartz mixed with other minerals detected, such as oxide magnetite, calcite, and coesite. These coal ash mixtures are classified as Class C ash and known as sub-bituminous coal with a higher concentration of calcium carbonate containing up to 30% of calcium compounds. In addition, the plant operation system also influences crystalline compounds. The recorded main crystalline contents in BFA mixtures modified with PVA include quartz, maghemite, berlinite, calcite, and uvarorite throughout all compositions. Similar crystalline compositions recorded for SLS modification with some other contents are observed, such as periclase, akemanite, and magnetite. The crystalline phase concentration of the BFA mixtures modified by PVA and SLS shows no significant difference from those of the unmodified mixtures. Furthermore, the alteration concentrations of PVA and SLS do not have a distinct impact on the crystalline compositions in those mixtures.

3.3 Morphological Surface of BFA Mixtures Modified with PVA and SLS

SEM images depict the surface morphologies of unmodified BFA and BA40 FA60 compositions modified with PVA and SLS at a 2.5% concentration (Fig. 2) cured for 28 days. For the unmodified specimen, the particles are angular, uneven, and irregular in shape, with rough, crooked surface appearance covered with dust on the surface (Fig. 2a). They tend to be comparatively inert due to the agglomerated size (Fig. 2b) and are often amalgamated due to less pozzolanic mechanism. Meanwhile, as BA content decreases from 100 to 40% in the PVA solution, the angular, uneven, and irregular shapes are changed by ball-shaped and spherical particle shapes (Fig. 2c, d).

Furthermore, some of the particles of PVA-modified BFA are significantly agglomerated and strongly attached, similar to a popcorn shape; this is due to the effect of PVA, which reacted between the coal ashes (Fig. 2c). A distinct morphological difference between the two chemicals can be observed as the BFA mixture is well-rounded at sizes ranging from 5 to 20 μm with fine thin wall and glassy spheres; furthermore, its surface appears to be very glossy and smooth in the SLS solution (Fig. 2e, f). The SLS-modified BFA is packed more closely and indicates a more

Table 3 Primary crystalline composition analyzed by XRD

BFA mixes	Crystalline compositions			
	BA100	BA60 FA40	BA40 FA60	FA100
Unmodified	Quartz (SiO_2), Calcite ($CaCO_3$)	Berlinite ($AlPO_4$), Quartz (SiO_2), Maghemite (Fe_2O_3)	Quartz (SiO_2), Coesite	Quartz low (SiO_2), Coesite
Modified with PVA (%)				
1.5	Quartz (SiO_2), Quartz low (SiO_2), Calcite ($CaCO_3$)	Quartz (SiO_2), Quartz low (SiO_2), Berlinite ($AlPO_4$)	Uvarovite, Quartz (SiO_2), Calcite ($CaCO_3$)	Uvarovite, Maghemite (Fe_2O_3), Quartz (SiO_2)
2.0	Quartz (SiO_2), Maghemite (Fe_2O_3)	Quartz (SiO_2), Maghemite (Fe_2O_3)	Quartz (SiO_2), Maghemite (Fe_2O_3)	Uvarovite. Quartz (SiO_2)
2.5	Quartz (SiO_2), Quartz low (SiO_2), Berlinite ($AlPO_4$)	Quartz (SiO_2), Calcite ($CaCO_3$)	Quartz (SiO_2), Quartz low (SiO_2), Maghemite (Fe_2O_3)	Quartz (SiO_2), Quartz low (SiO_2), Uvarovite
Modified with SLS (%)				
1.5	Maghemite (Fe_2O_3), Quartz (SiO_2), Calcite ($CaCO_3$)	Maghemite (Fe_2O_3), Quartz (SiO_2), Calcite ($CaCO_3$)	Uvarovite, Maghemite (Fe_2O_3), Quartz (SiO_2)	Quartz (SiO_2), Maghemite (Fe_2O_3), Akermanite, Calcite ($CaCO_3$)
2.0	Quartz (SiO_2), Quartz low (SiO_2), Berlinite ($AlPO_4$)	Quartz (SiO_2), Periclase (MgO), Berlinite ($AlPO_4$)	Uvarovite, Maghemite (Fe_2O_3), Quartz (SiO_2)	Quartz (SiO_2), Maghemite (Fe_2O_3), Calcite ($CaCO_3$)
2.5	Quartz (SiO_2), Calcite ($CaCO_3$), Magnetite (Fe_2O_4)	Quartz low (SiO_2), Calcite ($CaCO_3$)	Quartz (SiO_2), Quartz low (SiO_2), Periclase (MgO), Berlinite ($AlPO_4$)	Uvarovite, Quartz (SiO_2), Quartz low (SiO_2)

pozzolanic reaction, which can intensity its cementing characteristics and extend the plasticity and strength properties of soil [37].

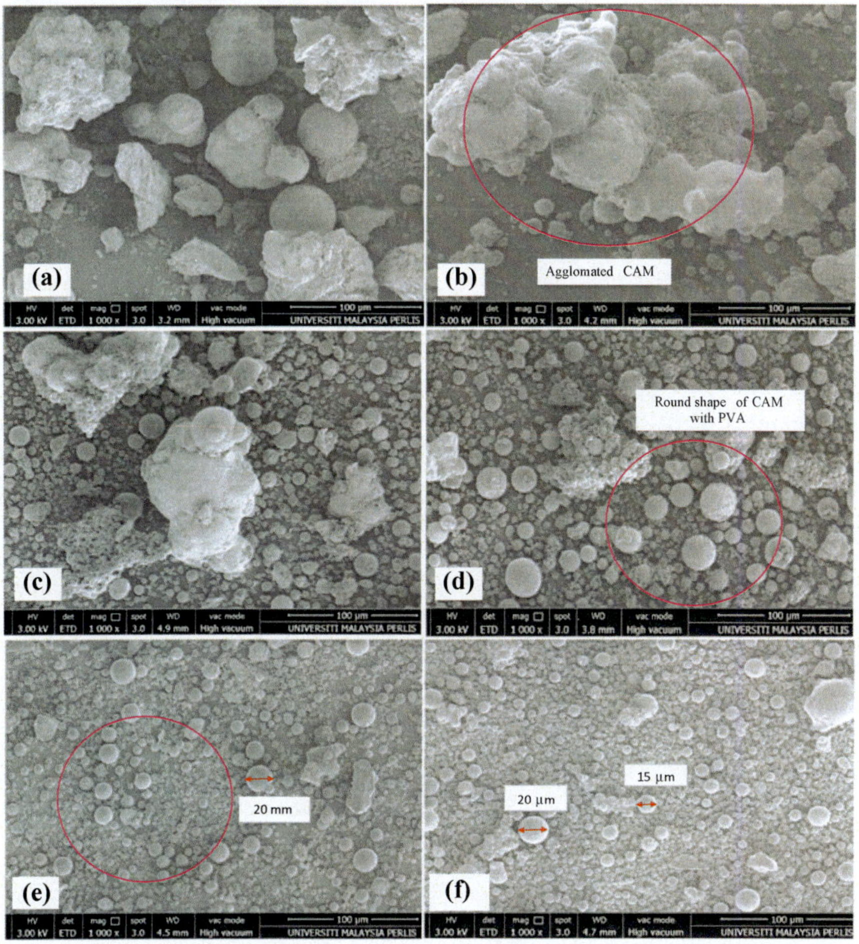

Fig. 2 Surface morphologies of BA40 FA60 composition captured by SEM **a** and **b** unmodified BFA; **c**, **d** PVA; **e** and **f** SLS modified at 2.5% concentration for 28-day curing days

3.4 Engineering Properties of BFA Mixtures Modified with PVA and SLS

Maximum Dry Density and Optimum Moisture Content. The highest MDD value of unmodified BFA is 1.62 g/cm^3 for FA100 specimens, as shown in Table 4. Meanwhile, for PVA-modified BA100, BA60 FA40, BA40 FA60, and FA100 compositions, the MDD values are comparatively increased from 1.29 to 1.53 g/cm^3. For SLS-modified BA40 FA60 mix, the MDD values are at a high range of 1.41–1.43 g/cm^3. When the composition of the FA increases in the BFA mix, it contributes a good-graded size distribution, which consequentially increases in MDD. The OMC values of PVA-modified BFA at concentrations of 1.5–2.5% increased within the range of 15.18–15.48% for BA40 FA40 compositions (Table 1). The OMC values observed are comparatively high for the same composition at 15.43%, 15.32%, and 15.20% at SLS concentrations of 1.5%, 2.0%, and 2.5%, respectively.

The increase of OMC values could be attributed to the tendency of BFA particles to achieve the cation exchange reaction for more modifiers' solution. Furthermore, with the application of compaction effort, the voids are filled by greater amounts of PVA and SLS solutions. In the application of soil stabilization with FA, the values of MDD and OMC largely depend on the fraction of FA in the mixture as well as the source of ash. As FA content increases in soil-FA mixtures, the MDD decreases, and OMC increases due to the generally lower density values for FA and higher OMC values [19].

Permeability of BFA mixtures modified with PVA and SLS. Figure 3 shows the trends of permeability for compacted unmodified BFA and BFA mixtures modified with PVA and SLS at concentrations of 1.5, 2.0, and 2.5%. Higher permeabilities for unmodified BA100, BA60 FA40, BA40 FA60, and FA100 mixtures are recorded within the range of 5.02×10^{-4} to 14.71×10^{-4} cm s^{-1}. However, the permeability of PVA-modified BFA gradually decreased as the contents of FA increased from 40 to 100% at a range of 4.51×10^{-4} to 12.36×10^{-4} cm s^{-1}. A significant pattern is observed for SLS-modified BFA with low permeability values ranging from 3.01×10^{-4} to 11.07×10^{-4} cm s^{-1}. A higher percentage concentration of PVA and SLS at 2.5% resulted in a compacted BFA mix with low permeability effect for both cases. This phenomenon can be explained by different coal ash grain structures as well as FA with a higher CaO content attached by crystallized, tiny spherical shape in an SLS concentration of 2.5%, which are responsible for reducing permeability over time (Fig. 3).

Table 4 MDD and OMC values for BFA mixes modified with PVA and SLS

BFA mix	MDD (g/cm^3)				OMC (%)			
	BA100	BA60 FA40	BA40 FA60	FA100	BA100	BA60 FA40	BA40 FA60	FA100
Unmodified	1.37	1.30	1.52	1.62	16.6	11.00	12.45	7.25
Modified with PVA (%)								
1.5	1.39	1.25	1.41	1.53	16.16	13.97	15.18	16.37
2.0	1.35	1.29	1.42	1.52	14.95	14.40	15.33	15.10
2.5	1.30	1.32	1.43	1.50	13.71	14.82	15.48	13.76
Modified with SLS (%)								
1.5	1.25	1.28	1.41	1.53	13.63	15.75	15.43	13.51
2.0	1.26	1.27	1.42	1.55	14.21	16.06	15.32	14.48
2.5	1.25	1.25	1.43	1.55	14.79	16.37	15.20	15.44

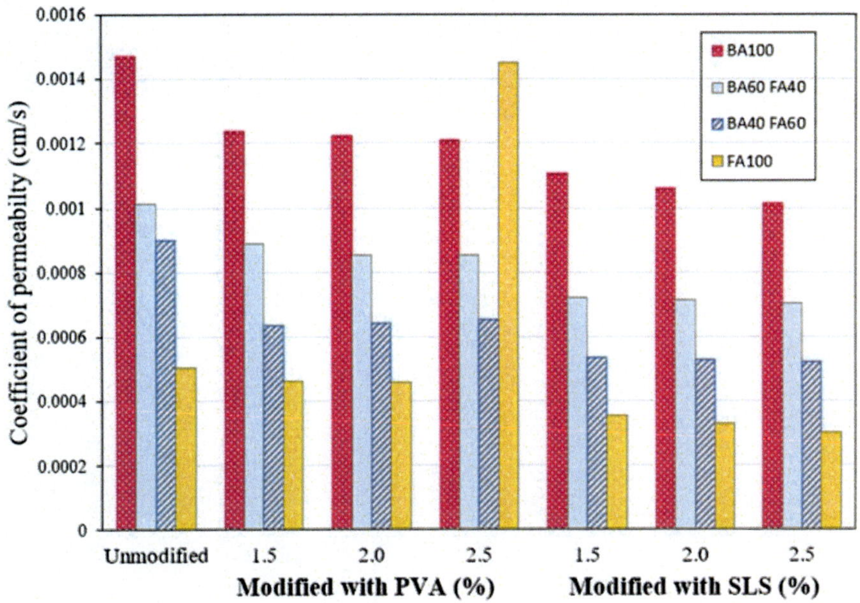

Fig. 3 Permeability of BFA mixes modified with PVA and SLS

Unconfined Compressive Strength of BFA Mixtures Modified with PVA and SLS. The UCS values for unmodified mixtures of BA60 FA40, BA40 FA60, and FA100 show an increase from 24.01 to 27.41 kPa at curing periods of 7 and 28 days, respectively, as depicted in Fig. 4. BA is classified as a non-cohesive material so the unmodified BA100 is broken during mixing, recording 0 kPa for the UCS value. The PVA-modified BA40 FA60 shows significant increases in UCS values of 34.04, 35.06, and 36.04 kPa for concentrations of 1.5%, 2.0%, and 2.5%, respectively, at a 28-day curing period (Fig. 4a). The results also show consistently higher UCS values of 43.23, 46.39, and 49.55 kPa for BA40 FA60 compositions by modification of SLS solutions with 1.5%, 2.0%, and 2.5% concentrations, respectively (Fig. 4b).

In this case, prominent UCS values can be identified by the fact that SLS content resulted in the improved bond mechanism of the BFA particle emulsion interface. Moreover, the adsorption mechanism of anionic SLS emulsion may have feasibly bonded to fine FA composition than to the BA particles. Thus, the SLS molecules can simply form an electrostatic affix with the FA particles. The retention on BFA typically transpire via the absorption of surfactant molecules on both internal and external pore surfaces or microscopic interlayer spaces. Further, this mechanism caused a higher cohesion of particles and enhanced UCS values. This condition could be due to the interlocking between the higher content of 60% FA particles originated from the chemical reactions. Additionally, BFA particles obtain higher UCS from the cementitious reaction of calcium content or via the hydrolysis process experienced by the particles.

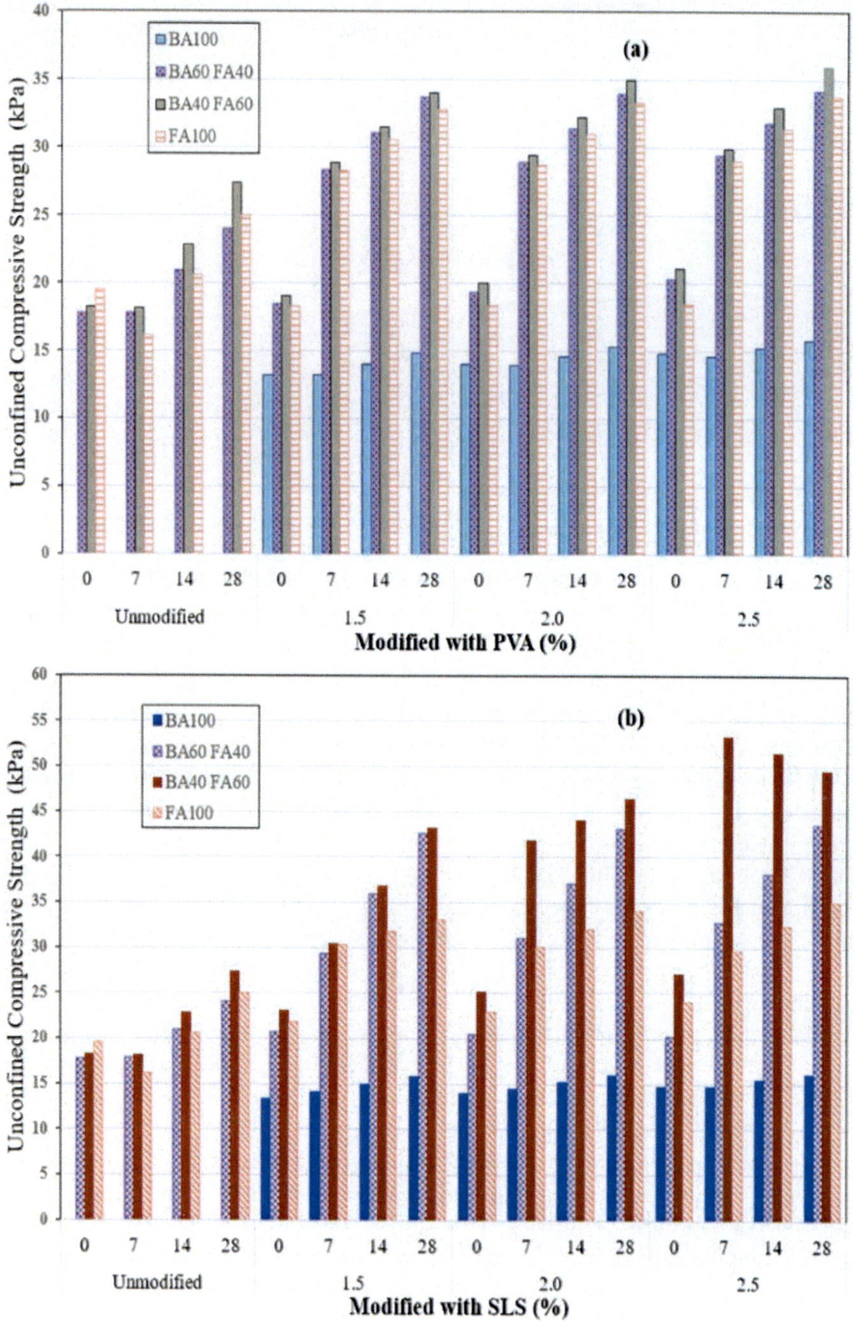

Fig. 4 Unconfined compressive strength for BFA mixtures modified with **a** PVA and **b** SLS

4 Conclusion

The generation of CCRs has increased over the years since the introduction of coal-fired plants in Malaysia to replace gas combustion thermal plants. The BA40 FA60 composition yields fine sandy to gravel distributions of 0.08–5.0 mm, with lower Gs of 2.45 compared with the typical value for conventional aggregates, which is 2.7. The modification of BFA mixtures with PVA and SLS concentrations ranging from 1.5–2.5% at mildly alkaline pH of 8.9–10.7 does not induce specific changes in the contents of main oxide and the phase of crystallinity compositions.

Furthermore, SEM shows distinct images of BA40 FA60 mix modified with SLS of 2.5% concentration cured at 28 days, which has a well-distributed, smooth, and fine spherical appearance; size ranges between 5–20 μm and less flocculation. The same composition in the same modifier shows high MDD and OMC values of 1.42 g/cm^3 and 15.32%, respectively. Modified BFA ashes mixed with SLS show lower permeability from 3.0×10^{-4} to 11.07×10^{-4} cm s^{-1} as compared to the counterpart of PVA. A prominent pattern of higher UCS values of SLS-modified BFA is observed, with values of 43.23–49.55 kPa for BA40 FA60 compositions at 1.5–2.5% SLS concentrations cured for 28 days. Such a result indicates that the non-cohesive and highly erosive BFA particles show improved strength, making them suitable for geotechnical applications, such as materials for bearing capacity enhancement to support retaining walls or the construction of low- to moderate-rise buildings.

Acknowledgements This study is fully funded by the Research Acculturation Grant Scheme (RAGS No: 9018-00078) from the Ministry of Higher Education, Malaysia. The authors acknowledged the Janamanjung Power Plant, Perak, for the permission and assistance of getting the bottom and fly ashes sample from the site.

References

1. International Energy Agency: Global Energy & CO_2 Status Report 2019 Paris. https://www.iea.org/reports/global-energy-co2-status-report-2019. Accessed online Jan 2021
2. Zahari, N.M., Mohamad, D., Arenandan, V., Beddu, S.: Study on prediction fly ash generation using statistical method study on prediction fly ash generation using statistical method. In: AIP Conference Proceedings, vol. 020038 (2018)
3. Xu, S., Malik, M.A., Qi, Z., Huang, B.T., Li, Q., Sarkar, M.: Influence of the PVA fibers and SiO_2 NPs on the structural properties of fly ash based sustainable geopolymer. Constr. Build. Mater. **164**, 238–245 (2018)
4. Marto, A., Tan, C.S.: Properties of coal bottom ash from power plants in Malaysia and its suitability as geotechnical engineering material. J. Teknologi **78**(8–5), 1–10 (2016)
5. Jorat, M.E., Aziz, M.A., Marto, A., Zaini, N., Jusoh, S.N., Manning, D.A.C.: Sequestering atmospheric CO_2 inorganically: a solution for Malaysia's CO_2 emission. Geosciences **8**, 483 (2018)
6. Mangi, S.A., Ibrahim, M.H.W., Jamaluddin, N., Arshad, M.F., Mudjanarko, S.W.: Recycling of coal ash in concrete as a partial cementitious resource. Resources **8**(2), 7–9 (2019)

7. Robert, C.G., Ayob, A., Muhammad Zaki, M.F., Razali, M.E., Lew, E.V., Hong, P.Y.: Characterization of fly and bottom ashes mixtures treated using sodium lauryl sulphate and polyvinyl alcohol. E3S Web Conf. **34**, 01018, (2018)
8. US Environmental Protection Agency EPA: https://www.epa.gov/coalash Frequent questions about the 2015 coal ash disposal rule, US EPA accessed online 5 Feb 2021 (2014)
9. Monika, F., Prayuda, H., Zega, B.C., Cahyati, M.D.: Flexural and compressive strength on no-fines concrete slab using variations of fly ash and superplasticizer. Int. J. Integr. Eng. **11**(9), 275–284 (2019)
10. Chang, H.L., Chun, C.M., Aksay, I.A., Shih, W.H.: Conversion of fly ash into mesoporous aluminosilicate. Ind. Eng. Chem. Res. **38**(3), 973–977 (1999)
11. Deb, T., Pal, S.K.: Effect of fly ash on geotechnical properties of local soil-fly ash mixed samples. Int. J. Res. Eng. Technol. **03**(05), 507–516 (2014)
12. Dharmalingam, U., Dhanasekaran, M., Balasubramanian, K., Kandasamy, R.: Surface treated fly ash filled modified epoxy composites. Polimeros **25**(6), 540–546 (2015)
13. Ramzi, N.I.R., Shahidan, S., Maarof, M.Z., Ali, N.: Physical and chemical properties of coal bottom ash (CBA) from Tanjung Bin Power Plant. IOP Conf. Series Mater. Sci. Eng. **160**(1), 012056 (2016)
14. ASTM C618: Standard specification for coal fly ash and raw or calcined natural pozzolan for use in concrete (2014)
15. Zabielska-Adamska, K.: Hydraulic conductivity of fly ash as a barrier material: some problems in determination. Environ. Earth Sci. **79**, 321 (2020)
16. Gimhan, P.G.S., Disanayaka, J.P.B., Nasvi, M.C.M.: Geotechnical engineering properties of fly ash and bottom ash: use as civil engineering construction material. Eng. J. Inst. Eng. Sri Lanka **51**(1), 49–57 (2018)
17. Elkhebu, A., Zainorabidin, A., Bakar, I.H., Huat, B.B.K., Abdeljouad, L., Dheyab, W.K.: Alkaline activation of clayey soil using potassium hydroxide & fly ash. Int. J. Integr. Eng. **10**(9), 99–104 (2018)
18. Andrade, L.B., Rocha, J.C., Cheriaf, M.: Influence of coal bottom ash as fine aggregate on fresh properties of concrete. Constr. Build. Mater. **23**(2), 609–614 (2009)
19. Bhatt, A., Priyadarshini, S., Acharath Mohanakrishnan, A., Abri, A., Sattler, M., Techapaphawit, S.: Physical, chemical, and geotechnical properties of coal fly ash: a global review. Case Stud. Constr. Mater. **11** (2019)
20. Alam, Q., Dezaire, T., Gauvin, F., Delsing, A.C.A., Brouwers, H.J.H.: Valorization of bottom ash fines by surface functionalization to reduce leaching of harmful contaminants. J. Environ. Manage. **271**, 110884 (2020)
21. Ju, C., Liu, Y., Yu, Z., Yang, Y.: Cement-lime-fly ash bound macadam pavement base material with enhanced early-age strength and suppressed drying shrinkage via incorporation of slag and gypsum. Adv. Civil Eng. **2019**, 10 (2019)
22. Jongphimai, D., Suksiripattanapong, C., Thumrongvut, J. Effect of Na_2SiO_3/NaOH ratio on compressive strength of lateritic soil mixed with bagasse ash geopolymer. IOP Conf. Series Mater. Sci. Eng. **600**(1) (2019)
23. Sun, Y., Zhang, P., Guo, W., Bao, J., Qu, C.: Effect of nano-$CaCO_3$ on the mechanical properties and durability of concrete incorporating fly ash. Adv. Mater. Sci. Eng. **7365862**, 10 (2020)
24. Nath, D.C.D., Bandyopadhyay, S., Yu, A., Blackburn, D., White, C.: High strength bio-composite films of poly(vinyl alcohol) reinforced with chemically modified-fly ash. J. Mater. Sci. **45**(5), 1354–1360 (2010)
25. Yang, Y.F., Gai, G.S., Cai, Z.F., Chen, Q.R.: Surface modification of purified fly ash and application in polymer. J. Hazard. Mater. **133**(1–3), 276–282 (2006)
26. van der Merwe, E.M., Mathebula, C.L., Prinsloo, L.C.: Characterization of the surface and physical properties of South African coal fly ash modified by sodium lauryl sulphate (SLS) for applications in PVC composites. Powder Technol. **266**, 70–78 (2014)
27. Rahman, J.A., Mohamed, R.M.S.R., Al-Gheethi, A.A.: Microbial activity in peat soil treated with ordinary portland cement and coal ashes. IOP Conf. Series Earth Environ. Sci. **140**(1), 012087 (2018)

28. Haweel, C.K., Saad, H.A.: Preparation of polyvinyl alcohol from local raw material. Iraqi J. Chem. Pet. Eng. **9**, 15–21 (2008)
29. Hue, N.V., Phong, N.N., Hanh, P.H.: Effects of sodium lauryl sulfate on co-electrodeposition of nickel and fly ash particles. Int. J. Innov. Eng. Technol. (IJIET)
30. ASTM D422-63: Standard test method for particle-size analysis of soils (2007)
31. ASTM D854-14: Standard test methods fot specific gravity of soils solids by water pycnometer (2000)
32. ASTM D698: Standard test methods for laboratory compaction characteristics of soil using standard effort (2003)
33. ASTM D5084-16a: Standard test methods for measurement of hydraulic conductivity of saturated porous materials using a flexible wall permeameter (2010)
34. ASTM D2166: Standard test method for unconfined compressive strength of cohesive soil (2013)
35. Abdullah, M.H., Abuelgasim, R., Rashid, A.S.A., Mohdyunus, N.Z.: Engineering properties of tanjung bin bottom ash. MATEC Web Conf. **250**, 01006 (2018)
36. Robert, C.G., Ayob, A., Zaki, M.F.M., Zahid, M.Z.A.M., Von, L.E.: Polyvinyl-alcohol-modified coal-ash mixtures for productive geotechnical application. J. Adv. Res. Mater. Sci. **10**(1), 12–25 (2015)
37. Ayob, A., Zahid, M.Z.A.M., Zaki, M.F.M., Yunus, N.M.: Physical, morphological and strength properties of Jana Manjung coal ash mixture for geotechnical applications. Adv. Environ. Biol. **8**(22), 25–30 (2014)

Correlation Study of Landfill Leachate Concentration and Resistivity Value for Prediction of Landfill Leachate in Shallow Groundwater

M. N. Kamarulzaman, M. H. Zawawi, A. Ahmad, M. A. Mazlan, N. H. Hassan, and M. A. Kamaruddin

Abstract Electrical resistivity imaging (ERI) is being applied to hydrogeological applications studies like survey of groundwater contamination. Monitoring of groundwater using this approach was more effective because it does not use wells for observation. Among the purposes of this research were to determine the leachate and sand characteristics, evaluate the resistivity for different types of permeating fluids and measure the resistivity correlation between water and leachate. Due to the infiltration of leachate, the bad impact at the landfill site has been declared will give contribution to groundwater contamination. Methods that have been implemented was soil box method. The purpose of this method is to measure the electrical resistivity of permeates liquid and sand sample. The moist fine sand was placed into an oven of 100 °C for 24 h to obtain the dried fine sand oven. Various materials have been tested such as leachate, tap water, and distilled water for 15, 30, and 45 ml with dried fine sand oven. Characteristics of leachates were determined from dissolved solid determination, turbidity measurement, colour measurement, and more. It can be summarized that the resistivity value of leachate is lower than that of tap water and distilled water. Also, from the comparison it can be concluded that leachate gives minimal impact for turbidity value as soil itself has a great effect on increasing the value of turbidity.

Keywords ERI · Groundwater · Leachate · Resistivity · Contamination

M. N. Kamarulzaman · M. H. Zawawi (✉) · N. H. Hassan
Department of Civil Engineering, Universiti Tenaga Nasional, 43000 Kajang, Malaysia
e-mail: MHafiz@uniten.edu.my

A. Ahmad · M. A. Mazlan
Department of Electrical and Electronic Engineering, Universiti Tenaga Nasional, 43000 Kajang, Malaysia

M. A. Kamaruddin
Environmental Technology Division, School of Industrial Technology, Universiti Sains Malaysia, 11800 Penang, Malaysia

N. Sabtu (ed.), *Proceedings of AWAM International Conference on Civil Engineering 2022 - Volume 3*, Lecture Notes in Civil Engineering 386,
https://doi.org/10.1007/978-981-99-6026-2_3

1 Introduction

Leachate is a form of wastewater which is produced in solid landfills, waste collection, and composting stations [1]. These liquid wastes which can be described as a liquid which passes through a landfill and has dissolved and extracted suspended matter are considered a major hazard to surface water, groundwater, and soil [2]. Once the leachate comes into contact with the environment, groundwater becomes polluted and causes adverse environmental effects [3]. Leachate deposit treatment is a primary environment protection and sustainable development measure [4]. Several factors have been identified affecting leachate composition, such as landfill age, local climate season, landfill waste depth, and most of all, waste material composition [5].

The volume of leachate leakage was increased as the main functional units of a landfill have deteriorated over time, resulted in even more serious adverse impacts on the ecosystem and human health [6]. The diffusion of leachate contaminants into groundwater was depended on the physical and chemical properties of MSW and the pattern of groundwater flow. Degradation of waste components in rainy seasons was enabled the absorption of leachate components with rainfall into groundwater [7]. Certain parameters such as the resistivity of soil, conductivity, total dissolved solids, and soil moisture were definitely changed. All these changes were resulted in many bad impacts especially for environmental concern like groundwater quality.

The technique of 2D resistivity imaging (RI) was used after leachate recirculation to track moisture movement and variations. Tests were conducted at different times using 56 electrodes, with an instrument manufactured by Advanced Geosciences Institute (AGI) that is programmable eight-channel, super-sting R8/IP instrument. Tests were repeated at the same location to observe the possible leachate flow after leachate recirculation.

A systematic variation in resistivity was observed with the addition of leachate through pipe H2 and used to evaluate variations in moisture and to determine the extent of leachate recirculation. It was evident from the study that a useful tool for monitoring periodic recirculation of leachate is the 2D RI technique.

Last but not least, after one day and 15 days of leachate recirculation, the field resistivity test showed that the resistivity value returned very close to the baseline value after 15 days. This means that no significant changes in the ionic properties of solid waste have occurred due to leachate recirculation, resulting in very little change in the resistivity value [8].

The soil's electrical resistivity is dependent on soil type, degree of saturation, ion concentration, and pore water temperature [9]. The properties that affect a soil or rock's resistivity also include porosity, water content, composition, pore water salinity, distribution of grain size, and particle shape and orientation.

When leachate from landfill that contains dissolved material as well as suspended material infiltrate into the ground, the soil will mix with this pollutant. This will give effect to some characteristics of soil ground like soil resistivity. Once the leachate comes into contact with the environment, groundwater becomes polluted and causes adverse environmental effects [3].

Overall, focus of this research is to analyse the resistivity correlation among dried fine sand oven and different types of permeating fluids. The result of soil sample and leachate parameters of this landfill, including leachate characteristics, will be observed and analysed using ERI to identify their contamination potential and to plan suitable remedial measure. The unavailability of the leachate's parameter can lead to an estimated error. Hence, error estimation will be carried out to reduce the inaccuracy in measurements and to mark the mistake in time ahead studies.

2 Materials and Method

2.1 Material

Leachate was obtained from disposal waste bin from local supermarket. Waste was placed in a container and sprayed by water, and then, the leachate came out and sampled into a bottle. A bag of fine sand was bought in local store for the research. Both materials were used is this study with additional material to fulfil the condition such as distilled water and tap water.

2.2 Method

All materials were prepared beforehand with sampled bottle of leachates from disposal waste, and fine sand have dried by using an oven at 100 °C for 24 h.

Various experiments have been conducted to determine the characteristics of leachates such as total solid determination, suspended solid determination, dissolved solid determination, colour measurement, and turbidity measurement.

Soil box method is used to determine the resistivity of the leachates.

2.2.1 Total Solid Determination

This experiment is used to determine the solid concentration in leachate sample by obtaining the total solids concentration (TS), total fixed solids concentration (TFS), and total volatile solids concentration (TVS).

The first step was collected a porcelain dish from desiccators and weighed the dish, $M1$. Next, transferred the dish to a steam bath. Poured out between 15 and 20 ml of aliquot of a sample using graduated cylinder and transferred them to the dish, $V1$. After there was no sign of liquid left, transferred the dish to an oven at 100 °C ± 5 °C for one hour.

After the hour has lapsed, weighed the sample, $M2$. After that, transferred the dish to a muffle furnace at 550 °C for 20 min. After the 20 min has lapsed, took the dish out from the furnace and cooled it by placing at a tripod for 5 min. Once the 5 min has lapsed, transferred to a desiccator to cool further for 30 min. Then, weighed the sample, $M3$. By using some calculations, the data required have obtained:

$$\text{Total solids concentration, TS, (mg/l)} = \frac{\text{Mass of solids}}{\text{Volume of sample}} = \frac{M2 - M1}{V1}$$

$$\text{Total fixed solids concentration, TFS, (mg/l)} = \frac{\text{Mass of solids after ignition}}{\text{Volume of sample}} = \frac{M3 - M1}{V1}$$

$$\text{Total volatile solids concentration, TVS, (mg/l)} = \text{TS} - \text{TFS}$$

2.2.2 Suspended Solid Determination

This experiment is used to determine the suspended solid concentration of the sample by obtaining the total suspended solids concentration (TSS), fixed suspended solids concentration (FSS), and volatile suspended solids (VSS).

Firstly, took a piece of filter paper and weighed, $M4$. Next, poured out 20–25 ml of sample and the volume of the sample, $V2$, was recorded. Next, transferred the filter paper to a Buchner funnel and filtered the sample. Slowly removed the paper from the funnel and placed it in an oven with temperature 100 °C for one hour.

After one hour, removed the filter from the oven and weighed, $M5$. Transferred the filter to a furnace at 550 °C for 20 min. After 20 min has lapsed, transferred the filter to desiccators for 30 min. Finally, after 30 min has lapsed, weighed the sample, $M6$. For discussion, discuss the results and compare it with water quality standard. By using some calculations, the data required have obtained:

$$\text{Total suspended solids concentration, TSS, (mg/l)} = \frac{\text{Mass of suspended solids}}{\text{Volume of sample}} = \frac{M5 - M4}{V2}$$

$$\text{Fixed suspended solids concentration, FSS, (mg/l)} = \frac{\text{Mass of remained suspended solids}}{\text{Volume of sample}} = \frac{M6 - M4}{V2}$$

$$\text{Volatile suspended solids concentration, VSS, (mg/l)} = \text{TSS} - \text{FSS}$$

2.2.3 Dissolved Solid Determination

The relationship between the total solids concentration (TS), total suspended solids concentration (TSS), and total dissolved solids concentration (TDS) can be given by the formula:

$$\text{Total dissolved solids concentration, TDS} = \text{TS} - \text{TSS}$$

$$\text{Fixed dissolved solids concentration, FDS} = \text{TFS} - \text{FSS}$$

$$\text{Volatile dissolved solids concentration, VDS} = \text{TVS} - \text{VSS}$$

2.2.4 Colour Measurement

Firstly, poured the sample into Nessleriser tube to the mark and covered the tube with the lid. Make sure to avoid any built up of air bubble in the tube and then dry the side of the tube with tissue paper.

Next, placed the tube in the comparator and the colour disc was rotated till the colour almost matches the sample and recorded the reading. After that, transferred an aliquot of sample into a sample vial. Switched on a spectrophotometer and chose colour measurement. Then, placed the distilled water and the reading has been cleared. Once it was ready, placed the sample cell and read. For further observation, discuss the colour reading of each sample and compare to water quality standard.

2.2.5 Turbidity Measurement

The first step was poured the sample into the vial provided. Next step was dried the vial and placed into the turbidimeter. Finally, the turbidity reading was recorded. For further observation, after the experiment, draw a graph comparison between samples.

2.2.6 Soil Box Method

For soil box method, various types of materials have been tested for resistivity values such as leachate, dried fine sand oven (DFSO), tap water, distilled water, dried fine sand oven with 15 ml of leachate, dried fine sand oven with 30 ml of leachate, dried fine sand oven with 45 ml of leachate, dried fine sand oven with 15 ml of tap water, dried fine sand oven with 30 ml of tap water, dried fine sand oven with 45 ml of tap water, dried fine sand oven with 15 ml of distilled water, dried fine sand oven with 30 ml of distilled water, and dried fine sand oven with 45 ml of distilled water.

3 Result and Discussion

3.1 *Leachate Characteristics*

Leachate characteristics have determined by experiments conducted in laboratory with various procedure and produced results (Table 1 and Fig. 1).

Comparison for colour and turbidity measurement between each type of samples shows that dry fine sand itself have much reaction with liquids. The value for DFSO with distilled water is higher than value for DFSO with leachate. The value continues to increase as the volume of sample increases. The collection of these data shows that type of permeating fluid used influences the resistivity significantly.

Table 1 Results of leachates characteristics test

Parameters	Results		
Total solid determination	Total solid (TS), mg/L	Total fixed solids (TFS), mg/L	Total volatile solids (TVS), mg/L
	1050	700	350
Suspended solid determination	Total suspended solids (TSS), mg/L	Fixed suspended solids (FSS), mg/L	Volatile suspended solids (VSS), mg/L
	700	400	300
Dissolved solid determination	Total dissolved solids (TDS), mg/L	Fixed dissolved solids (FDS), mg/L	Volatile dissolved solids (VDS), mg/L
	350	300	50

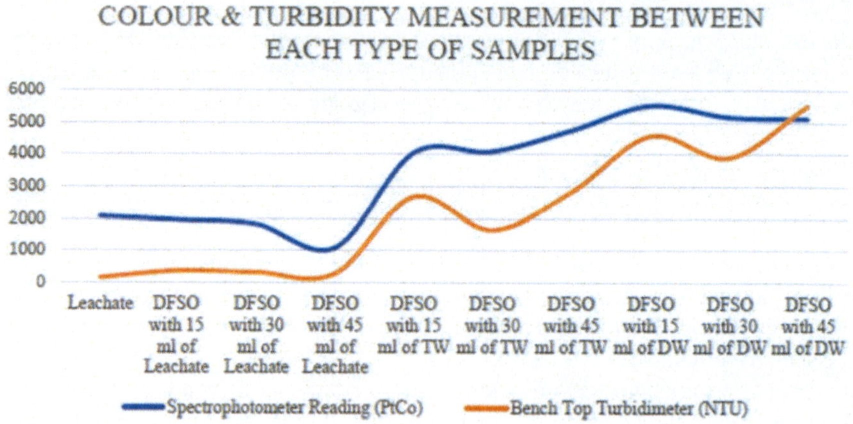

Fig. 1 Colour and turbidity for different amount of sample with dried fine sand oven (DFSO)

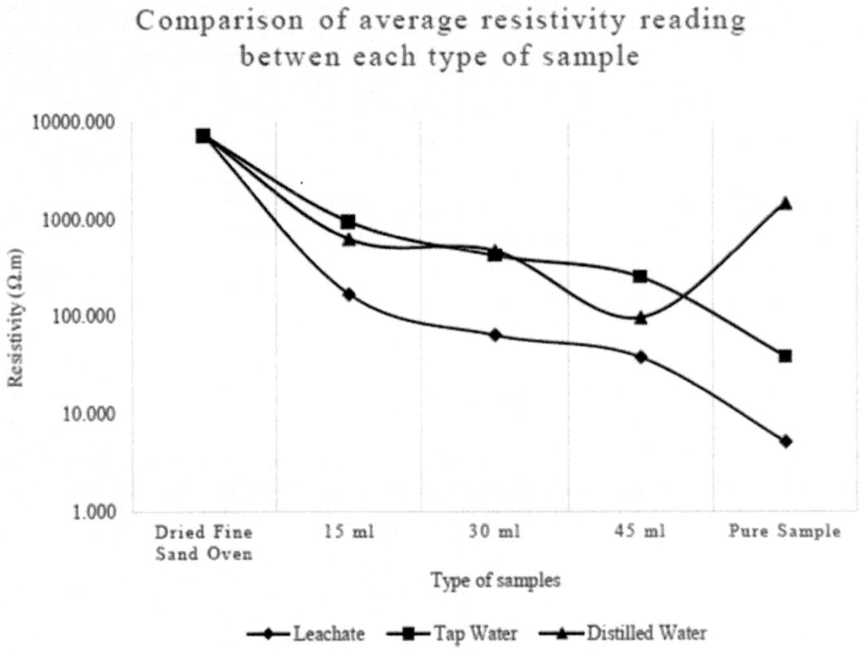

Fig. 2 Resistivity value with different amount of sample and dry fine sand

3.2 *Resistance Values*

Soil box been used in this study to measure the resistivity of the leachate itself, and fine dry sand mix with multiple type and amount of liquids. The following outcomes (Fig. 2) determined with 25 trials by 10 s interval range.

From the graph above, each value used is an average of 25 readings. Firstly, the average value of dried fine sand oven is 7218 Ω m. The pure sample refers to the average resistivity value for each type of samples such as leachate, tap water, and distilled water that are 5.098, 38.416, and 1160.4 Ω m.

Next, for mixtures of dried fine sand oven with 15, 30, and 45 ml of leachate, the average resistivity values for each type of composition are 167.480, 63.520, and 37.696 Ω m. After that, for mixtures of dried fine sand oven with 15, 30, and 45 ml of tap water, the average resistivity values for each type of composition are 939.320, 416.800, and 248.720 Ω m. On the other hand, for mixtures of dried fine sand oven with 15, 30, and 45 ml of distilled water, the average resistivity values for each type of composition are 619.360, 465.840, and 96.368 Ω m. The resistivity keeps decreasing when water content increases.

4 Conclusion

The electrical resistivity method is based on the fact that the electrical resistivity of soils is much higher than the electrical resistivity of leachates, tap water, and distilled water. Consequently, absorption of even a small amount of any of these permeates induces a shift in the electrical properties that can be easily identified by the use of the electrical resistivity method. To predict the landfill leachate in shallow groundwater, we can determine by observing the difference in resistivity value. The inclusion in the soil box of only a small amount of leachate containing dried fine sand oven results in a significant decrease in the resistivity. The resistivity with rising fluid content is showing a rapid decrease. The landfill leachate's resistivity is less than that of fresh groundwater. Due to an excess of ions or molecules in the groundwater, the electrical resistivity values of polluted area were lower than those of safe zones. It can be concluded from the findings of this research that the soil box method is a realistic option for electrical resistivity imaging (ERI) and boreholes to detect and analyse the imaging sub-surface structures and their properties.

References

1. Azadi, S., et al.: Photocatalytic treatment of landfill leachate using cascade photoreactor with immobilized W-C-codoped TiO_2 nanoparticles. J. Water Process Eng. **36**(April), 101307 (2020). https://doi.org/10.1016/j.jwpe.2020.101307
2. Collado, S., et al.: Effect of landfill leachate ageing on ultrafiltration performance and membrane fouling behaviour'. J. Water Process Eng. **36**(April) (2020). https://doi.org/10.1016/j.jwpe.2020.101291
3. Siddiquea, R., Kaur, G., Rajor, A.: Waste foundry sand and its leachate characteristics. Resour. Conserv. Recycl. **54**(12), 1027–1036 (2010). https://doi.org/10.1016/j.resconrec.2010.04.006
4. Yusoff, M.S., et al.: Floc behavior and removal mechanisms of cross-linked duriozibethinus seed starch as a natural flocculant for landfill leachate coagulation floccation treatment. Waste Manag. **74**, 362–372 (2018). https://doi.org/10.1016/j.wasman.2018.01.016
5. Kulikowska, D., Klimiuk, E.: The effect of landfill age on municipal leachate composition. Biores. Technol. **99**(13), 5981–5985 (2008). https://doi.org/10.1016/j.biortech.2007.10.015
6. Ya, X., et al.: Long-term dynamics of leachate production, leakage from hazardous waste landfill sites and the impact on groundwater quality and human health. Waste Manag. **82**, 156–166 (2018). https://doi.org/10.1016/j.wasman.2018.10.009
7. Mishra, S., et al.: Impact of Municipal Solid Waste Landfill leachate on groundwater quality in Varanasi, India. Groundwater Sustain. Develop. **9**(October) (2019). https://doi.org/10.1016/j.gsd.2019.100230
8. Manzur, S.R., et al.: Monitoring extent of moisture variations due to leachate recirculation in an ELR/bioreactor landfill using resistivity imaging. Waste Manag. **55**, 38–48 (2016). https://doi.org/10.1016/j.wasman.2016.02.035
9. Yoon, G.L., Park, J.B.: Sensitivity of leachate and fine contents on electrical resistivity variations of sandy soils. J. Hazard. Mater. **84**(2–3), 147–161 (2001). https://doi.org/10.1016/S0304-3894(01)00197-2

Centrifuge Modeling of Rockfalls with Different Slope Surface Materials using the MPS Method

Gaoyuan Lyu, Akihiko Hirook, Makoto Tobo, and Toshiyuki Ozaki

Abstract Rockfall, one of the common natural catastrophes, complicates the phenomena due to a mix of inherent predisposing variables and external factors such as slope surface material constants. In this study, aiming to investigate effects on the behavior of rockfall of parameters, a 50 G centrifuge experiment is implemented with two 45° inclined slope surfaces of different materials, two of which are different in hardness, and two aluminum cubes of different sizes, which are recorded by four high-speed cameras. The moving particle semi-implicit (MPS) method is a type of particle method that discretizes a continuum by using computation points as particles that move with the physical quantity with meshless and has attracted much attention because it is easier to model the object of analysis and can easily handle large deformation and fracture. From the perspective of verification and validation (V&V), the MPS method is employed to validate the results based on centrifuge experiments.

Keywords Rockfall · Centrifuge experiment · The MPS method

1 Introduction

It is currently difficult to predict the occurrence of rockfalls due to complicating factors, such as irregularity of slope shape and materials. In recent years, rockfall simulations have become popular for effective planning of countermeasures against rockfalls, since rockfall protection facilities are often designed based on experience and field judgment. By changing the initial angle of the falling rock, Renka Kanno

G. Lyu · A. Hirook (✉) · M. Tobo
School of Engineering, Kyushu Institute of Technology, Sensuicho 1-1, Tobata-Ku, Kitakyushu-Shi, Fukuoka 804-8550, Japan
e-mail: hirooka.akihiko032@mail.kyutech.jp

T. Ozaki
Foundation Design of O.H.L Department, Kyushu Electric Engineering Consultants Inc, Kiyogawa 2-13-6, Chuo-Ku, Fukuoka 810-0005, Japan

N. Sabtu (ed.), *Proceedings of AWAM International Conference on Civil Engineering 2022 - Volume 3*, Lecture Notes in Civil Engineering 386,
https://doi.org/10.1007/978-981-99-6026-2_4

et al. [1] employ DEM analyses to study the trajectory and mechanical energy of the rockfalls. Besides, in order to obtain the state of motion of rockfalls associated with fracturing, Kazuya Ito et al. [2] implement a centrifugal experiment of 100 G with rocks of different shapes (spherical and square) and materials (plaster + Toyoura sand and aluminum). However, due to the complex factors, such as irregularity of slope shape and materials, etc., the mechanisms of rockfalls motion had not yet been fully elucidated. In this study, Young's modulus and density are focused on discussing its effects on rockfall by experiments and numerical analysis.

2 The Centrifuge Experiment

2.1 The Condition of Experiment

The rockfall model (Fig. 1), which is adjusted to 1/50 with a centrifugal acceleration field of 50 G, has been made with two types of slopes (Fig. 1a, b). The conditions of centrifuge experiments are shown in Table 1. The rock is a cube of 1 and 2 cm on each side and is made of aluminum (Fig. 1c). The surface layer of the hard slope is made of acrylic plate, and the surface layer of the soft slope is made of a mixture of silty sand and Toyoura sand. Although Toyoura sand has adhered to the surface of the acrylic plates, the values of the physical properties of the surface layer shall be those of the acrylic plates. The physical properties of the rockfall model, hard model slope (acrylic plate), and soft model slope are shown in Table 2. As for the modulus of deformation and Poisson's ratio of the mixed sample, the modulus of deformation is calculated from the relationship between consolidation stress and strain in the over-consolidation state by conducting consolidation tests. The Poisson's ratio is calculated by transforming Eq. (1) using the volume compression coefficient m_{υ} and the modulus of deformation E obtained from the consolidation test,

$$K = \frac{E}{3(1 - 2\upsilon)}, \quad m_{\upsilon} = \frac{1}{K}, \tag{1}$$

where K is bulk modulus, and υ is the Poisson's ratio.

2.2 The Results of the Experiment

The speed is computed by taking the average of the monitor images obtained right before each of the two bottom cameras passed the edge of the slope and the previous one. The total kinetic energy of the falling rocks is the sum of the linear velocity energy and the rotational energy. The results are shown in Table 3.

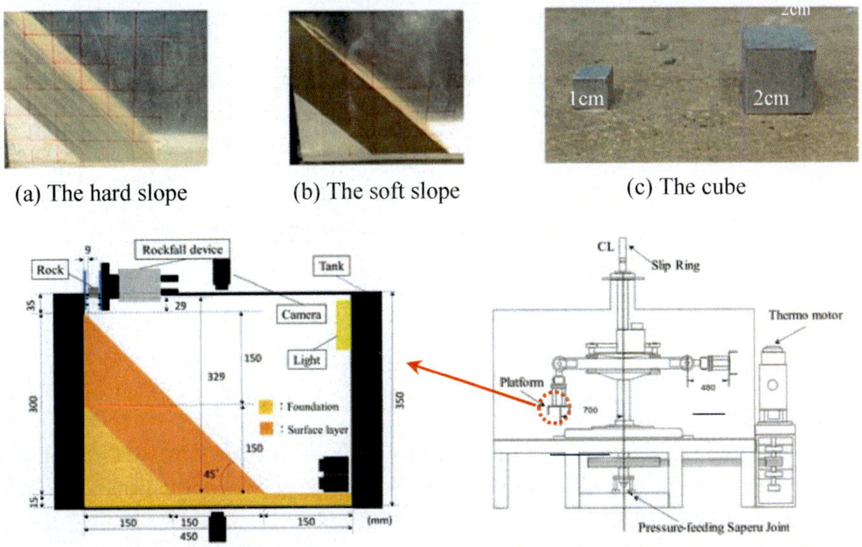

(a) The hard slope (b) The soft slope (c) The cube

(d) The rockfall model (e) Centrifuge device

Fig. 1 Rockfall model in the centrifuge experiment

Table 1 Condition of centrifuge experiment

Centrifugal acceleration	Model scale	Model slope inclination	Slope height	Hard slope	Soft slope	Rock
50 G	1/50	45°	300 mm	Acrylic	Sand	Aluminum

Table 2 Material constants in the experiment

Materials	Poisson's ratio	Young's modulus (Pa)	Density (kg/m^3)
Rockfall	0.34	6.83×10^{10}	2700
Hard slope	0.35	3.20×10^{9}	1200
Soft slope	0.33	3.99×10^{4}	2620

Table 3 Velocity and total kinetic energy at the bottom of the slope

Case	Velocity (m/s)	Total kinetic energy
Case HS	14.1	0.404
Case HB	4.02	0.262
Case SS	7.56	0.116
Case SB	3.96	0.254

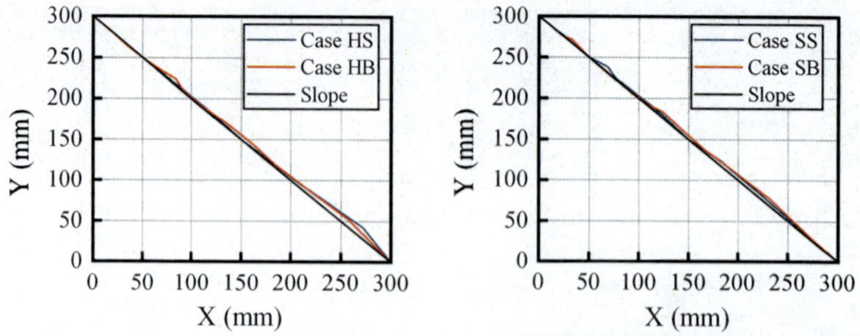

(a) Trajectory in the hard slope (b) Trajectory in the soft slope

Fig. 2 Side view of trajectory of rockfall in the experiment

In the experiment, case HS (hard surface with a small rock) takes 0.05 s longer than case SS (soft surface with a small rock) to reach the bottom of the slope, whereas case HB (hard surface with a big rock) and case SB (soft surface with a big rock) take the same time. From the side view in Fig. 2, on the hard surface, both case HS and case HB had a clear jump at the bottom of the slope (see Fig. 2a), although not in the first half of the, indicating the transition from rolling or sliding to jumping. On the other hand, on the soft surface, case SS and case SB almost no significant jumping occurred, and judging by the trajectory after the experiment, they rolled and slid until they left the slope (see Fig. 2b). Additionally, the soft slope consumed more energy due to a large number of collisions during rolling and sliding. Based on the above results, it is hypothesized that:

(a) The smaller the slope surface's Young's modulus, that is, the softer the slope surface, the more the falling rock will be to rotate.
(b) Hard slope leads to larger rockfall velocity and kinetic energy.

Then, the MPS method is employed to validate the results.

3 The MPS Method

3.1 Radius of Influence

In the MPS method [3, 4], particle interaction models are prepared for the differential operators of vector analysis such as gradient, divergence, and Laplacian, respectively, and the differential equations are discretized using these models. The particle interaction model is also defined by introducing the weight function W shown in Eq. (2):

Fig. 3 Radius of influence

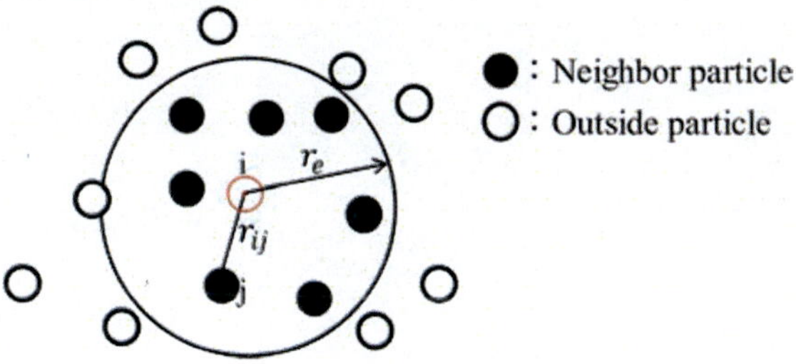

$$w(r_{ij}) = \begin{cases} \frac{r_e}{r_{ij}} - 1 & (0 \le r_{ij} \le r_e) \\ 0 & (r_e \le r_{ij}) \end{cases} \tag{2}$$

$$n_i = \sum_{j \neq i} w(|\boldsymbol{r}_j - \boldsymbol{r}_i|) \tag{3}$$

where r_{ij} is the distance between particle i and particle j, and r_e is the radius of influence of particle i. Besides, the weight function in Eq. (2) is used to express the particle number density n_i for particle i as shown in Eq. (3). Only particles within the radius of influence are involved in the calculation (see Fig. 3), for interaction assignment. Recommended by the author Seiichi Koshizuka here goes the radius of influence r_e of 2.1 l_0.

3.2 The Rockfall Model

The simulation time is set to 2.0 s, the time step to 1×10^{-7} s, the initial particle distance to 0.005 m, and the radius of influence to 2.1 referring to [4]. The rock is created using free particles (the small one is 64 particles, and the big one is 216 particles) as shown in Fig. 4, whereas the slope particles are created using free particles just in the upper first layer and wall particles in the second and succeeding layers. The rock is modeled by a cube made of aluminum. In this study, the rock's Poisson's ratio is adjusted to 0.00 in the analysis in order to maintain the cubic shape. According to the density and rock size of the centrifuge experiment, there are 4 series in simulation, and each series is analyzed by changing Young's modulus in 6 cases as shown in Table 5, a total of 24 cases of analysis results. (1200-1 ~ 6 with the small rock, 1200-1 ~ 6 with the big rock, 2620-1 ~ 6 with the small rock, and 2620-1 ~ 6 with the big rock).

The simulation conditions are shown in Table 4.

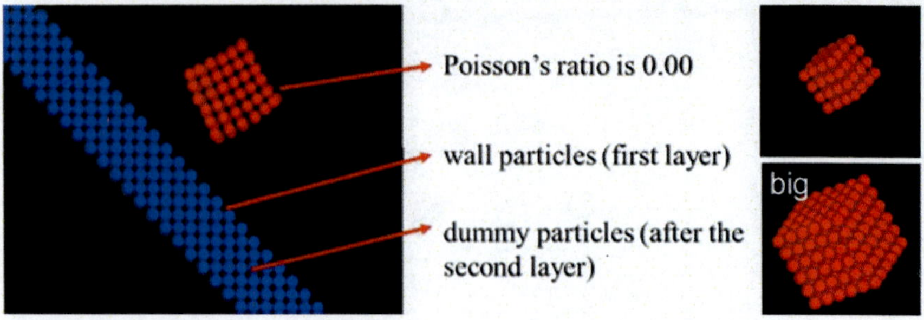

Fig. 4 Rockfall model in the MPS method

Table 4 Simulation condition

Simulation time (s)	2.0
Time increment (s)	1.0×10^{-7}
Initial distance (m)	0.005
Rock (particle type)	Free
Density of rock (kg/m^3)	2700
Young' s modulus of rock (Pa)	6.83×10^{10}
Poisson' s ratio of rock	0.00
Slope (particle type)	Wall, dummy
Density of slope (kg/m^3)	1200 or 2620
Young' s modulus of slope (Pa)	see Table. 5
Poisson' s ratio of slope	0.33 or 0.35

Table 5 Each series in simulation

Case	Young's modulus of slope	Poisson's ratio of slope
1	9.04×10^{10}	0.35
2	3.20×10^{9}	
3	3.00×10^{7}	
4	3.50×10^{6}	0.33
5	3.99×10^{4}	
6	3.99×10^{4}	

3.3 *The Results of Simulation*

3.3.1 Motion Process in the Hard Surface with a Small Rock

In this paper, only the motion of 1200-1 is shown (Fig. 5). It is observed that the rock collisions with the slope three times total in each case after the rock is dropped, at the top of the slope, and near the edge of the slope, respectively. Despite the number of collisions is the same, the behavior after the third collision in 1200-1 and 2 shows a jump with rotational motion and horizontal translational motion, whereas, after the third collision in 1200-3 to 6, there is no jump with rotational motion and horizontal translational motion.

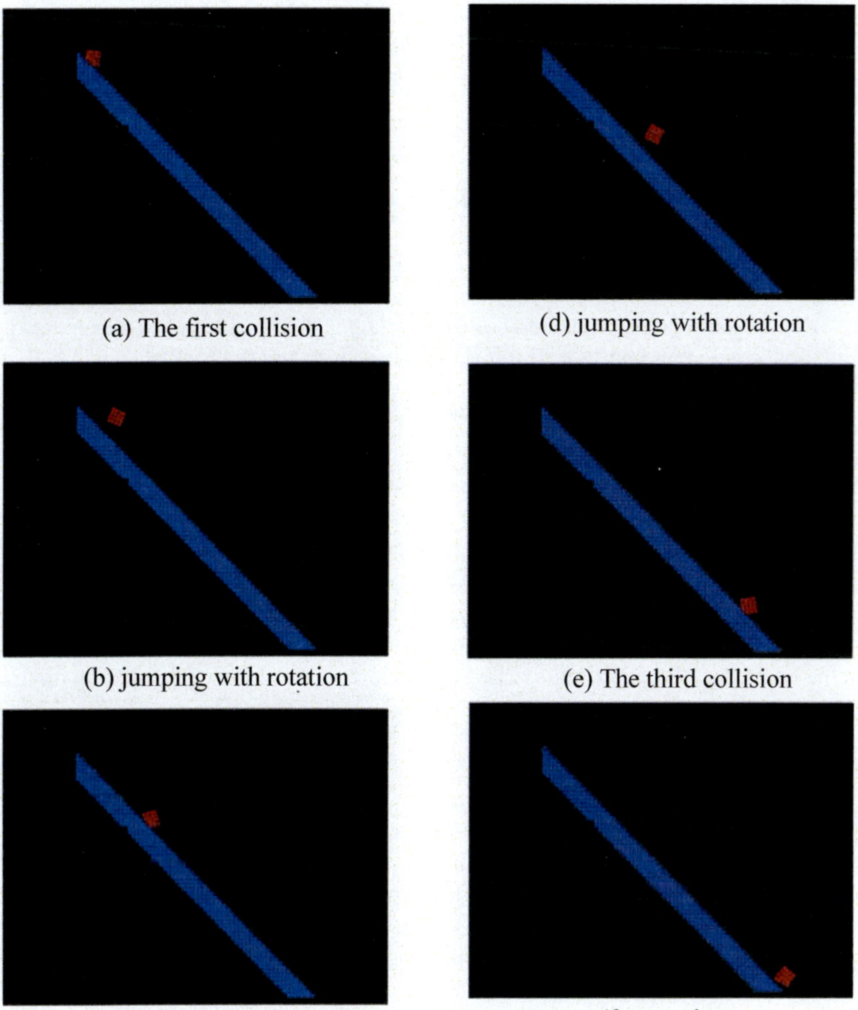

(a) The first collision

(b) jumping with rotation

(c) The second collision

(d) jumping with rotation

(e) The third collision

(f) over time

Fig. 5 Visualization of analysis results of case 1200-1 with a small rock

3.3.2 Motion Process in the Hard Surface with a Big Rock

The number of collisions is observed three times in 2620-1 and 2 and two times in 2620-3 to 2620-6. The reason is that the rock in cases 7 and 8 jumps with rotational motion and horizontal translational motion after the second collision, whereas in cases 2620-3 to 6 do not jump after the second collision and pass the edge of the slope while rotating. The jump by rotational motion and horizontal translational motion is observed in 2620-1 after the third collision, but not in 2620-2 after the third collision.

3.3.3 Motion Process in the Soft Surface with a Small Rock

The number of collisions is twice in 1200-1and three times from 1200-2 to 6. The reason for the difference in the number of collisions is the same as 2620-3 to 6. In 1200-1, the jump after the second collision is larger than in 1200-2 to 6 because of rotational motion and horizontal translational motion, and the rock passes the edge of the slope without colliding with it.

3.3.4 Motion Process in the Soft Surface with a Big Rock

There are two collisions between the rock and the slope in 2620-1, 2620-3 to 6, and three collisions in 2620-2, which is the only one that exhibited a variation in the number of collisions. The reason is the same as 1200. In 2620-3 to 6, after the second collision, the rocks do not show any jumping by rotational and translational motion in the horizontal direction and reached the bottom of the slope while rotating on the slope.

3.3.5 The Max. Jumping

In this study, the max. jumping is defined as the vertical distance from the center particle of the rock to the slope. Only 1200-3 in the second collision in Fig. 6a shows an abrupt drop, the rest are consistent with the pattern: As the Young' s modulus of slope surface grows, while the max. jumping shows a decreasing trend (see Fig. 6). The result is that when the rock collides with a soft slope surface, the surface absorbs the energy from the rock due to the large deformation of the magnitude.

3.3.6 The Rate of Decrease of the Velocity

In cases with large Young' s modulus such as case1 and case 2, the rock' s rate of decrease of the velocity before and after collisions becomes smaller or stays constant with the falling; the other hand, the cases with larger Young' s modulus are almost

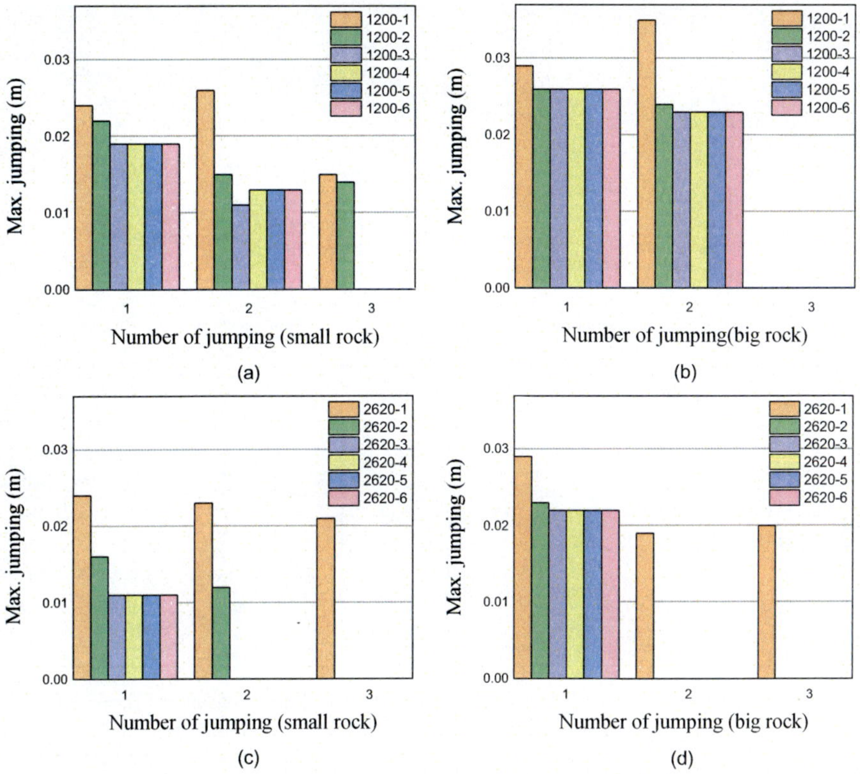

Fig. 6 The max. jumping

close to 100%, as shown in Fig. 7. This indicates that the slope with a large Young's modulus, that is, a soft slope, absorbs the energy of the rock and weakens the ability to jump

4 Conclusions

(a) From the results of the rockfall centrifuge experiments and numerical analysis, it is found that the smaller Young's modulus, which means the slope surface layer is more soft, the smaller the velocity and kinetic energy of the falling rock when it reaches the edge of the slope due to more losses from collisions.

(b) From the results of the visualization using the elastic model of the MPS method, it is confirmed that in the case where Young's modulus of the slope is large, there is a transformation from rolling to jumping, while in the case where Young's modulus is small, rocks are more likely to roll and slide.

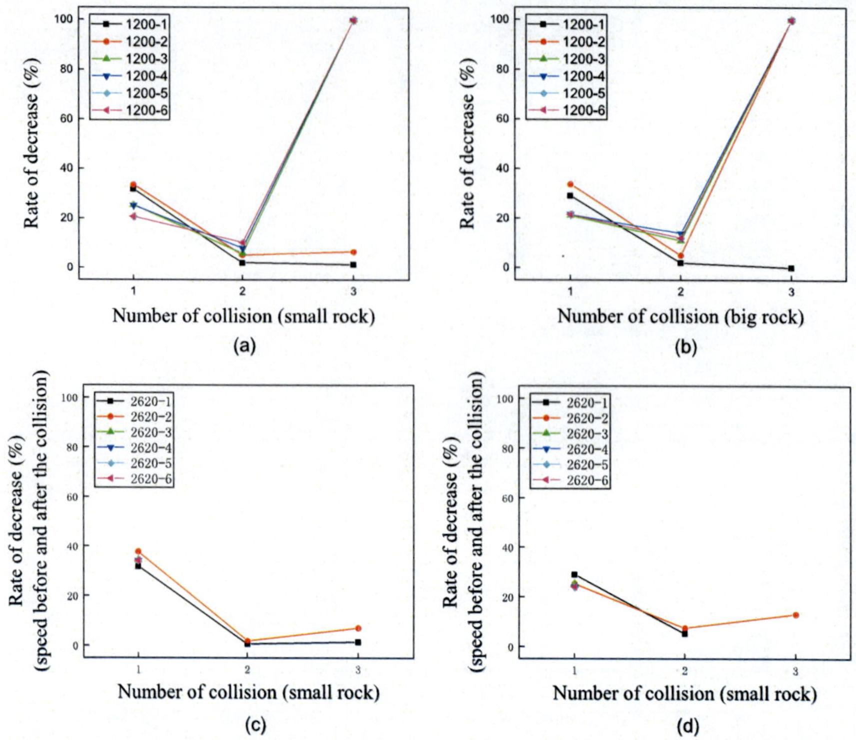

Fig. 7 Rate of decrease of the velocity before and after collision

References

1. Renka Kanno, F.: A basic study of optimal design method for rockfall countermeasures based on discrete element analysis. Proc. JSCE **73**(2), I-469–476 (2017)
2. Kazuya Itoh, F.: Centrifugal modelling of rockfall simulation associated with rock fracturing. Research reports of the Research Institute of Industrial Safety, pp. 17–26 (2004)
3. MooSeop Song, F.: Dynamic analysis of elastic solids by MPS method. Trans. Jpn. Soc. Mech. Eng. (A) **71**(701), 16–22 (2005)
4. Seiichi Koshizuka, F.: Ryushihou Nyumon. Maruzen Publishing Co., Ltd. [published in Japan] (2014)

Bearing Capacity of Deep Foundation Socketed into Granitic Rock

Yong Ping Oh and Mohd Ashraf Mohamad Ismail

Abstract With advancements in drilling techniques and improved coring tools, deep foundations socketed into granitic rock have been widely used to support the heavy loads imposed by the high-rise buildings and bridges. To optimize the pile design, designers are often required to estimate the pile's bearing capacity and subsequently validate the design parameters with pile load tests. Most of the time, the pile's bearing capacity is calculated based on empirical models suggested by various researchers around the world. However, these equations may not be suitable, as the data used to derive them were based on the pile load tests carried out at various location, likely with different construction methodologies. This paper aims to develop an empirical model to estimate the pile shaft resistance in granitic rock based on uniaxial compressive strength (UCS) by using data from twelve pile load test results performed in Southeast Asia and three pile load test results carried out in East Asia. Thirteen of the piles tested by using bi-directional static load test (BDSLT) method and remaining tested by conventional top-down static load test method (SLT). Comparison between the actual test data and estimated capacity by using existing empirical models is also presented. This paper also further demonstrates the maximum displacement of pile shaft segment in rock upon reaching failure. For pile shaft resistance not reaching its ultimate value, a suitable method is applied to predict the ultimate shaft resistance values. End bearing capacity obtained from the pile tests is also presented herein as reference.

Keywords Pile bearing capacity · Bi-directional static load test · Granitic rock · Uniaxial compressive strength · Sustainable construction

Y. P. Oh · M. A. M. Ismail (✉)
School of Civil Engineering, Universiti Sains Malaysia Engineering Campus, Nibong Tebal, 14300 Seberang Perai Selatan, Penang, Malaysia
e-mail: ceashraf@usm.my

Y. P. Oh
e-mail: jason@arges.com.sg

N. Sabtu (ed.), *Proceedings of AWAM International Conference on Civil Engineering 2022 - Volume 3*, Lecture Notes in Civil Engineering 386,
https://doi.org/10.1007/978-981-99-6026-2_5

1 Introduction

A deep foundation's bearing capacity is mainly composed of two components, namely the shaft friction and the end bearing. The capacity of these two components is affected by several factors such as type of geological formation, construction practices, and pile material. For pile socketed into rock, factors such as interface roughness and uniaxial compressive strength (UCS) affect the pile bearing capacity.

Since 1970s, researchers have proposed empirical models to estimate the pile shaft resistance from the uniaxial compressive strength (q_u) of rock. These empirical models were established based on pile load test data conducted at various location around the world with different construction methodologies. Hence, it is important to have locally established empirical models for pile designers to estimate the rock shaft resistance with a good degree of confidence. Two types of rocks commonly encountered are igneous rocks and sedimentary rocks. The weathering classification is specified in Annex B of BS EN ISO 14689 Approach 2.

Pile shaft friction capacity develops through shearing of the bond between concrete and the rock or sliding friction or combination of both. Nam and Vipulanandan [1] conducted a study on the excavation roughness produced by different type of drilling tools. Seidel and Collingwood [2] also suggested socket roughness factor for prediction of rock socket shaft resistance. However, it is still not common to quantify the surface roughness as the instrument required for this work is not commercially available.

Most of the pile socketed into rock are constructed by using bored piling machine with its dedicated rock coring tools such as rock auger, core barrel with bullet teeth, and core barrel with roller bits. The bullet teeth or roller bit is fixed on the core barrel with certain angle typically between 1 to 7°, which is adjusted based on the hardness of the rock during trial bore on site. Another method of rock coring is by using reverse circulation drilling (RCD), where the drilled material is returned to the surface by flushing the medium inside the drill pipes. Case 13–15 in this study was constructed by using reverse circulation drilling method in East Asia. While the rest was constructed by using normal bored piling machine with core barrel-roller bit in Southeast Asia. Corel barrel with roller bit and rock core sample is shown in Fig. 1.

Polymer and bentonite are the two types of stabilizing fluids widely used for the pile excavation. As the bentonite filter cake reduced the shear strength of the pile soil interface over time, most of the piles constructed in recent years are utilizing polymer as the support fluid as it does not have a detrimental effect on the interface shear resistance [3]. Cleaning of the pile toe by de-sanding method or air-lift method is crucial step to ensure the performance of the pile especially the pile is heavily relied on end bearing, particularly for large diameter pile. Inspection of the toe cleanliness shall also be carried out, be it using conventional "sounding" or "dipping" method or quantitative method like sediment probe with depth encoder to record the sediment thickness prior to the concreting works.

Fig. 1 Core barrel-roller bit and rock core sample by using core barrel-roller bit

2 Pile Test Method

Several test methods are available to test the pile socketed into granitic rock. Over 85% of the test pile results presented in this paper are tested with bi-directional static load test (BDSLT) or better known as "Osterberg Cell test (O-cell)" and two cases tested by conventional top-down static load test (SLT). One of the benefits of BDSLT is the sacrificial jack-like device can be pre-installed on the reinforcement cage and positioned near to the tip of the pile. As the jack expands, the end bearing provides reaction for the side shear and vice versa [4]. This method enables engineer to test the pile in full scale, loaded the pile in compression from the bottom until one of the two components (shaft friction or end bearing) reaches its ultimate capacity or until the maximum capacity of the jack installed in the pile. One of the benefits for pile tested with BDSLT method is the sacrificial jacks can be positioned within the rock socket or near to the toe of the excavation and test the rock socket capacity directly.

All the test piles are instrumented with telltale extensometers to measure the pile compression and pile displacement. Load distribution along the pile is estimated by attaching the vibrating wire strain gauges in pairs onto the reinforcement cages (Refer to Table 1 for the test pile summary).

Table 1 Test pile summary

Test ID	Diameter (mm)	Length (m)	Socket Length (m)	Test Method
Case 1	1100	39.09	2.10	BDSLT
Case 2	1100	35.29	2.30	BDSLT
Case 3	800	32.47	4.52	BDSLT
Case 4	1200	41.12	3.05	BDSLT
Case 5	800	22.70	1.60	SLT
Case 6	800	45.00	1.70	BDSLT
Case 7	1000	32.30	5.00	BDSLT
Case 8	1500	28.90	2.50	SLT
Case 9	1000	23.86	3.80	BDSLT
Case 10	1200	49.86	4.80	BDSLT
Case 11	1000	21.00	3.30	BDSLT
Case 12	1000	17.56	3.50	BDSLT
Case 13	2350	51.58	14.58	BDSLT
Case 14	2350	58.35	23.35	BDSLT
Case 15	2350	54.20	14.80	BDSLT

3 Pile Test Result

3.1 Shaft Resistance Capacity

To establish a more reliable empirical relations, the ultimate value of the shaft friction shall be adopted. Test pile program shall be designed with additional test capacity to fully mobilize the pile. Figure 2 shows the top section of bi-directional static load test (BDSLT) fully mobilized. Maximum displacement of rock segment at failure for four cases in this study shown in the Fig. 3 below is approximately at 10–15 mm. Ayithi and Ryan [5] reported that average additional capacity beyond 10 mm displacement is less than 5% of the total capacity. This finding agrees well with data presented in this paper. Hence, shear displacement of 10 mm can be considered as upper bound for maximum shaft resistance.

Chin [6] suggested a hyperbolic curve fitting method (Fig. 4) to estimate the ultimate capacity of the shaft resistance based on actual test data. This method usually fits well with actual test data, where plot of the settlement versus ratio of settlement and resistance gives a linear relationship. Therefore, for the cases where pile displacement is less than 10 mm, ultimate value is obtained through Chin's method.

Mobilized unit shaft resistance of moderately weathered granitic rock is presented in Table 2. The ultimate values for case 4 and case 10 are estimated based on Chin's method with the maximum displacement limited to 10 mm.

Fig. 2 Top section fully mobilized (Case 9)

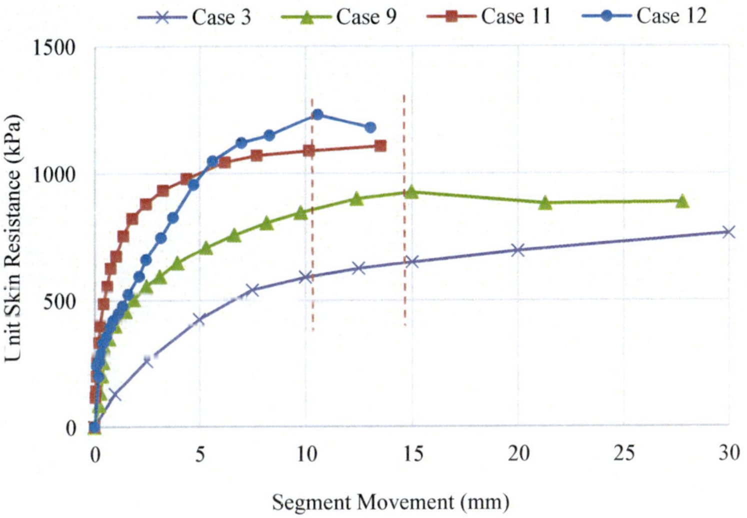

Fig. 3 Unit shaft friction versus segment movement

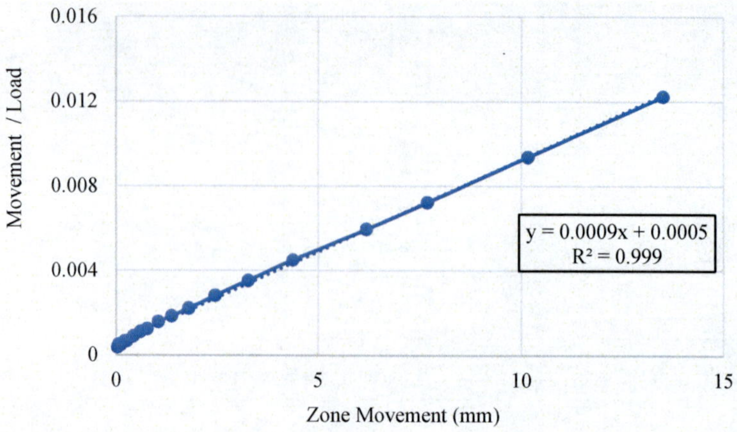

Fig. 4 Hyperbolic curve fit–Chin's method–Case 11

Table 2 Mobilized unit shaft resistance

Test ID	UCS	Unit shaft resistance	Displacement (mm)	Ultimate value
Case 1	46.4	839	> 10	839
Case 2	90.7	823	> 10	823
Case 3	93.8	712	> 10	712
Case 4	16.1	1255	4.8	1587
Case 5	88.0	1068	> 10	934
Case 6	35.5	653	> 10	653
Case 7	112.9	990	> 10	990
Case 8	129.7	932	> 10	932
Case 9	69.3	925	> 10	925
Case 10	95.9	1091	7.2	1176
Case 11	35.6	1105	> 10	1105
Case 12	131.2	1230	> 10	1230
Case 13	34.8	1685	> 10	1685
Case 14	15.7	1484	> 10	1484
Case 15	18.5	1799	> 10	1799

3.2 End Bearing Capacity

The test pile unit end bearing capacity and its respective toe displacement show no clear correlation (see Table 3 and Fig. 5). As discussed by Fellenius [7], the end bearing resistance does not exhibit an ultimate value. Larger displacement could be due to the combination of the pile material strength and cleanliness of the pile base.

Table 3 Mobilized unit end bearing

Test ID	UCS	Unit end bearing	Displacement (mm)
Case 1	46.4	13,519	4.1
Case 2	90.7	13,030	5.9
Case 3	93.8	13,455	2.8
Case 4	16.1	13,692	18.6
Case 5	88.0	12,214	18.0
Case 6	35.5	8814	17.0
Case 7	112.9	10,501	41.5
Case 8	129.7	8217	37.2
Case 9	69.3	12,244	25.7
Case 10	95.9	12,141	18.0
Case 11	35.6	11,502	2.0
Case 12	131.2	10,828	3.8
Case 13	34.8	20,477	30.5
Case 14	15.7	16,877	108.4
Case 15	18.5	25,556	80.4

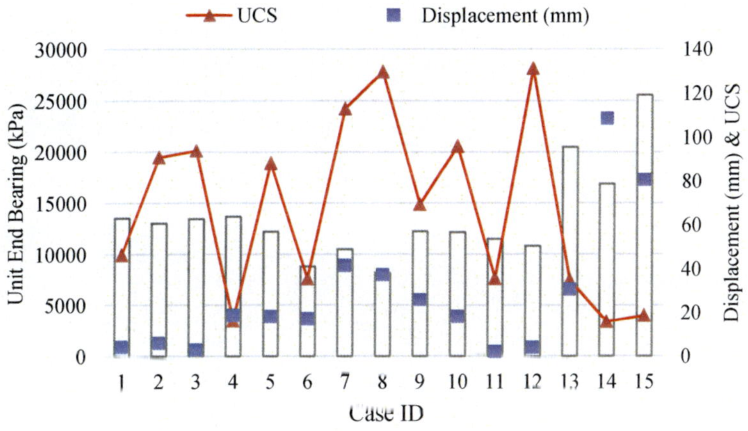

Fig. 5 Mobilized unit end bearing

4 Review on Empirical Models for Shaft Resistance

Several empirical models were proposed to estimate the rock shaft resistance based on the uniaxial compressive strength (UCS) value. All these models were based on Eq. 1.

$$\frac{f_s}{P_a} = B\left(\frac{q_u}{P_a}\right)^n \tag{1}$$

where fs is shaft resistance, q_u is uniaxial compressive strength, P_a is atmospheric pressure (101.325 kPa), and B and n is the empirical factor derived from load test data. Summary of the selected empirical models is stipulated in Table 4.

Fifteen pile test data result was compared with the empirical models suggested by various researchers. It is clearly seen that the empirical models predicted shaft resistance is deviated away from the actual test data from test pile drilled with core barrel-roller bit. Three of the test pile result with its rock socket cored by using reverse circulation drilling method (RCD) is higher than empirical models suggested by Horvath and Zhang (see Fig. 6).

Based on the actual test data collected, an empirical model is proposed to suit the local soil condition as well as the construction practices (see Fig. 7). The assessment

Table 4 Summary of selected empirical models

Empirical model	B	n	Remarks
Rosenberg and Journeaux [8]	1.09	0.52	Based on 6 data points
Horvath and Kenney [9]	0.65	0.5	Diameter > 400 mm
Rowe and Armitage [10]	1.42	0.5	Test did not reach failure
Zhang and Einstein [11]	0.63	0.5	Smooth socket
Kulhaway et al. [12]	1	0.5	Lower bound, $B = 0.63$

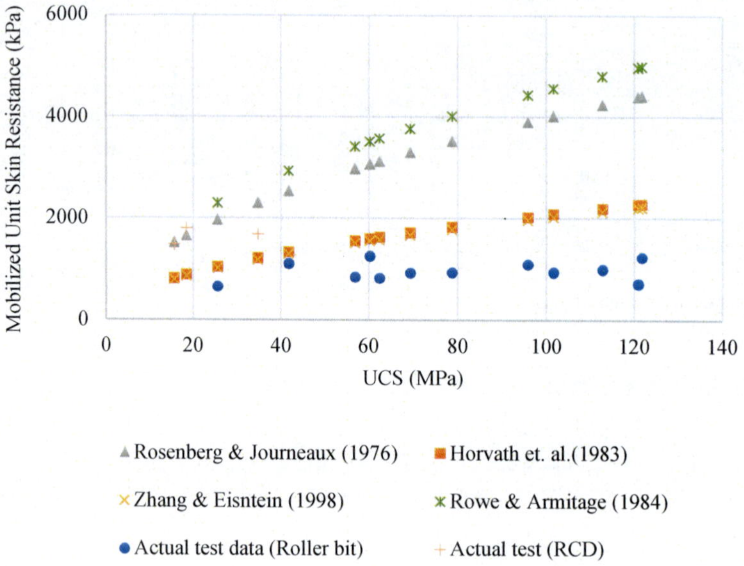

Fig. 6 Comparison of empirical models with actual test data

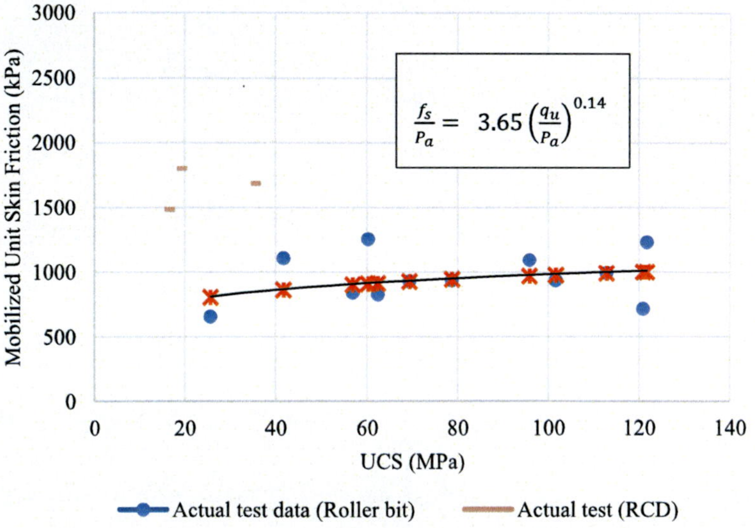

Fig. 7 Suggested empirical model based on actual test data

suggested the following:

$$\frac{f_s}{P_a} = 3.65\left(\frac{q_u}{P_a}\right)^{0.14} \tag{2}$$

Due to limited data for test pile with rock socket drilled by using reverse circulation drilling method, the three test piles data point are not included in the assessment. However, it shall be noted that the shaft resistance capacity is much higher than pile constructed with core barrel-roller bit (see Fig. 7).

5 Summary and Conclusions

A total of fifteen pile load test data was compiled, and a comparison was made against the empirical models. The data show that the existing available empirical models are not suitable for estimating the rock shaft friction in Southeast Asia region. Actual pile test data show that there is no significant increase in pile shaft resistance with the increase in UCS value. Hence, an empirical model is proposed specifically to estimate the pile shaft resistance in granitic rock socket and cored by core barrel with roller bits.

Bored pile socketed into rock can be constructed with different types of drilling tools and stabilizing fluids. The empirical equation shall be used with caution, and the adopted design parameters for construction shall be proven by full-scale pile load test.

References

1. Nam, M.S., Vipulanandan, C.: Roughness and unit side resistance of drilled shafts socketed in clay shale and limestone. J. Geotech. Environ. Eng. 1272–1279 (2008)
2. Seidel, J.P., Collingwood, B.: A new socket roughness factor for prediction of rock socket shaft resistance. Can. Geotech. J. **38**(1), 138–153 (2001)
3. Lam, C., Jefferis, S.A., Suckling, T.P., Troughton, V.M.: Effects of polymer and bentonite support fluids on the performance of bored piles. Soils Found. **55**(6), 1487–1500 (2015)
4. Schmertmann, J.H., Hayes, J.A.: The Osterberg cell and bored pile testing-a symbiosis. In: Proceedings of Third International Geotechnical Engineering Conference, pp. 139–166. Cairo University, Cairo Egypt (1997)
5. Ayithi, A., Ryan, W.G.: Unit side shear in rock-socketed bored piles. In: 10th International Conference on Stress Wave Theory and Testing Methods for Deep Foundations, pp. 97–127 (2019)
6. Chin, F.K.: Estimation of the ultimate load of piles from tests not carried to failure. In: Proceedings of the 2nd Southeast Asian Conference on Soil Engineering, pp. 81–92. University of Singapore, Singapore (1970)
7. Fellenius, B.H.: Capacity versus deformation analysis for design of footings and piled foundations. Geotech. Eng. J. **41**(2), 70–77 (2011)
8. Rosenberge, P., Journeaux, N.L.: Friction and end bearing tests on bedrock for high capacity socket design. Can. Geotech. J. **12**(3), 324–333 (1976)
9. Horvath, R.G., Kenny, T.C.: Shaft Resistance of Rock Socketed Drilled Piers, Symposium on Deep Foundations, pp. 182–214. ASCE, New York (1979)
10. Rowe, P.K., Armitage, H.H.: A design method for drilled piers in weak rock. Can. Geotech. J. **24**(1), 126–142 (1987)
11. Zhang, L., Einstein H.H.: Closure to "End bearing capacity of drilled shafts in rock". Geotech. Eng. J. **125**(12), 1109–1110 (1999)
12. Kulhaway, F.H., Prakoso, W.A., Akbas, S.O.: Evaluation of Capacity of Rock Foundation Sockets, pp. 1–8. American Rock Mechanics Association, Alexandria, VA (2015)
13. Lam, C., Jefferis, S.A., Suckling, T.P.: Construction techniques for bored piling in sand using polymer fluids. Proc. Inst. Civil Eng. Geotech. Eng. **167**(GE6), 565–573 (2014)
14. Zhang, L., Einstein, H.H.: End bearing capacity of drilled shaft in rock. Geotech. Eng. J. **124**(7), 574–584 (1998)
15. Leung, C.F.: Case studies of rock-socketed piles. Geotech. Eng. J. **27**(1), 51–67 (1996)
16. Rezazadeh, S., Eslami, A.: Empirical methods for determining shaft bearing capacity of semi-deep foundations socketed in rocks. J. Rock Mech. Geotech. Eng. 1140–1151 (2017)

Ground Improvement Techniques for Soft Soil

S. Sharmeelee and F. Ahmad

Abstract A large number of ground improvement methods are available that can be adopted to treat poor soil site conditions. Several of these methods have been in use for many decades, while others have been recently developed. The selection of the most appropriate ground improvement method can be a complex process that depends upon the consideration of available techniques and a number of performance-specific and site-specific factors. Ground improvement using prefabricated vertical drain (PVD) with surcharge loading is one of the effective systems to accelerate the removal of excess pore water pressure in saturated soft clay prone to excessive settlement. The main ground improvement techniques and the key components of surcharge preloading with PVD are discussed.

Keywords Ground improvement · Surcharge preload · Prefabricated vertical drain

1 Introduction

The need for ground improvement has been driven by the increasing need to use less favored sites for construction and to minimize the risk of failure or possible poor performance post-construction. Engineers are presented with several alternatives, where unsuitable or marginal soil conditions are encountered including eliminating the site with poor soil through relocation of the project to a more suitable site, use of deep foundation, designing the planned structure to accommodate the poor/marginal soils or improving the existing soil either in place or by removal, treatment and replacement of the in situ soil or completely abandoning the project [1, 2]. With the availability of a wide range of ground improvement technologies, less favored sites and unsuitable in situ soil can be improved to meet the project specification requirements thereby making the improved site an economically viable solution.

S. Sharmeelee (✉) · F. Ahmad
School of Civil Engineering, Universiti Sains Malaysia, Penang, Malaysia
e-mail: sharmeelee@gcu.com.my

N. Sabtu (ed.), *Proceedings of AWAM International Conference on Civil Engineering 2022 - Volume 3*, Lecture Notes in Civil Engineering 386,
https://doi.org/10.1007/978-981-99-6026-2_6

Development on highly compressible soil is typically saddled with the common geotechnical problems such as low bearing capacity, high settlement and longer treatment time. Designing foundations of structures on these areas sometimes involves soil improvement techniques as useful alternative that would limit settlements within criteria and at the same time allow for reasonable duration of construction to be completed within schedule. Geotechnical problems and failures may also be minimized or prevented if appropriate ground improvement is designed and implemented in the field with proper quality control/assurance measures.

Sites comprising soft soil with poor engineering properties may be improved with a variety of ground improvement techniques. Some of these techniques have been practiced for many years, with aspects such as design and construction being well established, while other techniques are comparatively new and are continually being developed and improved. The ground can be treated by several techniques which includes surcharge preloading with/without PVD, deep soil mixing (DSM), vacuum preloading, stone columns, compaction grouting and several other methods. The type, function and method of ground improvement methods which has been extracted from Mitchell [2], ASCE [1, 3], Munfakh and Wyllie [4], Terashi and Juran [5], Elias et al. [6] and Chu et al. [7] are given in Table 1.

2 Ground Improvement Techniques

In the present time, various treatment methods are available that may be adopted to improve the poor ground condition. For the selection of the most appropriate ground improvement method to treat soft soil, a complete assessment of the in situ site condition should be carried out. The selection of an appropriate method or the most suitable method for a specific project can only be made after the consideration of the performance and site-specific factors. The most important factors were originally listed by Mitchell [2] and subsequently improved by Holtz [8] and by Munfakh [9]:

1. The operational criteria for the development, e.g., stability requirements, allowable total or differential settlement, rate of settlement, seepage criteria, durability and maintenance requirements, etc.
2. The extent of the site to be treated, i.e., area, depth, volume, etc.
3. The initial soil properties.
4. Depth to groundwater table.
5. Availability of materials, e.g., sand, gravel, water, admixtures, reinforcing elements.
6. Availability of specialized equipment and skilled labor force.
7. Construction and environmental factors, e.g., site accessibility and constraints, waste disposal, erosion, water pollution and effects on adjacent facilities and structures.
8. Local experience and preferences.
9. Time available.

Table 1 Ground improvement methods

Category	Function	Methods
Densification	Increase density, bearing capacity and frictional strength; decrease compressibility, increase strength of cohesive soils	Vibro compaction Dynamic compaction Compaction grouting
Consolidation	Accelerate consolidation, reduce settlement, increase strength	Surcharge preload with/without drains Vacuum consolidation electro-osmosis
Load reduction	Reduce load on foundation soils, reduce settlement, increase slope stability	Geofoam Foamed concrete Lightweight fill
Reinforcement	Inclusion of reinforcing elements in soil to improve engineering characteristics; provide lateral stability	Mechanically stabilized earth Soil nail/anchor Piled embankment Stone column Sand compaction piles Jet grouting
Chemical treatment	Increase density, increase compressive and tensile strength, fill voids, form seepage cutoffs	Permeation grouting Jet grouting Compaction grouting Deep soil mix
Thermal stabilization	Increase shear strength, provide cutoff	Ground freezing/heating Vegetation on slopes
Miscellaneous	Remediate contaminated soils	Electrokinetic methods Chemical methods

10. Cost.

The various ground improvement techniques that are available include the following.

2.1 Soil Removal and Replacement (R&R)

Where the unsuitable or weak soil is found at shallow depths, R&R treatment can be adopted. The very soft compressible cohesive soils can be excavated and replaced with suitable compacted material such as compacted sand or suitable fill that provide a competent and less compressible foundation. This also serves to minimize excessive settlement and also to strengthen the soil bearing resistance on the fill. However, this method is commonly limited to depths not exceeding 3.0 m depth of unsuitable soil from the existing ground level. Subsequently, the replaced suitable soil shall be well-compacted in layers and followed by field density test for the verification of the quality of the compaction work.

2.2 Deep Soil Mixing (DSM)

Deep soil mixing is a ground improvement method which involves in situ mechanical mixing of the weak soil with a cementitious binder. This action of mixing materials such as cement, bentonite, etc., with the weak or unsuitable soil leads to an improved soil which appears more like a hard soil. The improved soil generally has improved strength, reduced permeability and compressibility in comparison to the original in situ soil.

2.3 Stone Columns

Vibration and compaction of loose granular soil results in its densification as the soil particles rearrange themselves into a closer-packed configuration with a lower void ratio. For fine, cohesive soils, the effect of vibration and compaction is low due to its considerably lower permeability. Therefore, alternative techniques including the installation of stone columns or piled embankments are commonly adopted to improve the engineering properties of cohesive soils. Stone columns installed in cohesive soils act to create a composite material with lower compressibility, higher strength and higher permeability than the in situ soil.

Stone columns can be founded on a stiff stratum or within the weak material as floating columns. Though the column length is limited by the installation length, columns that extend to the stiff layer are generally favored because of the better load-settlement behavior in comparison with the floating columns.

2.4 Deep Compaction

Compaction is a method of treatment adopted to densify the soil with the aim of improving its engineering properties. The densification of the soil will lead to a reduction in the void ratio of the soil with a corresponding increase in its stiffness and strength, thereby leading to an increase in the bearing capacity and stability while reducing the total and differential settlement of the ground under the imposed loading. This treatment is generally carried out by using mechanical compaction.

2.5 Vacuum Consolidation

In this method, vacuum pressure is applied to the ground. Prior to this, the ground has to be protected from pressure losses at the ground surface by installing a sealing membrane. A negative pressure is created in the permeable soil which is sealed beneath the membrane and along the PVD installed within the ground under treatment. In practice, vacuum pressures as high as 90 kPa may be achieved however typically a value of 80 kPa is considered in design [7].

PVD facilitates the distribution of balanced vacuum pressure along the treated soil depth while also discharging the extracted pore water up to the permeable soil layer at the ground level [7]. Flexible PVDs have generally replaced sand drains and granular piles predominantly due to its cheaper cost [10]. Static methods of PVD installation are generally preferable since they produce less disturbance to the surrounding soil though dynamic methods of PVD installation such as vibrating or drop hammer techniques are sometimes used [11]. Indraratna et al. [12] noted that band PVDs and circular PVDs perform similarly in areas, where only the embankment load is placed. In areas where vacuum preload is applied, circular PVDs were observed to produce faster consolidation rates, likely due to their ability to transfer vacuum pressures to great depths more effectively.

2.6 Prefabricated Vertical Drains (PVDs) with Preloading

Preloading is a successful ground improvement technique which involves the loading of the ground surface to induce the maximum settlement. Usually, a surcharge load equal to or greater than the expected load is imposed on the soft ground to accelerate consolidation. PVD is often used in conjunction with surcharge preload to minimize the anticipated post-construction settlement caused by primary consolidation due to the embankment fill and imposed foundation load. PVD is generally used to decrease the overall time required for the completion of primary consolidation settlement by shortening the drainage path length. PVD has largely replaced other drainage techniques Previously, sand drains were used for acceleration of consolidation of soft soils. With the increasing popularity of polymer-based materials and the change in requirements from construction, synthetic or polymer-based PVDs have replaced conventional drains due to several reasons namely its economic competitiveness, less disturbance to the soil, speed and simplicity of installation [13].

The PVD consists of a synthetic drainage element composed of a central drainage core made up of a high mechanical strength polypropylene wrapped in an outer non-woven geotextile filter. It is widely used for soft ground improvement due to its excellent physical, mechanical and hydraulic properties.

By using PVD in combination with surcharge preload, pore water pressure is generated during the consolidation process and this pore water is able to flow faster horizontally. Subsequently, it flows into the vertical drain and dissipate the water into the permeable drainage layer. Therefore, the PVD shortens the drainage path as well as reduces the time of consolidation resulting in the rapid gain in strength of the weak clay soil [14–18].

PVD is the most suitable to be installed in clay, silt, organic layers, clayey and silty sand. The installation of PVDs is achieved by pushing a steel mandrel into the ground to the pre-determined depth with a mast mounted on a back hoe. As the mandrel is removed, an anchor plate is attached to the wick material to hold it in place. Subsequently, the PVD is cut off just above the ground level. As the distance between the drains becomes closer, the surcharge time decreases.

Factors Influencing the Vertical Drain Efficiency

The performance of the PVD installed in soft soil is affected by several factors. One of the most important subsoil parameters that affects the rate of consolidation is the horizontal coefficient of consolidation, ch. Miura and Chai [19] mentioned that there is no satisfactory laboratory test method to determine the horizontal coefficient of consolidation, ch and therefore back analysis from field measurement is recommended to determine this parameter. Apart from that, other factors such as drain spacing, drain diameter, smear effect and well resistance also influence the vertical drain efficiency indirectly.

Well Resistance

In the event that the discharge capacity of the PVD is reached during the consolidation process of the soft clay, and the overall consolidation process will be delayed. The well resistance is the measure of the PVDs resistance to pore water flow into them or the well resistance is defined as the resistance of water flow in the vertical drain. As the drain length increases, the well resistance of vertical drain increases and retard the water flow in drainage channel causing a reduction in the consolidation rate of soft clay layer. In general, the well resistance is controlled by drain's discharge capacity, the soil's permeability and the maximum discharge length. The well resistance factor is usually used as measure to check for the discharge capacity of PVD [20]. It was highlighted by Holtz et al. [21] that a large vertical settlement may cause the PVD to fold thereby reducing its discharge capacity. However, based on the study by Mesri and Lo [22], if the available well resistance is more than 5, the well resistance will be insignificant therefore, the minimum required discharge capacity for a PVD should be 5.

Discharge Capacity

An appropriate discharge capacity of the PVD is required to analyze the well resistance factor. Indraratna and Redana [23] stated that the well resistance is always less significant than the smear effect. In order to measure the discharge capacity of the drain, it is important that the field condition be duplicated as closely as possible.

Miura and Chai [19] carried out a systematic laboratory test program to identify the factors influencing the discharge capacity. From their research, the factors contributing to the discharge capacity were determined as follows:

a. Lateral earth pressure: As lateral pressure increases, the reduction of cross-sectional area available for flow resulting in a lower discharge capacity.
b. Settlements: The drains tend to settle together with the ground during consolidation process, causing bending, folding or twisting of the drain.
c. Clogging: The effect of possible air bubbles trapped in the drainage path or the infiltration of soil slows down the pore pressure dissipation.

d. Time: Degradation in the drain material will possibly occur due to aging and possibly due to biological and chemical activities.

Indraratna and Redana [17] mentioned that Holtz et al. [8] reached a conclusion where if the initial discharge capacity is more than 100–150 m^3/year, the consolidation rates should not be heavily influenced by some reduction in the discharge capacity.

Smear Effect

Smear zones are developed during the installation of vertical drains, where the surrounding soil is compressed and remolded thereby causing a substantial reduction in the permeability and increasing the compressibility. The additional resistance which must be overcome by the excess water will retard the rate of consolidation. Therefore, the installation of vertical drains is usually carried out by the aid of a special mandrel designed so as to cause a minimum of disturbance on the surrounding soil.

Miura and Chai [19] found that the diameter of the smear zone (ds) and the hydraulic conductivity ratio (kh/ks) is necessary to classify the smear effect. The disturbed zone is typically related to the installation method, size and shape of the mandrel and the soil composition. While maintaining an adequate stiffness, a smaller cross section of mandrel is preferable to be used in order to minimize the smear effect. Barron [24] and Hansbo [18] modeled the smear zone by dividing the soil cylinder dewatered by the central drain into two zones. The smear zone is classified as the disturbed zone in the immediate vicinity of the drain and the undisturbed zone outside of this region.

Furthermore, the extent of the smear zone around a vertical drain was studied based on a large-scale consolidometer testing. Indraratna and Redana [23] have proved that the diameter of smear can be four to five times the diameter of central drain (mandrel), and the measured ratio of horizontal to vertical permeability approached unity at the drain soil interface. However, this will depend on the size of the particular drain used.

3 Conclusion

The selection of an appropriate method of ground improvement requires detailed knowledge of the available ground treatment methods and careful evaluation of critical factors especially if the design is optimized and the ground improvement is required to be tailored to meet specific project needs. The critical factors include the understanding of the functions of the proposed methods, the selection criteria based on the performance requirements, appropriate design procedures, implementation of the necessary measures for quality control and consideration of all relevant cost components and environmental factors.

Acknowledgements The authors would like to acknowledge USM Grant RUI/8014094 which has supported this research.

References

1. ASCE.: Soil Improvement-History, Capabilities, and Outlook. J.K Mitchell, Editor. American Society of Civil Engineers, New York, 182:6 (1978)
2. Mitchell, J.K.: Soil improvement-state of the art report. In Proc., 11th Int. Conf. on SMFE 4, 509–565) (1981)
3. Ground Improvement Ground Reinforcement Ground Treatment: Developments (1987–1997): Logan, Utah July 17-19, 1997 (Geotechnical Special Publication), ASCE
4. Munfakh, G.A., Wyllie, D.C.: Ground improvement engineering-issues and selection. ISRM. Int. Symp. Melbourne, Australia (2000)
5. Terashi, M., Juran, I.: Ground improvement-state of the art. ISRM. Int. Symp. ISRM (2000)
6. Elias, V., Welsh, J., Warren, J., Lukas, R., Collin, J.G., Berg, R.R.: Ground improvement methods, Volumes I and II, Publication No.'s FHWA NHI-06-019 and FHWA NHI-06-020, US Dept. of Transportation, Federal Highway Administration (2006)
7. Chu, J., Varaksin, S., Klotz, U., Menge, P.: State of the Art Report: Construction Processes. In: Proceedings 17th International Conference on Soil Mechanics and Geotechnical Engineering, Alexandria, Egypt, pp. 3006–3135 (2009)
8. Holtz, R.D.: "Treatment of Problem Foundations for Highway Embankments", Synthesis of Highway Practice 147, National Cooperative Highway Research Program, Transportation Research Board, 72 pp (1989)
9. Munfakh, G.A. Ground improvement engineering–the state of the US practice: part 1. Methods. Proceedings of the Institution of Civil Engineers-Ground Improvement, 1(4), 193–214 (1997)
10. Indraratna, B.: 2009 EH Davis Memorial Lecture: Recent advances in the application of vertical drains and vacuum preloading in soft soil stabilization. Aust. Geomech. J. AGS. **45**(2): 1–43 (2010)
11. Ghandeharioon, A., Indraratna, B., Rujikiatkamjorn, C.: Analysis of soil disturbance associated with mandrel-driven prefabricated vertical drains using an elliptical cavity expansion theory. Int. J. Geomech. **10**(2): 53–64 (2010)
12. Indraratna, B., Rujikiatkamjorn, C., Ameratunga, J., Boyle, P.: Performance and prediction of vacuum combined surcharge consolidation at Port of Brisbane. J. Geotech. Geoenviron. Eng. 137(11): 1009–1018 (2011)
13. Rixner, J.J., Kraemer, R.S., Smith, D.A.: Prefabricated Vertical Drains. Federal Highway Administration, Washington (I. D.C.Rep. No.FHWA/RD-86/168) (1986)
14. Bergado, T.D., Balasubramaniam, A.S., Fannin, J.R., Holtz, D.R.: Can. Geotech. J. **39**(2), 304–315 (2002)
15. Chai, J., Hong, Z., Shen, S.: Vacuum-drain consolidation induced pressure distribution and ground deformation. Geotext. Geomembr. **28**(6): 525–535 (2010)
16. Indraratna, B., Bamunawita, C., Khabbaz, H.: Numerical modeling of vacuum preloading and field applications. Can. Geotech. J. **41**(6): 1098–1110 (2004)
17. Indraratna, B., Redana, I.W.: Numerical modeling of vertical drains with smear and well resistance installed in soft clay. Can. Geotech. J. **37**(1): 132–145 (2000)
18. Hansbo, S.: Consolidation of fine grained soils by prefabricated drains. In: Proceedings of the 10th International Conference on Soil Mechanics and Foundation Engineering, vol. 3, Stockholm. pp. 677–682 (1981)
19. Miura, N., Chai, J.C.: Investigation of factors affecting vertical drain behavior. J. Geotech. Eng. ASCE. 125(1): 216–226 (1999)
20. Chen, C.S.: "Use of prefabricated vertical drain to expedite the consolidation settlement", Semin. Exhib. Building. Geosynthetics. **30** (2004)
21. Holtz, R.D., Jamiolkowski, M., Lancellotta, R. and Pedroni, R.: Prefabricated vertical drains: Design and performance, CIRIA ground engineering report, Butterworth-Heinemann Ltd., London (1991)
22. Mesri, G., Lo, D.O.K.: "Field performance of prefabricatedvertical drains." In: Proceeding International Conference on Geotechnical Engineering forCoastal Development: Theory and Practice on Soft Ground, CoastalDevelopment Institute of Technology, Tokyo, 231–236 (1991)

23. Indraratna, B., Redana, I.W.: Laboratory determination of smear zone due to verticaldrain installation. J. Geotech. Eng. ASCE. **124**(2), 180–184 (1998)
24. Barron, R.A.: Consolidation of fine grained soils by drain wells. Trans. ASCE **113**, 718–724 (1948)
25. Widodo, S., Ibrahim, A.: Estimation of primary compression index (Cc) using physical properties of Pontianak soft clay. Int. J. Eng. Res. Appl. (IJERA) **2**(5), 2232–2236 (2012)
26. Bhattacharya, A.K., Basack, A.: Geotechnical Conference, H-029 (2011) 2, vol. 2346, pp. 718–754. A.R. Barron, ASCE (1948)
27. Indraratna,B., Sathananthan, I., Rujikiatkamjorn, C., Balasubramaniam, S.A.: Analytical and Numerical Modeling of Soft SoilStabilized by Prefabricated Vertical DrainsIncorporating Vacuum reloading. Int. J. Geo. Mech. 10.1061 (2005)
28. Das, B.M.: Principles of Geotechnical Engineering (1995)
29. Terzaghi, K.: Theoretical Soil Mechanics. Willey, New York (1943)
30. Lau, K.W.K., Cowland, W.J.: Transport and Geoenvironment. Geotech. Spec. Publ. **103**, 140—161 (2000)
31. Kjellman, W.: Proceedings of 2nd ICSMFE, vol. 2, pp. 302–305 (1948)
32. Carrillo, N.: Simple two and three dimensional case in the theory of consolidation of soils. J. Math. Phys. **21**(1), 1–5 (1942)
33. Sukla, S.A., Kambekar, A.R.: Working of prefabricated vertical drain—a case study. IJIRSET **2003**(2) (2013)
34. Craig, R.F.: Craig's Soil Mechanics, pp. 227–274. Taylor & Francis (2004)
35. Hansbo, S.: Consolidation of clay by band-shaped prefabricated drains. Ground Eng. **12**(5), 16–25 (1979)
36. Holtz, R.D., Kovacs, W.D.: An Introduction to Geotechnical Engineering, pp. 283–368. Prentice-Hall, Inc. (1982)
37. Ng, K.S., Chew, Y.M., Lazim, N.I.A.: Prediction of consolidation characteristics from index properties. In: ICCEE 2018: International Conference of Civil and Environmental Engineering 2018 (2018)
38. Seah, T.H.: Design and construction of ground improvement works at Suvarnabhumi Airport. Geotech. Eng. (2006)
39. Skempton, A.W.: Notes on the compressibility of clays. Q. J. Geol. Soc. London **100**, 119–135 (1944)
40. Terzaghi, K., Peck, R.B.: Soil Mechanics in Engineering Practice, 2nd edn. Wiley, New York (1967)
41. Varaksin, S., Yee, K.: Challenges in ground improvement technologies for extreme conditions: concept and performance. In: Proceedings of 16th Southeast Asian Geotechnical Conference, Kuala Lumpur, pp. 101–115 (2007)
42. Vinod, P., Bindu, J.: Compression index of highly plastic clays-an empirical correlation. Indian Geotech. J. **40**(3), 174–180 (2010)
43. Whitlow, R.: Basic Soil Mechanics, pp. 402–450. Pearson Education Ltd. (2001)
44. Wroth, C.P., Wood, D.M.: The correlation of index properties with some basic engineering properties of soils. Can Geotech. J. **15**(2), 137–145 (1978)

Slope Stability of Sg Langat Under the Influence of Extreme Rainfall

Jeffery Nazrien Ng, Aizat Mohd Taib, Norinah Abd. Rahman, Wan Hanna Melini Wan Mohtar, Othman A. Karim, Muhamad Razuhanafi Mat Yazid, Safari Mat Desa, Suriyani Awang, and Mohd Syazwan Faisal Mohd

Abstract This study aims to determine the geotechnical characteristics and behavior of soils in Sg Langat with different angles of slopes, the effectiveness of groundwater situations, and the effect of heavy rainfall on slope failure through numerical modeling. Critical slope failures happen during or after rain. Landslides, a natural disaster associated with sliding, are one of the catastrophes that can lead to the loss of lives. Thus, a study and analysis about the effect of extreme rainwater on slope strength is done to estimate the stability of slopes according to the extreme rainfall. This study is substantially to probe the effect of rainfall infiltration on slope stability with a shear strength of unsaturated soil and to determine the factor of safety of the

J. N. Ng · A. Mohd Taib (✉) · N. Abd. Rahman · W. H. M. Wan Mohtar · O. A. Karim · M. R. Mat Yazid
Faculty of Engineering and Built Environment, Universiti Kebangsaan Malaysia, Bandar Baru Bangi, Malaysia
e-mail: amohdtaib@ukm.edu.my

N. Abd. Rahman
e-mail: norinah@ukm.edu.my

W. H. M. Wan Mohtar
e-mail: hanna@ukm.edu.my

O. A. Karim
e-mail: othman.karim@ukm.edu.my

M. R. Mat Yazid
e-mail: razuhanafi@ukm.edu.my

S. Mat Desa · S. Awang · M. S. F. Mohd
National Hydraulic Research Institute of Malaysia (NAHRIM), Ministry of Environment and Water, Seri Kembangan, Malaysia
e-mail: safari@nahrim.gov.my

S. Awang
e-mail: suriyani@nahrim.gov.my

M. S. F. Mohd
e-mail: syazwan@nahrim.gov.my

N. Sabtu (ed.), *Proceedings of AWAM International Conference on Civil Engineering 2022 - Volume 3*, Lecture Notes in Civil Engineering 386,
https://doi.org/10.1007/978-981-99-6026-2_7

unsaturated soil slope affected by rainfall intensity and different boundary conditions through numerical modeling. This analysis is performed by using GeoStudio software to imitate the soil characteristics and the behavior, with the given data of rainfall in a certain area. This review presents the numerical analysis of different slopes regarding former cases that are subordinated to extreme rainfall from different places. It is shown that Channel 2 has higher pore-water pressure than Channels 1 and 3, where phase 2 attained a higher pore-water pressure value of 78.08 kPa, whereas phase 1 reaches 67.95 kPa of pore-water pressure. In Channel 3, the factor of safety shown is the highest among all the channels which are 0.63, 0.88, and 1.11 for minimum, intermediate, and maximum values, respectively. The pros and cons of the numerical approaches are discussed and come with some general modeling recommendations.

Keywords Infiltration · Pore-water pressure · Rainfall · Factor of safety

1 Introduction

Industrialization, urbanization, and population growth have decreasingly led to ferocious land use. Numerous landslides frequently do during the construction stages of architecture and have caused significant damage to assets and people. According to the data of the Korea Meteorological Administration (2002) from 1993 to 2002, over 22.7% of the total death risk by natural hazards in Korea caused by slope failures including natural hazards and accidents during construction [1]. Regarding on this statistic result, slope failure has become one of the most serious natural hazards [2]. According to a previous occasion, landslides were happened at Kampung Batu 10, Jalan Cheras in Kajang, Selangor in May 25, 2017. It was reported that the incident happened after a heavy rainfall on the same day [3]. According to the Hulu Langat Drainage and Irrigation, S.V. Kalaiselvam Velayudan, the damage was expansive because of the high intensity of the rainfall and the slope protection done was eventually insufficient to withstand the landslides. This shows the influence of extreme rainfall that affects the slope strength that caused landslides and might be a trouble for the people living in the surrounding areas.

The probability of landslide to occur under pitch construction depends on numerous factors similar as geomorphology, soil type, foliage, and climate [4]. Other factors including non-homogeneous soil layers, pressure, cracks, dynamic lading or earth-shakes, and seepage inflow may also affect landslides [5]. Thus, the aim of this study is to determine the geotechnical characteristics and types of soils in Sg Langat with different slope angles, to study the effectiveness of groundwater situations on slope and to observe the effect of heavy rainfall on slope failure through numerical modeling.

2 Rainfall

Infiltration is defined as the inflow of water to the soil through the surface of soil. Heavy rainfall has caused a considerable amount of slope failures that eliminate lots lives of peoples and conduct significant losses on economics throughout the world. Extreme rainfall leads to slope failure are generally happened on unsaturated soils. Rainfall is known to be one of the main factors behind all the tragedies happened; nevertheless, the denseness of the soil with consociated soil properties has not admitted attention on the stability of slopes [6]. Due to infiltration, various types of slope stability resembling as erosion can contribute to changes, different mass flow movements depending on the slope of the anatomy, the characteristic soil curves and the shear strength involved.

Extrapolating the spatial and temporal of rainfall variability is important in gaining knowledge of water balance dynamics on various measurements for water resources management and planning [7]. Laboratory and in situ experiments had proven that a homogeneous slope under rainfall qualifications is vulnerable to bear against surface erosion or shallow landslides, whereas audits of landslides inspect that deep-seated failure are continually induced by rainfall in slopes with pre-existing crack or wimpish layer [8]. As an example, the landslides occur in South Auckland region of New Zealand exposed to severance stresses within cracks after rainfall [9]. Furthermore, the fluxes in groundwater hydrology could reduce the effective stress and the shear strength of soil that may result in rainfall-induced slope failures too [9, 10]. One solution to avoid these undesirable rainfall-induced slope failures is to determine and analyze the rainfall intensity to estimate rainfall patterns in the future.

The water infiltration characteristics consociated with irrigation conditioning which bring up to the instability of slopes [11]. The infiltration of water significantly reduces the shear stress of soil in a state of unsaturated and contributes to slope instability [12]. The active bearing between water content and soil suction, appertained to as the soil–water characteristic curve (SWCC), plays a central job in extrapolating the behavior of an unsaturated soil [13].

Soil–water characteristics curve are used to assess the hydraulic conductivity, shear stress, volume changes, and aqueous diffusion of unsaturated soils [14]. For resembling unsaturated soils, it is commonly practice to soil–water characteristics curve (SWCC) or soil–water retention curves (SWRC), depicting relationships between an unsaturated soil moisture content change and its degree of saturation changes with its total suction changes, which affects its shear strength behavior [15]. When the degree of saturation is near full saturation, menisci disappear, as some air dissolves in the pore-water and the remaining air travels in continuous pore-water by diffusion process.

Rainfall-induced slope failure is a common geohazard for tropical and sub-tropical areas, where residual soils are abundant [16–20]. Rainfall decreases the suction of the matrix and increases the weight of soil units, leading to a lower stability soil slope [12, 21] According to Zhang et al. [22], the infiltration rate depends on the initial soil–water content, and the presence of stratigraphic instability may influence the rainfall pattern [22]. Fredlund and Xing [23] described the characteristics of soil curves for soil defined as the relationship between water content and suction for soil [24] as water content is described as the quantity of water contained in the pore [23, 24].

2.1 *Shear Strength for Unsaturated Slope*

Previous studies have been conducted to determine the shear strength of unsaturated slopes [25]. Shear strength is one of the essential features of geotechnical engineering. Geotechnical problems involving shear strength include slope stability, shallow or deep foundation, cuts, dams, pavement design, and side stresses on retaining walls. The completed structure must be stable and could receive the maximum load when been applied to it. In design, the determination of shear strength is required. Sample selection also depends on the type of construction needed. For example, uninterrupted samples should be taken for soil supporting the foundation and disrupted samples of altered soil structures such as dams and roads.

3 Methodology

According to the research objective, the data needed such as soil properties and water table are collected to set up a geometry set up for model parameters. Figures 1 and 2 show the exact research location, whereas Fig. 3 shows the Sg Langat slope profile, together with the slope geometry of channels 1, 2, and 3. The water channels chosen are only focused on channels 1, 2, and 3 which are represented as CH1, CH2, and CH3, respectively. Next, the rainfall data is then collected based on the month of the highest rainfall intensity recorded every year (2008–2016) of the study area. Then, simulation is conducted using GeoStudio software: (a) to analyze groundwater table in different position which are high, intermediate, and low position, (b) to determine the strength and stability of soil in different slope angles, (c) to observe rainfall intensity in different occasions in average and typical days. Calculation analysis is then made to obtain the pore-water pressure, displacement of soil, and factor of safety (FOS) against lambda. Lambda can be characterized as the proportion of the interslice shear and interslice normal forces producing a similar variable of factor of safety for both moment and force equilibrium [26, 27]. As such, lambda is perceived as the width of gap in the middle of two slip surfaces. Through this agreement, factor of safety can be strongly controlled by lambda to differentiate between factor of safety on each slip surfaces. Lastly, a slope behavior model will be done.

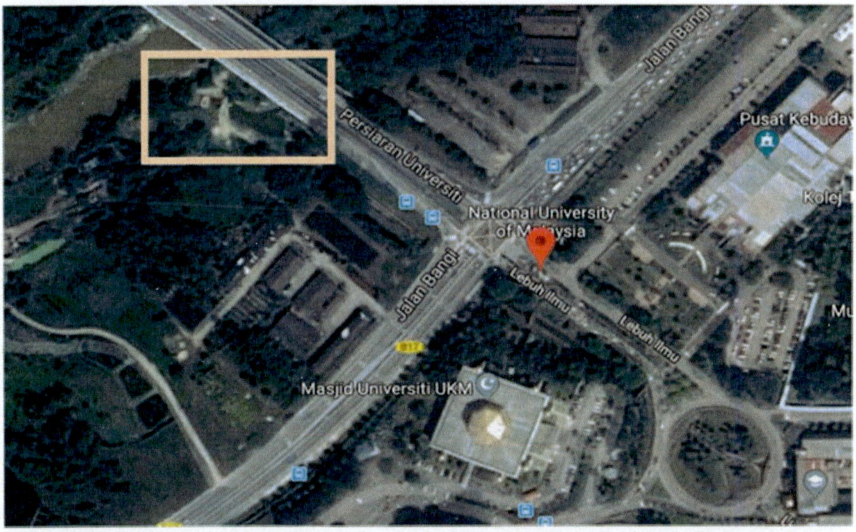

Fig. 1 Location of Sg Langat near UKM Bangi

Fig. 2 Location view of research site in Sg Langat

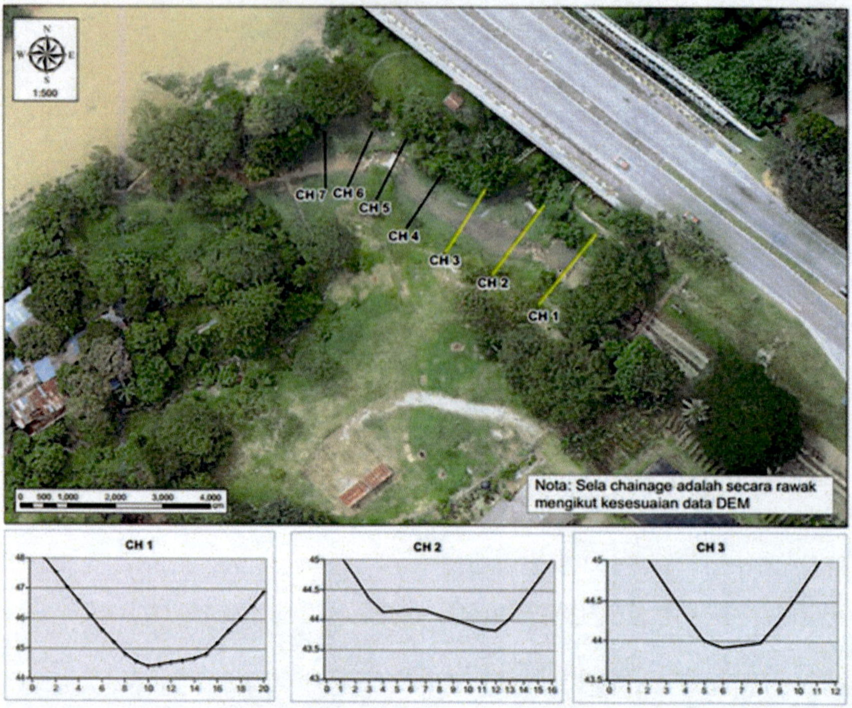

Fig. 3 Sg Langat slope profile

3.1 Soil Parameter for Slope Model

The model is utilized as plane strain model with 15-nodes mesh elements under the behavior of Mohr–Coulomb. After the slope geometry was drawn, the soil parameters were assigned onto the three channels given in Table 1 which includes the soil unit weight, permeability, and stiffness. The values of each soil properties were assumed and based on the suitability of the slope conditions. In addition, the general fixities are created regarding the slope model boundaries. General fixities consists an open flow for the top surface, lateral flow at both sides, and a close boundary for the bottom layer. These fixities are applied to ensure natural water flow taking place within the respected boundaries (Table 2).

3.2 Calculation and Estimation of Missing Rainfall Data

Regarding the obtained rainfall data, there are some data that found missing and errors on certain date on certain areas. Some of the data of certain stations are experiencing short breaks which can be led by certain reasons where instrumental

Table 1 Unsaturated soil parameter [28]

Channel	Bulk unit weight, γ (kN/m^3)	Permeability (m/day)	Stiffness, E (kN/m^2)	V (nu)	Strength, c_{ref} (kN/m^2)	Φ (phi)
1	18	0.001	10,000	0.35	5.00	28.00
2	18	0.001	3000	0.35	5.00	29.00
3	17	0.001	15,000	0.35	10.00	28.00

Table 2 SWCC and permeability function parameter

Soil type	θ_s	a	n	m	$k_a\left(\frac{\text{m}}{\text{s}}\right)$	p
Sand	0.45	10	1	1	10^{-4}	4

failures or absence of observer of the certain timeline might happen. Thus, it is necessary to compute the missing records to recover the rainfall data. These missing data is computed based on the observations of other stations nearby as possible to obtain the average and more accurate results. This estimation is carried out by interpolate the stations which consist of missing data with the other stations (index stations) to compare and recover the missing data. Thus, in this situation, simple arithmetic mean method is used to recover and calculate the missing data by using the formula:

$$P_x = \frac{1}{n}\sum_{i=1}^{i=n} P_i \tag{1}$$

For example, in Table 3, data error can be seen which is −999 on the date 01/06/2010 and 03/06/2010 at RTM Kajang. This shows that taking one of the dates as a calculation example: 02/06/2010, by using the simple arithmetic mean method. P_x represents the stations with missing data (Table 4).

$$P_x = \frac{1}{n}\sum_{i=1}^{i=n} P_i$$

$$P_x = \frac{1}{3}(P_1 + P_2 + P_3)$$

$$P_x = (0.4 + 6.1 + 7.6) = \mathbf{4.7}$$

$$P_x = \frac{1}{n}\sum_{i=1}^{i=n} P_i$$

$$P_x = \frac{1}{3}(P_1 + P_2 + P_3)$$

$$P_x = (0.5 + 0.5 + 1.5) = \mathbf{0.83}$$

Table 3 Missing rainfall data in the four gauging stations

Rainfall (mm)	RTM Kajang	Kg. Jenderam Hilir	Putrajaya	Kolam Sg. Merab
01/06/2010	0.4	6.1	−999	7.6
02/06/2010	5.6	41.8	2.1	8.1
03/06/2010	0.5	0.5	−999	1.5

Table 4 Recovered rainfall data in the four gauging stations

Rainfall (mm)	RTM Kajang	Kg. Jenderam Hilir	Putrajaya	Kolam Sg.Merab
01/06/2010	0.4	6.1	4.7	7.6
02/06/2010	5.6	41.8	2.1	8.1
03/06/2010	0.5	0.5	0.83	1.5

3.3 *Rainfall Data on Every Stations*

Kampung Jenderam Hilir

Kampung Jenderam Hilir is located near to Sungai Langat which is near to Dengkil area. According to the rainfall data on 26/02/2013, Kampung Jenderam Hilir experienced the highest rainfall intensity which is 152 mm (Figs. 4 and 5).

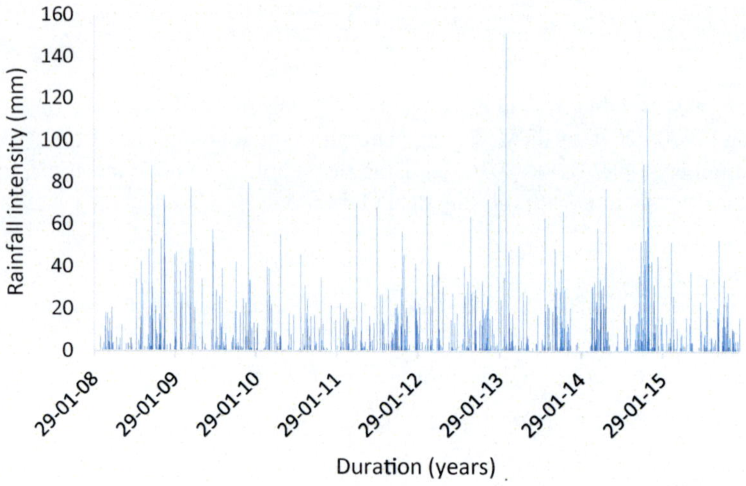

Fig. 4 Graph of rainfall intensity at Kg. Jenderam Hilir in 8 years [28]

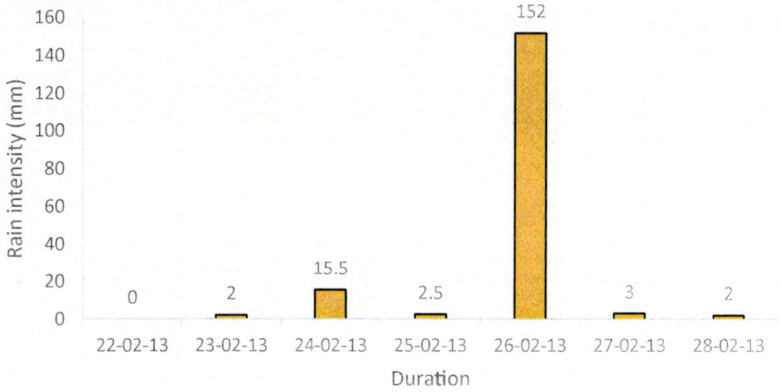

Fig. 5 Graph of rainfall depth on extreme day in 1 week interval in Kg. Jenderam Hilir

Puncak Niaga Putrajaya

Puncak Niaga Putrajaya is located near to Putrajaya Lake at Putrajaya. Based on its rainfall data, it is shown that during 28/11/2012, Puncak Niaga Putrajaya has the highest rainfall intensity which is 115.9 mm (Figs. 6 and 7).

Kolam Takungan Sungai Merab

Sungai Merab is located in between Dengkil and Bandar Baru Bangi. In 10/10/2013, Kolam Takungan Sg. Merab has the highest rainfall depth along the years which is 121 mm (Figs. 8 and 9).

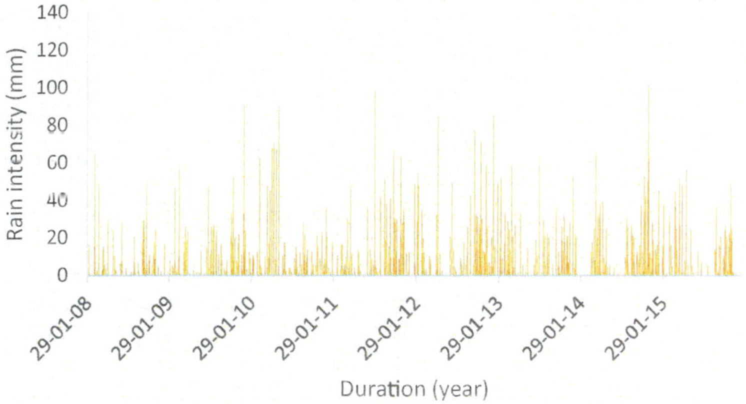

Fig. 6 Graph of rainfall intensity at Puncak Niaga Putrajaya in 8 years [28]

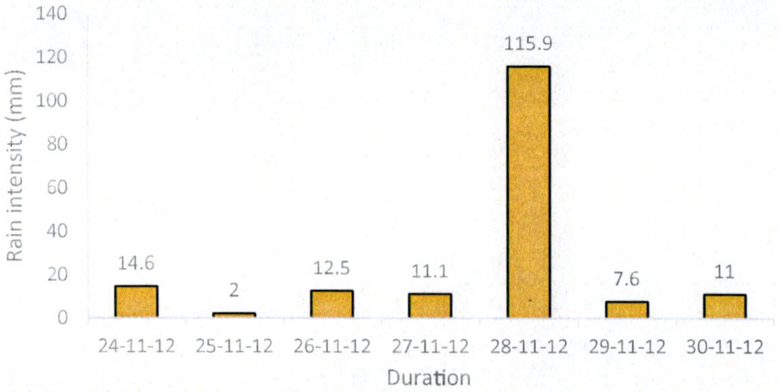

Fig. 7 Graph of rainfall depth on extreme day in 1 week interval in Puncak Niaga Putrajaya

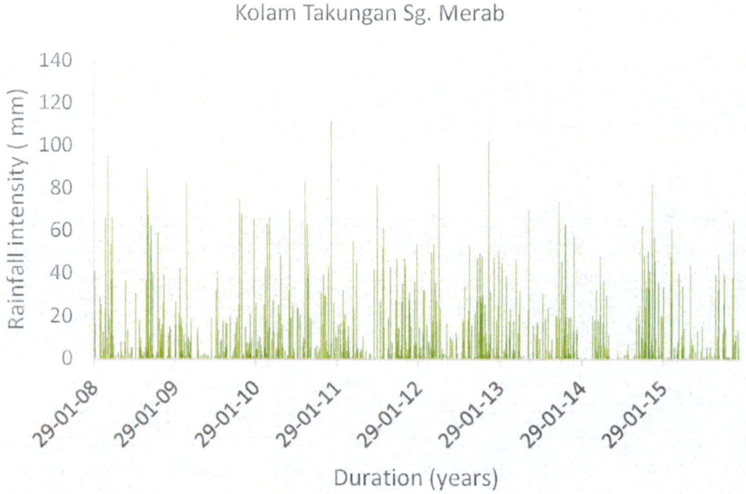

Fig. 8 Graph of rainfall intensity at Kolam Takungan Sg. Merab in 8 years [28]

RTM Kajang

RTM Kajang is located at Kajang near Jalan Cheras. According to rainfall data given, 9/4/2012 has the highest rainfall depth which is 142 mm on RTM Kajang area (Figs. 10 and 11).

As for conclusion, throughout rainfall data obtained from all of the four station, Kampung Jenderam Hilir has the most extreme rainfall depth which is 152 mm on 26/02/2013 compared to other all four stations (Fig. 12).

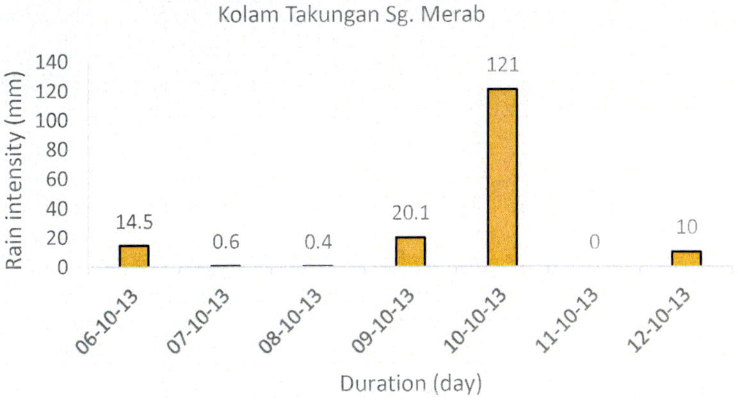

Fig. 9 Graph of rainfall depth on extreme day in 1 week interval in Kolam Takungan Sg. Merab

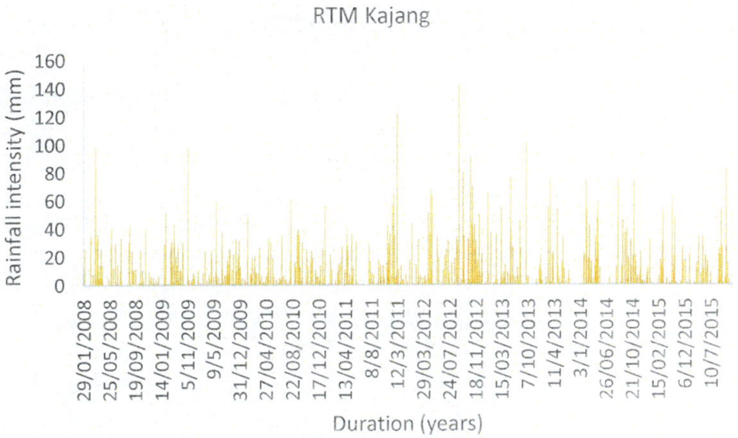

Fig. 10 Graph of rainfall depth at RTM Kajang in 8 years [28]

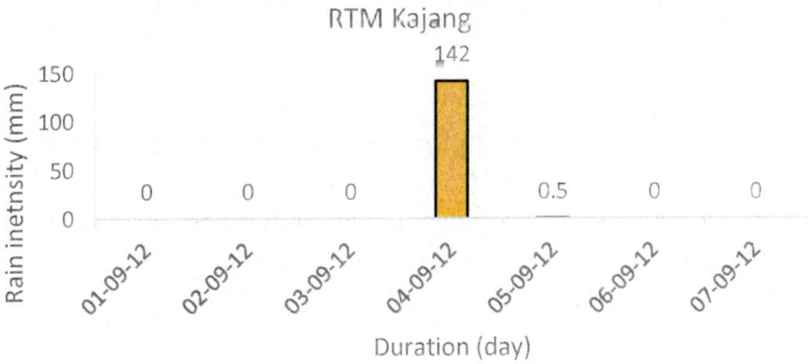

Fig. 11 Graph of rainfall intensity on extreme day in 1 week interval at RTM Kajang

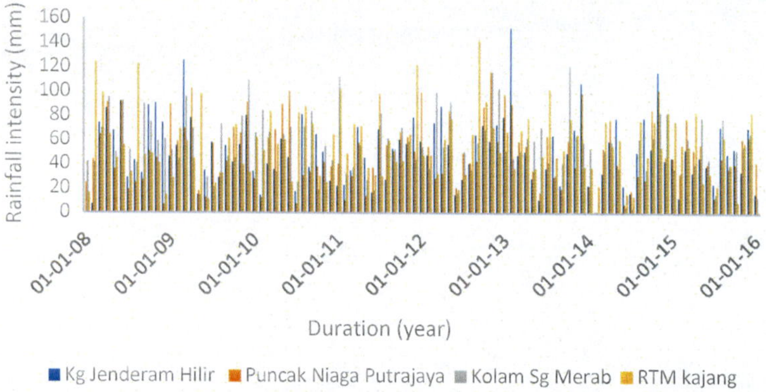

Fig. 12 Overall rainfall depth from 2008 to 2016 for all four stations [28]

3.4 Software Packages in Slope Modeling

Factor of safety is the proportion of a construction's outright strength (primary capacity) to a genuine applied burden in engineering. It communicates how much more strength a framework is than it should be for an actual applied load. SLOPE/W and SEEP/W in Geostudio are applied to register the component of wellbeing of slope and investigate both basic and complex issues for an assortment of slip surface shapes just as pore-water pressure conditions. Subsequently, Channels 1, 2, and 3 from Sg Langat profile have been decided to acquire their pore-water pressure conditions and factor of safety. In SEEP/W, the finite element technique is utilized to display simple saturated steady-state problems and analyze groundwater flow by using a comprehensive set of material models and boundary conditions. Then, the results from the SEEP/W is then moved to be analyzed in SLOPE/W. The SLOPE/W used the limit equilibrium method that models heterogeneous soil types, complex stratigraphic and slip surface geometry, and variable pore-water pressure conditions under the advancement of an enormous selection of soil models.

SEEP/W

In SEEP/W, the channels are drawn and designed while regions are attracted to the model geometry to separate between the upper layer and lower layer soils. Then, materials of models are characterized according to the soil hydraulic properties (e.g., hydraulic conductivity, hydraulic boundary, volumetric water content) and soil types before these model boundaries are appointed into the soil models. Then, at that point, the boundary conditions, e.g., seepage and pressure head are characterized and named onto the model to recognize the limits of each condition. The element mesh is additionally applied to the soil models. The cross section and shape utilized are both triangle and rectangular. Once the groundwater level is assigned, the groundwater flow analysis can be carried out. The rainfall data is applied onto the slope by means of seepage behavior. Then, at that point, the calculation of data can be settled. As the

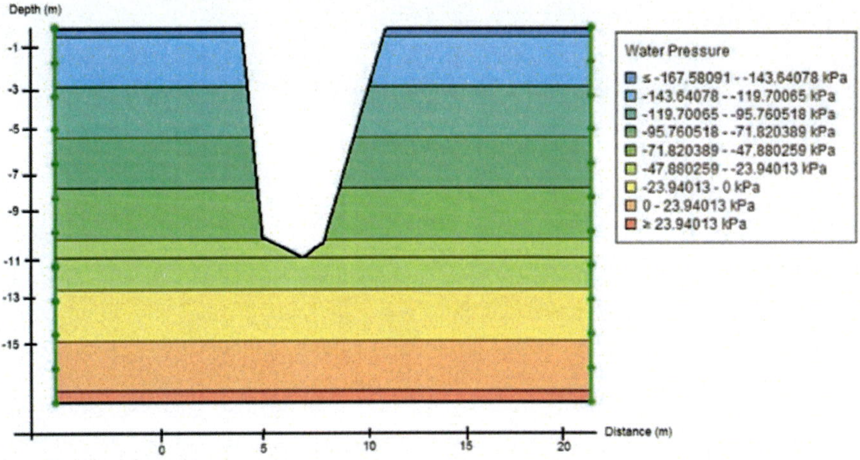

Fig. 13 Example of Channel 3 based on contour level for the pore-water pressure

result, the pore-water pressure of the slope can be acquired on a different contour level as shown in Fig. 13.

SLOPE/W

In SLOPE/W, comparative model geometry procedures are directed concerning the SEEP/W models. Then, materials of models are characterized by the soil mechanical boundaries including soil unit weight, design cohesion, and angle of friction, in this manner allocated to the soil models. The phreatic lines are drawn on the soil layer to produce the current pore-water pressure in the soil. Besides, entry and exit slip are determined on the model to show the critical slip surface. It is important to allow the visualization of the critical zones before solving the analysis to record the relevant factor of safety. The section will be at the most elevated surface of the incline (peak) and the ways out will be at the lower surface of the slope (toe). The computation is set, and the analysis would now be able to be solved. The information obtained shows the component of factor of safety which can be picked concerning the stability of slope and the situation of the slip surface.

4 Results and Analysis

In general, the construction phases are divided into two stages, namely the static (Phase 1) and consolidation (Phase 2). The static phase is used to generate initial conditions which determine the existing pore-water pressure at the site. The following consolidation phase is undertaken with the effect of phreatic level (9.81 kN/m) applied in the slope model. The durations for both phases used are different as the static phase is conducted immediately under zero hours, whereas the consolidation phase is simulated transient for two months. No water infiltration (i.e., rainfall) is applied at this stage. The results are obtained in a manner to represent the initial and changes of existing pore-water pressure underground.

4.1 Pore-Water Pressure

The outcomes of pore-water pressure are obtained from SEEP/W for all the analysis particular to the three channels. The pore-water pressure created can be observed against the depth of soil on various levels with values as displayed in Figs. 14, 15 and 16.

In Channel 1, initially both models started off with positive pore-water pressure at almost the same amount which is 1.49 kPa for phase 1 and 1.59 kPa for phase 2. The big difference that can be observed is when the depth moving from the surface toward 1 m below the surface of the ground. Phase 2 has a higher amount of pore-water pressure which reaches 62.65 kPa, whereas phase 1 only reaches to a value of 60.19 kPa of pore-water pressure. The pore-water pressure for both phases then slowly shifting to backwards, where pore-water pressure values changes are lower and the distance (depth) is larger. Lastly, only phase 1 reaches to the negative pore-water pressure with the value of −5.29 kPa.

In Channel 2, the situation happened to be the same as Channel 1, where both phases started off with almost the same value of pore-water pressure which are 0 kPa and 1.45 kPa for phase 1 and phase 2, respectively. When the depth moving from the surface toward 1 m below the surface of the ground, differences of pore-water pressure can be seen between two phases, where phase 2 attained a higher pore-water

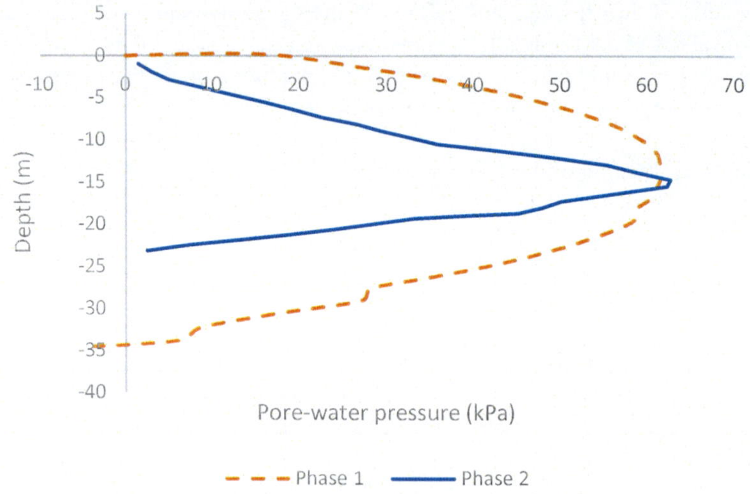

Fig. 14 Pore-water pressure measured against depth for Channel 1

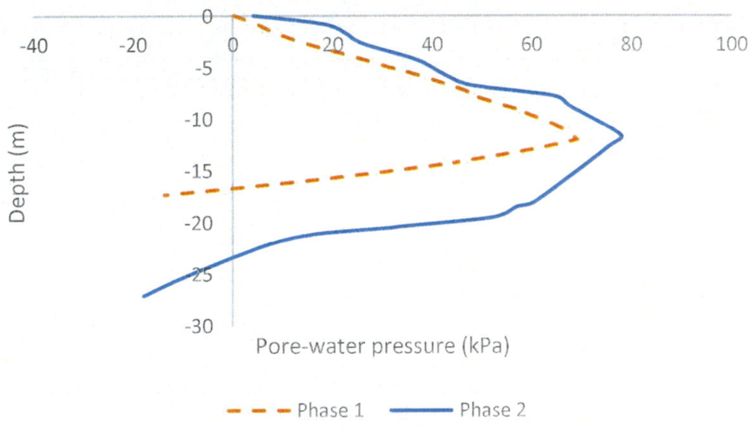

Fig. 15 Pore-water pressure measured against depth for Channel 2

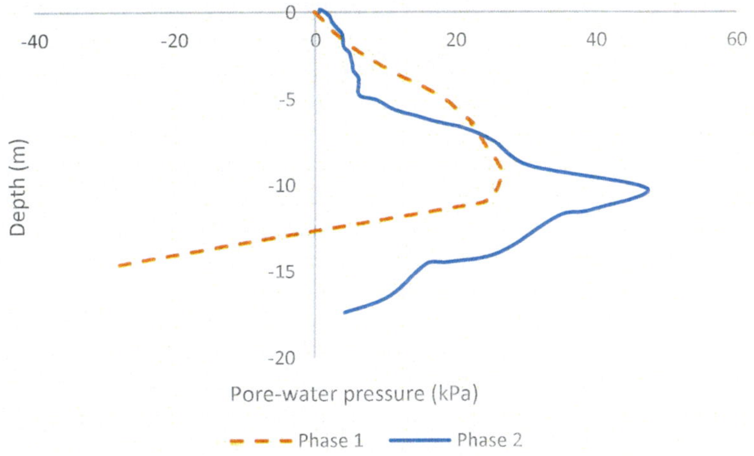

Fig. 16 Pore-water pressure measured against depth for Channel 3

pressure value which is 78.08 kPa, whereas phase 1 reaches 67.95 kPa of pore-water pressure. Same goes to Channel 2, both phases are then slowly shifting backwards as the depth increases until both phases reach the negative value of pore-water pressure which are −13.62 and −17.77 kPa.

In Channel 3, the big differences can be seen as phase 2 increases drastically when the depth moving from the surface below 4 m. The pore-water pressure of phase 2 reaches a high value of pore-water pressure which is 47.19 kPa compared to phase 1 which is only 26.6 kPa. However, as the distance (depth) increases, both phases slowly shift backwards and only phase 1 reaches a negative pore pressure value of −29.10 kPa.

It tends to be seen that Channel 2 has higher pore-water pressure followed by Channels 1 and 3. This is due to consolidation occurs where the soil is exposed to infiltration in phase 2 compared to phase 1 which is in static phase. Zero pore-water pressure demonstrates the area of groundwater level, henceforth, the zones at the channel openings with high pore-water pressure face the chance of soil disintegration through overflow. The dependability of the riverbank at these zones is inconsistent.

4.2 *Factor of Safety*

The limit equilibrium strategy isolates a potential sliding mass, characterized by a preliminary slip surface, into vertical slices. An iterative solution is utilized to decide the variable by which the shear strength, all things considered, should be decreased to such an extent that the sliding mass is exactly at the place of static balance (before failure happens) [26]. This reduction factor is referred to as the factor of safety. Equilibrium can be observed concerning to moment or force equilibrium. Generally, a factor of safety of 1 address that the stress is at the allowable limit. Factor of safety with under 1 most likely to experience failure though a factor of safety of more than 1 represents how much the pressure is within the allowable limit.

The factor of safety from Sg Langat channels can be acquired where it compares about the slip surfaces among entry and exit. The information is recorded into three classifications which are in the minimum, intermediate, and maximum values of factor of safety. Figures 17, 18 and 19 are introduced in chart of the factor of safety against lambda for minimum, intermediate and maximum values for all the channels.

Figure 17 distinctly shows the factor of safety determined at convergence with 0.38, 0.54, and 0.69, respectively, for minimum, intermediate, and maximum values. These values, however, encouraged slope instability as it neglects to reach at more than 1.0 factor of safety to the least. Whereas, in Fig. 18, on Channel 2, the factor of safety shows, respectively, of minimum, intermediate, and maximum values of 0.18,

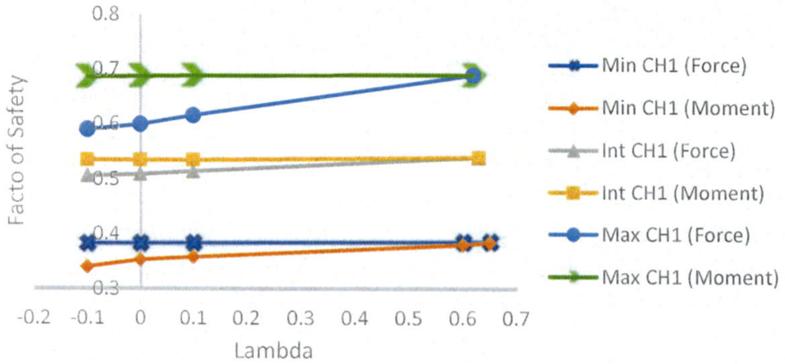

Fig. 17 Graph of factor of safety versus lambda for Channel 1

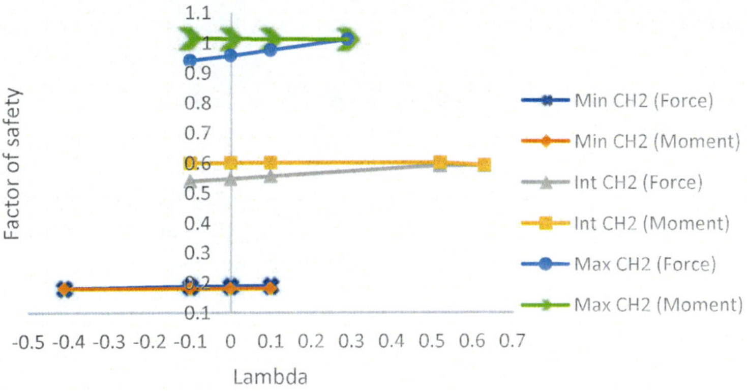

Fig. 18 Graph of factor of safety versus lambda for Channel 2

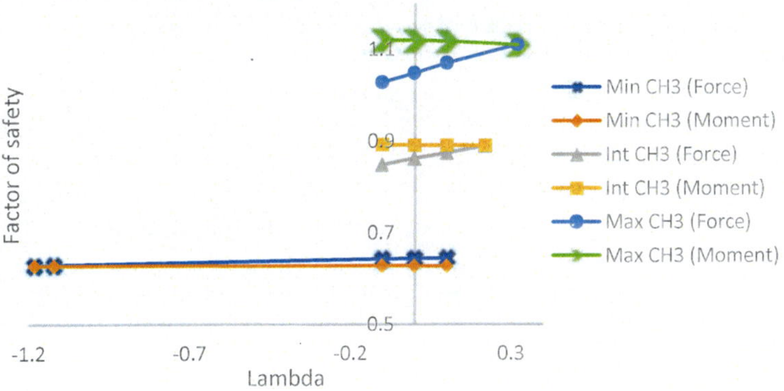

Fig. 19 Graph of factor of safety versus lambda for Channel 3

0.59, and 1.01. In Channel 3, the factor of safety shown are the highest among all the channel which are 0.63, 0.88, and 1.11 for minimum, intermediate, and maximum values, respectively. Contrasted with the past investigations which included a comparable circumstance [27, 29, 30], in spite of the fact that Figs. 18 and 19 exhibit a higher factor of safety, surpassed the 1.0 values, the factor of safety obtained are still extremely low to keep up with for long-term stability and eventually collapsed or fail. At these moments, bringing down shear strength through continuous infiltration or rigorous runoff on the slope surface can undoubtedly prove the initiating in slope failure [11]. It is, in this manner, suggested that the riverbank slope of Sg Langat be secured under both natural and built-up wall to protect the slope from additional failure in the future, caused by the impact from infiltration.

5 Conclusion

The outcomes obtained from the parametric review are introduced in the following segments where: (a) the impact of infiltration (rainfall intensity) toward the strength of soil, (b) the influence of slope angle during heavy rainfall, and (c) the precipitation and duration to cause shallow slope failure. Extreme rainfall occasions bring about the changes in pore-water pressure, where expanded pressure causes high water infiltration. The connection between infiltration and slope stability is recorded by greater rainfall infiltration making the slope to act insecurely and subsequently fail. It can be proven that rainfall controls the rate of decrease in factor of safety which compares to the minimum factor of safety under the influence on the duration of rainfall.

From the results of the analysis, Channel 2 has higher pore-water pressure compared to Channels 1 and 3, where phase 2 attained a higher pore-water pressure value which is 78.08 kPa, whereas phase 1 reaches 67.95 kPa of pore-water pressure. In Channel 3, the factor of safety shown are the highest among all the channels which are 0.63, 0.88, and 1.11 for minimum, intermediate, and maximum values, respectively. It tends to be seen that the pore-water pressure relies upon the intensity of rainfall and the length of duration of infiltration in the slope region. Pore-water pressure for conditions that get heavy rainfall is higher than conditions that get low rainfall or conditions that consequently do not get any rainfall. On the other side, the consistency of continuous rainfall shows an expansion in pore-water pressure and is higher than the conditions that receive rainfall or none. Variables of rainfall intensity and rainfall patterns has been a part in increasing the pore-water pressure.

At last, the result of simulation, utilizing GeoStudio PC programming can help in deciding the degree of slope stability and can study the level of water infiltration on slope. The utilization of SEEP/W programming in GeoStudio can capture the development of groundwater such as pore-water pressure effectively and precisely. Meanwhile, the utilization of SLOPE/W in GeoStudio factor of safety can be obtained rapidly and accurately. Thus, the stability of slopes can be determined.

Acknowledgements The author would like to thank the people involved in the research project and acknowledge the Universiti Kebangsaan Malaysia for financial assistance under grant GUP-2021-022 and Dana Padanan Antarabangsa (MyPair)-Natural Environment Research Council (NERC) with grant number; NEWTON/1/2018/TK01/UKM//2.

References

1. Kim, Y., Sohn, H.-G.: Disasters from 1948 to 2015 in Korea and Power-Law Distribution, pp. 77–97 (2018). https://doi.org/10.1007/978-981-10-4789-3_3
2. Duc Long, P., Dung, N.T.: Correction to Geotechnics for Sustainable Infrastructure Development (2020)

3. Sg Langat Landslip Worries Residents. https://www.nst.com.my/actionline/2017/06/248675/sg-langat-landslip-worries-residents. Accessed 20 Sep 2022
4. Jeong, S., Lee, K., Kim, J., Kim, Y.: Analysis of rainfall-induced landslide on unsaturated soil slopes. Sustainability **9**(7), 1–20 (2017). https://doi.org/10.3390/su9071280
5. Rahimi, A., Rahardjo, H., Leong, E.-C.: Effect of antecedent rainfall patterns on rainfall-induced slope failure. J. Geotech. Geoenvironmental Eng. **137**(5), 483–491 (2011). https://doi.org/10.1061/(asce)gt.1943-5606.0000451
6. Zhao, B. et al.: Effects of rainfall intensity and vegetation cover on erosion characteristics of a soil containing rock fragments slope. Adv. Civ. Eng. **2019** (2019). https://doi.org/10.1155/2019/7043428
7. Uhlenbrook, S.: Variability of rainfall in Peninsular Malaysia. Hydrol. Earth Syst. Sci. Discuss. **6**(4), 5471–5503 (2009). https://doi.org/10.5194/hessd-6-5471-2009
8. Zhang, G., Wang, R., Qian, J., Zhang, J., Qian, J.: Effect study of cracks on behavior of soil slope under rainfall conditions. Soils Found. **52**(4), 634–643 (2012). https://doi.org/10.1016/j.sandf.2012.07.005
9. Chatra, A S., Dodagoudar, G.R., Maji, V.B.: Numerical modelling of rainfall effects on the stability of soil slopes. Int. J. Geotech. Eng. **6362**, 0 (2017). https://doi.org/10.1080/19386362.2017.1359912
10. Taylor, P., Gofar, N., Lee, L.M.: Georisk: assessment and management of risk for engineered systems and geohazards extreme rainfall characteristics for surface slope stability in the Malaysian Peninsular (no. November 2014), pp. 37–41 (2008). https://doi.org/10.1080/17499510802072991
11. Hou, X., Vanapalli, S.K., Li, T.: Water infiltration characteristics in loess associated with irrigation activities and its influence on the slope stability in Heifangtai loess highland, China. Eng. Geol. **234**, 27–37 (2018). https://doi.org/10.1016/j.enggeo.2017.12.020
12. Aung, K.K., Rahardj, H., Leong, E.C., Toll, D.G., Rahardjo, A.H., Geotechnical, N.: Relationship between porosimetry measurement and soil-water characteristic curve for an unsaturated residual soil (no. Figure 1), pp. 401–416 (2001)
13. Jian, Z., Jian-lin, Y.U.: Influences affecting the soil-water characteristic curve, no. 8, pp. 797–804 (2005). https://doi.org/10.1631/jzus.2005.A0797
14. Thu, T.M., Rahardjo, H., Leong, E.C.: Elastoplastic model for unsaturated soil with incorporation of the soil-water characteristic curve (no. June 2017) (2011). https://doi.org/10.1139/t06-091
15. Egeli, I., Pulat, H.F.: Mechanism and modelling of shallow soil slope stability during high intensity and short duration rainfall. Sci. Iran. **18**(6), 1179–1187 (2011). https://doi.org/10.1016/j.scient.2011.09.010
16. Bordoni, M., Meisina, C., Valentino, R., Lu, N., Bittelli, M., Chersich, S.: (2015). https://doi.org/10.1016/j.enggeo.2015.04.006
17. Cho, S.E.: Effects of spatial variability of soil properties on slope stability. Eng. Geol. **92**(3–4), 97–109 (2007). https://doi.org/10.1016/j.enggeo.2007.03.006
18. Furuya, G., Suemine, A., Sassa, K., Komatsubara, T.: Relationship between groundwater flow estimated by soil temperature and slope failures caused by heavy rainfall, Shikoku Island, southwestern Japan, vol. 85, pp. 332–346 (2006). https://doi.org/10.1016/j.enggeo.2006.03.002
19. Zhai, Q., Rahardjo, H.: Determination of soil-water characteristic curve variables. Comput. Geotech. **42**, 37–43 (2012). https://doi.org/10.1016/j.compgeo.2011.11.010
20. Zhang, L.L., Zhang, J., Zhang, L.M., Tang, W.H.: Stability analysis of rainfallinduced slope failure: a review. Proc. Inst. Civ. Eng. Geotech. Eng. **164**(5), 299–316 (2011). https://doi.org/10.1680/geng.2011.164.5.299
21. Tran, A.T.P., Kim, A., Cho, G.: Numerical modeling on the stability of slope with foundation during rainfall Numerical modeling on the stability of slope with foundation during rainfall. (no. April) (2020). https://doi.org/10.12989/gae.2019.17.1.109
22. Zhang, X.: Test and application of hydrological models with a spatial modelling language (PCRaster) for the discharge simulation of a wetland dominated catchment in Northern Germany, p. 100 (2006)

23. Wang, S., Li, W., Chen, Z., Tian, G., Dai, G., Zhai, Q.: Results in Engineering Effect of Cr on the Performance of Fredlund and Xing (1994)'s Equation in Best Fitting Soil-Water Characteristic Curve Data, vol. 13, (no. February), pp. 8–10 (2022)
24. Likos, W.J., Lu, N., Godt, J.W.: Hysteresis and uncertainty in soil water-retention curve parameters. J. Geotech. Geoenvironmental Eng. **140**(4), 04013050 (2014). https://doi.org/10.1061/(asce)gt.1943-5606.0001071
25. Pham, H.Q.: An Engineering model of hysteresis for soil-water characteristic curves, pp. xxv (190, 96 leaves) (2001)
26. Wang, Y.J., Yin, J.H., Lee, C.F.: The influence of a non-associated flow rule on the calculation of the factor of safety of soil slopes. Int. J. Numer. Anal. Methods Geomech. **25**(13), 1351–1359 (2001). https://doi.org/10.1002/nag.177
27. Matsuura, S., Asano, S., Okamoto, T.: Relationship between rain and/or meltwater, pore-water pressure and displacement of a reactivated landslide. Eng. Geol. **101**(1–2), 49–59 (2008). https://doi.org/10.1016/j.enggeo.2008.03.007
28. Nazrien Ng, J., et al.: The effect of extreme rainfall events on riverbank slope behavior. Front. Environ. Sci. **10**, 1–14 (2022). https://doi.org/10.3389/fenvs.2022.859427
29. Chen, J., Ke, P., Zhang, G.: Slope stability analysis by strength reduction elasto- plastic FEM. Key Eng. Mater. **345–346**(6), 625–628 (2007). https://doi.org/10.4028/0-87849-440-5.625.
30. Merat, S., Djerbal, L., Bahar, R.: Numerical analysis of climate effect on slope stability (no. June), pp. 308–318 (2018). https://doi.org/10.1061/9780784481691.031

A Review on Simplified Image Analysis Method for Measuring LNAPL Saturation Under Groundwater Table Fluctuation

Doaa F. Almaliki and Harris Ramli

Abstract The leaking from surface spills and underground storage tanks of various light non-aqueous phase liquids (LNAPLs) caused hazardous contamination to the subsurface system, especially in case of groundwater table fluctuations. The toxicity of these compounds has made infeasible field studies and gets a replacement with laboratory studies. Researchers have recently become very interested in using image analysis techniques to measure the saturation migration of groundwater and LNAPLs. Over the last decade, the simplified image analysis method (SIAM) has become increasingly popular. SIAM has been proved to be a suitable and effective tool for characterization and measuring LNAPL migration in the subsurface system. This research introduces a review of the recent studies and published on the simplified image analysis method for LNAPL migration measurements. The experimental approaches in this study can be viewed as an important intermediary between column studies and tank studies. Besides discussion on the research efforts, recommendations for future research are provided.

Keywords Groundwater fluctuation · Saturation · LNAPL · Hazardous waste · SIAM

D. F. Almaliki · H. Ramli (✉)
School of Civil Engineering, Engineering Campus, Universiti Sains Malaysia, 14300 Nibong Tebal, Pulau Pinang, Malaysia
e-mail: cemhr@usm.my

D. F. Almaliki
e-mail: doaa.faisal@stu.edu.iq

D. F. Almaliki
Environment and Pollution Engineering Department, Basrah Engineering Technical College, Southern Technical University, Alzybair Street, Basrah, Iraq

N. Sabtu (ed.), *Proceedings of AWAM International Conference on Civil Engineering 2022 - Volume 3*, Lecture Notes in Civil Engineering 386,
https://doi.org/10.1007/978-981-99-6026-2_8

1 Introduction

The release of petroleum hydrocarbon liquids is a worldwide phenomenon that has been increasing in conjunction with rising oil consumption [1]. Rapid global population expansion [2], industrialization, and modern transportation have all increased demand for oil, increasing oil spills. An example of an oil incident that released oil between (2,100,000 and 2,400,000 barrels) is the Canada Atlantic Empress oil spill [3]. Pipeline ruptures, vandalism, human error, system failures, [4] collisions, and shipwrecks are the most common causes of oil spills. Crude oil spill expenses, cleaning costs, and rehabilitation costs are all examples of catastrophic economic and environmental disasters [5].

Environmental impacts of hydrocarbon product spills include water pollution (surface and groundwater) [6], seashore, and beach contamination [7]. Degradation of marine invertebrate habitat, individual oiling, and drowning, alteration of the food chain and toxicity [8, 9], mortality, and killing the marine birds [10–12] has all been identified. There has also been a decline in vegetation and marine mammals [13–15]. The slow response to oil leak disasters aggravates the severity of these risks. Fast reaction to oil spillage reduces its migration and lowers the potential effects [16–19].

Groundwater is an important element for humans. There is a significant need to reduce contaminated groundwater and use it as a source of drinking water in the future. Petroleum fuels are contaminating groundwater all over the world due to leakage from the underground storage tanks, leaking pipes, and unintentional spills. When these products are released on the ground, they migrate downward until reaching the water table. An example of petroleum fuels is the light non-aqueous phase liquids (LNAPLs). The LNAPLs cause reduction in the elevation of the water saturation if the spilled volume of LNAPL is high [20].

From previous LNAPL research to the current knowledge, there have been substantial improvements. The researchers provide that LNAPL is variably saturated in the underground system where the LNAPL can exist in free (mobile), entrapped, and residual forms. Understanding all the types of LNAPL distribution in the subsurface is essential for establishing a successful LNAPL remediation approach [21] as well as assessing the potential for environmental and human health harm [22].

Chevalier and Petersen [23] conducted an earlier literature evaluation of LNAPL flow, transport, and remedial techniques in 2D laboratory aquifer models. Later, Saleem et al. [24] investigated direct and indirect techniques for determining spilled hydrocarbon volume utilizing state-of-the-art technology. The authors in their research emphasized the significance of interdisciplinary collaboration between scientists and engineers. A review of multidimensional, multi-fluid, intermediate-scale experiments was demonstrated by Oostrom et al. [25]. Experimental aspects of LNAPL dissolution and increased remediation are covered in their review. The authors also studied the behavior of groundwater flow, saturation imaging techniques, and tracer detection in 2D laboratory experiments [26]. Recently, a critical review paper discussed the effect of ecohydrology, and ecohydrological interface on heavy metals and organic pollutants. It focused on the indispensable role of the ecohydrological interface in the removal of pollutants [27]. Another critical review focused on only two factors which are the ground-level fluctuation and the change in temperature and how it affects LNAPL spreading and degradation processes [28].

Many researchers used image analysis techniques to study the saturation distribution of groundwater and LNAPL in the subsurface system. Such methods are, the multispectral image analysis method (MIAM), light transmission visualization (LTV), ground penetration radar (GPR), and light reflection method (LRM). The widespread use of image analysis methods in a variety of research domains demonstrates that these methods are quite beneficial, particularly for examining the behavior of complicated types of pollutants and determining saturation. However, there is a lack of review papers on the investigation of groundwater and LNAPL using the simplified image analysis method. Therefore, this study will focus on using the simplified image analysis method to measure the saturation distribution of groundwater and LNAPL in the whole domain under fluctuated groundwater table. The next subjects will examine a general overview of groundwater movement and contamination by LNAPL as well as describe the main design of SIAM and some related studies that took into account this technique to measure the groundwater contamination saturation.

2 LNAPL Contamination on Subsurface

2.1 LNAPL Migration Behavior in the Subsurface

Light non-aqueous phase liquid (LNAPL) refers to the petroleum hydrocarbons liquid that is lighter than water. It is a type of groundwater constituent that has a lower density than water and is sparingly soluble. When LNAPL is released on the ground, it moves through the unsaturated zone before accumulating in the saturated zone. Capillary pressures produce a drop in the water percentage in the capillary fringe by keeping a residual fraction of the contaminant in the pore space during migration (see Fig. 1) [29–32]. The distribution of LNAPL in the subsurface during infiltration is complicated and varies on various parameters [33]. Dissolved and gaseous phases of LNAPL are commonly found in the ground as by-products of the pure phase. As a result, these various phases interact with the soil–water system, creating a variety of chemical and physical changes in the soil [34].

Several factors influence the migration of LNAPL in the subsurface, including the volume of NAPL released, the duration of the release, the surface of the infiltration, soil characteristics, the properties of LNAPL, and the subsurface flow conditions [35]. When LNAPL is released into the subsurface, it migrates down through the unsaturated zone as a separate liquid due to gravity's action. The vertical movement of the contaminant is accompanied by horizontal migration. This is attributed to the medium spatial variability and capillary forces effect [36]. During LNAPL

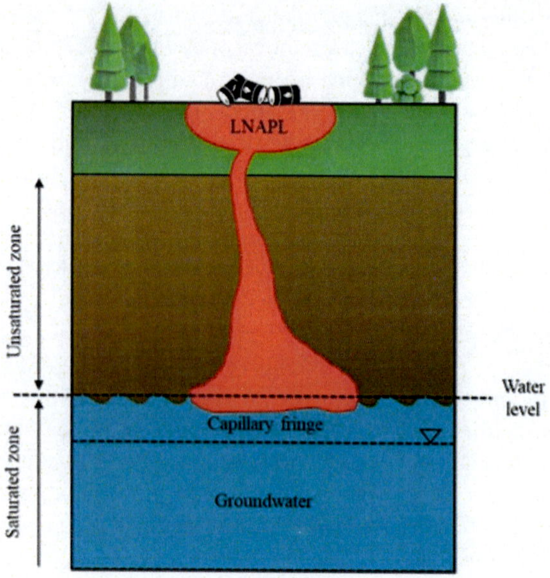

Fig. 1 LNAPL migration in the subsurface system [22]

vertical migration in the unsaturated zone, the residual liquid is trapped in pore spaces. Because of the surface tension effects, this trapping occurs. Some LNAPL will eventually reach the saturated zone where it will propagate horizontally through the capillary fringe when sufficient LNAPL volume is released [36].

2.2 *Effect of Groundwater Fluctuation on the Distribution of LNAPL*

Groundwater table variations can alter the distribution of the multiple LNAPL phases in the subsurface. Free LNAPL may become entrapped or entrapped LNAPL may become free LNAPL as the water table rises or decreases. Furthermore, as the groundwater level declines during the dry season, free LNAPL may become a residual form in the unsaturated zone, as the majority of free LNAPL migrates downward with the water table [37].

Water table oscillations may have a significant impact on LNAPL redistribution, mobility, and phase partitioning. Several in situ cleanup techniques rely on pollutant behavior within porous media. The chemical and the physical traits of the contaminants, the chemistry of the subsurface environment, the flow rate of groundwater, and other hydraulic characteristics are all factors that influence the water table fluctuation in the real world [38]. Most of the researchers concentrate on the dispersion and distribution of contaminants in groundwater using horizontal flow systems, however, there is still a little attention to the influence of the level of groundwater table on pollutants, where groundwater levels are unstable and fluctuate.

When the groundwater table drops, LNAPL follows, causing LNAPL redistribution and reformation of LNAPL components between phases. In the subsurface saturation zone, a portion of the LNAPL remains as discontinuous residual LNAPL ganglia. By successive water infiltration, the LNAPL's most volatile and soluble chemicals can volatilize or be leached, contributing to the dissolved plume's evolution. Mobile LNAPL is redistributed higher as the water table rises, leaving some immobile entrapped LNAPL droplets in the saturated zone [39–43].

During repeated drainage/imbibition phases, the influence of capillary pressure, as well as the trapping of fluids in the saturated and unsaturated zones, continuously changes the water flow paths [44]. Additional interactions between LNAPL and water (rainwater and groundwater) improve hysteresis effects, promoting the migration of pollutants (see Fig. 2) [45, 46]. As a result, seasonal or pump-induced groundwater level fluctuations commonly cause major pollution dispersion and redistribution through the water table fluctuation zone. It also lowers the average LNAPL saturation and mass of the mobile light NAPL that remains afterward.

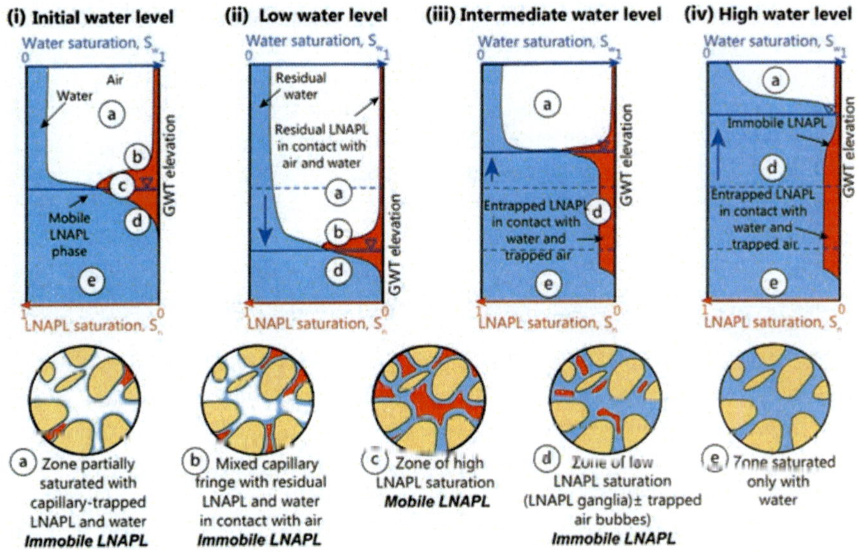

Fig. 2 Fluctuation of LNAPL (S_n) and water (S_w) saturation in soil under the influence of water table variations [47]

3 Simplified Image Analysis Method for Measuring Groundwater and LNAPL Saturation

Simplified image analysis method (SIAM) is a non-invasive and non-intrusive method for determining fluid saturations over a wide area. SIAM needs the least am out of equipment, is the cheapest, and produces no adverse radiation effects [48].

3.1 *Design and Advantages of SIAM*

Two types of digital cameras were used in the experiment using SIAM, which is Nikon D90 and Nikon D7000. Both of these digital cameras consist of two non-similar band-pass filters with wavelengths of 450 and 656 nm that are required to quantify the saturation distribution of LNAPL using SIAM. To collect photographs during the experiments, the cameras must be set in front of the experimental design (column or tank). As the only source of light in the dark room, two LED floodlights were placed. As a reference for white color, a GretagMacbeth white balance board was fixed next to the experimental design (see Fig. 3). Blue-dyed water (Brilliant blue FCF added to the solution at a 1:10,000 weight ratio) and red-dyed LNAPL were also used to enhance the visualization of LNAPL movement (Sudan III mixed with LNAPL Liquid at a 1:10,000 mass relation). The Nikon Camera Control Pro software was used to control the digital cameras to minimize displacement and vibrations. The aperture settings of the two lenses were modified using exposure periods of a few seconds to fully use the camera's dynamic range. The lenses' apertures were set to f-16 and the exposure length was 2.5 s for all of the photos. Nikon ViewNX 2.10.3 is required to convert all images from NEF (Nikon proprietary RAW version files) to Tagged Image File Format (TIFF). The TIFF images then are analyzed using MATLAB [49].

The theory that supports SIAM is the use of the Beer–Lambert equation of transmittance to develop a correlation between fluid saturations and average optical density, which can then be used to estimate fluid saturations of porous media across the entire experimental [48].

When a beam of monochromatic light I_o intersects with a block of absorbing matter perpendicular to a surface, its power is decreased to I_t as a result of absorption after traveling through a length b of the material, according to the Beer–Lambert law of transmittance:

$$D_i = \in bc = \log_{10} \frac{I_o}{I_t} \tag{1}$$

where D_i is the optical density, ε is the numerical constant, b is the length of the path, is the number of moles per liter of absorbing solution, I_o is the initial radiant power, and I_t is the transmitted power [50]. The average optical density (AOD) D_i

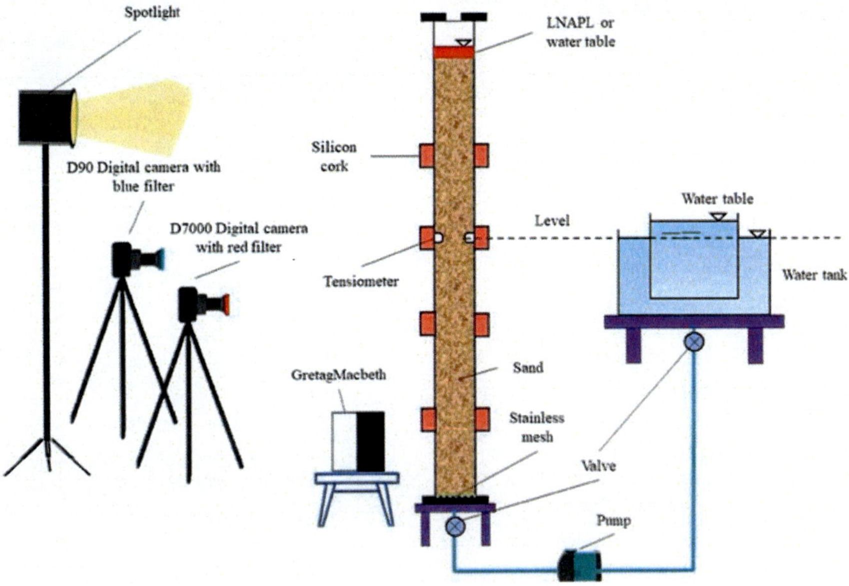

Fig. 3 Experimental setup [49]

for reflected light intensity in digital photographs is defined as:

$$D_i = \frac{1}{N}\sum_{j=1}^{N} d_{ji} = \frac{1}{N}\sum_{j=1}^{N}\left(-\log_{10}\left(\frac{I_{ji}^{r}}{I_{ji}^{0}}\right)\right) \tag{2}$$

From Eq. (2), a total number of pixels in the selected area is N, the optical density of the individual pixels for a given spectral band I is d_{ji}, I_{ji}^{r} is the reflected light intensity, as determined by individual pixel values, and I_{ji}^{0} is the intensity of reflected light using a brilliant white surface [51].

$$D_t = c.D_o \tag{3}$$

where D_0 is the optical density of a solution of unit concentration, and D_t is the optical density of a solution with a concentration of c. So, when two cameras with band-pass filters (wavelengths = I and j) are employed, and water and NAPL are combined with dyes whose predominant color wavelengths are both I and j, two sets of linear equations can be solved for S_w and S_o:

$$\left[\begin{pmatrix} D_i \\ D_j \end{pmatrix}_{m.n}\right] = \left[\begin{array}{c} \left(D_i^{10} - D_i^{00}\right).S^w + \left(D_i^{01} - D_i^{00}\right).S_o + D_i^{00} \\ \left(D_j^{10} - D_j^{00}\right).S^w + \left(D_j^{01} - D_j^{00}\right).S_o + D_j^{00} \end{array}\right] \tag{4}$$

Table 1 Summary of SIAM technique design

Technique	Simplified image analysis method (SIAM)
Method application	Measuring LNAPL and water saturation distribution in 1D column
Average optical density (AOD)	The AOD equations are unique to each mesh element (matrix)
Number of cameras	2 cameras
Number of filters	2 filters (450 and 656 nm)
Calibrated sample	Experimental domain (column)
Experiment calibration	3 non-linear points were used to generate a linear regression plane from these images ($S_w = 0\%$ $S_o = 0\%$ dry sand $S_w = 100\%$ $S_o = 0\%$ sand fully saturated with water $S_w = 0\%$ $S_o = 100\%$ sand saturated with LNAPL)
Advantages	Measuring the saturation in the whole domain

$[D_j]_{mn}$ is the average optical density of each mesh element for wavelengths i and j; $\left[D_i^{00}\right]_{mn}$ and $\left[D_j^{00}\right]_{mn}$ represent the average optical density of each mesh element of dry sand; $\left[D_i^{10}\right]_{mn}$ and $\left[D_i^{10}\right]_{mn}$, water-saturated sand, and $\left[D_i^{01}\right]_{mn}$ and $\left[D_i^{01}\right]_{mn}$ for LNAPL-saturated sand [48].

This technique requires two cameras with two different band-pass filters (wavelengths $\lambda = i$ and j) to obtain three calibration photos. Each camera must be calibrated based on the photo of the studied domain filled with dry sand, water-saturated sand, and LNAPL-saturated sand. The average optical densities for each mesh element of the examined domain are analyzed and differentiated to the corresponding ones for all three cases, as well as a correlation matrix. The next section will propose to summarize current knowledge on measuring groundwater and LNAPL distribution underwater table fluctuation using SIAM (Table 1).

3.2 *Measuring Groundwater and LNAPL Saturation Using SIAM*

Several researchers conducted their studies on the migration of groundwater and LNAPL using SIAM. Flores et al. [48] had a novel study on the simplified image analysis method. The simplified image analysis method had been used for analyzing the behavior of the two-phase air–water and three-phase air–water-LNAPL fluctuating groundwater systems. Under dynamic conditions, the 1D column porous soil (Toyoura sand) contaminated with LNAPLs showed a 4.7% difference in the drainage-imbibition three-phase air–water-LNAPL. Another study presented by Flores et al. [52] described the movement of LNAPL and groundwater in four drainage/imbibition cycles in a three-phase system in a 1D column containing fully saturated sand. At the same point of each step, the findings revealed distinct saturation levels, which improves the effect of the groundwater movement on the distribution of the water and the contaminant through rising and decreasing of the water table level. Similar research by Flores et al. [53] measured three-phase saturation of water and diesel in granular soil. SIAM was used to investigate the migration behavior of diesel in a 1D column filled with Toyoura sand under various precipitation conditions (6.6 and 63.2 mm/h). The study's findings revealed that the residual LNAPL saturation could be anticipated based on the maximum saturation. Further studies need to be done in a two-dimensional tank to measure the vertical and horizontal movements of LNAPL.

A one-dimensional column was used in conjunction with SIAM by Sudsaeng et al. [54]. They studied the migration of diesel in porous media under variable groundwater conditions. Toyoura sand and Silica #5 sand were used as porous media in this study. According to this study, diesel saturation is higher in Toyoura sand than in Silica #5 sand at the end of imbibition processes, and sand particle size has a major impact on diesel migration behavior. Water and LNAPL saturation distribution under the influence of groundwater fluctuation will be different in various sands, according to this study.

Only two published researches on the NAPL saturation migration in granular soil comparing different NAPLs with different densities and viscosity were explored by Flores et al. [55, 56]. The researchers established a logarithmic relationship between viscosity and residual saturation ratios of various NAPLs, which may be utilized to compute residual saturation values after imbibition. Even though LNAPLs has a lower density than water, their findings show that they can effectively be trapped below the water table. Flores et al. showed in the second investigation that the infiltrated LNAPL depth was significantly related to the viscosity of the pollutant ($R^2 = 0.84$). The variations in the depth of the LNAPL mobile fraction after both the drainage/imbibition steps were also directly related to interfacial tension values ($R^2 = 0.79$). The viscosity was also logarithmically connected to the residual saturation under the same conditions. The researches help to understand and anticipate the properties of various spilled contaminants on the ground.

In addition to studying the effect of density and viscosity of the contaminant on its migration, experiments on the influence of LNAPL volume on the contaminant migration were done by several researchers. The experiment was described by Alazaiza et al. [57] in laboratory 1D column tests. The effect of water infiltration on pooled light non-aqueous phase liquid (LNAPL) redistribution in porous media was investigated using SIAM. Low and high levels of LNAPL were determined in the experiments. The capillary fringe was found to be around 6–7 cm long, which corresponded to the capillary height calculated using empirical equations. This presented that LNAPL contamination migrated further into the capillary fringe zone when the volume of the contaminant was large. Also, it showed that injected water produced a greater hydrodynamic force on the entrapped LNAPL, causing it to move further into the capillary and saturation zones. Similarly, another study done by Alazaiza et al. [58] in one-dimensional column experiments to show the effect of the volume of LNAPL on the migration of the contaminant was carried out to see how varying LNAPL volumes affected the behavior of LANPL distribution in the capillary zone in the system. Various three diesel volumes (50, 100, and 150) ml were used in separate experiments. The research demonstrated that the larger volume of diesel (150 ml) leads to faster LNAPL migration. This means that the higher injected volume, the faster migration of the contaminant deeper into the capillary fringe which contributes to the distinction of groundwater quality. SIAM also was used by Alazaiza et al. [59] to expose groundwater and LNAPL migration under the effect of two diesel volumes (low and high) and two patterns of water table fluctuation. When the water table drained, the water saturation decreased, however, the LNAPL saturation rose. By increasing the diesel volume from 25 to 50 ml, the initial saturation of diesel was raised above the groundwater table and the capillary fringe height extended from 36 to 38 cm. It is relatively clear when comparing the two LNAPL volumes, that it was discovered that the LNAPL in the case of high LNAPL covered a larger area. It seems, therefore, relatively clear that increasing LNAPL volume while keeping the same groundwater fluctuation pattern shows a slight influence on LNAPL's overall behavior.

The emphasis of the research by Harris [49, 60] was to assess diesel migratory behavior in the subsurface as a result of precipitation in one 1D column and 2D tank tests. The impact of precipitation intensity was considered. The study showed that the higher the intensity, the deeper the LNAPL will migrate into the saturated zone as a result of precipitation. S-p relationships can also be used to track potential LNAPL spills and migration as a function of precipitation, according to the study's findings. The author also concluded that diesel saturation was higher in the first drainage than in the second drainage, implying that some residual diesel was retained beneath the water table. The author's findings also reveal that in a 2D tank, the higher the horizontal groundwater flow, the bigger the diesel contamination. The displacement of the LNAPL in vertical groundwater movements might be upward or follow above the soil surface. Continuously, Sudsaeng et al. [61] have used SIAM to evaluate the impact of ground-water fluctuation on diesel distribution in the subsurface in a one-dimensional column. They also used a 2D tank experiment to investigate the impacts of groundwater table fluctuation and lateral groundwater flow. Their findings show that the greater the contaminated area by diesel, the large rate of horizontal groundwater movement. In the case of the vertical flow of groundwater, the LNAPL moves upward to the soil surface.

Yimsiri et al. [62] completed a one-dimensional column experiment, in which they focused on the impact of different particle sizes on the saturation of water and LNAPL in the main domain under transient conditions. Also, they investigated two-phase experimental data to show the soil-liquid characteristic curves (SLCCs) of diesel/air and water/air systems. Scaling procedures were used to estimate the soil liquids characteristic curve of the diesel/water system. The authors showed that smaller particle sizes affect the distribution of the contamination. They cause lower pressure entrance, residual matric suction, and residual saturation degree. The coarser

Table 2 Summary of experimental research used SIAM

Ref	LNAPL type	Soil type	LNAPL volume	Design dimension	Software
[48]	Paraffin	Toyoura sand	28 g	1D (3.5 × 3.5 × 50) cm	MATLAB
[52]	Paraffin	Toyoura sand	13.8 g	1D (3.5 × 3.5 × 60) cm	MATLAB
[53]	Diesel	Toyoura sand	/	1D (3.5 × 3.5 × 50) cm	MATLAB
[54]	Diesel	Toyoura sand, silica #5	28 g	1D (3.5 × 3.5 × 50) cm	MATLAB
[55]	10 NAPL	Granular soils Toyoura sand	50 ml	1D (3.5 × 3.5 × 40) cm	MATLAB
[56]	10 NAPL	Toyoura sand	50 ml	1D (3.5 × 3.5 × 50) cm	MATLAB
[57]	Diesel	River sand	(25, 50) ml	1D (35 × 35 × 600) mm	MATLAB
[58]	Diesel	River sand	(50, 100, 150) ml	1D (30 × 30 × 500) mm	MATLAB
[59]	Diesel	River sand	(25, 50) ml	1D (35 × 35 × 600) mm	MATLAB
[49]	Diesel	Toyoura sand	(25, 50) ml	1D (35 × 35 × 500) mm, 2D(500 × 700 × 35) mm	MATLAB
[60]	Diesel	River sand, Toyoura sand	(25, 50) ml	1D (35 × 35 × 600) mm	MATLAB
[61]	Diesel	Toyoura sand	(1D = 28 g), (2D = 300 g)	1D (3.5 × 3.5 × 50) cm, 2D (80 × 50 × 7.5) cm	MATLAB
[62]	Diesel	Ottawa sand (#3821, #3820)	15 g	1D (3.5 × 3.5 × 110) cm	MATLAB
[63]	0	Quarry sand	0	1D (35 × 35 × 550) mm	MATLAB

particle size results in lower entrapped LNAPL saturation and increase the entrapped air saturation. Another study related to the effect of particle size on the migration of LNAPL under water table fluctuation was done by Harris et al. [63]. They emphasized using the SIAM to estimate water saturation and capillary rise to get the grain shape constant. The study included six one-dimensional columns were used to model capillary rise at constant groundwater levels. The findings of this study showed that coarser sand should be measured with values less than 25 mm^2, whereas finer sand should be measured with values greater than 25 mm^2. Table 2 gives a summary of LNAPL migration studies done by using SIAM.

4 Conclusion

This review evaluates the use of the simplified image analysis method (SIAM) for measuring the LNAPL and groundwater migration in the subsurface system under the influence of groundwater table fluctuation. As several researchers prefer doing their studies using non-destructive and non-intrusive techniques like the simplified image analysis method, using developed digital cameras, up-to-date toolbox like Image-Pro Plus software, and a new version of MATLAB is the best choice for image processing. The review concludes that SIAM is a good method in terms of groundwater contamination saturation measurements. It is a cost-effective method that doesn't require a lot of equipment, as well as it measures the contaminant saturation in the whole domain.

Studying the effect of groundwater fluctuation is still complex and often limited. According to worldwide predictions, the higher the groundwater table variations, the higher the LNAPL mobility. Higher precipitation and groundwater levels may cause the LNAPL compounds to be diluted and submerged in groundwater and connected streams. Groundwater table fluctuations in all studies show high influence on the migration of LNAPL in the system. The studies showed that the contaminant moves with the movement of the water table producing trapped and residual contaminants in the study region. The amplitude of LNAPL may influence the concentration of dissolved LNAPL and the moisture content of the soil.

Also, this paper shows that there are quite very few studies dealing with measuring the water and contaminant distribution in the subsurface system using a two-dimensional tank experiment, need, therefore, to be considered to study the movement behavior of the contaminant in term of groundwater fluctuations. Moreover, there is still a deficiency in the effect of both LNAPL and variation in the water table in conjunction with the particle size distribution in the main domain. Soil permeability is one of the key elements that determine the migration path and rainwater infiltration. It depends on the soil grain sizes and composition. The distribution of light non-aqueous phase liquid in layered homogeneous sand is still not covered by the simplified image analysis method. Furthermore, there are also very few studies that examined the effects of density and viscosity on the movement of the contaminant in the whole domain. For these issues, the complexity of LNAPL chemical, soil composition, and heterogeneity, and the groundwater table fluctuations amplitude in terms of groundwater table fluctuation needs to be studied. Finally, future laboratory simulations should take into account the most current technological advances in this field.

References

1. Chen, J., Zhang, W., Wan, Z., Li, S., Huang, T., Fei, Y.: Oil spills from global tankers: status review and future governance. J. Clean. Prod. **227**, 20–32 (2019)
2. Yekeen, S., Balogun, A., Aina, Y.: Early warning systems and geospatial tools: managing disasters for urban sustainability. In: Sustainable Cities and Communities, pp. 1–13. Springer International Publishing, Cham, Switzerland (2019)
3. Michel, J., Fingas, M.: Oil spills: causes, consequences, prevention, and countermeasures. In: Fossil Fuels. Research Planning, Inc., Columbia, SC, USA (2015)
4. Pelta, R., Carmon, N., Ben-Dor, E.: A machine learning approach to detect crude oil contamination in a real scenario using hyperspectral remote sensing. Int. J. Appl. Earth Obs. Geoinf. **82**, 101901 (2019)

5. Jiao, Z., Jia, G., Cai, Y.: A new approach to oil spill detection that combines deep learning with unmanned aerial vehicles. Comput. Ind. Eng. **135**, 1300–1311 (2019)
6. Nwachukwu, A.N., Osuagwu, J.C.: Effects of oil spillage on groundwater quality in Nigeria. Am. J. Eng. Res. AJER **3**, 271–274 (2014)
7. Mignucci-Giannoni, A.: Assessment and rehabilitation of wildlife affected by an oil spill in Puerto Rico. Environ. Pollut. **104**, 323–333 (1999)
8. Fingas, M.: The Basics of Oil Spill Cleanup. CRC Press, Boca Raton, FL, USA (2012)
9. National Research Council.: Oil in the Sea III: Inputs, Fates, and Effects. National Academies Press (US), Washington, DC, USA (2003)
10. Li, P., Cai, Q., Lin, W., Chen, B., Zhang, B.: Offshore oil spill response practices and emerging challenges. Mar. Pollut. Bull. **110**, 6–27 (2016)
11. Westerholm, D.A., Rauch, S.D., III, Kennedy, D.M., Basta, D.J.: Deepwater horizon oil spill: final programmatic damage assessment and restoration plan and final programmatic environmental impact statement. In: Natural Resources Science Plan 2011–2015. Springer, Berlin/Heidelberg, Germany (2011)
12. Piatt, J.F., Lensink, C.J., Butler, W., Nysewander, D.R.: Immediate impact of the 'Exxon Valdez' oil spill on marine birds. Auk **107**, 387–397 (1990)
13. Nevalainen, M., Helle, I., Vanhatalo, J.P.: Estimating the acute impacts of Arctic marine oil spills using expert elicitation. Mar. Pollut. Bull. **131**, 782–792 (2018)
14. Prabowo, A.R., Bae, D.M.: Environmental risk of maritime territory subjected to accidental phenomena: correlation of oil spill and ship grounding in the Exxon Valdez's case. Results Eng. **4**, 100035 (2019)
15. Amir-Heidari, P., Arneborg, L., Lindgren, J.F., Lindhe, A., Rosén, L., Raie, M., Axell, L., Hassellöv, I.-M.: A state-of-the-art model for spatial and stochastic oil spill risk assessment: a case study of oil spill from a shipwreck. Environ. Int. **126**, 309–320 (2019)
16. Grubesic, T.H., Nelson, J.R., Wei, R.: A strategic planning approach for protecting environmentally sensitive coastlines from oil spills: allocating response resources on a limited budget. Mar. Policy **108**, 103549 (2019)
17. Fan, C., Hsu, C.-J., Lin, J.-Y., Kuan, Y.-K., Yang, C.-C., Liu, J.-H., Yeh, J.-H.: Taiwan's legal framework for marine pollution control and responses to marine oil spills and its implementation on T.S. Taipei cargo shipwreck salvage. Mar. Pollut. Bull. **136**, 84–91 (2018)
18. Bullock, R.J., Perkins, R.A., Aggarwal, S.: In-situ burning with chemical herders for Arctic oil spill response: meta-analysis and review. Sci. Total Environ. **675**, 705–716 (2019)
19. Sardi, S.S., Qurban, M.A., Li, W., Kadinjappalli, K.P., Manikandan, K.P., Hariri, M.M., Al-Tawabini, B.S., Khalil, A.B., El-Askary, H.: Assessment of areas environmentally sensitive to oil spills in the western Arabian Gulf, Saudi Arabia, for planning and undertaking an effective response. Mar. Pollut. Bull. **150**, 110588 (2019)
20. Lenhard, R.J., Rayner, J.L., García-Rincón, J.: Testing an analytical model for predicting subsurface LNAPL distributions from current and historic fluid levels in monitoring wells: a preliminary test considering hysteresis. Water **11**, 2404 (2019). https://doi.org/10.3390/w11112404
21. US.EPA.: A decision-making framework for cleanup of sites impacted with light non-aqueous phase liquids (LNAPL). Us.Epa. 542-R-04-0, 86 (2005)
22. CRC CARE.: A practitioner's guide for the analysis, management and remediation of LNAPL cooperative research centre for contamination assessment and remediation of the environment, technical report series (2015)
23. Chevalier, L.R., Petersen, J.: Literature review of 2-D laboratory experiments in NAPL flow, transport, and remediation. Soil Sediment Contam. **8**, 149–167 (1999). https://doi.org/10.1080/10588339991339289
24. Saleem, M., Al-Suwaiyan, M.S., Aiban, S.A., Ishaq, A.M., Al-Malack, M.H., Hussain, M.: Estimation of spilled hydrocarbon volume—the state-of-the-art. Environ. Technol. **25**, 1077–1090 (2004). https://doi.org/10.1080/09593330.2004.9619401
25. Oostrom, M., Hofstee, C., Wietsma, T.W.: Behavior of a viscous LNAPL under variable water table conditions. Soil Sediment Contam. **15**, 543–564 (2006). https://doi.org/10.1080/15320380600958976

26. Oostrom, M., Dane, J.H., Wietsma, T.W.: A review of multidimensional, multifluid, intermediate-scale experiments: flow behavior, saturation imaging, and tracer detection and quantification. Vadose Zo. J. **6**, 610–637 (2007). https://doi.org/10.2136/vzj2006.0178
27. Zhao, S., Wang, J., Feng, S., Xiao, Z., Chen, C.: Effects of ecohydrological interfaces on migrations and transformations of pollutants: a critical review. Sci. Total Environ. **804**, 150140 (2022). https://doi.org/10.1016/j.scitotenv.2021.150140
28. Cavelan, A., Golfier, F., Colombano, S., Davarzani, H., Deparis, J., Faure, P.: A critical review of the influence of groundwater level fluctuations and temperature on LNAPL contaminations in the context of climate change. Sci. Total Environ. **806** (2022). https://doi.org/10.1016/j.scitotenv.2021.150412
29. Alazaiza, M.Y.D., Ngien, S.K., Bob, M.M., Kamaruddin, S.A., Ishak, W.M.F.: Non-aqueous phase liquids distribution in three-fluid phase systems in double-porosity soil media: experimental investigation using image analysis. Groundwater Sustain. Dev. **7**, 133–142 (2018). https://doi.org/10.1016/j.gsd.2018.04.002
30. Huang, Y., Wang, P., Fu, Z., Shen, H.: Experimental and numerical research on migration of LNAPL contaminants in fractured porous media. Hydrogeol. J. **28**, 1269–1284 (2020). https://doi.org/10.1007/s10040-020-02118-w
31. Badv, K., Mohammad Seyyedi, B., Nimtaj, A.: Numerical investigation of propagation of BTEX compounds in soil. Geotech. Geol. Eng. **38**, 3875–3890 (2020). https://doi.org/10.1007/s10706-020-01263-z
32. Foong, L.K., Rahman, N.A., Nazir, R., Sa'ari, R., Mustaffar, M.: Investigation of aqueous and non-aqueous phase liquid migration in double-porosity soil using digital image analysis. Chem. Eng. Trans. **63**, 685–690 (2018). https://doi.org/10.3303/CET1863115
33. Crawford, R.L., Alcock, J., Couvreur, J.F., Dunk, M., Fombarlet, C., Frieyro, O., Lethbridge, G., Mitchell, T., Molinari, M., Ruiz, H., Walden, T., Martin, D.E.: European oil industry guideline for risk-based assessment of contaminated sites (revised). CONCAWE Rep. (2003)
34. Alesse, B., Orlando, L., Palladini, L.: Non-invasive lab test in the monitoring of vadose zone contaminated by light non-aqueous phase liquid. Geophys. Prospect. **67**, 2161–2175 (2019). https://doi.org/10.1111/1365-2478.12809
35. Mercer, J.W., Cohen, R.M., Va, U.S.A.: Review Paper a Review of Immiscible Fluids in the Subsurface: Properties, Models, Characterization and Remediation [Nonaqueous phase liquids (NAPL's) have been discovered at numerous hazardous waste sites (e.g., Faust, 1985; Mercer et al., 19. J. Contam. Hydrol. 6, 107–163)] (1990)
36. Kacem, M., Esrael, D., Boeije, C.S., Benadda, B.: Multiphase flow model for NAPL infiltration in both the unsaturated and saturated zones. J. Environ. Eng. **145**, 04019072 (2019). https://doi.org/10.1061/(asce)ee.1943-7870.0001586
37. Azimi, R., Vaezihir, A., Lenhard, R.J., Majid Hassanizadeh, S.: Evaluation of LNAPL behavior in water table inter-fluctuate zone under groundwater drawdown condition. Water (Switzerland) **12** (2020). https://doi.org/10.3390/W12092337
38. Yang, Y.S., Li, P., Zhang, X., Li, M., Lu, Y., Xu, B., Yu, T.: Lab-based investigation of enhanced BTEX attenuation driven by groundwater table fluctuation. Chemosphere **169**, 678–684 (2017). https://doi.org/10.1016/j.chemosphere.2016.11.128
39. Charbeneau, R.: LNAPL Distribution and Recovery Model. Distribution and Recovery of Petroleum Hydrocarbon Liquids in Porous Media, vol. 1, p. 4760. API Publication (2007)
40. Jeong, J., Charbeneau, R.J.: An analytical model for predicting LNAPL distribution and recovery from multi-layered soils. J. Contam. Hydrol. **156**, 52–61 (2014). https://doi.org/10.1016/j.jconhyd.2013.09.008
41. Kechavarzi, C., Soga, K., Illangasekare, T.H.: Two-dimensional laboratory simulation of LNAPL infiltration and redistribution in the vadose zone (2005). J. Contam. Hydrol. **76**, 211–233 (2004). https://doi.org/10.1016/j.jconhyd
42. Kemblowski, M.W., Chiang, C.Y.: Hydrocarbon thickness fluctuations in monitoring wells. Groundwater **28**, 244–252 (1990). https://doi.org/10.1111/j.1745-6584.1990.tb02252.x
43. Lenhard, R.J., Johnson, T.G., Parker, J.C.: Experimental observations of nonaqueous phase liquid subsurface movement. J. Contam. Hydrol. **12**, 79–101 (1993). https://doi.org/10.1016/0169-7722(93)90016-L

44. Parker, J.C., Lenhard, R.J.: A model for hysteretic constitutive relations governing multiphase flow: 1. saturation-pressure relations. Water Resour. Res. **23**, 2187–2196 (1987). https://doi.org/10.1029/WR023i012p02187
45. Sookhak Lari, K., Johnston, C.D., Davis, G.B.: Gasoline multiphase and multicomponent partitioning in the vadose zone: dynamics and risk longevity. Vadose Zone J. **15**, 1–15
46. Van Geel, P.J., Sykes, J.F.: The importance of fluid entrapment, saturation hysteresis and residual saturations on the distribution of a lighter-than-water non-aqueous phase liquid in a variably saturated sand medium. J. Contam. Hydrol. **25**, 249–270 (1997). https://doi.org/10.1016/S0169-7722(96)00038-1
47. ITRC.: LNAPL site management: LCSM evolution, decision process, and remedial technologies. LNAPL-3. Interstate Technology & Regulatory Council, Washington, DC
48. Flores, G., Katsumi, T., Inui, T., Kamon, M.: A simplified image analysis method to study lnapl migration in porous media. Soils Found. **51**, 835–847 (2011). https://doi.org/10.3208/sandf.51.835
49. Ramli, M.H.: Dynamic Effects on Migration of Light Non-Aqueous Phase Liquids in Subsurface. Dissertation. Kyoto University (2014). https://doi.org/10.14989/doctor.k18487
50. Skoog, D.A., Holler, F.J., Crouch, S.R.: Principles of Instrumental Analysis. Thomsom Brooks/Cole, Belmont (2007)
51. Kechavarzi, C., Soga, K., Wiart, P.: Multispectral image analysis method to determine dynamic fluid saturation distribution in two-dimensional three-fluid phase two laboratory experiments. J. Contam. Hydrol. **46**(3–4), 265–293 (2000)
52. Flores, G., Katsumi, T., Eua-Apiwatch, S., Lautua, S., Inui, T.: Effects of repeated drainage and imbibition on the contamination behavior of a LNAPL and on its S-p relation. J. Geo-Eng. Sci. **3**, 15–30 (2016). https://doi.org/10.3233/jgs-150033
53. Flores, G., Katsumi, T., Inui, T., Ramli, H.: Characterization of LNAPL distribution in whole domains subject to precipitation by the simplified image analysis method. Coupled Phenom. Environ. Geotech. Theor. Exp. Res. Pract. Appl. Proc. Int. Symp. ISSMGE TC **215**, 573–577 (2013). https://doi.org/10.1201/b15004-76
54. Sudsaeng, S., Flores, G., Katsumi, T., Inui, T., Likitlersuang, S., Yimsiri, S.: Experimental study of diesel migration in P orous Media by Simplified Image Analysis Method. Geo-Environmental Engineering 2011 Kagawa National College of Technology, Takamatsu, Japan, May 21–22 2011
55. Flores, G., Tkatsumi, T., Inui, T., Takai, A.: Measurement of NAPL saturation distribution in whole domains by the simplified image analysis method. In: 18th International Conference on Soils Mechanics and Geotechnical Engineering (ICSMGE). Challenges Innov. Geotech. **4**, 3017–3020 (2013)
56. Flores, G., Katsumi, T., Eua-Apiwatch, S., Lautua, S., Inui, T.: Migration of different LNAPLs in subsurface under groundwater fluctuating conditions by the simplified image analysis method. J. Geo-Eng. Sci. **3**, 15–30 (2016). https://doi.org/10.3233/jgs-150033
57. Alazaiza, M.Y.D., Ramli, M.H., Copty, N.K., Ling, M.C.: Assessing the impact of water infiltration on LNAPL mobilization in sand column using simplified image analysis method. J. Contam. Hydrol. **238**, 103769 (2021). https://doi.org/10.1016/j.jconhyd.2021.103769
58. Alazaiza, M.Y.D., Al Maskari, T., Albahansawi, A., Amr, S.S.A., Abushammala, M.F.M., Aburas, M.: Diesel migration and distribution in capillary fringe using different spill volumes via image analysis. Fluids **6**, 1–11 (2021). https://doi.org/10.3390/fluids6050189
59. Alazaiza, M.Y.D., Ramli, M.H., Copty, N.K., Sheng, T.J., Aburas, M.M.: LNAPL saturation distribution under the influence of water table fluctuations using simplified image analysis method. Bull. Eng. Geol. Environ. **79**, 1543–1554 (2020). https://doi.org/10.1007/s10064-019-01655-3
60. Ramli, H., Flores, G.: The migration of LNAPL in subsurface affected by spill volume and precipitation. Int. Conf. Contam. Sites **2016**, 96–100 (2016)
61. Sudsaeng, S., Yimsiri, S., Flores, G., Katsumi, T., Inui, T., Likitlersuang, S.: Diesel migration in sand under groundwater movements. In: 5th Asia-Pacific Conference on Unsaturated Soils, vol. 2, pp. 493–498 (2012)

62. Yimsiri, S., Euaapiwatch, S., Flores, G., Katsumi, T., Likitlersuang, S.: Effects of water table fluctuation on diesel fuel migration in one-dimensional laboratory study. Eur. J. Environ. Civ. Eng. **22**, 359–385 (2018). https://doi.org/10.1080/19648189.2016.1197158
63. Ramli, H., Lee, Z.X., Azmi, M., Ahmad, F.: Capillary rise determination using simplified image analysis method. E3S Web of Conf. **195**, 03017 (2020). https://doi.org/10.1051/e3sconf/202019503017

Modeling of Monopile for Onshore Wind Turbines Tower by COMSOL Multiphysics

Ahmed B. Shaath and Harris Ramli

Abstract This paper presents a study of a wind tower foundation subjected to wind loading, accomplish by using COMSOL Multiphysics. The restriction of the program is discussed with the way, in which the soil and monopile are modeled. The aim of this paper is to examine multiple monopile structure comparisons, and the problem of geotechnical engineering loads via numerical models. The computer codes are briefly described, and the numerical model is made between the results obtained from the FEM methods. Numerical methods in geotechnical engineering loads applied to the model comprise various wind turbines tower, with two different soil types. Foundation displacement, stress distribution, and total displacement at critical sections underestimate design load/stress were compared. Generally, this research looks at the virtual loads on surface of monopile on top part of monopile and the body load of foundation by applied Mohr–Coulomb in numerical model as well as similar results are found using the program for these cases, except for the wind loads predicted. Additionally, the impact of the mechanical characteristics of the adjacent soft clay and loose sand soil on the load pattern was as well investigated, together with the angle of internal friction and the cohesion of soft clay and loose sand. The outcomes of this paper may possibly be used as standards and recommendations for accomplishing a satisfactory design of monopile for onshore wind turbines tower.

Keywords Wind turbines tower · Monopile · Soft soil · Renewable energy · Foundation design · COMSOL multiphysics · FEM methods

A. B. Shaath · H. Ramli (✉)
School of Civil Engineering, Universiti Sains Malaysia, Engineering Campus, 14300 Nibong Tebal, Pulau Pinang, Malaysia
e-mail: cemhr@usm.my

A. B. Shaath
e-mail: ahmedshaath@student.usm.my

N. Sabtu (ed.), *Proceedings of AWAM International Conference on Civil Engineering 2022 - Volume 3*, Lecture Notes in Civil Engineering 386,
https://doi.org/10.1007/978-981-99-6026-2_9

1 Introduction

Wind turbines have been playing a very vital role to production environmental and green energy to achieve the rising of electricity demands [1]. The stability of the foundation structure on soft soil is more valuable than placing a wind turbine next to a beach. The engineers involved in the design of the substructure consider any potential difficulties with serviceability on soft soil. Due to their significant role in the creation of green electricity, onshore wind turbines have been evolving into a primary source of producing energy. Up to 30% of the world's green energy capacity will be produced by onshore wind energy by 2021 [2]. Extreme settlement from soft soil could lead to structural failure of the foundations' overall system [3]. The monopile system has been the particularly reasonable from cost aspects for sea level depths less than 25 m [4]. The main objective for high use of this groundwork sort is to attract high loads from the water, wind, and dead loads as well as the water depths and the size of wind turbines tower increase, the radius of monopile needed to face the different type of forces from the ecosystem such as water and wind force turn into considerable higher [5]. Figure 1 shows an example of a monopile foundation used for onshore wind turbines [6].

The groundwork category regularly implemented for wind turbines tower is monopile, reporting at around 75% of wind turbines tower installed to date [7]. In the modeling and evaluation of the dynamic attributes of wind turbine structure, the groundwork was supposed rigid by several scholars such as Van der Tempel and Molenaar [8], Reinhard and Gideon [9]. The monopile is the preponderant sort owing to various rewards. The primary regard for a monopile foundation of onshore wind

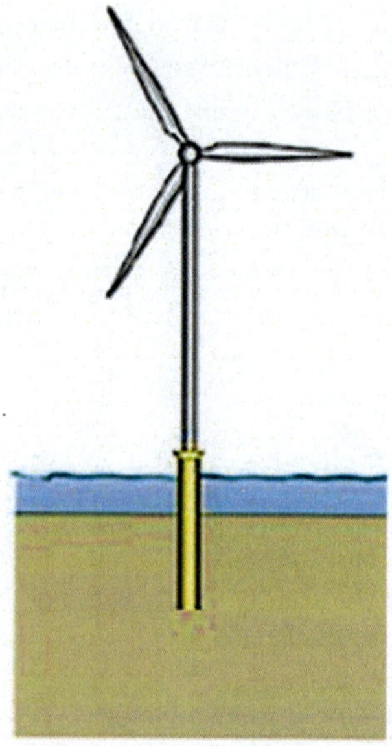

Fig. 1 Monopile foundation used for onshore wind turbine tower [6]

turbine comprises solidity, strength, and natural frequency together with the acceptable settlement of the structure throughout the process [10], naturally administered through permanent settlement in natural frequency and the serviceability limit state (SLS) [11].

The groundwork is surrounded mainly by soil with varying degrees of saturated [12]. Groundwork settlement in soft clay is a challenging phenomenon in front of geotechnical engineers worldwide [13]. During the lifespan of a groundwork system, monopile supported on soft clay, which has large amount of liquid could be achieved by high ratio of settlements due to their elevated compressibility and tiny strength of a material [14]. In insufficient geotechnical situations, traditional isolated foundations or even mat foundations might not extremely accomplish performance. The groundwork constructed on soft soil might possibly produce unsuitable damage or harm to the superstructure [15]. The conditions are able to turn out to be not as good as if there is a seismic behavior [16].

In challenging soft soils, monopile is commonly applied on force and onshore to carry on massive onshore wind turbine tower loads to strong layers of soil, relying on end bearing capacity and skin friction [17]. Additional, monopile is taken into account as a clarification to manage the settlement of onshore wind turbine towers built on weak ground [16]. The static forces from the onshore wind turbine tower, this monopile has been objected to further seismic forces from the adjacent soil during vibration actions [18]. If they are not considered to support these extra loads, the monopile could be unsuccessful. Hence, the reaction of monopile and adjacent ground during seismic events must be clearly recognized for the security purpose of monopile and the onshore wind turbine tower.

COMSOL Multiphysics computer software has been employed for the analysis of monopile subjected to applied loadings. The program is briefly introduced, and the basis of the program is described. Although the obtainability of many different geotechnical engineering software, it sounds that slight variance has been made to compare the results of the analysis carried out with alternative soil types. The main objective of this paper is to critically examine the above-mentioned software so could be used with more confidence when approaching a real case. The model was implemented in COMSOL Multiphysics and used to perform simulations of suction-assisted installation in soft soil.

2 Methodology

The finite element method (FEM) has been completed by applying COMSOL Multiphysics software to make sure the load and soil properties affect monopile foundations' behavior. The model was accomplished by taking into account soft soil. The monopile action on the soft clay under various geotechnical and loading cases is the fundamental purpose of this paper. The numerical simulation was examined for the six cases established on soft clay and six cases on loose sand. Twelve (12) cases of monopile were considered in this study. Each case varies in terms of loads.

The terminology of the numerical analysis studies is illustrated in Table 1. The first and second letter represents the soil condition and the third and fourth letter indicates the software used in the case study while the number indicates the estimated design load. To illustrate, 'CASE-SCCM 100' means Soft Clay (SC) modeling via COMSOL Multiphysics (CM) software has been tested to underestimate design load 100 kPa.

The strategic goal of this paper is to estimate the performance of the Monopile for onshore wind turbines tower under variable loading conditions compare with the above-mentioned software so they could be applied with more confidence when considering a real case. The FEM base numerical analysis has been implemented in this paper. The numerical analysis has been completed by applying the solid mechanic's approach in software. It has been promising with using software to model the monopile and soil together. The index properties of soft clay and loose sand soil defined in Sect. 3 were integrated into the case studies as cohesive material. The Mohr–Coulomb model involves a very limited set of parameters, namely: friction angle, cohesion, and dilation angle (the elastic range: Young's modulus and Poisson's ratio). Every one of these factors would be very simply defined in the research workshop, and hence, the Mohr–Coulomb is applied mostly got from the limitation of the prototype itself rather than from the uncertainty in its standardization. The soil is modeled as Mohr–Coulomb's material [19], with the following governing equations:

The function of normal and shear stress:

$$\tau_f = f(\sigma) \tag{1}$$

The linear function:

$$\tau_f = c + \sigma \tan \Phi \tag{2}$$

The linear function in terms of effective stress:

$$\tau_f = c' + \sigma' \tan \Phi' \tag{3}$$

Table 1 Terms of the considered cases in the FEM analysis

Estimate design load/stress (kPa)	Soil type	
	Soft clay	Loose sand
100	CASE-SCCM 100	CASE-LSCM 100
150	CASE-SCCM 150	CASE-LSCM 150
200	CASE-SCCM 200	CASE-LSCM 200
250	CASE-SCCM 250	CASE-LSCM 250
300	CASE-SCCM 300	CASE-LSCM 300
350	CASE-SCCM 350	CASE-LSCM 350

whereas τ_f is the shear stress, σ is the normal stress, c is defined as material cohesion, Φ is the frictional angle of the soil. c' and σ' are respective values in terms of effective stresses. The monopile foundation is assumed to be reinforced concrete. The concrete is assumed as linear elastic with normal standards of elasticity modulus and density Poisson's ratio.

3 Soil Properties

It is well known that soil index properties are not similar on all construction sites. The soil characteristics differ depending on the local earth science. The ground situation determines whether to select a shallow foundation or monopile for the wind turbine tower. Especially, the massive masses have been essential to carry on to the ground and if ground soil is soft, the majority of the engineers propose to set out for monopile that have been extremely suitable to strengthen the wind turbine tower from fallen and other foundation system challenges. The monopile could keep the stress impacts owing to wind forces and bending moments on the structure. Dense build of wind structure transfers enormous forces onto the earth through the monopile [20]. To bear such enormous weight, the bearing capacity of soil should be strong enough. Basically, poor soils properties will lead to low bearing capacity, and these soil types immediate would not be suitable for the construction of the groundwork for wind tower [21]. Hence, the index properties of soft clay and loose sand soil have been necessary to be enhanced. Table 2 gives, that the index properties of soft clay soil are used in this paper. The soft clay soil models had obtained from Bangkok MRT project and the loose sand had taken from the geotechnical laboratory at the Faculty of Engineering, Islamic University of Madinah [22, 23]. The specifics of the parameters applied in this paper are illustrated in Table 2. Figure 2 shows soil models of length (L) 8000 mm and diameter (d) 3500 mm were applied in this numerical.

Table 2 Index properties of soft clay and loose sand soil

Parameters	Soft clay	Loose sand
Grain size distribution		
• Sand	4%	92%
• Silt	31.7%	2%
• Clay	64.3%	0
• Gravels	0%	6%
Plasticity index (P.I.)	43%	0%
Liquid limit (L.L.)	118%	0%
Plasticity index (P.I.)	75%	0%
Natural moisture content (w)	122%	0%
Specific gravity (G_s)	2.75	2.52

Fig. 2 Soil geometry of onshore wind turbine tower

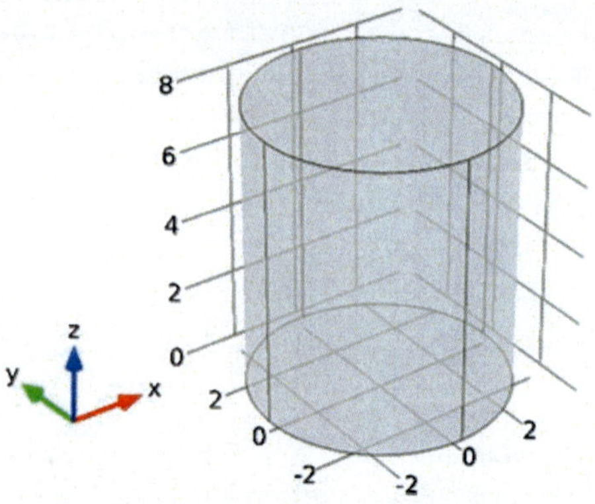

4 Numerical Analysis

This section goes through the required calculations to achieve the study detailed in the flowchart in Fig. 3 and the 11-step model would be defined in this section. The monopile of onshore wind turbines tower has been done to support the superstructures insensitive to monopile depth. In researching the reaction of the wind turbine tower, finite element method (FEM) is typically applied in structural design and analysis by Farooq and Shaath [23]. A reinforcement concrete monopile of length (L) 4000 mm and diameter (d) 500 mm was applied in this study. The specifics of the parameters applied in this paper are illustrated in Table 3.

The numerical modeling has been completed by utilizing the solid mechanic's module of COMSOL Multiphysics software. It was possible with the assistance of software to model the pile and soil together. The soil could be investigated and examined by applying several loads patterns [24]. The monopile is simpler than the gravity base and in this paper, the gravity force takes into account in the monopile. It is a cylindrical pipe with a constant large diameter that is drilled or driven into the soil. The Gravity point adds the force F_g, which is equal to ρ_g, where g is the gravity vector. Gravity enter the coordinates for the gravity vector g (SI unit: m/s^2). The default value is -g_const in the z component. g_const is the acceleration of gravity (a predefined physical constant) [24].

5 Evaluation Results

The outcomes achieved formed on the stress–strain study are shown in this section. Each soil type has been examined for all the six loads pattern. The outcomes are presented for total displacement, and stress or pressure distribution, and the models for each case against the above-mentioned parameters have been presented individually in the following subsections.

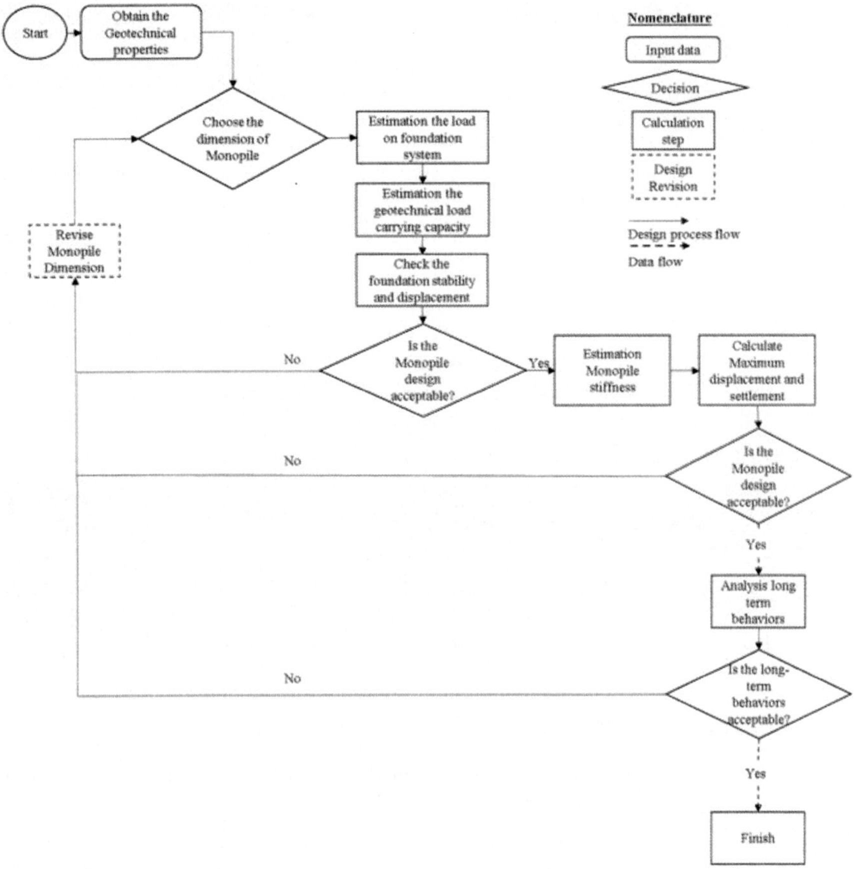

Fig. 3 Flowchart of the monopile modeling

Table 3 Adopted/calculated parameters for the numerical models

Description and standard symbol	Soft clay	Loose sand	Unit
Soil bulk modulus (K_{soil})	0.69	9	MPa
Soil dry density (ρ_d)	16.5	14.7	kN/m^3
Soil dilation angle (ψ)	0	0.007	rad
Soil friction angle (ϕ)	0.471	0.537	rad
Poisson's ratio (n)	0.495	0.250	–
Void ratio (e)	3.11	0.72	–
Cohesion (C)	10	10	kN/m^2

5.1 Soft Clay Cases Study

Total Displacement for Monopile

The outcomes of the displacement analysis for the estimated design load/stress from 100 to 350 kPa are shown in Fig. 4. Figure 4 shows the 3D total displacement, it could be noticed that the highest displacement turns out at the center of the monopile. Under the different loading conditions and same geotechnical properties, the maximum total displacement for the model was 36.39 mm when 350 kPa was applied, while it decreases to 0.27 mm for the minimum estimated design load. In CASE-SCCM 300 and CASE-SCCM 350, the monopile might collapse in order to the displacement of the monopile is 36.39 mm and 28.98 mm, respectively. According to Eurocode 7: Geotechnical Design Worked, the maximum allowable displacement is 25 mm more than the limit value is an unacceptable displacement that leads to the redesign of the monopile dimensions according to Fig. 3 [25]. Meanwhile, the minimum monopile displacement has been achieved in the bottom of soil simply in all load cases, the displacement bottom zone for all cases is zero and generally increased around the monopile foundation. Randolph [26] observed a scientific approach to determining the displacement of monopile should consider the building complex stress–strain record, that involves; the initial in situ situation, installation, and loading the following section discusses stress/pressure distribution for the monopile foundation for soft clay.

Stress or Pressure Distribution

The stress and characteristics of the soil items at different load pattern after hunt were extracted from the model to calculate a normal stress. The normal stress is a stress that happens once a unit is filled by an axial load. The amount of the normal stress for any section is basically the force (P) divided by the cross-sectional (A) area is defined by Eq. 4:

$$\sigma = \frac{P}{A} \tag{4}$$

The results have obtained from the FEM study based on stress (mean pressure) for CASE-SCCM 100, CASE-SCCM 150, CASE-SCCM 200, CASE-SCCM 250, CASE-SCCM 300, and CASE-SCCM 350 are presented in this section. The limit stress established at the intersection of monopile and soil while the pressure presented for the soft clay cases is shown in Fig. 4. The pressure "P" presented in Fig. 4 is defined by Eq. 5:

$$\pi = \frac{\sigma_{xx} + \sigma_{yy} + \sigma_{zz}}{3} \tag{5}$$

In these cases, π indicated for the pressure, σ_{xx}, σ_{yy}, and σ_{zz} are normal stresses. In these cases of principal stresses, this pressure is similar to the mean principal

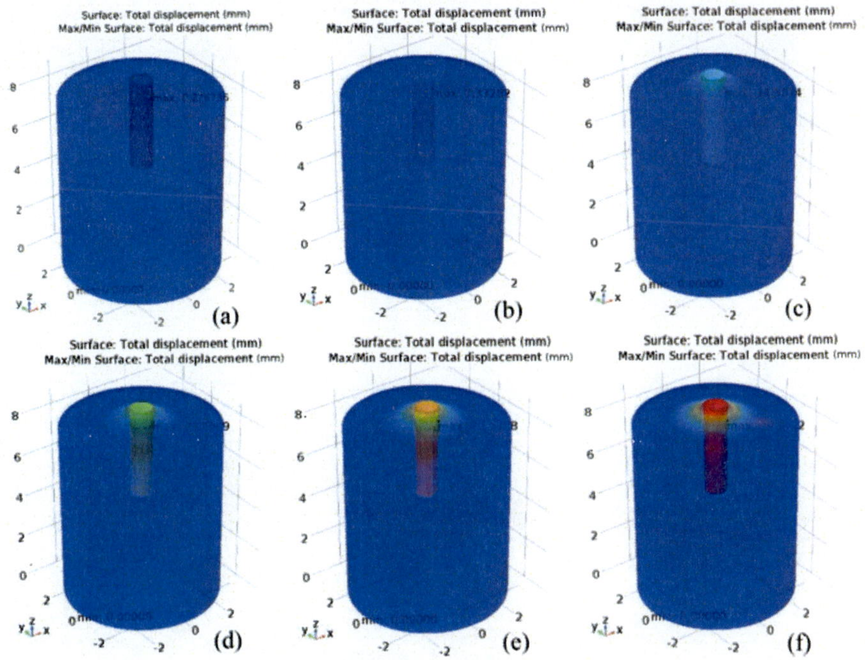

Fig. 4 Displacement for monopile **a** CASE-SCCM 100, **b** CASE-SCCM 150, **c** CASE-SCCM 200 **d** CASE-SCCM 250, **e** CASE-SCCM 300, and **f** CASE-SCCM 350

stress, used to indicate the stress path in the triaxial tests. In Fig. 5f, the maximum pressure rate and minimum pressure rate are 107.8 kPa, and – 208.9 kPa respectively while, the maximum rate located at the lower point of the reinforced concrete (RC) monopile foundation and minimum placed on surface of foundation. Due to being a strong material RC, has absorbed the majority of the pressure. Meanwhile in Fig. 5a is shown, the minimum pressure rate between – 34.19 and 0.8605 kPa. The rate of stress in the considered soft clay layer ranges between – 200 and 120 kPa for different estimate design load/stress (kPa) Meanwhile, Fig. 5a is shown, the minimum pressure rate between – 34.19 and 0.8605 kPa. The rate of stress in the considered soft clay layer ranges between – 200 and 120 kPa for different estimated design load/stress (kPa).

5.2 Loose Sand Cases Study

Total Displacement for Monopile

A successful groundwork model of the monopile should ensure efficient, safe, and economic design of the whole wind turbine foundation system. The numerical method

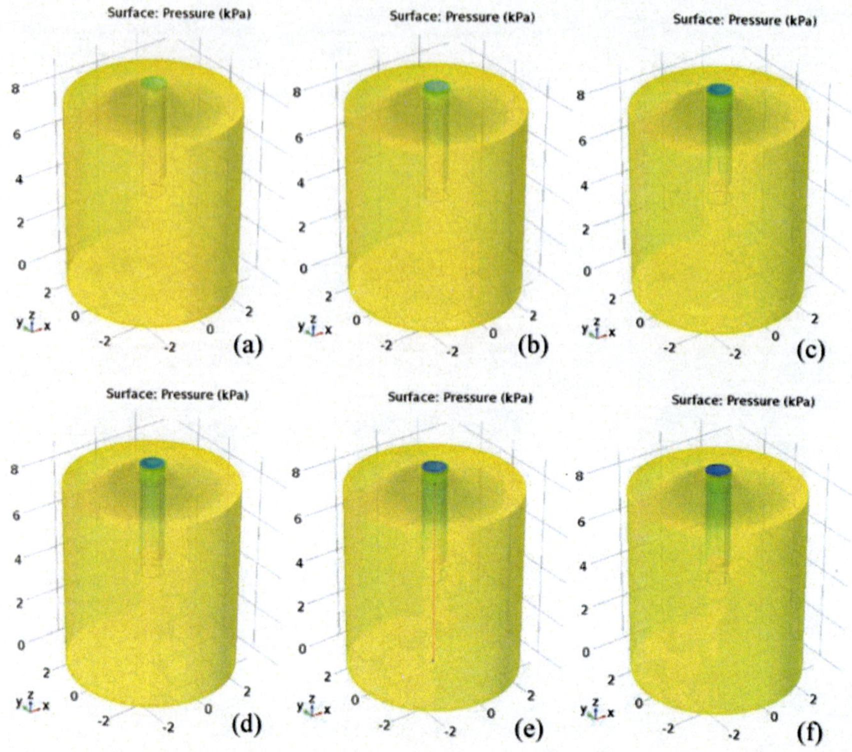

Fig. 5 Pressure distribution for monopile **a** CASE-SCCM 100, **b** CASE-SCCM 150, **c** CASE-SCCM 200, **d** CASE-SCCM 250, € CASE-SCCM 300, and **f** CASE-SCCM 350

is applied to obtain a forced response for different types of displacement. The optimistic model must merge the economic features of the overall safety model. The FEM model results for loose sand cases have been indicated in Fig. 6. The conclusions of the displacement result for the estimate design load/stress from 100 to 350 kPa are illustrated in previous plotted. The maximum displacement applied at 350 kPa is almost 5 mm that mean according to Eurocode 7: Geotechnical Design Worked, the displacement is acceptable [25]. The primary reason for a higher performance in the monopile system is soil properties. In Fig. 6a, the displacement might be neglected since it is considered a very tiny changeable that could not be influential on the foundation system performance. Although, the other cases CASE-LSCM 100, CASE-LSCM 150, CASE-LSCM 200, CASE-LSCM 250, and CASE-LSCM 300 have been examined, it considered very low displacement and to achieve the economic model the geotechnical engineers could redesign the model by changing the monopile dimensions to the small measures.

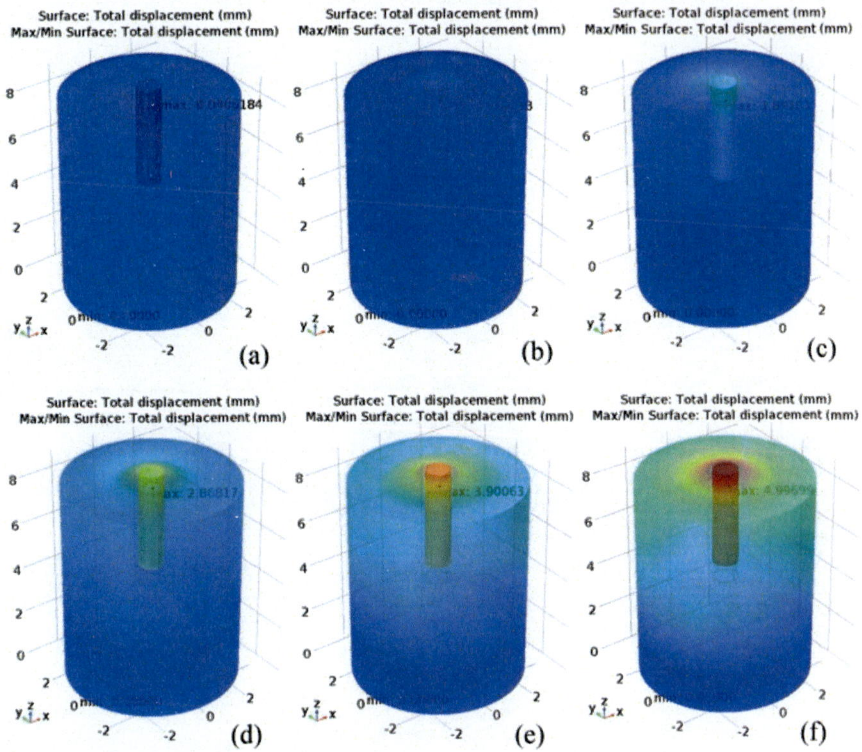

Fig. 6 Displacement for monopile **a** CASE-LSCM 100, **b** CASE-LSCM 150, **c** CASE-LSCM 200, **d** CASE-LSCM 250, **e** CASE-LSCM 300, and **f** CASE-LSCM 350

Stress or Pressure Distribution

Looking from an overall perspective, it is readily apparent that the pressure distribution for Monopile CASE-LSCM 100, CASE-LSCM 150, CASE-LSCM 200, CASE-LSCM 250, CASE-LSCM 300, and CASE LSCM 350 is very low comparing with cases study in soft clay. The FEM study outcomes for loose sand soil cases have been shown in Fig. 7. The location of maximum pressure was at the base edge of monopile foundation, while the peak value in all loose sand case was 55.3 kPa and it located in bearing capacity of CASE-LSCM 350. Like soft clay cases obvious stresses were noted throughout the lower point in monopile.

6 Discussion

COMSOL Multiphysics software have been compared for the displacement, and stress in this cases study. In all the twelve cases, the soil stresses have been very low, i.e., the stresses and pressure are well above the bearing capacity of the two-soil

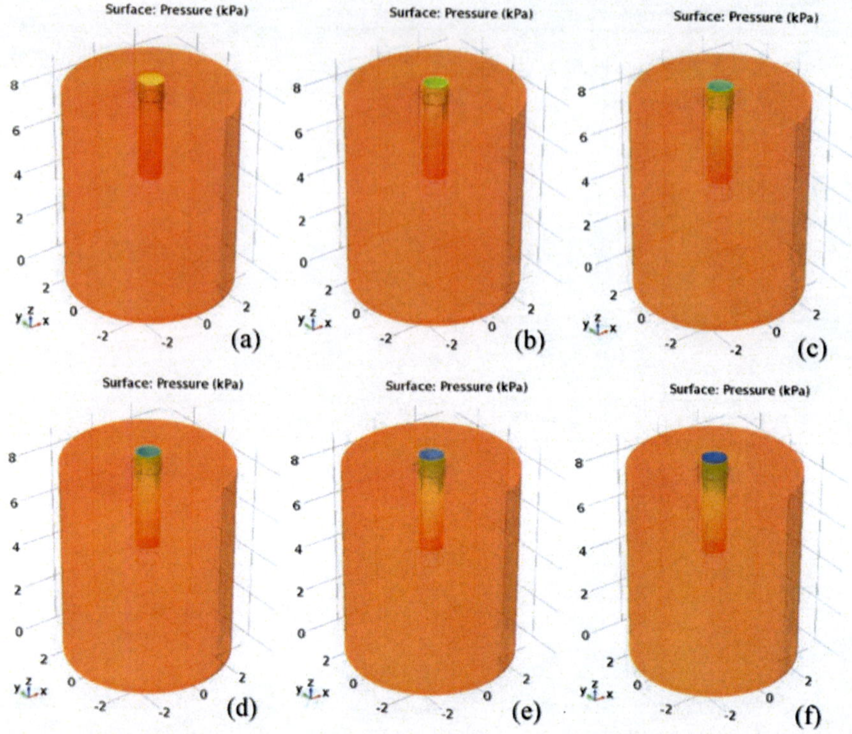

Fig. 7 Pressure distribution for monopile **a** CASE-LSCM 100, **b** CASE-LSCM 150, **c** CASE-LSCM 200, **d** CASE-LSCM 250, **e** CASE-LSCM 300, and **f** CASE-LSCM 350

type. The maximum displacement value studied from all the foundation cases under six various loading situations is plotted in Fig. 8a. The monopile system on loose sand soil has discovered the most excellent resistance compared to soft clay case on overall settlement of the groundwork system. The soft clay soil has indicated the highest displacement. The load settlement distribution has the same pattern for all cases. The primary reason for a higher displacement in soft clay soil cases is the poor of index properties.

The stresses at this location are playing a very significant role and controlling the overall protection of the Onshore Wind Turbines Tower. The highest stress established at layer point in the monopile system is selected for each case. These stresses are then plotted for all twelve cases against six estimate design load/stress in Fig. 8b. The stresses in the loose sand cases had been extremely small as compared to the soft clay cases. Therefore, stresses in Fig. 8b were presented on a natural logarithmic scale higher, the small stresses in the loose sand cases are due equivalent roughly a good soil index properties.

The monopile foundation system in soft clay soil (CASE-SCCM 350) has revealed the maximum displacement. The deformed figure of the groundwork system has been

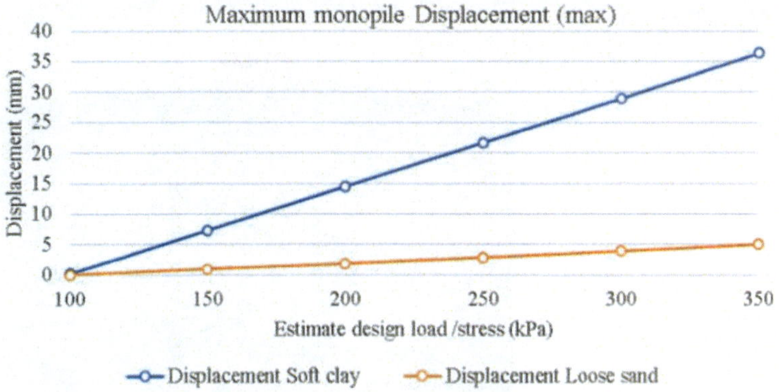

(a) Displacement line graph

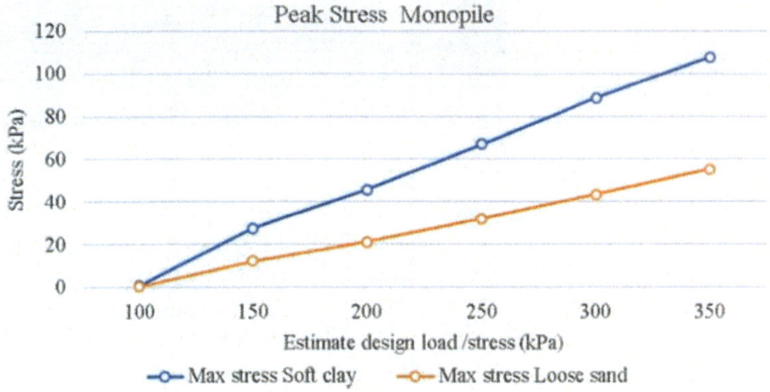

(b) Stresses line graph

Fig. 8 Summary of FEM results

described in Fig. 9. The disFig.ment (sagging) of the monopile and soil might be visible in Fig. 8.

7 Conclusions

The unique computer software was implemented for the monopile analysis. The software has been used in this paper COMSOL Multiphysics. This software was illustrated in brief, and then matched for the evaluation of a simple case. Preliminary monopile designs for soft clay and loose sand soils were established using linear elastic materials. The loose sand soils be able to use for heavily structures, where stresses at lower point of monopile is essential. The displacement might be governed

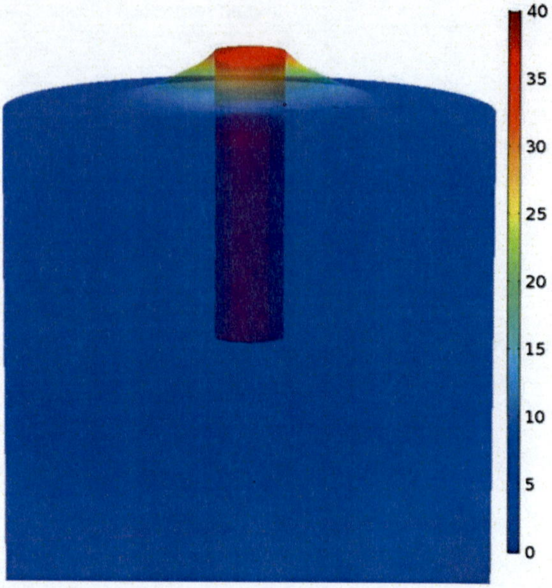

Fig. 9 Deformed shape for CASE-SCCM 350 (mm)

by soil treatment or redesign foundation dimension or type. This cases study was accomplished on soft clay and loose sand soil. On the flip side, consequences could be used in other analogous topographical regions. The methodology for forecasting the displacement of loading monopile systems using a finite element software code is offered by commercial companies such as COMSOL Multiphysics and PLAXIS LE. The design steps were confirmed by comparing the modeled estimate design load/ stress to the results of displacement behavior and Stress or pressure distribution.

The performance of monopile at COMSOL Multiphysics software has been evaluated the displacement, and stress in same time the under fluctuating groundwater conditions and rotation will be studied as future research work, and potential research with challenges and laboratory investigations will be planned to reinforce the numerical study.

Acknowledgements This research is done with the support from the Ministry of Higher Education (MOHE) and Malaysia International Scholarship (MIS). The authors would like to thank MOHE, MIS and Universiti Sains Malaysia for providing the sponsorship and scholarship.

Conflicts of Interest The author(s) declare(s) that there is no conflict of interest regarding the publication of this paper.

References

1. Seguro, J.V., Lambert, T.W.: Modern estimation of the parameters of the Weibull wind speed distribution for wind energy analysis. J. Wind Eng. Ind. Aerodyn. **85**, 75–84 (2000). https://doi.org/10.1016/S0167-6105(99)00122-1
2. GWEC: GWEC | Global wind report (2021)
3. Roy, S., Bhalla, S.K.: Role of geotechnical properties of soil on civil engineering structures. Resour. Environ. 103–109 (2017). https://doi.org/10.5923/j.re.20170704.03
4. Wang, X., Zeng, X., Yang, X., Li, J.: Feasibility study of offshore wind turbines with hybrid monopile foundation based on centrifuge modeling. Appl. Energy **209**, 127–139 (2018) (no. October 2017). https://doi.org/10.1016/j.apenergy.2017.10.107
5. Zhang, Q., Liu, S., Zhang, S., Zhang, J., Wang, K.: Simplified non-linear non-linear approaches for response of a single pile and pile groups considering progressive deformation of pile–soil system. Soils Found. **56**(3), 473–484 (2016). https://doi.org/10.1016/j.sandf.2016.04.013
6. Kaynia, A.M.: Seismic considerations in design of off shore wind turbines. Soil Dyn. Earthq. Eng. **124**, 399–407 (2019). https://doi.org/10.1016/j.soildyn.2018.04.038
7. Murphy, G., Igoe, D., Doherty, P., Gavin, K.: 3D FEM approach for laterally loaded monopile design. Comput. Geotech. **100**, 76–83 (2018). https://doi.org/10.1016/j.compgeo.2018.03.013
8. Van der Tempel, D., Molenaar, J.: A review of the principles for modern power generation, onshore and offshore. Wind Eng. **25**(4), 211–222 (2002). https://doi.org/10.1260/030952402321039412
9. Reinhard, H., Gideon, P.: Structural stability of concrete wind turbines and solar chimney towers exposed to dynamic wind action. J. Wind Eng. Ind. Aerodyn. **95**(9–11), 1079–1096 (2007). https://doi.org/10.1016/j.jweia.2007.01.028
10. Koulin, G., Sewell, I., Shaw, B.A.: Faceted monopile design suitable for mass production and upscaling. In: 1st International Conference on Structural Integrity Faceted, vol. 114, pp. 385–392 (2015). https://doi.org/10.1016/j.proeng.2015.08.083
11. Arany, L., Bhattacharya, S., Macdonald, J.H.G., Hogan, S.J.: A critical review of serviceability limit state requirements for monopile foundations of offshore wind turbines. In: In Offshore Technology Conference. OnePetro (2015). https://doi.org/10.4043/25874-MS
12. Briaud, J.-L.: Geotechnical Engineering Unsaturated and Saturated Soils, 1st edn. Wiley, New Jersey (2013)
13. Ravichandran, N., Shrestha, S.: Performance- and cost-based robust design optimization procedure for typical foundations for wind turbine. Int. J. Geotech. Eng. **14**(4), 395–408 (2020). https://doi.org/10.1080/19386362.2018.1428387
14. Hoang, L.T., Matsumoto, T.: Long-term behavior of piled raft foundation models supported by jacked-in piles on saturated clay. Soils Found. **60**(1), 198–217 (2020). https://doi.org/10.1016/j.sandf.2020.02.005
15. Das, B.M., Sobhan, K.: Principles of Geotechnical Engineering (2012)
16. Garala, T.K., Madabhushi, G.S.P.: Seismic behaviour of soft clay and its influence on the response of friction pile foundations. Bull. Earthq. Eng. **17**(4), 1919–1939 (2019). https://doi.org/10.1007/s10518-018-0508-4
17. Hazzar, L., Hussien, M.N., Karray, M.: Influence of vertical loads on lateral response of pile foundations in sands and clays. J. Rock Mech. Geotech. Eng. **9**(2), 291–304 (2017). https://doi.org/10.1016/j.jrmge.2016.09.002
18. Zaaijer, M.B.: Foundation modelling to assess dynamic behaviour of offshore wind turbines. Appl. Ocean Res. **28**, 45–57 (2006). https://doi.org/10.1016/j.apor.2006.03.004
19. Krabbenhøft, K., Lyamin, A.V., Sloan, S.W.: Three-dimensional Mohr—Coulomb limit analysis using semidefinite programming. Commun. Numer. Methods Eng. **24**(11), 1107–1119 (2008). https://doi.org/10.1002/cnm.1018
20. Zhang, Q., Zhang, Z.: A simplified nonlinear approach for single pile settlement analysis. Can. Geotech. J. **49**, 1256–1266 (2012). https://doi.org/10.1139/T11-110
21. Mohamed, W., Austrell, P.: A comparative study of three onshore wind turbine foundation solutions. Comput. Geotech. **94**, 46–57 (2018). https://doi.org/10.1016/j.compgeo.2017.08.022

22. Surarak, C., Likitlersuang, S., Wanatowski, D., Balasubramaniam, A., Oh, E., Guan, H.: Stiffness and strength parameters for hardening soil model of soft and stiff Bangkok clays. Soils Found. **52**(4), 682–697 (2012). https://doi.org/10.1016/j.sandf.2012.07.009
23. Farooq, Q.U., Shaath, A.B.: The appraisal of hybrid foundations on loose granular soil of western Saudi Arabia. Proc. Eng. Technol. Innov. (2020). https://doi.org/10.46604/peti.2020.5515
24. COMSOL.: COMSOL Multiphysics Reference Manual, Version 5.3. COMSOL AB (2018)
25. Bond, A.J., Schuppener, B., Scarpelli, G., Orr, T.L.L.: Eurocode 7: Geotechnical Design Worked Examples (2013). https://doi.org/10.2788/3398
26. Randolph, M.F.: Science and empiricism in pile foundation design. Geotechnique **53**(10), 847–875 (2003). https://doi.org/10.1680/geot.2003.53.10.847

Review of Natural Fiber Application for Sustainable Ground Improvement

Noor S. Al-Hassnawi, Fauziah Ahmad, Mohammed Y. Fattah, and Mastura Azmi

Abstract Ground modification refers to a variety of processes for enhancing the engineering characteristics of the ground. It works in several soil types, including collapsible, expansive, soft, and mechanically deficient soils. Strips, bars, sheets, membranes, and fibers are examples of reinforcing materials that can help to strengthen the ground. Natural fiber materials, rather than synthetic ones, have recently gained appeal in sustainable geotechnics as a soil reinforcement strategy. Cost, mass availability, and environmental friendliness are all advantages of natural fibers versus synthetic fibers. Natural fibers, on the other hand, differ fundamentally from synthetic fibers, and fiber-reinforced soil's behavior is influenced by both physical and features. In the present review, the brief history, characterization, and properties of natural fiber material are extensively discussed. Review also attempts to explore the advantage and disadvantage of natural fiber material. The degradation mechanism and the treatment method are also presented. As well as the soil reinforcement application, review on the application of geosynthetic and natural fiber as reinforcement in the foundation bed under different loading conditions is presented. Based on the detailed literature review, this paper lay out identified research gaps and present future direction for research.

N. S. Al-Hassnawi · F. Ahmad · M. Azmi (✉)
School of Civil Engineering, Universiti Sains Malaysia, 14300 Nibong Tebal, Malaysia
e-mail: cemastura@usm.my

N. S. Al-Hassnawi
e-mail: nooralhassnawi@student.usm.my

F. Ahmad
e-mail: cefahmad@usm.my

N. S. Al-Hassnawi
Roads and Transportation Engineering Department, College of Engineering, University of Al-Qadisiyah, Al-Qadisiyah, Iraq

M. Y. Fattah
Civil Engineering Department, College of Engineering, University of Technology, Baghdad, Iraq

N. Sabtu (ed.), *Proceedings of AWAM International Conference on Civil Engineering 2022 - Volume 3*, Lecture Notes in Civil Engineering 386,
https://doi.org/10.1007/978-981-99-6026-2_10

Keywords Soil reinforcement · Geosynthetics · Natural fibers · Synthetic material · Sustainable geotechnics

1 Introduction

The demand for structures built on weak/marginal soils has increased as a result of rapid urbanization, imposing a great deal of stress on engineers to enhance ground. Many researchers have attempted to use recycled materials for stabilization of soil and ecofriendly biological mechanisms; however, such new strategies still required to be dealt with the practical applications in the field and lowering costs due to their limited in situ applicability [1]. Chemical stabilization, mechanical stabilization, and soil reinforcement are some of the most often utilized ground stabilization methods. The concept of soil reinforcing by inclusions has been employed in traditional geotechnical infrastructures to improve soil strength in embankment projects [2]. Planar reinforcements like geotextiles, piles, sheets, and geomembrane, as well as discrete random inclusions like synthetic fibers and tire chips, are examples of these inclusions [3].

These synthetic reinforcing materials are often known as reinforcing materials, and on the other hand, these synthetic reinforcing materials are petroleum by-products that are both exhaustible and nonrenewable in nature [4]. As a result of the current infrastructure demand, engineers have been advised to look for alternative materials and use environmentally friendly and sustainable reinforcing agents. A variety of soil management procedures have been developed and utilized in the past to improve the low strength soils prior to development. The recommended improvement approaches can be divided into two categories: (i) mechanical stabilization technique and (ii) chemical stabilization technique. The synthetic reinforcing application is one of the mechanical approaches [5].

The mechanical and chemical stabilization techniques are associated with major environmental problems, e.g., temperature rising caused by massive carbon dioxide emissions, high energy costs, pollution of the environment (air, land, and water), and nonrenewable resource depletion, and inflow of heavy and dangerous compounds into the geo-environment [6]. As a result of the global focus on a healthy future, ecofriendly solutions in the geotechnical and geo-environmental engineering fields are highly preferred. Engineers' current goals are to apply ground renovation techniques to improve existing weaker ground and soils while maintaining land use sustainability [7].

2 Natural Fiber Materials

2.1 Brief History

In 1969, Vidal was the first to introduce the concept and theory of fiber-based soil reinforcement, discovering that reinforcing soil boosts the shear resistance of soil [8]. Since Vidal's invention in 1969, utilizing the soil reinforcement concept, around 4000 applications have been carried out in over 37 countries [2]. Previous research has shown that several reinforcing materials have been employed as geosynthetics to enable soil fiber reinforcement, vary from low polymeric materials to high tensile strength metallic sheets [9].

Traditional synthetic fibers were mostly made from petroleum, which is a nonrenewable, finite resource. As a result, their processing flexibility, minimal cost and large specific stiffness, and geosynthetic products have become increasingly popular [10]. Global capacity for such plastic composites expanded dramatically from a total of 0.36 million metric tons in 2007 to approximately 2.33 million metric tons in 2013, with a forecast of 3.45 million metric tons by 2020 [11]. Furthermore, because corroded steel is particularly toxic to the natural at the end of its lifetime, inserting steel bars as ground reinforcement has already been recognized as a non-environmentally friendly technique [12]. Soil reinforcement with natural fibers has recently gained traction as one of the advanced technologies for sustainable soil strengthening methods in geotechnical engineering [13] because of its advantages like preserving the environment, abundant resources, minimum consumption of energy, cost reasonability, and a lot of potential over other founded materials.

2.2 Characterization and Properties of Natural Fibers

Plant, animal, and mineral fibers are examples of natural fibers. Plant fibers have been projected as geotechnical goals in terms of natural fibers, due to their widespread availability and applicability [14] Cellulose, hemicelluloses, lignin, pectin, and waxy compounds make up lignocellulosic natural fibers from plants. Lignocellulosic fibers are another name for these fibers. The stem's stiffness is provided by lignin, a complex hydrocarbon polymer. It preserves fibers from biological attack. The reinforcing ability of fiber is determined by the crystallinity of cellulose. Hemicellulose is a hydrophilic polymer that forms the cementing matrix. Pectin gives fibers their flexibility [15]. Figure 1 shows a typical fibril matrix structure schematic (Cellulose strands embedded in a matrix of hem cellulose and lignin).

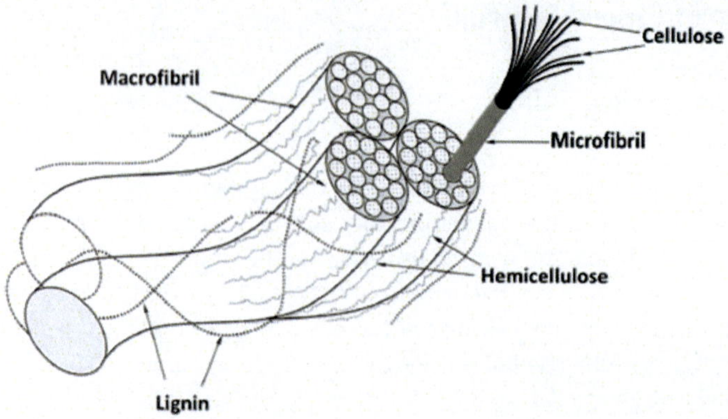

Fig. 1 Fibril matrix structure of plant fiber [16]

2.3 *Advantage and Disadvantage of Natural Fibers*

Natural fibers have a number of advantages, including being relatively inexpensive, simple to produce, and renewable. They have a lower carbon footprint and a higher rate of recovery at the end of their lives [17]. Although natural fibers are less expensive than synthetic fibers, there is a risk of biodegradation in the long run [18]. Due to a lack of adequate interfacial adhesion, a low melting point, and poor moisture resistance, natural-fiber-reinforced composites become less appealing. These factors necessitate pretreatment techniques to clean and chemically change the surface, halt moisture absorption, and increase surface roughness [19]. Natural fiber geotextiles, despite their relative advantages, are rarely used in engineering projects due to their low tensile strengths and sensitivity to biological, chemical, and physical degradation [20]. Natural fibers' use in civil engineering applications is restricted, owing to their proclivity to degrade under various environmental conditions [21].

2.4 *Degradation Mechanism of Natural Fibers*

Natural fibers breakdown more quickly than synthetic fibers in natural surroundings, resulting in less pollution. Natural fiber biodegradation, on the other hand, may limit the life of reinforcing projects. Natural fibers are treated to protect them from hazardous chemicals, allowing them to operate better in naturalistic environments [22]. Natural fibers are mostly made up of biodegradable components such as cellulose, hemicellulose, and lignin. These primary components are subjected to a number of degradation mechanisms, such as temperature deterioration and water uptake, which effect on the fiber's lifespan. Based on the structure and chemistry of the plant fiber, different species have different levels of endurance. The hydrophobic characteristic of natural fibers with a high lignin content restricts water uptake and, as

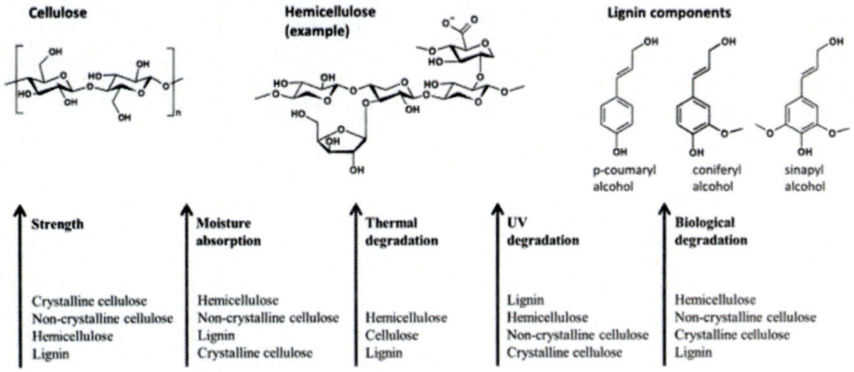

Fig. 2 Plant fibers' primary chemical compositions and their qualities, listed in descending order [23]

a result, microbial development [23]. Figure 2 shows a summary of plant components and their sensitivity to deterioration.

Natural fiber degradation in the subsurface environment, which lowers the reinforcing action of fiber, is generally regarded as a problem that must be avoided. Many treatments have been applied to fibers to change their undesirable characteristics and ensure their long-term durability as it will also be discussed in the next section.

2.5 *Treatment Methods of Natural Fibers*

Natural fibers can be made more durable by chemically altering the hydroxyl groups in the cellulose polymer or applying various coatings to the fiber's surface. Because of their degradability, cellulose-based materials such as natural fibers require protection from damaging agents. Because of water absorption, soil organisms, and minerals, the protection may require a special condition to use in soil. Natural fibers can be altered since they contain hydroxyl groups from cellulose and lignin. In addition, coating the fibers with chemical substances decreases water absorption while also protecting them from bacterial and fungal attack. The hydroxyl groups may assist hydrogen bonding within the cellulose molecules. Many therapies have been used to improve the undesirable characteristics of fibers [22] and are briefly summarized in Table 1.

Table 1 Summary of previous research on fiber treatments

Geo-natural materials	Recommended treatment	Prime target	Reference
Oil palm empty fruit bunches	(ABS) acrylonitrile protecting fibers from water sorption increased the reinforcing ability (synthetic monomer produced from propylene and ammonia), butadiene (petroleum hydrocarbon obtained from butane) and styrene	Protecting fibers from water sorption increased the reinforcing ability	[22]
Sisal fiber	Portuguese gum rosin dissolved with turpentine	Enhanced stiffness of the fibers The coated fibers indicated lower failure elongation than uncoated fibers in tensile strength test The crystalline nature of rosin endowed the fibers with enhanced stiffness	[24]
Areca leaf sheaths	(CCA) water soluble (fixed)—copper chromium arsenic composition	Extend the durability, subsequently decrease areca leaf sheaths biodegradation, and enhance lifespan	[25]
Waste weed, water hyacinth (WH), fibers (jute and coir)	Nanoparticle of ferric hydroxide and aluminum hydroxide	Increases the tensile strength of the material by increasing the surface roughness of the fiber	[26]
Coir fiber	Kerosene	Kerosene reduces moisture intake of coir by up to 170%	[27]

3 Soil Reinforcement Application

Soil reinforcement is a process that involves inserting materials with desired properties into the soil to improve its engineering features and behavior. The fundamental purpose of soil reinforcement is to increase the stability, shear, and bearing capacity of the soil mass, and thereby reduce soil settlement [2]. Physical techniques of reinforcement, mechanical methods of reinforcement, and chemical methods of reinforcement are the three basic types of soil reinforcement procedures. Figure 3 depicts the various soil reinforcing methods and their subcategories.

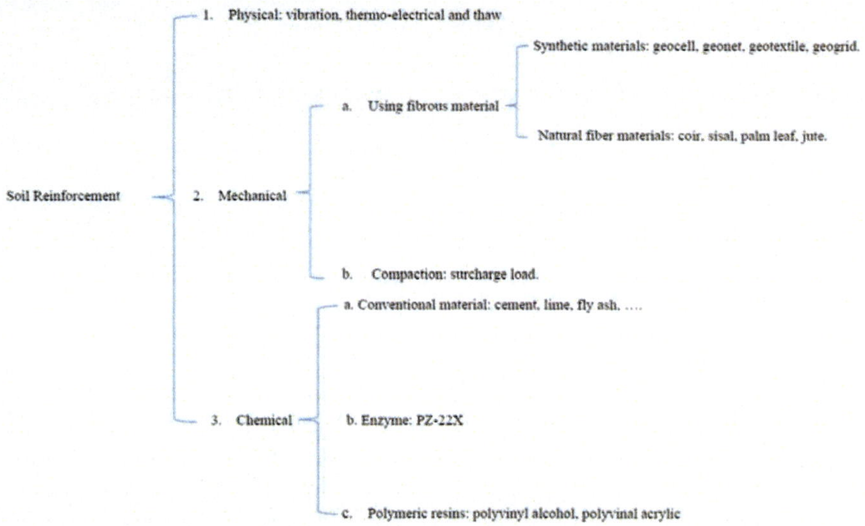

Fig. 3 Soil reinforcement methods [16]

Geosynthetics are widely applied for filtration, drainage, waterproofing, separation, and reinforcing, among other things. Geosynthetics-reinforced soil foundations can increase the foundation's carrying capacity and reduce footing settlement, and they have been frequently used to cure and enhance soft soil foundations.

Geosynthetics is a broad term that refers to a variety of items such as geocell, geotextiles, geogrids, geofoams, geomembranes, geocells, geonets, and geocomposites that are made primarily of polymeric materials (e.g., polystyrene, polypropylene, polyethylene, polyamide, polyvinyl chloride, and polyester) that are utilized in soil, rock, and/or other civil engineering projects [28].

Natural fiber-based products (jute, coir, cotton, bamboo, and so on) are also being used in interaction with soil, rock, and/or other civil engineering-related materials, particularly in temporary/short term civil engineering applications. Due to their biodegradable qualities, such products, known as geonaturals, have a relatively short lifespan when utilized with the soil and so have fewer field applications than geosynthetics. As a result, it must be treated with some coated materials to protect it from the surrounding conditions and thus extend its lifespan and durability.

Based on their application method, there are two main forms of fiber-reinforced soil: oriented distributed fiber-reinforcing soil symbols as (ODFS) and randomly distributed fiber-reinforcing soil symbols as (RDFS) [14, 22]. Figure 4 illustrates the schematics for the two type of fiber-reinforced soil in detail [16, 29].

RDFS known as ground improvement method where fibers with appropriate qualities and amounts are spread and compacted in situ. On the other hand, ODFS is a reinforcement method in which natural fibers are inserted in horizontal, vertical or both directions using planner/cellular systems. In general, ODFS allows natural fibers to be mechanically enhanced through changes such as, combining binding, or

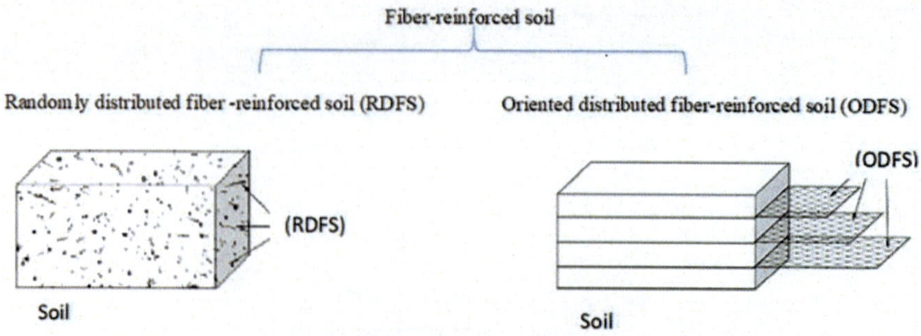

Fig. 4 Illustration and categorization of fiber-reinforced mechanism of soil [16]

punching, depending on the application's needs. The method of ODFS is comparable to classic geosynthetic procedures, in which materials such as geogrids, geocells, geomats, geotextiles, and other geosynthetic materials were inserting in problematic soil planes. Despite the fact that the ODFS technique mobilizes extra frictional strength along fiber-reinforced planes, unreinforced area must rely on their own strength to survive. Failure planes can still be formed through weaker unreinforced zones [30].

3.1 Studies on Cellular Reinforcement Under Static Load

On a soft clay bed reinforced with three different types of 3D-cellular confinement systems, a plate load test was performed. Geogrid cells, commercial geocells, and bamboo cells are all cellular system. Due to its higher stiffness, higher tensile strength, and higher surface roughness values, the foundation bed reinforced with bamboo cells had the highest bearing capacity [31].

Investigated the use of coir geocells as a viable reinforcement material for shallow foundations, with the goal of enhancing soil load-carrying capacity. The introduction of coir geocells greatly improved the performance characteristics of the foundation, according to the findings. This is due to the restricting effect provided by the geocell's three-dimensional structure and the interlocking of dirt between the geocell's openings. In addition, the reinforcing arrangement and configuration have a massive effect on the reinforced soil's performance characteristics [32].

A series of large-scale steel square footing experiments on sand reinforced with two types of reinforcement methods were explored. There are two types of geocell reinforcement: full geocell reinforcement and geocell with an opening reinforcement as shown in Fig. 5. The application of opening geocell reinforcement and full geocell reinforcing methods considerably improves the footing load-carrying capacity, reduces footing settlement, and reduces surface heave, according to the findings. Footing bearing pressure increases as the number of geocell layers grows

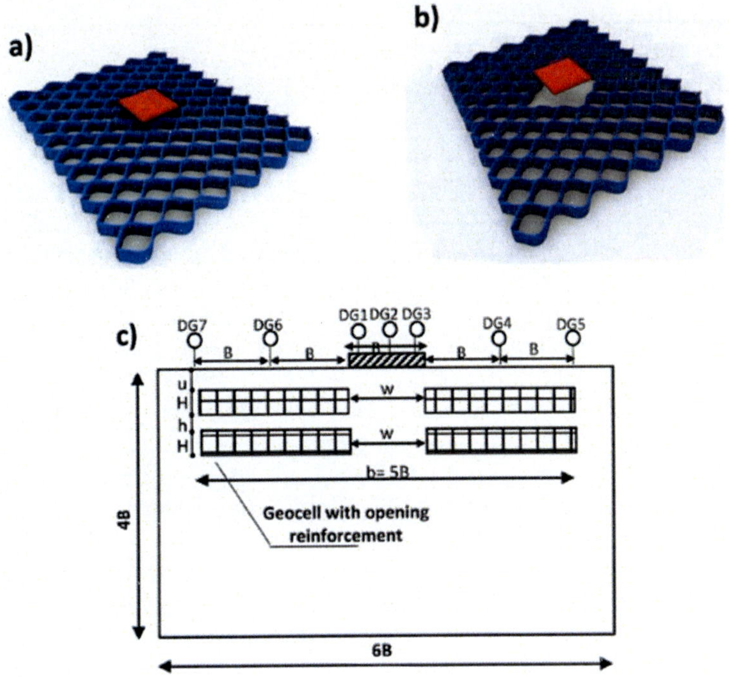

Fig. 5 Schematic view of geocell reinforcement: **a** full geocell reinforcement, **b** geocell with an opening reinforcement, and **c** contributing parameters for square footing and geocell arrangement [33]

from one to two in both the opening geocell reinforcement and full geocell reinforcement techniques, while footing settlement, surface heave, and the difference in performance between the two methods mattress diminish [33].

To better understand the soil reinforcement mechanism, a numbers of model plate load tests were carried out on unreinforced soil and soil reinforced with coir geocells. The load-bearing capacity of the soil bed improved up to three times with the addition of coir geocells, and there was a significant decrease in settlement in the weak soil layer underneath. The study also compares soil reinforced with natural coir cell to soil reinforced with conventional HDPE geocell in terms of performance. Furthermore, the impact of the lateral resistance effect and vertical load dispersion effect absorbed by coir cells in reinforcing the soil bed is demonstrated analytically in this article [34].

3.2 Studies on Cellular Reinforcement Under Dynamic Load

Most of the research has focused on the behavior of cellular-reinforced bases when subjected to static loading. The performance of cellular-reinforced bases under repeated stress has only been studied in a few cases.

Using a single geocell and a loading plate, evaluate the behavior of reinforced bases under repeated loads. Under repeated loads, single geocell reinforcement reduced plastic deformation and enhanced percent elastic deformation, according to the results. It only required ten cycles to reach an elastic deformation of 80% or higher. After 150 loading cycles, the elastic deformation had exceeded 95% of the total deformation. The contribution of geocell can be acknowledged for this advantageous behavior [35].

By conducting load tests in a large size tank, researchers evaluated the cyclic behavior of sand reinforced with four geosynthetic products: polymeric-woven geotextile, uni-axial geogrids of two different capacities, and a coir geotextile. The dynamic response was measured using the 'Coefficient of Uniform Compression', or CUC (Cu). The results showed that geosynthetic products with greater mobilization tensile strength improved Cu values more effectively [36].

Strip footings supported on unreinforced and reinforced sand beds with geocell were tested in a laboratory model under a mix of static and recurring loads. The results show that the reinforcement reduces the settlement magnitude, the geocell acts as a settlement retardant, and allows for larger loads or increased cycling. As the geocell's height and width were raised, the reinforcement's efficacy in lowering maximum footing settlement declined. Geocells significantly reduced plastic deformation during repeated loading compared to static loading, with the decrease being greatest in the presence of additional reinforcement and the loading rate was highest. It is hypothesized that a rapidly loaded polymeric geocell's more resilient stiffness attracts load to itself, shielding the soil from some of the more problematic stress states and, as a result, reducing deformation [37].

Large-scale cubical triaxial experiments were used to evaluate the behavior of reinforced and unreinforced subballast under cyclic load. The number of cycles, confining pressure, and frequency of train-caused vibration all had an effect on subballast stabilization, according to the laboratory findings. The geocells altered the subballast behavior under cyclic loading, especially at low confining pressure and high frequency, according to the results. The geocell's enhanced confining pressure lowered the vertical and volumetric strains [38].

Geocells and geocells with added basal geogrid were used to perform cyclic plate load testing on the reinforced soft clay beds. The elastic uniform compression coefficient (Cu) was computed. In the presence of geocell reinforcement, the Cu value increased. The largest rise in the Cu value was found in the case of the clay bed reinforced with a combination of geocell and geogrid, according to the findings. In addition to the reduction in settling, the results showed an improvement in strain modulus, bearing capacity, and stiffness [39].

The cyclic loading of geocell-reinforced subballast was examined. To model the geocell-reinforced subballast, a cyclic loading with a periodic and positive full-sine waveform was used, which is similar to the load applied in the laboratory. The effect of subballast strength and geocell stiffness on mobilizing tensile strength in the geocell mattress was also evaluated using parametric research. The largest mobilized tensile stress was discovered to occur on the subballast with the lowest degree of stiffness. Furthermore, the results demonstrated that increasing geocell stiffness reduced lateral displacement even more, and that a geocell with a low stiffness performs very well in comparison with a geocell with a higher stiffness [40].

Using a unique track process modeling device and cyclic loading frequencies ranging from 10 to 30 Hz, study the load-deformation behavior of geocell-reinforced subballast exposed to cyclic loading. The geocell reduced the lateral and vertical deformation of subballast assemblies at any given frequency, according to the findings [41].

Investigated the influence of geocell confinement, thickness and strength of base course, and also subgrade strength on permanent and resilient deformations of recycled asphalt pavement (RAP) bases by performing cyclic plate loading tests on geocell-reinforced recycled asphalt pavement (RAP) bases with three different thicknesses above subgrades that are weak and moderate. When compared to the performance of unreinforced bases, geocell confinement improved the performance of stabilized RAP bases by limiting the permanent surface deformations and increasing resilient deformations and therefore increasing the percentages of resilient deformation [42].

Performed geosynthetic-reinforced soil beds supporting machine foundation in large-scale field tests, the dynamic response is studied in terms of reduction in resonant amplitude, peak particle velocity (PPV), and enhancement in dynamic characteristics of the soil. The results revealed that the displacement amplitude of vibration significantly reduced in the presence of geosynthetics. The greatest reduction is seen in the presence of geocell reinforcement in comparison with the other conditions [1].

4 Conclusion

Soil reinforcement, mechanical, and chemical stabilization, are all prevalent ground stabilization procedures; however, due to their high cost and environmental impacts, these techniques are becoming less popular as society becomes more concerned about sustainable products and materials. Using geosynthetics as a soil reinforcement technology is believed to be a long-term and cost-effective solution. The use of geosynthetics in pavement engineering and construction has yielded several benefits in terms of reduced settlements, increased bearing capacity, and so on. By reducing the amount of construction material used, the geosynthetic inclusion allows for easier construction and lower construction costs.

Despite all this, the manufacture of these synthetic materials is in a way that is not environmentally friendly, as their manufacture is accompanied by a lot of emissions, in addition to that they are plastic materials that do not expire at the end of the desired benefit, and since most of the world's interest at present has become in the search for economic alternatives at the same time environmentally friendly. Therefore, the searchers began to research to replace the geosynthetic materials with natural materials that are less expensive and environmentally friendly.

Natural materials are widely available materials in various parts of the world, and therefore, the cost of their manufacture is low, as their manufacture and formation to use as soil reinforcement materials require hand work and thus will contribute to providing job opportunities for many workers at a lower cost than the geosynthetic industry. Although natural fibers are environmentally friendly and economical, they have a negative characteristic which is degrade over time. Furthermore, altering the negative features of fiber materials with various treatments might improve fiber–soil behavior and fiber lifetime. The treatment procedure should, however, be chosen based on the fiber material and application goal.

Most studies focused on the behavior of cellular-reinforced bases under static loading. Limited studies conducted to investigate the performance of natural fibers as cellular-reinforced bases under static loading. Cellular-reinforced bases used for pavement applications are subjected to repeated loading. Limited studies conducted to investigate the performance of cellular-reinforced bases under repeated loading, and there is no research on natural fibers as cellular-reinforced bases under dynamic loading, however.

Acknowledgements This project is supported by the funding from University Sains Malaysia (USM) Grant RUI/8014094.

References

1. Venkateswarlu, H., Ujjawal, K.N., Hegde, A.: Laboratory and numerical investigation of machine foundations reinforced with geogrids and geocells. Geotext. Geomembr. **46**(6), 882–896 (2018). https://doi.org/10.1016/j.geotexmem.2018.08.006
2. Hejazi, S.M., Sheikhzadeh, M., Abtahi, S.M., Zadhoush, A.: A simple review of soil reinforcement by using natural and synthetic fibers. Constr. Build. Mater. **30**, 100–116 (2012)
3. Bordoloi, S., Patwa, D., Hussain, R., Garg, A., Sreedeep, S.: Nano-particle coated natural fiber impregnated soil as a sustainable reinforcement material. In: Urbanization Challenges in Emerging Economies: Resilience and Sustainability of Infrastructure, pp. 435–444. American Society of Civil Engineers Reston, VA (2018)
4. Wambua, P., Ivens, J., Verpoest, I.: Natural fibres can they replace glass in fibre reinforced plastics? Compos. Sci. Technol. **63**(9), 1259–1264 (2003)
5. Sharman, A.K., Sivapullaiah, P.V.: Ground granulated blast furnace slag amended fly ash as an expansive soil stabilizer. Soils Found. **56**, 205–212 (2016)
6. Al-Swaidani, A., Hammoud, I., Mezia, A.: Effect of adding natural pozzolana on geotechnical properties of lime-stabilized clayey soil. Rock Mech. Geotech. Eng. **8**, 714–725 (2016)
7. Fagone, M., Loccarini, F., Ranocchiai, G.: Strength evaluation of jute fabric for the reinforcement of rammed earth structures. Compos. Part B **113**, 1–13 (2017)
8. Lekha, B.M., Goutham, S., Shankar, A.U.R.: Evaluation of lateritic soil stabilized with Arecanut coir for low volume pavements. Transp. Geotech. **2**, 20–29 (2015)

9. Bordoloi, S., Hussain, R., Garg, A., Sreedeep, S., Zhou, W.H.: Infiltration characteristics of natural fiber reinforced soil. Transp. Geotech. **12**, 37–44 (2017)
10. Pujari, S., Ramakrishna, A., Padal, K.T.B.: Comparison of ANN and regression analysis for predicting the water absorption behaviour of jute and banana fiber reinforced epoxy composites. Mater. Today Proc. **4**, 1626–1633 (2017)
11. Bordoloi, S., Garg, A., Sekharan, S.: A review of physio-biochemical properties of natural fibers and their application in soil reinforcement. Adv. Civ. Eng. Mater. **6**, 323–359 (2017)
12. Gaw, B., Zamora, S., Albano, L.D., Tao, M.: Soil Reinforcement with Natural Fibers for Low-Income Housing Communities. Bachelor's Thesis, Worcester Polytechinic Institute, Worcester, MA, USA21 (2011)
13. Ibrahim, M.M., Dufresne, A., El-Zawawy, W.K., Agblevor, F.A.: Banana fibers and microfibrils as lignocellulosic reinforcements in polymer composites. Carbohydr. Polym. **81**, 811–819 (2010)
14. John, M.J., Thomas, S.: Biofibres and biocomposites. Carbohydr. Polym. **71**, 343–364 (2008)
15. Hansen, C.M., Bjorkman, A.A.: The ultrastructure of wood from a solubility parameter point of view. Holzforschung **52**(4), 335 (1998)
16. Gowthaman, S., Nakashima, K., Kawasaki, S.: A state-of-the-art review on soil reinforcement technology using natural plant fiber materials: past findings, present trends and future directions. Materials **11**, 553 (2018). https://doi.org/10.3390/ma11040553
17. Manthey, N.W., Cardona, F., Aravinthan, T., Wang, H., Cooney, T.: Natural Fibre Composites with Epoxidised Vegetable Oil (EVO) Resins: A Review. Southern Region Engineering Conference, University of Southern Queensland (2010)
18. Ramesh, H.N., Manoj Krishna, K.V., Meena.: Performance of coated coir fibers on the compressive strength behavior of reinforced soil. Int. J. Earth Sci. Eng. **4**(6), 26–29 (2011)
19. Kalia, S., Kaith, B.S., Kaur, I.: Pretreatments of natural fibers and their application as reinforcing material in polymer composites—a review. Polym. Eng. Sci. 1253–1272 (2009)
20. Saha, P., Roy, D., Manna, S., Adhikari, B., Sen, R., Roy, S.: Durability of transesterified jute geotextiles. Geotext. Geomembr. **35**, 69–75 (2012)
21. Azwa, Z.N., Yousif, B.F., Manalo, A.C., Karunasena, W.: A review on the degradability of polymeric composites based on natural fibres. Mater. Des. **47**, 424–442 (2013)
22. Bateni, F., Ahmad, F., Yahya, A.S., Azmi, M.: Performance of oil palm empty fruit bunch fibres coated with acrylonitrile butadiene styrene. Constr. Build. Mater. **25**(4), 1824–1829 (2011). https://doi.org/10.1016/j.conbuildmat.2010.11.080
23. Prambauer, M., Wendeler, C., Weitzenböck, J., Burgstaller, C.: Biodegradable geotextiles—an overview of existing and potential materials. Geotext. Geomembr. **47**, 48–59 (2019)
24. Kafodya, I., Okonta, F.: Effect of fibre surface coating on the mechanical properties of natural fibre-reinforced soil. Int. J. Geotech. Eng. **00**(00), 1–11 (2018). https://doi.org/10.1080/19386362.2018.1542557
25. Kolathayar, S., Aravind, C.A., Tg, S.: Model tests and analytical studies on performance of areca leaf cells as cellular confinement in soil. Geomech. Geoeng. 1–12 (2019). https://doi.org/10.1080/17486025.2019.1664774
26. Bordoloi, S., Leung, A.K., Gadi, V.K., Hussain, R., Garg, A., Sekharan, S.: Water retention and desiccation potential of lignocellulose-based fiber-reinforced soil. J. Geotech. Geoenviron. Eng. **145**(11), 06019013 (2019)
27. Torio-Kaimo, L.V., Diego, A.M.S., Alcantara, M.T.M.: Unconfined compressive strength of clay reinforced with kerosene-treated coir fiber. Int. J. GEOMATE **18**(69), 97–103 (2020). https://doi.org/10.21660/2020.69.9295
28. Shukla, S.K.: Geosynthetics and ground engineering: sustainability considerations. Int. J. Geosynthetics Ground Eng. **7**(1), 1–3 (2021). https://doi.org/10.1007/s40891-021-00256-z
29. Sharma, V., Vinayak, H.K., Marwaha, B.M.: Enhancing compressive strength of soil using natural fibers. Constr. Build. Mater. **93**, 943–949 (2015)
30. Maher, M.H., Gray, D.H.: Static response of sands reinforced with randomly distributed fibers. J. Geotech. Eng. **116**, 1661–1677 (1990)

31. Hegde, A., Sitharam, T.G.: Experiment and 3D-numerical studies on soft clay bed reinforced with different types of cellular confinement systems. Transp. Geotech. **10**, 73–84 (2017). https://doi.org/10.1016/j.trgeo.2017.01.001
32. Lal, D., Sankar, N., Chandrakaran, S.: Behaviour of square footing on sand reinforced with coir geocell. Arab. J. Geosci. **10**(15), 1–8 (2017)
33. Shadmand, A., Ghazavi, M., Ganjian, N.: Load-settlement characteristics of large-scale square footing on sand reinforced with opening geocell reinforcement. Geotext. Geomembr. **46**(3), 319–326 (2018). https://doi.org/10.1016/j.geotexmem.2018.01.001
34. Kolathayar, S., Narasimhan, S., Kamaludeen, R., Sitharam, T.G.: Performance of footing on clay bed reinforced with coir cell networks. Int. J. Geomech. **20**(8), 04020106 (2020). https://doi.org/10.1061/(asce)gm.1943-5622.0001719
35. Pokharel, S.K., Han, J., Leshchinsky, D., Parsons, R.L., Halahmi, I.: Behavior of Geocell-Reinforced Granular Bases Under Static and Repeated Loads, pp. 409–416 (2009). https://doi.org/10.1061/41023(337)52
36. Goud, A.P.K.: Behaviour of Geosynthetic Reinforced Sand Bed Under Cyclic Load. Indian Geotechnical Conference (2011)
37. Moghaddas Tafreshi, S.N., Dawson, A.R.: A comparison of static and cyclic loading responses of foundations on geocell-reinforced sand. Geotext. Geomembr. **32**, 55–68 (2012). https://doi.org/10.1016/j.geotexmem.2011.12.003
38. Indraratna, B., Biabani, M.M., Nimbalkar, S.: Behavior of geocell-reinforced subballast subjected to cyclic loading in plane-strain condition. J. Geotech. Geoenviron. Eng. **141**(1), 04014081 (2015). https://doi.org/10.1061/(asce)gt.1943-5606.0001199
39. Hegde, A., Sitharam, T.G.: Behaviour of geocell reinforced soft clay bed subjected to incremental cyclic loading. Geomech. Eng. **10**(4), 405–422 (2016). https://doi.org/10.12989/gae.2016.10.4.405
40. Biabani, M.M., Indraratna, B., Ngo, N.T.: Modelling of geocell-reinforced subballast subjected to cyclic loading. Geotext. Geomembr. **44**(4), 489–503 (2016). https://doi.org/10.1016/j.geotexmem.2016.02.001
41. Ngo, N.T., Indraratna, B., Rujikiatkamjorn, C., Mahdi Biabani, M.: Experimental and discrete element modeling of geocell-stabilized subballast subjected to cyclic loading. J. Geotech. Geoenviron. Eng. **142**(4), 04015100 (2016). https://doi.org/10.1061/(asce)gt.1943-5606.0001431
42. Thakur, J.K., Han, J., Parsons, R.L.: Factors influencing deformations of geocell-reinforced recycled asphalt pavement bases under cyclic loading. J. Mater. Civ. Eng. **29**(3), 04016240 (2017). https://doi.org/10.1061/(asce)mt.1943-5533.0001760

Physical Properties of Soil and Its Correlation with River Bank Erosion and Soil Erodibility: Sungai Pusu Case Study

Nur Aqilah Mohd Rosli and Saerahany Legori Ibrahim

Abstract Riverbank erosion is a complex process where the extent of riverbank erosion is influenced by many factors including soil properties. In this study, soil samples were collected from different sections of riverbanks at Sungai Pusu. Several physical soil properties such as the median grain size, percentage of sand, silt and clay, plasticity index, and specific gravity were measured for each soil samples. Riverbank erosion rates were measured at the site periodically using erosion pin method. Linear regression was conducted to identify the correlation between the erosion rate and the physical soil properties using the Statistical Package for the Social Science (SPSS) software. It was found that the percentage of sand content had significantly influenced the erosion rate, while the other parameters showed weak correlation and no correlation with the erosion rate. Soil erodibility coefficient, kd, was calculated using empirical equation, and the value ranged between 0.0553 and 0.1023 cm^3/N s. The erodibility of the soil samples can be categorized as "Moderately resistant" based on the τ_c versus k_d plot produced by previous study.

Keywords Bank erosion · Soil properties · Soil erodibility · Sungai Pusu

1 Introduction

Over the past decades, scientific research and studies on riverbank erosion have increased dramatically. Numerous case studies had been conducted to measure the rate of erosion in various locations worldwide. Riverbank erosion as defined by Chatterjee and Mistri [1] is the wearing away of materials from the riverbank. It is a geological process in which soil and rock are removed from the earth's crust into the receiving water body. Riverbank erosion consists of two primary process groups which are fluvial erosion and mass failure [2]. Fluvial erosion or bank scour

N. A. M. Rosli · S. L. Ibrahim (✉)
Department of Civil Engineering, Kulliyyah of Engineering, International Islamic University Malaysia (IIUM), 53100 Gombak, Selangor, Malaysia
e-mail: saerahany@iium.edu.my

N. Sabtu (ed.), *Proceedings of AWAM International Conference on Civil Engineering 2022 - Volume 3*, Lecture Notes in Civil Engineering 386,
https://doi.org/10.1007/978-981-99-6026-2_11

is the removal of earth's material from the bank which is caused by the direct action of flowing water. In general, the rate of fluvial erosion depends on the action of hydraulic force and the composition of stream bank's material [3]. On the other hand, mass failure is the movement and collapse of bank material into the stream due to gravitational instability. The two processes are frequently linked where fluvial erosion is the precursor of mass failure [4].

The degree of riverbank erosion was influenced by various factors including soil physical properties. Soil physical properties has a significant impact on the bank erosion process [5]. The physical properties of soil that influence its erodibility factor are aggregate size, particle size distribution, bulk density, water content, and temperature [6]. Abidin et al. [7] added that erodibility varies with soil texture, soil structure, stability, shear strength, soil depth, infiltration capacity, soil organic matter, and chemical constituent. Soil structure or the composition of soils significantly influences the rate of erosion [8]. The soil structure varies according to the proportion of clay, sand, and silt particles. Soil structure affects the soil resistance against erosion. Silty soil is more susceptible to grain-by-grain erosion as it is easily detached and eroded due to its fine texture and weak bond between particles. On the contrary, soil with high percentage of clay is less vulnerable to grain-by-grain erosion owing to its cohesive properties which cause the particles to stick together [4]. However, it can be rapidly eroded by mass movement as it is more likely to be poorly drained.

The study focused on the bank erosion of Sungai Pusu where considerable amount of erosion has been spotted along the riverbank. The aim of this study is to identify the effects of bank material physical properties on riverbank erosion by finding the correlation between several physical soil properties and the rates of riverbank erosion. The soil erodibility for each sample was also determined using existing equation and was categorized using a plot of τ_c versus k_d produced by previous researcher.

2 Materials and Methods

2.1 *Study Area*

Sungai Pusu, which flows through International Islamic University Malaysia campus before joining Sungai Gombak, has a drainage area of 12.4 km^2 and a total length of around 4.1 km. Figure 1 depicts the Sungai Pusu flow channel as well as the fieldwork locations.

A total of four sections of Sungai Pusu were selected as the study area where it is deemed suitable for field measurements and data collection. Table 1 presents the coordinates of all the selected bank sections of Sungai Pusu.

Sungai Pusu is presently classified as a Class IV river by Malaysian Water Quality Standards, which indicates that it is in the worst possible state [9]. Sand mining activities taking place upstream of Sungai Pusu and its major tributaries, Anak Pusu

Fig. 1 Sampling locations at Sungai Pusu

Table 1 Coordinates of section

Section no	Coordinates
1	3° 15′ 01″N 101° 43′ 52″E
2	3° 15′ 00″N 101° 43′ 55″E
3	3° 14′ 59″N 101° 44′ 05″E
4	3° 14′ 59″N 101° 44′ 07″E

and Batang Pusu, are one of the key factors impacting river water quality. Riverbank erosion can also be seen at various places along the river.

2.2 *Field Works*

Figure 2 shows one of the sections of the riverbank of Sungai Pusu where the soil samples were collected. Ten soil samples were collected from four different sections of Sungai Pusu using hand auger at various depth of the riverbank. The soil samples were properly labeled and taken to the laboratory. Sieve analysis and hydrometer test were used to determine the median particle size (d_{50}), clay percentage, sand percentage, and silt percentage where the test was conducted according to MS1056:2005, Part 2, Clause 10.5. Atterberg limits test was also conducted to measure the plastic limit and liquid limit which are then used to measure the plasticity index of the soil.

Table 2 presents the physical soil properties measured from the soil samples collected. The soil samples were classified according to British Soil Classification System which is detailed in BS5930 Soil Investigation.

Fig. 2 Sungai Pusu riverbank section

Table 2 Results of physical soil properties

Sample	PI (%)	ρ_s	d_{50} (mm)	Clay (%)	Silt (%)	Sand (%)	Gravel (%)	Soil classification
1	13	2.64	0.057	12	38	46	4	Clay of intermediate plasticity
2	4	2.69	0.071	12	18	52	18	Sandy silt
3	3	2.68	0.034	9	18	52	21	Sandy silt
4	9	2.68	0.170	5	32	52	11	Clay of intermediate plasticity
5	21	2.67	0.132	7	33	59	1	Clay of intermediate plasticity
6	12	2.7	0.186	2	22	76	0	Very silty sand
7	13	2.69	0.239	17		77	6	Very silty sand
8	14	2.69	0.333	4	20	70	6	Very silty sand
9	24	2.63	0.060	11	41	47	1	Clay of high plasticity
10	20	2.66	0.155	11	24	64	1	Clay of intermediate plasticity

Fig. 3 Erosion pins installed at the riverbank

The method that was used for measurement of riverbank erosion rate is erosion pins. The diameter of the erosion pins is 6 mm, and the length is approximately 60 cm. The pins were installed at the bank in a grid pattern along the river section. Figure 3 shows the erosion pins installed at the study area.

The pin diameter was selected to be as small as possible to avoid material disturbance and minimize public visibility. The pins were inserted horizontally into the bank until only 10 cm of the pin length remain visible. To facilitate pin measurement, the end of the pins was labeled by numbers. Erosion pin exposures were measured weekly and after every rainfall and peak flow event.

2.3 *Soil Erodibility Determination*

Soil erodibility coefficient will be determined using the equation by Julian and Torres [9] which is a correlation between critical shear stress and clay-silt fraction as shown below.

$$\tau_c = 0.1 + 0.1779\ (SC\%) + 0.0028\ (SC\%)^2 - 2.343 \times 10^{-5}\ (SC\%)^3 \quad (1)$$

where τ_c is the critical shear stress and SC% is the combined percentage of clay and silt.

After calculating the critical shear stress, the soil erodibility coefficient will be determined using the empirical correlation found by Hanson and Simon [10].

$$k_d = 0.2\tau_c^{-0.5} \tag{2}$$

where k_d is the soil erodibility coefficient (cm^3/N m).

3 Results and Discussion

Linear regression was used in this study to evaluate the effect of each variable on riverbank erosion rate. Each variable was regressed individually against the erosion rate using the Statistical Package for Social Science (SPSS) software. The coefficient of determination (R^2) and p-value were used to identify the significance of the relationship between the variable and erosion rate.

The median particle size (d_{50}) of the soil samples collected range from 0.0343 to 0.333 mm. Figure 4 suggests that there was a moderate correlation between erosion rate and d_{50} in this study area which demonstrates that erosion rates increase with increasing particle grain size.

Clay content in the soil samples collected varied from 2 to 12%. Figure 5 shows the variation of erosion rate and percentage of clay content. The coefficient of determination (R^2) value indicates that there was a weak correlation between erosion rate and clay content. Most of previous studies analyze the relationship between percentage of clay content with erosion rate indirectly by studying the relationship between clay content with critical shear stress [9, 11, 12]. In the paper by Carlson and Enger [11], Mazumdar and Talukdar [5], strong correlation was found between clay content and

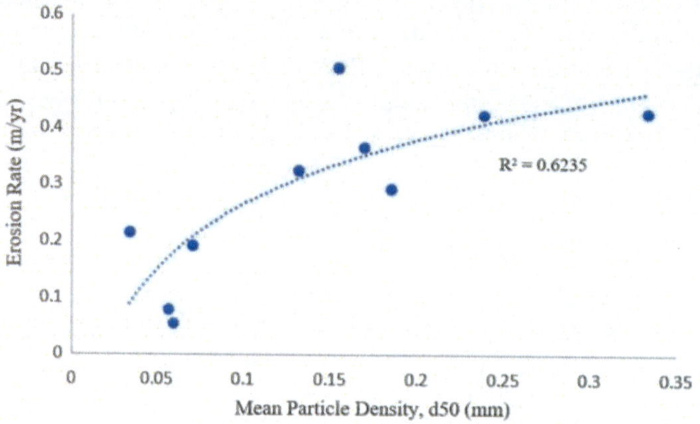

Fig. 4 Erosion rate versus median grain size

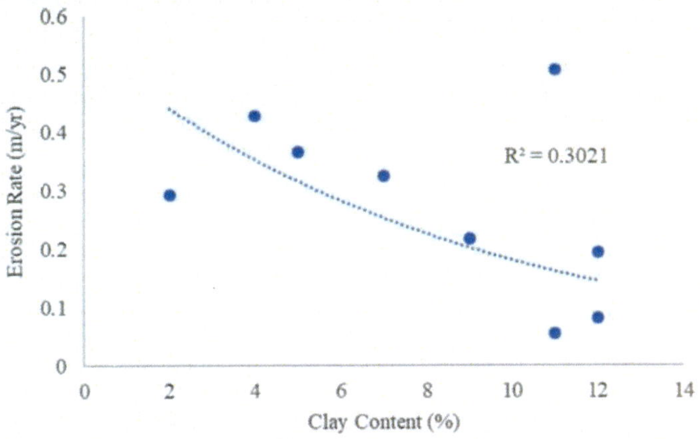

Fig. 5 Erosion rate versus clay content

critical shear stress. However, Briaud [13] and Kimiaghalam et al. [12] reported a weak correlation between clay content and critical shear stress in their study. In the study by Kimiaghalam et al. [12], the author suggests that the in situ conditions have greater impact on the erodibility of cohesive soil than its inherent properties.

Sand content in the soil samples collected in this study ranged between 46 and 77% which suggests that the bank material in this study area has a considerably high percentage of sand particles. According to previous researchers, riverbank which consists of high percentage of sand is more prone to fluvial erosion [13–15]. Figure 6 shows the variation between erosion rate and the percentage of sand content. The coefficient of correlation (R^2) is 0.56 which shows that the erosion rate is moderately correlated to percentage sand content. The *p*-value obtained also indicates that the relationship between sand content and erosion rate is significant. A very strong positive correlation between percentage of sand content and erosion rate was also demonstrated by past study [10] which shows that percentage of sand content highly affects the bank erosion rate.

Silt content in the samples collected varied between 17 and 41%. Figure 7 demonstrates that the erosion rate has a weak correlation with percentage of silt content. However, it can be seen from the graph that the erosion rate decreases with increasing silt content.

Figure 8 shows that there was no correlation between erosion rate and the plasticity index. Results from this graph indicates that plasticity index is not a significant predictor for erosion rate in this study area.

Figure 9 suggests that there was a moderate correlation between erosion rate and specific gravity. The general trend of the graph indicates that an increase in specific gravity results in higher erosion rate which is in accordance with the results that has been reported in [5].

Table 3 presents the summary of the R^2 and *p*-value for the relationship between erosion rate and the soil physical properties. The results of linear regression showed

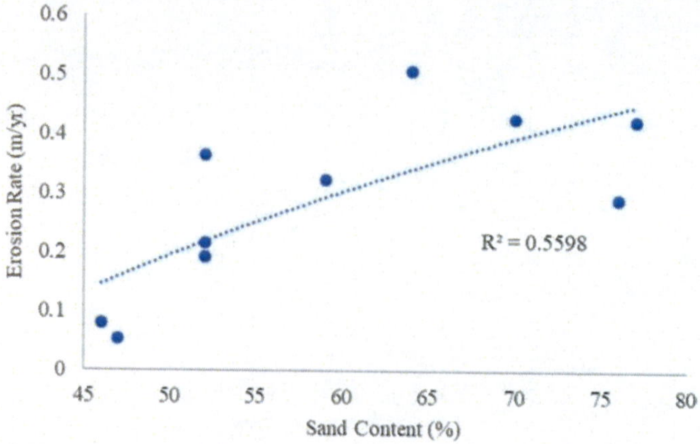

Fig. 6 Erosion rate versus sand content

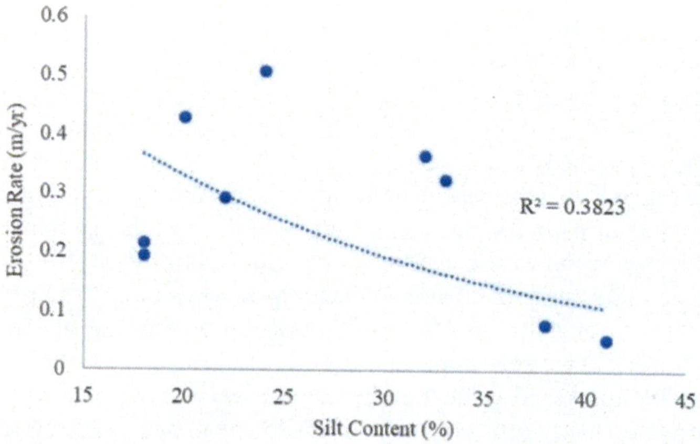

Fig. 7 Erosion rate versus silt content

that mean particle density, d_{50}, has the highest coefficient of determination (R^2) value followed by percentage sand content and specific gravity of particle. The significance of the relationship between the erosion rate and the soil physical properties was measured using the *p*-value. A *p*-value of lower than 0.05 indicates statistically significant relationship between the variables. Mean particle density, percentage sand content, and specific gravity of particle have a *p*-value of lower than 0.05 and thus can be considered statistically significant.

Critical shear stress (τ_c) was calculated using Eq. (1), and soil erodibility coefficient, k_d, was determined subsequently by using Eq. (2). Soil erodibility coefficient, k_d, ranged between 0.0553 and 0.1023 cm^3/N s as given in Table 4.

A plot of τ_c versus k_d produced by Julian and Torres [9] as shown in Fig. 10 was used to identify the soil erodibility category. The erodibility coefficient for all

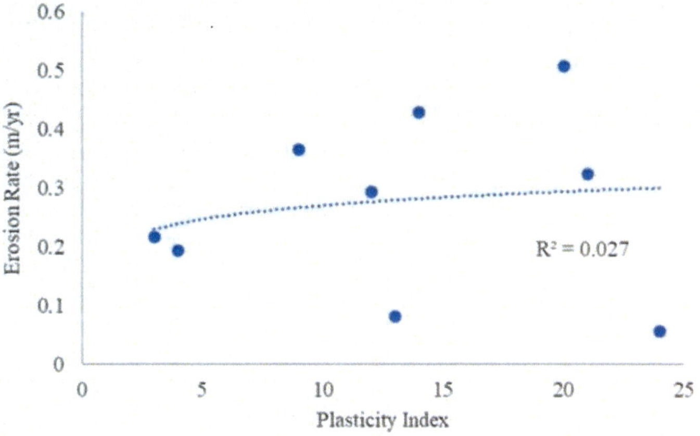

Fig. 8 Erosion rate versus plasticity index

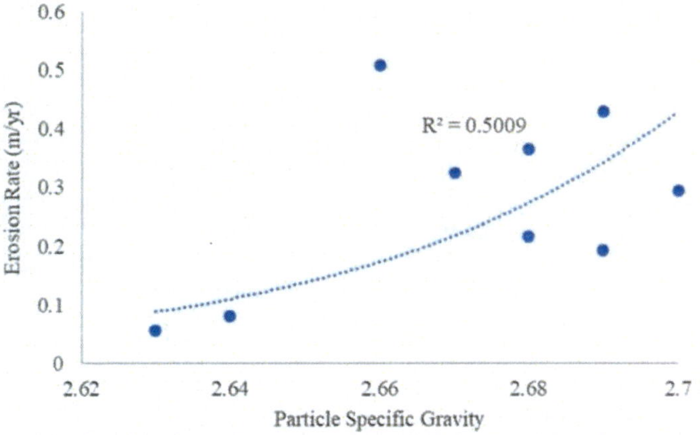

Fig. 9 Erosion rate versus specific gravity

samples in this study falls into the same category which is "Moderately Resistant". However, based on the data collected, the erosion rate was observed to be higher, if the erodibility coefficient has high value. The highest k_d value was found to be from sample 7 which has the lowest percentage of silt and clay. Meanwhile, samples 1 and 9 which composed of higher percentage of silt and clay have lower value of k_d and erosion rate. Clay content plays significant role in resisting erosion due to the cohesive properties of the material. The results from this study were in agreement with previous studies conducted by Kimiaghalam et al. [12] and Constantine et al. [16].

Fig. 10 τ_c versus k_d [9]

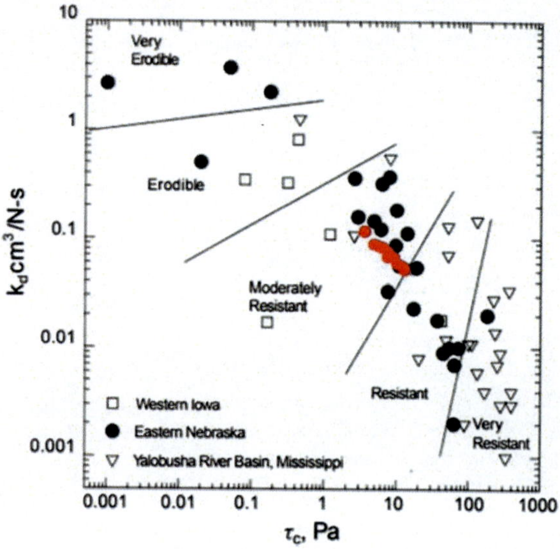

Table 3 Summary of R^2 and p-value of the soil properties

	Coefficient of determination, R^2	p-value
Mean particle diameter	0.624	0.007
Clay content	0.302	0.125
Sand content	0.560	0.013
Silt content	0.382	0.076
Plasticity index	0.027	0.673
Particle specific gravity	0.501	0.033

Table 4 Critical shear stress and soil erodibility coefficient

Sample	% SC	τ_c (N/m^2)	k_d (cm^3/N s)	Erodibility category	Erosion rate, u_b (m/yr)
1	50	13.07	0.055	Moderately resistant	0.080
2	30	7.33	0.074	Moderately resistant	0.193
3	27	6.48	0.079	Moderately resistant	0.216
4	37	9.33	0.065	Moderately resistant	0.365
5	40	10.20	0.063	Moderately resistant	0.324
6	24	5.66	0.084	Moderately resistant	0.292
7	17	3.82	0.102	Moderately resistant	0.424
8	24	5.66	0.084	Moderately resistant	0.428
9	52	13.63	0.054	Moderately resistant	0.054
10	35	8.75	0.068	Moderately resistant	0.507

4 Conclusion

In this study, soil samples were collected from the bank of Sungai Pusu, and for each sample, several physical soil properties were measured. The results from particle size distribution reported that the soil samples collected at the sites consists of high percentage of sand particles. The section of the river with higher percentage of sand is more erodible compared to bank material with lower percentage of sand. It can be seen from the regression analysis that percentage sand content, mean particle density, and specific gravity of particle showed moderate correlation with riverbank erosion rate. Percentage clay content and percentage silt content showed weak correlation with riverbank erosion rate. Meanwhile, plasticity index showed no correlation with the erosion rate. Soil erodibility coefficient, k_d, calculated using the equation ranged between 0.0553 and 0.1023 cm^3/N s. The soil erodibility for this study area can be categorized as "Moderately resistant". Overall, the results were in good agreement other study such as [5, 12]. Other soil parameters can be studied further to develop a predictive model to estimate bank erosion rate using soil properties.

Acknowledgements This research was financially supported by IIUM Research Acculturation Grant Scheme (IRAGS) Research Project IRAGS18-025-0026 entitled "River Meander Migration Prediction: Evaluation of Existing Techniques and Establishment of New Empirical Model".

References

1. Chatterjee, S., Mistri, B.: Impact of river bank erosion on human life: a case study in Shantipur Block, Nadia District, West Bengal. Int. J. Humanit. Soc. Sci. Invention **2**(8), 108–111 (2013)
2. Duan, J.G.: Analytical approach to calculate rate of bank erosion. J. Hydraul. Eng. **131**(11), 980–990 (2005)
3. Rinaldi, M., Darby, S.E.: Modelling River Bank Erosion Process and Mass Failure Mechanisms: Progress Towards Fully Coupled Simulations. Gravel Beds Rivers VI from Process Understanding to River Restoration, pp. 231–258 (2007)
4. Watson, A.J., Basher, L.R.: Stream Bank Erosion: A Review of Processes of Bank Failure, Measurement and Assessment Techniques, and Modelling Approaches. A Report Prepared for Stakeholders of the Motueka Integrated Catchment Management Programme and the Raglan Fine Sediment Study. Landcare Research, Hamilton, New Zealand (2006)
5. Mazumdar, N., Talukdar, B.: Role of physical properties of soil in river bank erosion assessment: a case study in lower Assam Region of River Brahmaputra of India. Am. J. Eng. Res. **7**(9), 197–205 (2018)
6. Ibrahim, S.L., Ariffin, J., Abdullah, J., Muhamad, N.S.: Jet erosion device (JED)—measurement of soil erodibility coefficients. Jurnal Teknologi (Sci. Eng.) 63–67 (2016)
7. Abidin, R.Z., Sulaiman, M.S., Yusoff, N.: Erosion risk assessment: a case study of the Langat River bank in Malaysia. Int. Soil Water Conserv. Res. **5**(1), 26–35 (2017)
8. Pimentel, D.: Soil erosion: a food and environmental threat. Environ. Dev. Sustain. **8**, 119–137 (2006)
9. Julian, J.P., Torres, R.: Hydraulic erosion of cohesive riverbanks. Geomorphology **76**, 193–206 (2006)
10. Hanson, G.J., Simon, A.: Erodibility of cohesive streambeds in the loess area of the midwestern USA. Hydrol. Process. **15**, 23–38 (2001)

11. Carlson, E.J., Enger, P.F.: Studies of Tractive Forces of Cohesive Soils in Earth Canals. US Department of the Interior Bureau of Reclamation, Denver, USA, Report No. Hyd-504 (1962)
12. Kimiaghalam, N., Clark, S.P., Ahmari, H.: An experimental study on the effects of physical, mechanical, and electrochemical properties of natural cohesive soils on critical shear stress and erosion rate. Int. J. Sedim. Res. **31**(1), 1–15 (2016)
13. Briaud, J.L.: Erodibility of fine grained soils and new soil test. In: Proceedings of the Erosion of Soils and Scour of Foundations, pp. 135. ASCE, GSP (2005)
14. Bhowmik, M., Das, N., Das, C., Ahmed, I.: Bank Material Characteristics and Its Impact on River Bank Erosion, West Tripura district, Tripura, North-East India, vol. 115, issue 8, pp. 1571–1576 (2018)
15. Hooke, J.: An analysis of the processes of river bank erosion. J. Hydrol. **42**, 39–62 (1979)
16. Constantine, C.R., Dunne, T., Hanson, G.J.: Examining the physical meaning of the bank erosion coefficient used in meander migration modeling. Geomorphology **106**(3–4), 242–252 (2009)

Quick Review of the Approaches of Landslide Risk Assessment

Mohamad Firdaus Mahamad Yusob, Fauziah Ahmad, Mohd Fadzil Ain, and Mastura Azmi

Abstract Landslide prevention and mitigation have a long history. Retaining walls and ground anchors are popular methods for preventing slope failure, as these prevention methods increase the factor of safety against failure. These methods are widely utilised and proven to be effective all around the world. The methods are however very costly, leading to a restricted application only for wide slopes. A landslide early warning system (LEWS) with real-time monitoring system is required to identify appropriate moment for preventive measures in order to handle landslide emergencies. However, depending on the type of landslide, different techniques are required. The approach is frequently used to help determine a set of crucial thresholds that can be used to determine when alerts can be sent, and it is based on LEWS' own theoretical basis. This paper presents a quick review on the three general approaches used in LEWS, namely statistical approach, physical-based approach and monitoring-based approach. This quick review will attempt in discovering more on the strengths and advantages rather than diving into the limitations of each approach.

Keywords Landslide early warning system · Statistical approach · Physical-based approach · Monitoring-based approach

1 Introduction

Due to global climate change, the intensity and frequency of heavy rainfall events in a landslide-prone area have increased, resulting in significant landslides and high-intensity rainfall. Aside from the presence of weathered material on a slope, the

M. F. M. Yusob · M. F. Ain
Electrical and Electronic Engineering, Universiti Sains Malaysia, Penang, Malaysia

F. Ahmad · M. Azmi (✉)
Civil Engineering, Universiti Sains Malaysia, Penang, Malaysia
e-mail: cemastura@usm.my

N. Sabtu (ed.), *Proceedings of AWAM International Conference on Civil Engineering 2022 - Volume 3*, Lecture Notes in Civil Engineering 386,
https://doi.org/10.1007/978-981-99-6026-2_12

frequent high-intensity rainfall is a well-known criterion for causing landslides [1–8]. In many countries, the growing number of urban or residential settlements inhabiting mountainous environments has resulted in a vast amount of damage caused by landslides [9]. Affected areas and accompanying annual recovery expenditures have increased dramatically in Korea since roughly 2000, while a locality in Amboori, India, experienced its most devastating landslide incident in 2001 [10–12]. Thus, an early warning system is a necessary prerequisite for places prone to landslides, as it is sometimes referred to as the low-cost option for mitigating the losses and casualties associated with natural hazards [12–16]. However, in order to develop an efficient and optimal landslide early warning system (LEWS), the important first step is to acquire knowledge on risks that the landslides impose. Numerous techniques, plans, frameworks and tactics have been devised and used in response to the social and physical environmental characteristics of the study areas [16]. Geological and meteorological elements, covered area, landslide type, data availability and dependability, local authorities and policy- and decision-makers' concerns are among them. Based on a substantial quantity of literature from prior functional examples or suggested LEWS, the majority of the research has determined that each LEWS requires its unique risk assessment approach according to the area's physical and social characteristics mentioned earlier [17]. The approach is commonly used to help determine a series of crucial thresholds that can be used to determine when warnings and alerts can be given, and it is based on LEWS' own theoretical basis. Park et al. [12] concluded that the methodologies can be divided into three categories: statistical based, physical based and monitoring based. The LEWS that use the same type of risks assessment approaches generally have similar features, advantages, and disadvantages in numerous components of the system. Table 1 displays the common attributes of LEWS based on a statistical-, physical- and monitoring-based approaches, which was partly adapted from Park et al. [12]. This work includes a monitoring-based approach that use contact-sensing methods with ground instrumentation [14, 18], and remote-sensing methods with GB-InSAR [17]. The reason for this inclusion is to also compare statistical- and physical-based approach with the monitoring-based approach with other approaches previously done by Park et al. [12].

2 Discussion on Different Approaches, Its Advantages and Disadvantages

2.1 *Basic of Statistical Approach*

The method employs statistically studied correlations where the final results were obtained using, almost exclusively, historical landslide data and rainfall data, which later translated into rainfall thresholds. The rainfall threshold usually comprises intensity-duration curves, among various data included. It has been one of the most

Table 1 Common attributes of LEWS based on statistical-, physical- and monitoring-based approaches [12]

Approach		Statistical	Physical based	Monitoring based
Covered area		Regional scale	Local scale	Regional and local scale
Type		Debris flow, shallow landslide	Shallow landslide, debris flow	Rockslide, shallow landslide, debris flow
Input data		Rainfall data	Rainfall data; hydrological and geotechnical data; and other physical attributes	Measurements from sensor units or monitoring system
Evaluation method		Rainfall thresholds deriving from analysis of rainfall data	A variety of models for analysis purposes	Wireless sensor units, or ground-based synthetic aperture radar interferometry (GB-InSAR)
Time period for risks assessment		A few days ahead	1–24 h ahead	A few days—2 h or 3 h ahead
Resolution of alert level	Update frequency	Twice per day–every 6 min	Daily, twice a day, hourly	60 s, 5 min, 10 min, or 8 h–24 h
	Spatial	Ranging from 1 km grid mesh to regional-scale area	Whole area	Whole area, or 2 × 2 m at 1 km range

popular methods for developing an operational LEWS in recent years. Mandal and Mondal [19] described this method as a method in landslide studies that employ numerical models and procedures that incorporate multiple important variables and inventories of landslide before producing a susceptibility map of a specific area with extreme certainty by utilising statistical, mathematical and GIS software. On the other hand, the quality of the evaluation is contingent upon the inventory's accuracy and completeness [9, 20, 21]. Ordinary statistical models, for instance, logistic regression and linear regression, are only relevant to binary-dependent variables, such as the existence (one) or absence (zero) of a landslide. Nevertheless, the likely landslides in valleys, for example, correspond to multivalent dependent variables with values between zero and one, which will necessitate the employment of statistical models with several classifications [20]. According to Park et al. [12], the evaluation process for landslide early warning uses susceptibility maps to examine the spatial element of landslide hazard. In order to create the map, a statistical approach is applied to various geoproperty variables [22]. Commonly, the statistical approach used is rainfall thresholds that use recorded and forecasted rainfall data to determine landslide warning levels which are considered to be cost-effective, convenient and the preferred tools for years now. Park et al. [12] and other researchers [15, 23–25] have studied various landslide early warning systems that apply statistical approach in order to evaluate different managed area ranging from nation-size area to province-, city- and lastly district-size area. This approach involves the data collection of input factors in real-time and risk assessment calculations. Despite this, the approach has issued relatively high warning level updates over a wide area due to its relatively light requirements [12]. Initially, the influence of different geoproperties (geotechnical, hydraulic, geology and topography properties), that have not been clearly integrated into this method might offer answers for more thorough spatial discrimination in landslide forecasting. Moreover, the accuracy of data obtained in real-time from rainfall is critical to the reliability of an LEWS based on a statistical approach [12]. Hence, because the data sources are dispersed, the statistical approach may not be particularly suitable for regional-scale areas. The geographical variability in rainfall throughout these locations are not adequately captured, particularly localised strong rainfall effects. Alternatively, by using satellite-based precipitation tracking or forecasting techniques as data sources, it may provide higher-resolution rainfall data on a regional scale at reduced update rates.

2.2 *Fundamental of Physical-Based Approach*

The approach represents the techniques for developing rationales for landslide early warning systems based on analytic interpretations of the physical process underlying landslide events. Marin et al. [26] stated that under normal circumstances, it is the input parameters determine the accuracy of a physical-based landslide risk assessment. Numerous models have been researched and proposed to explain landslide-triggering physics by integrating hydrological and geomechanical theories. A lot of these models have been widely used in the landslide research throughout the world [26–30]. The theories usually range from rainfall drainage into porous soil surface to groundwater flow to changes in underground water state up until to the shear failure due to the loss of mechanical balance between stress and strength of materials. Although, numerical method is the more preferred method to derive the solutions, analytically solved equations have been utilised in many cases of simpler models. While different model designs and implementation research of intermediate prototype systems have been suggested, the system's application has been ignored and the work has not yet proceeded to the degree that solely physical-based LEWS are in use. This approach has been applied in areas such as targeted slopes, valleys and watersheds, as certain level of detail and advancement in investigations and supervisions of landslides are demanded [12]. However, the LEWS has undergone some simplifications in terms of physical-based models, and this kind of implementation is rarely practiced [31]. In contrary to the statistical approach, physical-based approach models incorporate particular characteristics ranging from topographical, geological, meteorological and hydrological aspects to geotechnical issues. This will make it difficult to apply the physically dependent approach to a wide field. In addition, unless intentionally simplified, a coupled hydro-geomechanical models simulating landslide-triggering and hydraulic processes typically require a significant amount of computational time to determine whether a threshold associated with a particular type of safety factor is exceeded [12, 25]. This time-consuming assessment cycle which defines the physical-based approach needs to be refined in order to determine near-future risks with fairly long assessment cycles at a site. This is due to its computational intensiveness. Given the assurance of the precision of anticipated rainfall data is at a certain standard, the nowcasting ability is, according to Park et al. [12], as essential as the forecasting ability. This ensure nowcasting to offer means in identifying current risk levels and also to initiate suitable plans against the imminent landslides. With higher update rate of the alert level than the statistical approach, it can be invaluable in countries with localised extreme heavy rainfall that causes rapid increase of danger level over short period of time. This supported by Park et al. [12] as they stated that Korea can benefit as it has similar weather conditions.

2.3 Monitoring-Based Approach

The monitoring-based approach is not a new methodology that needs introduction. Even though, the approach may not receive similar preference as the statistical approach or even the physical-based approach, this approach should be considered for future LEWS. It is impractical and expensive to develop a LEWS appropriate for all landslide types. Due to the variation of precursors and monitored parameters, the LEWS is designed specifically for the designated area, whether it is local-scale or regional-scale area [14]. Though, Uchimura et al. [18] previously have focused more on the small-scale slope disaster. Their work observed the behaviour of the soil implementing a minimal number of measurement points on a slope using low-cost and advanced wireless sensor network, so that the residents occupying near the landslide-prone areas can use it. In order to monitor the parameters, a number of at least three measurement points are set up. Numerous LEWS utilize a wireless sensor network (WSN), which comprises extensometers, thermometer, rain gauge and camera, while others are using ground-based synthetic aperture radar interferometry (GB-InSAR) and also microelectromechanical systems (MEMS) tilt sensor as reported by Uchimura et al. [18], Intrieri et al. [14] and Casagli et al. [17]. The benefit from using the tilt sensor as proposed by Uchimura et al. [18] is that the installation and maintenance become straightforward and low-cost, due to absence of long wire from an extensometer. This means that the transfer of measurement data was mostly done through Wi-Fi or radio communication. However, the downside of using the tilt sensor is that it incapable of determining the slope surface displacement directly [18]. Additionally, the power supply is a key component of an LEWS that must be considered, since disruptions in monitoring leading to a shortage of electricity may be extremely detrimental at times of high landslide activity [14]. Another option of power source is the AAA alkaline batteries, which performed for more than a year at the site remarkably well, according to Uchimura et al. [18]. From the monitoring activities, the researcher can determine whether to decide where it is required to install new sensor units by analysing the information gathered by LEWS monitoring units in terms of the displacement velocity of the soil. This is convenient for planning the expansion of the covered area in the future.

Intrieri et al. [14] stated that the presence of rain events throughout this campaign had no apparent effect on landslide behaviour, at least in the near term. Therefore, most monitoring-based approach LEWS decided to exclude rainfall thresholds attributed to a tenuous link between them and probable failure and insisted on using displacement velocity as thresholds in issuing warning levels. If thresholds using records of rainfall and other additional parameters, then the number of thresholds could be considered as too many and thus would increase the frequency of false alarm produced. This may have more serious implications than the landslide itself, as it will undermine the population's credibility, and the LEWS would be less conservative [14]. Moreover, a large number of thresholds defined would defeat the purpose of a low-cost, simple and straightforward design that many of the studies of monitoring-based LEWS are aiming for [18]. Intrieri et al. [14] have stated the solution for counter-measuring the false alarm, which is to have experts as the last resort to produce a much-thought judgement and decision in issuance of warning levels. Uchimura et al. [18] suggested a criteria for the judgment before issuing the warning, which was based on the combination of data obtained by the monitoring system of LEWS and the complementary sensors, such as volumetric water content sensors. In addition to experts' decision, redundancy and data averaging were used by most studies to minimise the chance of false alarms. In a study by Intrieri et al. [14], they considered ground water content calculations and weather forecasts before issuing warning levels. As for the frequency of warning update, the approach may be related to the frequency of measurements update obtained from the monitoring system. According to Uchimura et al. [18], Intrieri et al. [14] and Casagli et al. [17], the frequency ranging from 60 s to a maximum of 10 min, while the lowest warning level, the monitoring system would produce measurements in the range of 8–24 h cycle [17]. In terms of warning levels, Intrieri et al. [14] stated that further deriving or defining more alert levels would almost certainly necessitate the development of additional thresholds. In this case, the previously mentioned high frequency of false alarms would become a problem. Moreover, lowering the alert levels would be more economically viable [14].

Aside from the normal framework of monitoring-based approach by Uchimura et al. [18], Intrieri et al. [14] and Casagli et al. [17] choose to combine the landslide displacement data with rainfall data to obtain the spatial information. This is for purpose of detecting acceleration on the surface which can determine imminent event of a structural failure of slope. They stated that monitoring-based approach through in situ instrumentation is insufficient in representing the whole area information. They proposed a remote sensing to obtain a two-dimensional deformation maps using the GB-InSAR. According to Casagli et al. [17], GB-InSAR is able to obtain displacement evidence in the observed area, enabling information to be collected from hazardous and rapidly moving areas where no in situ equipment can be mounted.

In the end, the monitoring-based LEWS was trying to achieve a LEWS that is low cost and simple, and in addition to that, the monitoring activities can be maintained for a long time. With less thresholds defined, this approach can be lightweight as the statistical approach, which means that the system is simple and easy to use. Other criteria added from Casagli et al. [17] is data reliability and quick availability. All the studies seem to describe monitoring-based approach as the suitable approach for landslides with high slope acceleration, especially the study by Casagli et al. [17].

3 Conclusion

In summary, the statistical approach is a practical approach for implementing an LEWS and addressing landslide hazards on a regional scale using relatively straightforward evaluation techniques for generating early alerts. Physical arguments and geographically accurate discriminations derived from complex analytical techniques enable the physical-based approach to provide early alerts. Given the complementary nature of the two approaches, various earlier research have presented a method for assessing landslide risks at a regional scale that incorporates and utilises both the statistical and physical-based methodologies. However, this review lacks the findings of a method combining the three approaches or the combination of either one mentioned before with the monitoring-based approach.

Moreover, the physical-based assessment has the advantage of determining higher alert levels with a greater degree of certainty related to physical explanations determined from comprehensive investigations of landslide-triggering occurrences. Meanwhile, due to the frequent updates to the alert level, the statistical approach exhibits a wide range of temporal success in landslide early detection. Lastly, the monitoring-based approach is found to be low-cost approach, simple, and has the potential of a long-term successful investment. With less thresholds defined, this approach can be as simple as the statistical approach. Moreover, it offers data reliability, fast availability and suitable approach for landslides with high slope acceleration.

Acknowledgements The authors would like to thank the Ministry of Education Malaysia and Universiti Sains Malaysia for the support provided by the Malaysia Research University Network–Long Term Research Grant Scheme (MRUN-LRGS) under the program title: River Basin Approach for Integrated Risk Assessment and Reduction of Water-Related Hazards (program grant number: LRGS/1/2016/UTM/01/1/3), under the project title: Intelligent Green Energy Landslide Real-Time Alerting System in the Tropics (grant number: 203/PAWAM/6770006; and grant number: 203/PAWAM/6776001). The authors acknowledge the support from the School of Civil Engineering and the School of Electrical and Electronics Engineering, USM.

References

1. Bíl, M., Andrášik, R., Zahradníček, P., Kubeček, J., Sedoník, J., Štěpánek, P.: Total water content thresholds for shallow landslides. Outer Western Carpathians. Landslides **13**(2), 337–347 (2016). https://doi.org/10.1007/s10346-015-0570-9
2. Fell, R., Corominas, J., Bonnard, C., Cascini, L., Leroi, E., Savage, W.Z.: Guidelines for landslide susceptibility, hazard and risk zoning for land use planning. Eng. Geol. **102**(3), 85–98 (2008). https://doi.org/10.1016/j.enggeo.2008.03.022
3. Guzzetti, F., Peruccacci, S., Rossi, M., Stark, C.P.: Rainfall thresholds for the initiation of landslides in central and southern Europe. Meteorol. Atmos. Phys. **98**(3), 239–267 (2007). https://doi.org/10.1007/s00703-007-0262-7
4. Montrasio, L., Valentino, R.: Experimental analysis and modelling of shallow landslides. Landslides **4**(3), 291–296 (2007). https://doi.org/10.1007/s10346-007-0082-3
5. Petley, D.: Global patterns of loss of life from landslides. Geology **40**(10), 927–930 (2012). https://doi.org/10.1130/G33217.1
6. Segoni, S., Piciullo, L., Gariano, S.L.: A review of the recent literature on rainfall thresholds for landslide occurrence. Landslides **15**(8), 1483–1501 (2018). https://doi.org/10.1007/s10346-018-0966-4
7. Smith, D.M., Oommen, T., Bowman, L.J., Gierke, J.S., Vitton, S.J.: Hazard assessment of rainfall-induced landslides: a case study of San Vicente volcano in central El Salvador. Nat. Hazards **75**(3), 2291–2310 (2015). https://doi.org/10.1007/s11069-014-1422-y
8. Thiery, Y., Terrier, M., Colas, B., Fressard, M., Maquaire, O., Grandjean, G., Gourdier, S.: Improvement of landslide hazard assessments for regulatory zoning in France: STATE–OF–THE-ART perspectives and considerations. Int. J. Disaster Risk Reduction **47**, 101562 (2020). https://doi.org/10.1016/j.ijdrr.2020.101562
9. Guzzetti, F., Gariano, S.L., Peruccacci, S., Brunetti, M.T., Marchesini, I., Rossi, M., Melillo, M.: Geographical landslide early warning systems. Earth Sci. Rev. **200**, 102973 (2020). https://doi.org/10.1016/j.earscirev.2019.102973
10. Naidu, S., Sajinkumar, K.S., Oommen, T., Anuja, V.J., Samuel, R.A., Muraleedharan, C.: Early warning system for shallow landslides using rainfall threshold and slope stability analysis. Geosci. Front. **9**(6), 1871–1882 (2018). https://doi.org/10.1016/j.gsf.2017.10.008
11. Park, J., Lee, S., Son, D.: Safety of elevation from superficial fascial plane versus traditional deep fascial plane for flap elevation in a porcine model. Arch Hand Microsurg. **23**(2), 99–109 (2018). https://doi.org/10.12790/ahm.2018.23.2.99
12. Park, J.Y., Lee, S.R., Lee, D.H., Kim, Y.T., Lee, J.S.: A regional-scale landslide early warning methodology applying statistical and physically based approaches in sequence. Eng. Geol. **260**(October 2018), 105193 (2019). https://doi.org/10.1016/j.enggeo.2019.105193
13. Chen, M., Jiang, Q.: An early warning system integrating time-of-failure analysis and alert procedure for slope failures. Eng. Geol. **272**(April), 105629 (2020). https://doi.org/10.1016/j.enggeo.2020.105629
14. Intrieri, E., Gigli, G., Mugnai, F., Fanti, R., Casagli, N.: Design and implementation of a landslide early warning system. Eng. Geol. **147–148**, 124–136 (2012). https://doi.org/10.1016/j.enggeo.2012.07.017
15. Osanai, N., Shimizu, T., Kuramoto, K., Kojima, S., Noro, T.: Japanese early-warning for debris flows and slope failures using rainfall indices with radial basis function network. Landslides **7**(3), 325–338 (2010). https://doi.org/10.1007/s10346-010-0229-5
16. Pecoraro, G., Calvello, M., Piciullo, L.: Monitoring strategies for local landslide early warning systems. Landslides **16**(2), 213–231 (2019). https://doi.org/10.1007/s10346-018-1068-z
17. Casagli, N., Catani, F., Del Ventisette, C., Luzi, G.: Monitoring, prediction, and early warning using ground-based radar interferometry. Landslides **7**(3), 291–301 (2010). https://doi.org/10.1007/s10346-010-0215-y
18. Uchimura, T., Towhata, I., Anh, T.T.L., Fukuda, J., Bautista, C.J.B., Wang, L., Seko, I., Uchida, T., Matsuoka, A., Ito, Y., Onda, Y., Iwagami, S., Kim, M.S., Sakai, N.: Simple monitoring

method for precaution of landslides watching tilting and water contents on slopes surface. Landslides **7**(3), 351–357 (2010). https://doi.org/10.1007/s10346-009-0178-z

19. Mandal, S., Mondal, S.: Concept on landslides and landslide susceptibility BT. In: Mandal S., Mondal S. (eds.) Statistical Approaches for Landslide Susceptibility Assessment and Prediction, pp. 1–39. Springer International Publishing (2019). https://doi.org/10.1007/978-3-319-93897-4_1
20. Du, J., Glade, T., Woldai, T., Chai, B., Zeng, B.: Landslide susceptibility assessment based on an incomplete landslide inventory in the Jilong Valley, Tibet, Chinese Himalayas. Eng. Geol. **270**, 105572 (2020). https://doi.org/10.1016/j.enggeo.2020.105572
21. Reichenbach, P., Rossi, M., Malamud, B.D., Mihir, M., Guzzetti, F.: A review of statistically-based landslide susceptibility models. Earth Sci. Rev. **180**, 60–91 (2018). https://doi.org/10.1016/j.earscirev.2018.03.001
22. Piciullo, L., Dahl, M.-P., Devoli, G., Colleuille, H., Calvello, M.: Adapting the EDuMaP method to test the performance of the Norwegian early warning system for weather-induced landslides. Nat. Hazard. **17**(6), 817–831 (2017). https://doi.org/10.5194/nhess-17-817-2017
23. Calvello, M., d'Orsi, R.N., Piciullo, L., Paes, N., Magalhaes, M., Lacerda, W.A.: The Rio de Janeiro early warning system for rainfall-induced landslides: analysis of performance for the years 2010–2013. Int. J. Disaster Risk Reduction **12**, 3–15 (2014). https://doi.org/10.1016/j.ijdrr.2014.10.005
24. Jakob, M., Owen, T., Simpson, T.: A regional real-time debris-flow warning system for the district of North Vancouver, Canada. Landslides **9**(2), 165–178 (2012). https://doi.org/10.1007/s10346-011-0282-8
25. Segoni, S., Battistini, A., Rossi, G., Rosi, A., Lagomarsino, D., Catani, F., Moretti, S., Casagli, N.: Technical note: an operational landslide early warning system at regional scale based on space-time-variable rainfall thresholds. Nat. Hazard. **15**(4), 853–861 (2015). https://doi.org/10.5194/nhess-15-853-2015
26. Marin, R.J., Velásquez, M.F., Sánchez, O.: Applicability and performance of deterministic and probabilistic physically based landslide modeling in a data-scarce environment of the Colombian Andes. J. South Am. Earth Sci. **108**, 103175 (2021). https://doi.org/10.1016/j.jsames.2021.103175
27. Aristizábal, E., Vélez, J.I., Martínez, H.E., Jaboyedoff, M.: SHIA_Landslide: a distributed conceptual and physically based model to forecast the temporal and spatial occurrence of shallow landslides triggered by rainfall in tropical and mountainous basins. Landslides **13**(3), 497–517 (2016). https://doi.org/10.1007/s10346-015-0580-7
28. Mergili, M., Marchesini, I., Alvioli, M., Metz, M., Schneider-Muntau, B., Rossi, M., Guzzetti, F.: A strategy for GIS-based 3-D slope stability modelling over large areas. Geosci. Model Dev. **7**(6), 2969–2982 (2014). https://doi.org/10.5194/gmd-7-2969-2014
29. Mergili, M., Marchesini, I., Rossi, M., Guzzetti, F., Fellin, W.: Spatially distributed three-dimensional slope stability modelling in a raster GIS. Geomorphology **206**, 178–195 (2014). https://doi.org/10.1016/j.geomorph.2013.10.008
30. Reid, M.E., Christian, S.B., Brien, D.L., Henderson, S.: Scoops3D—Software to Analyze Three-Dimensional Slope Stability Throughout a Digital Landscape. US Geological Survey Techniques and Methods, p. 14 (2015)
31. Montrasio, L., Valentino, R., Corina, A., Rossi, L., Rudari, R.: A prototype system for space–time assessment of rainfall-induced shallow landslides in Italy. Nat. Hazards **74**(2), 1263–1290 (2014). https://doi.org/10.1007/s11069-014-1239-8

Evaluation of Slope Stability Due to Earthquake and Rainfall Occurrences

Goh Chin Sin, Mastura Azmi, and Mazdak Ghasemi Tootkaboni

Abstract Malaysia is a tropical country with high humidity, heavy rainfall, and unfrequent earthquake, while tremors can exist due to neighbouring regions earthquakes such as Sumatra. The 2004 Sumatera earthquake caused 68 death in Penang, Langkawi, and Kedah. The combined effect of rainfall and earthquake is unknown for the existing slope structures in Penang Island. The research presents the results on slope stability after the impact of rainfall and earthquake using the GeoStudio software. In the first step, the single effect of rainfall shows that longer duration and higher intensity adversely affect slope stability. However, the safety factor was lower when both effects of rainfall and earthquake were considered in the analysis. The combined effect was assessed through two factors of FOS and deformation since the seismic waves imposed as secondary damage to the hill. The results show that higher deformation is recorded for slope geometry with a bigger angle and higher height. The slope is classified as a failed candidate when the factor of safety is less than 1.0. The slope geometry in Sungai Ara was relatively unstable, and the factor of safety (FOS) has not complied with Penang safety guidelines for hill site development 2020. Various slope reinforcement procedures are mainly discussed in relation to Sungai Ara's slope to improve the slope stability and lead to a higher factor of safety. The research proposes several reinforcement approaches, including retaining walls, soil nailing, and a drainage system. The results demonstrated that the factor of safety improved, and seismic waves deformation reduced, which proved the capability of reinforcement approaches in reducing the effect of rainfall and earthquake.

Keywords Slope stability · Rainfall intensity · Rainfall duration · Earthquake

G. C. Sin · M. Azmi (✉) · M. G. Tootkaboni
School of Civil Engineering, Engineering Campus, Universiti Sains Malaysia, 14300 Nibong Tebal, Penang, Malaysia
e-mail: cemastura@usm.my

N. Sabtu (ed.), *Proceedings of AWAM International Conference on Civil Engineering 2022 - Volume 3*, Lecture Notes in Civil Engineering 386,
https://doi.org/10.1007/978-981-99-6026-2_13

1 Introduction

In today's globalised world, property developing and building construction are common to fulfil the community's demand. However, the lack of land resources had become an issue due to rapid population growth. As a result, many construction projects had to be carried out on hillslope, which might be followed by the downward movement of soil and rock under the influence of gravity. This phenomenon is defined as one of the common types of slope failure known as landslide hazard. Generally, natural disasters or human activities induce an unexpected landslide that leads to destructive outcomes such as damage to properties, financial loss, and fatalities. The landslide hazard can happen in different forms, such as mudflows, debris flow, and rockfalls [1]. Therefore, the phenomenon of slope failure can happen unexpectedly without proper preventive measures and analysis. It is vital to study the slope failure mechanism and the effect of driving force or resisting force acted on slope faces through slope stability analysis; indeed, understanding the slope failure mechanism and the causes of slope instability is critical to prevent landslides from happening [2]. However, slope instability is a complicated issue that relates to various geotechnical principles, which needed site inspection and professional engineers' experiences. For instance, the determination of the equilibrium of shear stress and shear strength is the key variable to avoid the evolution of slope failure and landslide. If the shear stress applied is larger than soil shear strength, the landslide has a greater chance to take place.

Theoretically, two factors contribute to slope failures, which are internal and triggering factors. The internal factor includes the slope geometry, strength of soil materials, and groundwater [3]. In contrast, the triggering factor stems from natural phenomena such as volcanic eruptions, earthquakes, and heavy rainfall. According to prior inspections, earthquake-induced landslide and rainfall-induced landslide are two common types of slope failure [4]. Earthquake-induced landslide is related to the magnitude of an earthquake, distance from the epicentre, and geological condition [5]. In general, an earthquake involves a series of ground movements, and the ground acceleration will act as dynamic loading, which increases the driving force and sliding force along the potential slip surface [6]. Subsequently, the sliding surface will transform into plastic state and reduce the sliding resistance that leads to slope failure. On the other hand, a landslide can also trigger by rainfall; indeed, the water will infiltrate into the ground and increase the groundwater table level [7]. The groundwater level fluctuation will lead to a drop in soil matric suction, increase pore water pressure, saturation, and weaken the soil shear strength [8]. However, the formation of landslides depends on certain rainfall intensity, duration, and geological conditions. For instance, loose soil and weathered rock are more prone to rainfall-induced landslide.

This research seeks to assess the potential of slope failure in one of the Malaysian territory, Penang Island. Malaysia is recognised as a tropical country by equatorial climate with high temperatures and heavy rainfall. Generally, Malaysia undergoes two types of monsoons with different types of rainfall patterns: Southwest Monsoon and Northeast Monsoon. According to the Malaysian Meteorological Department, the weather is hot and dry during Southwest Monsoon, leading to drought and haze issues [9]. On the other hand, the local areas experienced a higher rainfall intensity during Northeast Monsoon from October to March. For instance, seven episodes of heavy rainfall occurred in November and December 2019 during Northeast Monsoon 2019/2020, resulting in flooding. Flood disasters had become part of Malaysian life during Northeast Monsoon. Besides, the occurrence of extreme weather events such as Typhoon Damrey caused slope failure in Penang Island due to the overflowing water carrying the soil down the hillslope [10]. In other words, the change in rainfall intensity might be a possible cause of slope failure. Thus, the variation in rainfall patterns should be given attention when evaluating slope stability. Besides, Malaysia was categorised as tectonically stable since it is located far from plate tectonics boundaries and the ring of fire region [11]. Hence, natural disasters like volcanic eruptions and major earthquakes caused by plate tectonics movement seldom take place. Malaysia is part of the Sundaland Block, covering a major part of Southeast Asia. It is mainly surrounded by several plate tectonic, including Australian Plates and the Philippines Plate. Malaysia experienced a significant change in direction and rate of tectonic motion after the 2004 mega-earthquake [9]. This statement had been proven by observation using Global Positioning System (GPS): a co-seismic displacement of 17 cm in Langkawi Island, Malaysia [12]. As a result, researchers must understand and consider the impact of seismic waves when assessing slope stability since Malaysia is no longer earthquake-free. Comprehensive studies on the impact of earthquakes and rainfall on slope stability should be carried out promptly by considering the effects of both the triggering factors. The ground shaking induced by seismic waves, combined with rainfall patterns in the Penang area, might have a greater impact on the stability of the slope and result in irreversible disaster. The study's main objectives are (i) determining the effect of rainfall on slope stability, and (ii) determining the combined effect of earthquake and rainfall on slope stability. The study scope mainly focused on slope structure modelling and using GeoStudio software, including SEEP/W, QUAKE/W, and SLOPE/W. The deformation of slope structure can be analysed with consideration of ground shaking induced by earthquake and rainfall intensity through the software.

1.1 Study Area

Penang Island was selected as the study location. One slope geometry in Sungai Ara was chosen (5° 19′ 33.2"N 100° 15′ 39.4"E). The selected site was developed as housing areas, and the soil investigation reports were done and provided by the local consultant (Figs. 1 and 2).

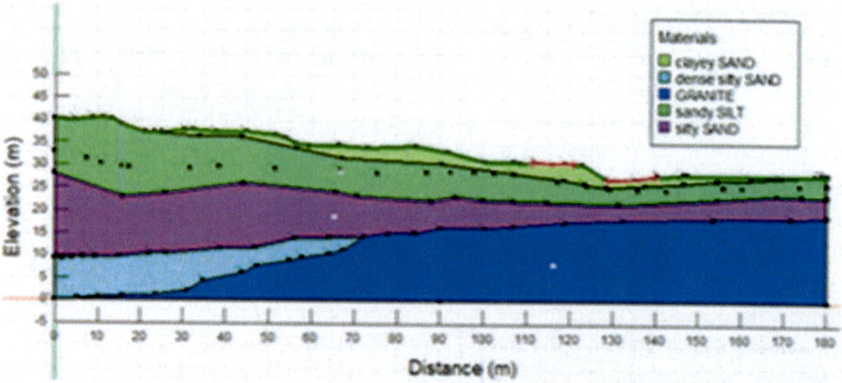

Fig. 1 Slope geometry in Tanjung Bungah

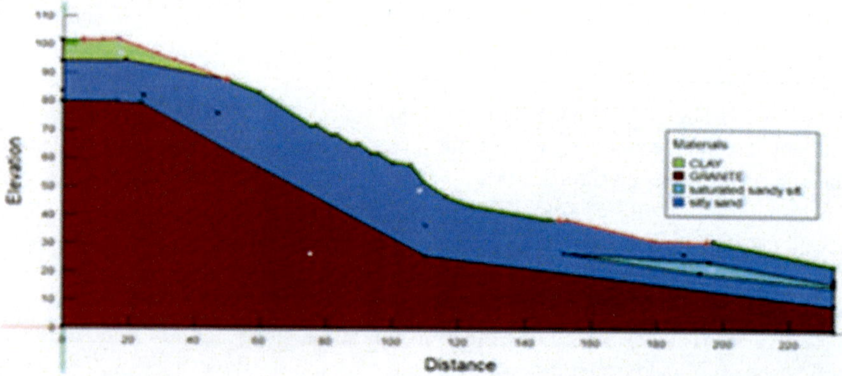

Fig. 2 Slope geometry in Sungai Ara

2 Methodology

This research develops an innovative methodology to optimise the study of slope stability and determination slope failure due to rainfall and earthquake effects. The study was mainly conducted using GeoStudio software, including SEEP/W, QUAKE/W, and SLOPE/W. Besides, the deformation of slope structure analysed through slope structure modelling. The methodology processes were summarised in the form of a flowchart in Fig. 3.

In the first stage, a review was done on slope stability affected by the factor of rainfall which adopted different rainfall intensities and rainfall durations to evaluate the stability of the slope. The slope stability is accessed by using the parameter of the safety factor. Next, the slope stability affected by the combined effects of rainfall and earthquake was discussed. In this stage, the deformation induced by seismic waves and factors of safety (FOS) were studied. Notably, the ground profile that

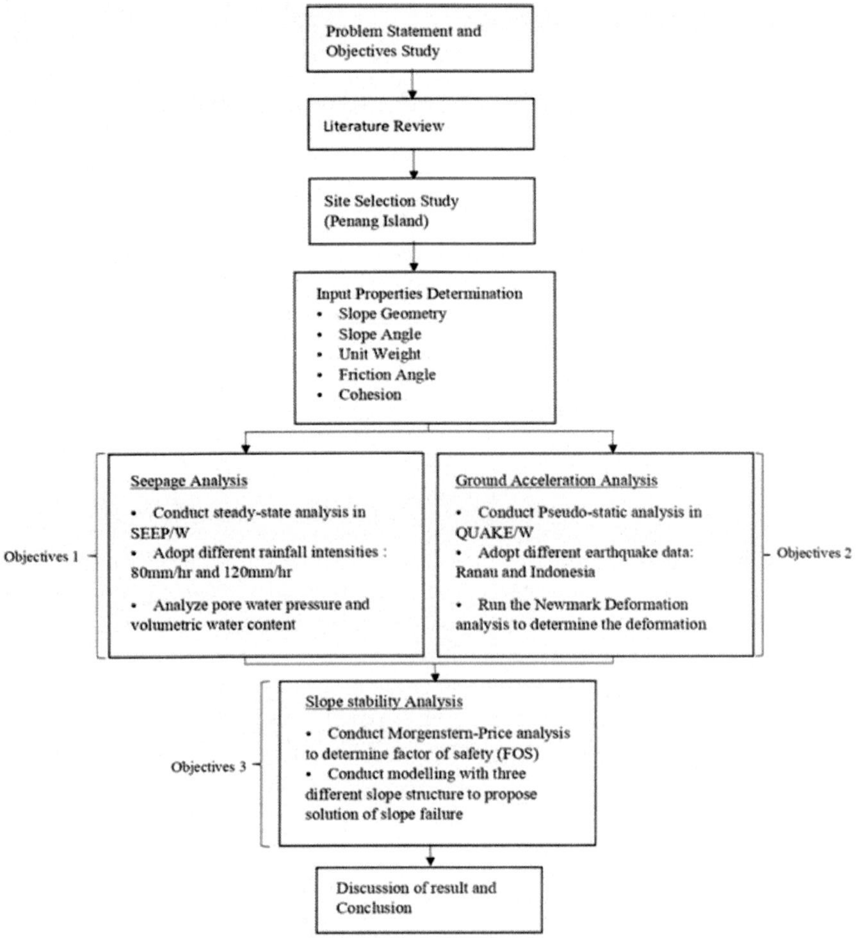

Fig. 3 Flowchart for the methodology

should be determined before modelling can be carried out. Data collection could be the most time-consuming step because a broad subset of undefined variables could be described and initially considered the input data. Undoubtedly, slope failure has stuck to key details, the data such as slope geometry, slope angle, and soil behaviours like unit weight, friction angle, and cohesion. Data records covering a wide range of inputs were collected from soil bore log reports at the study area in the first step. After data collection, the modelling of slope structure starts from seepage analysis. Next, ground acceleration analysis is modelled in the QUAKE/W software. Further, slope stability analysis was conducted in SLOPE/W, which take advantage of SEEP/W and QUAKE/W analysis. The stability evaluations help to create better insights regarding stability reinforcements.

2.1 Input Properties for Slope

Generally, the bore log report contains a variety of details like type of soil and *N*-number. However, extra information such as unit weight, cohesion, and friction angle can be calculated by using data provided in the bore log report. The friction angle and cohesion were calculated through Eqs. 1 and 2, where $(N_1)_{60}$ = value of N_{60} corrected to a standard value of $\sigma'o$ (100 kN/m^2) and K = Constant (3.5–6.5 kN/m^2). Friction angle was calculated based on Peck et al. [13], while cohesion was calculated based on SPT *N*-value. The data sets utilised in this investigation were chosen based on field experience and information found in the prior investigations, presented in Tables 1 and 2.

$$\varphi' = 27.1 + 0.3(N_1)_{60} - 0.00054\left[(N_1)_{60}\right]^2 \tag{1}$$

$$\mathrm{Cu} = KN_{60} \tag{2}$$

The modelling of this slope structure will start from seepage analysis which carries out in SEEP/W with steady-state analysis. The study implemented two different extreme rainfall intensities. The first one is extracted from a previous study of Zaki [14] with reference to historical rainfall data by the Department of Irrigation and Drainage (DID), which analyse and obtain the highest hourly rainfall intensity as 80 mm/h. The second one is extracted from reported cases in the annual report 2019 by the Malaysian Meteorological Department, which stated the extreme rainfall data as 120 mm/h.

Table 1 Soil parameters for each soil layer in Sungai Ara site

Characteristics	Clayey sand	Sandy silt	Silty sand	Dense silty sand
Unit weight (kN/m^3)	18	18.4	20	21
Cohesion	3.2	38.4	3.2	3.2
Friction angle	30.0	29.7	32.3	40.8
Poisson ratio	0.28	0.3	0.33	0.33
Gmax value	81,600	96,500	109,000	109,000

Table 2 Typical water content and conductivity for each type of soil

Soil types	Saturated water content	Residual water content	Conductivity (m/s)
Clayey sand	0.35	0.045	8.25×10^{-5}
Sandy silt	0.43	0.034	6.94×10^{-7}
Silty sand	0.38	0.057	4.06×10^{-5}
Dense silty sand	0.38	0.057	4.06×10^{-5}
Sandy clay	0.38	0.068	3.33×10^{-7}

2.2 Seepage Analysis

Seepage analysis needs to be carried out in SEEP/W that divides analysis into two categories: steady-state and transient analysis [15]. The water pressures and flow rates remain constant over time in steady-state category. Notably, the analysis does not consider the time frame, which leads to a simplifier process. In contrast, transient analysis is accompanied by more complicated steps due to the fact that the pressure conditions might change over time; indeed, the procedure provides more realistic outcomes in modelling [16]. The paper intended to study the seepage flow over time that convinced us to select transient analysis. In SEEP/W, the process starts by assigning material to create the slope geometry and cross-sectional. In the next step, boundary conditions need to be determined, which can be described in terms of the head (H) or total flux (Q) [17]. In this study, boundary conditions followed flux (Q) terms and conditions set as shown in Fig. 4. The flux value for the boundary of the slope surface will be the rainfall intensity extracted from historical data. Also, the flux value for impenetrable layer and side border condition above groundwater table level is 0 $m^3/s/m^2$. The total head is used to define the groundwater table condition for the soil below the groundwater table level. Besides, the volume water content and hydraulic conductivity are two parameters that need to be determined to conduct the seepage analysis [18]. For the volume water content function, the data point function was chosen, and the sample function was used as the estimation method by specifying the type of soil and saturated water content. For hydraulic conductivity function, the data point chooses to develop the graph of hydraulic conductivity over matric suction; however, several methods can be generated to estimate the unsaturated hydraulic conductivity function. In this study, the most widely used soil water retention curve (SWRC) model known as Van Genuchten is mainly used as the estimation method. The saturated water content, residual water content, and conductivity were extracted through the Van Genuchten model, which provides the typical values according to different soil types. After defining the volumetric water content function and saturated conductivity, SEEP/W adopt the parameters to predict the hydraulic conductivity function. Table 2 presented all the information needed to develop the volumetric water content and hydraulic conductivity. From the other aspects, two different rainfall data is used as flux when modelling, which is 80 mm/h and 120 mm/h, respectively. Besides, different durations of rainfall (1 and 6 h) were modelled to determine the effect of rainfall duration on the stability of both selected slopes.

Next, ground acceleration analysis will be modelled in the software of QUAKE/W. The acceleration data for each time step of the earthquake happened in Pulau Bangkaru, Indonesia, was used as input. Further, initial static was conducted to analyse the in situ condition of stress in the ground before the seismic waves. The equivalent linear dynamic analysis is used to analyse the seismic waves considering the time step of the earthquake. Then, the Newmark deformation is run in the software to determine the slope deformation caused by the ground shaking.

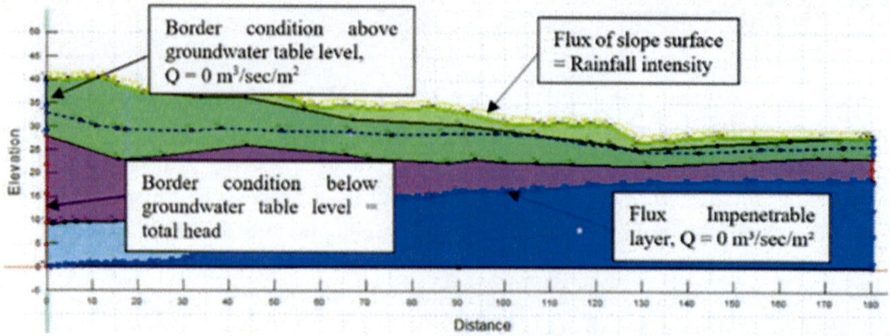

Fig. 4 Boundary conditions for Tanjung Bungah slice in SEEP/W

2.3 Ground Acceleration Analysis

The process starts with analysing the in situ stress condition in the ground. The study augmented QUAKE/W as another valuable software of GeoStudio, which can stimulate dynamic forces and help SLOPE/W to calculate accelerations and possible permanent deformations in every trial slip surface. The type of analysis used in QUAKE/W is initial static, which considers the initial pore water pressure condition referring to the pore water pressure condition obtained in SEEP/W examination [19]. Besides equivalent linear dynamic analysis is used to evaluate the slope stability during earthquake shaking which takes the initial static stress as the initial stress condition, the equivalent linear dynamic analysis begins with a given G value then proceeds to go through the whole ground motion record, identifying peak shear strains at each Gauss numerical integration point in each element. The shear modulus is then adjusted using a defined reduction function (G_{max} reduction function), and the process is repeated. The iterative approach is repeated until the appropriate G modifications are within the specified threshold. Notably, historical horizontal earthquake records need to be defined before the equivalent linear dynamic analysis can be run in QUAKE/W and stimulate the seismic conditions in the slope area. An earthquake with a magnitude of Mw 8.6 struck Pulau Bangkaru, Indonesia (latitude 2.09, longitude 97.11) on March 2005 was used as the input ground motion. Azmi stated that the peak ground acceleration of 0.15 g with 2500 years return period would be imposed on the slope after interpreting the Probabilistic Seismic Hazard Analysis map for Penang Island [8]. Further, the peak ground acceleration input for the earthquake record modified based on the findings. Figure 5 shows the modified earthquake record with peak ground acceleration of 0.15 g at about the 66-s mark in the 400 s of shaking.

A few parameters need to be determined for each soil property, including Poisson's ratio, damping function, G reduction function, and G_{max} constant (maximum shear modulus). G_{max} can be defined as a constant related to shear wave velocity or as a function of effective confining stress. Following equation used to calculate the values of G_{max} for each soil layer, where v_s is shear wave velocity for each soil type and ρ

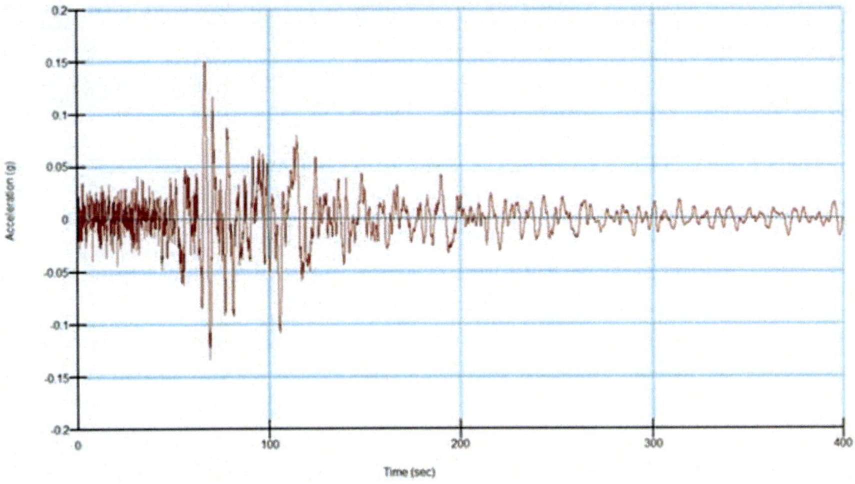

Fig. 5 Horizontal earthquake records import in QUAKE/W

is the soil bulk density.

$$G_{\max} = v_s^2 \cdot \rho \tag{3}$$

The G-reduction function is another parameter that needs to be defined to indicate response to cyclic shear stress. In the equivalent linear model, the G-reduction function can be described as a ratio of $G_{\max}$. In this study, an equivalent linear model was used in defining all types of soil properties. The Poisson's ratio and $G_{\max}$ are set based on Tables 1 and 2 depending on the location of slopes. In the next step, the G-reduction function and damping function were developed through the generated method in the QUAKE/W [20]. Boundary conditions of the slope are important and need to be defined before starting the analysis. The left and right boundaries had to be fixed in x-direction in the initial static analysis and in the y-direction in dynamic analysis. Eventually, the analysis carried out in SEEP/W and QUAKE/W can be used in determining the slope stability. SLOPE/W mainly conduct limit equilibrium analysis to determine the factor of safety (FOS). Two slope structures conducted the modelling with different geometry located at Tanjung Bungah and Sungai Ara. Further, three slope reinforcement methods were suggested to improve slope stability for the safety factor that do not satisfy the guideline.

2.4 Slope Stability Analysis

Slope stability analysis conducted in SLOPE/W, which mainly uses limit equilibrium methods such as Morgenstern Price, Bishop, Janbu, and Spencer to analyse the stability of a slope structure [21]. The structure divides into a few vertical slices by using the limit equilibrium approach, which expresses the equilibrium of the sliding mass in terms of force or moment equilibrium. In this study, the Morgenstern-Price approach was selected due to the analysis characteristics. The method satisfies both force and moment equilibrium acting to the sliding mass. Besides, it considers the interslice shear and normal force. In SLOPE/W, the interslice function selection is on half-sine. In other aspects, the slip surfaces can be selected manually by using the entry and exit function. The soil model goes through Mohr–Coulomb, and the analysis is conducted according to the soil parameters presented in Tables 1 and 2. The factor of safety (FOS) is the outcome of the slope stability study. Besides, Newmark deformation analysis can be conducted to analyse the permanent deformation of slope due to shaking in SLOPE/W. In general, the combination of inertia forces and initial static force will overcome the shear resistance and cause permanent deformation. The Newmark method will double integrate the acceleration with time and calculate the displacement of the slope, which gives access to both parameters, including the factor of safety (FOS), and deformation [22].

3 Results and Discussion

The study considered all probable theories in three stages. In the first stage, the single effect of rainfall intensities and durations was adopted to evaluate slope stability using the safety factor parameter. Next, the combined effects of rainfall and earthquake were assessed through the deformation induced by seismic waves and factor of safety (FOS). Lastly, several recommendations and slope reinforcements in reducing the impact of rainfall and seismic waves were analysed.

3.1 Single Effect of Rainfall

In the study's first phase, the impact of various rainfall duration and intensities was considered in Tanjung Bungah, Penang Island, to evaluate the stability of slopes. Firstly, the pore water pressure was analysed as one of the characteristics affected by rainfall. Figure 6 shows the pore water pressure distribution for the slope in Tanjung Bungah.

As shown, the highest point of pore water pressure is located near the bedrock due to surface tension effects above the groundwater table. It can conclude that the pore water pressure is proportional to the depth, and the pore pressure above the water table is negative. The location near the bedrock possesses the greatest depth, which transforms it into a good candidate for the most considerable pore water pressure by the range of 300–350 kPa. At the same time, negative pore water pressure exists above the groundwater level that corresponds to the pore pressure condition in unsaturated soil conditions due to a suction effect induced by the surface tension. The groundwater

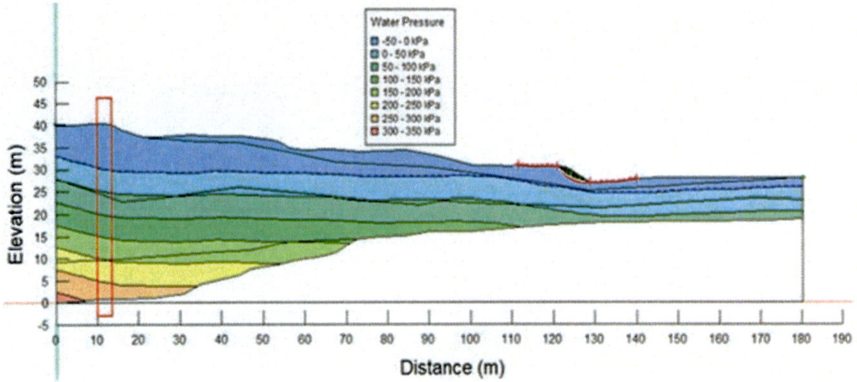

Fig. 6 Pore water pressure condition of Tanjung Bungah's slope before rainfall

level for Tanjung Bungah's slope was around 7.0 m below the ground surface level or 33.0 m above the bedrock layer. The distribution of pore water pressure varies with depth of slope for the highlighted part in Fig. 6.

Case study 1: Effect of different duration of rainfall. In the first case, the Tanjung Bungah slope was used to simulate the effect of different rainfall duration by the specified rainfall intensity (80 mm/h). Figures 7 and 8 show the factor of safety for Tanjung Bungah slice after 1 and 6 h of 80 mm/h rainfall.

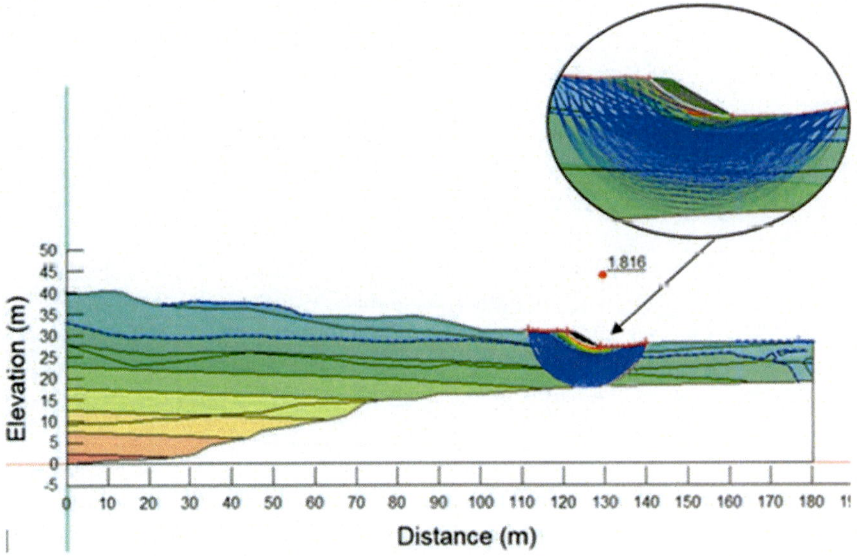

Fig. 7 Factor of safety (FOS) after 1 h of 80 mm/h rainfall

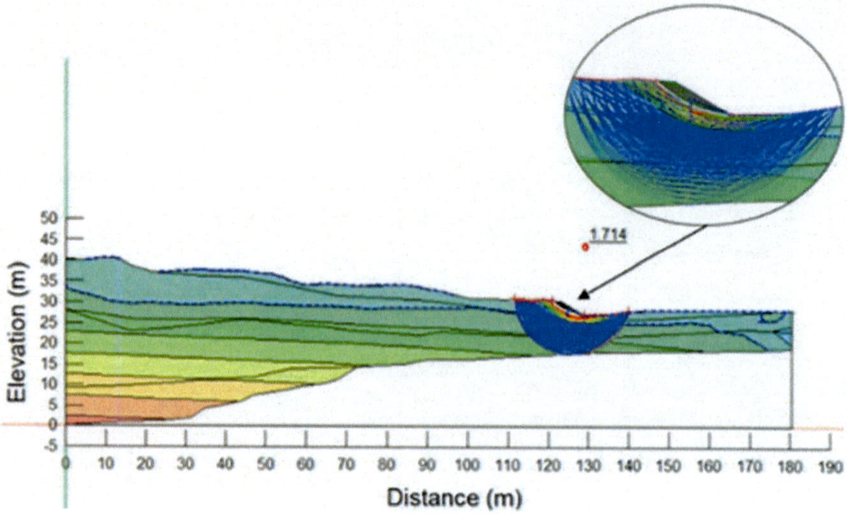

Fig. 8 Factor of safety (FOS) after 6 h of 80 mm/h rainfall

As indicated, the slope's safety factor reached to 1.816 after 1 h of 80 mm/h rainfall, while the factor decreased to 1.714 after 6 h. The safety factors for both rainfall durations were acceptable according to the Penang safety guideline for hill site development 2020. It can conclude that longer rainfall duration with the same amount of rainfall intensity led to a lower factor of safety (FOS). Figure 9 displays the pattern of the safety factors in accordance with rainfall duration.

As shown, the factor of safety remained unchanged until the crucial point. The FOS reflected a downward trend after 5.5 h of continuous rainfall, which was attributed to the rise of the groundwater table. As a result, longer rainfall duration caused a rise in soil permeability and a reduction in soil matric suction, increasing the likelihood of a landslide event. Further, several rainfall durations were chosen to stimulate and analyse the safety factor for the Tanjung Bungah's slope and study the relationship between the factor of safety and different durations of rainfall. A linear equation (Eq. 4) was generated to evaluate the relationship between the two parameters, where y = Factor of safety and x = duration of rainfall. It can be interpreted that safety factors were inversely proportional to rainfall duration from Fig. 10. Besides, the linear best fit line showed an R^2 value of 0.8225, representing the linear model that could evaluate the data variation around its mean.

$$y = -0.0323x + 1.8847 \tag{4}$$

Case study 2: Effects of different rainfall intensities. In this phase, the Tanjung Bungah slope was used to model the impact of different rainfall intensities with an identical duration of 6 h. The rainfall intensities were 80 and 120 mm/h. Figure 11 shows the safety factor after 6 h of 120 mm/h rainfall.

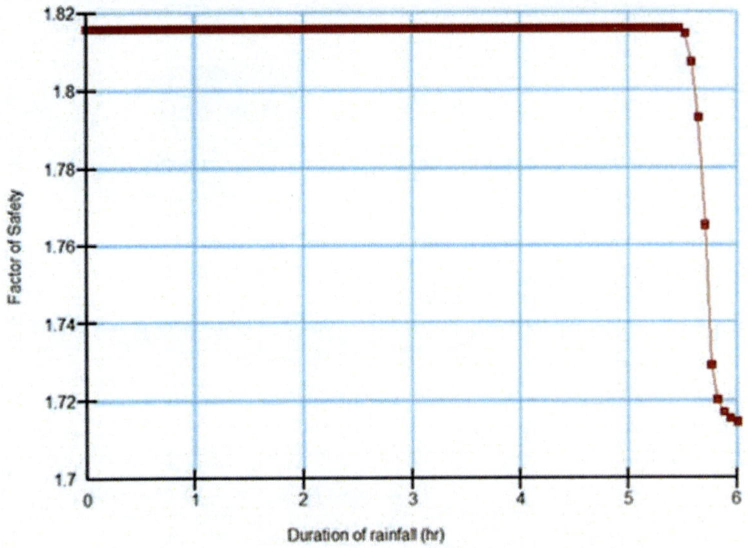

Fig. 9 Factor of safety (FOS) varies with the duration of rainfall

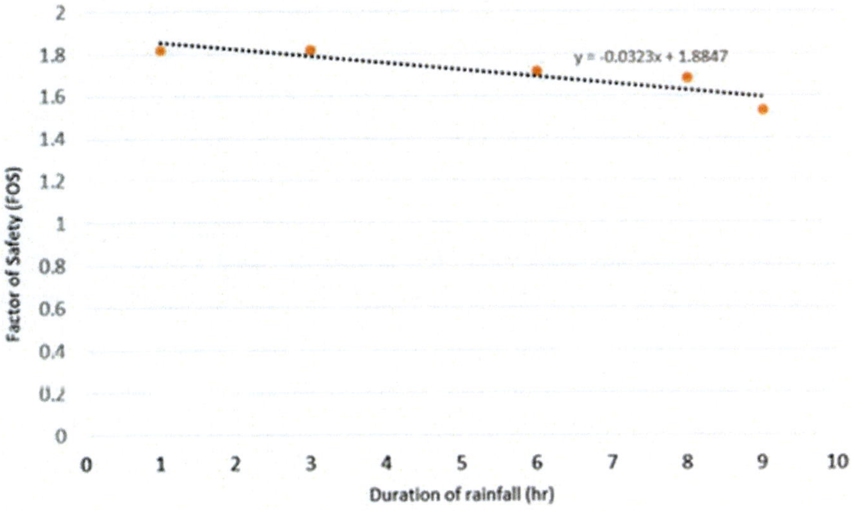

Fig. 10 Factor of safety versus duration of rainfall

The comparison between Figs. 8 and 11 indicated that the FOS for 6 h of 80 mm/h is 1.714 while the number reached 1.674 for 6 h of 120 mm/h rainfall. Thus, a higher rainfall intensity in the fixed duration contributed to a lower FOS due to the high level of groundwater table and decrease in matric suctions. The higher intensity rainfall increased pore water pressures and led to a decrease in shear strength which

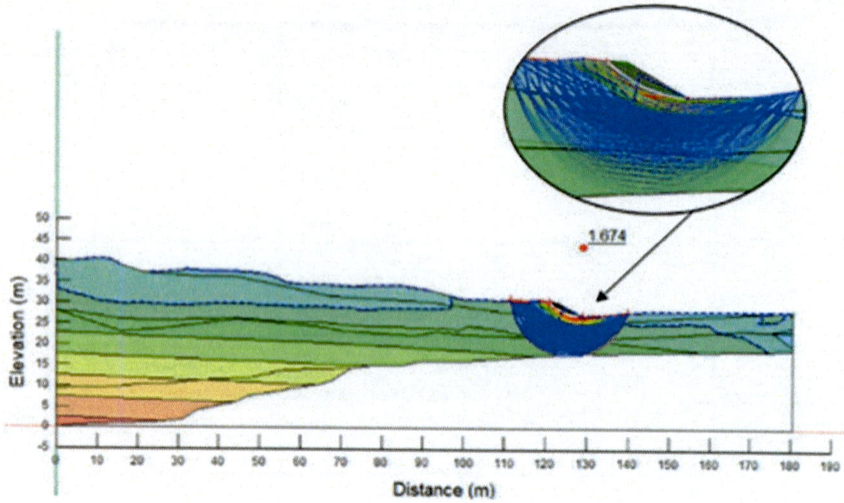

Fig. 11 Factor of safety (FOS) for Tanjung Bungah slope after 6 h of 120 mm/h rainfall

caused a drop in the factor of safety. Results indicate that the safety factor for the slope in Tanjung Bungah caused by 80 and 120 mm/h rainfall is tolerable since they are greater than 1.2.

3.2 Slope Stability Induced by Earthquake and Rainfall

In this section, the combined effects of earthquake and rainfall are considered to comprehend and evaluate the stability function of the slope. Firstly, QUAKE/W processed the input earthquake data with modified peak ground acceleration which was generated from the Sungai Ara slope. After 6 h of 80 mm/h rainfall, the pore water pressure condition was taken into consideration. Figure 12 shows the factor of safety (FOS) obtained for the critical slip surface under the influence of seismic waves. It is indicated that the modified PGA of 0.15 g had the lowest safety factor by 0.85 in 70 s, while the highest factor of safety was 2.36 at about 66 s. The deformation will occur when the safety factor is less than 1.0.

It can be interpreted from Fig. 13 that the average acceleration of 0.07 g resulted in FOS that was lower than 1.0 and subsequently produced deformation. This crucial point is known as yield acceleration value or variance that depends on the earthquake record and will lead to different deformation values. Further, Figs. 14 and 15 demonstrate the average acceleration and velocity against time. The integration is done on the area under the graph in Fig. 14. Generally, integration of the area under the yield acceleration values provides the velocity. Eventually, deformation of the slope can be obtained after integration of the area under the graph, as shown in Fig. 15, and

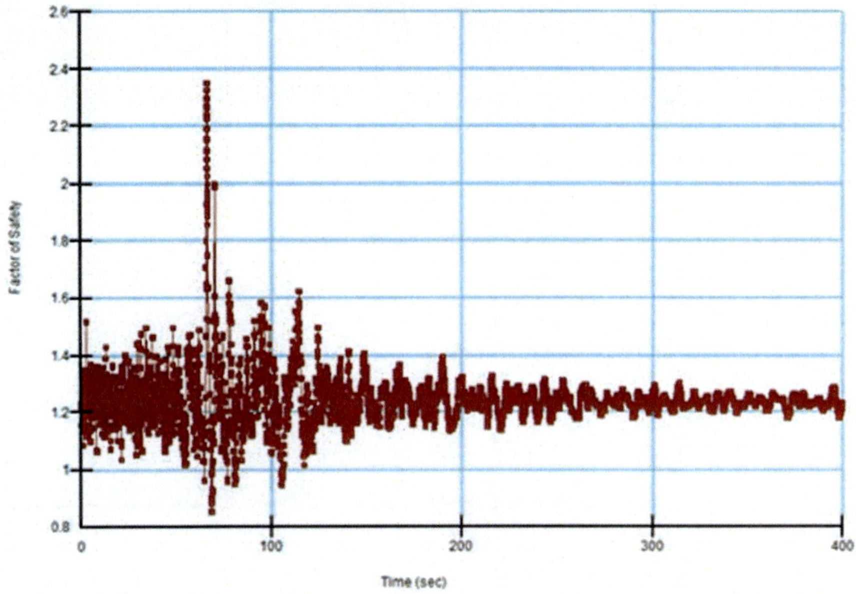

Fig. 12 Factor of safety vs. time for Indonesia earthquake (modified PGA 0.15 g)

the total deformation caused by the seismic wave is indicated as 0.46 m, as shown in Fig. 16.

Case Study 3: Effect of different rainfall intensities with same earthquake record for Tanjung Bungah slope. This case study stimulated different rainfall intensities with the same duration in the Tanjung Bungah slope. Firstly, the natural slope geometry is stimulated with 80 mm/h and 120 mm/h rainfall intensity. Then, the earthquake record with modified peak ground acceleration of 0.15 g imposed on the slope for stability evaluation. Figure 18 indicates that the slope with 6 h of 120 mm/h rainfall and ground shaking induced by the earthquake demonstrated a lower factor of safety (1.502). In comparison, the slope with 80 mm/h rainfall and the same earthquake rooord had a higher factor of safety (1.576) (Fig. 17). It can be concluded that the higher intensity negatively affects slope stability. Surprisingly, the ground shaking induced by the earthquake did not significantly affect the factor of safety, which can be verified through the comparison of Figs. 8, 11, 17, and 18. For instance, the safety factor of 1.714 obtained for the Tanjung Bungah slope affected by 6 h of 80 mm/h rainfall, which has not changed significantly when considering the combined effect of rainfall and earthquake.

The results indicated there is no deformation due to the earthquake record with modified 0.15 g peak ground acceleration, but a higher modified peak ground acceleration of 0.45 g will impose a deformation; indeed, zero deformation might be due to the slope geometry, which has a lower slope height and smaller slope angle. Thus, it can be concluded that under 0.15 g of peak ground acceleration which is 2500 years

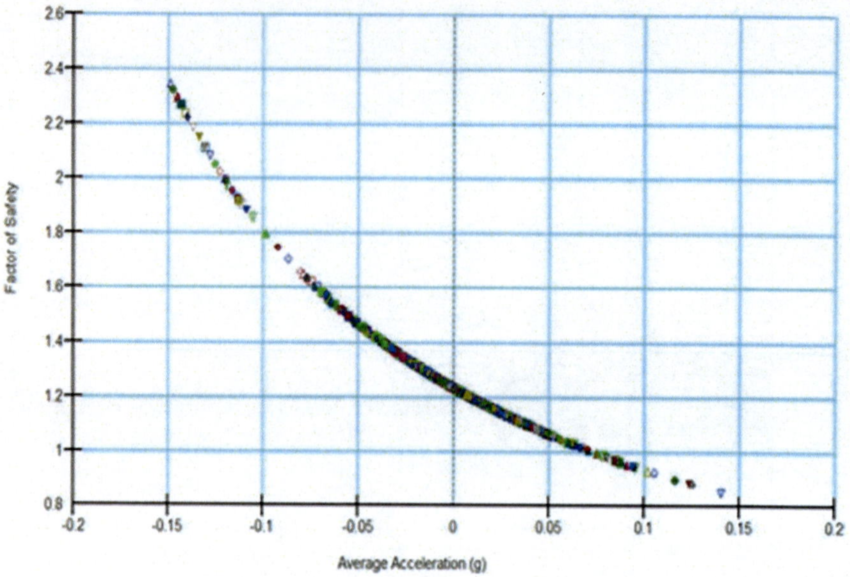

Fig. 13 Factor of safety versus average acceleration

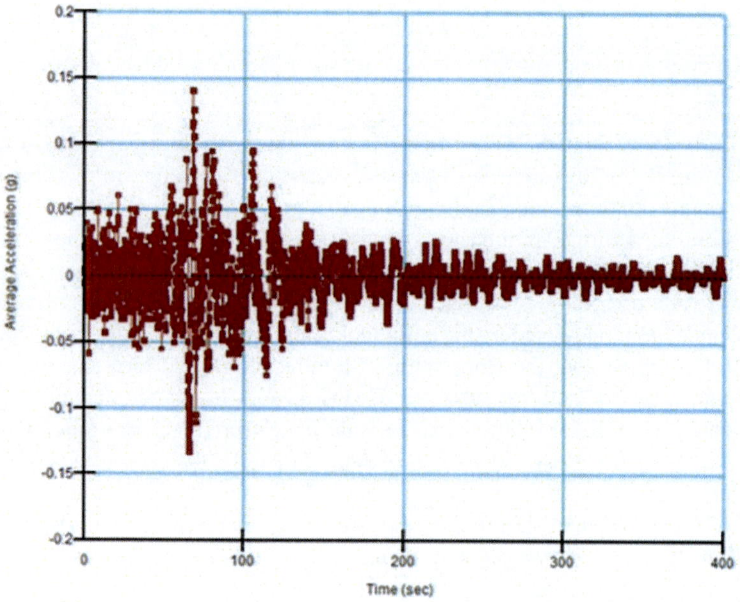

Fig. 14 Average acceleration versus time

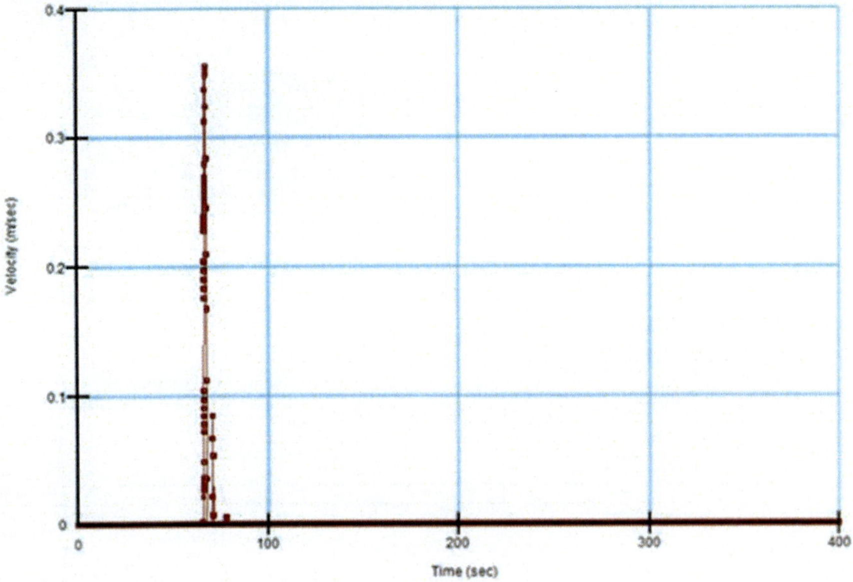

Fig. 15 Velocity versus time

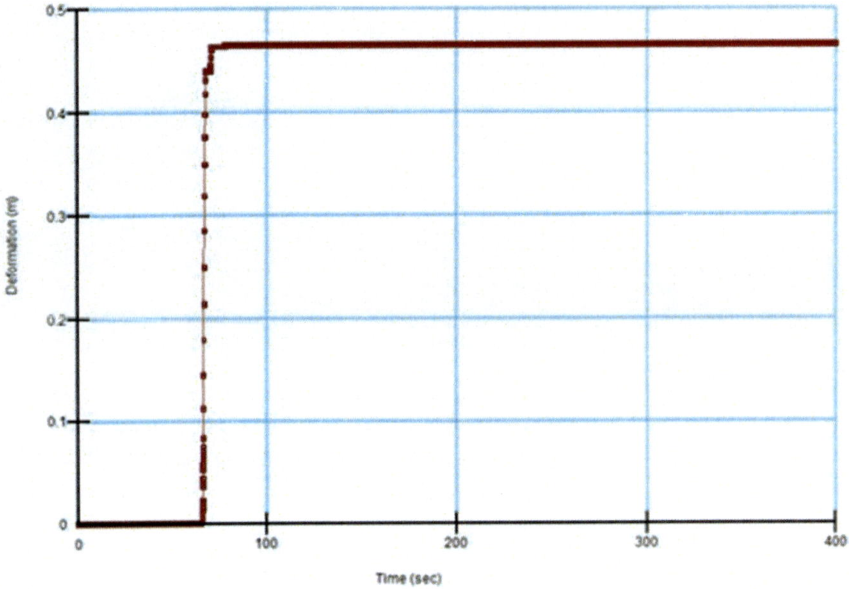

Fig. 16 Deformation versus time

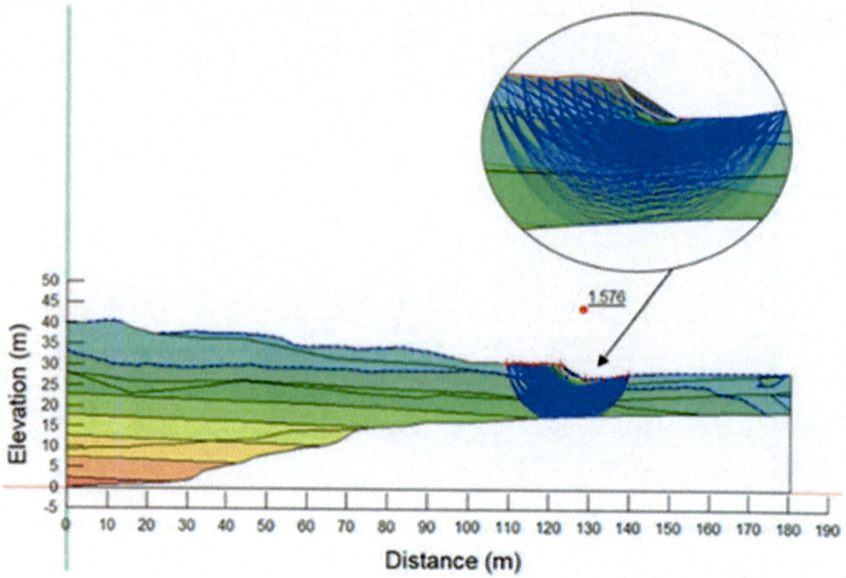

Fig. 17 Critical factor of safety for slope in Tanjung Bungah affected by 80 mm/h rainfall intensity and seismic waves

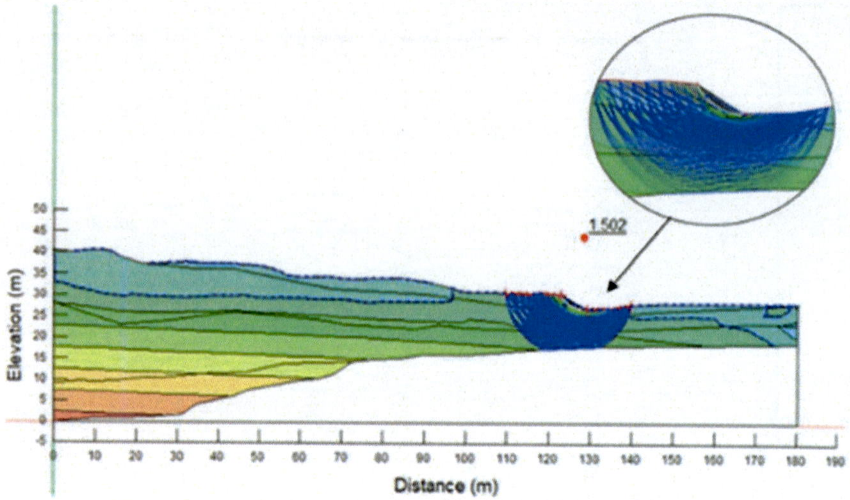

Fig. 18 Critical factor of safety for slope in Tanjung Bungah affected by 120 mm/h rainfall intensity and seismic waves

return period for the slope located in Penang Island, the selected slope located in Tanjung Bungah is safe.

4 Conclusion

This research is conducted to determine the effect of rainfall and earthquake on slope stability under different kinds of scenarios. In the first phase, rainfall's impact was discussed in terms of different intensities and duration through GeoStudio software. For case study 1, 1 h and 6 h of 80 mm/h rainfall were imposed to the slope surface, which resulted in the rise of groundwater table level and the soil permeability. It concluded that the extended rainfall duration would decrease the factor of safety and more likely to trigger landslide events. For case study 2, 80 mm/h and 120 mm/h with the duration of 6 h were discussed in terms of safety (FOS). It is indicated that higher rainfall intensities increase groundwater table level which cause a reduction in the soil shear strength and the soil matric suction. As a result, higher rainfall intensity can be more risky for triggering a landslide compared to lower rainfall intensity by the same duration.

In the second phase, the research determines the combined effect of earthquake and rainfall on slope stability, mainly accessed through the Newmark deformation method. The slope stability is expressed in terms of FOS and deformation. For case study 3, the combined effect of 80 mm/h and modified earthquake record yielded a higher safety factor than the combined effect of 120 mm/h and the same earthquake record for the Tanjung Bungah slope. The comparison between cases 3 and 2 shows that the effect of the earthquake caused a negative trend for the factor of safety (FOS). Thus, higher rainfall intensity combined with the same earthquake might impact slope stability. At the same time, zero deformation was recorded for the slope geometry with a smaller slope angle and lower slope height. It can be concluded that under 0.15 g of peak ground acceleration which is 2500 years return period for slope located in Penang Island, the selected slope located in Tanjung Bungah is safe from slope failure either caused by the single factor of rainfall or combined factors of rainfall and earthquake. In case study 4, the combined effect of 80 and 120 mm/h with the same earthquake as case study 3 was adopted in the analysis for the Sungai Ara slope. The outcome displayed a lower factor of safety compared to case study 3, which may stem from the slope geometry with greater slope angle and higher slope height. The results indicated that it is crucial to implement reinforcements to improve slope stability and effective slope improvement measures that are needed to improve the slope stability.

Acknowledgements The authors would like to thank Universiti Sains Malaysia under project grant RUI 1001/PAWAM /8014135 for the support provided for this project.

References

1. Highland, L., Bobrowsky, P.T.: The Landslide Handbook: A Guide to Understanding Landslides. US Geological Survey Reston (2008)
2. Ishii, Y., et al.: Evaluation of slope stability by finite element method using observed displacement of landslide. Landslides **9**(3), 335–348 (2012)
3. Burns, S.F.: Urban landslides: challenges for forensic engineering geologists and engineers. In: Engineering Geology for Society and Territory, vol. 5, pp. 3–9, Springer (2015)
4. Chen, Y.-L., et al.: Stability evaluation of slope subjected to seismic effect combined with consequent rainfall. Eng. Geol. **266**, 105461 (2020)
5. Higaki, D., Abe, S.: Classification of the geomorphology, geology and movement types of earthquake landslides. In: Earthquake-Induced Landslides, pp. 37–44. Springer (2013)
6. Kumar, A., et al.: Assessment of landslide hazards induced by extreme rainfall event in Jammu and Kashmir Himalaya, northwest India. Geomorphology **284**, 72–87 (2017)
7. Huat, L.T., Ali, F., Ibrahim, A.S.: An investigation on one of the rainfall-induced landslides in Malaysia. Electron. J. Geotech. Eng. **17**, 435–449 (2012)
8. Azmi, M.B.: Study on Slope Stability of Penang Island Considering Earthquake and Rainfall Effects (2014)
9. Marto, A., et al.: Seismic impact in peninsular Malaysia. In: The 5th International Geotechnical Symposium-Incheon (2013)
10. Ahmad, F., et al.: Slope failure due to effect of Damrey Typhoon in Penang Island. Lowland Technol. Int. **22**(1, June), 35–39 (2020)
11. Simons, W., et al.: A decade of GPS in Southeast Asia: resolving Sundaland motion and boundaries. J. Geophys. Res. Solid Earth **112**(B6) (2007)
12. Vigny, C., et al.: Insight into the 2004 Sumatra-Andaman earthquake from GPS measurements in southeast Asia. Nature **436**(7048), 201–206 (2005)
13. Peck, R.B., Hanson, W.E., Thornburn, T.H.: Foundation Engineering, 2nd ed., John Wiley and Sons, New York, NY (1974)
14. Zaki, H.: A study on the effect of extreme rainfall intensity on slope stability using physical based modelling. Universiti Sains Malaysia, Thesis (2019)
15. Broaddus, M.R.: Performing a Steady-State Seepage Analysis Using SEEP/W: A Primer for Engineering Students (2015)
16. Mishal, U.R.: Stability analysis of an earth dam using GEO-SLOPE model under different soil conditions. Eng. Technol. J. **36**(5 Part A) (2018)
17. Arshad, I., Babar, M.M., Javed, N.: Numerical analysis of seepage and slope stability in an earthen dam by using geo-slope software. PSM Bio. Res. **2**(1), 13–20 (2017)
18. Dorairaj, D., Osman, N.: Present practices and emerging opportunities in bioengineering for slope stabilization in Malaysia: an overview. PeerJ **9**, e10477 (2021)
19. Chiaro, G., et al.: The Takanodai landslide, Kumamoto, Japan: insights from post-earthquake field observations, laboratory tests, and numerical analyses. In: Geotechnical Earthquake Engineering and Soil Dynamics V: Slope Stability and Landslides, Laboratory Testing, and In Situ Testing, pp. 98–111. American Society of Civil Engineers Reston, VA (2018)
20. Chiaro, G., Chew, K., Kim, J.: Numerical Analyses of the Earthquake-Induced Takanodai Landslide, Kumamoto, Japan (2018)
21. GEO-SLOPE, I.: Stability Modeling with Slope/w. GEO-SLOPE International Ltd. (2012)
22. Sumanth, H., Govindaraju, L.: A comprehensive seismic slope stability assessment of earth dams. J. Adv. Geotech. Eng. **3**(1, 2, 3) (2021)

Sustainable Infrastructure

Comparison of Microstructural Properties of One- and Two-Part Fly Ash Geopolymer Concretes

Nurulhuda Nadziri, Idawati Ismail, Amirul Syahmi bin Ardi, Nurul Hadirah binti Damni, and Annisa Jamali

Abstract Geopolymerization is a chemical process involving reacting raw aluminosilicate minerals with alkali activators such as sodium hydroxide and potassium hydroxide to produce a geopolymer binder. In most cases, anhydrous activator is typically made by dilution with water to form a liquid activator before being blended with other raw components such as sand and coarse aggregates. This traditional approach, also known as two-part mixing, can be hazardous due to the corrosive nature of the liquid activator, making it difficult to mix huge volumes of concrete. In recent years, a one-part mix or "just add water" geopolymer concrete has been proposed as a simpler mixing method to minimize mixing time and improve the perception of geopolymer concretes. Similar to the preparation of typical Portland cement concretes, a one-part mix is performed in which all of the dry ingredients, including the solid activator, are initially combined together before water is eventually added to the mix. As a result, the aim of the research is to compare these two geopolymer concrete types, as well as how these mixing processes affect the mechanical strength and microstructural properties of fly ash geopolymer concretes. This study also compares the effects of the activators used, which included sodium hydroxide (NaOH) and potassium hydroxide (KOH). Microstructural investigation was performed through Fourier transform infrared (FTIR) and scanning electron microscopy-energy-dispersive X-ray spectroscopy (SEM–EDX). It was observed that one-part geopolymer concrete using NaOH as activator shows higher mechanical strength. The fly ash geopolymer binder study reveals that sodium aluminosilicate hydrate (N-A-S–H) gels and potassium aluminosilicate hydrate (K-A-S–H) gels

N. Nadziri · I. Ismail (✉) · A. Syahmi bin Ardi · N. Hadirah binti Damni
Department of Civil Engineering, Faculty of Engineering, Universiti Malaysia Sarawak, 94300 Kota Samarahan, Sarawak, Malaysia
e-mail: iidawati@unimas.my

N. Nadziri
Centre for Foundation and General Studies, i-CATS University College, Jalan Stampin Timur, 93350 Kuching, Sarawak, Malaysia

A. Jamali
Department of Mechanical Engineering, Faculty of Engineering, Universiti Malaysia Sarawak, 94300 Kota Samarahan, Sarawak, Malaysia

N. Sabtu (ed.), *Proceedings of AWAM International Conference on Civil Engineering 2022 - Volume 3*, Lecture Notes in Civil Engineering 386,
https://doi.org/10.1007/978-981-99-6026-2_14

are the binding gels formed in one-part and two-part mixing methods at different alkali activators with the formation of N-A-S–H gels which is faster with NaOH activator than with KOH. It may be inferred that one-part geopolymer concretes perform better than two-part geopolymer concrete mixtures, and that NaOH-activated concretes exhibit more desired characteristics than KOH-activated concretes. It can be concluded that one-part geopolymer concretes perform better compared to two-part mix of geopolymer concretes and NaOH-activated concretes gives desirable properties compared than KOH.

Keywords One-part geopolymer · Two-part geopolymer · Fly ash

1 Introduction

Geopolymer is created by the alkaline activation of low-cost materials or industrial waste, such as fly ash, slag, and rice husk. Fly ash geopolymer is a cement-free building material that not only protects the environment [1, 2], but also often outperforms OPC in terms of performance and durability [3–5]. Corrosive hydroxides such as sodium hydroxide (NaOH) are frequently utilized in the manufacture of these geopolymer concretes. Traditionally, these concretes are mixed using anhydrous pallets of NaOH diluted with raw precursors and other materials such as gravel and sand [6, 7]. This technique is referred to as two-part mixing, and it has been successfully used on a large scale [8]. However, it is time consuming and requires additional safety measures owing to the activator solution's corrosive nature. Meanwhile, one-part fly ash geopolymer concrete, or simply "just add water," is made by mixing all of the dry components together, including the solid anhydrous activator, and then adding water [9, 10]. Duxson and Provis [11] and Thomas et al. [12] started the development of one-part geopolymer many years ago. At the moment, a one-part geopolymers remain commercially viable in comparison with two-part geopolymers [8].

The type of activator solution used to activate the fly ash is critical to the reaction progression. When fly ash is activated with solutions of NaOH and KOH in the presence of sodium silicate and potassium silicate, it has no crystalline phase other than the crystalline fly ash. However, when activated with NaOH solution, the fly ash will include some hydroxysodalite in addition to the minerals found in the fly ash. When fly ash is activated with KOH solution, potassium carbonate and potassium bicarbonate are produced [13]. In comparison with the KOH 18 M solution, the NaOH 12 M solution provides more strength and quicker activation. The residual hydroxide ion concentration in KOH when 18 M is used weakens the alkali cement [13]. The presence of an alkaline solution has a significant effect on the microstructure of a two-part fly ash geopolymer. In comparison with the KOH solution, the NaOH solution has a greater solubility of silica and alumina at the same concentration.

The microstructure of the two-part fly ash geopolymer is denser than that of OPC, resulting in much less chloride transport and porosity. The denser microstructure results from a greater reactivity with fly ash as a result of the alkaline activator's higher alkali concentration. García-Lodeiro et al. [14] also discovered that one-part and two-part fly ash geopolymers performed better than OPC concretes. Additionally, the performance of one- and two-part geopolymers is dependent on the amount of silica in the alkaline activator [13]. Criado et al. [15] concur stating that a product with a greater silica content will have a higher mechanical strength [16].

The difference between a one-part and two-part fly ash geopolymer lies in the amount of silica and aluminum present, as well as the rate of release. Hajimohammadi et al. [17] demonstrated that a one-part affluent-silica geopolymer paste with a low silica concentration produced a crystalline phase with a greater silica content than a pure geothermal silica geopolymer paste [17]. Additionally, one of the advantages of geopolymer-based concrete over OPC concrete is its resilience to heat. Both one-part and two-part fly ash geopolymers outperform OPC concretes in terms of heat resistance [18, 19].

The purpose of this study is to compare the mechanical characteristics and microstructures of one-part and two-part fly ash geopolymer concretes as a function of activator. In order to make a one-part fly ash geopolymer concrete, anhydrous activator (NaOH and KOH) pallets are mixed together with other ingredients. For a two-part mix, the alkali solution must be prepared and left for equilibrium for 24 h. Compressive strength testing, functional group determination testing, and microstructural observation will be performed on both techniques to determine the strength development of the fly ash geopolymer concrete after 7, 14, and 28 days. However, the microstructures of one- and two-part geopolymers are not extensively characterized. As a result, this study examines the mechanical and microstructural characteristics of fly ash geopolymer concrete as a function of the mixing technique, namely one-part and two-part mixing.

2 Method and Materials

2.1 Fly Ash

The fly ash utilized in this study was provided by Sejingkat Power Plant, which is located in Kuching, Sarawak. As indicated in Table 1, the mineral composition of Class F fly ash is composed of the following elements.

2.2 Alkaline Activator

In this study, two different types of activators were applied. It is the activators NaOH and KOH that are responsible for the reaction. NaOH is accessible in the form of

Table 1 Chemical composition of fly ash

Element	Composition
CaO	4.91
SiO_2	55.90
Al_2O_3	21.80
Fe_2O_3	6.62
MgO	2.00
SO_3	0.32
K_2O	2.20
Others	3.91
LOI	2.34

powder, whereas KOH is supplied in the form of pellets. When making the two-part geopolymer, the activator was made by dissolving the powder and the pellets in a solution of tap water. The amount of activator (solids) present in a solution is proportional to the concentration of the solution. The activator will be employed at a concentration of 12 M, which is the recommended concentration. The alkaline activator solution was produced first, then allowed to sit for 20 min before being combined with the fly ash. Therefore, heat is generated during the interaction between fly ash and activator before the mixing process takes place at room temperature, as shown in the diagram.

2.3 *Aggregate*

The coarse aggregate utilized in this study is a normal grade river sand, and the fine aggregate is a normal grade river gravel, according to the manufacturer. The maximum aggregate size that may be utilized is 20 mm.

2.4 *Samples and Curing Method*

The concretes were 100 mm cubes in volume when they were produced. Table 2 summarizes the mix proportions for the fly ash geopolymer concretes that were produced as part of this study. The samples were dried for 24 h at 60 °C in an oven. After 24 h of drying in the oven, samples will be removed from the mold and stored at room temperature in sealed plastic bags until testing can be performed on them. The samples for the chloride migration and permeability experiments were produced in cylindrical molds measuring 100 mm in height by 50 mm in diameter.

Table 2 Mix proportion design for geopolymer concrete

Fly ash proportion (wt%)		100
Total binder (kg/m^3)		400
Coarse aggregate (kg/m^3)		1150
Fine aggregate (kg/m^3)		640
Activator (molarity)	NaOH	12
	KOH	0
Liquid-to-binder ratio		

2.5 Testing Method

The sample is subjected to three distinct types of testing. The purpose of the compressive strength test was to determine the mechanical strength of the sample in accordance with British Standard, Testing Concrete-Part 116: Compressive strength of concrete cube was determined using BS1881: Part 116: 1983 [20], functional groups were identified using Fourier transform infrared spectroscopy (FTIR), and microstructures were identified using scanning electron microscopy (SEM). Compressive strengths of the concretes were evaluated after 7, 14, and 28 days of curing, while FTIR and SEM analysis were performed on paste samples of similar mixtures (without aggregates).

3 Results and Analysis

3.1 Compressive Strength

The compressive strength of fly ash geopolymer concretes is plotted in Fig. 1 as a function of the mixing technique and activator. The type of alkali activator used, as well as the one-type and two-type geopolymer, impacted the mechanical performance of fly ash geopolymer.

The compressive strength of fly ash geopolymer concretes using NaOH as an alkali activator is higher than that of fly ash geopolymer concretes with KOH. The type of alkali activated cement used greatly influences the performance of fly ash geopolymer concrete. In comparison to KOH, NaOH is more often used in the manufacture of geopolymer concrete [18]. The rate of dissolution is one of the reasons why NaOH is more concentrated than KOH [21].

The compressive strength of a fly ash geopolymer including NaOH is enhanced owing to the compact and denser solids, but the compressive strength of a fly ash geopolymer containing KOH is decreased due to increased porosity. NaOH forms a more crystalline structure than KOH [19]. Additionally, the production of oligomers in NaOH solution contributes to the increase in compressive strength when compared to KOH, because NaOH has a greater capacity for dissolving alumina silicate minerals

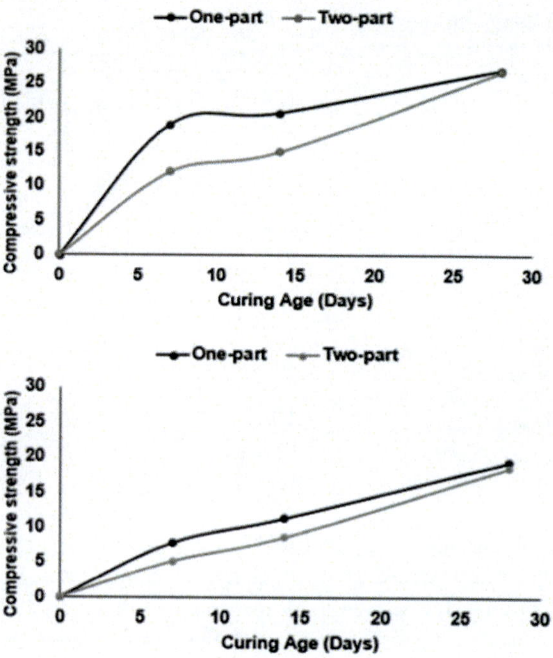

Fig. 1 Compressive strength development as a function of mixing technique and activator; **a** NaOH, **b** KOH

[19]. The compressive strength of geopolymer concrete will be improved as a result of its well-structured crystalline structure.

One-part fly ash geopolymer concretes showed better strength than two-part fly ash geopolymer concretes. Hajimohammadi et al. [17] show that decreasing the quantity of silica in a one-part fly ash geopolymer enhanced the formation of high-silica crystalline phases. As a result, the initial strength of a one-part fly ash geopolymer is greater than that of a two-part fly ash geopolymer. A geopolymer composed entirely of fly ash was shown to have a higher starting strength but a slower rate of strength growth [22].

3.2 *Fourier Transform Infrared Spectroscopy (FTIR)*

The infrared spectra of the unreacted precursors material utilized in this investigation are shown in Fig. 2 with their associated wavenumbers. Wavenumbers between 400 and 4000 cm^{-1} are detected. It is observed that dehydration had a significant effect on a number of significant alterations in the FTIR spectrum. In alkali-activated paste, the main bands are about 3604, 1622, 1072, 783, and 462 cm^{-1}.

Figure 3 illustrates the FTIR analysis of a one-part geopolymer binder, whereas Fig. 4 illustrates the study of a two-part geopolymer binder. After geopolymerization, all main bands are reduced except the O–H stretching band. The reactivity of fly

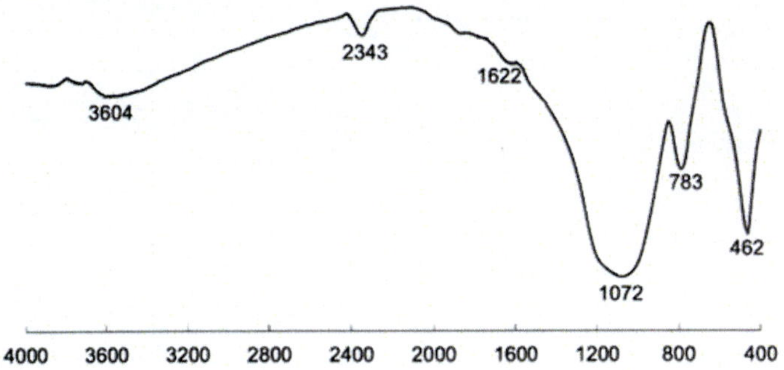

Fig. 2 FTIR spectra of the unreacted material used to synthesize alkali activated fly ash binders

ash with various activators is demonstrated by the peak at 980–1100 cm^{-1}, which corresponds to the development of a binding gel (T–O–T bond) in amorphous glasses. The production of binding gels varies according to the alkali activators used; sodium aluminosilicate hydrate gel (N–A–S–H gel) forms in reaction with NaOH, whereas potassium aluminosilicate hydrate gel (K-A-S–H gel) forms in response with KOH. Table 3 summarizes the band assignments from these FTIR analyses for both one-part and two-part fly ash geopolymer.

The peak number in each band region explained the structural group's bending and stretching. 400–500 cm^{-1} and 750–850 cm^{-1} band areas imply bending vibrations and symmetric stretching of the Si–O–Si bonds, respectively. While the band region between 980 and 1100 cm^{-1} indicates the asymmetric stretching of T–O–T and the development of binding gel. Stretching and bending vibrations of the O–C–O bond at 1410–1420 cm^{-1}, stretching and bending vibrations of the O–H band at 1600–3700 cm^{-1} band area with water losses (O–H).

From 7 to 56 days, Fig. 3 depicts the micro development of the binding gel in a one-part fly ash geopolymer. The shift in number is significant in the FTIR spectrum because it indicates the dehydration of the materials [23]. Around 3440, 1650, 1420, 1000, and 450 cm^{-1} are the main bands. The wide peak at about 3440 cm^{-1} is attributable to hydrogen-bonded O–H stretching and free water in both types of activator (Fig. 3a, b) [23, 24]. The development of N–A–S–H and K–A–S–H gel is described as a peak in the band region between 800 and 1200. At 7 days, the peak in the pastes with NaOH activator is sharper than those in the KOH activator. It demonstrates that the development of N–A–S–H gel is significantly more than that of K–A–S–H gel after seven days. Then, at 14 days, N–A–S–H and K–A–S–H gels appear to form equally, whereas N–A–S–H gel forms more rapidly than K–A–S–H gel at 28 days. After 56 days of curing, the peak's sharpness on intensity is comparable. Essentially, the steeper the peak, the more N–A–S–H and K–A–S–H gel is produced [4]. The wide peaks for K–A–S–H gels are displayed in comparison to N–A–S–H gels. This is due to the porous nature of the materials and their irregular forms [23], which are also discussed in the next sections (SEM–EDX). Additionally, the wider peaks have a relatively low mechanical strength, as already stated in the preceding section (compressive strength).

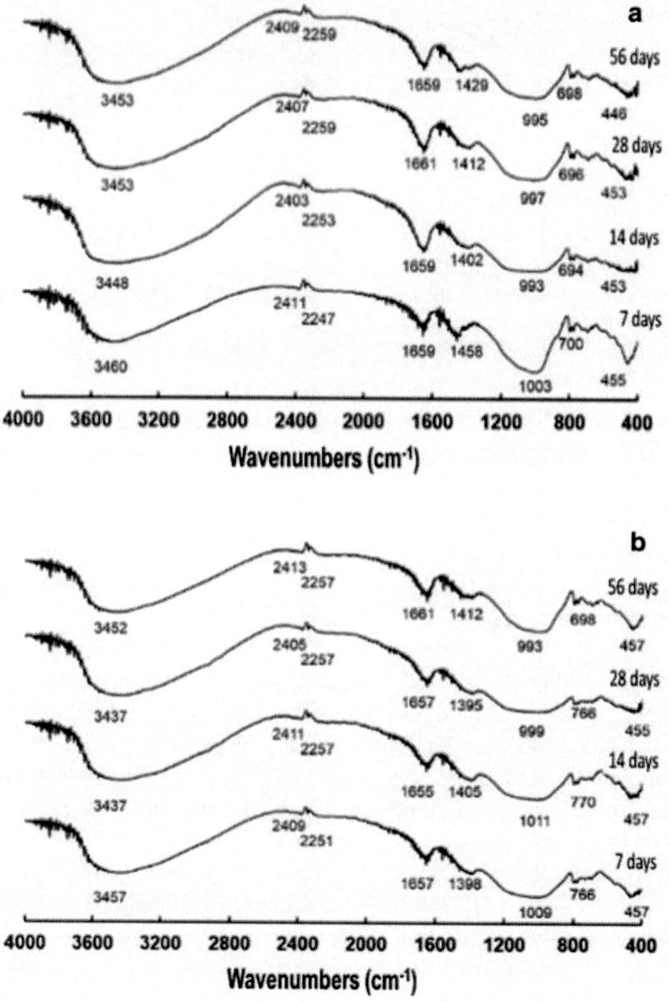

Fig. 3 FTIR analysis for one-part fly ash geopolymer paste as a function of alkali activator

From 7 to 56 days, Fig. 4 illustrates the microevolution of the binding gel in a two-part fly ash geopolymer. Following geopolymerization, all main bands are reduced except the O–H stretching band. The development of N–A–S–H and K–A–S–H gel is indicated by a peak in the band region between 800 and 1200. As the binding gel's microevolution progresses, the peak's sharpness decreases. In general, the higher the peak, the more N–A–S–H and K–A–S–H gel is produced.

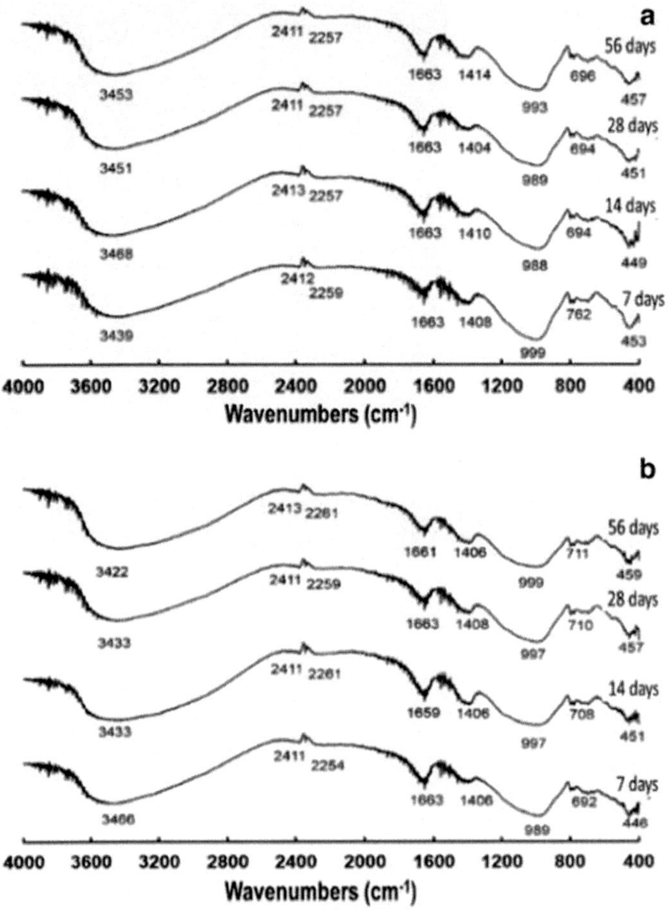

Fig. 4 FTIR analysis for two-part fly ash geopolymer paste as a function of alkali activator

Table 3 FTIR band assignment for one-part fly ash geopolymer paste as a function of alkali activator

Wavenumbers (cm^{-1})	Assignments
400–500	Bending vibrations of Si–O–Si and O–Si–O bonds
750–850	Symmetric stretching of Si–O–Si
980–1100	Asymmetric stretching of Si–O–T (T = Si or Al) (formation of binding gel)
1410–1420	Stretching of O–C–O
1600–1650	Stretching of O–H
2850–3700	Bending vibration of H–O–H

Fig. 5 Micrograph of raw fly ash

3.3 Scanning Electron Microscopy with Energy-Dispersive X-ray Spectroscopy (SEM–EDX)

The microstructure of the raw fly ash utilized in the experiment is shown in Fig. 5, which is consisting of spherical particles and predominantly glassy (amorphous) fly ash in nature with very small particles. Reduced friction between aggregate and fly ash particles improves the workability of concrete [23, 25]. Additionally, the small particle size acts as a filler for voids in geopolymer concrete, resulting in denser and more durable geopolymer materials [26].

3.4 Microstructure of One-Part Fly Ash Geopolymer Binder

The microstructure of fly ash geopolymer binder was depicted in Fig. 6a utilizing NaOH as an activator, whereas KOH. When activated with an alkaline activator, fly ash particles undergo transformation. The changes in the spherical shape of fly ash particles following the alkaline attack can be observed by comparing the morphology of raw fly ash particles in Fig. 6. According to van Deventer [27], the consolidation of sodium silicate gel with other particles of produced gel during precipitation results in a denser and more compacted cementitious matrix. Additionally, when NaOH is used as the activator, the microstructure is more compact and denser than when KOH is used [28].

The fly ash particles in Fig. 6b demonstrate that particles are not completely reacted when the spherical shape of unreacted particles is seen. Additionally, there is an empty hollow left by reacting particles, which increases the number of voids and contributes to the material's low mechanical strength. At 14 days, the hollow was filled with reaction product and the amount of reacted fly ash particles increased. While the diameter of the particles grows and they become compacted after 28 days,

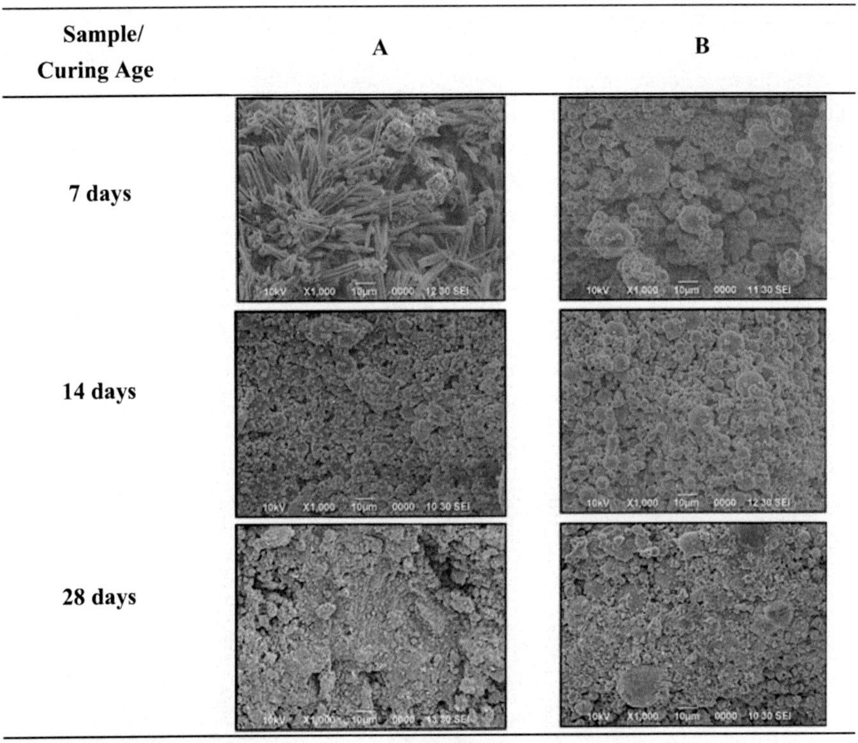

Fig. 6 Micrograph of one-part fly ash geopolymer paste as a function of curing age and activator: **a** NaOH and **b** KOH

they do not yet form a dense matrix like the products demonstrated when the NaOH activator is used. This is one of the reasons why geopolymers activated with NaOH has better mechanical strength than geopolymers activated with KOH. As a result of the slower rate of reaction of KOH as an activator, the particles require more time to be completely consumed.

3.5 Microstructure of Two-Part Fly Ash Geopolymer Binder

The microstructure of the particles in a two-part fly ash geopolymer binder including NaOH is smaller than the particle size in a one-part fly ash geopolymer binder. At 7 days after curing, SEM images reveal that the particles are not completely reacted, as evidenced by the spherical form of unreacted particles in Fig. 7a. At 14 days, the reaction product began to multiply and create a dense matrix. It therefore permits the compressive strength to be increased in comparison with 7 days. While at 28 days, the production of compacted matrix was detected. It is defined that fly ash particles are consumed completely after 28 days but require additional time to stabilize the structure due to the denser matrix.

The microstructure of a two-part fly ash geopolymer binder activated with KOH exhibits inconsistent characteristics due to the sample's physical properties during the experiment, which are damp and quickly crack. It then resulted in a lack of strength during compressive strength testing. The visualization of reacted and unreacted particles after seven days of curing is shown in Fig. 7b. At 14 days, the matrix becomes denser. Even after 28 days of cure, the spherical shape of unreacted particles may be detected under SEM.

The compressive strength of fly ash geopolymer concrete was evaluated at the 7th, 14th, and 28th days of concrete age. It can be shown that one-part fly ash geopolymer concrete has a higher compressive strength than two-part fly ash geopolymer concrete. However, the strength of a two-part fly ash geopolymer reached a satisfactory level by the 28th day.

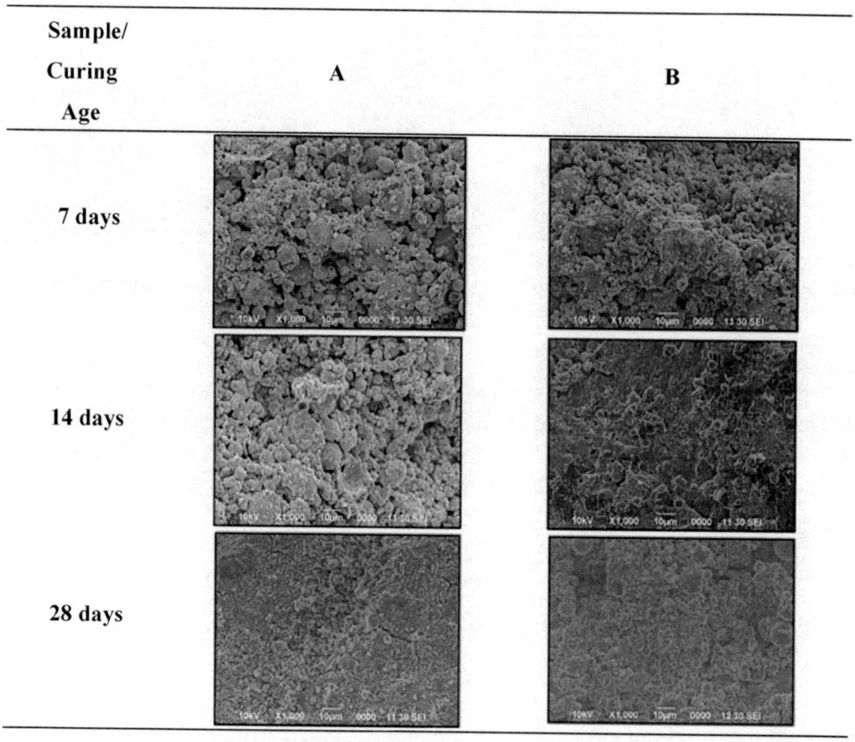

Fig. 7 Micrograph of two-part fly ash geopolymer paste as a function of curing age and activator: **a** NaOH and **b** KOH

4 Conclusions

The compressive strength of one-part fly ash geopolymer concrete is greater than that of two-part fly ash geopolymer concrete. The highest compressive strength and lowest volume of permeable voids are found in one-part fly ash geopolymer concrete. In the concrete mixer, NaOH pellets were combined with fly ash, gravel, and sand. The sodium ions dissolving all silica and alumina solids in the fly ash and the heat released during the hydration of the fly ash and NaOH pellets resulted in high compressive strength and lower porosity.

The fly ash geopolymer binder study reveals that sodium aluminosilicate hydrate (N–A–S–H) gels and potassium aluminosilicate hydrate (K–A–S–H) gels are the binding gels formed in one-part and two-part mixing methods at different alkali activators. Because of the higher presence of Si and Al dissolute when fly ash reacted with the alkali activator, the N–A–S–H and K–A–S–H gels have a higher amount in a one-part compared to two-part fly ash geopolymer binder. The ion dissociates the Si–O–Si and Al–O–Al bonds, allowing Si and Al to enter the liquid phase. Furthermore, the microstructural characteristics of one-part and two-part alkali activators were affected during hydration. The SEM image in one-part shows that the particles of NaOH-activated fly ash have a more compacted matrix than KOH, and the same is true in the second part. The mechanical strength of the fly ash geopolymer binder will be affected by these microstructural characteristics. As a result, the denser matrix form in particles with higher Si and Al values will have a higher compressive strength.

Acknowledgements The authors wish to thank Universiti Malaysia Sarawak for research funding through Cross Disciplinary Research Grant (No: F02/CDRG/1841/2019).

References

1. Gartner, E., Hirao, H.: A review of alternative approaches to the reduction of CO_2 emissions associated with the manufacture of the binder phase in concrete. Cem. Concr. Res. **78**, 126–142 (2015)
2. Flatt, R.J., Roussel, N., Cheeseman, C.R.: Concrete: an eco material that needs to be improved. J. Eur. Ceram. Soc. (2012)
3. Al Bakri, A.M.M., Kamarudin, H., Binhussain, M., Nizar, I.K., Zarina, Y., Ash, F.: Comparison of geopolymer fly ash and OPC to the strength of concrete. Comput. Theor. Nanosci. 1–4 (2013)
4. Ismail, I., Bernal, S.A., Provis, J.L., Nicolas, R.S., Brice, D.G., Kilcullen, A.R., Hamdan, S., van Deventer, J.S.J.: Influence of fly ash on the water and chloride permeability of alkali-activated slag mortars and concretes. Constr. Build. Mater. **48**, 1187–1201 (2013)
5. Aiken, T.A., Sha, W., Kwasny, J., Soutsos, M.N.: Resistance of geopolymer and Portland cement based systems to silage effluent attack. Cem. Concr. Res. **92**, 56–65 (2017)
6. Singh, N., Vyas, S., Pathak, R.P., Sharma, P., Mahure, N.V., Gupta, S.L.: Effect of aggressive chemical environment on durability of green geopolymer concrete. Certif. Int. J. Eng. Innov. Technol. **9001**(4), 2277–3754 (2008)
7. Kumaravel, S., Girija, K.: Acid and salt resistance of geopolymer concrete with varying concentration NaOH. J. Eng. Res. Stud. 3–5 (2013)

8. Glasby, T., et al.: EFC Geopolymer Concrete Aircraft Pavements at Brisbane West Wellcamp Airport (2015)
9. Bong, S.H., Xia, M., Nematollahi, B., Shi, C.: Ambient temperature cured 'just-add-water' geopolymer for 3D concrete printing applications. Cem. Concr. Compos. (2021)
10. Nematollahi, B., Bong, S.H., Xia, M., Sanjayan, J.: Digital Fabrication of 'Just-Add-Water' Geopolymers: Effects of Curing Condition and Print-Time Interval. In: RILEM Bookseries (2020)
11. Duxson, P., Provis, J.L.: Designing precursors for geopolymer cements. J. Am. Ceram. Soc. (2008)
12. Thomas, R., Ye, H., Radlinska, A., Peethamparan, S.: Alkali-activated slag cement concrete. Concr. Int. (2016)
13. Palomo, A., Grutzeck, M.W., Blanco, M.T.: Alkali-activated fly ashes: a cement for the future. Cem. Concr. Res. **29**(8), 1323–1329 (1999)
14. García-Lodeiro, I., Palomo, A., Fernández-Jiménez, A.: Alkali-aggregate reaction in activated fly ash systems. Cem. Concr. Res. **37**(2), 175–183 (2007)
15. Criado, M., Fernández-Jiménez, A., Palomo, A.: Alkali activation of fly ash: effect of the SiO2/Na2O ratio: part I: FTIR study. Microporous Mesoporous Mater. **106**(1), 180–191 (2007)
16. Criado, M., Fernandez-Jimenez, A., Palomo, A.: NoAlkali activation of fly ash: effect of the SiO2/Na2O ratio: part I: FTIR study. Microporous Mesoporous Mater. **106**, 180–191 (2007)
17. Hajimohammadi, A., Provis, J.L., van Deventer, J.S.J.: The effect of silica availability on the mechanism of geopolymerisation. Cem. Concr. Res. **41**(3), 210–216 (2011)
18. Sarker, P.K., Kelly, S., Yao, Z.: Effect of fire exposure on cracking, spalling and residual strength of fly ash geopolymer concrete. Mater. Des. **63**, 584–592 (2014)
19. Sarker, P.K., McBeath, S.: Fire endurance of steel reinforced fly ash geopolymer concrete elements. Constr. Build. Mater. **90**, 91–98 (2015)
20. BS1881: Part 116: 1983.: Method for Determination of Compressive Strength of Concrete Cube. Chiswick High Road London: BSI (1983)
21. Duxson, P., Fernández-Jiménez, A., Provis, J.L., Lukey, G.C., Palomo, A., Van Deventer, J.S.J.: Geopolymer technology: the current state of the art. J. Mater. Sci. (2007)
22. Nematollahi, B., Sanjayan, J., Shaikh, F.U.A.: Synthesis of heat and ambient cured one-part geopolymer mixes with different grades of sodium silicate. Ceram. Int. **41**(4), 5696–5704 (2015)
23. Nadziri, N., Ismail, I., Hamdan, S.: Binding gel characterization of alkali-activated binders based on palm oil fuel ash (POFA) and fly ash. J. Sustain. Cem. Mater. (2018)
24. Clark, S.M., Colas, B., Kunz, M., Speziale, S., Monteiro, P.J.M.: Effect of pressure on the crystal structure of ettringite. Cem. Concr. Res. **38**(1), 19–26 (2008)
25. Pacheco-Torgal, F., Castro-Gomes, J., Jalali, S.: Alkali-activated binders: a review. Part 1. Historical background, terminology, reaction mechanisms and hydration products. Constr. Build. Mater. **22**, 1305–1314 (2008)
26. Zerfu, K., Ekaputri, J.J.: Review on alkali-activated fly ash based geopolymer concrete. Mater. Sci. Forum **841**(1), 162–169 (2016)
27. van Deventer, J.S.J.: Geopolymers. Woodhead Publishing Series in Civil and Structural Engineering, University of Melbourne, Australia (2009)
28. Abdul Rahim, R.H., Rahmiati, T., Azizli, K.A., Man, Z., Nuruddin, M.F., Ismail, L.: Comparison of using NaOH and KOH activated fly ash-based geopolymer on the mechanical properties. Mater. Sci. Forum **803**, 179–184 (2014)

Strength Performance of Hybrid Fiber-Reinforced Concrete Containing Manufactured Polypropylene and Ring-Shaped Polyethylene Terephthalate Waste Fibers

Faisal Sheikh Khalid, Abdullah Nabil Abdullah Al-Jaberi, Shahiron Shahidan, Mohd Irwan Juki, and Syafiqa Ayob

Abstract Fibers are basically utilized as substitutes for ordinary support in nonprimary applications to control early warm withdrawal and drying shrinkage breaking. However, polypropylene (PP), lamellar or straight fibers are weak interfacial bond strength of surface during fiber bridge stress. Therefore, to solve the limitations of traditional straight or irregularly PP fibers, ring-shaped poly-ethylene terephthalate fibers were promoted as combination fiber with manufactured PP fiber in concrete. This study aims to determine the slump flow, density, compressive strength, tensile strength and water absorption of hybrid fiber-reinforced concrete (FRC) containing manufactured PP and ring-shaped PET fiber. An experiment was conducted in fresh and mechanical properties of PP and ring-shaped PET fibers with the percentage of 0%, 0.5%, 1.0% and 1.5% for both fibers. The results of fresh FRC properties showed maximum values for mix control samples, while the minimum values for slump flow test was recorded with 1.5% of PP and 1.5% of ring-shaped PET which is 3.2% less than control sample. Meanwhile, density after 28 days curing was increased 5.8% than control samples with 0% of PP and 1.0% of ring-shaped PET. However, 1.0% of both PP and ring-shaped PET fibers achieved 6 and 22% increment for FRC compressive strength and splitting tensile strength, respectively. In conclusion, the fresh properties of concrete were decreased with incorporation of PP and ring-shaped PET fibers while hardened properties were improved.

F. S. Khalid (✉) · A. N. A. Al-Jaberi · S. Shahidan · M. I. Juki · S. Ayob
Faculty of Civil Engineering and Built Environment, Universiti Tun Hussein Onn Malaysia, 86400 Parit Raja, Johor, Malaysia
e-mail: faisalsh@uthm.edu.my

A. N. A. Al-Jaberi
e-mail: gf200031@siswa.uthm.edu.my

S. Shahidan
e-mail: shahiron@uthm.edu.my

M. I. Juki
e-mail: irwan@uthm.edu.my

N. Sabtu (ed.), *Proceedings of AWAM International Conference on Civil Engineering 2022 - Volume 3*, Lecture Notes in Civil Engineering 386,
https://doi.org/10.1007/978-981-99-6026-2_15

Keywords Synthetic fiber · Fiber-reinforced concrete · Plastic bottle

1 Introduction

Fibers are basically utilized as substitutes for ordinary support in nonprimary applications to control early warm withdrawal and drying shrinkage breaking. These advantages have expanded the use of fibers in structural elements, especially those with low qualities, like foundations, walls, and slabs on grade. The utilization of fibers as a main part of structural elements of designed structures is consistently expanding. Fibers are added to develop the failure attributes and structures behavior through the ability of the fibers to cracks on bridges. Accordingly, numerous researches have been improved to analyze different types of fiber and shapes, especially in researching the exhibition of plastic fiber-reinforced concrete [1–3]. The reason of addition of manufactured plastic fiber was concrete properties such as weakness in tension and its brittleness under different types of loadings [4]. These required new adaption of various and convenient fibers such as poly propylene fiber that play main role in increasing the ductility, improve postcracking behavior of concrete and decrease shrinkage and thermal cracks [5].

In the research field of structural engineering, the reformed PET has started to be embraced in the concrete. Researches have joined manufactured straight PP fiber into concrete [1, 6]. Authors claimed that high volume from fibers with greater length have the ability to interlocking fiber bridges of concrete due to fibers can be embedded between aggregates compared with smallest volume with short mm-long fibers [1, 6]. Nevertheless, manufactured straight PP fibers exhibit limited performance due to fiber balling [6, 7]. This will cause the reduction of strength of concrete. In addition, manufactured PP fiber weak interfacial bond strength of surface during fiber bridge stress, particularly in fibers with irregular and lamellar shapes [7, 8].

Ring-shaped fibers are principally modeled to prepare fiber yielding (tensile rupture) instead of fiber pullout (fiber force slipped), which is an essential benefit over irregularly or straight PET fiber. In this way, choosing PET waste products as reformed materials is suitable from the point of view of the applications of civil engineering. Combination of manufactured PP fiber and ring-shaped PET (RPET) waste fiber was the fundamental factor that leads to bridges of fiber during tensile pressure [3]. The higher number of fibers in concrete that will raise fiber interlocking instrument among matrix concrete and fiber is required. Moreover, this study will have used concrete of Self-Compacting Concrete (SCC) combination showed adequate result on strength and workability of concrete [9].

The failure behavior of the postpeak, which consist of the cracking sound of fiber, was credited to the fiber of gradual rupture. Most of the fiber consolidated into the concrete failed through rupture, which is produced by the tensile pressure exposed to the tensile load. The purpose of the PET fiber in ring formed is to retain the ductile pressure at the vital fiber-bridging cross area [6]. The fiber-bridging system of PET fibers of ring shaped in FRC was much obvious compared to that of PET fibers in irregularly shaped. The PET fibers in ring shape offered particular highest tensile strength by yielding up to the fiber cracks. The structure of the PET fibers in ring-shaped is also dissimilar from that of the manufactured synthetic fibers that require interfacial connection strength [3].

Therefore, this research intended to prove the improvement of hybrid-manufactured polypropylene PP and RPET fibers in terms of fresh and hardened-state of concrete on mixture design according to SCC. The workability flow, strength and durability were examined to determine optimum combination of PP and RPET fibers concrete.

2 Materials and Methods

Portland cement Type I was used in this study. The superplasticizer used which conforms to ASTM D638 [10]. Size of coarse aggregates reached 14 mm. Meanwhile, fine aggregates measured was 5 mm. Waste PET bottles were also used in this study as shown in (Fig. 1).

RPET with widths or cross-sectional diameters of 60 ± 5 mm was used as shown in Fig. 1. The experiment also used manufactured synthetic macro-fibers, as shown in Fig. 2. The manufactured synthetic macro-fiber has aspect ratio ($Lf/\mathrm{d}f$) of 45 with tensile strength of 425 N/mm^2.

The absolute volume method was used to calculate the 1 m^3 of concrete as shown in Table 1. The water/cement ratio used in the concrete mix design is 0.45.

Slump flow test has been conducted in accordance with ASTM C 1611/C1611M-09b [11] to measure the filling capability of SCC. Slump flow was determined by measuring the diameter of the concrete spread in two perpendicular directions. The hardened state concrete compressive strength test was calculated in accordance with EN 12390-4:2012 [12]. Meanwhile, splitting tensile concrete was conducted at 28 days of age concrete in accordance with EN 12390-6:2009 [13]. Water absorption tests were per- formed following with EN 12390-8:2019 [14].

A total of 192 cube specimens were prepared for density, compressive strength and water absorption tests. The cube size used was 100 mm × 100 mm. A total of 48

Fig. 1 The dimension of ring waste bottles

Fig. 2 Manufactured PP fiber with 45 mm length

Table 1 Mixtures proportion for 1 m^3 of concrete

Mix designation	Volume of Fi BER (%)		Cement (kg/m^3)	Coarse aggregate (kg/m^3)	Fine aggregate (kg/m^3)	Water (kg/m^3)	Super-plasticizer (kg/m^3)
	PP	Ring PET					
Control	0	0					
PP0-R0.5	0	0.5					
PP0-R1.0		1.0					
PP0-R1.5		1.5					
PP0.5-R0		0					
PP0.5-R0.5	0.5	0.5					
PP0.5-R1.0		1.0					
PP0.5-R1.5		1.5	345	805	980	155.25	4.7
PP1.0-R0		0					
PP1.0-R0.5	1.0	0.5					
PP1.0-R1.0		1.0					
PP1.0-R1.5		1.5					
PP1.5-R0		0					
PP1.5-R0.5	1.5	0.5					
PP1.5-R1.0		1.0					
PP1.5-R1.5		1.5					

cylinders with a diameter of 100 mm and a height of 150 mm were prepared for the splitting tensile strength test. The molds were filled to the top with compaction. The concrete in the cylinder molds are screened to ensure that the top surface is even. All the specimens were stored in a water tank to cure the specimens for 7 and 28 days.

The average of the three test results for each concrete batch was considered as the final result.

3 Result and Discussion

Test was carried out to determine the workability, strength and durability of concrete containing waste RPET and PP manufactured fiber materials. Laboratory testing was conducted such as slump, density, compressive strength, tensile strength and water absorption.

3.1 Slump Flow

Slump flow values has been obtained are between 550 and 800 mm. The filling ability test was measured with respect to slump flow and T50 slump flow time. The slump flow diameter of the normal concrete of 0.45 water binder ratio was 591 mm as shown in Fig. 3.

It can be observed that mixes with RPET 1.0 with 0% PP fiber concrete resulted in higher slump flow numbers compare to other fibers concrete mixes. The minimum of slump flow number was measured at 1.50% of fiber RPET and 1.5 PP fiber. The slump flow outcomes in this study indicates that by increasing the amount of PP the slump flow decreases, this is similar to the outcomes obtained by Alalwany et al. [15] by replacing 0, 0.45, and 0.9 kg/m^3 of PP fiber and the results of slump test were 550, 500, and 430 mm respectively. Meanwhile, another study by Umasabor and Daniel

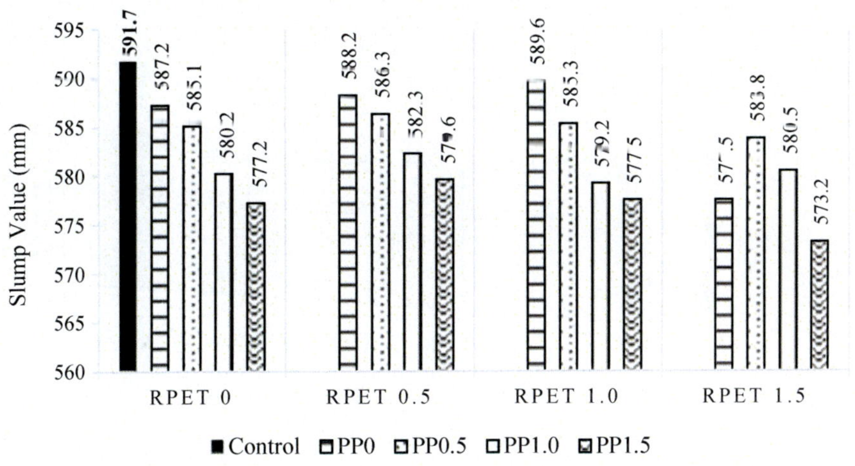

Fig. 3 Slump flow of fiber reinforced concrete

[16] examined the replacement of PET in concrete and the slump test results through replacing 0, 5, 10, and 15% of PET the results indicates that the increasing of PET percentage decreases the slump test value.

3.2 Density

The density of concrete was determined before the compressive strength test was conducted. The weight of the concrete was divided by the volume of the sample to deter mine its density. Figure 4 shows the density of fiber reinforced concrete.

The control sample of concrete without polypropylene and RPET exhibits the density value of 2261 kg/m^3. While the concrete with 1.0% polypropylene content has the lowest value of 2210 kg/m^3 of density. In term of polypropylene fiber content, the value of density decreases with the increase of polypropylene fiber content [17].

Current study agrees with a study conducted by Hasan et al. [18] which found that increasing polypropylene volume in concrete decreases the concrete density. However, Hasan et al. [18] showed the decrement of concrete density examined the effect of polypropylene fiber on the strength of concrete by replacing different volumes of PP from 0.06 to 2.16. The study results which found that the concrete without polypropylene fiber exhibits the highest density value of 2433 kg/m^3, while the concrete with 2.16% polypropylene fiber content has the lowest value 1969 kg/m^3 of density.

In terms of replaced PET, this study showed a positive effect of RPET in increasing concrete density by increasing RPET volume. According to Fig. 4, the density of concrete with 1.5 RPET and 0 PP volumes achieved the highest density volume with 2393 kg/m^3 which is 5.8% greater than control sample density.

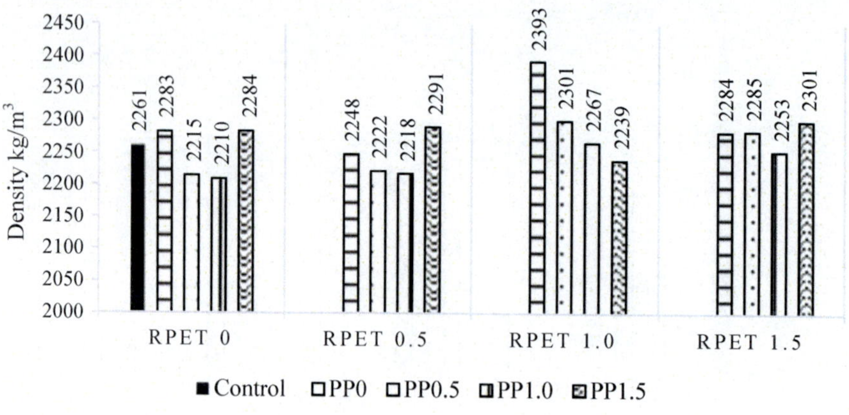

Fig. 4 Density of fiber reinforced concrete

3.3 Compressive Strength

Figures 5 and 6 show that this experiment indicates that RPET with PP fiber concrete shows increase in compressive strength compared to normal concrete with increases 5.0% for 7 days and 8.0% for 28 days for fiber content of 1.0%, respectively, compared to normal concrete. However, a pattern of strength shows decreases of 2% of 7 days and 8% of 28 days strength for fiber content of 0% PP with 1.0% RPET compared to normal concrete. For 0% RPET with 1.5 PP, the compressive strength develops from 21.27 MPa at 7th day and 36.00 MPa at 28th days. Finally, the strength development of mix 1.5 PP with 1.5 RPET increased 23.21 MPa at 7th days and finally 1.0 PP with 1.0 RPET the compressive strength increased to 37.01 at 28th days.

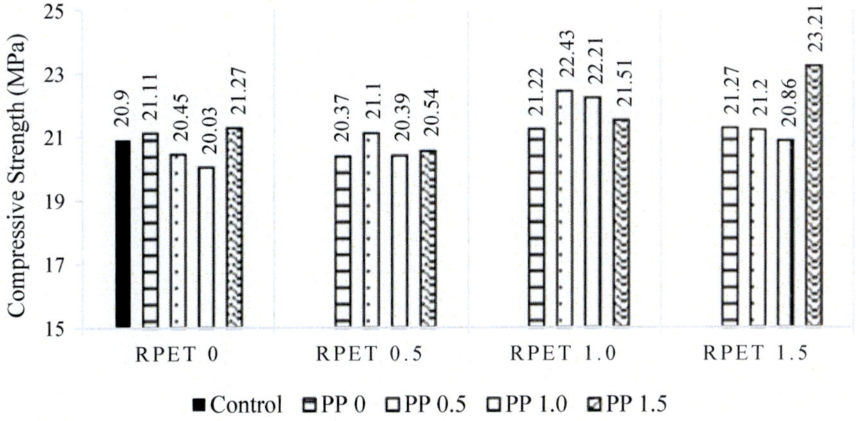

Fig. 5 Compressive strength of fiber reinforced concrete at 7 days

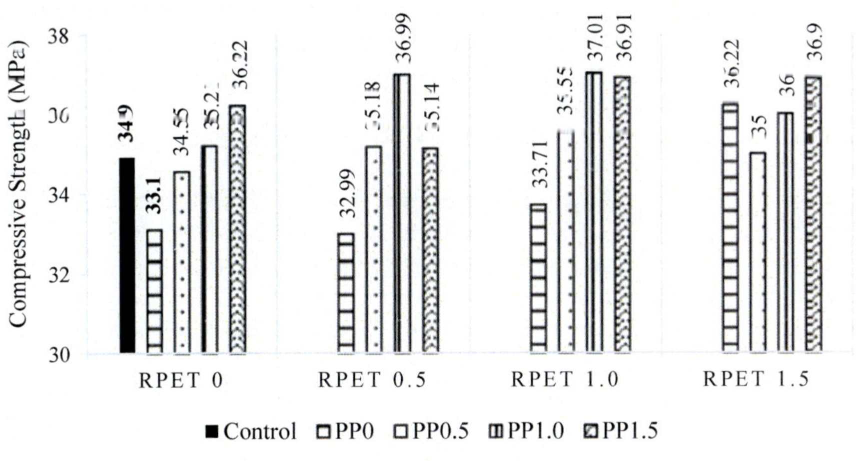

Fig. 6 Compressive strength of fiber reinforced concrete at 28 days

The results obtained from this current study agree with results of previous studies that using PP fiber in concrete increases concrete compressive strength. However, polypropylene fiber effects negatively in concrete compressive strength when the replaced values are greater than 1.0%. According to Hasan et al. [18] the results of compressive strength test show the highest value with replacing 0.72% of PP fiber in concrete after 28 days curing.

Another study carried by Alalwany et al. [15] reported that replacing 0.45, and 0.90% of polypropylene fiber in SCC showed compressive strength 25.2, and 33.5 MPa respectively after 28 days of curing. It can be observed that the highest compressive strength value is achieved by replacing 0.90% of PP in the concrete mixtures.

Umasabor and Daniel [16] claimed that an increasing the days of curing and PET volume increases the compressive strength of the blended concrete up to 5% of PET addition. However, the author mentioned that volume more than 5% PET, the compressive strength of the blended concrete reduces.

The initial increase of compressive strength may be due to the PET having proper bonding with the fine aggregate [19]. The decrease of the compressive strength beyond 5% may be due to the decrease in adhesive strength between the surface of the waste plastic and the cement paste [20].

3.4 Splitting Tensile

Splitting tensile strength test result for the RPET and PP fiber concrete were done by using a compression machine load, results are summarized in Fig. 7. This experiment indicated that each batch of RPET and PP fiber concrete have the optimum tensile strength which is depends on the fiber content of concrete. This experiment indicates that RPET with PP fiber concrete shows increase in splitting tensile test compared to normal concrete with maximum result of increases 30.0% for fiber content of 1.0% respectively compared to normal concrete. However, a pattern of strength shows the minimum increases of 4% strength for fiber content of 0% PP with 0.5% RPET compared to normal concrete.

According to Adnan and Dawood [21] the tensile strength of concrete is only 10% of its compressive strength. It shows that an addition of fibers to a concrete mixture is significant to the tensile properties of concrete. This due to the fibers act as crack arresters in the concrete matrix prohibiting the propagation of concrete cracks. Therefore, tensile strength is increased due to a bridging mechanism of polypropylene fibers, and after volume fraction 0.36% the extra fiber in the concrete causes a reduction in the bond strength between concrete ingredients so results in quick failure as compared to concrete with less volumes of fibers.

Another study carried by Hasan et al. [18] resulted that the tensile strength starts to increase with the increasing of the volume fraction of fiber content thus reaches the maximum value of 4 MPa at fiber content about 0.36% which is about 16% with a comparison to the tensile strength of reference mix. However, it has been reported

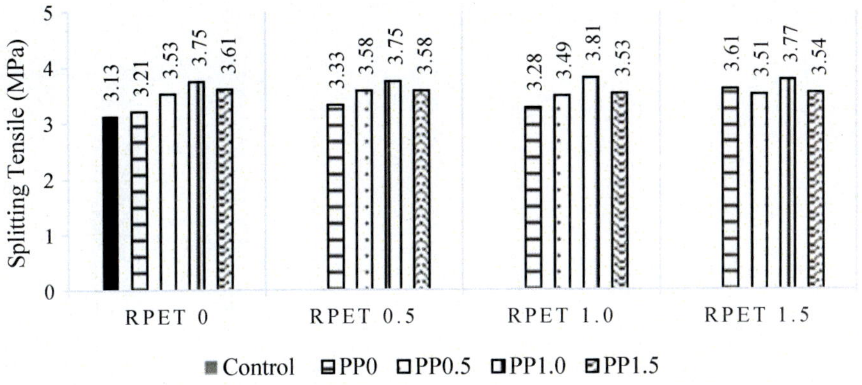

Fig. 7 Splitting tensile strength of fiber reinforced concrete

by Umasabor and Daniel [16] that replacing PET in concrete decreases splitting tensile strength and the results showed the maximum tensile strength of concrete with replacing 0% of PET after 28 days curing. Increasing the curing duration decreases the flexural strength except at day 3 curing duration where the opposite happens. Baboo et al. [20] attributed the decrease in the flexural strength to a decrease in adhesive strength between the surface of waste plastic and the cement paste.

3.5 Water Absorption

Figure 8 summarizes average water absorption percentages for tested samples, it can be observed that the minimum value was recorded with the control samples 2.23% and increasing PET volumes increases water absorption significantly. Furthermore, increasing PP volumes gradually increases water absorption. Water absorption was increased 23% from control samples to the maximum value recorded 2.74% with replacement of 1.5 PP and 1.5 RPET volumes.

Water absorption results obtained by this study correspond with results obtained by Hama and Hilal [22] which found that increasing PET volume increases water absorption significantly. Hama and Hilal [22] replaced PET volumes from 1 to 9% and recorded an increment of water absorption maximum value 3.26% with replacement of 9% PET in concrete specimens.

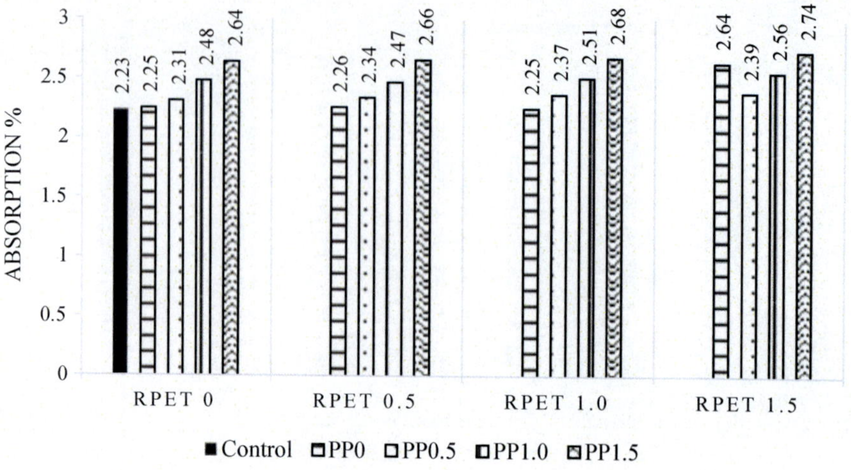

Fig. 8 Water absorption of fiber reinforced concrete

4 Conclusion

FRC including PP and RPET and the results were compared to that mixes with RPET 1.0 with 0 PP fiber concrete resulted in higher slump flow numbers compare to other fibers concrete mixes. The minimum of slump flow number was measured at 1.50% of fiber RPET and 1.5 PP fiber.

Density showed that decreasing PP volumes and increasing RPET volumes causes and increment in concrete density. However, compressive strength results showed that the maximum compressive strength after 28 days curing was recorded by replacing 1.0% of PP and 1.0% of RPET. Furthermore, splitting tensile strength test results showed that replacing 1.0% of PP and increasing RPET volumes scores higher splitting strength values. Finally, water absorption test results indicated that increasing PP and RPET volumes increases water absorption percentages gradually.

References

1. Foti, D.: Use of recycled waste pet bottles fibers for the reinforcement of concrete. Compos. Struct. **96**, 396–404 (2013). https://doi.org/10.1016/j.compstruct.2012.09.019
2. Fraternali, F., Ciancia, V., Chechile, R., Rizzano, G., Feo, L., Incarnato, L.: Experimental study of the thermo-mechanical properties of recycled PET fiber-reinforced concrete. Compos. Struct. **93**(9), 2368–2374 (2011)
3. Khalid, F.S., Irwan, J.M., Wan Ibrahim, M.H., Othman, N., Shahidan, S.: Splitting tensile and pullout behavior of synthetic wastes as fiber-reinforced concrete. Constr. Build. Mater. **171**, 54–64 (2018)

4. Irwan, J.M., Asyraf, R.M., Othman, N., Koh, H.B., Aeslina, A.K., Annas, M.M.K., Faisal, S.K.: Deflection behaviour of irregular-shaped Polyethylene Terephthalate fibre reinforced concrete beam. Adv. Mater. Res. **911**, 438–442 (2014)
5. Sadrmomtazi, A., Tahmouresi, B.: Effect of Fiber on Mechanical Properties and Toughnessof Self-Compacting Concrete Exposed to High Temperatures, vol. 1, issue 2, pp. 153–166 (2017)
6. Azmi, N.B., Khalid, F.S., Irwan, J.M., Mazenan, P.N., Zahir, Z., Shahidan, S.: Performance of composite sand cement brick containing recycle concrete aggregate and waste polyethylene terephthalate with different mix design ratio. IOP Conf. Ser. Earth Environ. Sci. **140**(111), 012129 (2018)
7. Irwan Juki, M., Muhamad, K., Annas Mahamad, M.K., Boon, K.H., Othman, N., Kadir, A.A., Asyraf Roslan, M., Khalid, F.S.: Development of concrete mix design nomograph containing polyethylene terephtalate (PET) as fine aggregate. Adv. Mater. Res. **701**, 12–16 (2013)
8. Al-Hadithi, A.I., Abdulrahman, M.B., Al-Rawi, M.I.: Flexural behaviour of reinforced concrete beams containing waste plastic fibers. IOP Conf. Ser. Mater. Sci. Eng. **737**(1) (2020)
9. Rahmani, E., Dehestani, M., Beygi, M.H.A., Allahyari, H., Nikbin, I.M.: On the mechanical properties of concrete containing waste PET particles. Constr. Build. Mater. **47**, 1302–1308 (2013)
10. ASTM D638.: Standard Test Method for Tensile Properties of Plastic. The American Society for Testing and Materials Standard (ASTM), West Conshohocken, Pennsylvania, USA
11. ASTM C1611/C1611M-14.: Standard Test Method for Slump Flow of Self-Consolidating Concrete. The American Society for Testing and Materials Standard (ASTM), West Conshohocken, Pennsylvania, USA
12. BS EN 12390-4:2012.: Testing Hardened Concrete—Shape, Dimensions and Other Requirements for Specimens and Moulds. British Standards Institution, London
13. BS EN 12390-6:2009.: Testing Hardened Concrete—Tensile Splitting Strength of Test Specimens. British Standards Institution, London
14. BS EN 12390-8:2019.: Testing Hardened Concrete—Depth of Penetration of Water Under Pressure. British Standards Institution, London
15. Alalwany, A., Mohamad, N., Abdul Samad, A.A., Goh, W.I.: Mechanical and fresh state properties of medium strength self-compacting concrete (SCC) containing polypropylene fibers. MATEC Web Conf. **10301011** (2017)
16. Umasabor, R., Daniel, S.: The effect of using polyethylene terephthalate as an additive on the flexural and compressive strength of concrete. Heliyon **6**(8), e04700 (2020)
17. Ding, Y., Wang, Q., Pacheco-Torgal, F., Zhang, Y.: Hybrid effect of basalt fiber textile and macro polypropylene fiber on flexural load-bearing capacity and toughness of two-way concrete slabs. Constr. Build. Mater. **261**, 119881 (2020)
18. Hasan, A., Maroof, N., Ibrahim, Y.: Effects of polypropylene fiber content on strength and workability properties of concrete. Polytechnic J. **9**(1), 7–12 (2019)
19. Khilesh, S.: Study of strength property of concrete using waste plastics and steel fibers. Int. J. Eng. Sci. **3**(5), 9–11 (2014)
20. Baboo, R., Rushad, T., Bhavesh, K., Duggal, K.: Study of waste plastic mix concrete with plasticizer. Int. Sch. Res. Netw. **20**, 2–5 (2012)
21. Adnan, H.M., Dawood, A.O.: Case studies in construction materials strength behavior of reinforced concrete beam using re-cycle of PET wastes as synthetic fibers. Case Stud. Constr. Mater. **13**, e00367 (2020)
22. Hama, S., Hilal, N.: Fresh properties of self-compacting concrete with plastic waste as partial replacement of sand. Int. J. Sustain. Built Environ. **6**(2), 299–308 (2017)

Thermo-Mechanical Properties of Concrete Mortar with Cenosphere

S. Beddu, N. A. N. Basri, Z. Itam, N. L. Mohd Kamal, T. S. Manan, Z. Che Muda, D. Mohamad, N. Sivakumar, Agusril, and K. X. Lee

Abstract Materials used in modern constructions received attention from researchers and engineers recently due to their impact on climate change, power consumption and operational cost. The usage of recycled materials gives extra advantages towards environment. It was found that power plant produced various types of waste product that can be utilize in many applications. For instance, Cenosphere produced as waste product from power plant have ability promotes

S. Beddu (✉) · Z. Itam · N. L. Mohd Kamal · Z. Che Muda · D. Mohamad · Agusril · K. X. Lee
Department of Civil Engineering, College of Engineering, Universiti Tenaga Nasional, 43000 Kajang, Selangor, Malaysia
e-mail: drsalmiabeddu@gmail.com

Z. Itam
e-mail: iZarina@uniten.edu.my

N. L. Mohd Kamal
e-mail: Yana_Kamal@uniten.edu.my

Z. Che Muda
e-mail: zakariachemuda@gmail.com

D. Mohamad
e-mail: Daud@uniten.edu.my

Agusril
e-mail: Agusril@uniten.edu.my

T. S. Manan
Institute of Tropical Biodiversity and Sustainable Development, Universiti Malaysia Terengganu (UMT), Kuala Terengganu, Terengganu, Malaysia
e-mail: tehsabariah@umt.edu.my

N. Sivakumar
Department of Civil Engineering, SSN College of Engineering, Old Mahabalipuram Road, SSN Nagar, Kalavakkam 603110, India
e-mail: sivakumarn@ssn.edu.in

N. A. N. Basri
College of Graduate Studies, Universiti Tenaga Nasional, 43000 Kajang, Selangor, Malaysia
e-mail: amalinabasri798@gmail.com

N. Sabtu (ed.), *Proceedings of AWAM International Conference on Civil Engineering 2022 - Volume 3*, Lecture Notes in Civil Engineering 386,
https://doi.org/10.1007/978-981-99-6026-2_16

thermo-mechanical properties of mortar to certain extents. Synthetic fibers such as Polypropylene (PP) was added to increase mechanical properties of mortar as well as thermal value. The aim of this study is to evaluate the effect of cenosphere and PP fiber added to concrete mortar on its mechanical properties and thermal conductivity at different compositions (0, 5, 10, 15 and 20%). Five separate mixes were tested in this report. The findings for mechanical properties tests suggested that the usage of cenosphere as sand replacement has significantly reduce the mechanical strength as well as density of samples. Twenty percentage of cenosphere caused the compressive and tensile strength together with density reduced to lowest value which are 8.6 MPa, 8.20 MPa and 1416 kg/m^3, respectively, as compared to other percentages. This study also revealed that k value reduced as low as 0.62 W/m K due to incorporation of cenospheres and mixtures. It can be concluded that higher percentage of cenosphere used in mortar samples, thermo-mechanical value decreases.

Keywords Cenosphere · Mechanical properties · Thermal properties · Compressive strength · Tensile strength

1 Introduction

Industrial pollution is characterized as pollution that originates directly from industry. This form of contamination is one of the most common sources of pollution around the world. Industrial pollution can impact air quality and can enter the soil, causing widespread environmental problems. The World Health Organization estimates that outdoor air pollution alone accounts for around 2% of all heart and lung diseases, about 5% of all lung cancers, and about 1% of all chest infections. Power plant is one of example that caused industrial pollution. It has ability to contaminate air during coal combustion process in order to generate electricity by releasing poisonous gaseous (carbon monoxide, carbon dioxide, nitrogen oxides and sulfur dioxide, among others), particulate matter and liquids from multiple sources, which concentrate in the atmosphere [1]. Fly ash (FA), bottom ash (BA), boiler slag (BS) and flue gas desulphurization (FGD) materials, which are commonly known as coal combustion by-products (CCPs) are examples of particulate matter produced during generation of electricity. The number of dangerous by product produced increasing over years as Malaysia using coal as main sources to generate electricity.

As mentioned above, fly ash (FA) is one of particles that produced during combustion process and considered as waste from industrial. Generally, FA composed about 75% from total ash produced. It is made up from silicates and unburned carbon [2]. Even though FA contributes large percentage of waste materials produced, it has other components that deposited in it and can be used in many applications. Different components and applications existed in FA was suggested by [3] in Fig. 1.

Cenosphere is one of particles that deposited in FA. The word "Cenosphere" extracted from two Greek words which are kenos (hollow) and sphaira (sphere). The color cenospheres are vary from grey to almost white [5]. This hollow spheres

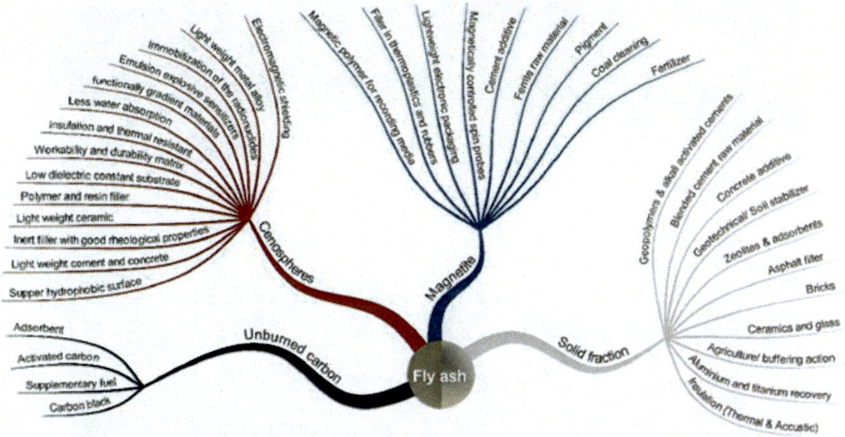

Fig. 1 Different components and applications exist in fly ash [4]

as shown in Fig. 2 mainly consisting of silica and alumina, are filled with air or gases, and are by-products of the combustion of pulverized coal at the thermal power plants. These ash particles get their hollow spherical shape as a result of cooling and solidifying around a trapped gas (generally CO_2 and N_2bubble) from the molten droplets of inorganic coal residue [6]. During coal burning process in thermal power plants, FA containing cenosphere produce high composition of alumina (Al_2O_3) and silica (SiO_2). Throughout the complicated process in the chemical, physical transformation, and high temperatures of 1500–1750 °C of producing electrical power, the composition of element in fly ash and cenosphere produced are varied depending on the origin of coal. Besides hollow in shape and light weight, it also has low density, low thermal conductivity, good insulation, good flowability, chemically inert, good resistance to thermal shock as well as high mechanical strength [6–9]. With these properties, Cenosphere has advantages that can be used in many applications.

According to [10], the inclusion of cenosphere promotes good mechanical properties of concrete as well as reducing thermal conductivity. For instance, DeSouza et al. [11] showed that the replacement of sand of 33% by cenosphere slightly reduce the compressive strength to 80 MPa as compared to control mix which about 85 MPa. Patel et al. [4] added that 50% of sand replacement with cenosphere reduce compressive strength to 22.6 MPa which is lower than 0% sand replacement (35.11 MPa).

The basic term for classifying thermal insulation materials is the thermal conductivity coefficient. The Kelvin meter per watt W/m K unit of thermal conductivity coefficient can be defined as the amount of heat that passes through a given unit area and thickness in unit time when the temperature difference between two surfaces is 1 K. Because of the lower unit weight, the thermal conductivity of lightweight aggregate concrete is approximately 50% lower than standard concrete. Standard

Fig. 2 Spherical shape of cenosphere [6]

concrete has a thermal conductivity of 0.62–3.3 W/m K [12], whereas lightweight concrete has a thermal conductivity of 0.4–1.89 W/m K [13].

2 Design of Experiment Setup

The purpose of the study is to explore the behavior of building materials using cenosphere as cement replacement. Various percentages of cenosphere with constant values of cement, coarse aggregate, fine aggregate and water will be used in the mix design of building materials to observe its performance against mechanical as well as thermal properties. In order to obtain more accurate results, one of the most critical conditions for experimental investigation as part of any study program is the practice of strict quality control steps. Therefore, by applying the existing code of standards and requirements for the collection of concrete producing products and mixing proportions, spurious findings and false results can be avoided. Compressive and tensile strength determined by using Compression Testing Machine and Universal Testing Machine respectively. On the other hands, as for thermal properties, the test will be conducted by using FOX 50 instrumentation. Five different types of mixes were prepared with vary composition of cenosphere while percentage of fibrillated Polypropylene remain same for all mixtures. Water and admixture were measured in percentage by weight proportion of cement used (Fig. 3).

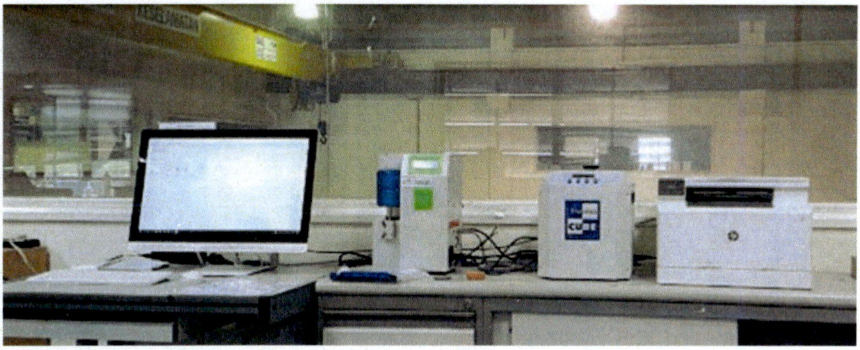

Fig. 3 Fox 50 instrument

Fig. 4 Ordinary portland cement (OPC)

2.1 Materials and Method

2.1.1 Ordinary Portland Cement (OPC)

Ordinary Portland Cement from Tasek Corporation Berhad was used in this study. The selection was based on the fact that with this specific gravity, the concrete properties of the mix could be observed under normal hydration process (Fig. 4).

2.1.2 Cenosphere

Cenosphere is lightweight, inert and hollow sphere that are produced as a by-product from coal combustion process. Different percentage of cenosphere (0, 5, 10, 15, 20) were used as sand replacement in this study (Fig. 5).

Fig. 5 Cenosphere

2.1.3 Superplasticizer (SP)

Superplasticizer commonly known as super Ps usually used to maintain high workability while at the same time maintaining strength. Many trials mixes were done and proportions of water and super plasticizers to determine the optimum dosage.

2.1.4 Polypropylene

Polypropylene fibers are synthetically inactive, hydrophobic and stable in soluble condition of cement that are cut into 19 mm each. In this study, PP was added in CM mixtures as it would reduce the k values. However, PP fibers need to be mixed thoroughly in the wet cement paste (Fig. 6).

2.2 Design Mix

The mixing of mortar was carried out using the 1:3 mix according to the standard specifications given by ASTM C91. This has to do with the fact that mortar of type M is the suitable mortar type for plastering works. The 1:3 mix simply means that the mixture of cement consists of 1 part of cement and 3 parts of sand. In order to find the perfect mix design, several trials had been done. Then, it was concluded that the w/c ratio needed to be used was 0.5% with 2% of SP as the water reducer. At the same time, the fiber volume fractions were added individually as follows (Table 1).

Five types of samples with different cenosphere composition were used in this study. Density of each sample taken by using Oven Dried method was taken after

Fig. 6 Polypropylene (PP)

Table 1 Percentage of cenosphere and polypropylene

Mixing code	Cenosphere, Cf (%)	PP fibrillated fibres (%)
CM	0	0.5
Cf-5	5	0.5
Cf-10	10	0.5
Cf-15	15	0.5
Cf-20	20	0.5

24 h of curing by measuring weight samples and divided by its volume. Compressive test was done to determine the ability of CM to carry loads before failure by following ASTM C109 standard. Referring to ASTM C496 standard, tensile strength of CM was deduced from splitting tensile test. Both compressive and splitting tensile strength was measured by using Universal Testing Machine. Thermal conductivity was measured by using FOX 50 Heat Flow Meter as per ASTM C1783 standard.

3 Results and Discussions

3.1 Compressive Strength

To determine the compressive strength of cube specimen, a load was applied until failure by using compression testing machine. Compressive strength test was carried out at 3, 7 and 28 days age. The results obtained are presented in Fig. 7. By observing

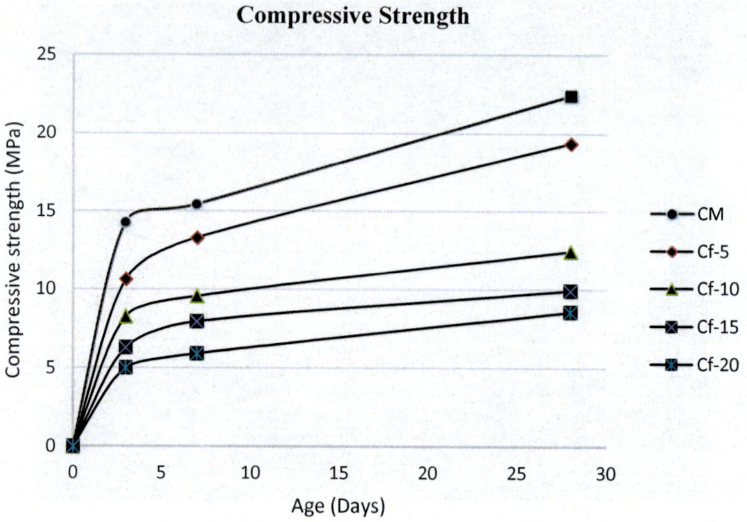

Fig. 7 Compressive strength (MPa)

the Fig. 7 below, a significant increase in strength can be noticed for CM and Cf-5 comparing to 7 days results. As for the rest of mixtures series, a slight increase in strength can be observed. The data presented showed that the inclusion of Cf-5 caused the compressive strength of mortar decrease by 14% compared with control mix. Compressive strength value reduces up to 44%, 56%, and 62% when 10%, 15% and 20% of cenosphere was used to replace sand respectively compared to control mix. The continuously decrement of compressive strength was also noticed when more cenosphere was added. These experimental results are consistent with previous researches where a decrease in strength was reported due to incorporating cenosphere with concrete mortar. The lower the density of concrete mortar, the higher the strength loss for the concrete mortar [14].

The relationship between percentage of Cenosphere and compressive strength at 3, 7 and 28 days shown below:

3 days;

$$f_c = -2.265C_f + 15.641$$

$$R^2 = 0.9363$$

where f_c = Compressive strength (MPa), and C_f = Dosage of Cenosphere (%)

7 days;

$$f_c = -2.519C_f + 17.903$$

$$R^2 = 0.9867$$

where f_c = Compressive strength (MPa), and C_f = Dosage of Cenosphere (%)

28 days;

$$f_c = -3.717C_f + 25.689$$

$$R^2 = 0.9408$$

where f_c = Compressive strength (MPa), and C_f = Dosage of Cenosphere (%).

3.2 Splitting Tensile Strength

Splitting tensile strength test was carried out at 28 days to determine tensile strength. Three cylinders of size 50 × 100 was tested for concrete mortar and each mixtures series. Indirect splitting tension load was applied along the vertical axis of the test sample and was increased until failure occurred when the cylinder split into two halves. This test was run to determine the strength between cenosphere and mortar. The data obtained are shown in Fig. 8. Broadly speaking, the test results showed that the introduction of wastes enhanced tensile strength for concrete mortar. An obvious decrease in strength can be noticed when cenosphere was incorporated with concrete mortar comparing to concrete mortar strength. In addition, Cf-5 caused the lowest decrease in tensile strength by 22% whereas the control mix of the concrete mortar. The tensile strength value reducing when higher percentage of sand replaced with Cenosphere added. Cf-20 demonstrate lowest tensile value as compared to other which is 8.19 MPa or equvalent to 37.95% reduction.

3.3 Density

Density of mortar was an important factor that has a significant effect on its thermal conductivity and therefore it had to be measured. The testing condition for thermal conductivity was oven-dried oven-dried densities were determined and presented in Fig. 9. It was found that the inclusion of fibers caused a reduction in oven-dried density for all cenosphere series. When 20% of sand replaced with cenosphere, mortar's density shows highest reduction percentage which is 30.6% as compared to Control Mix. The decrease in oven dried density was 26% when Cf-15 was added which is slightly higher than Cf-10 and Cf-5 (21.14% and 11.7%) respectively. From results obtained, its clearly shown that inclusion of cenosphere as sand replacement reduce the density of mortar. It is due to properties of cenosphere its self which

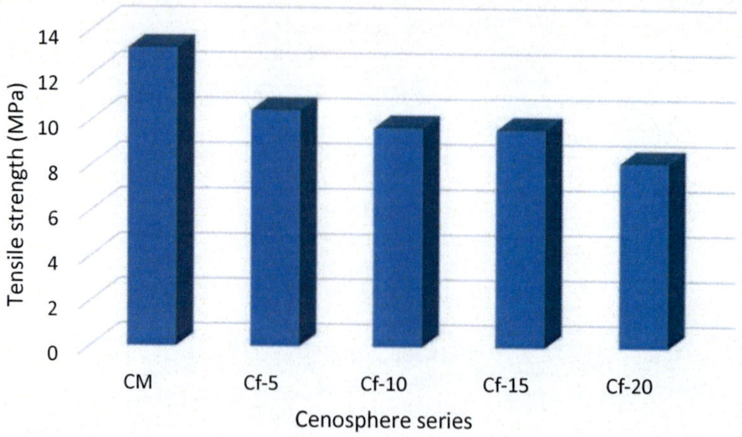

Fig. 8 Splitting tensile strength (MPa)

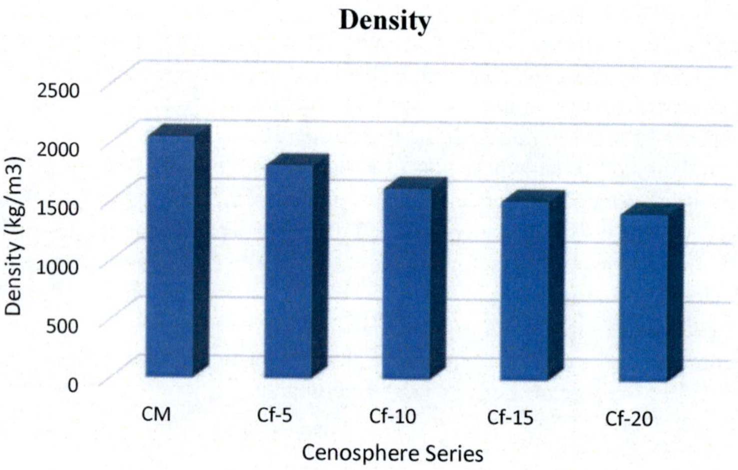

Fig. 9 Density

hollow in shape and this lead low in density. Hence, reducing the overall density of mortar.

3.4 Thermal Conductivity

Thermal conductivity for mortar was measured using heat flow meter apparatus FOX50. The test method was done in accordance with ASTM C1784-13 and ISO 8301. During the test, rubber was used to help in eliminating errors due to air gap between the specimen's surface and apparatus' plate. Moreover, rubber also helped in preventing poor contact between the specimen's surface and apparatus' plate as well and hence excluding the thermal contact resistance. The obtained k values were corrected after the test using the following formula (Fig. 10).

$$\boldsymbol{\lambda}_{\text{accurate}} = \left(\boldsymbol{d}_{\text{sample+rubber}} - \boldsymbol{d}_{\text{rubber}}\right) / \left[\left(\boldsymbol{d}_{\text{sample+rubber}} / \boldsymbol{\lambda}_{\text{sample+rubber}} - \left(\boldsymbol{d}_{\text{rubber}} / \boldsymbol{\lambda}_{\text{rubber}}\right)\right)\right]$$

The thermal conductivity test was carried out at the mean temperature of 27 °C and temperature difference between the upper and lower plates was 20 °C. By observing the figure below, it can be noticed that in general, inclusion of cenosphere will caused k values drop. However, the reduction in k values was not considerable. The lowest k values obtained was for Cf-20 mixtures series which caused a 64.54% decrease in thermal conductivity compared to the CM. The thermal conductivity of all mixture's series is 1.8 W/mK for CM, 1.04 W/mK for Cf-5, 0.95 W/mK for Cf-10, 0.70 W/mK for Cf-15 and 0.62 W/mK for Cf-20 respectively. This can be attributed to many reasons for instance, the increase in cenosphere dosage was causing the density to drop more and the lower the density hence lowering the k values.

4 Conclusions

Based on this study, it can be concluded that the utilization of cenosphere as sand replacement influenced thermo-mechanical properties of mortar. The results for each property also show that:

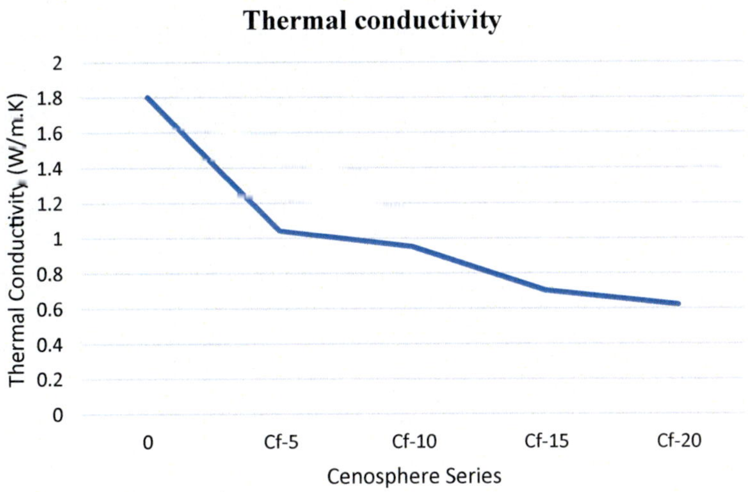

Fig. 10 Thermal conductivity (W/m K)

- Compressive strength of mortar with 20% of cenosphere recorded lowest compressive value which is 8.6 MPa while 5% gives 19.3 MPa reading.
- 5% of cenosphere reduced the mortar's tensile strength by 22% but there is no significant reduction when higher percentage of cenosphere was included.
- Density of mortar with 20% of cenosphere produce 1416 kg/m^3 which is much lower than normal value (2400 kg/m^3).
- Mortar with 20% of cenosphere recorded lowest thermal conductivity value which is 0.62 W/m K as compared to other ratios.

Acknowledgements The authors are thankful to Universiti Tenaga Nasional, Malaysia for providing financial support DLTK Grant (20238004DLTK) and TNB Seeding Fund (U-TG-RD-20-21) for this study.

References

1. Barreira, A., Patierno, M., Ruiz Bautista, C.: Impacts of pollution on our health and the planet: the case of coal power plants. Environmental **1**(28), 10 (2017)
2. Goodarzi.: Characteristics and composition of fly ash from Canadian coal-fired power plants. Fuel **85**(10–11), 1418–1427 (2006)
3. Ranjbar, N., Kuenzel, C.: Cenospheres: a review. Fuel **207**, 1–12 (2017)
4. Patel, S.K., Majhi, R.K., Satpathy, H.P., Nayak, A.N.: Durability and microstructural properties of lightweight concrete manufactured with fly ash cenosphere and sintered fly ash aggregate. Constr. Build. Mater. **226**, 579–590 (2019)
5. Manocha, L.M., Ram, K.A., Manocha, S.M.: Separation of cenospheres from fly ashes by floatation method. Eurasian Chem. J. **13**(1–2), 89–95 (2011)
6. Kolay, P.K., Singh, D.N.: Physical, chemical, mineralogical, and thermal properties of cenospheres from an ash Lagoon. Cem. Concr. Res. **31**(4), 539–542 (2001)
7. Yoriya, S., Tepsri, P.: Separation process and microstructure-chemical composition relationship of cenospheres from lignite fly ash produced from coal-fired power plant in Thailand. Appl. Sci. **10**(16), 1–21 (2020)
8. Yoriya, S., Intana, T., Tepsri, P.: Separation of cenospheres from lignite fly ash using acetone-water mixture. Appl. Sci. **9**(18) (2019)
9. Danish, A., Mosaberpanah, M.A.: formation mechanism and applications of cenospheres: a review. J. Mater. Sci. **55**(11), 4539–4557 (2020)
10. Satpathy, H.P., Patel, S.K., Nayak, A.N.: Development of sustainable lightweight concrete using fly ash cenosphere and sintered fly ash aggregate. Constr. Build. Mater. **202**, 636–655 (2019)
11. De Souza, F.B., Montedo, O.R.K., Grassi, R.L., Antunes, E.G.P.: Lightweight high-strength concrete with the use of waste cenosphere as fine aggregate. Rev. Mater. **24**(4) (2019)
12. Kim, J.J., Youm, K.S., Reda Taha, M.M.: Extracting concrete thermal characteristics from temperature time history of Rc column exposed to standard fire. Sci. World J. (2014)
13. Yun, T.S., Jeong, Y.J., Han, T.S., Youm, K.S.: Evaluation of thermal conductivity for thermally insulated concretes. Energy Build. **61**, 125–132 (2013)
14. Senthamarai Kannan, K., Andal, L., Shanmugasundaram, M.: An investigation on strength development of cement with cenosphere and silica fume as pozzolanic replacement. Adv. Mater. Sci. Eng. **2016** (2016)

Influence of Moisture Content and Position on the Tensile Strength of Four Air-Dried Bamboo Species

Dinie Awalluddin, Mohd Azreen Mohd Ariffin, Yusof Ahmad, and Nor Fazlin Zamri

Abstract Bamboo is one of the fastest growing plants globally, and it has been used for numerous applications, including in the construction industry. It also has been known that bamboo has an excellent mechanical property, especially in tensile. However, each species has different mechanical properties, and generally, moisture content and the bamboo position have a greater impact on the variation of bamboo properties along with the culm height. Thus, this study focuses on the effect of four bamboo species' moisture content and position: *Dendrocalamus asper, Bambusa vulgaris, Gigantochloa scortechinii,* and *Shizostachyum grande.* The test on the bamboo was conducted after the treatment and air-dried conditioning were completed. The moisture content test and tensile strength test were conducted for each species for 1, 5, and 7 months. The results show that the *D. asper* and *B. vulgaris* have the highest tensile strength value, followed by *G. scortechinii* and *S. grande.* Moreover, it was observed that the tensile strength of bamboo increased from the bottom to the top part of the bamboo. The moisture content of each species also differs, and the value also increased from the bottom to the top part of the bamboo. The moisture content decreased from 1 to 7 months while the tensile strength increased. It was concluded that the tensile strength of bamboo differs from each species and position, as well as the change in moisture content affects the tensile strength.

Keywords Bamboo · Mechanical properties · Tensile strength · Moisture content

1 Introduction

Bamboo is one of the fastest growing plants on the planet, with species ranging in height from a few centimeters to many meters. Some giant bamboo species can achieve their maximum height of up to 30 m in 2 to 4 months. In Malaysia, there are

D. Awalluddin · **M. A. M. Ariffin** (✉) · Y. Ahmad · N. F. Zamri
Faculty of Civil Engineering, Universiti Teknologi Malaysia, 81310 Skudai, Johor Bahru, Malaysia
e-mail: mohdazreen@utm.my

N. Sabtu (ed.), *Proceedings of AWAM International Conference on Civil Engineering 2022 - Volume 3*, Lecture Notes in Civil Engineering 386,
https://doi.org/10.1007/978-981-99-6026-2_17

many bamboo species such as *Dendrocalamus asper* (Buluh Betung*), Gigantochloa scortechinii* (Buluh Semantan), *Shizostacyum grande* (Buluh Semeliang), *Bambusa vulgaris* (Buluh Minyak), and others. It was found that the tensile strength of some species of bamboo is relatively high and can reach 581 MPa [1]. In fact, the ratio of tensile strength to the specific weight of bamboo is six times greater than that of steel [2].

However, bamboo must be treated to enhance its durability and resistance to fungus attack. The durability of untreated bamboo is often influenced by its age, species, and storage method [3]. It is expected that untreated bamboo will last 10 to 15 years if stored and used properly [4]. The starch content of untreated bamboo culm attracts insects and fungi, which will nest inside the bamboo culm. Meanwhile, treated bamboo has a lifespan of more than 3–4 years when placed outdoors, but a bamboo structure has a lifespan of more than 50 years when placed under a roof and is not in direct contact with rain or sunlight.

There are numerous aspects that affect the quality and viability of bamboo as a broad and economically beneficial alternative to wood or steel. Physical and mechanical properties are among these factors. Due to inadequate basic information on the properties of different bamboo species, only a few different types of local bamboo have been used commercially in the industry [5]. Factors such as age, moisture content, and treatment are very crucial in determining bamboo's mechanical properties. The strength and lifespan of bamboo also depend on the moisture content. Similar to timber, moisture content is a key factor for its utilization whether as for research or as a structural element purpose. Higher moisture content in the bamboo eventually will attract the fungi and borer insects which will speed up the rotting process of the bamboo itself. Naturally, bamboo also has higher moisture content, just like other plants. The distribution of moisture content in bamboo are depended on whether they are raw or processed. Raw bamboo or fresh felling bamboo might have 100% of moisture on a dry weight basis [6]. As a fibrous material, bamboo can change dimensionally when they gain or lose moisture. The anisotropic material properties in bamboo affected the changes in their dimensions, weight, and strength, especially in tensile. These properties also can cause the bamboo to deform when the bamboo possesses very low moisture content.

The moisture content is influenced by age, species, and harvesting season [7]. However, it is also known that bamboo in the green state has a higher moisture content, and the value differs within other culms and in relation to age, species, and season. The season has a greater effect on the moisture content of the bamboo. Minimum water content is usually at the end of the dry season, while the maximum water content is in the rainy season. The bamboo stem can increase the water content capacity during the rainy season. In addition, the moisture variation due to the season is higher than the difference between the bottom and top parts and between the bamboo species. Even at a similar locality, the water content of bamboo also may vary. Thus, it is important to know the strength of different bamboo species, positions, and the effect of moisture content along the bamboo culm height. The behavior of moisture absorbing of bamboo along the period also needs to be highlighted. Thus, this study aims to investigate the influence of moisture content of the bamboo as well as the effect of position on the tensile strength along the bamboo height.

2 Materials and Procedures

As shown in Fig. 1, the bamboo species involved in this study are (a) *D. asper*, (b) *B. vulgaris*, (c) *G. scortechinii,* and (d) *S. grande.* These bamboos were selected because it is the most common species in Malaysia. The bamboos were obtained and treated at Bamboo Jungle Adventure, Perak, Malaysia. The treatment aims to enhance their durability and resistance to fungus attack. The combination of boric acid and borax in a 1:1.5 ratio produces an alkaline salt, which was obtained as a ready-to-use powder. This powder was poured into a rectangular tank and mixed with water, and the bamboo was soaked in the tank for about a week before being dried. The bamboo was then received in air-dried conditions and stored in a closed shed before testing. Then, the bamboo was cut into the desired dimension according to the tests involved, such as moisture content and tensile strength test. The tests periods were selected at 1, 5, and 7 months in an air-dried condition.

2.1 Moisture Content Test

The moisture content of bamboo specimens was carried out before completing the tensile strength test. The specimens for moisture content in this research were taken near to the extracted tensile specimens. The ISO-22157-1 (2004) procedure was followed for the tensile strength test [6]. The specimens were prepared like prism

Fig. 1 Harvested of 3-m length with different species of bamboo **a** *Dendrocalamus asper*, **b** *Bambusa vulgaris*, **c** *Gigantochloa scortechinii* and **d** *Shizostachyum grande*

Fig. 2 Moisture content test specimens

form, approximately 25 mm high, 25 mm width, and thick as the wall thickness shown in Fig. 2.

The moisture content value can be calculated by using Eq. 1. The initial weight of the specimens was recorded as m_i. Then, using a hot-air oven, the specimens were dried at a temperature of 103 ± 2 °C for 24 h. The test specimens were weighed after 24 h and drying continued after that. The weighing process was carried out and recorded every 2 h until the successive determination of the mass did not exceed 0.01 g. The final reading of weighing was recorded as m_o.

$$\text{Moisture Content, }(\%) = \frac{m_\text{i} - m_o}{m_o} \times 100 \tag{1}$$

2.2 *Tensile Strength Test*

The tensile strength test was conducted following the ISO-22157-1 (2004) procedure [8]. The bamboo specimen was cut into a dog-bone shape, and the tests were conducted in an air-dried condition. Figure 3 shows the completed bamboo test specimen for the tensile strength test. The bamboo test specimen had an overall length of 260 mm and a specimen grip width of 20 mm. The test specimen's effective length was 65 mm, its width was 10 mm, and its thickness was determined by the thickness of the bamboo itself. For each species, nine specimens were taken from the internode parts, which consisted of three samples at the top part, three samples at the middle part and three samples at the bottom part of the bamboo. All the measurements were taken by using Vernier caliper.

The tensile strength test was conducted by using a Universal Testing machine with a capacity of 1000 kN. First, all the bamboo dimensions, such as width and thickness, were recorded before performing the test. Then, the specimens were loaded

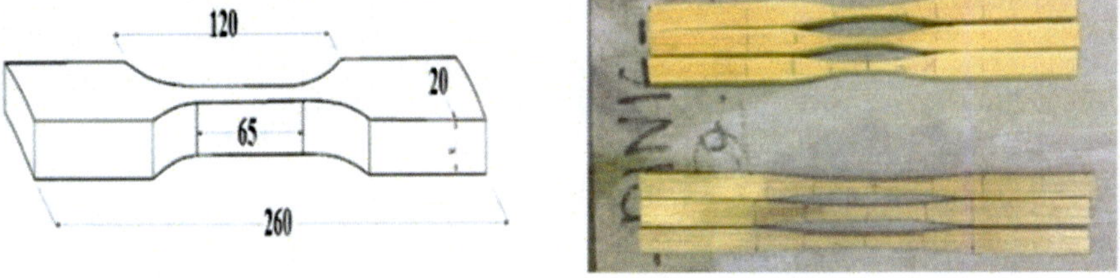

Fig. 3 Tensile strength test specimen

onto the tensile strength testing equipment, and both ends were tightened flat on the provided grips. Following that, pulling forces of 0.05 mm/s were applied with continuous movement until the maximum load was reached before the sample failed. An operational test set up for the tensile strength test is given in Fig. 4.

The tensile strength of bamboo specimens can be calculated in general by dividing the maximum load (P) with the area at the effective length, (A). The formula can be referred to the following Eq. 2.

$$\text{Tensile Strength, } \sigma = \frac{P}{A} \tag{2}$$

Fig. 4 Tensile strength test setup

3 Results and Discussions

3.1 Initial Moisture Content and Tensile Strength

Table 1 shows the results of the initial moisture content and tensile strength of bamboo species at a dried condition of 1 month. The results indicate that *D. asper* and *B. vulgaris* have the highest tensile strength among the other species, followed *by Gigantochloa scotechinii* and *S. grande*. Figure 5 shows average tensile strength trend of *B. vulgaris* where the tensile strength was observed to increase from the bottom to the top part of the bamboo culm. *G. scortechinii* and *S. grande* showed a strong incremental trend in tensile strength values. Even though *D. asper* did not show a consistent increase in tensile strength, the data can still be considered because the specimen on the top part of the bamboo culm gave a greater value than the specimen on the bottom part.

The increase in tensile strength along the height of the culm observed in this study for all species is believed to result from a simple mechanism. Fibers material that exists in the form of the sclerenchymatous sheath around the vessel in the culm is basically constant or increases along with the height of the culm [9, 10]. However, as height increases, the diameter and culm thickness decrease, reducing the area over which the stress is determined. As the wall thickness of the bamboo getting decreased at the top part of the bamboo, the condensation of the vascular bundle caused the attributed to increasing fibers density, thus, increasing the strength. Furthermore, since the fibers primarily convey the tensile strength, the apparent tensile strength of bamboo has increased. Thus, the fibers volume fraction increased along with the height of the culm [11].

Table 1 Average moisture content and tensile strength of bamboo at 1 month

Species	Part	Average moisture content (%)	Average tensile strength (MPa)
Dendrocalamus asper	Top	17.40	219.68
	Middle	20.25	198.35
	Bottom	22.30	214.06
Bambusa vulgaris	Top	16.25	219.08
	Middle	18.60	212.38
	Bottom	22.45	211.76
Gigantochloa scortechinii	Top	17.10	175.98
	Middle	18.20	149.29
	Bottom	20.19	147.36
Schizostachyum grande	Top	25.00	138.72
	Middle	26.79	107.83
	Bottom	27.14	105.83

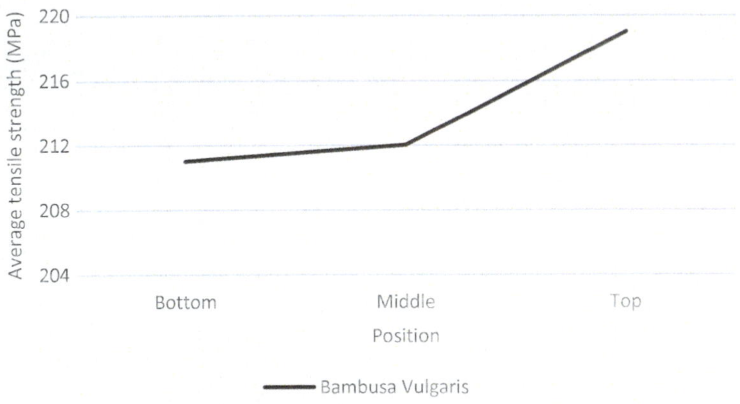

Fig. 5 Average tensile strength of *Bambusa vulgaris* species along with the height

Clearly seen that the moisture content also decreased from the bottom to the top part of the bamboo. The result on the change of moisture content obtained was also agreed upon in the previous study. Anokye et al. [5] conducted a study on the variation of the moisture content of green bamboo. They reported that the moisture content decreased from the bottom to the top part of the *B. vulgaris* species. Generally, water is transported from the bottom to the top part of bamboo [5]. This is because the bottom part has a smaller vascular bundle amount. However, the vascular bundle size at the bottom part is large, and the water is concentrated at the bottom, thus, contributing to decreasing trend of moisture content from the bottom to the top parts. Nevertheless, the moisture content variation along the culm did not differ significantly across the bottom, middle, and top parts.

3.2 *Moisture Content Reduction and Tensile Strength Development*

Figure 6 shows the trend of the average tensile strength graph from 1 to 7 months interval, while Table 2 shows the full data of bamboo specimens average moisture content and tensile strength at 5 and 7 months of closed shed storage. The tensile strength of all species was found to increase with time. The increment of tensile strength was associated with the moisture content of the specimens, which decreased at a different part of bamboo culm from 5 to 7 months. The moisture content of all bamboo species had been reduced gradually from 1 to 7 months. *Schizostachyum grande* shows the highest average moisture content followed by *D. asper*, while *B. vulgaris* and *G. scortechinii* possess only a small difference in average moisture content value.

Bamboo drying is a relatively slow process, which may take several days to several months, depending on the bamboo species, its thickness, and the drying

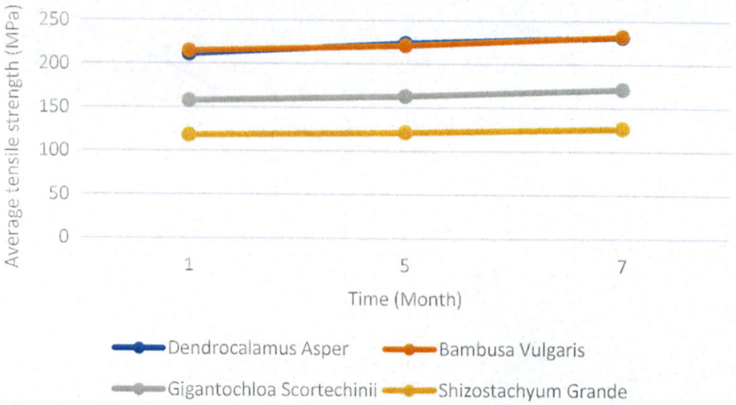

Fig. 6 Average tensile strength of bamboo (top part) against the time

Table 2 Average moisture content and tensile strength at 5 and 7 months

Species	Part	Average moisture content (%)	Average tensile strength (MPa)	Average moisture content (%)	Average tensile strength (MPa)
		5 months		7 months	
Dendrocalamus asper	Top	15.85	225.47	13.94	232.80
	Middle	17.91	222.75	16.36	228.59
	Bottom	18.44	222.48	17.23	232.32
Bambusa vulgaris	Top	14.01	222.31	13.86	231.67
	Middle	15.10	220.07	14.28	233.98
	Bottom	19.20	221.14	17.05	230.63
Gigantochloa scortechinii	Top	15.60	182.77	14.40	187.67
	Middle	16.95	154.10	15.30	163.97
	Bottom	18.09	151.48	16.15	162.44
Schizostachyum grande	Top	16.87	142.21	15.30	149.20
	Middle	17.98	110.61	16.84	114.93
	Bottom	19.63	109.20	18.80	113.01

methods used. In addition, bamboo is also a 'hygroscopic material', which means that its moisture content changes in response to its surroundings' temperature and relative humidity [12]. In constant temperature and relative humidity conditions, the bamboo will eventually reach a constant moisture content or known as the "equilibrium moisture content". However, in practice, such stability of conditions rarely occurs, and therefore a true equilibrium moisture content is never reached. It is also worth noticing that the strength of bamboo is weaker with the increasing moisture content. Further reduction of moisture content below optimum point will cause the

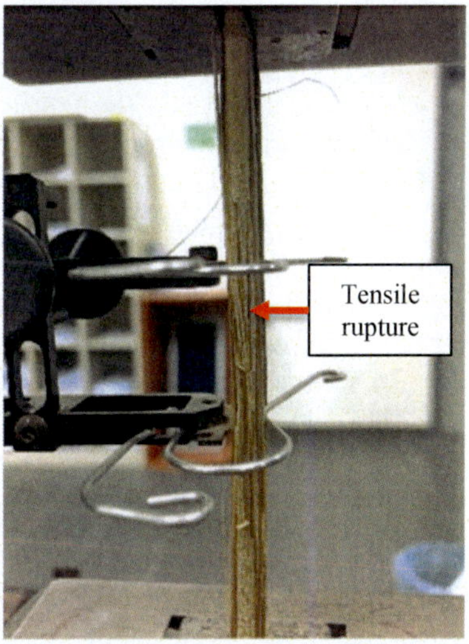

Fig. 7 Bamboo failure in tensile strength test

reduction of strength as the bamboo will start to show brittle characteristics due to loss of water [9].

3.3 Mode of Failure

Figure 7 shows the failure mode where the bamboo specimen experienced a tensile rupture along the gage length of the specimen. The bamboo specimen failure showed 'good' failure in the way of being mostly unaffected by the gripping process of the testing machine. Uneven grip pressure occurred when grip pressure was applied to the grip region because the bamboo itself had various material properties across the grip area, leading the apparent gripping pressure to that area to be uneven. It resulted in a single longitudinal splitting associated with a higher grip pressure area.

4 Conclusion

Understanding the factor that affects the mechanical properties of bamboo is essential because the material has numerous applications in the building industry. For example, it was known that moisture content is one factor that needs to be considered when using bamboo as a construction material. Therefore, the main purpose of this study was to present the effect of moisture content on the tensile strength of different bamboo species. Based on the extensive experimental testing conducted on tensile strength, it was found that *D. asper* and *B. vulgaris* have the highest tensile strength, with the average value of above 200 MPa. However, *G. scortechinii* and *Shizostacyum grande* species show the average tensile strength value below 200 MPa.

Moreover, the tensile strength for all species shows an incremental trend from the bottom part to the top parts. This trend may be attributed to moisture change and the variation of fibers distribution along with the height of bamboo. The study also has identified that the drying process of bamboo was relatively slow in an ambient condition due to the hygroscopic behavior of bamboo, and this was proved by the moisture content values recorded along the given periods. However, the tensile strength of bamboo was considered excellent when the moisture value was below 20%. Therefore, it can be concluded that bamboo also can be used as a construction material. It was also recommended to apply a protective layer such as a water repellent agent to cater to the hygroscopic behavior of bamboo. Thus, future works need to be done to widen the application of bamboo in the construction industry.

References

1. Shao, Z.P., Fang, C.H., Huang, S.X., Tian, G.L.: Tensile properties of Moso bamboo (*Phyllostachys pubescens*) and its components with respect to its fiber-reinforced composite structure. Wood Sci. Technol. **44**, 655–666 (2010)
2. Dewi, S.M., Nuralinah, D.: The recent research on bamboo reinforced concrete. MATEC Web Conf. **103** (2017)
3. Ghavami, K.: Bamboo: low cost and energy saving construction materials. In: International Conference on Modern Bamboo Structures, no. January (2008)
4. Jayanetti, D.L., Follet, P.R.: Bamboo in construction. In: Modern Bamboo Structures, 1st Ed., London, vol. 15, pp. 1–120 (2008)
5. Anokye, R., Kalong, R.M., Bakar, E.S., Ratnasingam, J., Jawaid, M., Awang, K.: Variations in moisture content affect the shrinkage of *Gigantochloa scortechinii* and *Bambusa vulgaris* at different heights of the Bamboo Culm. BioResources **9**, 7484–7493 (2014)
6. Li, X.: Physical, chemical, and mechanical properties of bamboo and its utilization potential for fiberboard manufacturing (2004)
7. Suriani, E.: A study of the physical-mechanical properties of bamboo in Indonesia. In: Proceedings of the Built Environment, Science and Technology International Conference, pp. 154–162 (2020)
8. I. O. for Standardization, ISO 22157-1:2004 (2004)
9. Janssen, J.J.: Designing and building with bamboo, no. 20. Inbar Technical Report 20, Beijing, China, 1–211 (2000)
10. Grosser, D., Liese, W.: On the anatomy of Asian bamboos, with special reference to their vascular bundles. Wood Sci. Technol. **5**, 290–312 (1971)
11. Amada, S., Munekata, T., Nagase, Y., Ichikawa, Y., Kirigai, A., Zhifei, Y.: The mechanical structures of bamboos in viewpoint of functionally gradient and composite materials. J. Compos. Mater. **30**, 800–819 (1996)
12. Okhio, C.B., Waning, J.E., Mekonnen, Y.T.: An experimental investigation of the effects of moisture content on the mechanical properties of bamboo and cane. Cyber J. Multidiscip. J. Sci. Technol. J. Sel. Areas Bioeng. November, 7–14 (2011)

The Aflaj Systems in Sultanate of Oman: Its Traditional-Engineer Construction and Operation

Ahmed S. Al-Marshoudi and Jasni Sulong

Abstract The ancient aflaj (singular falaj) system in Sultanate of Oman is not only an irrigation system conveys rainfall accumulated water-table (from aquifer) to the settlement areas, but also it hold a unique construction method since millennium. It is believed that the location of Oman, in an arid or semi-arid climate surrounding with harsh desert environment, has led the settlers to develop high experience in the art of water exploitation. Much were written in descripting these systems using Iranian qanat system. Because of a confusion found in describing the construction of the three main component (mother-well, shift and tunnel), this paper used author own knowledge in the aim to clarify such confusion. This knowledge has been gained and collected from the continuous contact with local and the use of wide range of literatures in the purpose to offer contribution in systematic manner with regard to the traditional engineer construction process. The finding clearly showed the Omani construction method for the aflaj hold its own way with accordance to the experiences of the local people.

Keywords Aflaj · Tradition · Construction · Physical

1 Introduction

While Sultanate of Oman is located in arid and semi-arid climatic zone, characterized by holding its own what is known as hydrogeology environment [1]. There are two main factors closely link recharge of the aquifers through the intense rain with slopes of hill mountain. The existence of hard rocks (ophiolite) capable to store water for longer period [2]. Furthermore, it is important to appreciate that the establishment of settlement within such environment is not so simple. First, the problem of finding a suitable site for cultivation is sometime more difficult to resolve than that of finding

A. S. Al-Marshoudi · J. Sulong (✉)
Section of Islamic studies, School of Humanities, Universiti Sains Malaysia, 11800 Pulau Penang, Malaysia
e-mail: jasni@usm.my

N. Sabtu (ed.), *Proceedings of AWAM International Conference on Civil Engineering 2022 - Volume 3*, Lecture Notes in Civil Engineering 386,
https://doi.org/10.1007/978-981-99-6026-2_18

water [3]. Second, the location of a traditional agricultural settlement is principally determined by the availability of good soil which in Arabia sometime more difficult to find than water [4]. The earth's fresh water sources normally divided into two main categories. Sources that obtained from rivers and lakes, sometimes referred to as surface water, and sources collected from groundwater. While surface water projects were often complex and large scale and hence resulted in more centralized societies, most groundwater activation systems remained quite localized servicing smaller areas. However, most groundwater activation systems remained quite localized servicing smaller areas. However, most of the groundwater techniques require to develop a social structural framework in order to form manpower skill, with a division of labor. One of the ancient technologies used to lift/collect groundwater from aquifer known as aflaj in Oman and qanat Iran. Such idea, of drawing groundwater to the surface through the use of natural conditions, has received extensive emphasis in the literature. This is attributed to the fact that such an idea is sustainable and is reliable for a long period and considered to be ancient technology from which modern life still benefits [5].

Given the significantly of water sustainability techniques and the possibility to preserve the most ancient water catchment system, this paper has two main purposes. First, it attempts to produce a literature review in order to contribute to a clearer debate about the issue of groundwater techniques. The second this paper explain a situation by which hold the most ancient physical water supply; aflaj system in Oman.

2 Background to the Problem

While substantial coverage of the water drawing technique in many countries worldwide, much emphasis placed toward the system in Iran (qanat) and Oman (aflaj). For example, Lightfoot [6] argued that qanats of Yemen are of ghayl type (known as miyan) are not remarkable in size/number, compared to those found in Iran and Oman. This is attributed mainly to the hydro-geological environment factor which normally reflect the aflaj/qanat construction process. First, it is well known that a unique traditional-construction process, which uses the physical-topographical characteristics of the site, has been carried out by local expertise and normally conducted manually; in Iran known as muqannis, and in Oman they are known as awamir [3, 7]. The idea is that the physical structure of the area has been designated in such a way to withdraw/collect water from aquifer without the use of mechanical devices or expansive fuel. This is achieved by traditional-constructing process of long tunnel (normally several kilometers long) starting from the main water collected source (locally known modari chah in Iran and mother-well in Oman. Second, despite the fact that several researchers who analyzed the aflaj in Oman attempted to use similar descriptive pattern used for the Iranian qanat, problem in understanding the function of the three main components attached to aflaj system (the mother well, shafts and the tunnels) are found. In addition, is the branches among these components or is it operated apart?

The earliest who described the Iranian qanat construction method Cressey [8] and English [7] showed a series of ventilated shafts are dug first and then the tunnel is constructed to connect the bottoms of these shafts. However, Wulff [9] indicated that a series of vertical shafts will have to be sunk from the surface to the conduit, since he thinks that the tunnel is constructed first (see Fig. 2). He further revealed the function of these shafts, as earth and rock are excavated from the tunnel face and winched to the surface through the shafts. Then Wilkinson [3] and Stanger [1] followed almost similar descriptive pattern provided different views in understanding the whole system components. One view provided by Stanger [1] described the whole construction as a hydraulic principle of sub-horizontal 'line sink-wells' with an active length of hundreds of meters including several shafts in addition to the mother well. On another occasion he stated, "…it is a sub-horizontal line consisting of adits (branches)" Similarly, Wilkinson (1983: 177) viewed the falaj as a tunneled horizontal well, characteristically up to ten kilometers long and drawing water from aquifers around twenty meters. To clarify this confusion, below is an attempt combining these literatures with author own local contact with people of Oman and literatures written in Arabic language.

3 Construction Idea

There are three types of aflaj in Oman known as daudi, ghaili and ayni. Although, they all extract water through rainfall accumulation, they greatly differ with respect to the physical water extraction process. It is whether from direct supply from the aquifer or surface of the oasis. For instance, the daudi and ayni type extracts water from the upper part of the aquifer in the mountains while ghaili from the surface of wadi (oasis). Since the daudi extraction process considered of being more sophisticated and complex, it received extensive discussion in the literatures. The earliest writers about daudi ancient construction idea were provided by Wilkinson [3] and Al-Abri [10], as the former used field work sampling and analytical techniques while Al-Abri [10] local experience used a descriptive logistic framework, followed by Stanger [1] who looked deeply into the matter. Below is a review of the construction steps as demonstrated by these writers.

3.1 Water Source Survey

In literature, our understanding of the aflaj construction process can be divided into two main views. Those who view the construction through investigation and primary measurement in digging the mother well [3, 10, 11]. Those who think that the whole construction can be considered as a hydraulic principle known as a sub-horizontal. Here a line sink first of a length of hundreds meters included several shafts in addition to mother well, as the upstream of the shafts is known as the "mother well" [1].

To provide more clarification to this issue, it is important to define each of the components involved in the process separately.

Mother well. This component considered as the first important element of the construction process. Wulff [9] argued that it is the head of the water source that is constructed in order to reach down to the water-table. However, there are different views show how this well first constructed. Wilkinson [3] argued the first requisite for building a qanat is to locate a suitable aquifer and sink a mother well to it. Al Abri [10] and Al-Rawas et al. [11] indicated the first thing is to survey and investigate the area through the use certain engineering measurement, thinking, imagination and then dug a well to find out the water depth. Furthermore, Al-Abri demonstrated certain facts from building such a well. (1) provide an idea of water depth (2) from this distance to the cultivated area then can be planned (3) allow workers to work out the land graduate.

Shafts and tunnel construction. The tunnel has been constructed in such a way to tap/collect water from aquifer and bring it to the surface for distribution/uses. In addition, a series of vertical shafts were constructed or sunk from the surface of the tunnel away from the water-table or mother well. These provided several functions: removing all construction material from the tunnel; providing ventilation and later maintenance. Again, different view has been found with respect to these important components. Wilkinson [3] pointed out that first need to locate a suitable supply of water then a line of the qanat is surveyed and a tunnel is dug back to this source at a graduate. Al-Abri [10], as earlier indicated, demonstrated in more details how this carried out and can be summarized as follow. Once the engineering measurement is accomplished, they start to work on the land surface to form an open channel, starting from the extreme end of the water source near the village cultivated field working backward. While, operating this channel, two things needed to be considered; graduate level and the base of the channel must be levelled and strong to allow water to flow without causing erosion. Once this channel reached three meters, they start putting roofing until it disappears completely from the surface and become a tunnel. Then it come another operation, that is once they complete tunnel roofing and completely disappear to about five meters, then ventilation is needed to allow air inside the tunnel, thus they invented small wells or shaft that constructed along the tunnel with ten meters in between. Wilkinson [3] argued that as tunnels are dugs, shafts are sunk for ventilation and maintenance.

Branches (sa'ad). This is another important component, which is considered as a supportive element of the falaj system, locally known as saad (adits). It is also referred as smaller tunnels branching off from the main tunnel to collect additional water supply for the main tunnel. In other words, these are commonly found only in the falaj system, by which constructed in the middle or away from the mother well, to allow extra water from the surrounding oasis to flow into the tunnel. This can be seen clearly looking at the falaj daris located in the city of Nizwa map layout (see Fig. 1), as it consists of two branches which are exactly like an arm spread from the main tunnel, they run parallel to the tunnel constructed horizontally.

Summary. Based on the above reviewed, the following can be considered as most likely account of how the construction of the aflaj system in Oman was carried out.

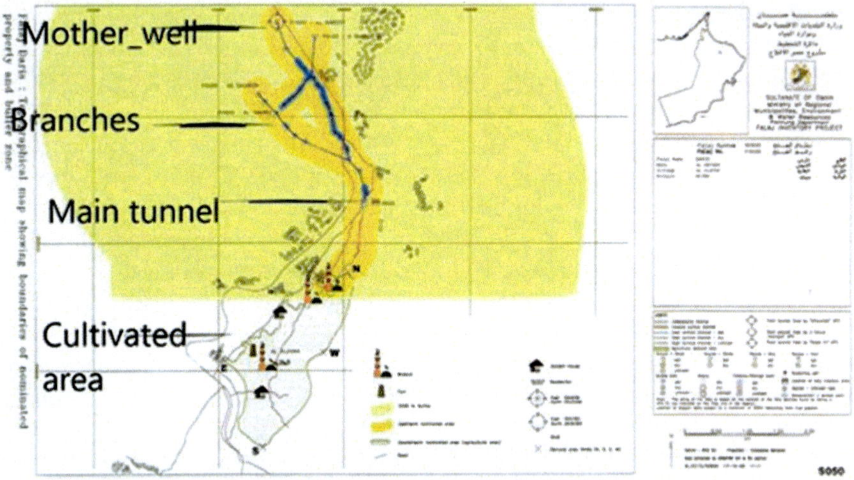

Fig. 1 Falaj daris in Nizwa layout showing mother-well and two branches

Fig. 2 Falaj al khatmeen tunnel in the village of Burkat-al-moz

The first task is to establish a depth of the water table around identified site. After this and knowing the point where the water needs to come out of the mountain, it is important to establish the slope by which the canal/tunnel needs to have in order to enter the aquifer at the right point. How this is conducted. First, the waters table is measured through digging a well at the point where the tunnel is expected to enter the aquifer to extract the water. This well is sometimes called the mother well. Second, once the gradient has been established; constructors start to work on the land surface by initiated an open channel, starting from the extreme end of the proposed system, near the village and the cultivated fields. Third, they need to work backwards. The next stage starts once the channel enters the hills and the points are identified where the water is anticipated to come out of the ground. At this point the channel slowly turns into a tunnel following the estimated slope. Once the channel in this way reaches a depth of three meters a roof is put over the channel and it then becomes a tunnel (see Fig. 2).

Once the tunnel is five meters underground, ventilation is needed to allow air inside the tunnel and also as a means to get rid of the dirt from the tunnel construction, and later on access for maintenance. Thus, smaller wells or shafts are built along the tunnel with a 10-m interval and connecting the tunnel with the surface of the ground.

According to Fig. 3 there are two newly constructed shift over approximately ten-meter interval. Note, authors own visit to the site found several of these shifts which were newly repaired using cement and stones (this is for falaj al-malki in the town of Izki). Starting from the second shift, where author and falaj administrated agent stand, to the first shift approximately ten-meter in distance.

4 Conclusion

A picture drawn with respect to the ancient aflaj construction method in Oman. First, the physical network consists of several linked components: mother well, shafts, tunnels and branches. These have been designated to convey/collect water from the aquifer to the surface naturally without the use of any mechanical devices. This is only achieved if the indicated components constructed over particular way with consideration of the topography of the area. However, several researchers who described the Omani aflaj followed an almost similar descriptive way which obtained from Iranian qanat construction processes. Because of this, confusion about a thorough understanding with respect to function of these components has been observed. As a result, different views have been wrote reflecting the overall operation aflaj system. Such confusion can be attributed to the fact that there might lack of close contact with local expertise, also the richness of the local words used in the construction was problematic due to lack of careful interpretation of the mother language of these terminologies. In conclusion aflaj system in Oman considered as sustainable system which the modern society can copy and incorporated with the water project construction process, as water is collected and transported in underground tunnels, and it flows continuously, day and night, throughout the year out put the use of any mechanical

Fig. 3 Fala al-malki cement protected with stone shift

devices. In addition, there is very unique feature when defining aflaj, as it provides a sense that falaj is not just to supply water, but it holds social organization structure. This provides a researchable opponent for future research.

References

1. Stanger, G.: The Hydrology of the Oman Mountains (1985)
2. Dewandel, B., et al.: A conceptual hydrogeological model of ophiolite hard-rock aquifers in Oman based on a multiscale and a multidisciplinary approach. Hydrogeol. J. **13**(5), 708–726 (2005)
3. Wilkinson, J.C.: Water and tribal settlement in South-east Arabia: a study of the Aflaj of Oman. Clarendon Press, Oxford [England] (1977)

4. Costa, P.W., Wilkinson, T.J.: Traditional methods of water supply. J. Oman Studies **9**, 35–40 (1987)
5. Sutton, S.: The falaj—a traditional co-operative system of water management. Waterlines **2**(3), 8–12(5) (1984)
6. Lightfoot, D.R.: The origin and diffusion of Qanats in Arabia: new evidence from the northern and southern Peninsula. Geogr. J. **166**(3), 215–226 (2000)
7. English, P.W.: The origin and spread of Qanats in the old world. Proc. Am. Philos. Soc. **112**(3), 170–181 (1968)
8. Cressey, G.B.: Qanats, Karez, and Foggaras. Geogr. Rev. **48**(1), 27–44 (1958)
9. Wulff, H.E.: The Qanats of Iran. Sci. Am. **218**(4), 94–107 (1968)
10. Al-Abri, B.B.S.: Explanations of some Aflaj of Oman. Golden Press, Ruwi, Sultanate of Oman (1980)
11. Al-Rawas, A.: Construction problems in traditional irrigation systems: a case study from the Afalj of Oman. In: Goosen, M. F. A., Shayya, W. H. (Eds.), Water Management Purification and Construction in Arid Climates. Technomic Press, US (2000)

Environmental Benefits of Sustainable Green Buildings (SGBs): A Project Life Cycle Perspective

Sharifah Akmam Syed Zakaria and Farah Hussain Saaed Ahram

Abstract Green building technology is one of several sustainability practices and technologies that have been developed with the goal of achieving environmental sustainability. The benefits of establishing sustainable green buildings (SGBs) are numerous and mirror various parts of the building's life cycle. One of the most significant variables that benefit from the development of SGBs is the environmental factor. There is various research in the area of SGBs but less focus is given on the environmental benefits of SGBs during their life cycle stages. This study intends to identify the environmental benefits of SGBs throughout their life cycle, based on the results of the in-depth semi-structured interviews with 25 professionals in the field of the built environment of Jordan in order to accelerate the adoption of this new building technology. The result of this research was obtained by the content analysis of data using Atlas ti software. The outcomes of this study shows that SGBs have several environmental advantages throughout their life cycle. The main environmental benefit in the design stage was related to "Innovative environmental solutions," while the major environmental benefit in the construction stage was regarding "Conservation of resources and environment," followed by the operational stage that was linked to "Decrease harmful environmental impacts of buildings". Lastly, the most significant environmental benefit in the maintenance stage was concerning "Preserving the environment." The importance of this research rests in the significant addition it will make to the literature in this field, and the relevance of green building technology adoption as an important topic in the construction industry.

Keywords Green buildings · Environmental benefits · Life cycle · Built environment

S. A. S. Zakaria (✉) · F. H. S. Ahram
School of Civil Engineering, Universiti Sains Malaysia, Engineering Campus, 14300 Nibong Tebal, Penang, Malaysia
e-mail: akmam@usm.my

N. Sabtu (ed.), *Proceedings of AWAM International Conference on Civil Engineering 2022 - Volume 3*, Lecture Notes in Civil Engineering 386,
https://doi.org/10.1007/978-981-99-6026-2_19

1 Introduction

The potential or aptitude of something to maintain or sustain established values of human well-being, social justice, and environmental integrity across time can be defined as sustainability [1, 2]. Green buildings are considered to synonymous with environmentally friendly architecture because they aim to lessen the whole impact of the built environment on human health and the natural environment [3]. The advantages of implementing sustainable green building (SGBs) measure vary and reflect many aspects during the life cycle of green building projects.

These advantages involve material savings from reuse and recycling, waste conversion into valuable forms, safe working conditions, cost savings, trash treatment, less usage of transport, reduced energy consumption during the manufacturing process, having a higher quality construction, cheaper life cycle costs, and a safer construction procedure are all advantages for adopting of the SGBs technology [4]. From a technological point of view [5], it is discovered that green building pieces of work are basically not similar to those that are traditional in nature as it specifically calls for the use of tools and approaches that are related to building construction activities with the focus to attain sustainability. Besides, it also calls for a comprehensive act of documenting, as well as monitoring, especially if the project's aim is to achieve an environmental certificate.

In order to reduce the possibilities of hazards and increase the project completion possibilities at affordable prices, the specific aspects of green construction include changes to the practice of project management in dynamic ways. Most of the changes indicate an urgent call for on-site selection, interdisciplinary teamwork, construction methods, construction network, and subsystems early during the project period. Despite a limited number of relevant studies and based on an exhaustive evaluation of the current eorks of literature, it is important to investigate the significant environmental benefits of adopting SGB technology throughout the life cycle of green building projects. As a result, this topic has been identified as an area that needs to be investigated in a more comprehensive manner. In this study, the benefits of SGBs are presented based on the life cycle of green building projects in Jordan according to the environmental pillar of sustainability. Therefore, this research will answer this question:

What are the most significant environmental benefits achieved from SGBs during the life cycle stages of green building projects in Jordan?

2 Literature Review

2.1 Benefits Achieved from SGBs During the Design Stage

Innovative environmental solutions. Green building technology adoption, as it is termed in the construction industry, allows designers to create high-performance buildings, utilizing the most up-to-date design technology and knowledge to assure that operating energy requirements are minimized or zero [6]. The sort of response and feedback received during the planning, decision-making, and design stages of a green building project will influence the project's actual performance, which will be helpful in closing the gap between actual and expected performance or resolving any potential challenges [7].

As green building is considered as a new technology, it includes creative features which can also increase the performance of building project, some team members believe that the innovation aspect is an influence or crucial factor in the decision-making of green building technology [8]. The necessity of considering sustainability aspects at the design stage of building projects could fulfill the requirement of long-term solutions that provide well-being while reducing natural resource demands such as land usage, wildlife, water, atmosphere, and energy. If a project is well-designed and sustainable requirements are incorporated early in the process, the likelihood of reducing negative consequences is increased, and the cost of implementing green technology can be decreased [9].

2.2 Benefits Achieved from SGBs During the Construction Stage

Preserving the environment. Fundamentally, sustainable building construction has numerous advantages for the environment, economy, and society [10]. Furthermore, the real estate and construction industries are major contributors to the global economy and labor market [11]. Sustainable building, on the other hand, is used by both the government and society to assist the accomplishment of sustainability goals in building projects because they are committed to conserving the environment, increasing life quality, and fostering community sustainability [12]. Thus, the construction phase of building projects generates the physical appearance that satisfies the concept, enables the realization, and implements the planning aspects of green technology in the design stage [13].

Therefore, green building materials, which mostly consist of recycled and reused materials, sustainable product production, locally available materials, and the use of green resources will reduce the effect of environmental damage and generate healthy green construction that can be sustainable for both the building occupants and the environment. Accordingly, building materials with the least amount of environmental impact might be chosen for their utility in a country's long-term sustainable growth.

Environmental-friendly construction products, as well as the recycling and reuse of wastes and the production of green construction materials, are great ways to improve a building's function while reducing its environmental and human health impact [14].

2.3 Benefits Achieved from SGBs During the Operation Stage

Environmental sustainability. Building project managers and owners can implement a green building strategy to minimize energy consumption, increase environmental performance, and lessen the environmental effect of their activities. As a result, the facility of green building can lower its operational costs, boost occupant productivity, and reduce health concerns while also being ecologically friendly [15, 16].

Sustainable technologies that concentrate on enhancing operational efficiencies, decreasing waste, and improving the structure's environmental quality are considered green building components [12]. The advantages of green buildings are straightforward to comprehend. For example, building owners and managers can gain financial benefits while taking care of the environment and society by following green design standards. Firstly, green buildings, offer lower operating expenses than traditionally constructed buildings. Additionally, sustainable design tends to decrease wide-range environmental implications like water use, ozone depletion, and the use of raw materials. Furthermore, considering and incorporating biodiversity aspects into the built environment increases the planet's ability to mitigate climate change, increase air quality, flood prevention, and people's overall health and welfare [17]. Finally, workers who work in well-lit, safe, and pleasant workplaces are more productive and satisfied, which are critical in the present labor market [18].

2.4 Benefits Achieved from SGBs During the Maintenance Stage

Preserving environmental sustainability. Maintenance is described as a collection of technical and organizational procedures aimed at keeping or restoring an item to a state where it can fulfill its intended purpose. As a result, the primary goal of any upkeep system is to achieve maximum system availability at the lowest possible cost while maintaining excellent service quality in order to improve building services [19].

Green maintenance has created a way to make maintenance more ecologically friendly by eliminating all relevant wastes, implementing energy efficiency, reducing emissions, and reducing water usage. On the other hand, the tendency of to employ renewable materials also serves the selection of new materials for sustainable maintenance [20].

3 Research Methodology

The qualitative technique was chosen for this study because it appears to be the most appropriate way to address the study issues and their measurements. Qualitative research is distinguished by its goals, which are concerned with comprehending some aspects of life, and by its techniques, which in general produce words rather than numbers as data for analysis, with the goal of measuring something. These strategies are designed to answer inquiries regarding a phenomenon's "how,' "what," or "why," rather than "how much" or "how many" [21]. Thus, the major goal of this strategy is to collect detailed information on the current condition of the environmental benefits achieved by SGBs during the life cycle stages.

By using semi-structured interviews, the researcher will be able to obtain as much detail and in-depth information on the subject as possible from the individual interviewed [22]. Respondents will have the best chance to react to the questions, which are limited in number and well-structured to elicit ideas and opinions from their topic expertise. At the same time, the researcher can dig further into the participants' responses, looking for more information on any difficulties that arise throughout the interview [23].

Because doing face-to-face interviews proved challenging for certain participants because of the COVID-19 pandemic, some interviews were conducted virtually using the online meeting program "Zoom." On the other hand, various face-to-face interviews were used with other interviewees whose circumstances were suited for face-to-face interviews while keeping the safety recommendation in mind (e.g. wearing the face mask and gloves and leaving the safety social distance) [24]. The interviews lasted somewhere between 40 and 50 min. Furthermore, the interviews were held in Arabic. In addition, the interview sessions were taped in order to ensure accurate results during the analysis stage's coding method.

3.1 Population Definition

Green building professionals, architects, project managers, civil engineers, contractors, academia, as well as end-users of green buildings in Jordan were selected for the semi-structured in-depth interviews. A minimal amount of sample is usually required for qualitative techniques [25]. The method used a purposive sampling technique to identify research participants who are proficient and well-informed with the study issues [26]. The interviews covered different samples of stakeholders because each participant has the experience or full interaction in a particular stage of SGB life cycle. The summary of the stakeholders as the research participants in this study is shown in Table 1.

Table 1 Data summary of the stakeholders in this study

Stakeholders' description	Codes of stakeholders	Numbers	Years of experience in the field
Consultant	C1, C2, C3, C4	4	At least 3 years of experience
Architect	A1, A2, A3, A4, A5, A6	6	At least 4 years of experience
Project manager	P1, P2, P3	3	At least 3 years of experience
Civil engineer	Ci1, Ci2, Ci3	3	At least 3 years of experience
Contractor	Co1, Co2, Co3	3	At least 3 years of experience
Academia	Ac1, Ac2, Ac3	3	At least 6 years of experience
End user	E1, E2, E3	3	At least 2 years of using the SGBs
Total number of stakeholders		25	

4 Findings and Discussion

This section presents the findings of data analysis on the benefits of SGBs throughout their life cycles in accordance with the basic qualities required for environmental sustainability, by conducting interviews with 25 professionals in the field of the built environment.

Qualitative analysis techniques particularly in-depth interviews can be used to include content analysis, which permits systematic coding of data by grouping information into categories and to identify patterns that cannot be found by merely listening to tapes or reading texts, and analyzing speech [27]. Content analysis has been defined as a valuable qualitative analysis technique that may create a summary or overview of a data collection in a systematic and comprehensive manner [28]. Coding and other data analysis procedures were carried out using Atlas ti software, which allows for the systematic organization, storage, and modification of all data, subjects, categories, findings, and research notes [29].

The analysis of the interviews revealed that each stage of the life cycle in SGB projects has a main environmental benefit achieved from these buildings during their life cycle. Table 2 shows the summary of qualitative data analysis and the results of the interviews.

Based on the results if this study, it is discovered that, during the life cycle stages (design, construction, operation, and maintenance) of SGB projects contributed to the accomplishment of environmental benefits. Discussions on the results of the qualitative data analysis are as follows:

- The attainment of "Innovative environmental solutions" benefits, was found as the most important environmental benefit achieved in the design stage of SGBs. The results demonstrated that this benefit was repeated 18 times by different interviewees throughout the interview sessions. The participants pointed out that the green and sustainable design of buildings contributes to creating creative

Table 2 The summary of qualitative data analysis and results of the interviews

Life cycle stage	Codes related to the environmental benefits	Number of code repetition	Some examples of quotations from interviews
Design stage	Innovation environmental solutions	18	"Finding solutions to the problems that the designer encounters during the design lead to obtaining various innovations in this field, this contributes to its development…" (Architect.A3)
Construction stage	Conservation of resources and environment	15	"…Building techniques for green buildings contribute to preserving the environment and reducing environmental pollutants. For example, the spread of dust and dirt is reduced through the use of a mesh with fine holes…" (Consultant.Co2)
Operation stage	Decrease harmful environmental impacts of buildings	21	"Green buildings contribute to preserving the environment and reducing harmful emissions such as carbon dioxide…" (Project manager. P1)
Maintenance stage	Preserving environmental sustainability	14	"…In general, GBs do not need substantial maintenance due to breakdowns and destruction, and that maintains their efficacy in achieving cost savings…" (End User, E1)

solutions and exploiting natural sources of energy in the best way through design solutions related to the orientation of the building with the places and spaces of windows. The participants also clarified the relationship between creativity in sustainable green design and its reflection on environmental practices in general so that it contributes to preserving the environment and renewable resources and reducing climate change.

- The results also found that the most important benefit achieved by SGBs during the construction stage was considered as "Conservation of resources and environment". The repetition of this code was 15 done times by different interviewees. The participants pointed out that the sustainable technologies adopted during the construction phase of green buildings contribute to reducing harmful emissions from construction, reducing harmful waste to the environment, and getting rid of these wastes resulting from construction activities in environmentally friendly ways. Moreover, green building techniques contribute to preserving the environment of construction projects and their surroundings limiting the spread of dust and dirt as well. In addition, the utilization of green resources, recycled and reused materials, sustainable products, and locally accessible materials, will also contribute to preserving the environment at this stage. Overall, the result of green technology adoption at this stage is that it contributes towards achieving sustainability and the trend towards environmentally friendly cities, and will also contribute to handling the issues of climate change and global warming.

- "Decrease harmful environmental impacts of buildings", was the highest code repeated by the interviewees, with 21 times repetition. This shows that the environmental benefits were also attained from SGBs during the operation stage. Construction stakeholders made it clear that the systems used in sustainable green buildings, which are based on preserving non-renewable resources, exploiting clean alternative energy, reducing emissions, and others, will invariably contribute to preserving the environment and resources, spreading a culture of sustainability, and reducing environmental problems such as waste disposal and pollution during the operation phase of SGBs.
- Finally, the environmental benefit achieved from SGBs during the maintenance stage was repeated 14 times as a code related to "Preserving the environmental sustainability". Environmentally friendly green buildings, according to the participants, rarely require extensive maintenance due to malfunctions or defects, as they only require routine maintenance procedures on a regular basis, making them highly efficient for a long time and achieving the goal for which they were designed, namely, environmental sustainability and the use of clean renewable energy.

Overall, as there is increasing demand for building comfort and promotions to minimize energy use, there is every need to strategize new ways of energy conservation and renewable energy generation to be used in making the built environment more comfortable and to achieve environmental sustainability. As a result, sustainable buildings have become the key guiding principle for the practice of construction and urban planning, for energy resource efficiency, occupant comfort, and restriction of environmental effects [11, 30]. A life-cycle approach to buildings is being encouraged, which highlights previously overlooked aspects of building management [31, 32]. The choice of a building's life cycle is characterized by various project aims, and the requirements of many stakeholders (client, designer, contractor, etc.) who are all involved in the life cycle of building projects [33].

5 Conclusion

Conclusively, the results of this study showed that SGBs throughout their life cycle have a number of environmental benefits. Thus, it is important to emphasize on the advantages or benefits of adopting green building technology as a topic for further research and studies in the built environment. The findings also revealed that, the most significant environmental benefit in the design stage was "Innovative environmental solutions", the most significant environmental benefit in the construction stage was "Conservation of resources and environment", the most significant environmental benefit in the operational stage was "Decrease harmful environmental impacts of buildings", and the most significant environmental benefit in the maintenance stage was "Environmental sustainability". As the environmental component of sustainability was only considered in this SGB study, it is recommended that other dimensions of sustainability, such as economic and social factors be included for future investigations in the adoption of green technology in building projects.

Acknowledgements Acknowledgement to Ministry of Higher Education Malaysia for Malaysia Research University Network—Long Term Research Grant Scheme (MRUN-LGRS) under the program title: River Basin Approach for Integrated Risk Assessment and Reduction of Water-Related Hazards (Program grant number: R.J.130000.7851.4L896). The authors also acknowledge the support from the School of Civil Engineering, Universiti Sains Malaysia.

References

1. Roostaie, S., Nawari, N., Kibert, C.J.: Sustainability and resilience: a review of definitions, relationships, and their integration into a combined building assessment framework. Build. Environ. **154**, 132–144 (2019)
2. Kaveh, A., Dadras Eslamlou, A.: Optimization of building components with sustainability aspects in BIM environment. Stud. Comput. Intell. **900**, 289–308 (2020). https://doi.org/10.1007/978-3-030-45473-9_13
3. Alawneh, R., Mohamed Ghazali, F.E., Ali, H., Asif, M.: Assessing the contribution of water and energy efficiency in green buildings to achieve United Nations Sustainable Development Goals in Jordan. Build. Environ. **146**, 119–132 (2018). https://doi.org/10.1016/j.buildenv.2018.09.043
4. Baloi, D.: Sustainable construction: challenges and opportunities. Assoc. Res. Constr. Manag. **1**(September), 289–97 (2003) [Online]. Available: www.csir.co.za/akani
5. Robichaud, L.B., Anantatmula, V.S.: Greening project management practices for sustainable construction. J. Manag. Eng. **27**(1), 48–57 (2011). https://doi.org/10.1061/(asce)me.1943-5479.0000030
6. Zakaria, S.A.S.: Behavioral economics factors in the decision-making of green building technology for sustainable infrastructure governance. Behav. Econ., 67
7. Zakaria, S.A.S., Brewer, G., Gajendran, T.: Understanding decision making: future direction for industrialised building system adoption in the Malaysian construction industry (2011)
8. Akmam Syed Zakaria, S., Gajendran, T., Rose, T., Brewer, G.: "Contextual, structural and behavioural factors influencing the adoption of industrialised building systems: a review. Archit. Eng. Des. Manag. **14**(1–2), 3–26 (2018)
9. Bragança, L., Vieira, S.M., Andrade, J.B.: Early stage design decisions: the way to achieve sustainable buildings at lower costs. Sci. World J. **2014** (2014). https://doi.org/10.1155/2014/365364
10. Djokoto, S.D., Dadzie, J., Ohemeng-Ababio, E.: Barriers to sustainable construction in the Ghanaian construction industry: consultants perspectives. J. Sustain. Dev. **7**(1), 134–143 (2014). https://doi.org/10.5539/jsd.v7n1p134
11. Goubran, S.: On the role of construction in achieving the SDGs. J. Sustain. Res. **1**(2) (2019). https://doi.org/10.20900/jsr20190020
12. Nelms, C., Russell, A.D., Lence, B.J.: Assessing the performance of sustainable technologies for building projects. Can. J. Civ. Eng. **32**(1), 114–128 (2005). https://doi.org/10.1139/l04-102
13. Akmam, S., Zakaria, S., Brewer, G.: Understanding decision making: future direction for industrialised building system adoption in the Malaysian construction industry. Construction (2002)

14. Iqbal, A., Swapnil, F.J.: Sustainable green materials for building construction
15. Kibert, C.J.: Green buildings: an overview of progress. J. L. Use Environ. Law **19**(2), 491–502 (2004)
16. Hui, S.C.M.: HVAC design and operation for green buildings. Building (June), 11–13 (2001) [Online]. Available: https://www.researchgate.net/publication/228606734_HVAC_design_and_operation_for_green_buildings
17. Opoku, A.: Biodiversity and the built environment: Implications for the Sustainable Development Goals (SDGs). Resour. Conserv. Recycl. **141**, 1–7 (2019)
18. von Paumgartten, P.: Building green. Consult. Specif. Eng., p. 12 (2002)
19. Aldairi, J., Khan, M.K., Munive-Hernandez, J.E.: Knowledge-based Lean Six Sigma maintenance system for sustainable buildings. Int. J. Lean Six Sigma **8**(1), 109–130 (2017). https://doi.org/10.1108/IJLSS-09-2015-0035
20. Ajukumar, V.N., Gandhi, O.P.: Evaluation of green maintenance initiatives in design and development of mechanical systems using an integrated approach. J. Clean. Prod. **51**, 34–46 (2013)
21. McCusker, K., Gunaydin, S.: Research using qualitative, quantitative or mixed methods and choice based on the research. Perfus. (United Kingdom) **30**(7), 537–542 (2015). https://doi.org/10.1177/0267659114559116
22. Lewis, S.: Qualitative inquiry and research design: choosing among five approaches. **16**(4) (2015)
23. Salmons, J.: Qualitative Online Interviews: Strategies, Design, and Skills. Sage Publications (2014)
24. Chiesa-Estomba C.M. et al.: Systematic review of international guidelines for tracheostomy in COVID-19 patients. Oral Oncol. **108**(May), 104844 (2020). https://doi.org/10.1016/j.oraloncology.2020.104844
25. Marshall, B., Cardon, P., Poddar, A., Fontenot, R.: Does sample size matter in qualitative research? A review of qualitative interviews in is research. J. Comput. Inf. Syst. **54**(1), 11–22 (2013). https://doi.org/10.1080/08874417.2013.11645667
26. Rai, N., Thapa, B.: A study on purposive sampling method in research. Kathmandu Sch. Law, pp. 1–12 (2015) [Online]. Available: http://www.academia.edu/28087388
27. Stemler, S.E.: Content analysis. In: Emerging Trends in the Social and Behavioral Sciences: An Interdisciplinary, Searchable, and Linkable Resource, pp. 1–14 (2015)
28. Roberts, C.W.: Content analysis. In: International Encyclopedia of the Social & Behavioral Sciences, 2nd ed., no. 2008, pp. 769–773 (2015).https://doi.org/10.1016/B978-0-08-097086-8.44010-9
29. Hwang, S.: Utilizing qualitative data analysis software: a review of Atlas.ti. Soc. Sci. Comput. Rev. **26**(4), 519–527 (2008). https://doi.org/10.1177/0894439307312485
30. Mateus, R., Bragança, L.: Sustainability assessment and rating of buildings: developing the methodology SBToolPT–H. Build. Environ. **46**(10), 1962–1971 (2011)
31. Mora, R., Bitsuamlak, G., Horvat, M.: Integrated life-cycle design of building enclosures. Build. Environ. **46**(7), 1469–1479 (2011). https://doi.org/10.1016/j.buildenv.2011.01.018
32. Nowogońska, B.: The life cycle of a building as a technical object. Period. Polytech. Civ. Eng. **60**(3), 331–336 (2016). https://doi.org/10.3311/PPci.7932
33. Guo, H.L., Li, H., Skitmore, M.: Life-cycle management of construction projects based on virtual prototyping technology. J. Manag. Eng. **26**(1), 41–47 (2010)

Light-Rail Transit (LRT) in Penang: LRT's Route Selection and the Impact Toward Neighborhood in Penang

Siti Nadia Kamarudin, Ernawati Mustafa Kamal, and Muneera Esa

Abstract Penang now is preparing and getting ready to achieve the state development vision under Penang's 2030 agenda, *A Family-Focused Green and Smart State that Inspires the Nation.* Penang's mobility and connectivity infrastructure are to undergo a major overhaul through public and private sector investments. One of the focuses is to strengthen mobility, connectivity, and digital infrastructure by improving public transportation infrastructure. This study was carried out to investigate the criteria that have been considered in selecting Penang Light-Rail Transit (LRT) route and the impacts on the neighborhood. An in-depth interview session with professional experts who have been involved directly with the project has been carried out to shed some light on this issue which gained a popular topic of discussion ever since it was announced by the Penang State Government. The findings show that four criteria were taken into consideration during the routing selection process: accessibility, finances, environment, and population. A point to highlight, the safety aspect is the least to be considered in Penang Light-Rail Transit (LRT) routing. But when it comes to impact, the proposed LRT routes become one of the environmental pollution issues such as noise and vibration. Even though there are drawbacks related to the proposed infrastructure causing major disruptions in the state, however, the LRT corridor's potential benefits may offset these disadvantages. The findings from this study will definitely stimulate further discussion and debate.

Keywords LRT · Penang · Route selection criteria · Environmental impact

S. N. Kamarudin
Aspen Group, Aspen House, 300 Macalister Road, 10450 Penang, Malaysia

E. M. Kamal · M. Esa (✉)
School of Housing, Building and Planning, Universiti Sains Malaysia, 11800 Penang, Malaysia
e-mail: muneera_esa@usm.my

N. Sabtu (ed.), *Proceedings of AWAM International Conference on Civil Engineering 2022 - Volume 3*, Lecture Notes in Civil Engineering 386,
https://doi.org/10.1007/978-981-99-6026-2_20

1 Introduction

Light-Rail Transit (LRT) is one of the good alternatives for public transport especially in congested areas like Penang Island. This alternative helps solve traffic issues, preserves safety, reliability, efficiency, and affordability for decent people as well as reduces air pollution. In other words, this kind of transportation system helps to improve the quality of life and allows individuals, institutions, corporations, regions, and nations to interact and assume economic, social, cultural, or political activities [1]. However, an appropriate route selection is crucial to consider because of a notable impact on the Penangites economically, environmentally, and socially. The development of a railway project can be considered as a complex process that requires a comprehensive decision to set the network design to satisfy the main objectives, list of criteria required for the project, and integrating with existing development [2]. The network design refers to the route selection and the placement of the train station.

In reaching Penang 2030 vision, Penang State Government is now actively preparing themselves to achieve A Family-Focused Green and Smart State that Inspires the Nation [3]. There are four focuses highlighted in the vision namely (i) increase livability to enhance quality of life, (ii) upgrade the economy to raise household income, (iii) empowers people to strengthen civic participation, and (iv) invest in the built environment to improve resilience. Referring to the fourth focus, the main concern is to strengthen the existing mobility, connectivity, and digital infrastructure by improving public transportation in the infrastructure sector. This includes the light-rail train and high-speed ferry services. The key initiatives mentioned are to prioritize the construction of the light rail and initiate both high-speed ferry and water taxis. This is because they are targeting the usage of public transport from 20% of 8 trips to 50% and reachable directly to Penang International Airport.

Penang Transportation Master Plan (PTMP) is a strategic plan for transportation set by the Penang State Government. As mentioned in PTMP, the transportation system plan comprises Light-Rail Transit (LRT), bus rapid transit (BRT), monorail, tram, ferry, taxi, e-hailing, and water taxi [4]. The execution of PTMP shows the Penang State Government's commitment to ensuring a better public transportation service, as well as solving the congestion issue, which is expected to become worse due to massive development in Penang. It is included the introduction of the LRT service in Penang. This effort was well received by many people, especially Penangites. A petition objecting Penang LRT project with a total of 23,218 voters shows that only 2.23%, which is equivalent to 540 people who voted to object to the effort [5].

In 2013, the Penang State Government approved the allocation budget of RM27 billion under PTMP to mitigate the state's transport and traffic issues [6]. An analysis of the infrastructure development discovers that many investments in previous projects led to low return values and also will negatively influence the environment. This analysis includes the execution stage and operating stage [2]. Hence, as an approval authority and advisor to the Transport Minister, Land Public Transport

Agency (APAD) plays a vital role in reviewing the proposed Penang LRT route selection to ensure the project meets the main objectives and offers a high return value. This includes the suitability of locations to coordinate train stations to resolve the traffic congestion issues at main places as well as possess minimum impact toward the environment, social, and economy.

2 Background of the Issue

Setting a route alignment is an intricate process where many criteria and concerns need to be carefully considered. Since the announcement of the Penang LRT project approval, many issues regarding the project have arisen. This includes the objection of proposed route selection that triggers social stigma, especially among Penangites. On June 28, 2019, a demonstration protest against the proposed LRT alignment which will cross over the Sungai Nibong Besar Mosque or its adjoining Muslim cemetery was said to be insensitive [4]. On the other hand, many Penangites expressed their frustration and demanded a solution to traffic congestion issues. As reported by New Straits Times dated February 20, 2020, Penangite publicly issued a letter expressing her frustration and demanded to speed up the LRT project instead of keeping revising the PTMP [5]. Therefore, a deep understanding of the project is essential to get an overview of why the whole process is timely and many revisions on project planning are required. Since no previous research focuses on this matter, this research is expected to give insight into the criteria for selecting the route for the Penang Light-Rail Transit (LRT) network and the impact on the neighborhood.

3 Methods

This research aims to investigate the criteria for selecting the Penang LRT route and its impact on the neighborhood. To achieve the aim, a qualitative approach has been used in this research to obtain the information. Since the information obtained is expected to be insightful to respond to the issues arising in the Penang LRT project, the criteria of informants were chosen based on their experience, knowledge, and direct participation in the project. This method is similar to research by Yusoff et al. (cited in [7]) to obtain information regarding the LRT project in Klang Valley. The selection of informants also complied with the guidelines as proposed by Steinberg (cited in [8]). In a project, stakeholders are the key players who are directly involved in the decision-making process. Representatives from each stakeholder who met the criteria were invited for an interview session. The main purpose was to collect a different perspective on the matter from each stakeholders. The informants were chosen among experts who have working experience of 10 years and above and are involved in this project to share their point of view on the Penang LRT route selection. The stakeholders involved are:

(a) Client: Bahagian Perancang Ekonomi Negeri (BPEN), Penang State Government
(b) Planning Consultants: AJC Planning Consultants Sdn Bhd (AJC)–PTMP
(c) Project Delivery Partners (PDP): SRS Consortium Sdn Bhd
(d) Authority: Land Public Transport Agency (APAD), Ministry of Transport
(e) End Users: Public/Citizen/Politician/Non-Government Organization (NGO).

The interview was a structured interview with a focus on investigating the criteria for selecting the Penang LRT route and the impact of route selection on the environment, economy, and social perspective. A questionnaire was given to each of the informants to rank the criteria based on the Penang LRT project. Justifications and opinions from each informant were asked and recorded as additional information to this research.

4 Results and Discussion

Due to private and confidential restrictions, the client needs to decline the invitation as well as the authority. On the other hand, three parties accepted the invitation: a planning consultant, project delivery partner, and two representatives on behalf of the end-user to share their point of view regarding the Penang LRT project. The background information of respondents present in Table 1 as follows:

4.1 Criteria in Selecting the Route for Penang LRT Network

According to criteria analysis in route selection for transportation routes, there are five main criteria to be considered in deciding LRT route selection: accessibility, safety, environment, economy, and population. To investigate the criteria related to the Penang LRT project, all informants were asked to rate the main criteria based on

Table 1 Background of informants

Informant	Representing	Profession/ Designation	Years of experience	Background
A	Project delivery partner	Project manager	15	Civil engineering and environmental
B	End-user	Non-government organization (NGO)	35	Banker and government officer
C	End-user	Politician	10	Built environment
D	Planning consultant	Project manager	19	Environmental science

a 5-point Likert scale which most related to the Penang LRT Project as well as their opinion. The results of analysis shows in Table 2.

According to informants A and D who are involved directly during the planning stage, all related factors have been considered to ensure the project complies with the requirements as setting up by the local authorities. This includes Agensi Pengangkutan Awam Darat (APAD), Local Authority; Majlis Bandaraya Pulau Pinang (MBPP), Department of Environment (DOE), Jabatan Kerja Raya (JKR), and other relevant authorities. In deciding the LRT route, railway multi-criteria analysis has been used as a guideline throughout the process. Both informant A and D explained the main concerns in deciding the routing for Penang LRT. First, the LRT shall connect these six main places: (1) George Town (Komtar), (2) Universiti Sains Malaysia (USM), (3) Sungai Nibong Bus Terminal, (4) Subterranean Penang International Convention and Exhibition Center (SPICE), (5) Free Industrial Trade Zone (FTZ), (6) Penang International Airport. Second, the station must easily be accessible

Table 2 Percentage of scoring bands on the 5-point Likert scale

	NI (%)	SNI (%)	N (%)	SI (%)	VI (%)
Accessibility					
Travel time	0	0	0	0	100
Mobility to job center	0	0	0	0	100
Mobility to public infrastructure	0	0	0	0	100
Mobility to high-density residents	0	0	0	0	100
Transit oriented development	0	0	0	25	75
Land use	0	0	50	0	50
Safety					
Black points number	0	50	25	25	0
Intersections numbers	0	0	25	50	25
Alignment	0	0	25	25	50
Environment					
Noise vibration	0	0	0	25	75
Aesthetic aspect	0	50	0	0	50
Impact on sensitive areas	0	0	0	0	100
Financial					
Path length	0	25	0	25	50
Construction cost	0	0	0	0	100
Land acquisition	0	0	0	0	100
Population					
Population density	0	0	25	0	75
Traffic count/traffic congestion	0	0	0	25	75

NI Not important, *SNI* Somewhat not important, *N* Neutral, *SI* Somewhat important, *VI* Very important for all people. Third, the project can integrate with existing transport systems and development. Fourth, the project should impose minimum social and environmental impacts. Fifth requires minimum land acquisition. Sixth, the constructability. Lastly, the LRT should offer optimum journey speed. On top of that, both informants also highlighted that each of the main concerns must be analyzed on what the project can offer based on four criteria: accessibility, environment, economy, and population.

According to informants A and D who are involved directly during the planning stage, all related factors have been considered to ensure the project complies with the requirements as setting up by the local authorities. This includes Agensi Pengangkutan Awam Darat (APAD), Local Authority; Majlis Bandaraya Pulau Pinang (MBPP), Department of Environment (DOE), Jabatan Kerja Raya (JKR), and other relevant authorities. In deciding the LRT route, railway multi-criteria analysis has been used as a guideline throughout the process. Both informant A and D explained the main concerns in deciding the routing for Penang LRT. First, the LRT shall connect these six main places: (1) George Town (Komtar), (2) Universiti Sains Malaysia (USM), (3) Sungai Nibong Bus Terminal, (4) Subterranean Penang International Convention and Exhibition Center (SPICE), (5) Free Industrial Trade Zone (FTZ), (6) Penang International Airport. Second, the station must easily be accessible for all people. Third, the project can integrate with existing transport systems and development. Fourth, the project should impose minimum social and environmental impacts. Fifth requires minimum land acquisition. Sixth, the constructability. Lastly, the LRT should offer optimum journey speed. On top of that, both informants also highlighted that each of the main concerns must be analyzed on what the project can offer based on four criteria: accessibility, environment, economy, and population.

According to Informant A, there are ten routing options for Penang LRT were outlined. There are: Option A, Option B, Option B1, Option B1a, Option B2, Option B3, Option B4, Option B4a, Option C, and Option C1. Each of the abovementioned options has been evaluated carefully to identify the best option based on the criteria met. **During the Railway Scheme stage, five options (Option A, Option B2, Option B3, Option C, and Option C1) were eliminated as they did not fulfill the basic criteria** for following reasons (See Appendix: Alignment Options):

(a) Option A: Did not link the majority of main places (serve only Komtar and Penang International Airport), had low ridership and high land acquisition cost (Jalan Jelutong area)
(b) Option B2: Had low ridership and did not link USM
(c) Option B3: Did not link to FTZ
(d) Option C: Did not link to USM and Sg. Nibong Bus Terminal
(e) Option C1: Had the lowest ridership among all options and did not link three of the main places (USM, Sg. Nibong Bus Terminal, and FTZ).

Among the four short-listed options (Option B, Option B1, Option B1a, Option B4), Option B4 was selected as the preferred option for the following reasons:

(a) Linking all main places including SPICE and FTZ
(b) Relatively high ridership

(c) Lower social impact and lower construction risk compared to Option B.

After the Railway Scheme submission, the preferred Option B4 was further reviewed to address concerns on noise and vibration, public safety, and traffic congestion at the short stretch of road in front of SJK (C) Kwang Hwa, SRJK (C) Shih Chung, and SMK Sg Nibong. Two additional options were identified, namely Option B4a (through Queensbay Mall) and Option B4b (through Bukit Gambir). Option B4 was determined as the most suitable option as Option B4a and Option B4b would require extensive land acquisition and incur significant social impacts. Furthermore, an additional 2 km alignment length for Option B4a would require additional cost on the construction, operating, and maintenance.

In April 2017, the Option B4 alignment near the Komtar area was further improved to avoid disturbing the Prangin Market and sluice gate on Prangin Canal. The Komtar station was shifted to an open area between Jalan Magazine and Lebuh Tek Soon, while the terminus of the alignment was shifted to Jalan Magazine. The alignment at the Penang International Airport has also been improvised to allow the integration of LRT with existing transport as well as reduce the need for land acquisition.

As an Environment Consultant of the project, Informant D further explained that Penang Airport Station, which was shifted closer to the Penang International Airport terminal. Alignment was shifted further away from Wesley Methodist International School and SJK (C) Kwang Hwa, which are noise-sensitive receptors. The alignment which previously traversed through Kg Baruh, Kg Padang, Kg Seronok, and Kg Binjai was realigned to travel along with the Sg. Tiram river corridor, south of the Penang International Airport. SPICE Station was relocated to Jalan Tun Dr. Awang, between Taman Tunas Muda and Taman Sunway Tunas, as the alignment was rerouted from Persiaran Mahsuri to Jalan Tun Dr. Awang. The alignment at IPG Kampus Pulau Pinang and Sri Poomarathammal Temple was revised to traverse behind the temple instead of going in front of it, and to avoid cutting through the IPG campus and sports field. Alignment in the vicinity of Masjid Jamek Sg Nibong Besar was revised to traverse straight along with Jalan Sultan Azlan Shah instead of turning into Kg Selamat and Taman Lip Sin. The alignment and Sg Nibong Station were relocated to Jalan Kampung Relau instead of being located along with Jalan Sultan Azlan Shah.

There are two underground options have been considered for the construction of the Komtar station and part of the alignment between Komtar and East Jelutong using two different methods, which were (i) tunnel boring machine (TBM) method and (ii) cut-and-cover method. However, both tunneling methods would pose risks and constraints. The TBM method will disrupt heritage buildings in the Komtar area and significantly incur costs to elevate the structure option. For the cut-and-cover method, the risk is very high as it will affect the environment in urban areas, relocate underground utilities, and damage the nearby heritage buildings. Considering the significant constraints and risks of the tunneling methods, the elevated LRT structure option was considered the most preferred and cost-effective option.

4.2 *Impact Toward Neighborhood*

As the end user, informants B and C mentioned this project is significant to them. However, the criteria should be carefully considered to ensure the public transport can be functioning efficiently where the users can fully utilize the services. The informants also stressed the environmental aspects that should be considered to avoid extreme noise and vibration especially at sensitive areas such as mosques, schools, and hospitals. Informant C also mentioned the route selection that proposed alignment went through the Youth Park and Sungai Ara Linear Park has arisen as an issue. This is because the public will lose their leisure space at the liveliest Sungai Ara Linear Park.

5 Results and Discussion

Penang is one of the heritage cities listed under the United Nations Educational, Scientific, and Cultural Organization (UNESCO). Thus, the route selection process took an empirical study to be carefully decided, to ensure the route chosen will possess a minimum impact on the neighborhood, especially at sensitive areas [2, 10]. Experts who have been involved directly since the planning stage sum up four main criteria that were taken into consideration during the route selection process which are accessibility [11], environment [12], financial [13], and population [14]. From these criteria, accessibility received the highest vote followed by financial, environmental, and population which is consistent with Vu et al., findings [4, 5, 15]. While safety is the least to be considered criteria in Penang LRT route selection. As a developed state, the existing building and public space were matured, developed, and fully utilized such as community parks, mosques, schools, commercial areas, and high-density residential areas. To cope with the rising issue regarding the Penang LRT project, the experts briefly explained that route selection requires a complex process to satisfy the main objectives and the criteria [6]. According to the previous findings, the numbers of population, flow of the people transferring, and expected travel time could be considered [16]. Such as, the first decade of LRT establishment in other regions shows the impact on density, property value, transit use, and other positive effects [17].

The expert emphasizes seven main concerns in evaluating route selection. The route must link six main places, stations setup must be easily accessible by all people, integrate with existing transport systems and development, minimum impact toward social and environmental, minimum land acquisition, constructability and offer optimum travel time [18, 19]. Ten route designs were proposed as an alternative to evaluating each of the proposed routes which offered the most satisfying criteria and promoted the least impact on the environment, economy, and social aspect. Option B4 was chosen as the most preferred route as it is linking all main places, has relatively high ridership, and offers lower social impact and construction risk. The finding of this investigation is expected to provide insightful information and respond to the issues of constant revision on PTMP and went through sensitive areas. Therefore, these criteria are multifaceted that derived from both local and federal objectives to satisfy needs of the region [18]. The set of criteria proposed is in line with the objectives to develop public transportation and maximize the usage of rail-based public transport to serve the community [20]. Based on the criteria it shows that the LRT route selection is timely as it needs an empirical study to ensure the project meets all the criteria.

Appendix: Alignment Option

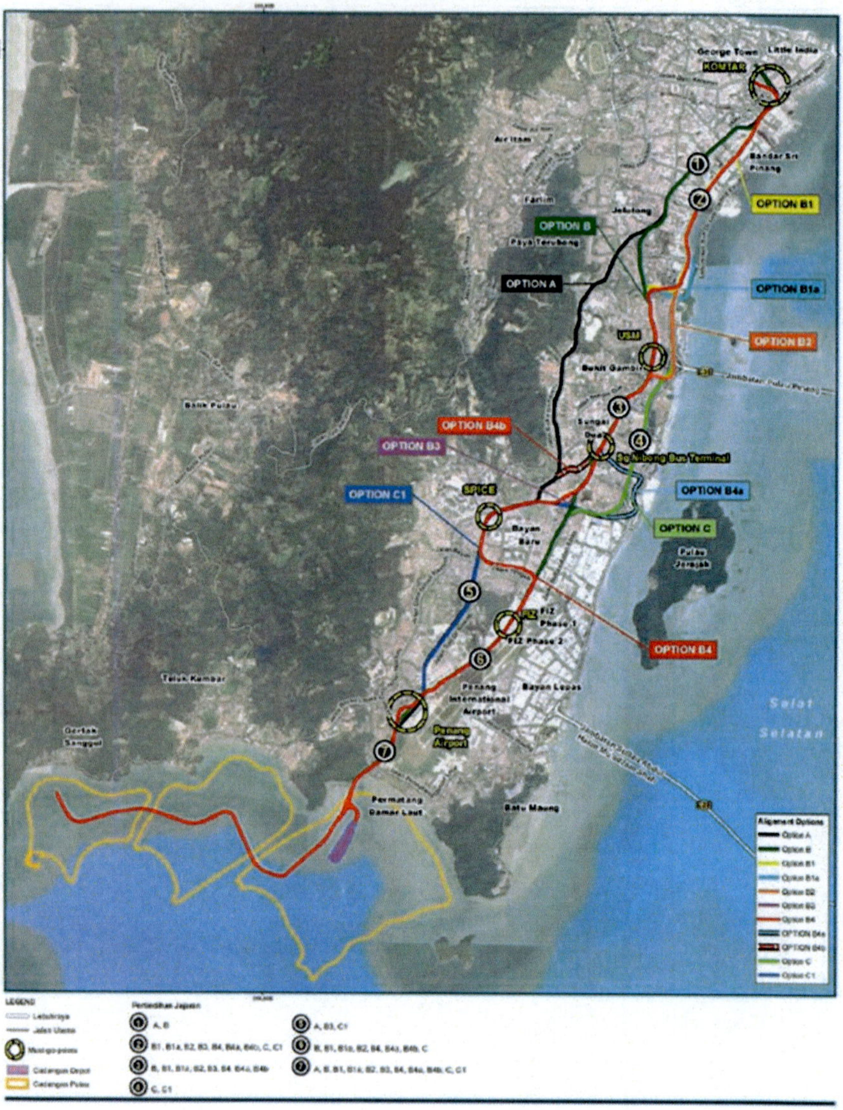

References

1. Rodrigue, J.P., Comtois, C., Slack, B.: The geography of transport systems (2016)
2. Kosijer, M., Marković, M., Belošević, I.: Multicriteria decision-making in railway route planning and design. Gradjevinar **64**(3), 195–205 (2012). https://doi.org/10.14256/jce.645.2011
3. The Penang2030: a family-focused and smart state that inspires the nation (2019) [Online]. Available: https://www.penang2030.com/files/ThePenang2030Guide_FirstEdition2019_eBook_pdf
4. PTMP e-Booklet (July ed.). (2020). Penang transportation master plan
5. Md Nasir, S.S.: 2.23 peratus daripada 23,218 maklum balas bantah projek LRT. Berita Harian, Nov. 27, 2019 (2019)
6. Mok, O.: Penang govt says LRT alignment won't cross over mosque, cemetery grounds. Malay Mail, Jun. 28, 2019 (2019)
7. Yusoff, I., Ng, B.K., Azizan, S.A.: Towards sustainable transport policy framework: a rail-based transit system in Klang Valley, Malaysia. PLoS ONE **16**(3 March) (2021). https://doi.org/10.1371/journal.pone.0248519
8. Libakova, N.M., Sertakova, E.A.: The method of expert interview as an effective research procedure of studying the indigenous peoples of the North. J. Sib. Fed. Univ. Humanit. Soc. Sci. **1**(8), 114–129 (2015). https://doi.org/10.17516/1997-1370-2015-8-1-114-129
9. Melaka and George Town, Historic cities of the straits of Malacca. United Nations Educational, Scientific and Cultural Organization. https://whc.unesco.org/en/list/1223. Accessed 26 Aug 2021
10. Cao, J., Cao, X.: The impacts of LRT, neighbourhood characteristics, and self-selection on auto ownership: evidence from Minneapolis-St. Paul. Urban Studies **51**(10), 2068–2087 (2014)
11. Koopmans, C., Rietveld, P., Huijg, A.: An accessibility approach to railways and municipal population growth, 1840–1930. J. Transp. Geogr. **25**, 98–104 (2012)
12. Cuthill, N., Cao, M., Liu, Y., Gao, X., Zhang, Y.: The association between urban public transport infrastructure and social equity and spatial accessibility within the urban environment: an investigation of Tramlink in London. Sustainability **11**(5), 1229 (2019)
13. Lee, D.J.: A multi-criteria approach for prioritizing advanced public transport modes (APTM) considering urban types in Korea. Transp. Res. Part A: Policy Practice **111**, 148–161 (2018)
14. Abdullah, J., Rashid, K.A., Shah, M.I.B., Ling, O.H.L., Majid, R.A., Ngah, R.: Land use development and ridership at Kelana Jaya lrt line, Malaysia. Planning Malaysia, p 18 (2020)
15. Vu, H.T., Norman, R., Pham, N.M., Pham, H.M., Nguyen, H.T., Nguyen, Q.N., Do, L.D., Huxley, R.R., Lee, C.M., Hoang, T.M., Reid, C.M.: Access route selection for percutaneous coronary intervention among Vietnamese patients: implications for in hospital costs and outcomes. The Lancet Region Health-Western Pacific **9**, 100116 (2021)
16. Holleczek, T., The Anh, D., Yin, S., Jin, Y., Antonatos, S., Goh, H.L., Low, S., Shi-Nash, A.: Traffic measurement and route recommendation system for mass rapid transit (mrt). In: Proceedings of the 21th ACM SIGKDD International Conference on Knowledge Discovery and Data Mining, pp. 1859–1868 (2015, August)
17. Dueker, K.J., Bianco, M.J.: Light-rail-transit impacts in Portland: the first ten years. Transp. Res. Rec. **1685**(1), 171–180 (1999)
18. Banai, R.: Public transportation decision-making: a case analysis of the Memphis light rail corridor and route selection with analytic hierarchy process. J. Public Transp. **9**(2), 1, 8 (2006)
19. Novales, M., Bertrand, D., Fontaine, L.: A proposed new approach to light rail safety management in Spain and other countries. Saf. Sci. **118**, 740–751 (2019)
20. Ludin, A., Latip, S.N.H.M.: Using multi-criteria analysis to identify suitable light rail transit route. Map Asia (2006)

The Relation Between Destructive and Non-destructive Test of Concrete Incorporated with Dredged Sediment as Fine Aggregate Replacement

Siti Aliyyah Masjuki, Nurul Ibrah Mat Gheni, Altamashuddinkhan Nadimalla, Siti Asmahani Saad, Nadiah Md. Husain, Siti Noratikah Che Deraman, Saerahany Legori Ibrahim, Nur Khairiyah Basri, and Shuhairy Norhisham

Abstract Accumulation of dredged sediment due to uncontrolled land clearing, construction works, agricultural activities, river control works, soil mining and dam construction at the upstream has led to an environmental issue in Malaysia. Sungai Pusu and its tributaries which is located at Gombak, Selangor Darul Ehsan, has been identified as the affected river that transports the excessive sediment to the stream. Part of Sungai Pusu area is situated within the International Islamic University Malaysia

S. A. Masjuki (✉) · N. I. M. Gheni · A. Nadimalla · S. A. Saad · N. Md. Husain · S. N. C. Deraman · S. L. Ibrahim · N. K. Basri
Department of Civil Engineering, International Islamic University Malaysia (IIUM), Kuala Lumpur, Malaysia
e-mail: aliyyah@iium.edu.my

N. I. M. Gheni
e-mail: nurul.ibrah@gmail.com

A. Nadimalla
e-mail: altamashk1987@gmail.com

S. A. Saad
e-mail: asmahanisaad@iium.edu.my

N. Md. Husain
e-mail: drnadiah@iium.edu.my

S. N. C. Deraman
e-mail: snoratikah@iium.edu.my

S. L. Ibrahim
e-mail: saerahany@iium.edu.my

N. K. Basri
e-mail: nurkhairiyahbasri@iium.edu.my

S. Norhisham
Department of Civil Engineering, Universiti Tenaga Nasional (UNITEN), Selangor, Malaysia
e-mail: shuhairy@uniten.edu.my

N. Sabtu (ed.), *Proceedings of AWAM International Conference on Civil Engineering 2022 - Volume 3*, Lecture Notes in Civil Engineering 386,
https://doi.org/10.1007/978-981-99-6026-2_21

(IIUM) boundary. Due to the construction work at the upstream of the river have contributed to the accumulation of dredged sediments resulting in the disturbance of the river ecosystem. This study has been conducted to analyze the properties of dredged sediment as a replacement for fine aggregate in concrete. The dredged sediment that was used was taken near IIUM Gombak. Basically, the quality and strength of the dredged sediments were investigated using destructive and non-destructive test analysis. For destructive test, compressive strength test was carried out, meanwhile for non-destructive test, we conducted rebound hammer and ultrasonic pulse velocity test followed by the analysis. The compression test reveals that all of the samples have achieved strength greater than 30 MPa which is the minimum strength specified in this test. At the same time, the rebound hammer test result showed that the estimated compressive strength for all samples has also achieved the specified strength with the difference of all estimated and actual compressive strength which is less than 10%. The result followed by UPV test yields a fast pulse velocity indicating a good quality concrete for the samples. All of the results showed that the replacement of 25% of dredged sediment produced the best result on the 28th day of curing compared to other mixture. Overall, all of the tests that were done on this project show that the fine aggregate can potentially be replaced by the dredged sediment in the production of the concrete.

Keywords Dredged sediment · Destructive test · Non-destructive test · Fine aggregate replacement

1 Introduction

Malaysia is one of the countries that are classified as a rising industrialized country and one of the largest export nations in the world. The construction industry is one of the factors that help in enhancing the development and economic sector in Malaysia. This industry is an important element for Malaysia because it is widely connected with other parts of the industries such as metal products, concrete manufacturing and electronics components. Recently, there is an increasing demand of concrete consumption due to rising rate of building and infrastructure construction in the country. In general, concrete is widely used as a construction material due to its strength, durability and low maintenance. It is a mixture of various materials including cement, sand, aggregate, and water. The aggregates used in concrete manufacturing are obtained from natural sources such as rivers and quarries. Considering the extinction of some of the raw materials, therefore the usage of these natural aggregates should be optimized by searching for other alternative sources.

There are many research and development made to improve sustainable techniques and materials for the construction industries, especially in concrete technology. To implement sustainable construction, there levers of practices have been introduced in the construction industry which are reduction, recycling and reuse. Continued rapid growth through land development, urbanization and industrialization invariably had a negative impact to the environment, as well as scarcity of natural resource [1]. Therefore, the idea is to encourage the usage of new materials in the construction industries to maintain the sources of nature for the future generation. Waste materials that is mainly derived from the environment are highly encouraged to use in the construction sector as it can save costs and preserve the nature. Some alternatives involve waste materials as substitution materials in construction industry such as brick manufacturing, pavement as well as land development.

Due to uncontrolled land development at the upstream of the river have contributed to the accumulation of dredged sediments resulting in the disturbance of the river ecosystem. The sediments are a by-product that can be produced excessively when the construction works are not supervised by the rightful bodies thus leading it to serious environmental issues such as deposition of the sediment from the river to the nearest lake and pond. This problem can already be spotted at IIUM where the excessive sediments were transported from Sungai Pusu and its tributaries to the lake and pond in IIUM particularly at Maryam Lake and Mahallah Bilal pond. To tackle this problem, the excessive dredged sediments can actually be used in producing the concrete by replacing the fine aggregates. The sediments can be revolutionized as green concrete material thus promoting the sustainability concept that was encouraged by the government.

The main objective for this project is to analyze the properties of dredged sediments near Gate 3 IIUM as a fine aggregate replacement in green concrete production, to investigate the quality of dredged sediment in concrete as partial and full fine aggregate replacement by non-destructive test analysis using rebound hammer (RH) and ultrasonic pulse velocity (UPV), and to determine the strength of concrete incorporated dredged sediment as partial and full fine aggregate replacement under compressive strength test for destructive test analysis.

The experimental data obtained were analyzed for the destructive and non-destructive test analysis. The destructive test that will be used is the compressive strength test which will give the data on the hardened properties of the concrete. Meanwhile, the non-destructive test that will be conducted is the rebound hammer (RH) and ultrasonic pulse velocity (UPV) method that will give data corresponding with the destructive test analysis. Both destructive and non-destructive tests will be carried out on the partial and full concrete incorporated dredged sediments.

2 Material Selection and Preparation

In concrete production, materials selection needs to be performed carefully in order to achieve the performance required for the concrete. The selection of materials is important since the concrete needs to perform several functions such as carrying the prescribed loads, satisfy serviceability and durability requirement and many more. The most basic constituents of concrete are coarse and fine aggregates, cement and water. Brief details of the materials are discussed below:

For this experiment, Ordinary Portland Cement type 1 that can be obtained in IIUM concrete laboratory was used as binder of concrete mix. The cement was packaged in airtight condition by supplier because it can easily harden if it reacts with water or high humidity environment. The OPC that was used followed in accordance to MS 522: Part 1: 2003.

Aggregates compensate for 60–80% of the concrete volume. The main aggregates used are sand, gravel/aggregates/crushed stone. All aggregates should practically be free from silt and/or organic matter. The aggregate size used in concrete varies in cross-section from tens of millimeters to fragments smaller than one- tenth of a millimeter. Aggregate from deposits that include a wide variety of sizes, from the biggest to the smallest, is often used when producing low-grade concrete; this is considered an all-in or pit-run aggregate. Thus, to manufacture a high quality concrete, aggregate is produced in at least two size classes; the major distinction is between the coarse aggregate, which consists of materials at least 5 mm or typically between 9.5 and 37.5 mm in size and the fine aggregate that have size not greater than 4 mm (also referred to as sand).

The coarse and fine aggregate together with dredged sediment will undergo sieving process before mixing with cement and water. The sieve procedure is done according to the ASTM C 33—BS 882:1992 standard. The fine aggregate will be sieved using 7 standard sieves with openings from 150 to 9.5 mm, while the coarse aggregate will used 13 sieves with openings from 1.18 to 100 mm. In this experiment, the coarse aggregate that was used is in the range of 5–20 mm according to BS 812-103.1:1985. In addition, fine aggregate that was used passed through 5 mm size of sieve pan and was stored in a clean tray.

A sediment may be used as a replacement for cement raw materials, judging from the chemical composition. Dredged sediments portray the advantages of a renewable resource and are readily accessible in enormous quantities, easily transportable via waterways in accordance with the needs of the cement industry.

According to Achour et al. [1], several experiments have been performed during the past two decades to verify the usage of sediments in different applications: road building, brick manufacturing, mortar processing, concrete production and cement production. The substitution of traditional aggregates with the dredged sediments can bring many advantages such as improved management of dredged sediments, easy resources access and prevent harm that was caused by the resource as it is used in its original environment.

Throughout Europe, dredging managers are urged to search for an economically sustainable solution to handle harbor and channel management by-product dredging. Some of the possible methods therefore are to use the discarded dredged sediments as material in concrete construction. The majority of dredged sediments (DS) consist of clays, silts and sands [2].

The dredged sediment for this project is taken near Gate 3 IIUM. Sieving process is done on the material to ensure that the size will be below 5mm as it will act as replacement for fine aggregate. Since the dredged sediment obtained from this location is a newly formed material, the percentage of it that will be incorporated into the concrete are 25, 50, 75 and 100% to allow ample results and data on its

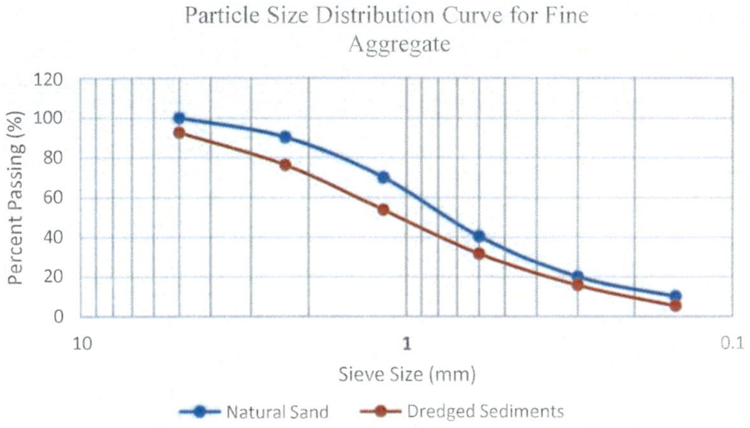

Fig. 1 Particle size distribution curve for fine aggregate

properties. A sieve analysis is done on dredged sediment and natural sand to compare the particle size distribution between both of the material as the fine aggregate.

In Fig. 1, the particle size distribution curve for natural sand and dredged sediments was plotted together in order to easily see the difference of the sizes between both of the fine aggregates. According to Fig. 1, the natural sand particle distribution is curvier than the dredged sediments and the percentage of sand passing through 5-mm sieve is 100% for the natural sand and 92.4% for dredged sediments. This difference of percentage shows that the dredged sediments most probably have sands that are not categorized in fine aggregates sizes or other things such as tree branches that stuck in the sample when it was taken. As Fig. 1 indicates, the particle size distribution curve for dredged sediment is lower than natural sand as the sieve size decreased to 0.15 mm which means that the percent passing of dredged sediments is lower than the percent passing for natural sand. This meant that the dredged sediment contained lesser amounts of fine particles compared to natural sand. Nevertheless, both of the fine aggregates are categorized as well graded since the curve is steep and most of the particles are in a very narrow size range specified.

3 Mixed Design Proportion

The selection of mix proportion is the solely essential method of choosing the correct ingredients for concrete and deciding their amounts, with the goal of manufacturing as efficiently as feasible a concrete with optimum efficiencies, such as strength, durability and greater consistency. Different percentages of cement, water, fine aggregates and coarse aggregates were combined to manufacture concrete with the necessary workability and strength fit for construction. The suggested replacement percentage of dredged sediment to fine aggregates is 0% as control mixture, 25, 50, 75 and 100%.

Table 1 Weight of the concrete material

Percentage of DS (%)	Cement (kg)	Water (kg)	Natural sand (kg)	Dredged sediment (kg)	Coarse aggregate (kg)
0	7.43	3.65	10.75	0	26.37
25	7.43	3.65	8.07	2.69	26.37
50	7.43	3.65	5.38	5.38	26.37
75	7.43	3.65	2.69	8.07	26.37
100	7.43	3.65	0	10.75	26.37

All the specifications and mix design calculation steps will be done in accordance with Department of Environment (DOE) method. Concrete mix shall be configured to have greater mean strength than the specified characteristics strength.

From the calculation that was done using the mix design, the proportion of dredged sediments to fine aggregates was adjusted according to the percentage of incorporation which are 0, 25, 50, 75 and 100%. Each trial mix of 0.0205 m^2 will be consisted of 6 cubes for testing at the 7th and 28th days of curing. The weight of the materials is shown in Table 1 for each batch of incorporated dredged sediments.

4 Result and Discussion

In this section, all the result tests were calculated for the average value from three cubes for each of dredged sediment percentage of replacement in concrete. Every test result was analyzed to get the optimum percentage of dredged sediment replacement in concrete.

4.1 Slump Test

The optimized value of slump that is specifically mentioned and used in mix design is 30–60 mm. Figure 2 illustrates the slump value for all of the fresh concrete with various percentage of dredged sediment. From Table 2, the result shows that the slump for concrete incorporated dredged sediment increased as the percentage of the dredged sediment in the concrete is increased; therefore, the workability is also increased. In Fig. 2, it can be seen that all of the concrete has a value ranging in 30–60 mm except for the concrete with 100% dredged sediment. Even though same water-to-cement ratio is used for all the mixture, concrete that was incorporated with 100% of dredged sediment has the highest slump measured which is 70 mm. For this reason, concrete incorporated with 100% dredged sediment is considered to be unusable since having a high slump can reduced the strength and durability of the concrete.

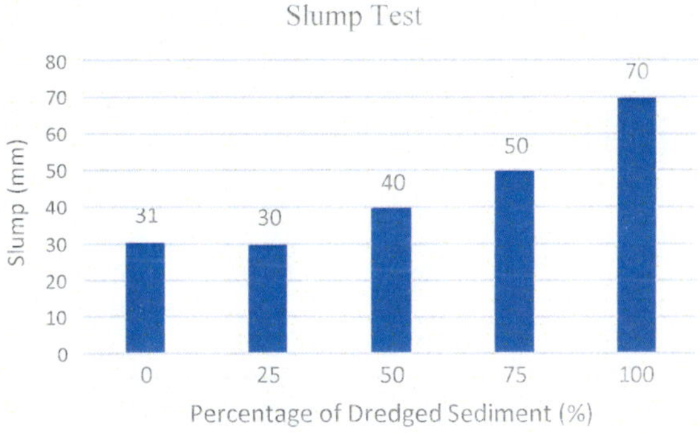

Fig. 2 Slump test graph for fresh concrete

Table 2 Slump test

Percentage of DS (%)	0	25	50	75	100
Slump (mm)	31	30	40	50	70

On the other hand, concrete incorporated with 25% of dredged sediment has the lowest value of slump which are 30 mm and only have a difference of 1 mm with control mixture which have a value of 31 mm. The data also show that utilization of 50% and 75% of dredged sediment gave quite a high value of slump which are 40 mm and 50 mm, respectively, compared to the control mixture. This can be attributed to the physical properties of the dredged sediments that is quite similar to the natural sand in which its round and smooth surface texture giving it the advantage of filling the concrete much easier and improving the workability. The dredged sediments also have finer particles compared to the natural sands thus the concrete can flow much easier when a higher volume of it is used in the concrete.

4.2 *Compressive Strength Test*

The minimum strength that each batches of concrete needed to achieve by 28th day is at least 30 MPa as specified in mix design. Table 3 contains data for the average strength of each percentage of dredged sediment, whereas Fig. 3 illustrates the difference of the strength on day 7 and 28.

Table 3 Compressive strength on 7th and 28th day of curing

Percentage of DS (%)		0	25	50	75	100
Average strength (MPa)	7 day	30.07	36.11	37.14	33.39	32.74
	28 day	33.71	47.83	47.29	45.76	45.01

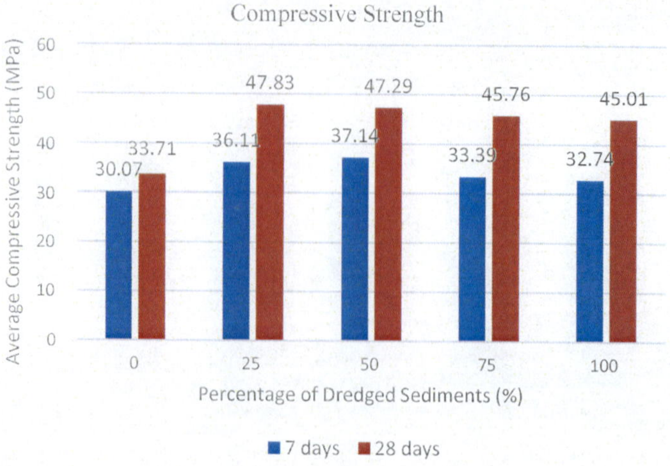

Fig. 3 Concrete compressive strength graph on 7th and 28th day of curing

From Fig. 3, it can be seen that all batches have already surpassed the strength required even before 28th day of curing. From all of the concrete that were incorporated with the dredged sediments, a 25% replacement has the best result on the day 28 and any further replacement has yield a lower compressive strength. According to Chu and Yao [3], a consensus has been reached that the dredged sediment adversely effects the concrete strength, and the higher the DS volume, the lower the concrete strength. This situation can occur due to the presence of materials like silt or clay. The higher the percentage of dredged sediment that was used, the higher the presence of silt in it. It is obvious that the existence of fines in dredged sediment can cause some detrimental effects on concrete due to their higher surface area and increasing water demand of concrete specimens. Concretes having a large percentage of dredged sediment will have high water absorption ratio thus lowering the strengths. In the research that was conducted by Norpadzilhatun et al. (2016), the result shows that only 10–40% of silt in the mix can achieve the required strength of 13.5 MPa on day 7. The concrete also shows a decreasing strength when the fine aggregate was replaced by dredged sediment from Kuala Perlis due to the increasing percentage of silt in the mix. These data support the notion that the higher the presence of the silt, the lower the strength of the concrete.

The result for 100% replacement of dredged sediment had shown poor indication; however, it still has better result compared with the mixture with natural sand (control mixture) which indicates that any percentage of replacement can be done without using any admixture to achieve the specified strength. An experiment by Manap et al. [2] also shows similar result where dredged sediment from Sungai Bebar, Pahang fully replaced the fine aggregate, and the result shows that the concrete has achieved a compressive strength (30.6 MPa) that is slightly above the normal strength for C20 mix (20 MPa) on day 28.

The increasing strength in all of the concrete incorporated with the dredged sediment compared to control mixture can be attributed to the quality of the dredged sediment. The surface texture of dredged sediment which is finer has filled in the gaps of the granular skeleton and strengthen the inner particle thus improving the compaction of the mixture since the cement content and w/c ratio were similar. To summarize all the data that has been presented for the compressive strength test, 25% of concrete incorporated with dredged sediment yields the best compressive strength; thus, it can be promoted and advocated as an acceptable replacement for fine aggregate in concrete production.

4.3 Rebound Hammer Test

In this project, three cubes for each percentage of dredged sediment replacement is marked with sixteen squares on one side of the cube to indicate the position where the rebound hammer will apply the load.

In Table 4, the result of average rebound number and estimated compressive strength obtained from the digital rebound hammer on 7th and 28th day of curing is given. According to Table 4 shown below, incorporation of 50% dredged sediment in the concrete has the highest value of rebound number while the lowest value of rebound number is obtained from concrete with fully incorporated dredged sediment for day 7. On the other hand, day 28 recorded the highest value of rebound number on the concrete incorporated with 25% of dredged sediment and the lowest rebound number is on the control mixture.

Figure 4 visualized the estimated compressive strength from Table 4. In the Fig. 4, the estimated compressive strength obtained from the digital hammer shows that all of the mixture has obtained the specified strength needed for the mix which is 30 MPa. The highest compressive strength on the 28th day is 44.8 MPa followed by 44.6 MPa, 42.8 MPa, 42.3 MPa and 32.8 MPa on the concrete that was incorporated with 25%, 50%, 75%, 100% and 0% of dredged sediment, respectively.

Table 4 Rebound number for 7th day of curing

Percentage of DS (%)	7 day		28 day	
	Average rebound number (Rn)	Estimated compressive strength (MPa)	Average rebound number (Rn)	Estimated compressive strength (MPa)
0	31.0	29.8	32.6	32.8
25	33.0	33.4	39.8	44.8
50	33.8	34.7	39.5	44.6
75	31.4	30.2	38.5	42.8
100	30.5	29.6	38.3	42.3

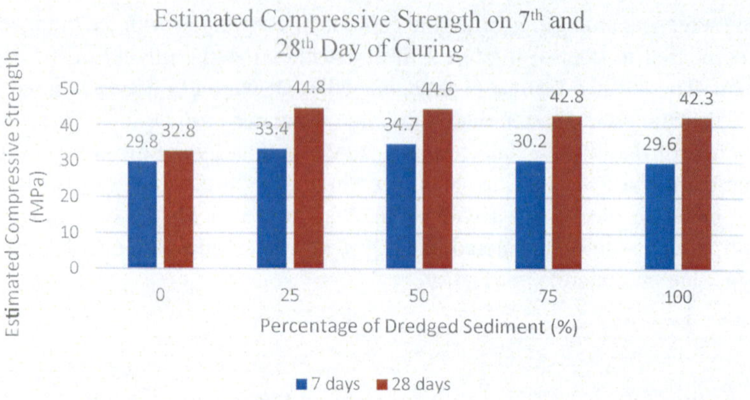

Fig. 4 Estimated compressive strength based on rebound hammer test

4.4 Correlation Between Rebound Hammer Test and Compressive Strength Test

A correlation has been set up and illustrated with data obtained from samples on 7th and 28th day of curing used for concrete strength and rebound hammer tests. When rebound hammer test on each cube of different proportion of dredged sediment was completed, the load was applied up to failure. The outcomes were plotted in the Fig. 5 which is given. Also, the R^2 value for samples on 7th day of curing is found to be 0.781 and its equation is $y = 1.7767x - 22.857$ (where y = compressive strength; x = rebound number). R^2 value of samples on 28th day of curing is found to be 0.9989 and its equation is $y = 1.976x - 30.656$. Both of the equations further affirmed that the relationship between compressive strength test and rebound hammer test is acceptable since both of the R^2 (correlation coefficient) was close to 1.

Tables 5 and 6 above further help in emphasizing on the acceptable relationship between both tests. The percentage of error is calculated between the estimated compressive strength obtained from the digital rebound hammer and the compressive strength from the compressive strength test. The highest percentage of error is produced by concrete incorporated with 75 and 100% of dredged sediments on 7th day of curing with the value of 9.6% which is still in the acceptable range of error. Both tests show a significant increase in the strength since the percentage of error on 28th day has become lower with the highest error value of 6.5% only. However, it should be noted that all of the estimated compressive strength is still lower than the actual compressive strength. This may be due to the influence of several factors such as the smoothness of the surface and moisture content of the sample. Although the hammer was not very accurate in predicting the actual strength, the observation still shows the reliability of the hammer in predicting an increase in strength as time varies.

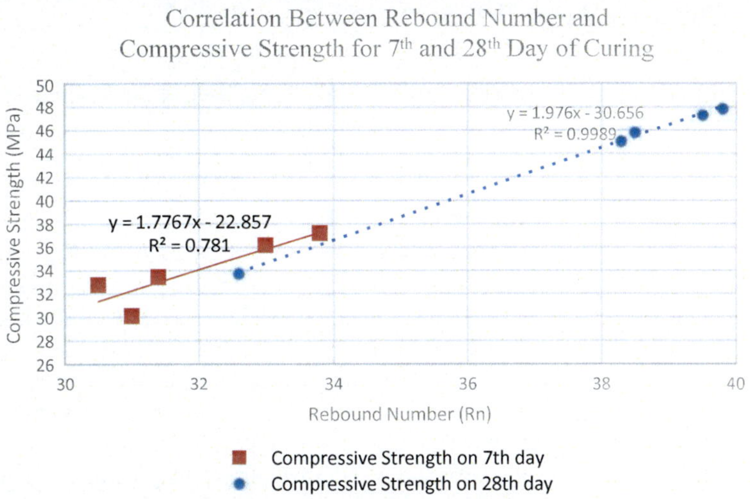

Fig. 5 Correlation between rebound number and compressive strength for 7th day and 28th of curing

Table 5 Percentage of error between estimated compressive strength and compressive strength test for 7th day of curing

Percentage of DS (%)	Estimated compressive strength (MPa)	Compressive strength test (MPa)	Percentage of error (%)
0	29.8	30.07	0.9
25	33.4	36.11	7.5
50	34.7	37.14	6.6
75	30.2	33.39	9.6
100	29.6	32.74	9.6

Table 6 Percentage of error between estimated compressive strength and compressive strength test for 28th day of curing

Percentage of DS (%)	Estimated compressive strength (MPa)	Compressive strength test (MPa)	Percentage of error (%)
0	32.8	33.71	2.7
25	44.8	47.83	6.3
50	44.6	47.29	5.7
75	42.8	45.76	6.5
100	42.3	45.01	6.0

4.5 Ultrasonic Pulse Velocity (UPV) Test

In this project, three cubes will be tested on two sides for each of the concrete samples with various dredged sediment proportion. Then, the result of both sides of each cube will be summed and recorded as the average transmit time on 7th and 28th day of curing as shown in Table 7. On the 7th day of curing, the highest transmit time that was recorded is by the control mixture with the value of 41.1 μs, while the lowest one is for the 50% dredged sediment with 37.0 μs. On the other hand, the highest and lowest transmit time for 28th day of curing is recorded by control mixture and 25% dredged sediment with the value of 39.3 μs and 35.2 μs respectively. It should be noted that, the higher the time taken to transmit the pulse, the lower the value of pulse velocity thus the result of it can be seen in Table 7. The tabulated data below were calculated using formula $V = L/T$ with V = pulse velocity, L = length of cube (150 mm) and T = transmit time. According to the data, 50% dredged sediment has the highest UPV, and control mixture has the lowest UPV for day 7 with value of 4.05 km/s and 3.65 km/s, respectively, while the highest and lowest UPV for day 28 is 4.26 km/s and 3.82 km/s for 25% dredged sediment and control mixture, respectively.

According to Latif and Rasoul [4], the quality of concrete can be divided into five ranges based on the pulse velocity which is excellent concrete with pulse velocity above 4.5 km/s, good concrete with pulse velocity in between 3.5 and 4.5 km/s, doubtful concrete (3.0–3.5 km/s), poor concrete (2.0–3.0 km/s) and lastly is very poor concrete with pulse velocity below 2.0 km/s. From the data that were collected from the experiment, all of the concrete with different proportion of dredged sediment belong to the good concrete quality since all pulse velocity is in the range of 3.5–4.5 km/s as illustrated in Figure 6.

Table 7 Average transmit time and UPV on 7th and 28th day of curing

Percentage of DS (%)	7 day		28 day	
	Average transmit time (μs)	Average pulse velocity (km/s)	Average transmit time (μs)	Average pulse velocity (km/s)
0	41.1	3.65	39.3	3.82
25	37.3	4.03	35.2	4.26
50	37.0	4.05	36.4	4.12
75	37.8	3.97	37.4	4.02
100	37.4	4.01	37.6	3.99

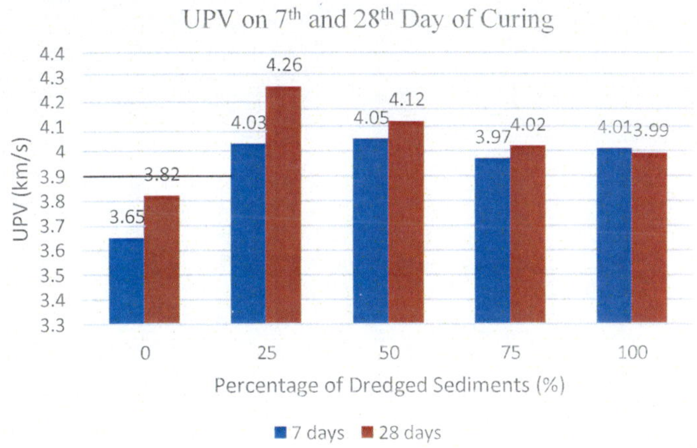

Fig. 6 UPV on 7th and 28th day of curing

4.6 *Correlation Between Ultrasonic Pulse Velocity (UPV) Test and Compressive Strength Test*

A correlation between concrete strength and ultrasonic pulse velocity (UPV) test has been set up and illustrated with data obtained from samples on 7th and 28th day of curing. The linear regression was applied to establish the relationship of compressive strength and UPV. Figure 7 shows the compressive strength–UPV relationship at the age of 7 and 28 days for various proportion of dredged sediments incorporated in the concrete. From the data, it has been observed that for concrete on day 7, the pulse velocity ranges from 3.50 to 4.10 km/s for which the values of compressive strength ranges from 30 to 38 MPa and for concrete on day 28, the pulse velocity ranges from 3.80 to 4.30 km/s for which the values of compressive strength ranges from 30 to 48 MPa. The correlation equations for the graph for 7th day is $y = 14.259x - 22.318$ and for 28th day of curing is $y = 30.964x - 81.238$, where y represents the compressive strength and x represents the ultrasonic pulse velocity. Their corresponding coefficients of determination R^2 are 0.707 and 0.7508, indicating a good relevance between data points and the regression curves since all the concrete produced using dredged sediment presented higher velocity compared to concrete with natural sand only. The incorporation of dredged sediment that is smoother and finer in size compared to natural sand can enhance the interlocking between cement paste and aggregate. This in turn makes the concrete become less porous thus speeding up the ultrasonic pulse velocity.

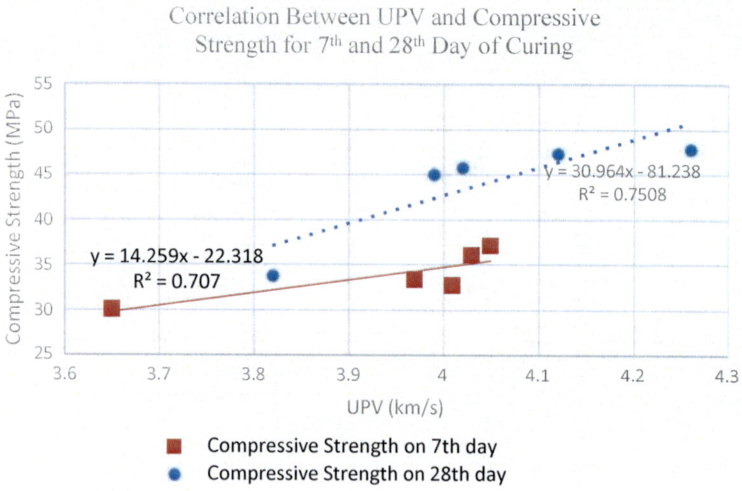

Fig. 7 Correlation between UPV and compressive strength for 7th and 28th day of curing

5 Conclusion

A study was undertaken to investigate the destructive and non-destructive analysis on concrete incorporated dredged sediment as fine aggregate replacement. The concrete produced with different replacement percentage of dredged sediment was used in the experiments. In this project, the ability of dredged sediment to act as substitute for fine aggregates incorporated in concrete is explored. From the sieve analysis that was done, the particle size distribution curve for dredged sediments is almost similar with the natural sand which means that dredged sediment can be potentially used as the replacement for fine aggregate. Besides that, the slump test reveals that a higher replacement of dredged sediment in the concrete can lead to higher workability but the mix will become less durable. Other tests that were conducted on the hardened concrete such as compressive strength test, rebound hammer test and ultrasonic pulse velocity test reveal that the mix with 25% of dredged sediment incorporated in it yields the best result on the 28th day of curing compared with other mixtures. All of the samples successfully achieved the strength that were specified in the experiment which is 30 MPa on day 28 when tested on the compression testing machine. The estimated compressive strength from rebound hammer test has also achieved the specified strength with the difference of all estimated and actual compressive strength is less than 10%. The correlation between compressive strength and UPV test has been established in which the UPV ranges for all of the concrete mixtures fall into the good quality concrete. Overall, the results that were collected in this project suggest that dredged sediment seems to be a good alternative for the fine aggregate in concrete production.

References

1. Achour, R., Zentar, R., Abriak, N.E., Rivard, P., Gregoire, P.: Durability study of concrete incorporating dredged sediments. Case Stud. Constr. Mater. **11**, e00244 (2019). https://doi.org/10.1016/j.cscm.2019.e00244
2. Manap, N., Govindasamy, G., Mohamed, S., Kasim, N., Musa, S.M.S., Yahya, M.Y.: Performance of dredged sediment as a replacement for fine aggregates in concrete mixture: case study at Sungai Pekan, Pahang. In MATEC Web of Conferences, vol. 266, p. 01017. EDP Sciences (2019)
3. Chu, S.H., Yao, J.J.: A strength model for concrete made with marine dredged sediment. J. Clean. Prod. **274**, 122673 (2020). https://doi.org/10.1016/j.jclepro.2020.122673
4. Latif, A.M.A., Rasoul, Z.M.R.A.: Correlation between the compressive strength of concrete and ultrasonic pulse velocity; investigation and interpretation. J. Kerbala Univ. **7**(3), 17–29 (2009). https://doi.org/10.13140/RG.2.2.15333.86241
5. Dang, T.A., Kamali-Bernard, S., Prince, W.A.: Design of new blended cement based on marine dredged sediment. Constr. Build. Mater. **41**, 602–611 (2013). https://doi.org/10.1016/j.conbuildmat.2012.11.088
6. Datir, U.D., Chauhan, N.L., Shah, D.R.: Correlation between pull-out force, pulse velocity and compressive strength of concrete. Indian Concr. J. **69**(12), 703–705 (1995)
7. Deepak, W.S., Naidu, T.G.: Effect on compressive strength of concrete using sea sand as a partial replacement for fine aggregate. Int. J. Res. Eng. Technol. **4**(6), 180–183 (2015). https://doi.org/10.15623/ijret.2015.0406030
8. Gal, E., Kryvoruk, R.: Properties of concrete. In: Computational Modelling of Concrete Structures (2010). https://doi.org/10.1201/b10546-17
9. Gholizadeh, S.: A review of non-destructive testing methods of composite materials. Procedia Struct. Integrity **1**, 50–57 (2016). https://doi.org/10.1016/j.prostr.2016.02.008
10. Hamidian, M., Shariati, M., Arabnejad, M.M.K., Sinaei, H.: Assessment of high strength and light weight aggregate concrete properties using ultrasonic pulse velocity technique. Int. J. Phys. Sci. **6**(22), 5261–5266 (2011). https://doi.org/10.5897/IJPS11.1081
11. Hussin, J.M., Abdul Rahman, I., Memon, A.H.: The way forward in sustainable construction: issues and challenges. Int. J. Adv. Appl. Sci. **2**(1) (2013). https://doi.org/10.11591/ijaas.v2i1.1321
12. Bin Ibrahim, A.R., Roy, M.H., Ahmed, Z., Imtiaz, G.: An investigation of the status of the Malaysian construction industry. Benchmarking **17**(2), 294–308 (2010).https://doi.org/10.1108/14635771011036357
13. Ravindra V, Buraka AK. Mechanical Properties of Concrete with Marine Sand as Partial Replacement of Fine Aggregate. Dr. V. Ravindra Int. Journal of Engineering Research and Applications. 2016;6(2):8–11. www.ijcra.com
14. Lambert, P.P. (n.d.).: Reinforced concrete—history, properties & durability. 1–6
15. Limeira, J., Etxeberria, M., Agulló, L., Molina, D.: Mechanical and durability properties of concrete made with dredged marine sand. Constr. Build. Mater. **25**(11), 4165–4174 (2011). https://doi.org/10.1016/j.conbuildmat.2011.04.053
16. Limeira, J., Agullo, L., Etxeberria, M.: Dredged marine sand in concrete: an experimental section of a harbor pavement. Constr. Build. Mater. **24**(6), 863–870 (2010). https://doi.org/10.1016/j.conbuildmat.2009.12.011
17. Limeira, J., Agulló, L., Etxeberria, M.: Dredged marine sand as construction material. Eur. J. Environ. Civ. Eng. **16**(8), 906–918 (2012). https://doi.org/10.1080/19648189.2012.676376
18. Manap, N., Polis, S., Sandirasegaran, K., Masrom, M.A.N., Chen, G.K., Yahya, M.Y.: Strength of concrete made from dredged sediments. **3**, 87–92 (2016)
19. Millrath, K., Kozlova, S., Shimanovich, S., Meyer, C.: Beneficial use of dredge material program (2006)
20. Zainul Abidin, N.: Sustainable construction in Malaysia—developers' awareness. Proc. World Acad. Sci., Eng. Technol. **5**(2), 122–129 (2010)

21. Ozer-Erdogan, P., Basar, H.M., Erden, I., Tolun, L.: Beneficial use of marine dredged materials as a fine aggregate in ready-mixed concrete: Turkey example. Constr. Build. Mater. **124**, 690–704 (2016). https://doi.org/10.1016/j.conbuildmat.2016.07.144

Transportation and Traffic Engineering

Review on the Main Characteristics of Freeway Merging Section

Wafaa Kh. Luaibi, Lee Vien Leong, and Hamid Athab Al-Jameel

Abstract The merging section is considered an essential part of the freeway segment, and it is one of the causes of traffic congestion along the freeway. This study reviews the main characteristics of merging sections and the methods to reduce traffic congestion at the merging section. Several significant factors affecting the merging section were identified such as freeway capacity, length of merging section, speed, traffic volume, and traffic composition. The effect of these factors has been thoroughly reviewed in this study to investigate their influences on traffic operations at merging sections. The effectiveness of ramp metering (RM) and gap metering (GM) was also reviewed in this study. The results indicated a significant impact of some traffic characteristics at merging sections, such as the lane-changing behavior, heavy vehicles, and RM effect. Combined with RM, the gap metering mechanism can reduce the overall delay at merging sections up to 27% compared to RM alone.

Keywords Driver behavior · Merging sections · Motorway capacity · Ramp metering

1 Introduction

Congestion is a serious problem since it reduces highway network performance, lengthens travel times, and increases vehicle emissions by increasing fuel consumption. Furthermore, distraction during a crowded period might increase the risk of

W. Kh. Luaibi (✉) · L. V. Leong
School of Civil Engineering, Universiti Sains Malaysia, Engineering Campus, 14300 Nibong Tebal, Penang, Malaysia
e-mail: wafaa.leabi@student.usm.my

L. V. Leong
e-mail: celeong@usm.my

H. A. Al-Jameel
Civil Engineering Department, University of Kufa, Kufa, Iraq
e-mail: hamid.aljameel@uokufa.edu.iq

N. Sabtu (ed.), *Proceedings of AWAM International Conference on Civil Engineering 2022 - Volume 3*, Lecture Notes in Civil Engineering 386,
https://doi.org/10.1007/978-981-99-6026-2_22

an accident, as all vehicles attempt to move through the congested section of the road. For example, for economic, urban, and environmental reasons, geometrical solutions to traffic congestion have proven ineffective. As a result, since the 1960s, the tendency has shifted to employing coordinated and integrated monitoring and control techniques to regulate link traffic and maximize capacity [1].

Due to the rise in vehicle ownership and changes in lifestyle or additional modes of transportation supplied by a system, traffic levels may exceed infrastructure capacity during peak hours, thus producing congestion. Many methods to alleviate traffic congestion on highways might be proposed to manage the repercussions of congestion; however, geometrical solutions may not be suitable in all situations due to economic and environmental factors [1].

Congestion on freeways leads to significant delays and, as a result, high societal costs. Local congestion is caused by infrastructure bottlenecks such as merging sections and work zones. This can expand to other areas of the road network. When traffic demand goes beyond a merging section's capacity, an operational bottleneck will form, causing a delay to build on the near-side lane of a significant line road. The backlog then expands laterally to adjacent lanes, causing a reduction in traffic flow, which is a capacity drop [2]. Several studies have found that traffic flow is unequally distributed throughout a freeway's available lanes [1–3]. However, the drop in one lane capacity affects the capacity for all other lanes.

The breakdown phenomenon is described as the alteration from free flow to synchronized movement in terms of freeway capacity and breakdown occurrences. Breakdown happens when speeds on a highway fall below a threshold value of 55 mph (88.5 km/h) when density exceeds the boundary of the levels of service D. Three methods based on speed decline, occupancy rise, and volume–occupancy correlation have been developed to identify breakdowns [3].

Figure 1 depicts a merging section in Navi Mumbai, India [4]. In this city, many factors are evaluated due to merging two-lane ramps on a two-lane main road such as vehicle merging mechanism and speed decrease at the merging section for various vehicle types, vehicle compositions, and flow levels.

Fig. 1 Merge section in Navi Mumbai, India [4]

Many traffic management techniques have been implemented to improve traffic flow, such as ramp control, route control with variable speed restrictions, and driver route assistance [1]. RM techniques have been recognized as an active method for regulating and managing highway traffic in the recent decade [1]. Therefore, this study reviews the main characteristics of merging sections and factors affecting the merging area, such as freeway capacity, length of joining the branch, speed, traffic volume, and traffic composition. The effect of these factors and the effectiveness of RM and gap metering were also reviewed in this study to investigate their influences on traffic operations at merging sections.

2 Freeway Capacity

The capacity of basic freeway sections is a function of the free flow speed (FFS), which is dependent on lane width, right-side clearance, and ramp density [5]. According to the HCM 2000, capacity breakdown of essential freeway section is defined as the "Maximum flow linked with the occurrence of some sort of breakdown, resulting in lower speeds and greater densities following the breakdown event." Maximum flow rates are used in HCM's capacity estimate model for basic highway segments. However, the maximum flow rate does not always correspond to the most significant, resulting in traffic congestion. Highway capacity studies have utilized average pre-breakdown flow, maximum pre-breakdown flow, and average discharge flow rates to account for this unpredictability. By analyzing both density and flow fluctuation, the unpredictability of capacity could be determined. According to the researchers, the variability of highway flow is smaller than that of critical density; hence, flow is a significant indicator of freeway capacity than critical density [6].

There are two types of capacity: the capacity before a breakdown (pre-breakdown capacity) and capacity in congested conditions (queue discharge flow). Field measurements have revealed that the queue discharge is between 10 and 30% lower than the pre-breakdown capacity and this discrepancy indicates a capacity drop. Multiple definitions of the pre-breakdown capacity were studied in previous studies, the majority of which are concerned with the occurrence of the breakdown phenomena. The flow rate before the breakdown is defined as capacity. The maximum pre-breakdown flow (top 5- or 15-min flow before the breakdown) and the maximum discharge flow were considered. Capacity is stochastic, according to several experts [3].

Moreover, at merge bottleneck sites, the ramp flow ratio has a detrimental impact on capacity. This means that, even if the upstream per-lane demand stays constant, more traffic from the on-ramp will lower merging capacity. Additionally, higher ramp demands result in a more significant loss in capacity [3].

By examining data from numerous sites around the United States, analyses were conducted to determine the influence of the geometric and operational factors on merging capacity and the likelihood of failure. The number of mainline lanes and lane reductions and extensions considered possible geometric elements that may impact merging throughput. Flow parameters like FFS and ramp flow rates, in addition to RM, were also evaluated as potential capacity limiting issues. Survival analysis was used to carry out the approach on numerous merging bottlenecks around the United States [6].

The ramp flow, number of lanes, existence of the lane drop, and existence of RM were all used to calculate the likelihood of breakdown and capacity. The average pre-breakdown flow rate per lane negatively relates to the number of lanes. Pre-breakdown flow rates were also 150 to 300 vehphpl greater at locations with RM than at those without them. The number of the lanes and the existence of RM were shown to own a statistically significant influence on the likelihood of a breakdown in survival analysis. At three-lane and four-lane unmetered locations, respectively, capacity was 2043 and 1952 vehphpl, defined as the 15th percentile breakdown of a probability flow rate. At three-lane metered sites, capacity was 2316 vehphpl, while at four-lane metered sites, capacity was 2011 vehphpl. At two of the four locations studied, more significant ramp flows were associated with a higher breakdown likelihood and lower capacity, implying that more research is needed to understand this relationship fully [6].

3 Merge Section Influence

At the merging section, speeds on separate lanes gradually decreased from the medium side lane to the lane next to the merging point. Most vehicles evade traveling in lanes closer to merging points in the major traffic stream under low and moderate flow conditions. This is done to decrease the likelihood of a collision and prevent competition for the same amount of road space. The speed versus flow relationship calculated for the basic highway segment and the merging section reveals that the merging operation reduces capacity by about 8% [7].

The influence of heavy vehicles moving on the expressway on the capacity of the kinematic merging segment was studied. It was discovered that an increase in the percentage of heavy vehicles resulted in a 1% reduction in capacity. A traffic simulation technique was used to test the lane closure idea. Kumar et al. discovered that closing on-ramp lanes, or closing both on-ramp and major line lanes, increased over 3% [7]. The authors similarly utilized a simulation model to examine the benefits of installing a variable message sign ahead of the merging point to direct truck drivers to move lanes to prevent collisions with merging vehicles. They found that by relocating 10% of the heavy trucks, the merging section's throughput might be increased by 1%. On expressways, merging portions are one of the most common causes of traffic accidents. Figure 2 depicts the merging of sections of the Delhi expressway [7].

The integrated model may reasonably replicate all necessary behavior of single drivers in merging regions, including regular car following, close-following, cooperative lane shifting, courtesy yielding, and gap acceptance. This indicates high activities of lane changing, which has a significant impact on merging section capacity.

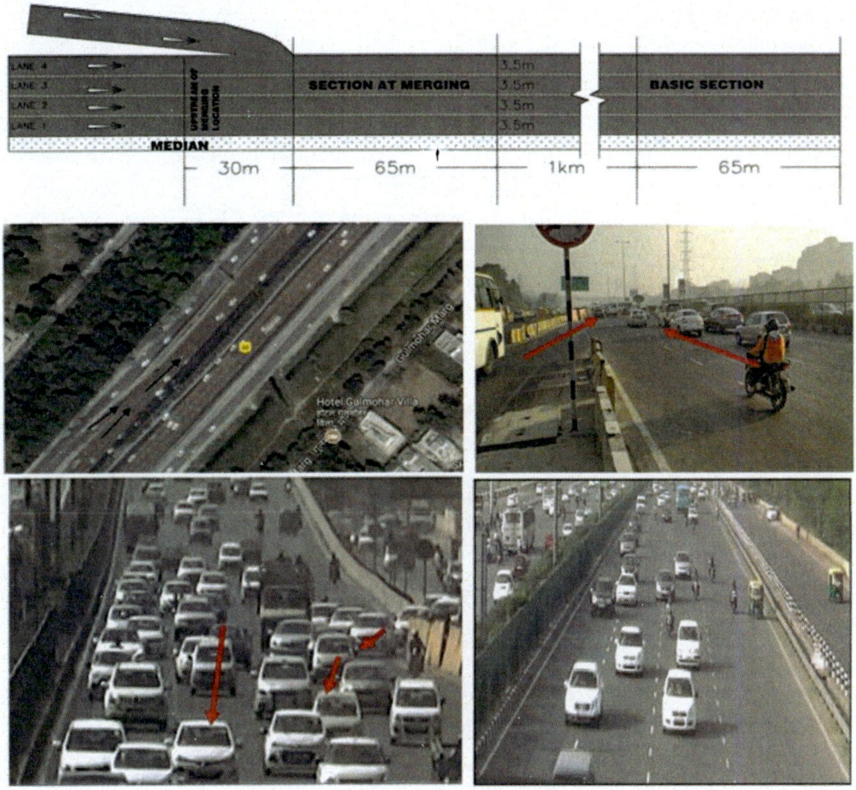

Fig. 2 Merging sections on expressway Delhi [7]

According to sensitivity testing on various merging durations, increased length may minimize merging failures. It may be interpreted that with longer merging distances, more merging traffic could effectively take off the following gaps, which has consequences for the geometric design of the acceleration lane [8]. There is no widely accepted benchmarking technique at the moment, and this framework has the benefit of allowing the concept and suggested methodology to be applied to a range of microsimulation models employing detector data sets.

Queues may occur downstream of the ramp, upstream of the ramp on the highway, on the existing on-ramp, or both at an on-ramp merge, depending on the prevailing needs. For example, when an acceleration lane exists after the ramp merge, the queue will be automatically represented as happening directly upstream of the lane drop [9]. However, for a diverge section, the queue is mainly due to downstream mainline congestion, despite the sufficient capacity on both the upstream and off-ramp lanes.

4 Effect of RM

There are two types of RM either local, used for the isolated on-ramp, or area-wide (coordinated) [10]. The main aim of using a RM system is to prevent congestion by avoiding downstream traffic from exceeding the overall capacity of the freeway by storing some traffic on the slip road (on-ramp). The motivation of RM operation is the occupancy value which reflects whether traffic operation is in a congested state or otherwise. Occupancy could be defined as the percentage of vehicles occupied by a traffic loop detector embedded in the road pavement. The main importance of using occupancy lies in measuring it from the loop detector. Moreover, occupancy can describe the traffic conditions (i.e., normal or congestion) in the same way as density [11].

The ramp length (storage length) and selection of suitable algorithms to control and update traffic signals timing are significant factors to get a successful RM system in mitigating traffic congestion for merging sections [12]. Unlimited ramp length leads to successful RM in reducing traffic congestion [13–15]. However, limited length may reduce the efficiency of RM.

Regarding the suitable algorithms used to control local RM, the main widely used methods are Demand–Capacity (D–C) and the ALINEA [16] ALINEA mainly depends on keeping occupancy levels downstream of the merging area close to the critical occupancy. However, several types of ALINEA algorithms have been developed based on the positions where flow and occupancy were determined, such as FL-ALINEA (based on downstream flow), UP-ALINEA (based on upstream occupancy), and UF-ALINEA (based on upstream flow) [14]. For traffic signal timing of RM, a 20-s green time signal is used in most sites in the UK [17].

It was found that the critical occupancy is 23%, and the optimum location for the loop detector for measuring occupancy is 300 m according to their simulation model [12]. On the other hand, RM impacts drivers' route-choice behavior and may be used as the dynamic assignment mechanism to encourage the usage of corridor networks [18].

In addition, the RM system, which is meant to avoid or decrease traffic congestion on highways, might have good and negative consequences for traffic on the nearby road network. On-ramp delays might encourage vehicles to choose a road link route rather than a highway route, resulting in extra traffic on the road subnetwork. Reduced recurring congestion on highways, on the other hand, may encourage more vehicles to take a highway path, resulting in fewer traffic difficulties on the road network.

RM basically minimizes recurring traffic congestion in both space and time while increasing the highway's average speed and somewhat increasing the supplied demand. As a result, the number of people diverting from the highway to the parallel arterial reduces, and traffic situations on the parallel arterial improve [10]. Under all assessment criteria, ALINEA was shown to enhance the no-control scenario and be superior to all other control techniques with which it was tested in all implementations. In the absence of events, ALINEA was comparable to coordinated RM.

ALINEA was used to enhance traffic circumstances on the highway and in the surrounding streets [18].

5 RM System

One of the RM systems which can be applied at the merging section is the Cooperative Intelligent Transportation Systems (C-ITS). This system enables data interchange between cars and infrastructure. A C-ITS application for in-vehicle use increases traffic efficiency around a merging area. By giving targeted lane-changing advice to a selection of vehicles, that application balances traffic distribution over an available lane of a highway. In the lane-changing advice framework, there are various vehicle classifications. Design a feedback-feed forward control rule based on a linear quadratic regulator using a multi-class multi-lane macroscopic traffic flow model (LQR). The microtraffic simulator is used to test the proposed system's performance. The findings show that the lane-changing warning system may reduce traffic shockwaves and considerably reduce congestion. In addition to providing significant travel time savings of up to almost 21% around merging portions, the system substantially lowers the variation of travel time losses in the system [2].

Furthermore, the fast conceptualization and commercialization of connected autonomous vehicles (CAV) have resulted in the issue of mixed traffic, or traffic containing both CAVs and traditional human-operated vehicles. Therefore, the cooperative decision-making for mixed traffic (CDMMT) technique is devised using discrete optimization to allow RM to know cooperative and non-cooperative behaviors in a traffic stream appropriately. The CDMMT mechanism is the bi-level optimization method incorporating state-constrained optimum control-based trajectory design issues into a sequencing problem. A bi-level dynamic programming-based solution technique is proposed to address the problem efficiently. The suggested modeling mechanism and solution technique can ensure system-efficient solutions for deterministic judgments. For model validation and mixed traffic analysis, a microsimulation environment is created. The results demonstrate that RM can be smoother in a hybrid traffic environment than in a scenario with 100% human-operated vehicles. The section throughput improves by around 18% when there is a significant CAV penetration. Traffic throughput can be enhanced by 10–15% using the suggested CDMMT method. The proposed approaches serve as the foundation for traffic analysis and cooperative control at the RM sections in a mixed traffic environment.

To reduce traffic conflicts and increase merging efficiency for mixed traffic, the deterministic CDMMT ramp merging mechanism was devised. Human-operated vehicles are thought to be non-cooperative, although CAV control logic may be made to be cooperative. Deterministic automobile following and RM utilizes simulated longitudinal and lateral traffic choices. A suggested method is a bi-level optimization problem, including state-constrained optimum control-based trajectory design issues in the merge sequencing problem. A dynamic programming-based solution was created to successfully tackle this problem, and system-efficient solutions are assured. The findings demonstrate that ramp merging is smoother in mixed traffic than in pure human-operated vehicles. Simulation findings demonstrate that when 100% CAVs are used, the section throughput rises by around 18%. A microscopic simulation was used to validate the CDMMT ramp merging mechanism under various traffic flow circumstances, traffic mixtures, and mainline-to-ramp ratios. Furthermore, findings show that the suggested technique reduces traffic conflicts, provides smoother acceleration/deceleration, and increases throughput. Findings also show that when more people cooperate, the system's efficiency improves. The CDMMT framework may be modified to the various driving activities (e.g., lane shifting, diverging, and crossing) and traffic circumstances; however, it is particular to ramp merging. Figure 3 illustrates flow–density diagrams [19].

6 Gap Metering

Freeway merging sections are significant elements that can reactivate peak-hour traffic jams regularly. A unique vehicle gap management approach is added to the current intelligent transportation system (ITS) toolboxes for freeway merging control as a new active traffic management (ATM) strategy known as "Gap Metering." The suggested technique can be considered a non-stopping major line variant of the RM [9]. It uses signals to tell major lines through cars to leave enough space for the merging vehicles.

In gap metering, as soon as the traffic volume on the ramp reaches a certain threshold, the lane control algorithm will show signs that say "Lane closed." There was a substantial decrease in overall VHT (vehicle hours travelled) with a noteworthy enhancement in the ramp flow. This control approach temporarily reduces central line capacity by one lane by separating significant line traffic from ramp traffic upstream of the merging location. This might result in a new bottleneck in the upstream traffic flow, causing even more significant congestion on existing congested roads. An ATDM approach is presented for freeway recurrent and non-recurrent merge control by controlling the gaps in mainline traffic flow. Gap metering systems are well-suited to assisting traffic flow during bottlenecks caused by heavy merging on a regular and non-recurrent basis [20].

VISSIM, a microsimulation software program, develops and implements detailed system design and control techniques. Individual driver behavior is developed in conjunction with varying stationary headway values to transition between gap-metered and standard cars. The proposed system was assessed using two VISSIM models, one for the I-894 corridor in Milwaukee, Wisconsin, and the other for the Riverside Drive stretch on I-35 northbound in Austin, Texas. The morning rush hour is quite congested on both routes. The I-894 corridor is used to evaluate system design parameters, while the I-35 corridor compares RM techniques. The I-894 findings show a 10–20% decrease in network latency across all circumstances. Then, the scenario is put to the test on the I-35 corridor, comparing it to ALINEA of the RM.

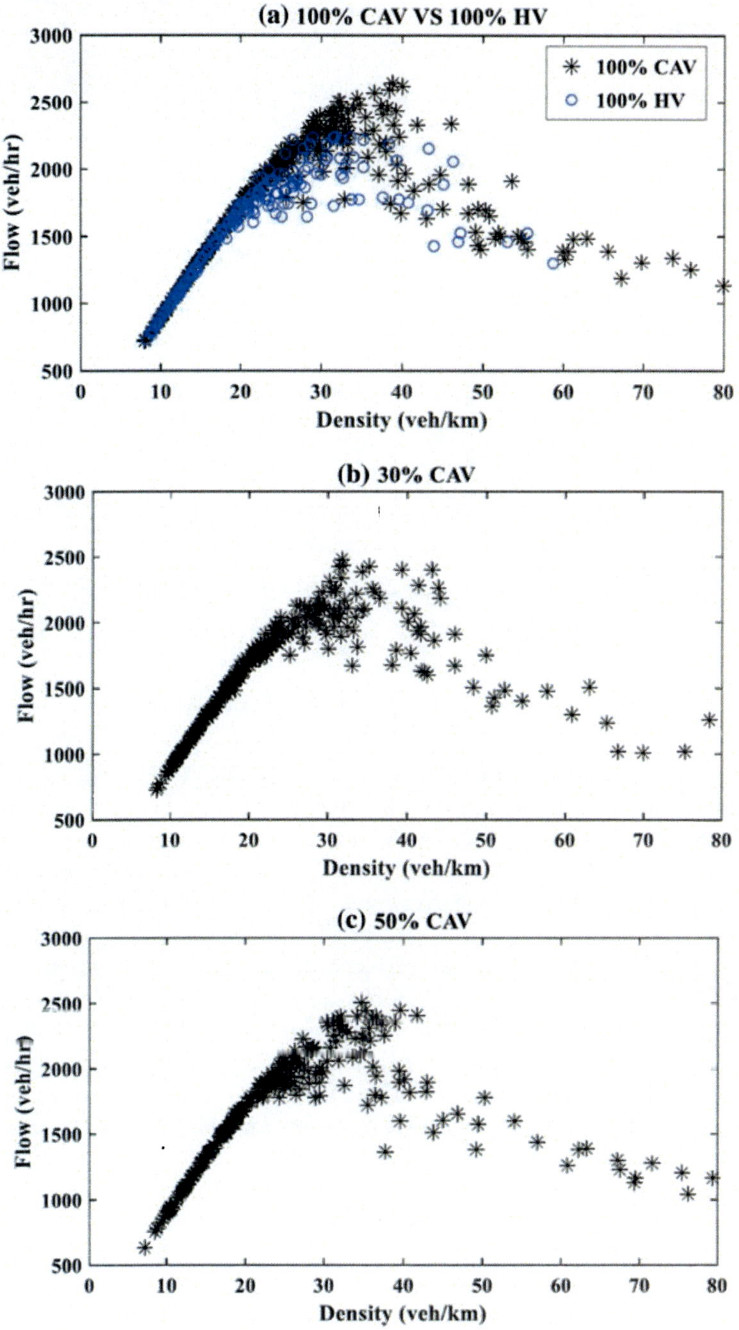

Fig. 3 Flow–density diagrams; **a** 100% CAV versus 100% HV; **b** 30% CAV; **c** 50% CAV; **d** 70% CAV

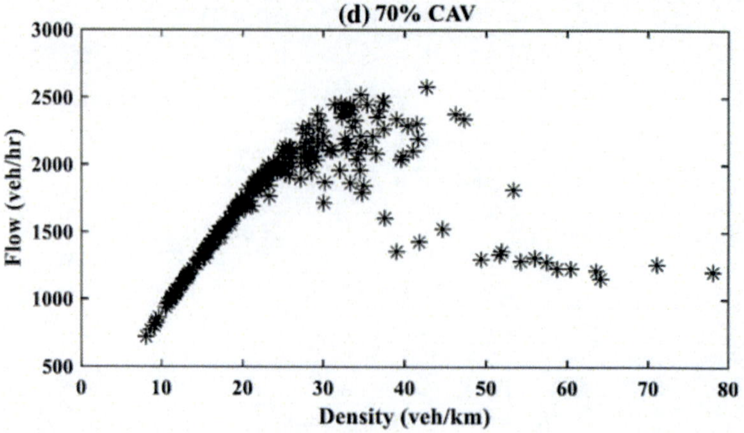

Fig. 3 (continued)

Gap metering methods alone or combined with RM can minimize overall delay by 17% and 27% more than RM alone at a 20% compliance rate [20].

7 Dynamic Traffic Assignment (DTA) Algorithm

The flow from an upstream connection is proportionate to its traffic demand, based on a detailed examination of current schemes. This method has been shown to capture essential aspects of a merging, such as a link capacity disparities and management of an on-ramp connection. This technique also produces a merging model that is computationally quick and simple to calibrate for dynamic traffic assignment (DTA) algorithms and other technologies used in advanced traffic management systems (ATMs). Network traffic flow models can capture system-wide components of traffic dynamics and are computationally effective for networks of actual scale which are desired by traffic engineers. The kinematic wave model is a good contender for these tasks because it provides an appropriate explanation of dynamic traffic events at the system level in terms of expansion and shock waves [21]. To utilize the kinematic wave model to simulate traffic dynamics on a network, first look at traffic dynamics at a merge, a diverge, or other network nodes [21].

The upstream section supply and downstream section demand are calculated first, and as a minimum, the boundary flow is used in a supply–demand technique. The principles of supply and demand are discussed here, however instead of "supply" and "demand," the words "sending flow" and "receiving flow" are used. The section demand is equal to its flow rate while traffic is under-critical (i.e., free flow); when traffic is over-critical (i.e., congested), the merging supply is equal to its flow capacity; and while traffic is over-critical (i.e., congested), the collection is similar to the merging flow capacity [21].

8 Simulation Models

One of the primary reasons for traffic congestion on highways is merging incoming vehicles. The disruptions in vehicle speed at the merge point could be decreased if the movements of cars joining the road could be organized in advance with the vehicle's movements currently traveling along the significant line motorway, reducing congestion and travel time. A high-reliability traffic microsimulation model is used to model these consequences. The outcomes are computed for a baseline scenario with manual merging, besides methods with 50, 60, 75, and 100% market penetration of vehicles with vehicle-to-vehicle (V2V) coordination and automated merge control abilities. The impact of vehicle location measurement flaws and V2V communication constraints (delays and packet losses) is also considered. The coordination of merging actions at freeway entry ramps, which are among the most major traffic bottlenecks on roadways, is one of the most significant potentials for V2V communication to contribute to traffic enhancements. This kind of collaboration comprises a diversity of message interactions that go beyond the basic safety messages (BSM) or cooperative awareness messages (CAM) sent by V2V collision awareness messages (CAM). Further communications are necessary to allow automobiles to request and negotiate the merging sequence [22].

Traffic simulation models may be utilized to evaluate several traffic engineering solutions, containing those that have yet to be built and tactics that have yet to be implemented. Each surrogate measure is predicated on the occurrence of a conflict event, which is a V2V contact in which one vehicle should take evasive action to prevent colliding. Time until collision, post-encroachment time, deceleration rate, maximum speed, and the speed difference is recommended as the best surrogate metrics. The intensity of the dispute may be measured using the time to collision, post-encroachment time, and deceleration rate. The maximum speed and the speed differential can be utilized to estimate the probable collision's severity. The post-processing tool would generate the statistics for different measurements and compare design options after the simulation model has been run for multiple iterations.

Microscopic simulation models show potential in collecting surrogate metrics of intersection safety. Microscopic traffic models classically replicate traffic systems vehicle by vehicle. Some simulations allow for even fewer time increments for more realistic behavioral analysis or an event-driven framework for greater computing efficiency. Microscopic simulations often involve comprehensive modeling of the traffic signal functioning, which is necessary for developing surrogate safety measures. On the other hand, all simulation models were built to assume that drivers behave safely based on their driver behavior characteristics. As a result, any derivation of alternate measures should consider this. Microscopic simulations must have the following properties: On a behavioral level, modeling of driver–vehicle interactions; access to comprehensive data from the simulation [APIs, output files, and open sources]. The ability to calibrate and choose model parameters, adjustments to a basis or outputs cost money, and user base, endurance, stability, and usefulness are general qualities [23].

9 Summary

The main points obtained from the above sections can be summarized in Table 1. This table describes the results of previous studies.

10 Conclusion

The main conclusions of this study are the following:

1. The lane-changing behavior has a significant effect to create a bottleneck or shockwave in traffic flow. Therefore, the lane-changing advisory system can significantly alleviate congestion.
2. RM lowers recurring traffic congestion in both space and time, while also increasing the average speed on a highway and somewhat increasing the serviced demand. As a result, the number of people diverting from the highway to the parallel arterial drops, and traffic conditions on the parallel arterial improve.
3. It was discovered that by shifting 10% of heavy cars, a merging section's throughput could be increased by 1%. On expressways, merging portions are one of the most common causes of traffic accidents. Because the variability of highway flow is smaller than that of a critical density, flow is a significant indicator of freeway capacity than the critical density.
4. Gap metering tactics alone or in combination with RM can reduce the overall delay by 17% and 27% more than RM alone at a 20% compliance rate.

Table 1 Main results of the previous studies

The author (s)	Results
Sharma et al. [2]	• A C-ITS application for in-vehicle use increases traffic efficiency around a merging area • The findings show that the lane-changing warning system may reduce traffic shockwaves and considerably reduce congestion. In addition to providing significant travel time savings of up to almost 21% around merging portions, the system substantially lowers the variation of travel time losses in the system
Asgharzadeh et al. [3]	• Field measurements reveal that queue discharge is between 10 and 30% lower than pre-breakdown capacity; this discrepancy is a capacity drop • At merge bottleneck sites, the ramp flow ratio has a detrimental impact on capacity. Additionally, higher ramp demands result in a more significant loss in capacity
Asgharzadeh and Kondyli [6]	• The average pre-breakdown flow rate per lane negatively relates to the number of lanes
Kumar et al. [7]	• The merging operation reduces the capacity of the basic highway segment by about 8%. Also, an increase in the percentage of heavy vehicles resulted in a 1% reduction in capacity. They found that by relocating 10% of the heavy trucks, the merging section's throughput might be increased by 1%
Wang [8]	• According to sensitivity testing on various merging durations, increased length may minimize merging failures
Van Aerde et al. [9]	• Added to the current ITS toolboxes for freeway merging control as a new ATM strategy "Gap Metering."
Yousif and Al-Obaedi [12]	• The critical occupancy is 23% and the optimum location for the loop detector for measuring occupancy is 300 m according to their simulation model
Carlson et al. [13], Smaragdis and Papageorgiou [14], Papamichail et al. [15]	• Length of ramp leads to successful RM in reducing traffic congestion the limited length may reduce the efficiency of RM
Zheng [17]	• For traffic signal timing of RM, a 20-s green time signal is used in most sites in the UK
Papageorgiou et al. [18]	• RM impacts drivers' route-choice behavior and may be used as the dynamic assignment mechanism to encourage the usage of corridor networks

(continued)

Table 1 (continued)

The author (s)	Results
Sun et al. [19]	• Traffic throughput can be enhanced by 10–15% using the suggested CDMMT method. Simulation findings demonstrate that when 100% CAVs are used, the section throughput rises by around 18%
Jin et al. [20]	• Gap metering methods alone or combined with RM can minimize overall delay by 17% and 27% more than RM alone at a 20% compliance rate
Gettman and Head [23]	• The maximum speed and the speed differential can be utilized to estimate the probable collision's severity

References

1. Abuamer, I.M., Celikoglu, H.B.: Local ramp metering strategy ALINEA: microscopic simulation-based evaluation study on Istanbul freeways. Transp. Res. Procedia **22**, 598–606 (2017). https://doi.org/10.1016/j.trpro.2017.03.050
2. Sharma, S., Papamichail, I., Nadi, A., Van Lint, H., Tavasszy, L., Snelder, M.A.: Multi-class lane-changing advisory system for freeway merging sections. IFAC-PapersOnLine **54**(2), 93–98 (2021). https://doi.org/10.1016/j.ifacol.2021.06.014
3. Asgharzadeh, M., Gubbala, P.S., Kondyli, A., Schrock, S.D.: Effect of on-ramp demand and flow distribution on capacity at merge bottleneck locations. Transp. Lett. **12**(8), 550–558 (2020). https://doi.org/10.1080/19427867.2019.1665774
4. Budhkar, A., Maji, A., Gandge, S.: Speed reduction on merging sections in mixed traffic: a case study. Transp. Res. Procedia **48**, 850–859 (2020). https://doi.org/10.1016/j.trpro.2020.08.096
5. Transportation Research Board (TRB).: Highway Capacity Manual (National Research Council, Washington, DC) (2000)
6. Asgharzadeh, M., Kondyli, A.: Effect of geometry and control on the probability of breakdown and capacity at freeway merges. J. Transp. Eng., Part A: Syst. **146**(7), 04020055 (2020). https://doi.org/10.1061/JTEPBS.0000381
7. Kumar, P., Arkatkar, S., Joshi, G.: Examining traffic flow parameters at merging section on high-speed urban roads in India. Curr. Sci. **117**(1) (2019). https://doi.org/10.18520/cs/v117/i1/94-103
8. Wang, J.: A merging model for motorway traffic Doctoral dissertation, University of Leeds (2006)
9. Van Aerde, M., Hellinga, B., Baker, M., Rakha, H.: Integration: an overview of traffic simulation features. Transp. Res. Records (1996)
10. Winyoopadit, S.: Development and comparative evaluation of ramp metering algorithms using microscopic traffic simulation. J. Transp. Syst. Eng. Inform. Technol. **7**(5), 51–62 (2007). https://doi.org/10.1016/S1570-6672(07)60039-7
11. Al-Jameel, H.A.E., Al-Jumaili, M.A.H.: Analysis of traffic stream characteristics using loop detector data. Jordan J. Civ. Eng. **10**(4) (2016)
12. Yousif, S., Al-Obaedi, J.: Modeling factors influencing the capacity of motorway merge sections controlled by ramp metering. Procedia Soc. Behav. Sci. **16**, 172–183 (2011). https://doi.org/10.1016/j.sbspro.2011.04.440
13. Carlson, R.C., Papamichail, I., Papageorgiou, M., Messmer, A.: Optimal main-stream traffic flow control of large-scale motorway networks. Transp. Res. Part C **18**, 193–212 (2010). https://doi.org/10.1016/j.trc.2009.05.014

14. Smaragdis, E., Papageorgiou, M.: Series of new local ramp metering strategies: emmanouil emaragdis and markos papageorgiou. Transp. Res. Record **1856**(1), 74–86 (2003). https://doi.org/10.3141/1856-08
15. Papamichail, I., Kotsialos, A., Margonis, I., Papageorgiou, M.: Coordinated ramp metering for freeway networks–a model-predictive hierarchical control approach. Transp. Res. Part C: Emerg. Technol. **18**(3), 311–331 (2010). https://doi.org/10.1016/j.trc.2008.11.002
16. Papamichail, I., Papageorgiou, M.: Traffic-responsive linked ramp-metering control. IEEE Trans. Intell. Transp. Syst. **9**(1), 111–121 (2008). https://doi.org/10.1109/TITS.2007.908724
17. Zheng, P.: A microscopic simulation model of merging operation at motorway on-ramps. Ph.D. Thesis, University of Southampton, UK (2003)
18. Papageorgiou, M., Hadj-Salem, H., Middelham, F.: ALINEA local ramp metering: summary of field results. Transp. Res. Rec. **1603**(1), 90–98 (1997). https://doi.org/10.3141/1603-12
19. Sun, Z., Huang, T., Zhang, P.: Cooperative decision-making for mixed traffic: a ramp merging example. Transp. Res. Part C: Emerg. Technol. **120**, 102764 (2020). https://doi.org/10.1016/j.trc.2020.102764
20. Jin, P.J., Fang, J., Jiang, X., DeGaspari, M., Walton, C.M.: Gap metering for active traffic control at freeway merging sections. J. Intell. Transp. Syst. **21**(1), 1–11 (2017). https://doi.org/10.1080/15472450.2016.1157021
21. Jin, W.L., Zhang, H.M.: On the distribution schemes for determining flows through a merge. Transp. Res. Part B: Methodol. **37**(6), 521–540 (2003). https://doi.org/10.1016/S0191-2615(02)00026-7
22. Chou, F.C., Shladover, S.E., Bansal, G.: Coordinated merge control based on V2V communication. In: 2016 IEEE Vehicular Networking Conference (VNC), pp. 1–8. IEEE (2016). https://doi.org/10.1109/VNC.2016.7835933
23. Gettman, D., Head, L.: Surrogate safety measures from traffic simulation models. Transp. Res. Rec. **1840**(1), 104–115 (2003). https://doi.org/10.3141/1840-12

Cooling Behaviour and Reusability of Hot Mix Asphalt

Aidiel Ashraf Abdul Razak, Norhidayah Abdul Hassan, Munzir Abdullah Zawawi, Mohd Zul Hanif Mahmud, Nordiana Mashros, and Azman Mohamed

Abstract Temperature is the most important element that can affect the performance of hot mix asphalt. There is a large amount of discarded asphalt mixture waste after construction due to the cooling phenomenon when exposed to surroundings. This study investigates the cooling rate of the AC14 mixture and the reusability of hot mix asphalt after reheating and compacting at various temperatures. In this study, the cooling behaviour of the loose asphalt mixture was monitored by measuring the surface temperature, the internal temperature of the mixture and the surrounding conditions. The cooling curve, cooling rate, and TAC were determined based on the results obtained. Other than that, for the reusability of hot mix asphalt, five different types of samples were prepared with a different number of reheating processes and compaction temperatures. The compacted samples were tested for volumetric properties and the Marshall stability test. The result determined the surface and internal TAC at 32 and 40 min, respectively. It was also found that reheating the cold mixture twice and compacting it at a considerably high temperature could achieve the required properties. Therefore, further studies should be conducted using large-scale samples to verify the finding.

Keywords Reusability of HMA · Cooling behaviour compaction time · Marshall stability · Compaction in various temperature

---

A. A. A. Razak · N. A. Hassan (✉) · M. A. Zawawi · M. Z. H. Mahmud · N. Mashros · A. Mohamed
Faculty of Civil Engineering, Universiti Teknologi Malaysia, Johor Bahru, Malaysia
e-mail: hnorhidayah@utm.my

M. Z. H. Mahmud
e-mail: mzhanif@utm.my

N. Mashros
e-mail: mnordiana@utm.my

A. Mohamed
e-mail: azmanmohamed.kl@utm.my

N. Sabtu (ed.), *Proceedings of AWAM International Conference on Civil Engineering 2022 - Volume 3*, Lecture Notes in Civil Engineering 386,
https://doi.org/10.1007/978-981-99-6026-2_23

1 Introduction

Asphalt pavement is known for its toughness and durability. Some states and agencies favour asphalt paving because of its durability and sustainability [1]. If correctly laid down, it can withstand for twenty to twenty-five years [2]. There are different types of asphalt mixture, such as cold mix asphalt (CMA), warm mix asphalt (WMA), and hot mix asphalt (HMA). HMA is manufactured at temperatures between 140 and 170 °C, and certain mixtures, including asphalt rubber and polymer-modified asphalt (PMA) mixtures, may be needed to achieve even higher mixing temperatures. These temperatures ensure that the asphalt coats the aggregate well and that the mixture has acceptable workability [3, 4]. Through the process of cooling, the temperature of HMA will decrease within some period until it becomes stable. This approach is generally linked to the time available for compaction, so the compaction work can be conducted to obtain the required density within the specific HMA cooling time. The rate of cooling of HMA is influenced by factors such as aggregate grading, lift thickness, and environmental factors such as wind, moisture, and surrounding temperature [5, 6]. These factors could increase the cooling rate and reduce the time available for compaction, leading to less durable pavement.

Many factors can make a poor condition for HMA paving. It is low-temperature surroundings, wind effects, and night-time construction. To get the optimum load bearing and weathering properties, the asphalt mixture must be compacted to some density range, and the time needed for HMA to attain acceptable compaction temperature to achieve considerable density will reduce as the rate of cooling goes up. The HMA compaction is usually conducted by using a roller. The roller operator can determine the most suitable time to begin the compaction by judging the depth of a heel imprint. This method will be used when the temperature is high enough. However, low temperature, high wind speed, and long paving time cause the rate of heat loss increases for the mix. During all these conditions, the ability to predict the temperature of the blend is more important because it limits the Time Available for Compaction (TAC). The control mechanisms cited from the locally used standards are generally the appropriate limits of transmission and laying completion temperature in local operation. No items are available to forecast these control components and to be directly connected to local circumstances. Therefore, the validation of these limitations involves study to be precisely connected to local circumstances, i.e. material properties, environmental conditions, and processes of compaction to satisfy local environmental conditions is significant for further study. Additionally, to avoid wastage in handling asphalt, the reusability of the asphalt after cooling was also evaluated for justification.

This study evaluates the cooling behaviour and reusability of the HMA compacted at various temperatures. The objectives are as follows:

1. To determine the cooling rate, time available for compaction, TAC, and reusability of dense graded asphalt under various surrounding conditions.
2. To evaluate the Marshall properties of dense graded asphalt compacted at various temperatures.

2 Literature Review

2.1 *Hot Mix Asphalt*

Hot mix asphalt (HMA) is used in local roads, highways, and airport pavement as a surface layer. HMA being laid on the surface of the road structural layer can provide high endurance for the pavement to receive the loads and stresses from the traffic [7]. HMA consists of an evenly blended and coated aggregate with bitumen. It comprises approximately 95% aggregate of different sizes, blended with approximately 5% bitumen as a binder. Both the aggregate and bitumen must be heated before mixing to dry the aggregates and achieve adequate bitumen fluidity for proper mixing and workability. Bitumen is composed of a complicated mixture of hydrocarbons, including calcium, iron, sulphur, and oxygen [8]. It is a solid, flexible, and weather–and the chemical-resistant binding medium used for many applications, perhaps most commonly to coat aggregate into a hard surface. Refined bitumen is created by distilling crude oil, whereas natural bitumen is extracted from bituminous sands and rocks that developed in nature over 360 million years [9].

On the other hand, the required aggregate must also be chosen for the prerequisite of a good combination when a decision recognizes a particular aggregate and measures its specific gradation and absorption [10]. HMA is graded by measuring the percentage of different aggregate sizes to achieve low void content within the mix. The shear strength relies primarily on the internal friction supplied by the aggregate. The cubical, rough-textured aggregate gives more shear resistance than the squared, smooth-textured aggregate. The internal friction allows the aggregate to interlock and produce a solid mass capable of withstanding the traffic load applied [11].

2.2 *Cooling Rate*

Usually, HMA is produced at the premix plant, then brought to the construction plant, and spread on the road surface with a temperature of 140–160 °C [3]. While the bitumen is in a liquid state, it can easily coat the aggregates and bind the material effectively. During the paving process of HMA, the cooling phenomenon must be well monitored, particularly when the compaction process is held, as it should provide sufficient time to compact. If the HMA cools to the surrounding temperature, it will become highly viscous, hard enough, and difficult to compact. These could lead to insufficient compaction to the pavement and low density and strength. The temperature of HMA decreases after it has been laid on the road surface. There are many ways for heat loss in HMA. The quantity of heat transmission is determined by the surface material and colour and the incoming radiation's wavelength [12]. The temperature distribution is influenced directly by the characteristics of the thermal environment to which it is exposed. Heat loss in asphalt mixture is based on three mechanisms: convection, conduction, and radiation [13].

2.3 Compaction

Compaction is one of the processes involved in road construction to produce a great-strength of asphalt mixture. Also, the compaction process causes the asphalt-concrete mix to be compressed, and its volume is reduced. As the density of the hot mix asphalt material increases, the air-void content of the mix decreases (they are inversely proportional to each other) [14]. As this compaction step directly affects the pavement life cycle, HMA layer compaction is one of the most important parameters to be monitored when paving new roads [15]. This process reduces the void in the HMA to meet the specified quality. Mix design is another important element in determining the efficient compaction process. Asphalt mix design is the procedure to determine the optimum percentage of bitumen, fine aggregates, and coarse aggregates to have a long-lasting pavement. Many factors must be considered during pavement construction, especially during the compaction process. The TAC is one of the major aspects of the compaction process. It can be characterized as the time of cooling and stiffness of the asphalt mix to the point that it can absorb the compaction energy added without causing the aggregate particles to travel [16]. TAC is the most important factor to ensure the pavement compaction is done adequately at the desired density to get stable and endurance pavement.

3 Methodology

This study focuses on preparing AC14 Marshall samples, the cooling behaviour of the AC14 mixture and the reusability of the AC14 mixture after cooling with the different reheating and compaction temperatures.

3.1 Preparation of AC14 Samples

This study used the conventional mix with bitumen penetration grade 60–70 and granite aggregates as stated in JKR/SPJ/2008-S4 specification [17]. The HMA of AC14 was prepared according to the JKR specification, as shown in Table 1.

3.2 Cooling Monitoring

For this monitoring, three samples (of 1200 g weight) were produced and exposed to the surrounding. This study used several tools during the cooling process monitoring, i.e. a stopwatch, anemometer, thermal imaging camera, and thermocouple. An

Table 1 Gradation limits for asphaltic concrete (AC14) [17]

Sieve size (mm)	Percentage passing by weight (%)
20.0	100
14.0	90–100
10.0	76–86
5.0	50–62
3.35	40–54
1.18	18–34
0.425	12–24
0.150	6–14
0.075	4–8

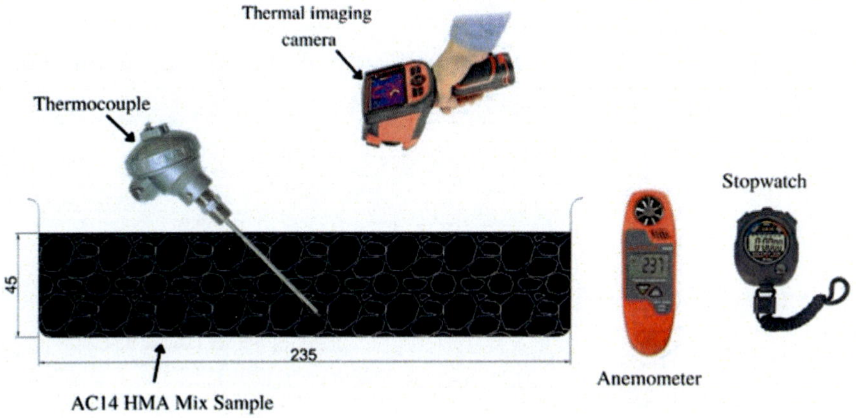

Fig. 1 Cross-section of sample and tools during monitoring

anemometer was used to measure the wind speed and thermal imaging. The set-up preparation is shown in Fig. 1.

3.3 Compacted Sample Preparation (Reusability)

After monitoring the cooling curve, the sample was reheated once and twice for the reusability evaluations. A total of 12 samples were produced. There are four types of samples: single-reheated and compacted at 135 °C, single-reheated and compacted at 150 °C, double-reheated and compacted at 135 °C, and double-reheated and compacted at 150 °C. The loose mix was cooled and reheated multiple times before compaction to determine the recyclability of the asphalt after cooling on-site to avoid wastage during construction. After compaction, the samples were determined for volumetric properties and Marshall stability tests.

3.4 Marshall Properties Test

The Marshall properties were evaluated in terms of void filled with asphalt (VFA), void in the total mix (VTM), stability, flow, and stiffness.

VFA is a percentage of void filled with bitumen. It can be calculated by using Eq. 1.

$$VFB = [1 - (G_{mm}{*}PA/G_{mb})] \times 100 \quad (1)$$

where

G_{mm} Theoretical Maximum Density.
PA Aggregate Percentage.
G_{mb} Aggregate Bulk Specific Gravity.

VTM is the percentage volume of void in the asphalt mixture. It can be calculated by using Eq. 2.

$$VTM = [1 - (G_{mb}/G_{mm})] \times 100, \quad (2)$$

where

G_{mb} Bulk Specific Gravity.
G_{mm} Theoretical Maximum Density.

4 Results and Discussion

This section presents the results of laboratory works, which include the cooling curves and surrounding conditions under monitoring for cooling behaviour monitoring, followed by the Marshall properties results of AC14. Samples were reheated once and compacted at 135 and 150 °C; samples were reheated twice and compacted at 135 and 150 °C to evaluate the reusability of hot mix asphalt.

4.1 Cooling Behaviour (Day Time)

Figure 2 shows the cooling curves of AC14 (average) monitored in the morning at approximately 11:00 am. The sample was monitored in a loose form after the mixing process. Based on the graph, the total time for the internal temperature (thermocouple) to achieve the cessation temperature is 40 min. While the total time for the surface temperature is at 32 min. Based on this figure, the rate of cooling for the surface temperature is higher than the internal temperature due to the direct exposure to the surrounding. The surface temperature drops rapidly for the first 6 min compared

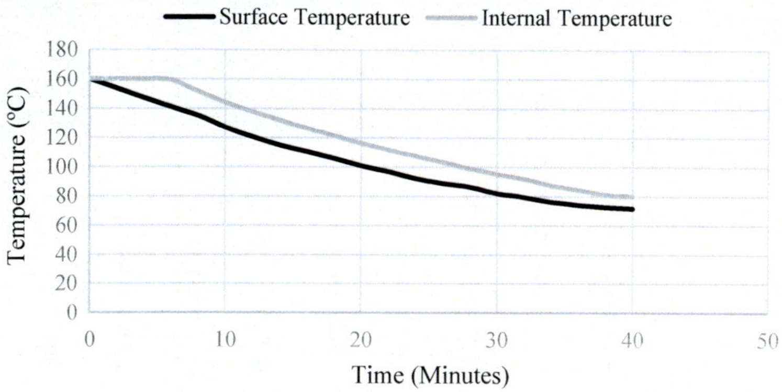

Fig. 2 Average cooling curves for internal and surface temperatures

Table 2 Cooling rate at 4 min interval for AC14

Time (min)	Cooling rate (°C/min)										TAC (min)
	4	8	12	16	20	24	28	32	34	36	
Internal temperature	0.0	2.1	3.4	2.9	2.6	2.2	2.2	1.7	1.9	1.1	40.0
Surface temperature	3.2	3.1	3.6	2.5	2.4	2.2	1.5	1.7	1.3	0.6	32.0

to the internal temperature. The curve is flat at the beginning of monitoring due to trapped heat within the mixture that maintains the temperature.

The cooling rate value is summarized in Table 2 and plotted in Fig. 3. The value was determined at the interval of 4 min during the monitoring. By comparing both surface and internal cooling rates, the sample's surface has a higher cooling rate than the sample's internal structure, particularly at the beginning of the exposure to the surrounding at approximately the first 12 min. However, after 12 min onwards, further monitoring shows a higher cooling rate for internal temperature as the internal heat starts to release to the surrounding until it reaches the cessation temperature.

4.2 *Humidity and Surrounding Temperature*

Figure 4 shows the surrounding temperature and percentage of humidity for an average of three samples. Based on the graph, the relative humidity was quite constant, ranging between 63 and 64%. While the temperature of the surrounding was measured at approximately 32 °C. Due to certain circumstances, exposure to surroundings has been conducted in the laboratory within a controlled environment.

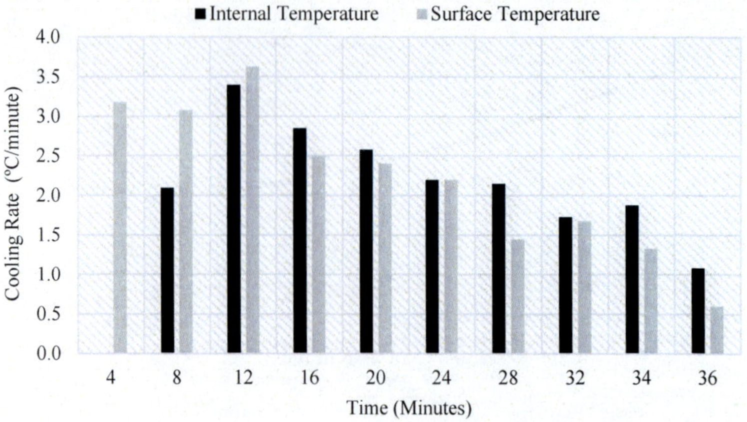

Fig. 3 Plot of cooling rate versus time

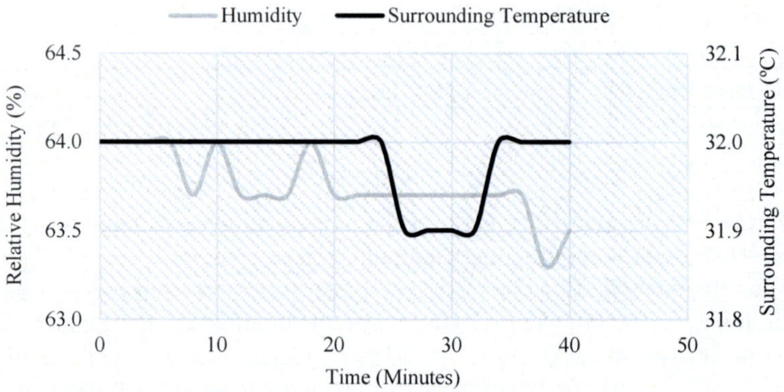

Fig. 4 Humidity and surrounding temperature versus time

4.3 Reusability of AC14 Mixture

Table 3 summarizes the result of the Marshall properties evaluated from five different types of sample preparation. The non-reheated samples are verified samples produced for the Marshall mix design. For 'single reheat', the cold samples were reheated once and compacted at two different temperatures; for 'double reheat', the cold samples were reheated twice and compacted at two specified temperatures. From the table, the single-reheated sample compacted at 135 °C complies with all the specified standards. The same observation was made for a double-reheated sample compacted at a higher temperature of 150 °C, where the results comply with the specification. This shows that by reheating the samples twice, more heat is needed to compact the material, possibly due to the binder's ageing. Other samples were found good in terms of stability and stiffness but slightly out of range for flow, VFB, and VTM, as shown in Table 3. This shows that the compaction temperature and the heating action affect the Marshall properties of the sample.

Figure 5 shows the plot of void in the total mix (VTM) and the void filled with bitumen (VFB) for all samples produced. Based on the JKR requirement for the VTM of the Marshall sample, the value must be within 3–5%. On the other hand, for the VFB, the acceptable value based on the specification is between 70 and 80%. From the graph, all the compacted samples comply with the specification except for the double-reheated samples compacted at 135 °C. It was also observed that the aged samples cooled faster for the second cooling attempt than the first attempt as the material became stiffer than the initial stage and less workable, potentially resulting in less compaction. However, as the compaction temperature increases, the total air voids reduce.

Figure 6 shows the stability and flow data measured for the different types of samples. For stability, all the values measured comply with the JKR standard of more than 8000 N. The single-reheating process seems to produce comparable samples with comparable stability to the control mixture, even though the samples were compacted at different temperatures. However, as the samples were reheated twice, compaction at a lower temperature seemed to reduce the stability at 12,033 N. Still,

Table 3 Marshall properties results

Types	Reheating (No Reheat (NR)/ Single Reheat (SR)/Double Reheat (DR))	Flow (mm)	Stiffness (N/mm)	Stability (N)	VTM (%)	VFB (%)	Specific gravity
Control (Compacted @150 °C)	NR	4.0	3455	13,844	4.8	70	2.310
Single reheat (Compacted @135 °C)	SR	3.1	4419	13,625	4.6	71	2.320
Single reheat (Compacted @150 °C)	SR	**4.4**	3076	13,673	4.8	70	2.310
Double reheat (Compacted @135 °C)	DR	4.0	3021	12,033	**5.2**	**69**	2.300
Double reheat (Compacted @150 °C)	DR	3.5	4099	14,293	4.4	72	2.320
JKR requirement	–	2.0–4.0	> 2000	> 8000	3.0–5.0	70–80	–

The bold values refer to values that don't comply with JKR requirements

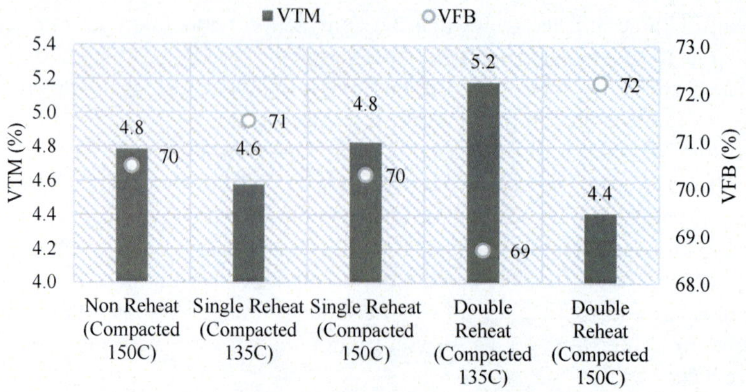

Fig. 5 Void in total mix (VTM) and void filled by bitumen (VFB)

the stability was improved to the highest at 14,293 N after compacting at a higher temperature. The increase in stability could be due to the aged binder after so much exposure to heat and better workability at higher compaction temperatures. On the other hand, the flow results show good agreement with the specification except for the sample that was reheated once and compacted at 150 °C as the value was slightly higher than the maximum limit at 4.4 mm.

Figure 7 shows the plot of the stiffness and density for all the prepared samples. Based on the JKR requirement, the stiffness must comply with the minimum limit of 2000 N/mm. Overall it can be summarized that the stiffness of all the prepared samples exceeds the minimum requirement. The reheating process does not negatively impact the stiffness value even after the double-reheating process, with the lowest observed at 3021 N/mm for samples compacted at 135 °C. The graph shows that the highest density is achieved for the double-reheat sample compacted at 150 °C, as reflected by the lowest air voids discovered for the sample at 4.4%. When the samples were compacted at 135 °C after the second reheating process, the mixture seemed to experience less workability during the compaction, as shown by the lowest density achieved at 5.2% total air voids.

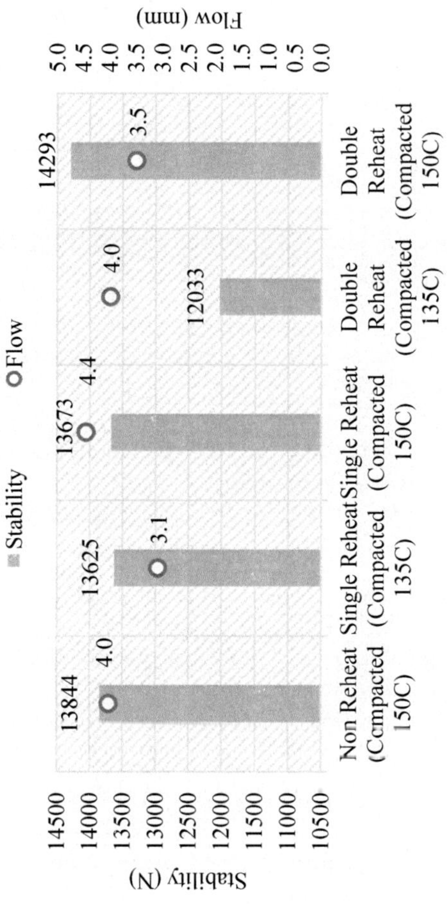

Fig. 6 Stability and flow

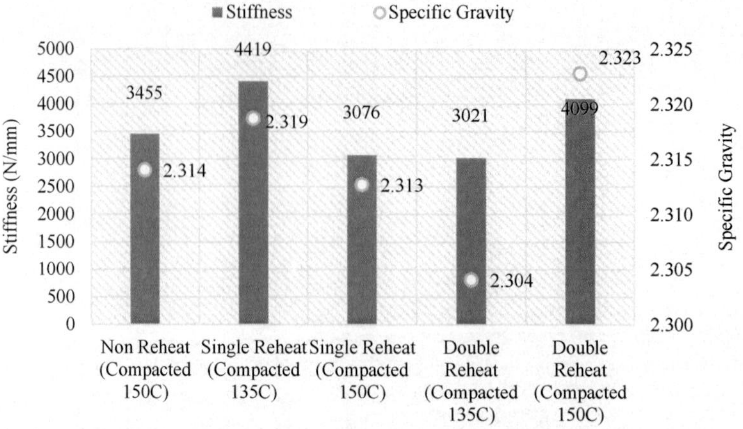

Fig. 7 Stiffness and specific gravity

5 Conclusion

This study focuses on determining the cooling behaviour of cooling rate, time available for compaction (TAC) for AC14 mix, and reusability of AC14 hot mix asphalt after cooling by using several tests such as volumetric properties and Marshall stability test. Overall, it was found that surrounding conditions affect the cooling rate of the mix. Other than that, the reusability of the AC14 mix should emphasize several factors that can affect the performance of the compacted asphalt mixture. The following conclusions are made based on the result obtained from this study:

1. The cooling rate and TAC are affected by the surrounding conditions, such as surrounding temperature and humidity.
2. The AC14 mix was found to have time for compaction (TAC) for internal and surface of 40 min and 32 min, respectively.
3. From the Marshall properties, HMA (AC14) reusability is applicable by reheating the cold mixture during the first and second attempts. A higher compaction temperature is preferred for the second reheat to improve workability.

For recommendation, further study should be conducted on evaluating other mechanical performances of the reheated samples, particularly in resilient modulus, dynamic creep, and rutting susceptibility.

Acknowledgements This work was supported/funded by the Malaysia Ministry of Higher Education under the Fundamental Research Grant Scheme (FRGS/1/2019/TK01/UTM/02/6).

References

1. Al-Saffar, Z.H., Yaacob, H., Katman, H.Y., Mohd Satar, M.K., Bilema, M., Putra Jaya, R., Eltwati, A.S., Radeef, H.R.: A review on the durability of recycled asphalt mixtures embraced with rejuvenators. Sustainability **13**(16), 8970 (2021). https://doi.org/10.3390/su13168970
2. Neal, B.P.P.: What is Asphalt Pavement? Pavemanpro (2019, July 4). https://www.pavemanpro.com/article/how_to_sealcoat_parking_lot_chapter_1/
3. Kim, Y., Lee, J., Baek, C., Yang, S., Kwon, S., Suh, Y.: Performance evaluation of warm- and hot-mix asphalt mixtures based on laboratory and accelerated pavement tests. Adv. Mater. Sci. Eng. **2012**, 1–9 (2012). https://doi.org/10.1155/2012/901658
4. Speight, J.G.: Asphalt paving. In: Asphalt Materials Science and Technology, pp. 409–435. (2016).https://doi.org/10.1016/b978-0-12-800273-5.00010-6
5. Badruddin, I., Hassan, N.A., Mahmud, M.Z., Ismail, S., Jaya, R.P., Oleiwi Aletba, S.R., Hainin, M.R.: Effect of environmental exposure on the cooling rate of asphalt mixtures. IOP Conf. Ser.: Earth Environ. Sci. **682**(1), 012072 (2021). https://doi.org/10.1088/1755-1315/682/1/012072
6. Ismail, S., Hassan, N.A., Yaacob, H., Hainin, M.R., Ismail, C.R., Mohamed, A., Satar, M.K.I.M.: Properties of dense-graded asphalt mixture compacted at different temperatures. IOP Conf. Ser.: Earth Environ. Sci. **220**, 012010 (2019). https://doi.org/10.1088/1755-1315/220/1/012010
7. Garg, N., Mounier, D.: Comparison of US (FAA) and French (DGAC) airport pavement HMA. Road Mater. Pavement Des. **14**(sup1), 242–261 (2013). https://doi.org/10.1080/14680629.2013.774760
8. Bitumen Definition.: Investopedia (2020, October 28). https://www.investopedia.com/terms/b/bitumen.asp
9. Infinity Galaxy Co.: What is Bitumen? Everything About Bitumen Definition and Types. Infinity Galaxy (2021, November 14). https://infinitygalaxy.org/bitumen/
10. Yuan, Q., Liu, Z., Zheng, K., & Ma, C. (2021). Chapter 7—Asphalt. In: Civil Engineering Materials (pp. 287–325). www.sciencedirect.com. https://doi.org/10.1016/B978-0-12-822865-4.00007-6
11. Bhasin, A., Masad, E., Little, D., Lytton, R.: Limits on adhesive bond energy for improved resistance of hot-mix asphalt to moisture damage. Transp. Res. Rec.: J. Transp. Res. Board **1970**(1), 2–13 (2006). https://doi.org/10.1177/0361198106197000101
12. Aletba, S.R.O., Abdul Hassan, N., Putra Jaya, R., Aminudin, E., Mahmud, M.Z.H., Mohamed, A., Hussein, A.A.: Thermal performance of cooling strategies for asphalt pavement: a state-of-the-art review. J. Traffic Transp. Eng. (Engl. Ed.) **8**(3), 356–373 (2021). https://doi.org/10.1016/j.jtte.2021.02.001
13. Basheer Sheeba, J., Krishnan Rohini, A.: Structural and thermal analysis of asphalt solar collector using finite element method. J. Energy **2014**, 1–9 (2014). https://doi.org/10.1155/2014/602087
14. Compacting hot mix asphalt pavements: part I. Roads Bridges (2021b, February 16). https://www.roadsbridges.com/compacting-hot-mix-asphalt-pavements-part-i
15. Fauchard, C., Li, B., Laguerre, L., Héritier, B., Benjelloun, N., Kadi, M.: Determination of the compaction of hot mix asphalt using high-frequency electromagnetic methods. NDT E Int. **60**, 40–51 (2013). https://doi.org/10.1016/j.ndteint.2013.07.004
16. O'Flaherty, C.A.: Highway Engineering, vol. 2. Edward Amold Publishers, London UK (1998). http://www.sciepub.com/reference/92272
17. Raya, J.K.: Standard Specification for Road Works. Section 4: Flexible Pavement. Cawangan Jalan, Jabatan Kerja Raya Malaysia (2008)

Road Crashes Among Food Delivery Riders (P-Hailing) During Pandemic in Kuala Lumpur

Ahmad Raqib Ab Ghani, Al Insyirah Abd Malek, Wan Azfizatul Az Zarah Wan Mohamad Yusoff, Nik Shahidah Afifi Md Taujuddin, and Kamarudin Ambak

Abstract Due to COVID-19 outbreak in Malaysia, a movement control order was implemented on March 18, 2020. As a result, many people ordered food through food delivery services, which may have come at a high cost due to the recent spike in food deliveries. This study aims to establish the causes and the most frequent mentioned elements that contribute to road crashes among riders. Road accidents are a severe concern since they result in death or injury, yet the number of fatalities among riders continue to rise. The study was done in the vicinity of Kuala Lumpur and the data was gathered through a survey using online Google Form questionnaire. The instrument was measured using a reliability test and exploratory factor analysis (EFA) to generate an empirical verification of the questionnaire's validity and reliability. Then, a descriptive analysis was undertaken for each variable to determine the primary factor. As a result, the critical fact resulted in cell phone use while travelling on the highway was a human factor component. The study's findings drew more attention to the issue and raised awareness about road safety.

Keywords Food delivery rider · P-hailing · Motorcycle accident · Exploratory factor analysis · Descriptive analysis

1 Introduction

Road transportation is a requirement that is highly beneficial to a country and its citizens, especially in enhancing access to the workplace, financial services, educational facilities, and health care facilities. After chronic diseases such as heart failure, stroke, and pulmonary infection and difficulty in breathing, injury caused by a car collision is one of the leading causes of death worldwide. It had also harmed the city's primary

A. R. A. Ghani (✉) · A. I. A. Malek · W. A. A. Z. W. M. Yusoff · N. S. A. M. Taujuddin · K. Ambak
Universiti Tun Hussein Onn Malaysia, Parit Raja, Johor, Malaysia
e-mail: ahmadraqib@uthm.edu.my

N. Sabtu (ed.), *Proceedings of AWAM International Conference on Civil Engineering 2022 - Volume 3*, Lecture Notes in Civil Engineering 386,
https://doi.org/10.1007/978-981-99-6026-2_24

mode of transportation, which is traffic accidents, including deadly accidents. From 1990 to 2011, total accident-related deaths in Malaysia grew every year, while total road-related deaths increased by 70% [1, 2]. According to police records gathered during Operation Selamat 16, the number of fatal road accident cases has increased somewhat, according to Deputy Inspector-General of Police Datuk Mazlan Mansor. According to him, there was a minor increase in fatalities this year when 206 fatal accident instances were reported compared to 197 incidents in 2019.

The number of deaths has also risen from 216 to 230. Deaths have also increased from 216 to 230. With 147 deaths, motorcyclists and pillion riders had the largest number of fatal and most significant statistic [3].

Motorcycle delivery services are well-known in Malaysia, and they operate in a variety of businesses, including food delivery, postal delivery, and small item delivery. Except those who work for registered business owners, no specific figures on the number of registered motorcyclists were available [4]. Food delivery personnel have been viewed as a vulnerable category among other types of motorists because they are forced to deliver orders to consumers within the period set by the managements. They were more likely to be exposed to the road while on the mission. Even though they are at a higher risk of being engaged in a collision, they are also at a higher hazard stage due to their longer exposure on the lane [5].

COVID-19, a novel coronavirus discovered in Wuhan, China, in December 2019, was declared a pandemic by the World Health Organization (WHO) in early March 2020. Most countries' most popular countermeasure to COVID-19's proliferation was a "lockdown," which limited people's daily activities to only the most essentials and prevented them from leaving their homes [6]. Meanwhile, to prevent the virus from spreading further, Malaysia had implemented a movement control order (MCO) that took effect on March 18, 2020. However, as more people become stranded at home due to the COVID-19 outbreak, the demand for food delivery services is increasing. Unfortunately, with motorcycle riders risking their lives to meet deadlines while attempting to increase commissions by increasing the number of pick-ups and deliveries, the recent increases in food delivery may have come at a high cost. A total of 150 fatal accident riders from two well-known food delivery firms were recorded three months after the MCO was enforced on March 18 to contain the COVID-19 outbreak [7].

As a result, this study aims to establish the causes and the most frequent mentioned elements that contribute to road crashes among riders. The paper's goals could be met, and an accurate instrument to assess each variable's reliability and validity on the tool could be created. Food delivery rider criteria and features, methodology, data analysis, findings, debate, and conclusion are all covered in this section.

2 Research Methodology

2.1 Sample of Study and Data Collection

This study was primarily conducted in Kuala Lumpur. The target participants for this study was all types of food delivery riders from various Kuala Lumpur-based companies. Because most of these meal delivery services were centred in Kuala Lumpur's big cities, "FoodPanda," "GrabFood," "DeliverEat," "UberEats," "Honestbee," "RunningManDelivery," "FoodTime," "Dahmakan," "Mammam," and "Shogun2U" were among the companies [8]. In Kuala Lumpur, it was believed that over 3500 motorcycle delivery riders, including food and parcel deliveries, were present [9]. As a result, the sample size needed for this study was $s = 346$. The sample size grew slower and remained relatively steady in slightly more than 380 cases [10].

After determining the target population, sampling frame, sampling method, and sample size, the next step was to collect data [11]. First, a "Google Form" was used to create the survey question. The information was then disseminated via social media platforms such as "Twitter," "Facebook," "Whatsapp," "Telegram," and "Instagram," as no personal interaction, including interviews and questionnaire delivery, was permitted during the outbreak.

2.2 The Instrument (Questionnaires)

A group of food delivery riders was investigated. The survey approach was used to construct many questionnaires. They were separated into three sections: A, B, and C. Section A comprised several questions based on the population randomly assigned to them based on their profile. Section B included several questions about human variables that contribute to traffic crashes among cyclists. Lastly, Section C was where the questions about environmental factors that cause road crashes among riders were asked. There will be 33 questions, with five questions in area A, 11 questions in section B, and 13 questions in section C.

Section A was, first and foremost, a demographic data survey that allowed for the gathering of background information. Gender, age range, work time hour, riding experience, and riding frequency were the five sorts of inquiries. The questions provided context to the collected survey data, allowing the respondents to be described and the data to be further analysed.

Second, the inquiries in this area revolved around what human variables contribute to motorcycle accidents. Human factors were researched and grouped into three types of behaviour: red-light running, cell phone use while riding, and speeding. As a result, multiple-choice survey questions required respondents to select only one option from various options. Because the response options were pre-determined, the respondents had a more straightforward time filling out the survey. They were also simple to employ in various ways, and they assisted in the generation of knowledge that was simple to analyse and provide mutually exclusive preferences.

Finally, to achieve the goal in this part, a type of dichotomous survey question was used. Many designed criteria were used to inquire about environmental factors that cause road crashes among riders: lighting, roadway surface, and the weather conditions. There are two possible answers to the dichotomous question. In surveys, dichotomous questions such as Agree/Disagree, Yes/No, True/False, or Fair/Unfair are frequently employed. They were used to distinguish between respondents' characteristics, experiences, and opinions. As a result, Yes/No is employed as a sort of response in each question.

2.3 Data Analysis

Reliability and Validity Test. Alpha Cronbach will express the instrument's reliability level based on the significant objectives mentioned. Alpha Cronbach's alpha value of 0.6 is considered high dependability and a good index value [12, 13]. In comparison, Alpha Cronbach's score was less than 0.6, indicating that it was faulty. Alpha Cronbach's coefficients of 0.60–0.80 are considered modest yet acceptable. In the 0.8 and 1.00 ranges, however, Alpha Cronbach was regarded to be quite good. Therefore, the level of reliability of the devices was assessed using Alpha Cronbach values for the instruments created [14].

On the other hand, validity refers to an empirical measure's ability to accurately detect the genuine meaning of the concepts under investigation [15]. According to its definition, validity is a tool that analyses whether data is suitable, honest, relevant, and valuable [16]. Exploratory Factor Analysis is used in this study to determine the instrument's validity (EFA). Furthermore, in this study, EFA served another purpose: identifying and combining multiple items from the questionnaire into structures under a single variable [17]. As proposed by Hair et al. [18], EFA was required to examine the organization of hidden dimensions among the observable variables displayed in the items of an instrument. As a result, in this study, using Exploratory Factor Analysis and Cronbach's Alpha would be the best technique to acquire empirical verification of the questionnaire's validity and reliability.

Descriptive Statistic In a quantitative research study, descriptive statistics were statistical representations of what was provided or the data disclosed understandably. The central tendency metrics were examples of the mean, median, mode, and variability sizes suited for standard deviation, variance, and minimum and maximum variables. Tables, graphs, and discussion threads were utilized in two of these measures to help people understand the significance of the data collected and processed.

3 Results and Discussion

3.1 *Reliability Analysis*

Statistical Analysis of the Reliability of the Pilot Study. For sections B and C of the questionnaire form, which was sent to 35 respondents in the circle of the riders, the statistical findings of the pilot study's reliability were examined using SPSS software. Cronbach's Alpha values were slightly low for part B (0.486) and extremely good for section C (0.913). As a result, the section C questions can be directly applied to the actual study without further adjustment. The questions in section B had been thoroughly evaluated, and if one of the questions was removed from the questionnaire form, fresh Cronbach Alpha values of 0.610 were created. The Cronbach Alpha value was higher than before, even though the fifth question related to one of the human variables, cell phone use while riding, was removed from SPSS. As a result, the question will no longer be required for the research.

Statistical Analysis of the Reliability of the Actual Study. After executing a pilot study, Cronbach's Alpha values were still required for the primary investigation, which had a total of 346 participants. As a result, the analysis was carried out in the same way as the pilot research. This study's reliability analysis was divided into two portions, B and C, referred to as human factors and environmental variables, respectively. Both sections' Cronbach Alpha ratings were moderate, at 0.781 and 0.856, respectively, indicating that both were good because they exceeded 0.6. In addition, the Cronbach's Alpha value for section B grew in this study, surpassing that of the previous one. Both portions were sufficient in demonstrating that all the cases mentioned above occurred among the motorcyclists, resulting in the road accident.

3.2 *Validity Analysis*

The validity test is carried out using factor analysis. It is also utilized to ensure that the quantifiable items employed in this study are acceptable. The two statistical measures employed to determine the capacity to do factor analysis were the Kaiser–Meyer–Olkin (KMO) sample adequacy measure and Bartlett's Test of Sphericity. Bartlett's Test of Sphericity must be significant (p 0.05) for the factor analysis to be considered acceptable, and the KMO index must be 0.60 as the minimum value for good factor analysis [19]. For the independent and dependent variables in this section, two-factor analyses were done separately.

The Kaiser–Meyer–Olkin (KMO) Test is a measure of how suited our data is for Factor Analysis. The test measures sampling adequacy for each variable in the model and for the complete model. The statistic is a measure of the proportion of variance among variables that might be common variance. The lower the proportion is, the more suited our data is to Factor Analysis. KMO return values are between

Table 1 KMO and Bartlett's test for independent variables

KMO and Bartlett's test		
Kaiser–Meyer–Olkin measure of sampling adequacy		0.650
Bartlett's test of Sphericity	Approx. chi-square	167.816
	Df	10
	Sig	0.000

0 and 1. A rule of thumb for interpreting the statistic by KMO is discussed. KMO values that are between 0.8 and 1 indicate that the sampling is adequate while values that are less than 0.6 indicate that the sampling is not adequate and remedial action should be taken. Some authors put this value at 0.5, so use our own judgement for values that are between 0.5 and 0.6, whereas values that are close to zero means that there are large partial correlations compared to the sum of correlations. In other words, there are widespread correlations which is a huge problem for factor analysis.

Bartlett's Test of Sphericity compares an observed correlation matrix to the identity matrix. Essentially, the test is used to check if there is a certain redundancy between the variables that we can summarize with a few numbers of factors. The null hypothesis of the test is that the variables are orthogonal, i.e. not correlated. The alternative hypothesis is that the variables are not orthogonal, they are correlated enough to where the correlation matrix diverges significantly from the identity matrix. This test is often performed before we use a data reduction technique such as principal component analysis or factor analysis to verify that a data reduction technique can compress the data in a meaningful way.

Independent Variables. The significance of the KMO and Bartlett's Tests for the independent variable is shown in Table 1. The KMO was 0.650, more significant than the previous reading of 0.60. According to Bartlett's test, the correlation matrix is factorable, which yielded a statistically significant result (p 0.00).

Dependent Variables. The significance of the KMO and Bartlett's Tests for the dependent variable is shown in Table 2. The KMO was 0.834, more significant than the previous value of 0.60. The correlation matrix is factorable according to Bartlett's test, which yielded a statistically significant result (p 0.00).

Table 2 KMO and Bartlett's test for dependent variables

KMO and Bartlett's test		
Kaiser–Meyer–Olkin measure of sampling adequacy		0.834
Bartlett's test of Sphericity	Approx. chi-square	2709.656
	df	253
	Sig	0.000

3.3 Descriptive Statistics

Tables 3 and 4 show the data collection of respondent responses, including the minimum and maximum data and the mean and standard deviation. The questionnaire form includes two variables: dependent and independent variables.

The independent variable was demographic data, whereas the dependent variable was human and environmental factors, divided into three groups. Red-light running (RLR) on the road, mobile phone use (MPU) while riding on the highway, and speeding (S) when riding on the road were all considered human factors. Meanwhile, lighting conditions (LC), roadway surface (RS), and weather conditions were also environmental factors (WC). Those variables for human and environmental aspects each have three and two options, respectively. According to Tables 1 and 2, the minimum and maximum variables should be in the range of 1–3.

Furthermore, the many options for each element were not equally distributed. The environmental factors, for example, employed a Yes/No format to ask respondents to choose an alternative. In terms of human aspects, the solutions presented for each type were not the same. However, the order of all answers suggested the same interpretation, with the first choice indicating that respondents were very interested in the elements presented. The second option stated that respondents were somewhat dedicated to the aspects supplied, implying that the respondent executed the deed only on a few occasions. Finally, the third option was the opposite of the first and second options, referring to people who were either disengaged or never engaged with the provided element.

The mean and standard deviation for each variable were displayed in Tables 1 and 2 by subtracting the most outstanding value from each variable and putting them together. As a result, the mean and standard deviation values are 1.6994 and 2.4711, respectively, and 0.46634 and 0.74479. As a result, the highest and lowest mean values correspond to cell phone use while riding on the road and illumination circumstances, respectively.

4 Conclusion

This study aimed to discover the elements that contribute to road crashes among cyclists and establish which factor was the most frequently mentioned. The questionnaire form includes two variables: dependent and independent variables. The independent variable was demographic data, whereas the dependent variable was human and environmental factors, divided into three groups. Red-light running on the road (RLR), mobile phone use while riding on the highway (MPU), and speeding while riding on the road were all considered human factors (S). Meanwhile, lighting conditions (LC), roadway surface (RS), and the weather conditions were environmental factors (WC). According to the study's findings, factors in the questionnaire

Table 3 Descriptive statistic and normally assessment for human factors

Variables			Descriptive statistics			
Human factors	Item label	Item	Min	Max	Mean	Standard deviation
Red light running behaviour on the road	RLR 1	1. Recalling the last five traffic lights you ride through, how many of them were red when you entered the intersection	1.00	3.00	2.1156	0.74479
	RLR 2	2. You are approaching an intersection with fewer cars, and the traffic has just turned red. Which of the following would you likely do?	1.00	3.00	2.4711	0.52808
	RLR 3	3. You are in a rush to deliver a customer's order and have been stopped by several red lights in a row. Then, you are approaching another intersection with yellow light for several seconds, but you know it is about to turn red. Which of the following would you likely do?	1.00	2.00	1.4595	0.49908
Mobile phone uses while riding on the road	MPU 1	4. Do you think it is dangerous to use your phone while riding?	1.00	3.00	2.2919	0.67998

(continued)

Table 3 (continued)

Variables			Descriptive statistics			
Human factors	Item label	Item	Min	Max	Mean	Standard deviation
	MPU 2	5. How frequently do you browse social media (e.g. Instagram, Twitter, Tiktok, and Facebook) on the mobile phone while riding a motorcycle?	1.00	3.00	2.3815	0.68871
	MPU 3	6. How constantly were you checking the assigned order or competing for ordering via the food delivery apps while riding on the road?	1.00	3.00	1.7486	0.73647
	MPU 4	7. How often you are answering or initiating a call while riding a motorcycle?	1.00	3.00	1.9133	0.65364
Speeding while riding on the road	S1	8. Do you speed regularly?	1.00	3.00	1.6965	0.58269
	S2	9. If you did speed, how often you were more likely to be involved in a crash while trying to deliver the ordered food?	1.00	3.00	1.9509	0.64193
	S3	10. Why would you be speeding on the road?	1.00	3.00	1.6618	0.65814

passed the reliability and validity tests, indicating that respondents may use the questionnaire. The respondents' data was then evaluated using descriptive analysis. It was discovered that using a cellphone while riding on the highway has become the leading cause of traffic crashes among food delivery riders in Kuala Lumpur.

According to the report's findings, the research location should be expanded because it was only conducted in Kuala Lumpur. Other Malaysian states, such as

Table 4 Descriptive statistic and normally assessment for environmental factors

Variables			Descriptive statistics			
Environmental factors	Item label	Item	Min	Max	Mean	Standard deviation
Lighting condition	LC 1	11. Early morning	1.00	1.00	1.6994	0.45917
	LC 2	12. Night-time with light	1.00	1.00	1.6503	0.47757
	LC 3	13. Daylight	1.00	1.00	1.5636	0.49666
	LC 4	14. Night-time without light	1.00	1.00	1.5636	0.49666
Roadway surface	RS 1	15. The road surface is dry	1.00	1.00	1.7197	0.44982
	RS 2	16. The road surface is wet	1.00	1.00	1.3468	0.47665
	RS 3	17. The road surface is flooded	1.00	1.00	1.6243	0.48501
	RS 4	18. The road surface is oily	1.00	1.00	1.5607	0.49702
	RS 5	19. The road surface is muddy	1.00	1.00	1.6821	0.46634
Weathers condition	W1	20. The weather is clear	1.00	1.00	1.6908	0.46285
	W2	21. The weather is windy	1.00	1.00	1.7601	0.42763
	W3	22. The weather is foggy	1.00	1.00	1.7139	0.45260
	W4	23. The weather is rainy	1.00	1.00	1.3179	0.46634

Johor, Pulau Pinang, and Kelantan, could be included in this study. In recent years, food delivery services have grown in popularity due to the continuous movement control order (MCO) imposed by the COVID-19 epidemic. A different research location means that the researchers will discover and learn more about the circumstances that contributed to the road accident. As a result, another researcher may develop various strategies and approaches to reduce the number of road accidents.

Acknowledgements Authors would like to thanks the Ministry of Higher Education (MOHE) that gave the funding of Fundamental Research Grant Scheme (FRGS) (FRGS/1/2020/TK02/UTHM/03/1), Research Management Centre (RMC) and Registrar Office, and Universiti Tun Hussein Onn Malaysia (UTHM) those gave financial support. Also, thanks to the Faculty of Civil and Environmental Engineering, UTHM, and Smart Driving Research Centre (SDRC) for support and providing facilities to accomplish the study.

References

1. Odero, W., Garner, P., Zwi, A.: Road traffic injuries in developing countries: a comprehensive review of epidemiological studies. Trop. Med. Int. Health **2**(5), 445–460 (1997). Available: https://doi.org/10.1111/j.1365-3156.1997.tb00167.x
2. Suriyawongpaisal, P., Kanchanasut, S.: Road traffic injuries in Thailand: trends, selected underlying determinants and status of intervention. Inj. Control Saf. Promot. **10**(1–2), 95–104 (2003). Available: https://doi.org/10.1076/icsp.10.1.95.14110. Accessed 29 June 2021
3. Transport Ministry releases guidelines for motorcycle delivery riders. The Star (2020). [Online]. Available: https://www.thestar.com.my/news/nation/2020/04/20/transport-ministryreleases-guidelines-for-motorcycle-delivery-riders. Accessed 05 Jan 2021
4. Yellappan, K., Mani, K.K.C., Tamrin, S.BMd.: How safe are standard certified motorcycle safety helmets? Malaysian postal delivery rider's scenario. Traffic Inj. Prev. **20**(6), 624–629 (2019)
5. Kulanthayan, S., See, L., Kaviyarasu, Y., Nor Afiah, M.: Prevalence and determinants of non-standard motorcycle safety helmets amongst food delivery workers in Selangor and Kuala Lumpur. Injury **43**(5), 653–659 (2012). Available: https://doi.org/10.1016/j.injury.2011.06.197
6. Katrakazas, C., Michelaraki, E., Sekadakis, M., Yannis, G.: A descriptive analysis of the effect of the COVID-19 pandemic on driving behaviour and road safety. Transp. Res. Interdiscip. Perspect. **7**, 100186 (2020)
7. Reporters, F.M.T.: A quick guide to essential services under a movement control order. Free Malaysia Today (FMT) (18-Mar-2020). [Online]. Available: https://www.freemalaysiatoday.com/category/nation/2020/03/18/a-quick-guide-to-essential-services-under-movement-control-order/. Accessed 29 June 2021
8. Chai Lau, T., Ng Ching Yat, D.: Online food delivery services: making food delivery the new normal. Sarawak Res. Soc. **1**(1), 62–63 (2021). Available: https://www.researchgate.net/publication/334050513_Online_Food_Delivery_Services_Making_Food_Delivery_the_New_Normal. Accessed 29 June 2021
9. Malaysian Road Safety Institute Pushes for Better Training Of Food Delivery Riders. Voice of America (2021). [Online]. Available: https://www.voanews.com/economy-business/malaysian-road-safety-institute-pushes-better-training-food-delivery-riders. Accessed 29 June 2021
10. Krejcie, R., Morgan, D.: Determining sample size for research activities. Educ. Psychol. Meas. **30**(3), 607–610 (1970). Available: https://doi.org/10.1177/001316447003000308
11. Taherdoost, H.: Sampling methods in research methodology; How to choose a sampling technique for research. SSRN Electron. J. (2016). Available: https://doi.org/10.2139/ssrn.3205035
12. Nunnally, J.: Psychometric theory. 25 years ago and now. Educ. Res. **4**(10), 7 (1975). Available: https://doi.org/10.2307/1175619
13. Watson, R.: SPSS survival manual by Julie Pallant, Open University Press, Buckingham, 2001, 286 pages, f16.99, ISBN 0 335 20890 8. J. Adv. Nurs. **36**(3), 478–478 (2001). Available: https://doi.org/10.1046/j.1365-2648.2001.2027c.x
14. Azhar Mat Daud, K., Zulkarnaen Khidzir, N., Rasdan Ismail, A., Aryani Abdullah, F.: Validity and reliability of an instrument to measure social media skills among small and medium entrepreneurs Pengkalan Datu River. Isdsnet.com (2021). [Online]. Available: https://isdsnet.com/ijds-v7n3-15.pdf. Accessed: 29 June 2021
15. Metwally, E.: Survey research methods. J. Organ. Change Manage. **25**(1), 186–188 (2012). Available: https://doi.org/10.1108/09534811211199655
16. bin Darusalam, G., & Hussin, S.: Metodologi penyelidikan dalam pendidikan
17. Chua, Y.P.: Goodreads.com (2021). [Online]. Available: https://www.goodreads.com/author/show/4783148.Chua_Yan_Piaw. Accessed: 29 June 2021

18. Hair, J.F., Black, B., Babin, B.J., Anderson, R.E.: Multivariate Data Analysis: Global Edition, 7th Edition. Pearson. Pearson.com (2021). [Online]. Available: https://www.pearson.com/uk/educators/higher-education-educators/program/Hair-Multivariate-Data-Analysis-Global-Edition-7th-Edition/PGM916641.html. Accessed: 29 June 2021
19. Pallant, J.F.: SPSS Survival Manual: A Step by Step Guide to Data Analysis Using SPSS for Windows Version 15. 2021 (2007)

Characterization of Penetration Grade Bitumen Blended with Petroleum Products for Cutback Production

Nurul Tasnim Che Noh, Norhidayah Abdul Hassan, Abdullah Farhan Nasralddin, Muhammad Naqiuddin Mohd Warid, Mohd Zul Hanif Mahmud, and Mohd Khairul Idham Mohd Satar

Abstract Cutback bitumen is a potential binder that can be used for producing cold mix asphalt. This study presents the characteristic of produced cutback bitumen according to different types of diluent used. The 60/70 penetration grade bitumen was mixed with different petroleum solvents (gasoline, kerosene, and diesel) to produce cutback bitumen for potential usage in asphalt pavement. The binder's viscosity and evaporation rate (curing time of solvent) were investigated to select the appropriate percentage of solvent that provides lower mixing and compacting temperature range. The percentage of diluent was varied at 30, 35, 40, and 45%. The viscosity was measured at a temperature ranging from 30 to 70 °C with an increment of 10 °C, while the evaporation rate was measured at an ambient temperature. The results show that the viscosity of each blended bitumen decreases as the percentage of diluent in the mix increases. The evaporation rate of the solvent is higher for gasoline and kerosene than diesel due to the volatility and chemical compositions of the solvents. From the data obtained, the most effective percentage of solvent for the cutback production is 40–45% by weight of the penetration grade bitumen, which establishes the lowest potential mixing and compacting temperature range for the asphalt.

Keywords Cutback · Viscosity · Evaporation rate · Petroleum solvent · Rapid curing · Medium curing · Slow curing

1 Introduction

Pavement performance is influenced by applied loads, thickness and durability of various pavement components, tire pressure, and mechanical properties of pavement materials. The quality of asphalt will directly influence the good performance of the

N. T. C. Noh · N. A. Hassan (✉) · A. F. Nasralddin · M. N. M. Warid · M. Z. H. Mahmud · M. K. I. M. Satar
Faculty of Civil Engineering, Universiti Teknologi Malaysia, Johor Bahru, Malaysia
e-mail: hnorhidayah@utm.my

N. Sabtu (ed.), *Proceedings of AWAM International Conference on Civil Engineering 2022 - Volume 3*, Lecture Notes in Civil Engineering 386,
https://doi.org/10.1007/978-981-99-6026-2_25

pavement [1]. The stiffness and resistance to distortion of asphalt allow it to withstand the pressure from the load, while the flexural strength prevents damage from occurring when pressures are applied. Hence, the paving quality is crucial because it shows how long the asphalt surfacing will hold up against the weather elements and repeated use over time. Several issues regarding the production of hot mix asphalt, such as environmental effects, have induced asphalt pavement with the latest technological advances to improve the quality of asphalt [2]. The emission of greenhouse gases has become one of the most priority awareness in improving the road construction industry. Hot mix asphalt production requires a highly energy-consuming process due to a higher mixing temperature. In addition, the production of hot mix asphalt also contributes to noise emissions that are not environmentally friendly. Therefore, one initiative to overcome this problem is by exploring the production of cold mix asphalt to achieve environmental and economic sustainability. Cold mix asphalt has several advantages as it requires no heat to manufacture or lay over the pavement, has a less environmental impact, is more cost-effective, and has less energy consumption [3]. The cold mixing technology can lower the mixing temperature of asphalt mixture, leading to approximately 50% of energy savings. Cold mix asphalt can be produced by mixing the aggregate with the cutback asphalts or emulsified asphalts. There are several cutbacks available in the market for the road industries, but research and detailed information regarding the properties and components composition for the production of cutbacks are lacking. Hence, further study is required to improve understanding regarding the cutback. This study establishes a mixing procedure for cutback bitumen as a potential binder for a cold mix asphalt and evaluates the viscosity and curing time of the produced cutback bitumen.

2 Literature Review

This literature provides an overview of the cutback bitumen from mixing the penetration grade bitumen with the petroleum solvent. The discussion includes the characterization of cutbacks, types of cutbacks, percentage of diluent, and their chemical composition.

2.1 Cutback Bitumen

Bitumen is a thick black crude oil consisting of complex hydrocarbons commonly used for asphalt pavement construction [1]. It also has highly viscous waterproof materials at ambient temperatures. The bitumen's chemical composition, particularly referring to the proportion of asphaltenes and maltenes, will determine the physical, rheological, and mechanical properties of bitumen [4]. Bitumen possesses unique characteristics that can change and adapt to temperature changes. It displays elastic behavior at low temperatures while presenting viscous behavior at high temperatures

[5]. However, it must be fluid enough at the application temperature to permit it to spray, mix, coat properly, or penetrate as the application requires [6]. The main requirements of the binder's performance are its stiffness at any temperature and recovery after loading release [7]. It also must possess curing or setting properties compatible with the construction type [6]. Hard binders will contribute to low rutting potential in the pavement, while soft binders contribute to more rutting potential but are low prone to fatigue and temperature cracking. In addition, the viscosity of bitumen displays a complex reaction under different conditions, which influences the mixing, laying, and compaction of asphalt mixtures and also pavement performance [8].

The cutback is a binder that is produced by combining the bitumen with a hydrocarbon solvent. The viscosity of the cutback bitumen is inversely proportional to the proportion of solvent added. Mixing bitumen with a solvent reduces its viscosity, allowing the mixture to be prepared at a lower temperature. As the solvent evaporates, bitumen binds the aggregate together [9]. As the petroleum solvent added increases, the viscosity of the modified bitumen decreases. The percent of petroleum content added into the bitumen can be increased until the targeted viscosity is achieved at the lowest mixing and compaction temperatures. The viscosity of the cutback as a binder becomes important due to its permeability characteristics and also for good workability. Good workability of the binder can provide proper performance for mixing and compacting the mixture. The viscosity of cutback bitumen depends on certain factors, such as types of petroleum solvents and their chemical compositions. In addition, the solvent's evaporation rate, either quickly or slowly, depends on the nature of the solvent, local weather conditions, and the amount of heating. Mixture stability will be gained through the evaporation of solvents at service [10]. Therefore, the solubility and evaporation rate greatly influence the workability, strength, and durability of cutback for asphalt mixtures [11]. Cutback has excellent wetting and penetrating ability due to its low surface tension [12]. It is often used as a primer for hot-applied asphalt. Besides, cutback tends to dry from the top-down, forming a skin that will reduce the drying rate.

Cutbacks are used because their viscosity is lower than penetration grade bitumen and can be applied in low-temperature applications. The higher the percentage of solvent, the lower the viscosity of the cutback produced. Cutback can be divided into three types based on the relative rate of evaporation as follows:

- Rapid-Curing (RC)—It is produced by adding a light diluent of high volatility to bitumen and is generally used for a tack coat.
- Medium-Curing (MC)—It is produced by adding a medium diluent of intermediate volatility to bitumen and is generally used for prime coat, stockpile patching mixtures, and road-mixing operations.
- Slow-Curing (SC)—It is produced by adding oils of low volatility to bitumen and is generally used for prime coat and stockpile patching mixtures.

The advantages of cutback possess a much higher residual bitumen percent, typically over 80%, compared with over 40–65% for bitumen emulsions. This results in more bitumen left on the roadway after the curing process. Hence, the application of cutback is widely utilized in road construction. Cutback agents reduce the bitumen's viscosity when applied as a primer to the pavement base course or substrate surface. Rapid curing cutback is usually recommended for surface dressing and patchwork. Medium curing cutback is suggested for premix with less quantity of the aggregates. Slow curing cutback is employed for premix with an appreciable quantity of the aggregates [13]. Besides, another high-volume cutback usage is within structures' waterproofing and damp-proofing [12].

2.2 *Types of Diluents*

The petroleum solvent used to modify the penetration grade bitumen can be divided into three types which are raping curing, medium curing, and slow curing. Petroleum solvent contains aliphatic hydrocarbons or hydrocarbons composed of hydrogen and carbon. The carbon atoms link together in chains of different lengths. The hydrocarbon molecules of different lengths have different properties and behaviors. As the number of carbon atoms in the hydrocarbon chain increases, the properties of the alkanes increases, such as autoignition tendencies, molecular weights, and melting and boiling points. The purpose of the diluent is to temporarily soften the binder so that the surfacing chip can be manipulated by traffic to form a strong aggregate mosaic. Two important properties of volatile solvents are the solubility parameter and evaporation rate, which will influence the viscosity value and the binder curing [14].

2.2.1 Gasoline

Gasoline is a complex mixture containing hundreds of different hydrocarbons in their basic molecular structure derived from petroleum distillation [15]. On the other hand, gasoline is one of the most commonly used fuels for automotive transportation worldwide [16]. Gasoline establishes flash point as low as − 40 °C. The gasoline hydrocarbons are typically composed of 4–12 carbon atoms with a boiling range between 30 and 210 °C. The rate of the volatility of gasoline is influenced by the molecular weight, where the light components were easy to evaporate, while the heavy hydrocarbon components possessing higher boiling points were not easy to volatilize at room temperature. Hence, the light components will evaporate into the air continuously with decreasing gasoline content. This situation resulted in the evaporation rate becoming slow by the time the gasoline weight tends to achieve stability at the end of the evaporation process [15].

2.2.2 Kerosene

Kerosene consists mainly of normal and open-chain alkanes, cycloalkanes, alkylbenzenes, and alkyl naphthalene. The boiling point range is generally from 145 to 300 °C.

2.2.3 Diesel

Diesel fuel contains hydrocarbons such as paraffinic, naphthenic, and aromatic hydrocarbons. With approximately 12–20 carbon atoms, the boiling point is between 170 and 360 °C. The flash point of diesel fuel is between 125 and 180 °C.

3 Methodology

This research consisted of two phases. Phase I involves sample preparation, while phase II is laboratory testing and characteristic analysis. All the laboratory tests were conducted at the Transportation Laboratory, Universiti Teknologi Malaysia.

3.1 Phase I: Preparation of Sample

The 100 g of pure liquid bitumen was prepared according to the number of samples required. The 60/70 PEN bitumen was heated at 70–80 °C while stirring for melting purposes. The different percentages of petroleum solvent (gasoline, kerosene, diesel), 30, 35, 40, and 45% by weight of the bitumen, were added into the bitumen. The mixture was then blended using the laboratory mixer for 10–15 min. The rotation speed of the impeller is constant at 80 rpm to ensure a uniform blend (Fig. 1).

3.2 Phase II: Testing and Characteristic Analysis

The cutback bitumens were evaluated for viscosity and curing time at ambient temperature. The viscosity of the mix was used as a reference to produce a cutback with an optimum ratio that can establish a lower range of temperature for mixing and compacting asphalt mixture. The type of solvent controls the curing time, while the amount determines the viscosity of the cutback bitumen.

Fig. 1 Sample preparation

3.2.1 Viscosity Test

This test was used to evaluate the workability during the mixing and construction of the produced cutbacks. The 11 g mixture was poured directly into the Brookfield viscometer for viscosity measurements at a selected temperature of 30–70 °C. Two replicates for each percentage were prepared to obtain the average viscosity value. The time spent for the rotation at each increment of 10 °C was 20 min. Torque is applied to the spindle through the viscometer motor to make it rotates at a constant speed of 20 revolutions per minute (RPM). The rotational viscometer displays the viscosity value as a digital readout.

3.2.2 Curing Time

The test evaluation was conducted to estimate the rate at which the diluent evaporates from cutback bitumen when applied as a coating. Two samples with a film thickness of 0.5 l/m^2 were prepared for each different percentage of solvent. The evaporation rate was observed from the samples by measuring the weight of the samples at the pre-selected time interval. The rate at which the diluent evaporates from cutback bitumen when applied to a surface is called curing. The diameter of the evaporating tray is 14 cm. The weight of the trays containing cutback bitumen was initially measured and continuously monitored for 48 h at a specified time interval. It is noted that curing time is affected by the product type, application rate, climatic conditions, humidity, wind, rain, existing surface condition, and the area's prevailing range of ambient temperatures during application [17]. However, only the surrounding temperature and relative humidity were monitored during the measurements using a thermohygrometer.

4 Results and Discussion

The laboratory testing results in determining the viscosity and curing time of the produced cutback bitumen are discussed in the following sections.

4.1 Viscosity Test

The viscosity of the produced bitumen blended with different percentages of solvents was determined at different temperatures. This measurement reflects the workability of the binders during the mixing and compaction processes. Table 1 summarizes the viscosity test results for different solvents at different temperatures. The testing was performed at various temperatures with various solvents, i.e., gasoline, kerosene, and diesel. The viscosity was evaluated at various temperatures starting from 30 to 70 °C with an increment of 10 °C at 30, 35, 40, and 45% diluent. Generally, this test aims to establish the targeted viscosity at 180 Centipoise and 270 Centipoise for the lowest mixing and compacting temperatures, respectively. Based on the table, the range for adequate mixing and compaction temperature for each solvent at each percentage can be estimated as in italic. The ideal temperature range for 30% gasoline must be above 70 °C, between 60 and 70 °C for 35% gasoline, and between 50 and 60 °C for 40 and 45% gasoline. Besides, for 30 and 35% kerosene, the targeted temperature range must be above 70 °C, and 40–50 °C for 40 and 45%. On the other hand, the temperature range between 50 and 60 °C is acceptable enough for 30, 35, 40, and 45% diesel to achieve the targeted viscosity. In addition, it is clearly shown that the viscosity of produced cutback using kerosene is lower with the increment of kerosene content compared to cutback using gasoline. It shows that the increment in the percentage of kerosene significantly reduces the bitumen's viscosity compared to gasoline and diesel. Initially, it was expected to be dominated by gasoline as it is more volatile than others. However, the result may be influenced by the chemical composition of the kerosene used in this test. Therefore, future studies could provide details on the chemical composition of the diluents used for further testing.

Figure 2 shows the viscosity versus temperature for different gasoline, kerosene, and diesel percentages. The data obtained from the testing shows a decreasing trend as the temperatures increase. Data from Table 1 can only generate the acceptable temperature range to establish the required viscosity. To determine the exact mixing and compaction temperature for each percentage of solvent, the best fit lines for each percentage were plotted. The dotted lines of 180 and 270 Centipoise become the reference in establishing the required temperature to achieve targeted workability. The intersection value between the linear line and the dotted line of mixing and

Table 1 Viscosity of different solvents at different temperatures

Types of solvent	Percentage of solvent added (%)	Temperature (°C)				
		30	40	50	60	70
Gasoline	30	10,200	2800	1000	500	*250*
	35	6550	1850	650	*300*	*150*
	40	1600	500	*200*	*100*	*100*
	45	1750	600	*250*	*150*	*100*
Kerosene	30	9950	2650	900	500	*300*
	35	6900	1950	700	350	*200*
	40	950	350	*150*	*< 100*	*< 100*
	45	550	*200*	*100*	*< 100*	*< 100*
Diesel	30	5200	1400	450	*150*	*100*
	35	2800	800	*250*	*100*	*< 100*
	40	2000	600	*200*	*100*	*100*
	45	1650	550	*250*	*150*	*150*

compaction determined the temperature value required for good workability. Then, the exact temperature for mixing and compaction could be specifically generated. On the other hand, the similar trend for 40 and 45% for gasoline and diesel can be seen as their viscosity value represents only slight differences. Detailed information extracted from the plots for each percentage regarding the temperature is shown in Fig. 2.

From the study, Fig. 3 established the mixing and compaction temperature for each type of solvent. For gasoline and diesel, there were only slight differences in the mixing and compaction temperature determined for 40 and 45% of solvent. Hence, considering the economic factor where the consumption of a greater volume of solvent would express almost the same result, 40% was classified as the ideal or acceptable percentage that can fulfill a similar requirement. However, the chosen percentages for kerosene could be up to 45% due to the lower temperature offered when that percentage was used. The addition of petroleum solvent has indicated a declination temperature that is quite different from the value of 40%.

The graph shows that the mixing and compaction temperature of the sample containing kerosene establishes the lowest mixing and compaction temperature with the increment of the percentage of solvent added. This could be attributable to the chemical composition of kerosene, which was not tested in this study.

4.2 Curing Time

The test method was undertaken to determine the curing rate at which the diluent evaporates from cutback bitumen when applied as a binder or coating, which reflects

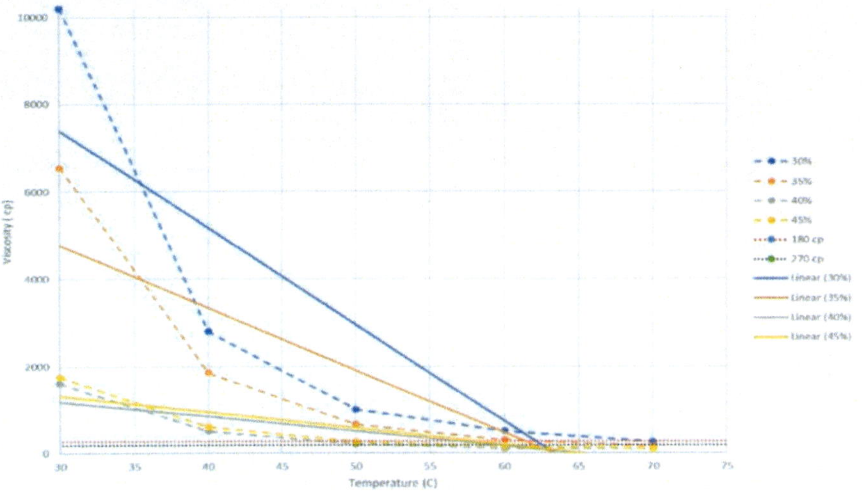

(a)

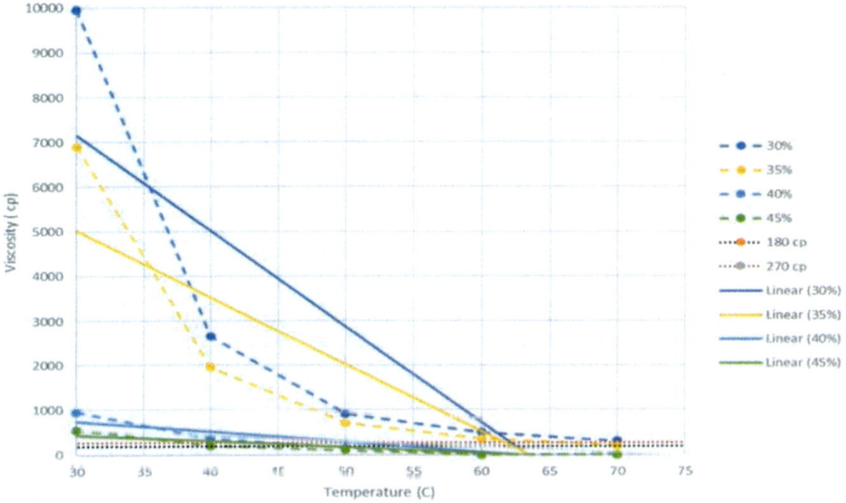

(b)

Fig. 2 Plot of viscosity versus temperature: **a** gasoline **b** kerosene **c** diesel

the application on the road surface. The data determine the classes of the selected petroleum products, whether it is rapid curing, medium curing, or slow curing. Figure 4 shows the data of ambient temperature and humidity monitoring for the curing process of the produced cutback conducted in the laboratory. The curing time monitoring was conducted at the ambient condition in the laboratory for 48 h. The figure provides an overview of the test's ambient temperature and relative humidity.

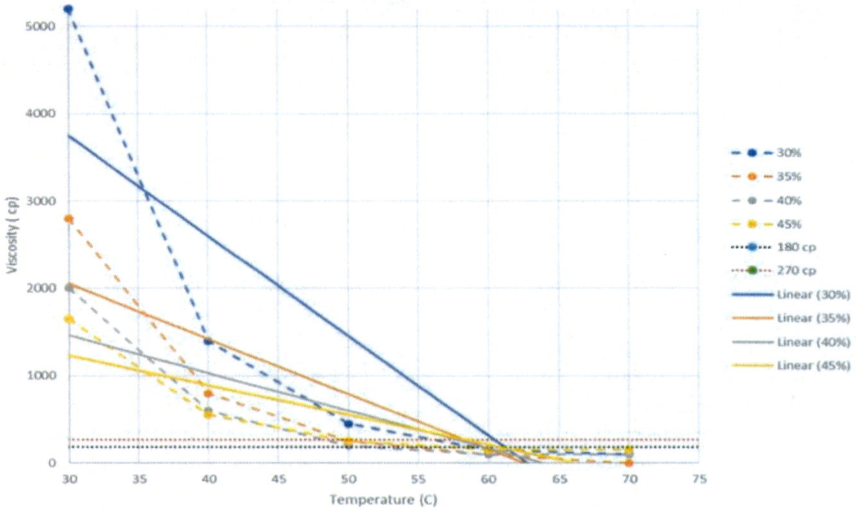

(c)

Fig. 2 (continued)

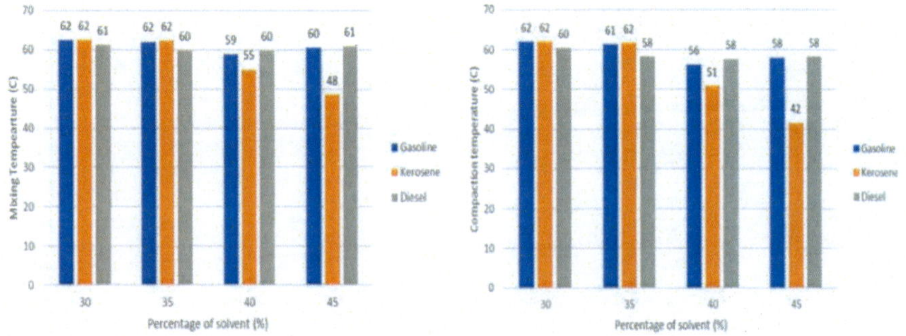

(a) **(b)**

Fig. 3 **a** Mixing and **b** compaction temperatures for different types of solvent

The ambient temperature range is between 27.2 and 32.2 °C, while the range for relative humidity is between 45 and 94%, averaged from a few days of monitoring. The aim of measuring the relative humidity and ambient temperature is to consider the factors that could influence the loss of solvent when the cutback is applied as a coating or in a cold mix. The obtained range will contribute to further understanding the factors that affect the percentage loss of solvent to the environment. In general, it can be observed that as the relative humidity increases, the ambient temperature decreases. Thus, it is clear that the percentage loss after curing would be higher whenever applied as a coating during road construction due to the greater temperature exposure, particularly the road surface temperature, which could be approximate

Fig. 4 Data of surrounding temperature and humidity monitoring for curing

60 °C. In addition, the loss rate of the solvent would potentially be quicker than being monitored in the laboratory at room temperature.

Figure 5 indicates the pattern of solvent loss to the environment for different types of solvent at different contents for 48 h of curing. The time range measured for the loss of solvent may vary for each percentage of petroleum solvent. It can be seen that the evaporation rate of the solvent is initially dominated by gasoline, then followed by kerosene and diesel at each percentage. The decreased rate of evaporation pattern for gasoline may be due to the decrease of light hydrocarbon, which almost achieves a stable state at room temperature. In addition, the heavier hydrocarbon residue components of the gasoline with higher boiling points were not easy to volatilize at room temperature [15]. However, by observing the graphical increment of kerosene, it can be concluded that the selected kerosene may possess a high chance of evaporation and high volatility compared to gasoline. This result may be relatable to the chemical composition of the kerosene. However, it should be noted that the samples were cured on different days. This is because of the constraint in the sample production, which might affect the percentage of loss measured for comparing the different types of samples due to potential changes in the surrounding conditions. Therefore further study should also consider this factor during the curing stage.

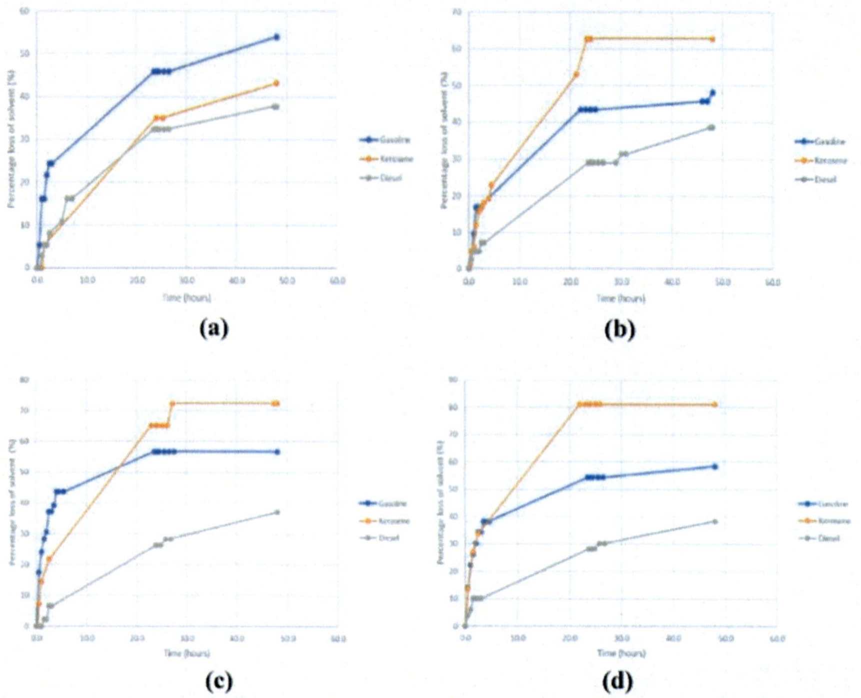

Fig. 5 Percentage loss for different types of solvent at different contents: **a** 30% **b** 35% **c** 40% **d** 45%

5 Conclusion

Performance evaluation of cutback properties is crucial in road construction. In this study, the behavior of cutback bitumen was investigated in terms of viscosity and curing time. The conclusions are as follows:

1. The increment of petroleum solvent content decreases the viscosity of the cutback and thus potentially decreases the asphalt mixing and compaction temperatures.
2. The evaporation rate depends on the type of petroleum solvent used, which is higher for kerosene and gasoline than diesel.
3. The performance of cutback with 40% gasoline and diesel was identified as an effective content of solvent to be added to the bitumen, based on the consideration made on the quantity of the solvent used and mixing and compaction temperatures requirements. However, for kerosene, the percentage can be chosen up to 45% to lower the temperature from 50 to 40 °C.

Acknowledgements This work was supported/funded by the University Teknologi Malaysia research grant (Q.J130000.2451.09G20).

References

1. Qasim, Z.I., Abed, A.H., Almomen, K.A.: Evaluation of mixing and compaction temperatures (MCT) for modified asphalt binders using zero shear viscosity and Cross-Williamson model. Case Stud. Constr. Mater. **11**, e00302 (2019). https://doi.org/10.1016/j.cscm.2019.e00302
2. Speight, J.G.: Asphalt binder. Science Direct. (2016)
3. Guo, M., Tan, Y., Zhou, S.: Multiscale test research on interfacial adhesion property of cold mix asphalt. Constr. Build. Mater. **68**, 769–776 (2014). https://doi.org/10.1016/j.conbuildmat.2014.06.031
4. González, O., Muñoz, M.E., Santamaría, A., García-Morales, M., Navarro, F.J., Partal, P.: Rheology and stability of bitumen/EVA blends. Eur. Polymer J. **40**(10), 2365–2372 (2004). https://doi.org/10.1016/j.eurpolymj.2004.06.001
5. Lu, X., Isacsson, U.: Effect of ageing on bitumen chemistry and rheology. Constr. Build. Mater. **16**(1), 15–22 (2002). https://doi.org/10.1016/S0950-0618(01)00033-2
6. Williams, E.G.: When to Use What Grade of Asphalt and Why, pp. 218–228. (1876)
7. Hafeez, J.H.: Influence of time and temperature on asphalt binders rheological properties. Life Sci. **5** (2013)
8. Yero, S.A., Hainin, M.R.: Viscosity characteristics of modified bitumen. Arpn. J. Sci. Technol. **2**(5), 500–503 (2012). https://www.researchgate.net/publication/299645857_Viscosity_Characteristics_of_Modified_Bitumen
9. Jain, S., Singh, B.: CMA: an overview. J. Clean. Prod. **280**, 124378 (2021). https://doi.org/10.1016/j.jclepro.2020.124378
10. Taylor, P., Simonen, M., Blomberg, T., Pellinen, T.: Physicochemical properties of bitumens modified with bioflux. Road Mater. Pavement Des. **14**(1), 36–48 (2013)
11. Wang, X., Chen, X., Dong, Q., Jahanzaib, A.: Material properties of porous asphalt pavement cold patch mixtures with different solvents. J. Mater. Civ. Eng. **32**(10), 06020015 (2020). https://doi.org/10.1061/(asce)mt.1943-5533.0003399
12. Dickson, W.J.: ASPHALT GS, pp. 28–32. (1966)
13. Mali, S.N., Patil, D.: Mix design of HMA and CMA for pothole repairs. In: Proceedings of 4th RIT Post Graduates Conference (RIT PG Con-18), vol. 2581, pp. 44–47. (2018)
14. Dong, Q., Gao, J., Chen, X., Wang, X.: Development of a turpentine cutback asphalt mixture for porous pavement pothole repair. J. Mater. Civ. Eng. **32**(3), 05020001 (2020). https://doi.org/10.1061/(asce)mt.1943-5533.0003075
15. Zhu, L., Chen, J., Liu, Y., Geng, R., Yu, J.: Experimental analysis of the evaporation process for gasoline. J. Loss Prev. Process Ind. **25**(6), 916–922 (2012). https://doi.org/10.1016/j.jlp.2012.05.002
16. Kalvakala, R.K.C., Pal, P., Kukkadapu, G., McNeely, M., Aggarwal, S.: Numerical study of PAHs and soot emissions from gasoline–methanol, gasoline–ethanol, and gasoline–n butanol blend surrogates. Energy Fuels **36**, 7032–7064 (2022).https://doi.org/10.1021/acs.energyfuels.2c00897
17. White, G.: Australian airport asphalt surface treatments. (2017). https://doi.org/10.14311/ee.2016.306

Uncovering the Risky Riding Behaviors Among Young Motorcyclist in Urban Areas

Siti Zun Nurain binti Mohd Ali and Intan Suhana binti Mohd Razelan

Abstract In Malaysia, more than 50% of road traffic crashes involve motorcyclists, with motorcyclists being the main contributor to the road crashes. Thus, the traffic safety of motorcyclists is critical and must be given priority. Road traffic crashes showed a significant increase every year in Malaysia according to Department of Road Transport statistics. Numerous researches on the relationship between risky riding behaviours and traffic accidents have been identified; however, research focuses on young motorcyclists' risky riding behaviors in urban areas were somewhat limited. To demonstrate that risky riding behaviors which played a significant role in the number of motorcycle crashes among motorcyclists, a motorcycle-riding behavior questionnaire was distributed to groups of young motorcyclists in several schools and higher education institutions to assess their level of understanding and perception of risky riding behavior. Two primary analyses, namely frequency and percentage, cross-tabulation, and test of independence (chi-squared), were adopted to assess their level of understanding on motorcyclists' risky riding behaviors and its correlation to crashes. Findings have proven that failing to keep proper side-to-side movement with another vehicle and riding while tailgating with another vehicle at an unsafe distance were highly contributed to the motorcycle crashes among the secondary school students. A greater comprehension of motorcyclists' risky riding behavior based on their self-reported behaviour and beliefs can influence motorcyclists to make positive changes in their riding style. Hence, this result will be beneficial for enforcement bodies to strategize their effort in curbing the crash issues involving riders.

Keywords Road traffic crashes · Risky riding behaviors · Urban areas

S. Z. N. M. Ali · I. S. M. Razelan (✉)
Faculty of Civil Engineering Technology, Universiti Malaysia Pahang, Lebuhraya Tun Razak, 26300 Gambang, Pahang, Malaysia
e-mail: intan@ump.edu.my

N. Sabtu (ed.), *Proceedings of AWAM International Conference on Civil Engineering 2022 - Volume 3*, Lecture Notes in Civil Engineering 386,
https://doi.org/10.1007/978-981-99-6026-2_26

1 Introduction

The road crash fatality rate in Malaysia is becoming the third highest in Asia, matching certain African countries. Furthermore, Malaysia was categorized as the growing country with the riskiest roads, behind just South Africa and Thailand, according to the World Health Organization figures from 2013 with a rate of death of 23 per 100,000 population. Every year, 7000–8000 individuals in the country die on the roadways, despite the country's estimated population of 30 million [1]. This might be due to the fact that motorcycles have become a popular method of transportation, particularly in many emerging and middle-income nations as they are the most prevalent mode of transportation in Asian countries, particularly among the low-income urban population.

2 Theoretical Background

Motorcycles have been a popular means of transportation in Malaysia due to their affordability in terms of purchase and maintenance [2]. It is also because of traffic, which makes riding a motorbike lot simpler and faster, especially in densely populated and crowded areas. In Malaysia, less developed states such as Kelantan and Perlis have a larger proportion of motorbike users than Selangor and Wilayah Persekutuan, which are more active in economic growth. Individuals begin to convert their method of transportation from motorcycle to passenger automobile as their wealth rises.

According to Fig. 1, the number of individuals who died due to motorcycle crashes is three times higher than other type of vehicle and six times more than pedestrian mortality and nearly 50 times greater than bus passenger death [4]. In addition, nearly 1.25 million people die in road traffic collisions every year according to the World Health Organization (2018). Road traffic injuries are the primary factor of mortality among young people between 16 and 19 years old [5]. In 2018, nearly 1.25 million individuals suffered in traffic accidents.

Furthermore, according to Blackman (2012), motorcycle riders had a 20–40 greater probability of dying on the road than car passengers [6]. Hence, according to Vlahogianni et al. [7] he uncovered that the number of young motorcyclists is rising due to wounded in traffic collisions and it has enhanced their likelihood of incurring road traffic injuries. Between 2000 and 2008, the proportion of a 15-year-old teenage motorcyclists injured in traffic collisions in Austria climbed from 6 to 32%. The Decade of Action for Road Safety (DARS) 2011–2020 is a United Nations-led worldwide campaign to increase road safety and decrease traffic accident and collision fatalities by 50%, particularly among those at high risk of road mortality, such as youngsters. The worldwide strategy is one of the important steps to meet the DARS 2011–2020 targets which involves road users' conduct [8]. The UN's Decade of Action for road safety seeks to cut road fatalities by 50% by 2020. The aim is

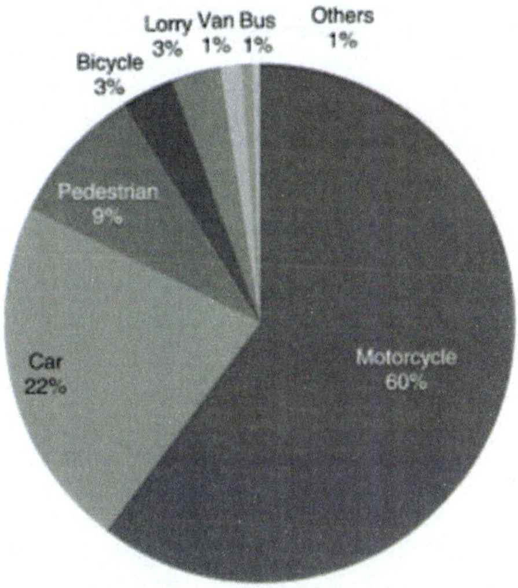

Fig. 1 Mortality distribution by mode of transport. *Source* PDRM [3]

reached by a series of initiatives that concentrate on changing the conduct of road users such as young motorcyclists.

According to Fig. 2, the riders aged 16–20 have had the highest significant number of road traffic deaths over the last ten years, with 1090 deaths in 2017. Several organizations, including Jabatan Keselamatan Jalan Raya (JKJR) and the Malaysian Road Safety Research Institute (MIROS), have made significant attempts to solve this issue. However, initiatives look insufficiently structured to handle the issue altogether. High-risk individuals such as young riders are often seen as the person who can contribute to a road traffic crash [4]. It is very crucial to be aware why these young riders are vulnerable to collisions, particularly on straight road portions. Figure 3 illustrates a trend in road traffic deaths among Malaysian motorcyclists, as stated in the Malaysian Institute of Road Safety Research (MIROS) status report.

Figure 3 shows the trend of road traffic crashes among motorcyclists in Malaysia from 2007 to 2017. As shown in Fig. 3, the highest number of road traffic death, which is 12,013 from 2007 until 2017, is the motorcyclist in age from 16 years until 20 years old.

According to the National Highway Traffic Safety Administration [11], 60% of fatal motorcycle crashes occurred in urban areas, and 40% of motorcycle crashes occurred in rural areas. Various factors could contribute to the increased rate of crashes in urban areas instead of rural areas. For instance, urban areas often have significantly more traffic than rural areas, increasing the rate of motorcycle crashes [12]. Urban areas also may contain more pedestrians and bicyclists, which could cause motorcyclists to abruptly break or swerve to avoid them. Road conditions, such as potholes, construction, and debris, may be more prevalent on urban streets, leading to more crashes. Emergency vehicles also tend to be more common, leading to increased accidents. Most fatal motorcycle crashes, whether in an urban or rural area, occurred on major, non-interstate roads [13].

RoadTrafficdeathswithinvariousageamongmotorcyclists(2007-2017)											
Year	200	200	200	201	201	201	201	201	201	201	201
0-5	11	10	3	3	3	2	5	7	7	7	7
6-10	10	10	6	9	8	5	4	6	5	7	7
11-15	23	24	36	34	35	38	29	32	36	44	41
16-20	97	114	105	125	126	103	96	113	93	116	10
21-25	93	103	75	104	104	111	113	89	82	100	98
26-30	79	68	53	67	70	78	85	52	64	65	61
31-35	63	49	67	49	48	57	58	40	57	57	53
36-40	37	38	63	40	37	47	46	39	48	45	42
41-45	36	36	62	35	35	45	48	36	44	41	38
46-50	34	35	36	36	35	41	32	36	42	41	38
>50	34	35	33	39	39	35	37	33	41	39	37

Fig. 2 Road traffic within various ages among riders in Malaysia from the year of 2007 to 2017. *Source* Khairul Amri Kamarudin [9]

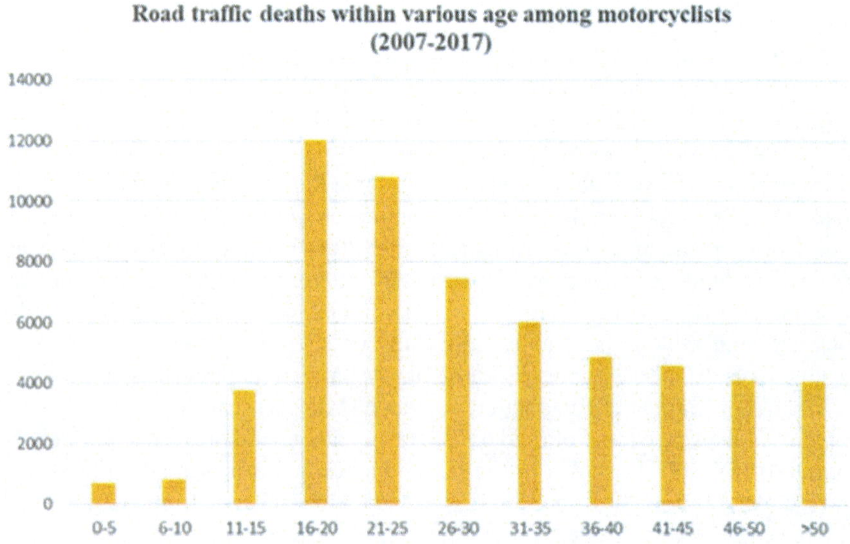

Fig. 3 Road traffic crashes trend among motorcyclists. *Source* Ref. [10]

Several factors such as speeding, violation of traffic laws, bad weather and faulty vehicles were among the most popular and well known road crash factors worldwide. A few contributor factors such as the human, road environment, and vehicle factors caused road traffic crashes.

Figure 4 shows that the highest contributing factor was the human factor (81.0%), on the other hand, the second-highest contributing factor followed by the environment factor (13.0%). Furthermore, the vehicle factor with (6.3%) was following quite near to the environment factor. According to Eboli et al. [15], the main contributing

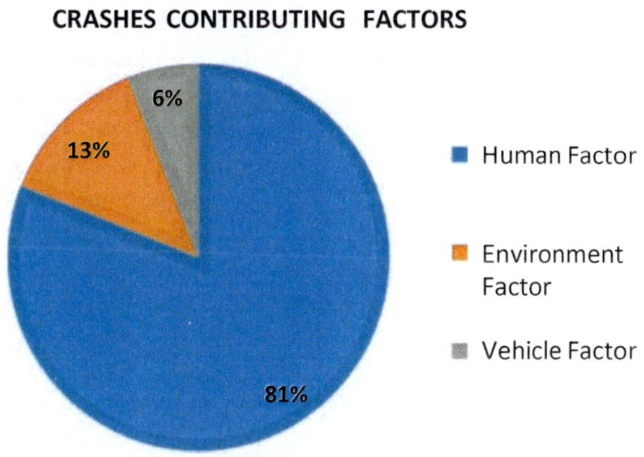

Fig. 4 Crashes contributing factors. *Source* Siti Zaharah [14]

factor of road traffic crashes among those four factors was the human factor. This is reinforced by Peden and World Health Organization [16], who claimed that road traffic crashes prevention focused on human factors which contributed approximately 90% of the entire road traffic collision intervention program.

Young motorists are regarded as a high-risk traffic category. Due to their relatively low riding experience that seemed insubstantial, younger generations are tend to be involved in a motorcycle crash more than the experienced riders [17]. Several previous studies in motorcycles crashes have suggested that high collision rate among young motorcyclist were most probably due to their lack of riding experiences. However, several studies have found that experienced young motorcyclists are still at significant risk of danger. Inexperience was not the only clarification for a collision; further research is needed to identify other elements distinguishing young motorists from other drivers [18].

3 Research Methodology

A questionnaire survey is extensively used to determine valuable insights into the respondents' subjective thoughts and beliefs, whereby it is hard to only extract them from the task's performance measures [19]. A questionnaire survey measures the risky riding behaviors and focuses on young motorcyclists' perception and attitudes based on their respondent background.

As shown in Fig. 5, the risky riding behaviors of young motorcyclists were assessed through the questionnaire risky riding behavior survey conducted online and face to face. Students were the targeted respondents for the survey. The questionnaire was designed and constructed based on the existing literature and similar research conducted in Malaysia and other countries with a high volume of motorcyclists and similar riding behaviors such as the Philippines, Taiwan, and Indonesia [20]. The questionnaire was divided into three sections. For the first section, demographic information for the respondents, including age, gender, years of riding, and type of motorcycle, was collected. Subsequently, the perceptions and attitudes towards risky riding behaviors were gathered in the second section. The third section was used to gather information about the risky riding behaviors of the respondents. In this section, respondents were required to select the risky riding behavior they most engaged in based on eight types of risky riding behaviors, as shown in Table 1. The study conducted by Goh et al. [21] used the same risky behaviors in which to assess the several factors that affecting the risky riding behaviors of riders in Malaysia.

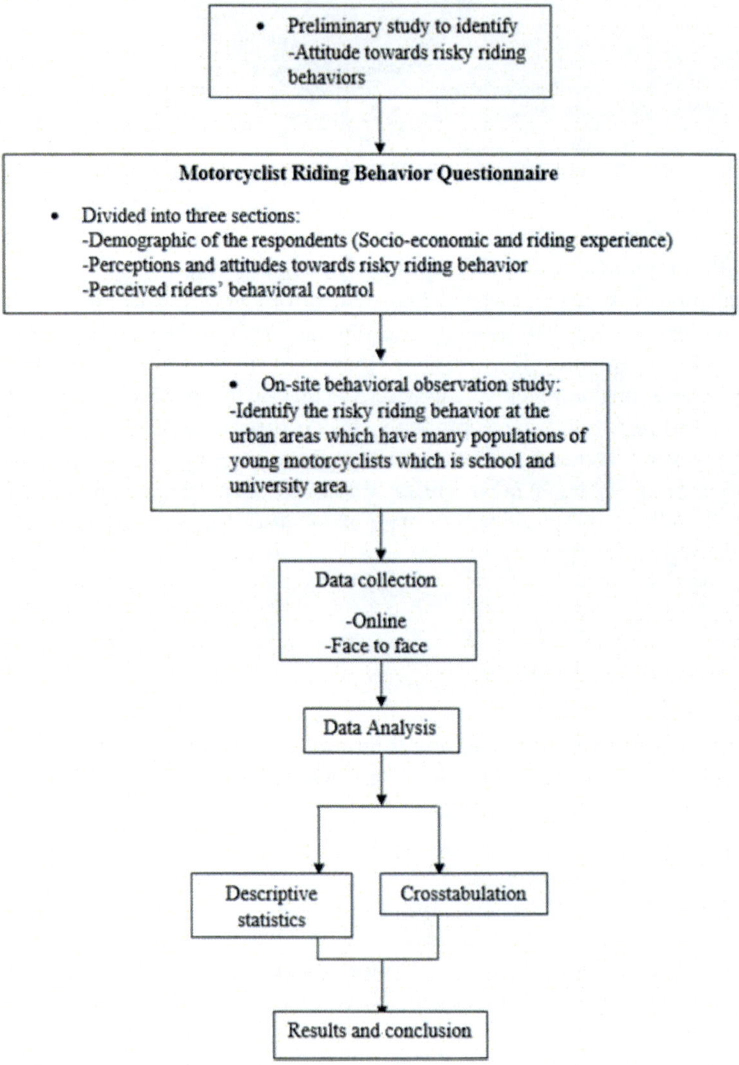

Fig. 5 Flowchart of study methodology

Table 1 Risky riding behavior factors

No	Risky riding behavior factors
1	Riding while breaking the speed limit
2	Riding without crash helmet
3	Riding while using phone
4	Riding without using signal
5	Riding while performing stunting acts
6	Riding against traffic
7	Riding while failing to keep proper side to side movement with another vehicle
8	Riding while tailgating with another vehicle with unsafe distance

In order to determine the most dynamic risky riding behaviors in the urban area, a behavioral observation study was conducted. Following a discussion held with the local authority, a school area in the urban area of Bandar Baru Bangi was chosen. The local authority considered this site a black spot due to the number of recorded crashes, especially riders. The types of risky riding behaviors were identified using a video camera in the school area. Figure 6 shows the site conditions of the school area of SMK Bandar Baru Bangi.

After conducting the questionnaire survey and pinpointing the hazardous riding behaviors within the school area, two main analytical approaches were employed: firstly, an assessment of frequency and percentage; secondly, cross-tabulation and utilization of descriptive statistics. The frequency of an occurrence is statistically defined as the number of times it occurs. One of the most common statistical

Fig. 6 School compound at Sekolah Menengah Kebangsaan Bandar Baru Bangi, Selangor

methods for analyzing and interpreting primary data is the frequency in percentages, which deals with the number of responses to a specific question. It is measured as a percentage of the total population chosen for the study.

Cross-tabulation analysis, also known as bivariate table analysis, is most often used to analyze categorical data. When performing cross-tabulation analysis, the chi-squared statistic is generally used to evaluate tests of independence in which the test assesses whether any relationship exists.

This study also used the statistical software for social sciences (SPSS) version 18.0. This program created a model (equation) and investigated the relationship between risk behavior characteristics and students' background. The risky riding behavior characteristics were the independent variable in this study, whereas the respondent background was the dependent variable. Because the number of students' backgrounds was predicted to fluctuate over time, the number of students' backgrounds was given as a dependent variable.

All procedures or techniques used for data were vital for the researcher to guarantee the intended study went successfully. Microsoft Excel 2010 (MS Excel) and Statistical Packages for Social Sciences (SPSS) software were used to analyze the data. Finally, the correlation of the relationship among the set of risk-riding behavior within the students in urban areas and the students' background.

4 Results

4.1 Observed Risky Riding Behavior in Urban Areas

A total of 2438 motorcyclists were recorded, and 59.4% or 1448 motorcyclists were observed to have performed risky riding behavior. Table 2 shows the frequency of the risk-riding behavior observed at the school area. The highest risk-riding behavior was "R8: Riding while failing to keep proper side to side movement with another vehicle" with a substantial value of 29.1%, followed by "R4: Riding without giving signal" with 26.4%, and the most minor recorded risky riding behavior was "R5: Riding while performing stunting acts" with only 1.74%.

4.2 Demographic Profile of Respondents

A total of 337 respondents from the students and staff of SMK Bandar Baru Bangi were participated in this survey, which was conducted from May 2021 to October 2021. All the students were notified about the aim of the survey who were hand out before asking for their consent and willingness to participate in this research. The completed questionnaires were screened and separated based on different types of risky riding behaviors according to the list in Table 2. Table 3 shows the demographic

Table 2 Frequency of risky riding behaviors observed in the school compound

Type	Description	Frequency (%)
R1	Riding without crash helmet	201 (14.5%)
R2	Riding while breaking the speed limit	110 (8.0%)
R3	Riding while performing stunting acts	24 (1.74%)
R4	Riding without giving signal	365 (26.4%)
R5	Riding while using phone	103 (7.5%)
R6	Riding against traffic	109 (7.9%)
R7	Riding while tailgating with another vehicle with unsafe distance	67 (4.85%)
R8	Riding while failing to keep proper side to side movement with another vehicle	403 (29.1%)
	Total	1382 (100.0%)

profile of the respondents. From the collected data, the age of respondents was categorized into four groups, as shown in Table 3. Most of the participants were in the age group of 16–20 years old (32%), followed by 21–25 years old (26.5%), followed by above 45 years old (13.5%). Lastly, only 7% of the respondents fell into the age group 31–35 years old.

Results of cross-tabulation between age groups (16–20 years old) with risky riding behavior are shown in Fig. 7. Since this study focuses on the young motorcyclist, all the data will specifically focus on the age group (16–20 years old). For this age group, "R8: Riding while failing to keep proper side to side movement with another vehicle"

Table 3 Demographic profile of respondents

Trip information		Frequency	Percentage (%)
Age group	16–20	201	59
	21–25	53	15.7
	26 30	23	6.8
	31 35	14	4.2
	36–45	19	5.6
	Above 45	27	8
Gender	Female	132	39.2
	Male	205	60.8
Riding license	Yes	285	84.6
	No	52	15.4
Riding experience	None	50	14.8
	< 2 years	122	36.2
	2–5 years	126	37.4
	> 5 years	39	11.6

engaged the most to the risky riding behavior, while "R3: Riding while performing stunting acts" engaged least in risky riding behavior. In addition, the chi-squared test for independence showed a significant association between age group and risky riding behavior ($X2 = 25.884$, df $= 17$, $p = 0.064$).

Subsequently, for the gender distribution of the respondents involved in the study, most of the respondents were male, with 59%, while female respondents were only 41%. Cross-tabulation result gender with risky riding behavior is shown in Fig. 8, and the results of the chi-squared test for independence showed that the gender has no significant association with risky riding behaviors ($X2 = 2.848$, df $= 4$, $p = 0.022$).

The majority of the respondents engaged more to "R8: Riding while failing to keep proper side to side movement with another vehicle" behavior has the highest number, which is 52%, while the "R4: Riding without signal" behavior engaged less to risky riding behavior turned out that the majority of the respondents had no valid riding license. The mean and standard deviation of having a valid riding license were 3.75 and 0.9225, respectively. Results of cross-tabulations between having a valid riding license and risky riding behavior are shown in Fig. 9. In addition, the results of the chi-squared test for independence showed that having a valid riding license has no significant association with risky riding behaviors ($X2 = 4.316$, df $= 3$, $p = 0.043$).

Regarding the riding experience, the majority of the 16–20 years old respondents had 2–5 years of experience (29%) of riding experience, while only (8%) of respondents had less than two years of riding experience. The mean and standard deviation of riding experience were 4.338 and 1.002 years, respectively. Results of cross-tabulation between riding experience and risky riding behavior are shown in Fig. 10. Generally, participants with riding experience of 2–5 years were more inclined to

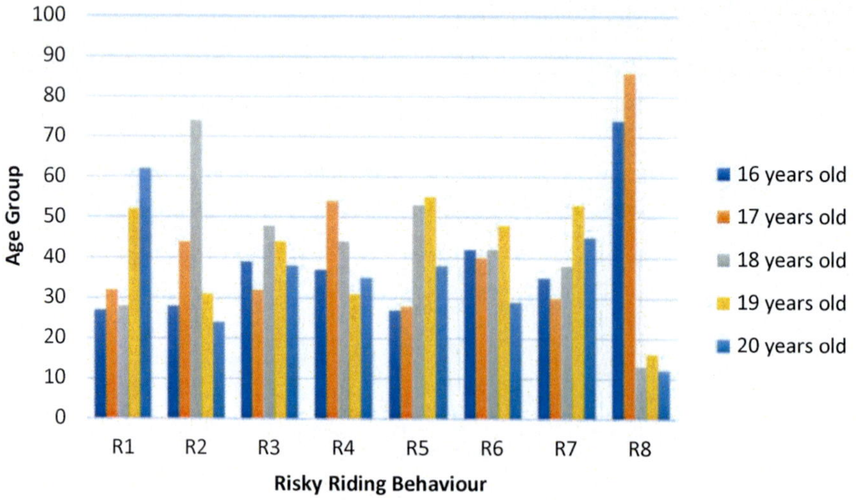

Fig. 7 Cross-tabulation between age group and risk riding behaviors

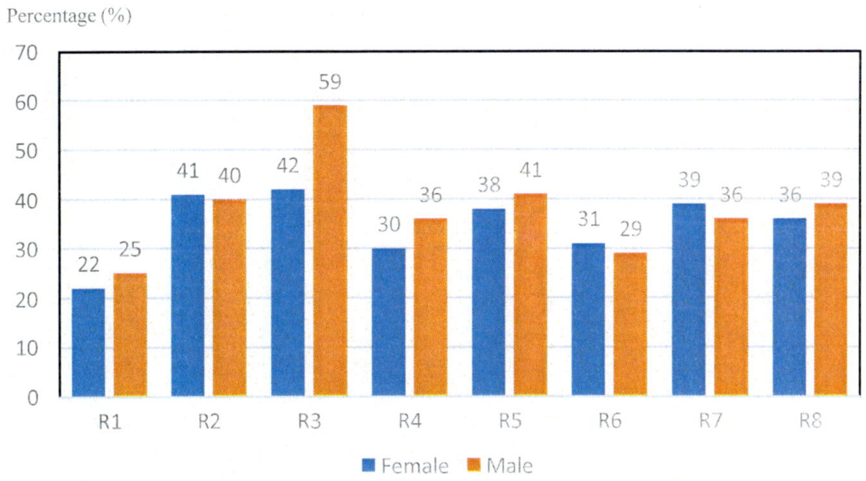

Fig. 8 Cross-tabulation between gender and risky riding behaviors

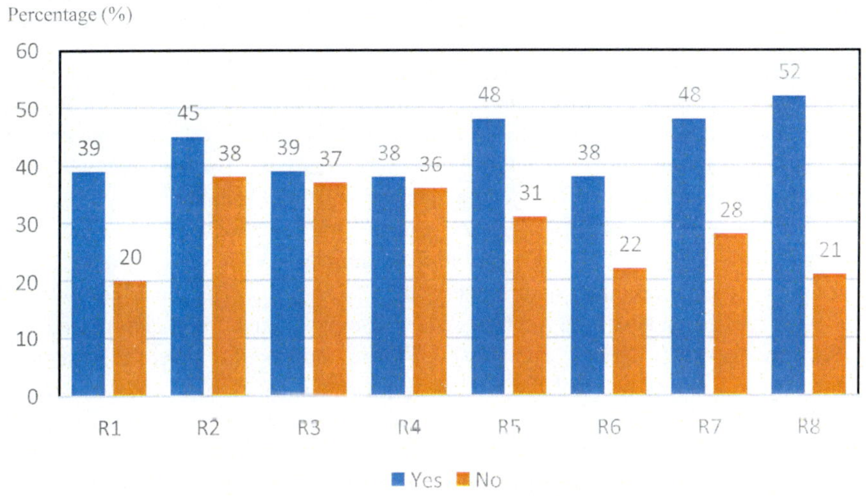

Fig. 9 Cross-tabulation between having riding license and risky riding behaviors

perform risky riding behaviors "R8: Riding while failing to keep proper side to side movement with another vehicle" behaviors and participants with riding experience of more than five years were more engaged to perform risky riding behaviors "R5: Riding while using a phone". Nevertheless, the results of the chi-squared test for independence showed that riding experience ($X2 = 10.897$, df $= 11$, $p = 0.045$) had no significant association with risky riding behaviors.

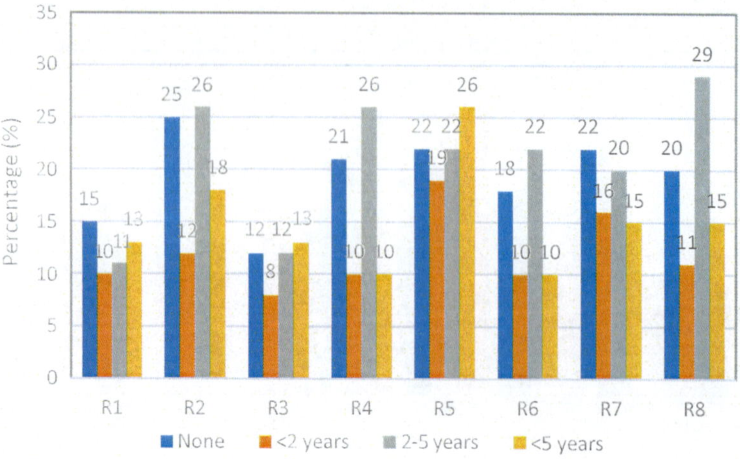

Fig. 10 Cross-tabulation between riding experience and risky riding behaviors

4.3 *Descriptive Analysis for All Variables*

The eight numerical variables were engaged from eight variables to the examination of the feature of numerical variables employed by descriptive analysis. In addition to that, the categories of unsafe riding activity were also assessed using descriptive statistics. The descriptive analysis reported here was based on the data gathered through questionnaire to assess general features of the variables. The acceptable range of skewness is between − 1.96 and + 1.96 which was regularly distributed by all eight numerical variables. In addition, the values of skewness of eight variables were presented and regularly distributed.

The curve line and histogram for secondary school students' risky riding behavior factors are shown in Fig. 11. The values of mean and skewness were represented. Generally, Fig. 11 shows that the risky riding behavior variables were skewed to the right and instead peaked. Result also shows that failing to keep proper distance with another vehicles has the highest mean values compared to the other risky riding behaviours performed by the young riders as shown in Fig. 11. Hence, the mean and skewness of the behavior were 3.02 and 0.449, respectively.

The histogram and curve line as depicted in Fig. 12 show the mean values and curve line for all types of risky riding behaviours performed by secondary school students between the age of 16 to 20 years old. The figure for the values of mean and skewness was represented. Generally, Fig. 12 shows that the risky riding behavior variables are skewed to the right and instead peaked. Scrutunizing the mean values for risky riding behaviours among school students age 16–20 years old, failing to keep safe distance with other vehicles seems to be most dangerous behaviours by this group of riders as shown in Fig. 12. Hence, the mean and skewness of the behavior were 1.66 and 0.082, respectively.

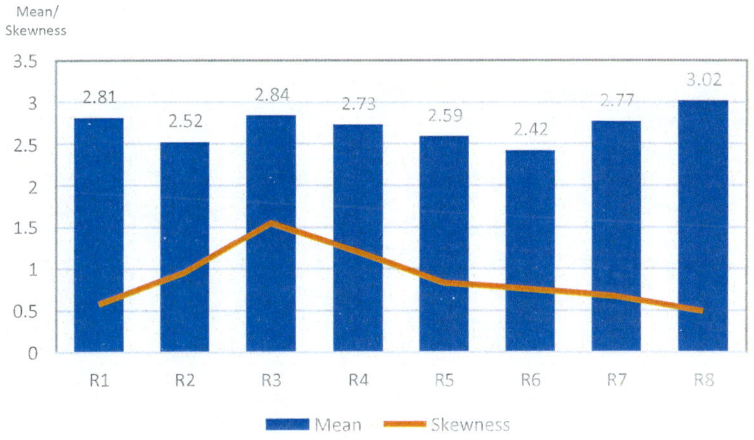

Fig. 11 Histogram of descriptive statistics between ages (16 and 20 years old) and risky riding behaviors

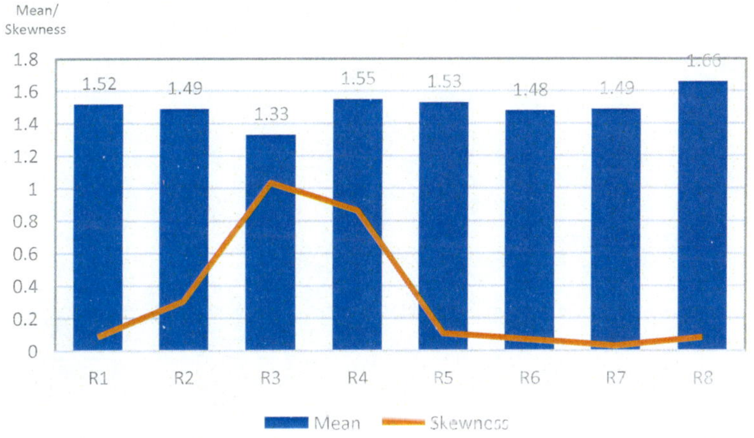

Fig. 12 Histogram of descriptive statistics between gender and risky riding behaviors

Figure 13 shows the curve line and histogram for the risky riding behavior factors among the young motorcyclist only. The figure for the values of mean and skewness presented in Fig. 13 showed that the risky riding behavior variables are skewed to the left and instead peaked. The negative skewed depicts that "R4: Riding without signal" behavior has the highest mean among all the variables shown in Fig. 13. Hence, the mean and skewness of the behavior were 1.48 and 0.124, respectively.

Figure 14 depicts the histogram's curve line for the risky riding behaviour among the young motorcyclist. The figure for the values of mean and skewness was represented. Generally, Fig. 14 shows that the risky riding behavior variables are skewed to the right and instead peaked. The results as in Fig. 14 also proved previous outcomes

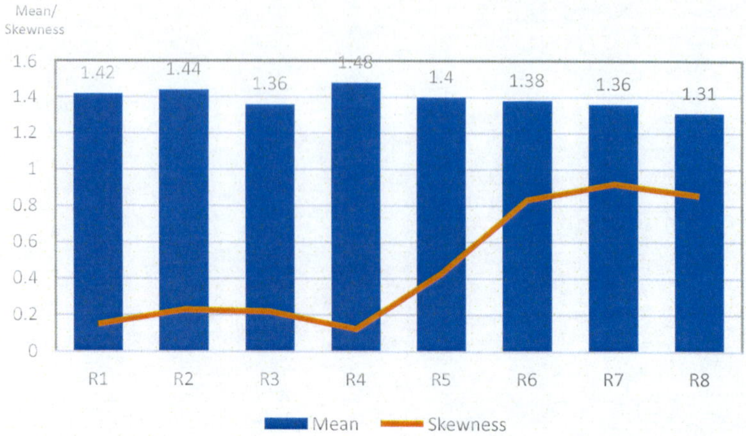

Fig. 13 Histogram of descriptive statistics between riding license and risky riding behaviours

that fail to keep safe distance from other vehicles was the riskiest riding behaviours performed by young motorcyclist. Hence, the mean and skewness of the behavior were 2.52 and 0.089, respectively.

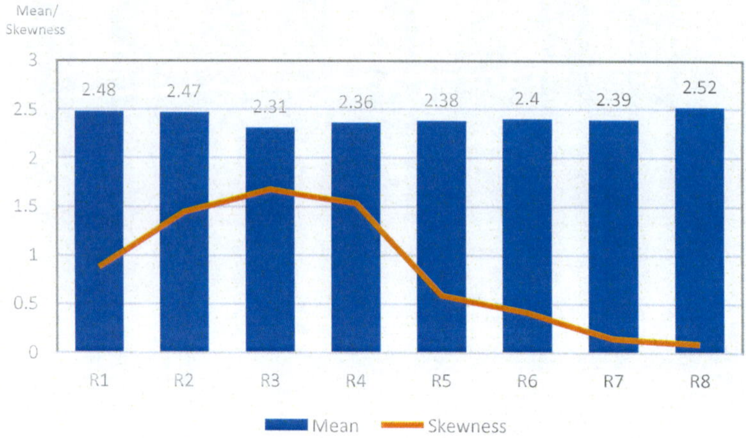

Fig. 14 Histogram of descriptive statistics between riding experience and risky riding behaviours

5 Discussion

Many studies on risky riding behaviors were either conducted using multivariate regression analysis or structural equation modeling. In this study, a more detailed study was conducted to understand the perceptions and attitudes of these young riders on the risky riding behavior in Malaysia using cross-tabulation and correlation analysis. Based on cross-tabulation analysis, young age group of 16–20 years old were associated with risky riding behaviors. Many studies have shown that adolescents tend to perform risky behavior than adults. They have a greater chance to ignore the consideration of risk perceptions in the decision-making process. A recent study by Cordellieri et al. [22] mentioned that the most vulnerable group regarding deaths due to road crashes is the young age group, aged between 15 and 29. They stated that the young age group is usually related to a higher risk of being involved in a road crash, such as speeding, drunk riding, and a lack of observation in urban areas.

According to Wong et al. [17], young drivers are more prone to traffic-law violations, speeding, and more often engaging in risky behavior. Furthermore, [23] mentioned those young riders who are poor in engaging in safe riding behavior. Therefore, in general, young riders more frequently perform risky riding behaviors than adults or older riders. In general, the cross-tabulation results between gender and risky riding behaviours revealed that male riders are more aggressive than female riders. This finding conforms with the conclusion by Goh et al. [23], and they mentioned that male riders are generally more likely to violate traffic regulations than female riders because of the negligence of potential risk, violation, and vehicle examination. In addition, Oltedal and Rundmo [24] concluded that male riders generally have higher scores in risky behaviour, accidents with damage, and accidents with personal injury than female riders.

Similarly, study by Wong et al. [17] found that nearly 60% of male motorcyclists displayed in fractions, notably in speeding, which is greater than the statistical data of female riders (36.3%). They found that male motorcyclists are more inclined to seek experiences and be impatient. Furthermore, a study by Khairul and Ibrahim [25] discovered that the percentage of female riders in the non-risk-taking category is higher than that of male riders. In other words, men motorcyclists are more prone to committing risky behaviors. However, the chi-squared test for independence obtained in this research demonstrated that gender had no significant connection with unsafe riding habits.

As for the riding experience, results of cross-tabulation between riding experience and risky riding behaviour showed that riding experience of between 2 and 5 years is more predominant in performing risky riding behaviours. This observation conforms with the conclusion drawn by McCartt et al. [26], who mentioned that novice or less experienced drivers have a lower tendency of not following the rules of overpassing another vehicle on perceived crash risk compared to medium-level experienced drivers. Additionally, a study conducted by Rolison et al. [27] revealed that the rate of road crashes significantly declined with older age groups and years of licensure. Moreover, the driving experience is related to driving skills. Riding skills concern performance limitations concerning riding tasks that affect riders' risk perception.

Overall, this study provided perception of the consequences of risk-riding behaviors on motorcycle crashes among these young riders. By concentrating on unsafe riding behaviors, these behaviors have been validated as factors that might result in immediate crash risk for young riders. Initially, eight risky riding behaviors were needed to demonstrate actual secondary school student behavior. According to cross-tabulation, several behaviors are firmly and inversely associated depending on the demography of the respondents. To sum up, the determined results based on the investigation were outlined below.

5.1 *The Selected Risk-Riding Behavior*

The following are the main conclusions that can be derived from the findings:

1. Riding while failing to maintain safe side-to-side movement with another vehicle was the most common cause of motorcycle crashes among students in urban areas.
2. Riding experience influenced the young motorcyclist's stunting behavior and failure to maintain correct side-to-side movement with another vehicle.
3. Riding while intoxicated and performing stunts are the behaviors that have had the most negligible impact on the incidence of motorcycle accidents.
4. The relationships of riding without a crash helmet, riding while exceeding the speed limit, riding while performing stunts, riding without giving a signal, riding while using a phone, and riding against traffic are the weakest, with a non-significant small correlation and a negative correlation relationship achieved.

6 Conclusion and Recommendation

This research attempts to find the risky riding conduct that would contribute most to motorcycle collisions and evaluate the view of the young motorcyclists of their understanding of risk attitudes they may do while riding the motorbike. Results of cross-tabulation among demographic characteristics such as age, gender, possessing a valid riding license, and riding experience on dangerous riding behaviors suggested that age and riding experience is substantially connected with risky riding behaviors. This study has effectively supplied a knowledge of the effects of risky riding conduct toward motorcycle crashes among young motorists in mostly metropolitan settings. By identifying a set of risk-riding behavior at the school area, the association between the set of risk-riding behavior among the students' background is based on the combination of risk-riding behaviors.

As most risky behaviours are due to the human factor, safety awareness within Malaysian motorcyclists should be applied on the policy-implementation level to generate an effective intervention. The findings of this study can provide authorities with a base to further understand factors affecting rider behaviour to improve road safety among Malaysian motorcyclists. In addition, findings from this study can streamline the authorities' operations in systematically capturing causes of motorcyclist crash incidents instead of evaluating every incident in an individualistic manner.

Further study can supplement this research by implementing more comprehensive structural equation modeling to determine the correlation among riders' demographic attributes and risk perception. Furthermore, secondary information such as crash data on frequency, type, and severity can be included in structural equation modeling to enhance the prediction capability of the model to anticipate better factors affecting Malaysian motorcyclists. The scalability of structural equation modeling can provide better insight for authorities to assist them in the decision-making process of generating an effective policy.

In a nutshell, to sum up, each risky riding behavior factor has a different effect on road safety among riders, especially the students in urban areas. Other than that, the estimated number of crashes can be a reference to authority to reduce the number of crashes in Malaysia.

References

1. Thomas, G.: Keeping death rate in check. New Straits Times NST Online 03 Sep 2019. https://www.nst.com.my/opinion/letters/2019/09/518489/keeping-death-rate-check. Accessed 04 Dec 2020
2. Idris, A., Hamid, H., Teik Hua, L.: Factors contributing to motorcycle accidents in Malaysia. IOP Conf. Ser.: Earth Environ. Sci. **357**, 012039 (2019). https://doi.org/10.1088/1755-1315/357/1/012039
3. PDRM: Laporan Tahunan PDRM 2009. In: Royal Malaysia Police (ed.) Royal Malaysia Police Annual Report. Kuala Lumpur, Malaysia (2009)
4. Abdul Manan, M.M., Várhelyi, A.: Motorcycle fatalities in Malaysia. IATSS Res. **36**(1), 30–39 (2012). https://doi.org/10.1016/j.iatssr.2012.02.005
5. WHO.: Global Status Report on Road Safety 2018: Summary; (WHO/NMH/NVI/18.20); World Health Organization, Geneva, Switzerland (2018). Available online: https://apps.who.int/iris/handle/10665/277370
6. Blackman, R.A.: The increased popularity of mopeds and motor scooters: Exploring usage patterns and safety outcomes. PhD Thesis, School of Psychology and Counselling, Queensland University of Technology (2012)
7. Vlahogianni, E.I., Yannis, G., Golias, J.C.: Overview of critical risk factors in Power-Two-Wheeler safety. Accid. Anal. Prev. **49**, 12–22 (2012). https://doi.org/10.1016/j.aap.2012.04.009
8. WHO.: Global Plan for the Decade of Action for Road Safety 2011–2020. WHO (2011). https://www.who.int/roadsafety/decade_of_action/plan/en/
9. Khairul Amri Kamarudin, M., et al.: Road traffic accident in Malaysia: trends, selected underlying, determinants and status intervention. Int. J. Eng. Technol. **7**(4.34), 112 (2018). https://doi.org/10.14419/ijet.v7i4.34.23839
10. MIROS Annual Report 2017. Malaysian Institute of Road safety Research, Asean Road Safety Centre (2017)

11. National Center for Statistics and Analysis.: Motorcycles: 2017 data (Up-dated, Traffic Safety Facts. Report No. DOT HS 812 785). National High- way Traffic Safety Administration, Washington, DC (2019)
12. Crump, B.: Where Do Most Motorcycle Accidents Happen? Motorcycle Accident Lawyers, Ben Crump, 2021. https://bencrump.com/motorcycle-accident-lawyer/where-do-most-motorcycle-accidents-happen/. Accessed 22 Nov 2021
13. Zepada, E.: Sobering Motorcycle Accident Statistics. Return of the Cafe Racers (2022). https://www.returnofthecaferacers.com/motorcycle-safety/sobering-motorcycle-accident-statistics/
14. Siti Zaharah, I.: Miros & its role in Asean—towards achieving fatality reduction. In: Malaysian Institute Of Road Safety And Research (MIROS) (2020). Accessed 16 Jan 2021
15. Eboli, L., Forciniti, C., Mazzulla, G.: Factors influencing accident severity: an analysis by road accident type. Transp. Res. Procedia **47**, 449–456 (2020). https://doi.org/10.1016/j.trpro.2020.03.120
16. Peden, M., World Health Organization.: World report on road traffic injury prevention: information kit. World Health Organization, Geneva, Switzerland (2004)
17. Wong, J.-T., Chung, Y.-S., Huang, S.-H.: Determinants behind young motorcyclists' risky riding behavior. Accid. Anal. Prev. **42**(1), 275–281 (2010). https://doi.org/10.1016/j.aap.2009.08.004
18. Dissanayake, S., Lu, J.: Analysis of severity of young driver crashes: sequential binary logistic regression modeling. Transp. Res. Record: J. Transp. Res. Board **1784**(1), 108–114 (2002). https://doi.org/10.3141/1784-14
19. Clark, I.A., Monk, A.M., Maguire, E.A.: Characterizing strategy use during the performance of hippocampal-dependent tasks. Front. Psychol. **11**, 2119 (2020). https://doi.org/10.3389/fpsyg.2020.02119
20. Idris, A., Hamid, H., Teik Hua, L.: Factors contributing to motorcycle accidents in Malaysia. IOP Conf. Ser.: Earth Environ. Sci. **357**, 012039 (2019). https://doi.org/10.1088/1755-1315/357/1/012039
21. Goh, W.C., Leong, L.V., Cheah, R.J.X.: Assessing significant factors affecting risky riding behaviors of motorcyclists. Appl. Sci. **10**(18), 6608 (2020). https://doi.org/10.3390/app10186608
22. Cordellieri, P., Sdoia, S., Ferlazzo, F., Sgalla, R., Giannini, A.M.: Driving attitudes, behaviours, risk perception and risk concern among young student car-drivers, motorcyclists and pedestrians in various EU countries. Transp. Res. Part F: Traffic Psychol. Behav. **65**, 56–67 (2019). https://doi.org/10.1016/j.trf.2019.07.012
23. Leong, L.V., Mohd Shafie, S.A., Gooi, P.K., Goh, W.C.: Assessing self-reported risky riding behavior of motorcyclists at unsignalized intersections for sustainable transportation. Sustainability **13**(16), 9144 (2021). https://doi.org/10.3390/su13169144
24. Oltedal, S., Rundmo, T.: The effects of personality and gender on risky driving behaviour and accident involvement. Saf. Sci. **44**(7), 621–628 (2006). https://doi.org/10.1016/j.ssci.2005.12.003
25. Khairul, M., Ibrahim, A.: A case study on risk-taking behaviours among motorcyclists in Klang Valley, Malaysia. In: Riding Behavior Study for Safer Road Design for Two-Wheelers View Project Driver Attitude Towards Motorcyclist in Malaysia View Project 2012. Accessed 09 Dec 2021
26. McCartt, A.T., Mayhew, D.R., Braitman, K.A., Ferguson, S.A., Simpson, H.M.: Effects of age and experience on young driver crashes: review of recent literature. Traffic Inj. Prev. **10**(3), 209–219 (2009). https://doi.org/10.1080/15389580802677807
27. Rolison, J.J., Regev, S., Moutari, S., Feeney, A.: What are the factors that contribute to road accidents? An assessment of law enforcement views, ordinary drivers' opinions, and road accident records. Accid. Anal. Prev. **115**, 11–24 (2018). https://doi.org/10.1016/j.aap.2018.02.025

How Drivers' Physiological Response of Blood Pressure on Unsignalized Conventional Roundabout?

Shahrul Jainudin, Jezan Md Diah, Suria Haron, and Sharifah Abdullah

Abstract A roundabout is one of the components of a critical road network, which consists of a circular intersection with a smaller island in the middle. In Malaysia, the traffic that moves through this critical network usually moves in a clockwise direction, meaning that the priority of traffic would be given towards the vehicle inside the roundabout to move first. The focus of this research was to measure the physiological behaviour of each of the participants. By focusing on the inner side of the transportation system, which is the driver itself, we could build an innovation that would contribute to a better transportation system. The youth participants were those whose age was set at 20–30 years old. One of the accessible physiological behaviours of drivers is blood pressure, measured by using an OMRON Digital Wrist Blood Pressure Monitor. By studying these data, we could determine the way they conducted their driving approach during a real-life situation. Although various variables could affect their driving experience in real life in this study, the focus will be only on the blood pressure of drivers before, during, and after driving, including their driving speed. Analysis by using the regression method to obtain the r-squared value has been applied to determine the correlation level between both of the variables. By comparing the data of blood pressure of genders between entering, driving through, and exiting the roundabout, the highest R^2 reading obtained was 0.8209 and 0.6779 for male and female drivers, respectively.

S. Jainudin (✉)
Faculty of Civil Engineering (FCE), Universiti Teknologi MARA (UiTM), F. T. Labuan, Malaysia
e-mail: shahruljainudin@gmail.com

J. M. Diah
Faculty of Civil Engineering (FCE), Malaysia Institute of Transport (MiTRANS), Universiti Teknologi MARA (UiTM), 40450 Cawangan Shah Alam, Malaysia

S. Haron
School of Civil Engineering, Universiti Teknologi MARA (UiTM), 13500 Cawangan Pulau Pinang, Permatang Pauh, Malaysia

S. Abdullah
Universiti Teknologi MARA (UiTM), 40540 Cawangan Shah Alam, Malaysia

N. Sabtu (ed.), *Proceedings of AWAM International Conference on Civil Engineering 2022 - Volume 3*, Lecture Notes in Civil Engineering 386,
https://doi.org/10.1007/978-981-99-6026-2_27

Keywords Driver behaviour · Blood pressure · Roundabout · Intersection · Youth driving experience · Speeding

1 Introduction

Drivers' behaviour can be defined as the way they act while controlling a vehicle in a course of motion. These behaviours can be affected by many factors, such as the health condition of drivers, weather, road conditions, other drivers on the road. The act of driving can be defined as the use of a person's cognitive, visual, and motor reactions to control a vehicle [1]. According to Batrakova and Gredasova [2], drivers' behaviour is one of the factors that contribute to the increased number of accidents, especially in the circular intersection area.

Physiological behaviour is one of the factors that use the mental capacity of a person and will affect their physical body response. In this research, the physiological behaviour of the driver can be used to determine whether the driver is in good or healthy condition for driving. By knowing the characteristics of drivers, we could study the manner of their driving on a road and find the problem with the road structure, etc. [3]. Blood pressure is one of the indicators in measuring physiological behaviour, and this response can change from time to time depending on the drivers' current condition [4]. According to the previous researchers, one of the ways to increase the safety level of drivers is by studying the way they behave during travel in a real-field situation. By learning from this analysis, we could access whether the drivers were in a suitable condition to drive a vehicle or not.

The same author also mentioned that blood pressure can be defined as arterial wall pressure that is caused by the blood flow in our cardiovascular system. Commonly, there are two types of blood pressure values, which are systolic and diastolic, where there are the highest and lowest readings of blood pressure, respectively. According to Mohan et al. [5], the normal values for systolic and diastolic blood pressures were 120/80 mmHg, respectively.

While in a critical intersection such as a roundabout, the probability of a car crash is higher due to a complicated decision that requires experience, and a high level of calculative instinct is required in order to go through it in a safe manner [2]. There were two types of intersection, which is signalized and unsignalized. According to the same author, the capacity of traffic can be one of the factors that contribute to the effectiveness of the roundabout. According to the Washington State Department of Transportation in 2021, by constructing a roundabout in a road network, it could increase the traffic capacity by up to 30–50% compared to a typical intersection.

Plus, a circular intersection is one of the road components where it is made of a few junctions and the traffic moves in one direction around a central island. According to United States Department of Transportation, 2021, roundabouts are one of the safety measures that have been used in reducing accidents that could cause serious injury or death. Besides improving the safety level of the road condition, the department also

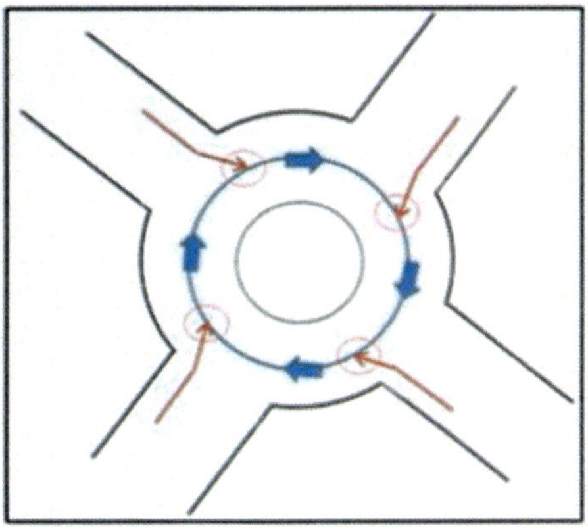

Fig. 1 Example of four-legged roundabout

stated that roundabouts aid in reducing driving speeds and promote the reduction of conflict points.

Next, a roundabout is a circular intersection which is usually made of one (1) or more legs where it is one of the critical road networks. The vehicle inside the roundabout will have the priority in movement, where there will only be one circular counterclockwise movement inside [6]. Figure 1 shows an example of the movement of a four-legged roundabout.

The objectives of this research were to identify the speed of vehicles and the driver's blood pressure among youth, to establish the trend pattern of driver's behaviour based on the driver's blood pressure among youths in gender difference, and lastly, to develop the relationships between drivers' blood pressure and speed among youths in the unsignalized conventional roundabout.

Various studies have proved that there is a correlation between the physiological behaviour of drivers and the way they drive on a course. The focus of the research itself, where the participants will be in the range of youths' age, and their blood pressure as the physiological indicator of behaviour during driving through the roundabout, was studied.

2 Materials and Method

In this research, there were several materials and equipment that were mainly used for data sampling.

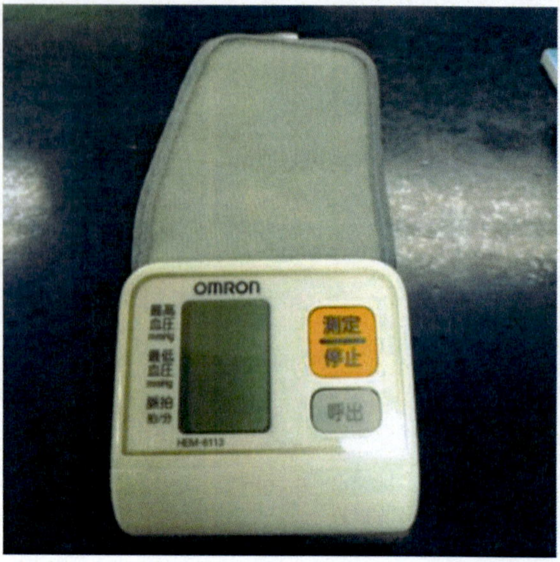

Fig. 2 Omron digital wrist blood pressure monitor

2.1 *Materials and Equipment*

The participants' blood pressure was measured in three phases, which were before, during, and after driving throughout the roundabout, where three readings were taken in each phase to increase the data accuracy. Blood pressure is measured in millimetres of mercury (mmHg). All readings taken will be described by classification by following the guidelines from the Ministry of Health Malaysia. Figure 2 shows the Omron Digital Wrist Blood Pressure Monitor, which has successfully sampled the systolic and diastolic blood pressure readings in mmHg units.

2.2 *Method*

By referring to the calculations using the Daniel Soper sample size calculator software in Fig. 3, 36 participants should have been selected, which focused on youth ages 20–30 years old. But due to the pandemic of COVID-19, a Movement Control Order (MCO) has been enforced in the Selangor area, which has caused the data sampling to be halted.

Besides using a Perodua Axia as the consistent variable for the participant, the participant drove through the driving route three times to get the average speed and blood pressure. The data were collected from 14 participants, the majority of whom were students at Universiti Teknologi Mara (UiTM) Shah Alam before the campus was closed due to COVID-19.

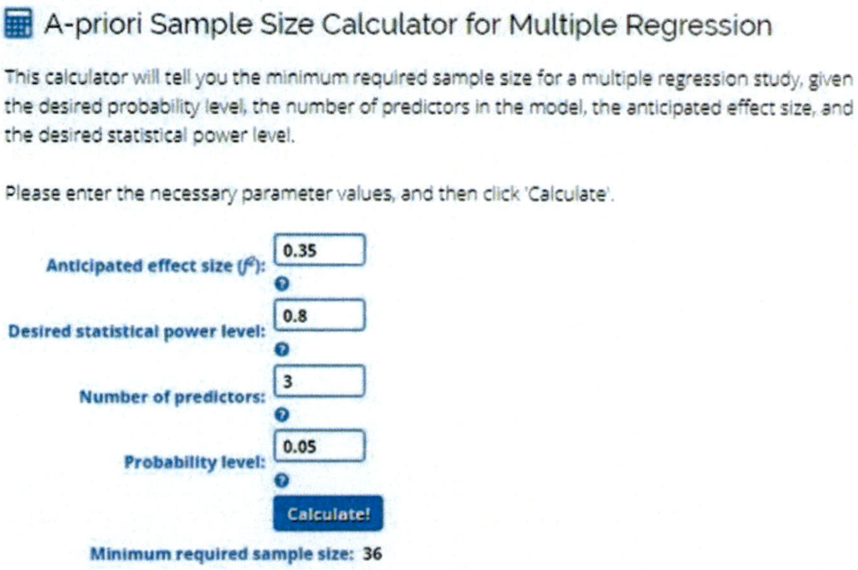

Fig. 3 Daniel Soper sample size calculator

2.3 *Research Location*

The location selected for this study was an unsignalized conventional roundabout at Bulatan Seksyen 2, Shah Alam, Selangor. Figure 4 shows the aerial map view taken by using Google Maps. The selected participants will drive the course which starts from the main gate of UiTM Shah Alam (No. 1), goes through the roundabout (No. 3), and goes back to the initial location as shown in Fig. 5.

2.4 *Research Limitations*

The limitations while conducting this research were that there were 36 participants, whose ages ranged from 20 to 30 years old. But due to the Movement Restrictions' Order (MCO) in Malaysia that has been implemented, the research has successfully sampled up to fourteen (14) participants. Next, the Omron Digital Wrist Blood Pressure Monitor will read the blood pressure of each participant before, during, and after driving throughout the roundabout. The study was only conducted during weekdays in non-peak hours, and the weather had to be bright and clear.

Fig. 4 Aerial map view of Bulatan Bestari Seksyen 2, Shah Alam, Selangor, taken by using Google Maps

Fig. 5 Aerial map view of the full driving route taken by using Google Maps

3 Results and Discussions

3.1 *Difference of Blood Pressure Between Gender Throughout the Route of Roundabout*

All the participants have a normal BMI and have gone through the full driving route three times to get the average data. Table 1 displays the average data for male participants, and Table 2 displays the average data for female participants systolic blood pressure (SBP) and diastolic blood pressure (DBP) while entering, driving through, and exiting the roundabout, as well as speed data.

3.2 *Discussions*

Learning human physiological behaviour is vital in such a way that the innovation of the vehicles can be done through the inner side of the transportation system, which is focused on the drivers themselves.

Figure 6 shows the average systolic blood pressure versus average speed before entering the roundabout for both genders.

Figure 7 shows the average systolic blood pressure versus average speed during in the roundabout for both genders.

Figure 8 shows the average systolic blood pressure versus average speed after entering the roundabout for both genders.

According to Md Diah et al. [7], there was a variable variability in human actions, which caused it to be hard to predict. From the data shown in Figs. 6, 7, and 8, all of the R^2 show a steadily positive relationship between blood pressure and speed. Compared with all the figures, the R^2 of female drivers is slightly lower than male drivers. It is one of the signs that female drivers have better composition and calmness during driving compared to male drivers, as stated by during their previous research. Besides that, it also proved that male drivers were more aggressive during driving compared to female drivers.

Referring to Fig. 7, it has the highest reading of R^2, which is 0.8209 and 0.6779 for male and female drivers, respectively. This R^2 reading of driving throughout the roundabout is the highest compared to the driving session before entering and exiting the roundabout. According to Refs. [8, 9], driving in the roundabout requires more cognitive and mental reactions where various factors need to be considered in order to cross the roundabout safely. This is one of the reasons that the reading of R^2 is higher compared to before and after exiting the roundabout.

Table 1 Average data for male participant's blood pressure and speed while entering, driving through, and exiting the roundabout

Participants 1, (age 25, male)	Entering roundabout	Driving through roundabout	Exiting roundabout
SBP, mmHg	109.67	106.33	109.33
DBP, mmHg	77.33	83.00	75.00
Speed, km/hr	42.33	41.67	48.00
Participants 2, 25 (male)	Entering roundabout	Driving through roundabout	Exiting roundabout
SBP, mmHg	114.33	127.67	113.67
DBP, mmHg	74.33	85.33	74.33
Speed, km/hr	44.00	45.00	49.33
Participants 3, (age 25, male)	Entering roundabout	Driving through roundabout	Exiting roundabout
SBP, mmHg	94.67	103.00	114.67
DBP, mmHg	66.67	66.00	65.57
Speed, km/hr	35.67	39.67	52.00
Participants 4, (age 25, male)	Entering roundabout	Driving through roundabout	Exiting roundabout
SBP, mmHg	104.33	118.67	96.00
DBP, mmHg	52.33	71.00	65.67
Speed, km/hr	46.33	40.33	50.00
Participants 5, (age 23, male)	Entering roundabout	Driving through roundabout	Exiting roundabout
SBP, mmHg	130.33	90.00	125.67
DBP, mmHg	86.33	65.00	98.67
Speed, km/hr	55.67	34.67	50.33
Participants 6, (age 23, male)	Entering roundabout	Driving through roundabout	Exiting roundabout
SBP, mmHg	95.33	97.67	88.67
DBP, mmHg	52.67	55.00	49.33
Speed, km/hr	45.00	37.67	42.00
Participants 7, (age 23, male)	Entering roundabout	Driving through roundabout	Exiting roundabout
SBP, mmHg	99.33	107.67	91.00
DBP, mmHg	62.33	69.00	47.00
Speed, km/hr	47.33	40.00	45.67

Table 2 Average data for female participant's blood pressure and speed while entering, driving through, and exiting the roundabout

Participants 8, (age 24, female)	Entering roundabout	Driving through roundabout	Exiting roundabout
SBP, mmHg	87.33	112.67	119.67
DBP, mmHg	57.00	64.67	76.33
Speed, km/hr	38.33	46.33	50.00
Participants 9, (age 23, female)	Entering roundabout	Driving through roundabout	Exiting roundabout
SBP, mmHg	105.67	80.00	83.00
DBP, mmHg	76.67	40.33	49.33
Speed, km/hr	40.33	35.67	45.00
Participants 10, (age 23, female)	Entering roundabout	Driving through roundabout	Exiting roundabout
SBP, mmHg	113.33	91.33	85.33
DBP, mmHg	70.67	62.33	58.33
Speed, km/hr	45.33	37.67	49.67
Participants 11, (age 23, female)	Entering roundabout	Driving through roundabout	Exiting roundabout
SBP, mmHg	105.00	104.67	113.00
DBP, mmHg	65.33	68.00	65.33
Speed, km/hr	44.67	42.67	53.00
Participants 12, (age 23, female)	Entering roundabout	Driving through roundabout	Exiting roundabout
SBP, mmHg	110.00	121.00	98.00
DBP, mmHg	78.67	88.00	73.67
Speed, km/hr	44.33	45.00	48.00
Participants 13, (age 24, female)	Entering roundabout	Driving through roundabout	Exiting roundabout
SBP, mmHg	115.67	118.67	88.67
DBP, mmHg	82.67	82.00	57.00
Speed, km/hr	48.33	39.33	43.67
Participants 14, (age 24, female)	Entering roundabout	Driving through roundabout	Exiting roundabout
SBP, mmHg	93.00	81.00	104.33
DBP, mmHg	62.33	53.67	81.67
Speed, km/hr	45.67	30.33	43.00

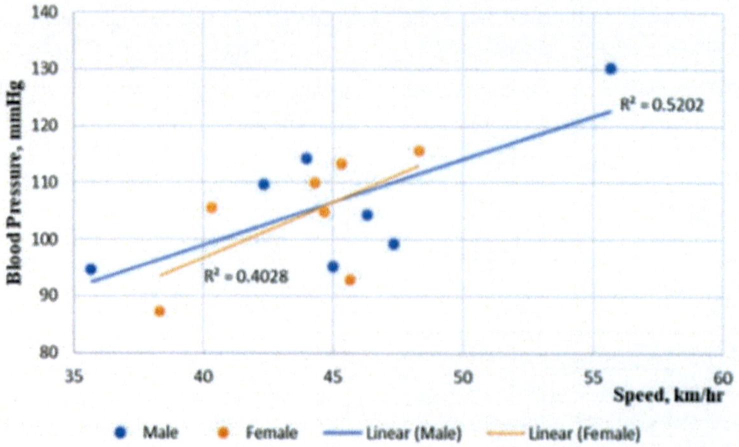

Fig. 6 Average systolic blood pressure versus average speed before entering the roundabout for both genders

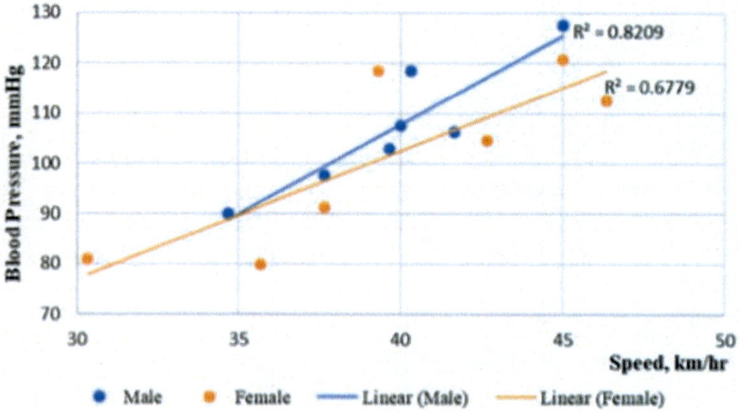

Fig. 7 Average systolic blood pressure versus average speed before entering the roundabout for both genders

4 Conclusion

Physiological studies have a lot of different factors that make it hard to get accurate information about how people act in a given situation. In this research, the physiological responses of drivers upon entering, throughout, and exiting the roundabout with differences in gender and age have been successfully investigated. Based on the results obtained, the highest R^2 reading in the difference of gender was 0.8209 and 0.6779 for male and female drivers, respectively. All these readings have shown a similar trend with a steadily positive gradient.

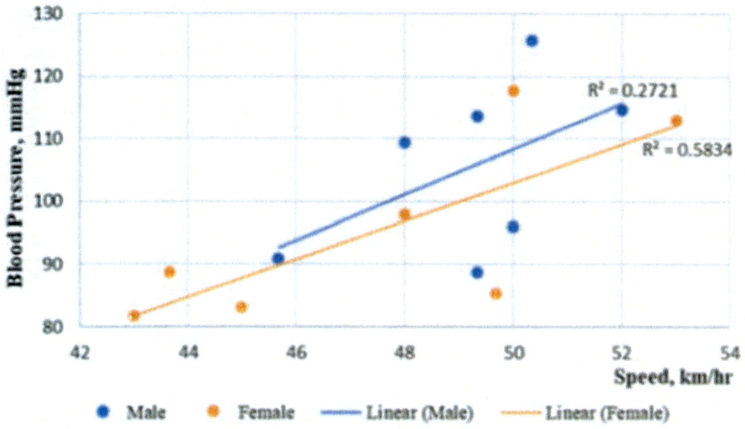

Fig. 8 Average systolic blood pressure versus average speed after exiting the roundabout for both genders

From the results obtained, the drivers' behaviour, such as speeding or aggressive driving between male and female drivers, could be the factor for the higher reading of blood pressure during driving [10]. Further studies are needed in order to gain a better understanding of the factors that influence the reading of blood pressure during driving, especially at the unsignalized conventional roundabout. By learning all these important characteristics of drivers, a better road layout and system could be constructed, and this would lead to a better road system.

This research has a lot of room for improvement, but the data that we have obtained so far are significant for future research. By having these data, the future accidents regarding roundabouts could be decreased by using various equipment, adding up to the current research progress such as electroencephalogram (EEG) and electrocardiogram (ECG) detection programmes conducted by previous researchers, such as [11, 12].

Acknowledgements The authors would like to thank the staff and organization of MiTRANS and the Faculty of Civil Engineering, Universiti Teknologi MARA (UiTM) for the significant contribution and full guidance throughout the process of completing this paper. Words of gratitude are also extended to all individuals and other organizations that have made this study possible.

References

1. Freydier, C., Berthelon, C., Bastien-Toniazzo, M.: Does early training improve driving skills of young novice French drivers? Accid. Anal. Prev. **96**, 228–236 (2016). https://doi.org/10.1016/j.aap.2016.07.026
2. Batrakova, A.G., Gredasova, O.: Influence of road conditions on traffic safety. Procedia Eng. **134**, 196–204 (2016). https://doi.org/10.1016/j.proeng.2016.01.060

3. Galovski, T.E., Blanchard, E.B., Malta, L.S., Freidenberg, B.M.: The psychophsiology of aggressive drivers: comparison to non-aggressive drivers and pre- to post-treatment change following a cognitive-behavioural treatment. Behav. Res. Ther. **41**(9), 1055–1067 (2003). https://doi.org/10.1016/S0005-7967(02)00242-5
4. Xu, J., Liu, J., Sun, X., Zhang, K., Qu, W., Ge, Y.: The relationship between driving skill and driving behavior: psychometric adaptation of the driver skill inventory in China. Accid. Anal. Prev. **120**(August 2017), 92–100 (2018). https://doi.org/10.1016/j.aap.2018.07.032
5. Mohan, B., Verma, A., Singh, K., Singh, K., Sharma, S., Bansal, R., et al.: Prevalence of sustained hypertension and obesity among urban and rural adolescents: a school-based, cross-sectional study in North India. BMJ Open **9**(9) (2019). https://doi.org/10.1136/bmjopen-2018-027134
6. WSDOT Public Transportation Division.: Public Transportation Mobility Report (2021). https://wsdot.wa.gov/sites/default/files/2021-11/2021-Public-Transportation-Mobility-Report.pdf
7. Md Diah, J., Abdul Rahman, M.Y., Adnan, M.A., Atan, I.: Weaving section flow model at the weaving area of Malaysian conventional roundabout. J. Transp. Eng. **136**(8), 782–792 (2010).https://doi.org/10.1061/(ASCE)TE.1943-5436.0000133
8. Frost, J.: Introduction to statistic. Frost. J. **18**(1) (2018). https://statisticbyjim.com/basic/correlations
9. Bou Samra, F., El Tomb, P., Hosni, M., Kassem, A., Rizk, R., Shayya, S., Assaad, S.: Traffic congestion and blood pressure elevation: a comparative cross-sectional study in Lebanon. J. Clin. Hypertens. **19**(12), 1366–1371 (2017).https://doi.org/10.1111/jch.13102
10. Wang, X., Liu, Y., Wang, F., Wang, J., Liu, L., Wang, J.: Feature extraction and dynamic identification of drivers' emotions. Transp. Res. F: Traffic Psychol. Behav. **62**, 175–191 (2019). https://doi.org/10.1016/j.trf.2019.01.002
11. da Silva, F.P., Santos, J.A., Meireles, A.: Road accident: driver behaviour, learning and driving task. Procedia—Soc. Behav. Sci. **162**(Panam), 300–309 (2014). https://doi.org/10.1016/j.sbspro.2014.12.211
12. Awais, M., Badruddin, N., Drieberg, M.: A hybrid approach to detect driver drowsiness utilizing physiological signals to improve system performance and Wearability. Sensors (Switzerland) **17**(9), 1–16 (2017). https://doi.org/10.3390/s17091991

Mobility Trend in Malaysia Throughout Restrictive Mobility Policies and National Immunization Programme Due to COVID-19 Pandemic

Surachai Airak and Nur Sabahiah Abdul Sukor

Abstract A study on the change in mobility amongst Malaysians during and post-restriction movement policies due to COVID-19 is needed, at the same time to understand the impact of the National Immunization Programme that was imposed by the government. A descriptive and comparative analysis was done for the percentage change in mobility of all states in Malaysia for retail and recreation, grocery and pharmacy, parks, transit stations, workplaces, and residential areas using Google Mobility Reports. Almost all the categories of mobility showed a decrease in percentage during 2020 due to the MCO, and the percentage for residential showed an increase as people were encouraged to just stay at home. In 2021, generally, the trend was an increase in all categories due to the ease of restriction and the proactive move by the government under the National Immunization Programme (NIP).

Keywords Mobility restriction · COVID-19 · Vaccination

1 Introduction

The first case of unknown cause was first detected in Wuhan, China, by the World Health Organization (WHO) on 31 December 2019 [1] which then became known as the Coronavirus Disease 2019 or "COVID-19". In just a mere 27 days, the virus spreads to 11 countries outside of China and the confirmed cases jumped to 2798 [2]. With the concern of the number of confirmed cases proliferating each day, the WHO declared the COVID-19 as a pandemic [3]. As of the time of writing this, by November 2021, a total of 269 million confirmed cases and nearly 5.3 million deaths have been reported globally [4].

Although the pandemic has caused an impact to almost all aspects and sectors globally [5–7], but one of the aspects that was hit most was mobility. This was

S. Airak · N. S. A. Sukor (✉)
School of Civil Engineering, Universiti Sains Malaysia, 14300 Nibong Tebal, Penang, Malaysia
e-mail: cesabahiah@usm.my

N. Sabtu (ed.), *Proceedings of AWAM International Conference on Civil Engineering 2022 - Volume 3*, Lecture Notes in Civil Engineering 386,
https://doi.org/10.1007/978-981-99-6026-2_28

because as the number of cases kept on increasing, governments around the world had to enforce lockdowns to curb the spread [8–10] which meant that activities on almost all sectors were brought to a halt.

In a study done to assess the impact that COVID-19 has caused to mobility, it was found that this pandemic has changed people's travel behaviour [11]. Most of the workers have shifted to working from home (WFH), and the frequency of people using the public transports has decreased significantly in Iran, Australia, and Norway with walking and bus which showed the most significant reductions in terms of transportation mode. Mobility reduction was also noted for multiple travel purposes, namely purchasing non-essential goods, visiting relatives, or hanging out with friends.

In the same study, it was also found that people's perceptions towards transportation has also changed. Usage of public transport was perceived as having higher probability of contracting the virus compared to non-public. This finding coincides with the finding of a study done by Przybylowski et al. [12] which showed that the public feels less safe in using the public transport during the pandemic compared to before the pandemic. Habib and Anik [13] in their study found that people are now shifting to private car, bicycle, and walking and avoiding public transport because of their fear of COVID-19.

Malaysia is no exception to the whole COVID-19 pandemic situation. The first case was detected in this region on the 25 January 2020 [14, 15] with the first death recorded on 17 March 2020 [16]. On 16 March 2020, then Prime Minister of Malaysia, Tan Sri Muhyiddin Yassin announced that a 14-day nationwide lockdown (Movement Control Order, MCO) would be implemented beginning 18 March 2020 to the 31 March 2020 [17], and since then, a series of total and partial lockdowns had been implemented. This MCO brought about new prohibitions such as travelling in and out of the countries, movement of persons and mass gatherings, and operation of educational institutions [18].

The introduction of the MCO has caused Malaysians' mobility to change drastically, especially their travelling purpose and mode of choice [19]. An analysis showed that mobility in Malaysia suffers from decline in almost all categories, especially mobility for public places [20]. However, workplaces showed small increases in mobility during late March to early April 2020. The finding was confirmed by Rajendran et al. [21] which in their study noted that there was a considerable decrease in human mobility due to the social restrictions done by the Malaysian government. Another study found that most of the individuals spent their time in home due to the lockdown and used personal motorized transport instead of public mode during the pandemic [22].

Therefore, with the impacts towards mobility mentioned above, this paper aims to portray the changes of mobility trend from the beginning of MCO until the ease of said restriction order. This is important as the community is ready to shift to the new normal, and with most of the population vaccinated, changes in mobility could be different than what was noted in previous studies.

2 Methodology

2.1 *Source of Data*

For this study, several sources were used to obtain the following data: percentage change in mobility; number of daily covid cases; percentage of vaccination. All data obtained were taken from 15 February 2020 to 11 December 2021. For the percentage change in mobility, data were obtained via the Google's COVID-19 Community Mobility Reports which can be accessed at https://www.google.com/covid19/mobility/ (accessed on December 12th 2021). The reports compare the change of everyday mobility for several locations with its baseline which is the median value for the corresponding day of the week from the 3 January 2020 until the 6 February 2020 [23].

The number of daily cases was retrieved from the Ministry of Health Malaysia's (MOH) official portal, COVIDNOW, accessible at https://covidnow.moh.gov.my/ (accessed on December 13th, 2021). The portal provides several statistics including Deaths, Hospital Admissions, and Tests Conducted to name a few. However, for the purpose of this study, only data on Confirmed COVID-19 Cases were retrieved. For percentage of vaccination, it was also retrieved from the MOH's COVIDNOW portal. The data were then cross-checked with several other sources such as The Special Committee for Ensuring Access to COVID-19 Vaccine Supply (JKJAV) official Facebook (https://www.facebook.com/jkjavmy) and official Twitter (https://twitter.com/jkjavmy?lang=en), in addition to its official website, accessible via https://www.vaksincovid.gov.my/.

2.2 *Categories of Mobility*

There is a total of six categories (refer Table 1) that were measured by the Google Mobility Reports, namely retail and recreation, supermarket and pharmacy, parks, public transport, workplaces, and residential.

2.3 *Analysis of Data*

For this study, descriptive analysis was used to briefly explain on the overall trend regarding the percentage change in mobility and the daily number of COVID-19 cases in the country.

Next, a comparative analysis was also done for each category of mobility, utilizing the mobility data for each state in Malaysia with the different stages of restrictive orders and vaccination rate (for 2021s data only). As compared to the previous studies which only focused on Malaysia as a whole [20, 21], this study will compare the

Table 1 Details of places for each category of mobility

Category	Details of places
Retail and recreation	Restaurants, cafés, shopping centres, theme parks, museums, libraries, and cinemas
Supermarket and pharmacy	Supermarkets, food warehouses, farmers markets, specialty food shops, and pharmacies
Parks	National parks, public beaches, marinas, dog parks, plazas, and public gardens
Public transport	Public transport hubs, such as underground, bus, and train stations
Workplaces	Places of work
Residential	Places of residence

mobility trend for all states (13 states and 3 federal territories) in Malaysia. This study will also analyze the trend for 2 years, namely 2020 and 2021, as compared to the previous studies which only looked at 2020.

3 Results

3.1 COVID-19 Situation in Malaysia

The number of daily cases with regard to different stages of Movement Control Order (MCO) for the year 2020 is shown in Fig. 1. During the Pre-MCO which was around February 2020 until early March 2020, there were less than ten COVID-19 cases reported each day. However, beginning the 15 March 2020, the daily cases sharply increased as the country reported above 100 new cases daily. The government then implemented MCO and Conditional Movement Control Order (CMCO). The cases showed a steady decline with the country reporting 117 cases in the beginning of MCO and only seven by the end of the CMCO. The government imposed the Recovery Movement Control Order (RMCO) due to the decreasing cases and ease the mobility for all sectors with strict Standard Operation Procedures (SOPs). Interstate travel was permitted, and people were allowed to meet and socialize with each other by adapting to the new norm. As expected, this resulted in a resurgence of daily COVID-19 instances, which increased from only two at the start of RMCO to 2525 at the end of that period.

Meanwhile, Fig. 2 shows the number of daily cases for different stages of restrictive movement policies in 2021. On 12 January 2021, the government declared a State of Emergency (SOE), which also endorsed the COVID-19 cases to private hospitals if necessary. SOE showed a decrease from 3309 to 1178 cases on 1 April 2021, but then increased tremendously and reached an all-time high with 17,150 new cases on the final day of SOE (1 August 2021). At the same time, the government also began the National COVID-19 Immunization Programme (NIP) to get as many

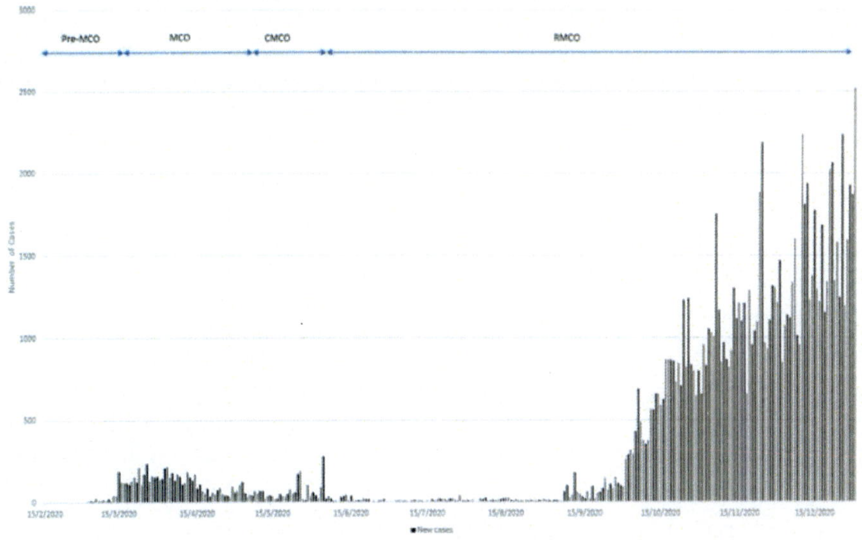

Fig. 1 Number of daily cases for different stages of MCO in 2020

Malaysians vaccinated as possible. The government then introduced a new plan called the National Recovery Plan (NRP), which aimed to help the country emerge from the COVID-19 economic fallout. However, interstate travel was banned, and the SOPs were similar to those for the MCO. Despite this, the cases kept on climbing, reaching a new all-time high on 26 August 2021, with 24,599 new cases reported. The cases then showed a steady decline, with 4626, new cases recorded on 11 December 2021, due to an increasing percentage of the population being vaccinated.

3.2 *Retail and Recreation*

The percentage change in mobility for 2020 and 2021 for retail and recreation purpose is shown in Figs. 3 and 4, respectively. All states showed a similar pattern throughout 2020. Once MCO was implemented, every state's mobility for this purpose showed a sharp decline, with most of the states noting a minimum of 60% decrease in mobility compared to its baseline. This is relevant as MCO banned all social gatherings and dine-ins were not allowed. Once CMCO was implemented, all states started to show an increase in mobility for this purpose until mid-September, with Kelantan reaching the highest increase in mobility (45%) on 30 July 2021. On the other hand, the lowest mobility change recorded in that period was by Wilayah Persekutuan (WP) Putrajaya, recording a decrease of almost 70% in mobility. Once the number of cases started to show a significant increase, the percentage mobility for retail and recreation for every state started decreasing again until the end of October, with Sabah having the largest decrease for a month. It is worth noting that during this period, RMCO was

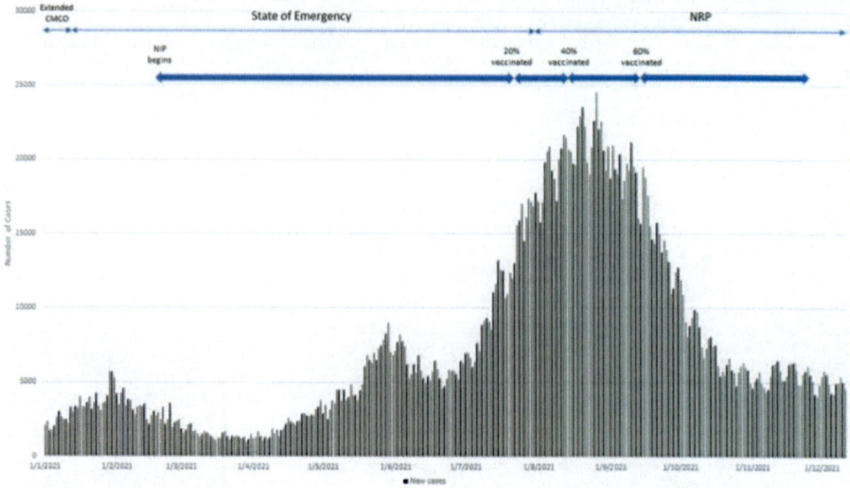

Fig. 2 Number of daily cases for different stages of restrictive movement policies in 2021

implemented, which means that restrictions were mostly relaxed, which caused the mobility to increase again alongside the number of daily cases.

In 2021, once the SOE was announced, almost all states showed a sharp decline in percentage mobility for retail and recreation except for Perlis, Kelantan, Melaka, and Terengganu which showed a rebound in percentage for one day, before decreasing again the next day after SOE was implemented. The mobility pattern then showed an increase with Kelantan dominating the highest percentage change in mobility for

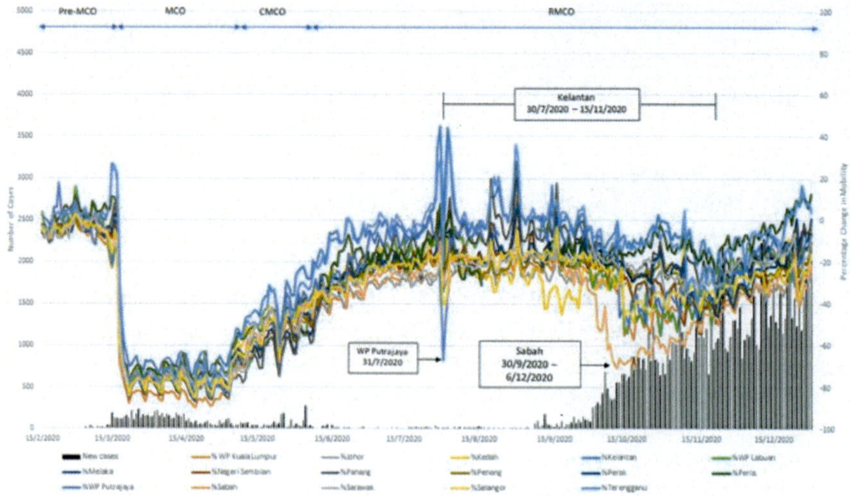

Fig. 3 Percentage change in mobility for retails and recreation in 2020

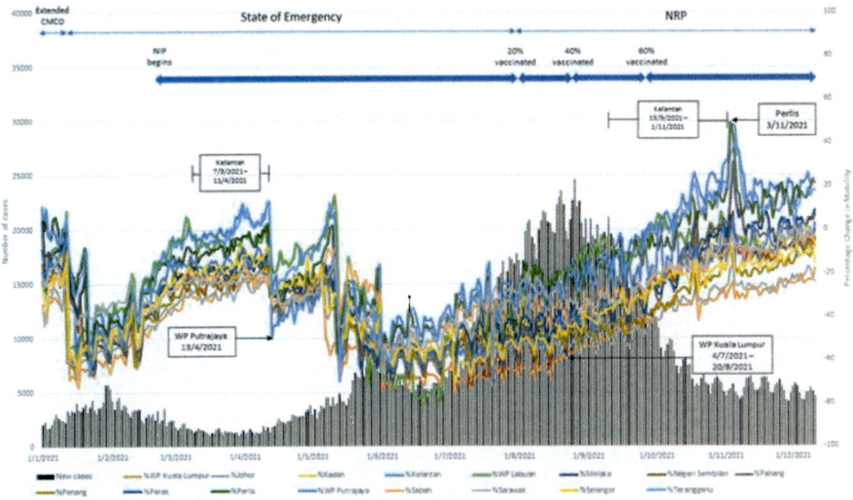

Fig. 4 Percentage change in mobility for retails and recreation in 2021

retail and recreation compared to other states from early March 2021 until April 2021, recording 10% mobility change at its highest change. After that period, on the other hand, WP Putrajaya recorded the highest decrease in mobility change, with the lowest mobility change being -50 percent during 13 April 2021. As the daily cases then started showing a gradual increase, the mobility for every state moved in the opposite direction, reaching its lowest level at the end of June 2021. Next, despite an increase in the number of daily cases afterwards, the change in percentage mobility kept on decreasing as the country was ramping up its NIP, which shows that Malaysians believe that vaccines help them protect against COVID-19. From early July 2021 until the end of August 2021, WP Kuala Lumpur recorded the lowest increase in mobility, meanwhile Kelantan recorded the highest which was then overtaken by Perlis on 3 November 2021 with an almost 50% change in mobility. The trend for those two states then decreased and joined the other states in their upward trend in mobility.

3.3 *Grocery and Pharmacy*

The percentage change in mobility for grocery and pharmacy purposes in 2020 and 2021 is depicted in Figs. 5 and 6. According to the findings, all states followed a similar trend pattern in 2020. In 2020, when the MCO was implemented, grocery and pharmacy mobility decreased significantly in comparison to pre-MCO levels. The mobility increased immediately following the enforcement and remained rather stable throughout the year 2020. This is because despite the restrictions on movement,

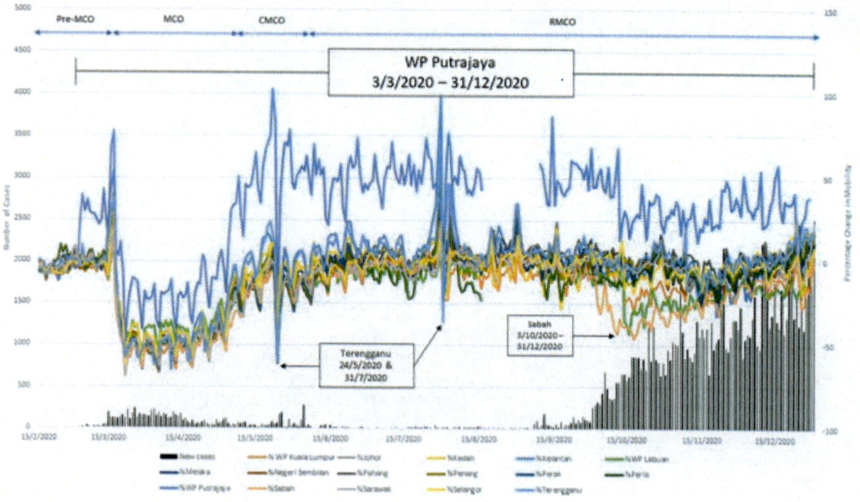

Fig. 5 Percentage change in mobility for grocery and pharmacy in 2020

people needed to go to stores to obtain daily essentials such as food and pharmaceuticals, which is why the grocery and pharmacy sectors saw such a trend line. Another interesting fact is that while Wilayah Persekutuan (WP) Putrajaya followed a similar trend pattern to the rest of Malaysia's states, their change was more significant, with the highest percentage mobility change for grocery and pharmacy almost throughout the entire year (3 March 2020 to 31 December 2020), with the highest point reached being over 100% change in late July 2020. There were occasional significant drops recorded by Terengganu on the 24 May 2020 and the 31 July 2020, ranging around − 30 to − 75%, respectively. After August 2020, Sabah was the state with the lowest percentage mobility, specifically from 3 October 2020 until the end of the year.

In 2021, WP Putrajaya again showed a considerable rise in mobility when compared to the rest of the states, with the biggest increase of 135% on 11 May 2021 compared to the baseline, surpassing Terengganu's 56% increase. However, the trend for all states in 2021 was similar to that of 2020, when all states experienced a decline in mobility following the implementation of the SOE. For the first time in over three months, WP Kuala Lumpur became the state with the lowest change in mobility. The mobility change for each state then decreased under the State of Emergency but climbed again until the end of 2021, with Kelantan recording two decreases on the 13 May 2021 and the 20 July 2021, totaling almost − 75 percent for both times.

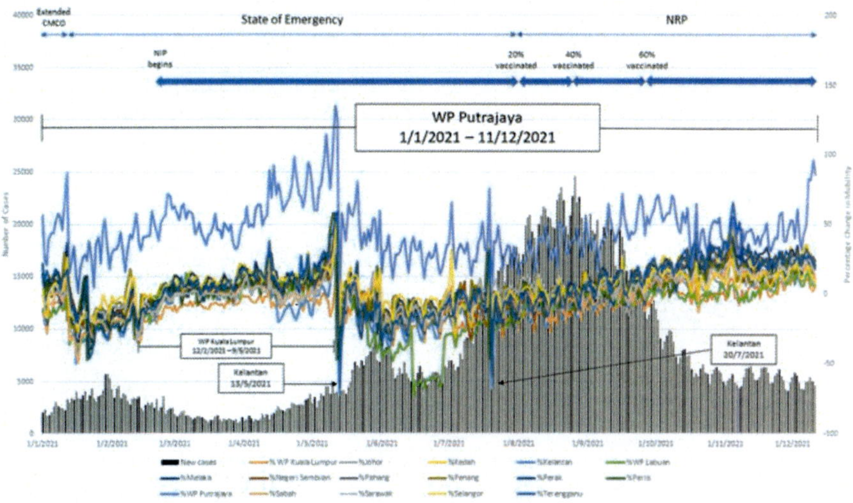

Fig. 6 Percentage change in mobility for grocery and pharmacy in 2021

3.4 *Parks*

Mobility to parks is depicted in Figs. 7 and 8 for 2020 and 2021, respectively. In 2020, during the Pre-MCO period, three separate states took turns leading the % change in mobility, namely WP Putrajaya (22 February 2020), WP Labuan (7 March 2020), and Terengganu (16 March 2020). After implementing MCO, the percentage mobility reduced dramatically. As the MCO became CMCO, the mobility of all states began to increase gradually. Kelantan experienced the greatest percentage shift in mobility (the 17 March 2020 to the 30 June 2020), while WP Putrajaya experienced the least, lingering at – 80% before being supplanted by Penang. Once the country joined RMCO, transportation to the parks increased significantly, with Terengganu recording the biggest percentage change in 2020 for this purpose (approximately + 175% rise) on 2 August 2020. Negeri Sembilan then peaked twice (end of August 2020 and mid-September 2020), with both peaks approaching + 125%. Terengganu then reported the greatest percentage change in mobility for parks through the end of the year, but overall, all states have shown an increase in mobility despite an increase in the number of instances. As a result of the implementation of RMCO, most social activities were reinstated but with rigorous SOPs, resulting in an increase in park usage for leisure activities.

As for 2021, a similar trend was found compared to 2020 as the percentage of mobility decreased once the government imposed an SOE. The mobility then showed a contrasting trend as compared to the number of daily COVID-19 cases because from the beginning of February 2021 until the end of May 2021, the percentage of mobility increased and then decreased in contrast with the number of cases which decreased and then increased. During this period, the highest increase was at WP

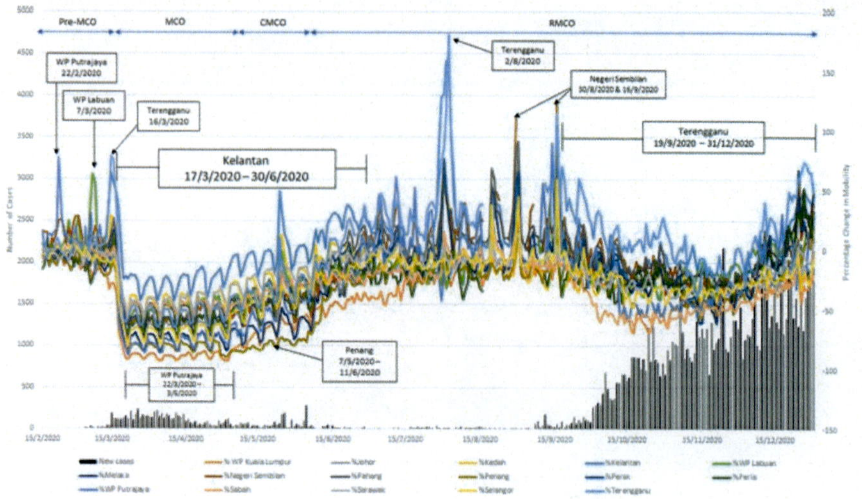

Fig. 7 Percentage change in mobility for parks in 2020

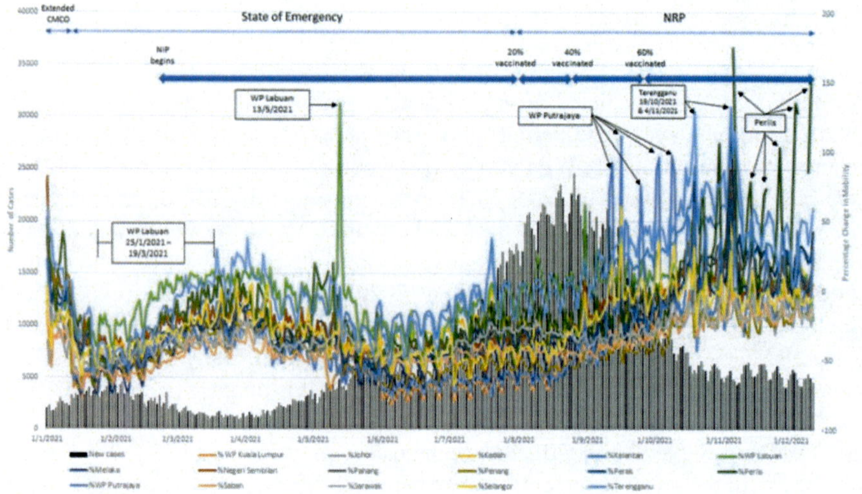

Fig. 8 Percentage change in mobility for parks in 2021

Labuan with a 134% increase in mobility for parks. This is because as the cases decreased, Malaysians felt safer to go out but did not feel so when the cases started showing signs of increasing again. Once the NIP initiative was launched, all states showed an increase in mobility to the parks. On mid-May 2021, WP Labuan recorded the highest increase of percentage mobility change (+ 130%) with Terengganu and WP Putrajaya having the highest overall positive percentage mobility change as compared to other states reaching the end of the year and Perlis having the highest positive change with + 111% at its peak.

3.5 Transit Stations

The trend for percentage change in mobility for this purpose is shown in Figs. 9 and 10 for year 2020 and 2021, respectively. In 2020, the mobility for transit stations showed a decrease as soon as the government imposed the MCO with all states recorded a decrease of between 76 and 41% for this purpose. However, right before it was implemented, Terengganu showed the highest increase in percentage change (almost + 100%). When MCO was imposed, all the public transportations were also put into halt which caused the dramatic decrease. As the country entered into a more relaxed CMCO, the mobility showed an increase for all states except Perlis which showed a somewhat stagnant trend with the decrease hovering around – 80 to – 90% change for 3 months. Once the country entered into RMCO and sectors were allowed to operate again, the mobility to transit stations increased but then showed a strong contrasting trend amongst several states at the end of July 2021. WP Kuala Lumpur and Selangor showed the highest decrease on 31 July 2021 (– 63% and –53%, respectively), but Terengganu showed the highest increase (+ 132%) for two periods (30 July 2020 and 3 August 2020). This is because although transit stations were opened again, WP Kuala Lumpur and WP Putrajaya are the centres of attention for all activities locally and internationally. The prospect of meeting strangers of unknown origins caused them to avoid public places like the transit stations. All states then showed a slow increase before decreasing again, with the highest percentage change recorded by Perak and Terengganu (around + 75% for both). However, the mobility trend surprisingly showed an increase until the end of the year although the number of daily COVID-19 cases was increasing as well, with the range of mobility changes from around – 80% (Sabah) to + 33% (Terengganu).

In 2021, similar to other categories of travel, the mobility to the transit stations decreased sharply when SOE was implemented with a change around – 50 to – 80% for all states. But after that, despite the SOE and also the NRP, the trend for all states had shown a steady increase regardless of the number of daily covid cases, with Kedah showing the highest increase in mobility (from – 16 to + 50%) throughout that period. On the other hand, WP Labuan recorded a significant decrease in mobility to transit stations beginning the 18 June 2021 until the end of that month before following the other states' trend of increasing. This could be attributed to the fact that although the restrictive movements were in place, most sectors were still allowed to open; therefore, the people were carrying out their activities and travelled publicly as usual. In addition to that, with the NIP in place, people were more relaxed with going out and carrying on their lives as usual.

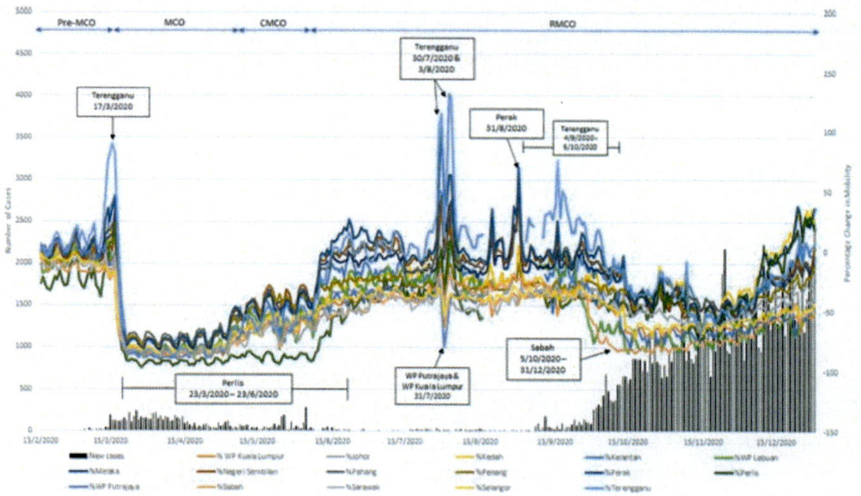

Fig. 9 Percentage change in mobility for transit stations in 2020

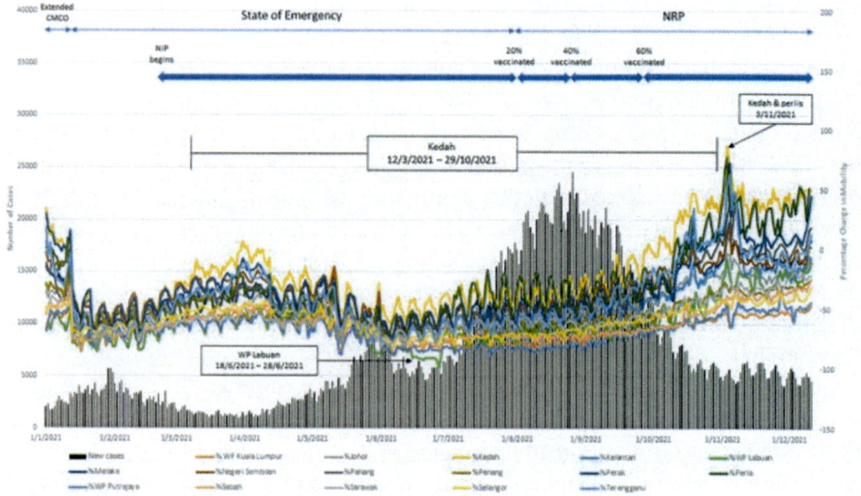

Fig. 10 Percentage change in mobility for transit stations in 2021

3.6 *Workplace*

Figures 11 and 12 show the percentage change in mobility regarding to workplace category for the years 2020 and 2021, respectively. During 2020, although the mobility for workplace dropped around – 40 to – 70% in the beginning though it was not as immediate as other categories. Before the MCO itself, Terengganu was already recording a low percentage change for workplace. During MCO, generally, all states

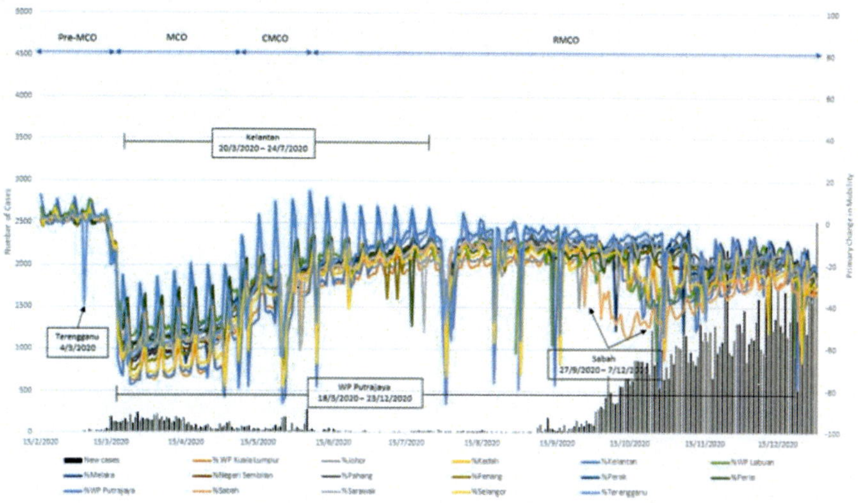

Fig. 11 Percentage change in mobility for workplace in 2020

showed a similar increasing and then stagnant trend with occasional drops for several states. Except for Kelantan and Terengganu, all other states in Malaysia showed a negative percentage change in mobility for workplace throughout the whole year as compared to its baseline, with WP Putrajaya having the highest negative percentage change of around 80% occasionally. This is because as the MCO was introduced, people were forced to stay at home and most companies were encouraged to implement a WFH strategy, hence causing such mobility trends. Therefore, although the number of daily cases was extremely low on some dates and several relaxations were allowed in CMCO and RMCO, the workers had started to adapt to WFH, causing an increase for a short time in the workplace mobility, but then it stayed somewhat stagnant throughout the year 2020.

In 2021, just like the other categories of mobility, overall, workplace also showed a slow increase in mobility but with occasional drops as well. However, similar to 2020, no states had a positive change in mobility for workplace, with the highest increase in mobility recorded for the whole year which was only + 2% in WP Labuan. At this stage, workers were now more comfortable working online. There were occasional drops, specifically for WP Putrajaya and Johor (− 60 to − 80%), and WP Kuala Lumpur took the place with the lowest change in percentage for mobility in workplace. Despite the generally negative change reported in all states, there was an increasing trend as well and this is because although people were encouraged to stay at home, essential sectors were allowed to operate (medical, groceries, municipalities, etc.); therefore, some Malaysians would still have to go to their workplace. In addition, the NIP was also on its way which meant that the people were more confident to go to their workplace once vaccinated.

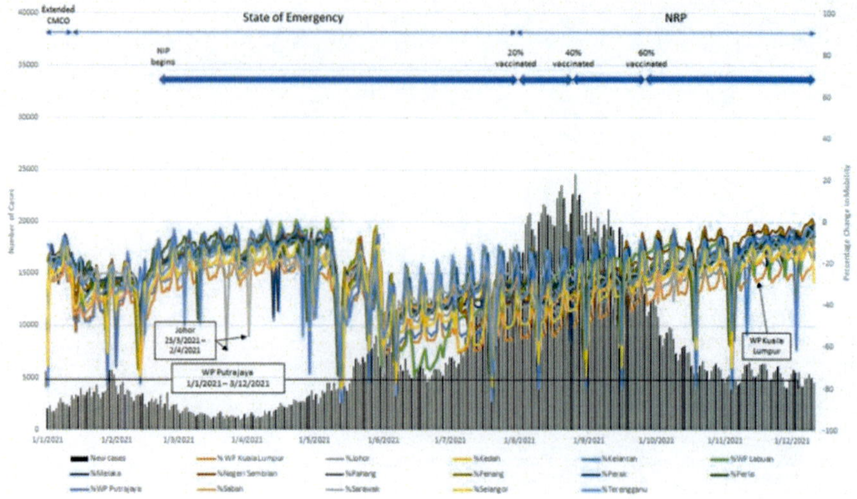

Fig. 12 Percentage change in mobility for workplace in 2021

3.7 *Residential*

Figures 13 and 14 show the percentage change in mobility for residential in 2020 and 2021, respectively. As expected, with the implementation of MCO, the mobility to residential places increased tremendously with all states recorded a sharp increase between 7 and 30%. Once the MCO moved to a more relaxed order (CMCO and RMCO), more sectors were opened and the trend showed a decrease in mobility for residential, and this is expected as during those stages, people started spending more time outside of their homes whether for work or recreational activities and stayed around − 10 to 10% of mobility from July 2020 to September 2020. Interestingly, the first half in the increase of COVID-19 cases beginning the 2 October 2020 reported a slight increase of mobility to residential places at 32% for Sabah, 26% for WP Putrajaya, and 18% for WP Labuan on 15 October 2020. However, the second half of the increase in cases showed a decreasing trend of mobility percentage for residential which means that people were spending more time outside and less time at home despite the increase in cases, with WP Putrajaya and WP Labuan leading the increase in mobility change.

Following the same pattern in 2020, the mobility change for residential increased in the beginning when the government imposed the State of Emergency (SOE), with Sarawak recording the lowest change around late January 2021 to early February 2021. Despite the rapid decrease and increase, no negative change in mobility was recorded throughout 2021. The highest increase for residential purpose throughout the year was by WP Putrajaya. Since Putrajaya is the administrative capital of Malaysia, therefore most of the residents there are white-collared workers which means that they are more capable to do WFH activity compared to other states. The

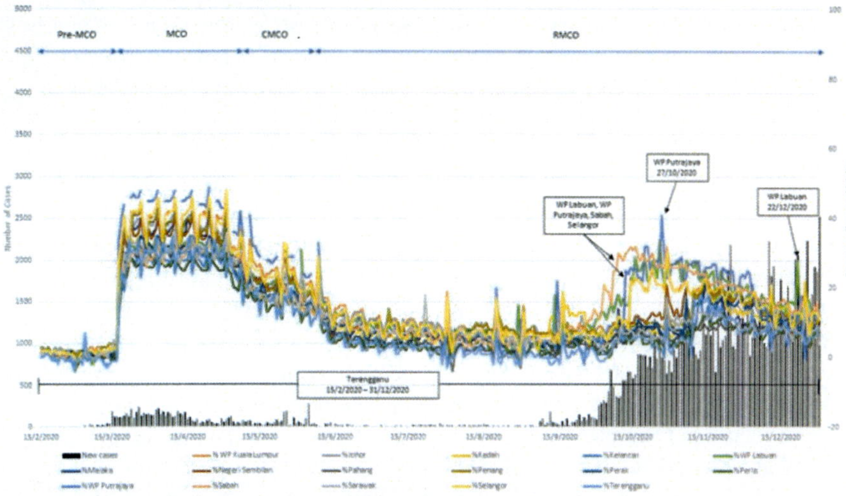

Fig. 13 Percentage change in mobility for residential in 2020

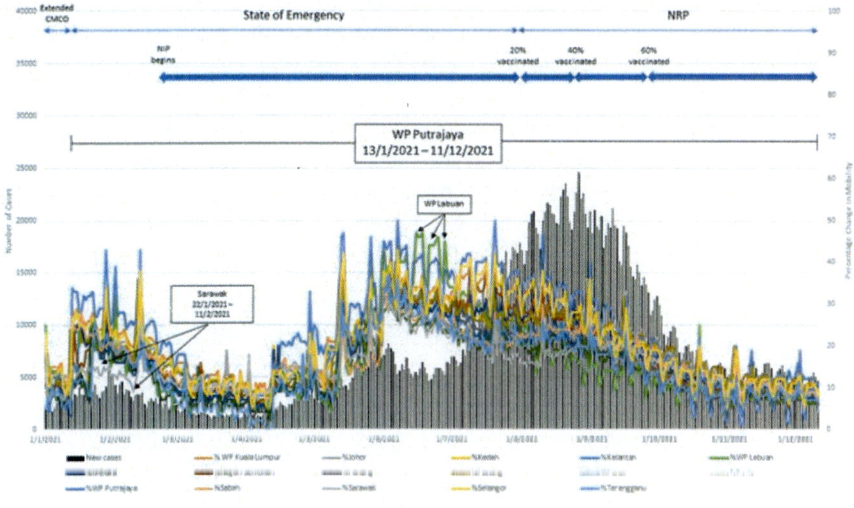

Fig. 14 Percentage change in mobility for residential in 2021

mobility trend for residential then decreased once NIP began in March 2021 but then rebounded on April 2021. Once the NIP was nearing its 20% population vaccinated, the mobility trend started to decrease steadily from + 50% for WP Putrajaya on 7 June 2021 to + 19% on 3 December 2021, showing that people were slowly carrying out their activities outdoors despite the increase in the number of daily cases.

WP Labuan recorded a sharp increase in the residential mobility for three different periods but then decreased alongside other states. Again, with NIP, people were more confident and felt safer in carrying out their activities after getting their vaccine shots as they believed that it would give them a form of protection towards COVID-19.

4 Conclusion

In conclusion, with the implementation of MCO in 2020, the trend in mobility for retail and recreation, grocery and pharmacy, parks, transit stations, and workplace decreased as those restrictive orders banned almost all sectors from operating and encouraged Malaysians to just stay at home. This is further reinforced by the residential mobility trend, which increased in percentage terms over that time period, indicating that people were actually remaining at home. However, after the restrictive movements (CMCO and RMCO) were relaxed, it was observed that the trend began to travel in the opposite way for all purposes as its MCO counterpart. In 2021, there was an overall upward trend for all mobility categories, as the government increased immunization rates, giving the majority of Malaysians the confidence to go about their daily lives. This is also due to the ease with which certain sectors and travel can be restricted during the NRP. This has aided the recovery of many industries that were closed in 2020, as Malaysians are now more comfortable and free to go about their daily lives under the "new normal" conditions. To avoid a repeat of the 2020 situation, the government should enforce tougher penalties on individuals who violate the COVID-19 SOPs, and the public should adhere to the SOPs scrupulously to avoid the formation of new clusters. It is worth noting that as of the time of this writing, a new COVID-19 variation, Omicron, has been discovered in Malaysia. As a result, additional research could be conducted to determine whether or whether this novel variety has caused or will induce changes in mobility.

References

1. World Health Organization (WHO).: Novel Coronavirus (2019-nCoV) Situation Report—1, vol. 205, pp. 1–19. (2020a)
2. World Health Organization (WHO).: Novel Coronavirus Situation Report 7 (2019-nCoV), pp. 1–7. WHO Bull (2020b)
3. WHO.: General's opening remarks at the media briefing on COVID-19 (2020). https://www.who.int/director-general/speeches/detail/who-director-general-s-opening-remarks-at-the-media-briefing-on-covid-19---11-march-2020. Accessed on 13 Dec 2021
4. World Health Organization (WHO).: COVID-19 Weekly Epidemiological Update, pp. 1–23. World Health Organization (2021)
5. Paulino, M., Dumas-Diniz, R., Brissos, S., Brites, R., Alho, L., Simões, M.R., Silva, C.F.: COVID-19 in Portugal: exploring the immediate psychological impact on the general population. Psychol. Heal. Med. **26**, 44–55 (2021). https://doi.org/10.1080/13548506.2020.1808236
6. Sarkodie, S.A., Owusu, P.A.: Impact of COVID-19 pandemic on waste management. Environ. Dev. Sustain. **23**, 7951–7960 (2021). https://doi.org/10.1007/s10668-020-00956-y

7. Wang, Q., Kang, W.: What are the impacts of Covid-19 on small businesses in the U.S.? Early evidence based on the largest 50 MSAS. Geogr. Rev. **111**, 528–557 (2021). https://doi.org/10.1080/00167428.2021.1927731
8. Haider, N., Osman, A.Y., Gadzekpo, A., Akipede, G.O., Asogun, D., Ansumana, R., Lessells, R.J., Khan, P., Hamid, M.M.A., Yeboah-Manu, D., Mboera, L., Shayo, E.H., Mmbaga, B.T., Urassa, M., Musoke, D., Kapata, N., Ferrand, R.A., Kapata, P.C., Stigler, F., Czypionka, T., Zumla, A., Kock, R., McCoy, D.: Lockdown measures in response to COVID-19 in nine sub-Saharan African countries. BMJ Glob. Heal. **5**, 1–10 (2020). https://doi.org/10.1136/bmjgh-2020-003319
9. Jang, S.Y., Hussain-Alkhateeb, L., Rivera Ramirez, T., Al-Aghbari, A.A., Chackalackal, D.J., Cardenas-Sanchez, R., Carrillo, M.A., Oh, I.H., Alfonso-Sierra, E.A., Oechsner, P., Kibiwott Kirui, B., Anto, M., Diaz-Monsalve, S., Kroeger, A.: Factors shaping the COVID-19 epidemic curve: a multi-country analysis. BMC Infect. Dis. **21**, 1–16 (2021). https://doi.org/10.1186/s12879-021-06714-3
10. Verma, B.K., Verma, M., Verma, V.K., Abdullah, R.B., Nath, D.C., Khan, H.T.A., Verma, A., Vishwakarma, R.K., Verma, V.: Global lockdown: an effective safeguard in responding to the threat of COVID-19. J. Eval. Clin. Pract. **26**, 1592–1598 (2020). https://doi.org/10.1111/jep.13483
11. Barbieri, D.M., Lou, B., Passavanti, M., Hui, C., Lessa, D.A., Maharaj, B., Banerjee, A., Wang, F., Chang, K., Naik, B., Yu, L., Liu, Z., Sikka, G., Tucker, A., Foroutan Mirhosseini, A., Naseri, S., Qiao, Y., Gupta, A., Abbas, M., Fang, K., Ghasemi, N., Peprah, P., Goswami, S., Hessami, A., Agarwal, N., Lam, L., Adomako, S.: A survey dataset to evaluate the changes in mobility and transportation due to COVID-19 travel restrictions in Australia, Brazil, China, Ghana, India, Iran, Italy, Norway, South Africa, United States. Data Br. **33**, 106459 (2020). https://doi.org/10.1016/j.dib.2020.106459
12. Przybylowski, A., Stelmak, S., Suchanek, M.: Mobility behaviour in view of the impact of the COVID-19 pandemic-public transport users in Gdansk case study. Sustainability **13**, 1–12 (2021). https://doi.org/10.3390/su13010364
13. Habib, M.A., Anik, M.A.H.: Impacts of COVID-19 on transport modes and mobility behavior: analysis of public discourse in Twitter. Transp. Res. Rec. J. Transp. Res. Board 036119812110299 (2021). https://doi.org/10.1177/03611981211029926
14. Hashim, J.H., Adman, M.A., Hashim, Z., Mohd Radi, M.F., Kwan, S.C.: COVID-19 epidemic in Malaysia: Epidemic progression, challenges, and response. Front. Public Heal. **9**, 1–19 (2021). https://doi.org/10.3389/fpubh.2021.560592
15. Shah, A.U.M., Safri, S.N.A., Thevadas, R., Noordin, N.K., Rahman, A.A., Sekawi, Z., Ideris, A., Sultan, M.T.H.: COVID-19 outbreak in Malaysia: Actions taken by the Malaysian government. Int. J. Infect. Dis. **97**, 108–116 (2020). https://doi.org/10.1016/j.ijid.2020.05.093
16. World Health Organization (WHO).: MALAYSIA: Strong Preparedness and Leadership for a Successful COVID-19 Response (2020c)
17. New Straits Times.: https://www.nst.com.my/news/nation/2020/03/575180/14-day-movement-control-order-begins-nationwide-wednesday. Last accessed on 14th Dec 2021
18. Prime Minister's Office.: https://www.pmo.gov.my/2020/03/movement-control-order/. Last accessed on 14 Dec 2021
19. Dias, C., Rahman, N.A., Abdullah, M.: Influence of COVID-19 mobility-restricting policies on individual travel behavior in Malaysia. Sustainability 1–18 (2021)
20. Ming, L.S., Jung, Y.P., Yong, N.K.: Movement Control Measures Against Covid-19: Mobility Changes in Penang and Malaysia (2020)

21. Rajendran, K., Ahmad, N., Singh, S., Heng, L., Ismail, R., Shaharudin, R., Aris, T., Sundram, B.M.: The effect of movement control order for various population mobility phases during COVID-19 in Malaysia 2019. 590–601 (2021)
22. Ali, M., de Azevedo, A.R.G., Marvila, M.T., Khan, M.I., Memon, A.M., Masood, F., Almahbashi, N.M.Y., Shad, M.K., Khan, M.A., Fediuk, R., Timokhin, R., Borovkov, A., Haq, I.U.: The influence of covid-19-induced daily activities on health parameters—a case study in malaysia. Sustainability 13 (2021).https://doi.org/10.3390/su13137465
23. Google.: Malaysia 12 December 2021 Mobility Changes (2021)

State-of-the-Art Review of Signalized Roundabouts: Evaluation, Analyses, and Gaps

Amani Abdallah Ali Assolie, Nur Sabahiah Abdul Sukor, and Ibrahim Khliefat

Abstract One of the easiest and most economical methods to increase traffic flow across the metropolitan network and relieve congestion at junctions is to implement signalized roundabouts. In order to provide the most current findings, identify any remaining study gaps, and suggest potential paths for future studies in the area, this article reviews recent literature on signalized roundabouts, including various traffic signal types, published between 2010 and 2021. By finding the publications that made the short list on Google Scholar and Scopus, the most recent literature review was utilized. Both Google Scholar ($n = 55$) and Scopus ($n = 156$) have found a total of 156 linked publications. The most thorough micro-simulation program utilized for signalized roundabouts was therefore thoroughly examined in this paper. We conclude by making a few projections about the future direction of this study.

Keywords State-of-the-Art · Sustainable transport · Signalized roundabout · Roundabout metering · Simulation software

A. A. A. Assolie (✉)
Department of Civil Engineering, Faculty of Engineering, Ajloun National University, P.O. Box 43, Ajloun 26810, Jordan
e-mail: AM.assolie@anu.edu.jo

N. S. A. Sukor
School of Civil Engineering, Universiti Sains Malaysia, 14300 Nibong Tebal, Penang, Malaysia
e-mail: cesabahiah@usm.my

I. Khliefat
Department of Civil Engineering, Faculty of Engineering, AL-Balqa Applied University, Salt, Jordan
e-mail: khliefat@bau.edu.jo

N. Sabtu (ed.), *Proceedings of AWAM International Conference on Civil Engineering 2022 - Volume 3*, Lecture Notes in Civil Engineering 386,
https://doi.org/10.1007/978-981-99-6026-2_29

1 Introduction

Constant population expansion and an increase in the number of vehicles on the road have had an impact on the environment and traffic congestion. As a result, several junction control system types have been researched for a long time to reduce the detrimental effects of traffic congestion [1–6].

For instance, roundabouts are a sort of junction where traffic moves continually in one direction around a center island. They are often utilized for clockwise ground traffic operations (in countries with right-hand traffic) [7–11]. Due to their capacity, cost-effectiveness, and safety, roundabouts are becoming more and more popular. By combining and diverging traffic flows at moderate speeds and small angles, roundabouts reduce the risk of crashes while also improving traffic flow at junctions [4, 11–15]. In terms of geometry, roundabouts have evolved to include single-lane, multi-lane, and more recently, new ones like turbo roundabouts, which function best when approach flows are balanced [9, 16–18]. Modern roundabouts are found to perform better than other types of intersections in terms of efficiency, safety, and environmental effect [3, 6, 11, 15, 19, 20]. In practice, it can be challenging to distinguish between the many types of roundabouts, and as a result, delays and queues would be considered negative effects (capacity, delay, and emissions). In reality, it can be challenging to distinguish between the many forms of roundabouts, making queuing and delays potentially unsafe (capacity, delay, and emissions) [19]. Numerous other elements, such as traffic volume, geometric characteristics, vehicle mix, and driver behavior, all have an impact on the roundabout's capacity [12].

Some roundabouts include signal control, which is useful in places with plenty of traffic and pedestrians. In this regard, the signalized roundabout approach offers a more effective way to alleviate congested conditions on roundabouts. One of the simplest and most affordable methods to improve traffic flow in city streets and relieve congestion at junctions is signalized roundabout control. Signalized roundabouts have a greater capacity than single-lane roundabouts. In fact, signalized roundabouts significantly cut delays and queue lengths on main roads (as could be observed in single-lane roundabouts) despite having a somewhat higher capacity (approximately 5%) [12]. Signalized roundabouts are designed to safely and quickly move traffic through a junction. Therefore, a right-of-way assignment plan must take into account the varied users. However, a number of elements, including geometric features, traffic volume associated with left-turning vehicles, storage area, and signal timing, have an impact on how well-signalized roundabouts work [18].

Additionally, research towards signalizing roundabouts was mainly focusing on adaptive traffic lights, and the issue of adaptive traffic lights has previously been the subject of high impact studies. This paper contains reviews of other studies on the widely used simulation technique for controlling traffic flow in urban traffic networks.

The remainder of the essay is structured as such: Sect. 2 of the article describes the study methods, and Sect. 3 describes the results. Section 4 identifies any gaps and

Table 1 Search parameters of the literature review

Target search	Search engine	Search keywords
R2010–2021	Scopus Google Scholars	• Signalized roundabouts • Roundabout metering • Simulation of signalized roundabouts • Adaptive roundabout • LOS at signalized roundabouts • Traffic flow at signalized roundabouts • Interface of signalized roundabouts

relevant research project ideas. Methodologies are discussed and analyzed in Sect. 5, which is followed by a list of references.

2 Research Methodology

This study investigates journal articles and conference proceedings on the subject of simulation for signalized roundabouts, published between 2010 and 2021. One of the main justifications for discussing the recently released research was technological innovation. A comprehensive search of databases (e.g., Scopus and Google Scholar) was carried out, using terms like "Signalized roundabouts," "Roundabout metering," "Adaptive roundabout," "LOS at signalized rounda-bouts," "Traffic flow at signalized rounda-bouts," and "Interface of signalized rounda-bouts."

The reference list of the chosen publications was expanded with additional works that were published within the time period mentioned. Table 1 accordingly provides a list of the search criteria used in this inquiry. The number of papers found between 2010 and 2021 in Scopus was 156 ($n = 156$) while that of Google Scholar was 55 ($n = 55$). The nominated papers were then divided into several types of micro-simulation tools throughout the search phase.

3 Finding

The results of the literature evaluation procedure with reference to the topic of signalized roundabouts are covered in this section.

The numerous forms of signalized roundabouts are described in subtopics 3.1 and 3.2 on method of signalized roundabouts, 3.3 on signal control system strategies, and in topic 3.4 on performance and operational analysis of signalized roundabouts.

Signalized roundabouts are found to be especially helpful in regions that are crowded with both pedestrians and mixed-vehicle traffic [18]. According to a prior research by Murat et al. [18], the best solution to reduce congestion at roundabouts is to specify the requirements for installing traffic signals at two-lane roundabouts using delay as a performance effectiveness metric. These requirements are applicable to two-lane roundabouts with uneven flow between the legs of roundabouts [21]. Understanding the goal of traffic signalized control and identifying the distinctive features of signalized roundabouts are prerequisites for conducting a state-of-the-art study.

3.1 *Different Types of Signalized Roundabouts*

Intersections with priority control roundabouts are autonomous. According to the notice at the intersections, motorists entering the roundabout must slow down, while circulatory vehicles already in the circulatory lanes receive priority. Despite the fact that roundabouts are by default equipped with yield control, traffic flows have been indicated by metering one or more entries or, if necessary, signaling the circulatory lane at each entry. Priority-controlled roundabouts frequently have high traffic demand and are challenging to accommodate, particularly as traffic loads increase. There are several subcategories of roundabout signal system use and three different forms of signalized roundabouts as depicted in Fig. 1 by Azhar and Svante [22]. Meanwhile, Table 2 lists the functions and purposes of signalized roundabouts.

In their study of signal control of roundabouts, Azhar and Svante [22] came to the following conclusions:

1. Alternative A1 should be avoided; B is the best option.
2. Due to capacity and traffic safety concerns, Alternative A2 should be situated at least 22 m from the roundabout.
3. If signalized crosswalks are required, alternative B2, which has a better capacity than B1, can be used.
4. Due to capacity issues during high pedestrian traffic of numerous approaches, Alternative C can be considered.

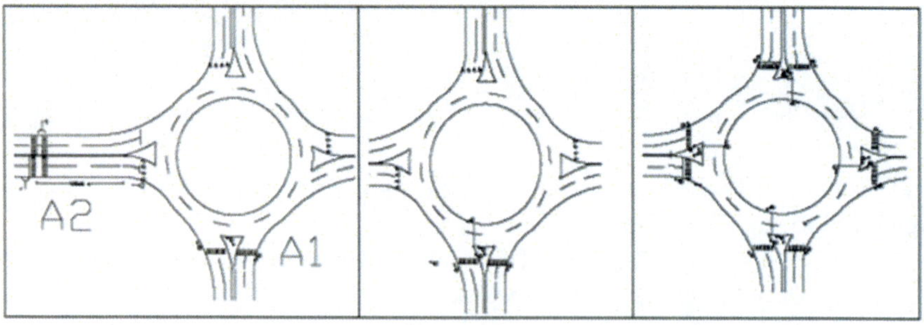

Fig. 1 Signal-controlled types A1, A2, B (in the middle), and C (to the right) [22]

Table 2 Purpose and functionality of the different types of signalized roundabouts [22]

Type of roundabout	Purpose and functionality
Type A Signal control of crosswalks (indirect control)	On the legs of the roundabout, there may be one or more crosswalks. Therefore, the roundabout movements are subjected to the basic principles of "Inside Priority"
Type A1 Off signal (near to the roundabout)	To avoid motorists from becoming confused if they receive a green signal just a few meters before a yield control, the signal does not display the green time if a crosswalk is adjacent to the roundabout
Type A2 On signal (far from the roundabout)	If a crosswalk is located at sufficient distance from the roundabout, the light may display enough green time
Type B Complete signal control of one or two legs (direct control)	On one or two of the roundabout's legs, the stop line in circulation is totally signalized, while the conflict between circulating traffic and the second legs is resolved by a giveaway in favor of the circulating streams
Type B1 (no gating)	Pedestrians can pass through signalized crosswalks in one leg
Type B2 (gating)	Pedestrians can pass through signalized crosswalks in two legs
Type C Fully-signalized roundabouts (direct control)	Circulating traffic streams at roundabouts equipped with type C signal control do not adhere to the typical "Inside priority" rules. During certain times of the day (off peak), these signalized roundabouts may operate as both unsignalized and signalized roundabouts (peak times). Allowing traffic to pass directly through the roundabout in one direction is a variation of signal-controlled roundabouts. While this design has a number of advantages, it does not allow for the signal to be switched off during specified times of the day

3.2 *Method of Signalized Roundabouts*

This article discusses the many signalization techniques, including full signalization and partial signalization that have been utilized at roundabouts, as reported in the literature.

3.2.1 Full Signalization

As a result of the excessive traffic volumes, high collision rates, and imbalanced traffic, some type of signalization is required to provide for proper gap construction and selection. Full signalization comes in various designs as discussed below:

(a) **Full Signalization with circulating lane**

This design encompasses a fully signalized roundabout concept: instead of simply one or two approaches, all approaches (external and circulating) are regulated by traffic lights [23–26]. This approach deviates from the traditional "Inside priority" rules, which state that signalized roundabouts may operate as both signalized and unsignalized roundabouts during off-peak hours (peak times); as a variation of signal-controlled roundabouts, it lets the traffic go past the intersection directly. The full signalization during peak hours is for dealing with unpredictable flows and for minimizing and balancing the queues and delay times on each of the approaches [26]. However, it does not allow the signal to be turned off during specific hours during off-peak hours [23–25].

(b) **Full signalization with left turn vehicles**

This design encompasses a fully signalized roundabout concept: instead of only one or two approaches, all approaches (both exterior and circulating) are under the direction of traffic lights [23–26]. This approach is significantly influenced by the amount of left-turning traffic, both during peak and off-peak periods. Otherwise, the signal duration could not be long enough to handle the volume of traffic, increasing to certain cars' delays. Additionally, signalized roundabouts with improper signal timing design may cause traffic congestion [18, 26, 27]. A signalized roundabout with left turn lanes is one that has been proposed; by eliminating heavy traffic movements from the signalized traffic stream, these slip lanes would only improve operation during peak hours [21, 26]. Therefore, by bringing down the delay to 9–35% at the fifth and sixth circle in Jordan [26] and to 20–80% in Michigan [21], a major increase in the roundabout's performance could be made.

(c) **Full signalization with entrance**

In this design, all approaches (external), as opposed to only one or two approaches, are under the control of traffic lights. This control strategy prevents competing traffic flows. The length of the green light may be adjusted based on how busy each entry is, while the traffic flow can be more evenly distributed [24, 28]. According to Sarrir [28], roundabouts should be fully signalized to increase capacity, while decreasing traffic jams and queue durations. Due to capacity issue, the queue length of certain approach becomes vital. So, signalization can change the flow balance to reduce queues and delays for some entries while increasing them for other entries. When a roundabout is operating at high levels of saturation, signals do help to decrease the overall delay and congestion by ensuring that all lanes and approaches are utilized to their fullest extent [28].

3.2.2 Partial Signalization

Traffic signals regulate partial signalization with one or more approaches and are classified into two categories based on the control of the internal and external approaches, namely, partial signalization and metered approach signalization, as described as follows:

(a) **Partial signalization**

Traffic lights manage a partially signalized roundabout with one or more approaches (internal and external) [26, 28]. In fact, when one or more entries are left under priority control, the flow of all traffic through the roundabout is frequently improved. If some of the arms are not signalized, one or more are left unmarked as giveaways. Having only three signal-controlled arms is preferable to having full signal control, which requires additional storage space for queuing within the roundabout. In such circumstances, no newly entering vehicles will be halted at the roundabout's initial stop line. It is now advised for safety reasons that vehicles not stop at the first stop line after approaching the roundabout. Only three arms must be signaled for perfect advancement. It has been observed that a successful progression occurs when freshly freed traffic traverses the entire or a considerable portion of the signalized roundabout without stopping at an internal stop line [12, 26].

According to a study conducted by Osei and colleagues, signalized roundabout intersections effectively increase capacity by up to 50% in some cases, significantly reduce delays and queue length, and maintain the benefits of roundabouts, including safety [12]. In contrast, a study conducted by Rashed and Imam found that the overall performance of Jordan's eighth roundabout would not change because the level of service remains at level F [26].

(b) **Metered approach**

Akcelik [29] conducted one of the most well-known studies on roundabout metering. It has two detectors, one on each approach—the dominant (controlling) way and the sub-dominant (metered) approach—and each is equipped with a single traffic light. It functions by turning red when the queue length on the controlling approach exceeds where Detector C is located. If Detector M on the metered approach detects queuing, the red light turns off. As such, the location of detectors on a metering roundabout significantly affects the timing of red signals, which in turn affects the duration of queues on regulated, metered, and other approaches.

Numerous studies have attempted to evaluate the effectiveness of metering roundabouts, concentrating on forecasting the queue lengths, which is essential for choosing where to deploy the detectors [30, 31].

An et al. [32] developed a Cuckoo Search/Local search Algorithm using variables like arrival volumes, conflicting volumes, detector distance, and phase time to study the connection between signal setting, detector position, and queuing formulations. The data were also subjected to a number of additional statistical tests to assure their appropriateness to their intended purposes. Strong model validations were carried out by comparing the model's output data to the AIMSUN model's output data.

The findings of An et al. demonstrated that phase time durations and vehicle queuing lengths on the regulated approach as well as queuing lengths on the metered approach are influenced by the queue detector's distance. According to a study on the Old Belair Road roundabout in Adelaide, South Australia [32], relocating the detector 209 m from the roundabout stop line reduces the total queue length (controlling + metered), resulting in longer phase green times and shorter intersection queuing lengths. According to a research [33], metering signals improve efficiency (by decreasing delays) and the capacity of roundabouts with unequal flows. When AN et al. [34] implemented the SIDRA INTERSECTION 6 model to a real-world scenario on Old Belair Road for the morning peak hour (8:00–9:00 AM), and they were able to enhance the roundabout's operating conditions and attain the level of service LOS C with a 23.2 s delay time on the dominant approach.

3.3 *Signal Control System Strategies*

3.3.1 Fixed-Time Signalized Control Strategy

Fixed-time techniques are mostly suitable for traffic signals, particularly when the flow of traffic is comparatively predictable, resulting in lower average delay and optimized network capacity. Urban regions have highly dynamic traffic systems, and thus, even a minor disruption from an accident, construction, etc. can drastically change traffic volumes and compromise the effectiveness of a traffic signal plan [35–37].

3.3.2 Actuated Signalized Control Strategies

Sensors are used in the actuated control schemes, and the actuated controller chooses the cycle duration based on previous data [35, 36].

3.3.3 Adaptive Signalized Control Strategy

The third signal approach is an adaptive control method that adapts to the traffic conditions of the moment in each phase. As a result, sensor technology was put to use. Initial sensor technologies could detect the presence of vehicles as they crossed the detector. Later, visual systems saw a widespread use. Notably, the adaptive strategies are a modified version of actuated control that forecasts the cycle length of current traffic circumstances using the available data. These sensors are installed on stop lines, right turn sensors, slip lane sensors, and advanced detection sensors, on every road that is a part of signalized network [35–38]. The existing traffic light system is designed with hard programmed delays, where the lights' transition time slots are established on a regular basis and do not rely on real-time traffic flow. This is one of the primary causes of traffic congestion [39]. The adaptive metering signal control system, which is a technology made specifically for roundabouts, was created to boost the efficacy of the roundabout metering system [38].

Based on current circumstances, Mahmood [40] analyzed the impact of main traffic parameters on the capacity of adaptive roundabouts, the results of which were found to be realistic and may serve as a guide for traffic planners. Meanwhile, the customized approach of traffic signal cycle design developed by Dmitry et al. [41] decreased the average transit delay by 28%.

Another study utilized the adaptive traffic signal at crowded roundabouts in Amman with long queues of over 90 m during peak hours [23]. In the second scenario, four signals were installed at the roundabout, whereby each had different cycle periods and was connected to the others. The *D* value of the level of service was defined by a 90-s cycle period. The third option involved putting adaptive signals on the roundabout, which connect traffic signals to ground detectors located at a predetermined distance ahead of the signal and depend on the length of the line of flow of autos passing the detector. To open and close the doors of approaches, adaptive signals were used. The first two lights were illuminated for 45 s, taking into account the Swefieh road and King Faisal's reference. It was then closed for 45 s to allow the signal connecting the fifth circle to Zahran Street to be opened.

3.4 *Performance and Operational Analysis of Signalized Roundabouts*

When the flow rate is less than capacity, traffic moves relatively smoothly; however, when the flow rate is at or near capacity, significant delays and congestion can occur. A capacity study is a quantitative evaluation of a road section's capacity to transport traffic under the existing highway traffic and control conditions. "Level of service" (LOS), which is a qualitative metric with a scale of A–F, was used to evaluate the quality of the flow [42]. The level of service (LOS) of a roadway indicates the performance of roadway when traffic loads are less than capacity. Numerical modeling and simulation modeling, along with a macroscopic and microscopic traffic flow models, were utilized to analyze the operation of a signalized roundabout. As a single vehicle unit is simulated using a microscopic traffic flow model, its dynamic variables indicate the position and speed of individual cars and offer an avenue for a thorough analysis. A macroscopic traffic flow model is a mathematical formulation of the interactions between many aspects of traffic flow, such as density, flow, mean speed of a stream of vehicles. This subtopic will contain the most recent research on signalized roundabout performance and operational analysis based on traffic patterns and geometric layouts.

3.4.1 Performance and Operational Analysis Based on Traffic Conditions

SIDRA has been utilized in certain research to analyze the effectiveness and operational analysis of roundabouts [15, 43, 44]. Delay time, degree of service, and queue length were assessed in a research by Yagoop and colleagues to determine traffic capacity for the Al-Mohafada roundabout. The results indicate that adding one lane for each approach causes the level of service to improve from F to C [15]. Additionally, AN and others [34] used the model to simulate a real-world scenario on Old Belair Road for the morning rush hour (8:30–9:00 AM), and they were successful in achieving the level of service LOS C with a 23.2 s delay time on the dominant approach. Martin [33] discovered that the delay value was 50 s, while the amount of left-turning traffic, the higher entry flow rates, and the lower improvement for low conflicting were the causes for worry.

In Malaysia, a study was conducted during the morning rush hour to see how much traffic could be moved by adding extra lanes and traffic signals at roundabouts. The findings demonstrated decreased congestion by increasing speed and delay from 57 to 20% and 2% to 5%, respectively [44], and Demir [19] discovered that the unsignalized roundabout performed better than the roundabout with a secondary signal. The 95th percentile queue was reduced to 82.2%, capacity rose to 67.8%, and the average delay decreased to 72.8% [19]. Given that the traffic demands do not exceed the recommended range for an effective application, the results of a study on signal metering in Malaysia using the SIDRA model gave good insights on the application of signal metering to improve roundabout performance by tacking the problem of unbalanced flow [45].

Through the use of an integrated modeling framework, the SIDRA technique highlights the consistency of capacity and performance study methodologies for roundabouts, sign-controlled intersections, and signalized intersections. The SIDRA program chooses two major crossings, analyzes the crossings, replaces them with roundabouts, and analyzes them once more. The findings indicate that the roundabout was in better condition than the intersections in the region of UKM and Bandar Baru Bangi. This study's next objective was to analyze and rationalize intersections and roundabouts, as well as the effects of doing so in a real-world setting, and to determine how this will affect the intersections and roundabouts in the area as a whole. In a case study comparing the performance of a hypothetical junction with that of the current traffic signal, it was found that there had been a considerable rationalization of the length of the queues, the level of service, and the cycle time, which resulted in a reduction in time. Using an intelligent traffic control system, the phasing sequences were altered to make the right lane turn longer and more functional [46].

A previous study [47] provided recommendations for selecting the appropriate type of control for an at-grade intersection. The scenarios explored in a case study of a four legs with two-lane roundabout were based on the traffic distribution on the approaches. Synchro and Sidra were used to evaluate the performance of this investigation. The findings showed that the roundabout outperformed a traditional signalized intersection in all traffic situations, including at a roundabout where the percentage of left turns was less than 20%, while the traffic flow was roughly 3500 cars per hour.

Additionally, AIMSUN was utilized in additional studies [25, 48] to model the extensive urban transportation networks. The performance metrics for the proposed hybrid roundabout were compared and analyzed for all statuses, including delay time, travel time, speed, density, fuel consumption, number of stops, queuing, and carbon emission. Additionally, comparisons of all statuses' traffic safety analyses were made. Results indicated that the proposed roundabout generally performed better on average than the current situation [3, 25, 48] at 40%, 41.8%, and 41% safer, and the accident risk was lower by 18.5%. Dmitry et al. [41] used AIMSUN to evaluate the performance or roundabouts (based on Webster's method), and they discovered that the adapted method of traffic signal cycle design could decrease the average transport delay by 28% [41].

A research by Hong [30] employed MUSIN to assess how queue detectors affected roundabout performance, and Hong discovered that the length of queue might be cut in half at the sites of the 380 m detector (for the controlling technique) and the 320 m detector (for the metering approach), respectively. According to a research by Hong [32], when the detector is moved to a location 209 m from the roundabout's stop line, longer phase of green times would result, reducing junction waiting durations.

Khliefat et al. [23] offered a C++-based solution for programming the adaptive traffic signals interfacing with VISSIM at roundabouts in Amman, Jordan, and VISSIM has been employed by many studies to minimize latency in various percentages. The unsignalized roundabout was very crowded during peak hours, with queues extending more than 90 m. The results with signalized roundabout indicated a considerable improvement in the LOS and a nearly 75% reduction in the delay at roundabout. In another study, CAKICI and MURAT [49] used VISSIM and concluded from the findings that the average vehicle delay at Kelleci Kardesler Intersection may be reduced to up to 45% with their proposed model compared with the current situation without traffic signal, while the delay was reduced by 36% [18] and by 25.7% [38]. Similar studies have been performed by other scholars as well [3, 50]. A study on delay was carried out [3] and the findings showed that delay was increased to 14 s, while cycle length was increased to 90 s after the installation traffic signal at a roundabout. In another study by wang et al. [51], delay was reduced by 9.3 s through the use of VISSIM.

A study in Spain [33] investigated the capacity and operational improvements of metering roundabouts. The goal of this study was to analyze the operational and capacity improvements of suburban roundabouts with metering signals using traffic micro-simulation called VISSIM for Spanish local conditions, with capacity estimates based on the Highway Capacity Manual for average delay. The findings showed the following: the improvement was smaller for low conflicting points, the increase of entering flow rates approached the capacity to criteria, and the value of delay was 50 s.

Another concern is the volume of left-turning traffic, which has a substantial impact on the design of signalized roundabouts. For optimal design, the number of cars and traffic composition must be considered. The timing of circulating flow signals should be determined by traffic loads, namely the proportion of heavy trucks. Variations in traffic volume, on the other hand, should be recorded during both peak and off-peak hours, and the trend of this fluctuation should be calculated. Otherwise, the signal period may not be sufficient for the traffic volume, causing additional delays for some vehicles. Additionally, traffic congestion may occur as a result of poor signal timing design at signalized roundabouts [18].

The high number of left-turn traffic also has been considered in the literatures. However, scholars appeared to have contradictory views pertaining to whether high percentage of heavy trucks, that is, percentage that is higher than planned, may clog the intersection. Notably, results from simulation model with VISSIM show large vehicles which may affect vehicle departure headways, resulting in gridlock [18, 27].

The use of analytical modeling improved traffic conditions at signalized roundabouts, for example, Al-Obaidi [17] used a statistical model to change and optimize the geometric elements and the traffic signal timing and Patnaik [52] was able to calculate the roundabout entrance capacity in heterogeneous traffic situations. In order to enhance operational performance, roundabouts in urban and suburban regions with heavy traffic volumes may be changed to signalized roundabouts [52].

Therefore, a study [53] suggested a novel capacity calculation approach for signalized roundabouts. This study, which was a case study, was conducted based on several current circumstances. Also, it encompassed an analysis of the impact of key elements on capacity, with the end result conforming to the reality; the study could serve as a guide for traffic planners.

Many studies demonstrated significant improvement of LOS after adjusting the signal time. For example, Al-Obaidi [17] reported significant improvement in LOS from F to B, while other studies found a considerable improvement from LOS D to better level [23, 34]. Two-level improvement was also reported from the use of signal time adjustment [26].

Some studies suggested the significant impact of infrastructure on vehicular traffic flow. For example, Al-Obaidi [17] estimated that improving the infrastructure will reduce congestion, particularly through the reduction of the average delay time per vehicle to less than one, and the reduction of delay time to nearly 43 times of that of current conditions. Furthermore, through the use of leading approach, delay time can be reduced to 23.2 s [23].

A study [33] examined operating performance and capacity at roundabouts, and the following findings were presented:

1. Follow-up headway should not be regarded a constant value because it grew as the conflicting traffic flow increased.
2. The improvement was smaller for low conflicting flows since the system was turned off most times.
3. The decrease in delay was considerable at all levels of entrance flow, and it was even greater when the entering flow rates were nearing the capacity requirement (delay equals to 50s).

Increased proportion of longer gaps and reduced follow-up time resulted in longer platoons approaching the roundabout at each available gap, resulting in capacity increases [33].

Others studies focusing on delays at roundabouts also concluded that installing traffic signals could reduce the average delays (Rashed & Imam) [26] and increase roundabout capacity [21], especially through the use of shorter cycle time [54].

3.4.2 Performance and Operational Analysis Based on Geometric Design of Roundabouts

A roundabout's geometric characteristics have a significant impact on its design. The turning radius, lane width, radius of the center island, and other geometric aspects or components should all match the specified specifications. When non-standardized parts are employed in intersection design, there may be flaws. On the other side, these design flaws may result in a reduction in intersection capacity [18].

Fortuijn et al. [55] studied the impacts of roundabout diameter on vehicle delay, as well as the effects of cycle duration on vehicle delay. They discovered that traffic signals are only beneficial to entrance sites with more than one lane using numerical calculations and experimental traffic engineering methodologies. Complete traffic management, on the other hand, barely increases the capacity of two-lane roundabouts with a modest diameter. With the aid of roundabout metering lights, it is possible to significantly improve the traffic flow.

According to a research [56] that determined the capacity and ideal cycle length, the capacity model's accuracy is poor without taking the roundabout's interior space constraints into account. In addition, the left-turn lane's capacity is significantly impacted by the center island's radius and cycle duration. The research also concluded that capacity could be optimized using appropriate signal cycle length.

Meanwhile, Ma et al. [50] proposed that an optimization model can be improved by concurrently considering lane markings and signal timings. If the roundabout is run according to the plans developed by the suggested optimization model, they calculated that the traditional signalized intersection will have shorter cycle durations, shorter average delays, and more capacity. The link between geometric qualities is increasingly essential, and yet studies, on this topic, are too few.

4 Gaps and Direction for Future Research

We noticed that less than half of the studies on signalized roundabouts used real-world data, and owing to the scarcity of standardized datasets [48, 57, 58], as well as the complexities of traffic flow data, research based on real-world data is more likely to have an impact in the area. Although there were studies that incorporate real-world data and real-time control, little or no mention was made on the findings and/or methodology being used or modified in the real world.

Many studies were designed to use real-time data (from sensors or cameras) to respond to traffic congestion and try to reduce the [3, 18, 19, 22, 23, 29–31, 33, 36, 40, 55, 59, 60]. In recent years, there has been a rapid advance in prediction algorithms, computing capacity, and the availability of real-time data. These facts, together with the developments in heuristic algorithms, can lead to proactive models that can now be effectively developed.

The literature reviews reveal that factors such as geometric circumstances, storage length, and a vehicle turning left at a roundabout all have an impact on geometric design. The turning radius, lane width, and radius of the center island have been popular subject matters as well. Notably, some academics excluded geometric characteristics in their researches [22, 26, 31, 48, 58, 61, 62]. In future, research on the impact of storage length and left turning within a roundabout in right-hand traffic movement (the anticlockwise circulation) may be required [12, 23, 59, 60].

The impact of pedestrian and driver behavior on the simulation models would be a significant field of research. In fact, pedestrians are a significant feature of heavy traffic at signalized roundabouts [22, 27, 36, 40, 60, 62]. It is therefore possible to study more impact of pedestrian at signalized roundabouts in the future. Furthermore, studies focusing on the influence of driving behavior at signalized roundabout may be also needed [26, 33, 50, 59, 62].

Moreover, only few researches were focusing on queue duration prediction even though it is crucial for determining the detector placements [30, 31], and so, this issue also needs to be explored more deeply.

5 Discussion and Conclusion

There are a number of benefits when comparing roundabouts with regular junctions with signalization. Along with the added advantages of decreasing the running costs and the speed control, the main advantages of signalized roundabouts include aesthetics, pedestrian safety, environmental impact, operational effectiveness, and traffic safety [3, 4, 6, 11, 15, 19, 20, 48, 63–66]. The majority of studies [66, 67] have largely concentrated on the effects of changing signalized junctions to single-lane roundabouts. In the USA, roundabouts have recently been built to replace signalized intersections at freeway ramp terminals as a safety measure. Accordingly, WSDOT [68] stated that the FHWA had done studies that proved that modern roundabouts could increase traffic capacity by 30–50% compared to the conventional intersection controls [68].

Roundabouts face a number of challenges because of the high traffic volumes [57] despite the fact that they could significantly and potentially improve junction's operational performance. These challenges include unbalanced flow and high capacity, which cause traffic congestion [12, 18, 21, 23, 24, 26, 69] and unrelenting congestion. Roundabouts are extremely beneficial when flows are balanced, but as flows rise, there may be operational challenges. The majority of roundabouts in Ghana's arterial routes frequently encounter peak-hour traffic jams, lengthy wait times, and excess truck rollover accidents [12, 18, 24]. Additionally, delays occur as a result of vehicles' misunderstanding brought on by wider circulating lanes at roundabout entrances. The entry capacities of approaches to a roundabout decreased as more vehicles were in the circulating lane [52], causing a bottleneck to appear at the entrance of the roundabout because vehicles need time to decide whether to enter the roundabout or to wait for another larger gap to appear [65, 70].

Vehicle movements at signalized roundabouts are significantly different from those at ordinary signalized intersections [2]. When automobiles turning left approach the opposing approach's entrance, vehicles on the opposite approach must yield, resulting in a loss of green time [59]. Rashed and Imam [26] compared roundabouts and signalized roundabouts based on roundabout efficacy and discovered that the average delay of signalized roundabouts was reduced by 9–35%, and the LOS was lowered by two levels. Additionally, others who investigated signalized roundabouts and signalized intersections discovered that signalized roundabouts had superior operational performance and safety [50, 54]. This is similar to Alozi and Hussein [48] who suggested that roundabouts are an important choice for reducing traffic accidents. The implementation of various signalization options at contemporary roundabouts may result in improvements in safety, the environment [48], operational performance, and pedestrian access at roundabouts, while resolving the issue of traffic congestion at roundabouts with imbalanced flows, according to the results of the literature review. In the latter case, signals alter the roundabout's inherent priority rule to produce delays that are more regularly spaced [24, 71]. Together, signals and detectors can track the length of queues in often aborted approaches and bias green times to reduce crucial queue characteristics like latency and length. Depending on the traffic flow management, operating duration, and the number of regulated approaches, the roundabout may be fully or partially signalized. We found that 71% of studies employed average delay, average trip time, queue length, and flow rate (or some function of these) for this purpose and that VISSIM used the majority of simulation methods in this study.

References

1. Szczuraszek, T., Klusek, R.: Influence on the type of intersection on road traffic safety in Poland. In: IOP Conference Series: Materials Science and Engineering. IOP Publishing (2019)
2. Pilko, H., Mandžuka, S., Barić, D.: Urban single-lane roundabouts: a new analytical approach using multi-criteria and simultaneous multi-objective optimization of geometry design, efficiency and safety. Transp. Res. Part C Emerg. Technol. **80**, 257–271 (2017)
3. Shah, R., et al.: 67. Traffic analysis evaluate and signalize the existing roundabout using Ptv Vissim software. In: 1st International Conference on Recent Advances in Civil and Earthquake Engineering (ICCEE-2021)–Oct 08, 2021
4. Owais, M., Abulwafa, O., Abbas, Y.A.: When to decide to convert a roundabout to a signalized intersection: simulation approach for case studies in Jeddah and Al-Madinah. Arab. J. Sci. Eng. **45**(10), 7897–7914 (2020)
5. Shatnawi, A., et al.: Analyzing the Effect of Driving Speed on the Performance of Roundabouts (2022)

6. Zhou, L., Zhang, L., Liu, C.S.: Comparing roundabouts and signalized intersections through multiple-model simulation. IEEE Trans. Intell. Transp. Syst. (2021)
7. Gustafson, J.: Uniformity of terminology for circular intersection designs. Transp. Res. Rec. **2672**(34), 63–72 (2018)
8. Alsaleh, N.M., Shbeeb, L.: Turbo, flower and conventional roundabouts in Jordan. In: Proceedings of 6th Annual International Conference on Architecture and Civil Engineering (ACE 2018), May 2018
9. Chong, E.E.M., Al-Mamari, R.N.S.: Characteristics of drivers' lane choice at large multi-lane roundabout. Int. J. Integr. Eng. **12**(9), 176–183 (2020)
10. Alkheder, S., Al-Rukaibi, F., Al-Faresi, A.: Driver behavior at Kuwait roundabouts and its performance evaluation. IATSS Res. **44**(4), 272–284 (2020)
11. Fakirah, M., et al.: Visible light communication-based traffic control of autonomous vehicles at multi-lane roundabouts. EURASIP J. Wirel. Commun. Netw. **2020**(1), 1–14 (2020)
12. Osei, K.K., et al.: Signalization options to improve capacity and delay at roundabouts through microsimulation approach: a case study on arterial roadways in Ghana. J. Traffic Transp. Eng. **8**(1), 70–82 (2021). (English Edition)
13. Kusuma, A., Koutsopoulos, H.N.: Critical gap analysis of dual lane roundabouts. Procedia Soc. Behav. Sci. **16**, 709–717 (2011)
14. Zhao, Y., Andrey, J., Deadman, P.: Whether conversion and weather matter to roundabout safety. J. Safety Res. **66**, 151–159 (2018)
15. Yagoop, N.K., Ahmad, N.Y., Abbood, A.K.N.: Improvements of some roundabout Intersections using capacity model Analysis. In: IOP Conference Series: Materials Science and Engineering. IOP Publishing (2020)
16. Žura, M.: Roundabout capacity estimation model considering driver behaviour on the exiting and entry flows. Promet-Traffic Transp. **34**(3), 397–405 (2022)
17. Al-Obaidi, M.K.: Improvement of the traffic management of deactivated Al-Faris Al-Arabi signalized roundabout in Baghdad City. In: IOP Conference Series: Materials Science and Engineering. IOP Publishing (2019)
18. Murat, Y.S., Cakici, Z., Tian, Z.: A signal timing assignment proposal for urban multi lane signalised roundabouts. Građevinar **71**(02), 113–124 (2019)
19. Demir, H.G., Demir, Y.K.: A Comparison of traffic flow performance of roundabouts and signalized intersections: a case study in Nigde. Open Transp. J. **14**(1) (2020)
20. Han, I.: Safety analysis of roundabouts and avoidance of conflicts for intersection-advanced driver assistance systems. Cogent Eng. **9**(1), 2112813 (2022)
21. Adegbaju, O.A.: Establishing Delay-Based Criteria for Installing Traffic Signals at Two-Lane Roundabouts (2018)
22. Azhar, A.-M., Svante, B.: Signal control of roundabouts. Procedia Soc. Behav. Sci. **16**, 729–738 (2011)
23. Khliefat, I., et al.: Improving the capacity and level of service at the sixth circle in Jordan by using traffic signals and roundabout metering approach. Open Transp. J. **15**(1) (2021)
24. Murat, Y.S., Guo, R.-J.: Signalized Roundabouts (2021)
25. User, Y., İlyas, S., Tinaztepe, G.: Performance evaluation of a hybrid roundabout using a microscopic simulation. Baltic J. Road Bridge Eng. **16**(3) (2021)
26. Rashed, O.: A functional and operational comparison between signalized and unsignalized roundabouts. Int. J. Eng. Res. Technol. **13**(6), 1448–1454 (2021)
27. Nikitin, N., Patskan, V., Savina, I.: Efficiency analysis of roundabout with traffic signals. Transp. Res. Procedia **20**, 443–449 (2017)
28. Sarir, E.M., Khan, E.M.T.: Evaluation of Capacity at Signalized Roundabout Using VISSIM (2020)
29. Akçelik, R.: Roundabout metering signals: capacity, performance and timing. Procedia Soc. Behav. Sci. **16**, 686–696 (2011)
30. An, H.K., Yue, W.L., Stazic, B.: Estimation of vehicle queuing lengths at metering roundabouts. J. Traffic Transp. Eng. **4**(6), 545–554 (2017). (English Edition)

31. An, H.K., Abdalla, A.N.: Prediction of queuing length at metering roundabout using adaptive neuro fuzzy inference system. Measur. Control **52**(5–6), 432–440 (2019)
32. An, H.K., Liu, Y., Kim, D.S.: Operational Optimization at Signalized Metering Roundabouts using Cuckoo Search/Local Search Algorithm (2020)
33. Martin-Gasulla, M., et al.: Capacity and operational improvements of metering roundabouts in Spain. Transp. Res. Procedia **15**, 295–307 (2016)
34. An, H.K., YUE, W.L., Stazic, B.: An analysis of a partially signalized roundabout using SIDRA 6 software. Asian Transp. Stud. **4**(2), 314–329 (2016)
35. Qadri, S.S.S.M., Gökçe, M.A., Öner, E.: State-of-art review of traffic signal control methods: challenges and opportunities. Eur. Transp. Res. Rev. **12**(1), 1–23 (2020)
36. Qadri, S.S.S.M., et al.: Analysis of various scenarios to mitigate congestion at a signalized roundabout using microsimulation. In: 2019 Innovations in Intelligent Systems and Applications Conference (ASYU). IEEE (2019)
37. Iscan, N.: Evaluation the Level of Service at a Roundabout: A Case Study on Al-Ibrahimeya Roundabout in Alexandria, Egypt (2019)
38. Duan, Y., et al.: Optimising total entry delay at roundabouts with unbalanced flow: a dynamic strategy for smart metering. IET Intel. Transp. Syst. **13**(3), 485–494 (2019)
39. Pratama, B., et al.: Adaptive traffic lights through traffic density calculation on road pattern. In: 2018 International Conference on Applied Science and Technology (iCAST). IEEE (2018)
40. Mahmood, K.: Improving Roundabout Performance Using Adaptive Metering Technique. Masters Thesis (2016). https://etd.ohiolink.edu/apexprod/rws_etd
41. Dmitry, S., Yuliya, S., Galina, Z.: Adaptation of the traffic signal control design method to a Hamburger roundabout. Archit. Eng. **6**(3), 70–76 (2021)
42. Manual, H.C.: Highway Capacity Manual, vol. 2, no. 1, Washington, DC (2000)
43. Macioszek, E., Akçelik, R.: A comparison of two roundabout capacity models. In: Proceedings of the 5th International Roundabout Conference, Transportation Research Board, Green Bay, WI, USA (2017)
44. Muhammad, S., Jamaludin, S.A.: Evaluating the performance of traffic flow using SIDRA for roundabouts in Ipoh, Perak. Jurnal Kejuruteraan **34**(3), 421–427 (2022)
45. Kabit, M.R.: An evaluation of signal metering applicationto mitigate roundabout's unbalanced flow conditions: a case study in Kuching. Malays. Constr. Res. J. **27**(1), (2019)
46. Mohammed, A.A.: Intersection analysis by aaSIDRA software for computer simulation to optimize the traffic flow. Eng. Technol. J. **31**(4 Part (A) Engineering) (2013)
47. Abdelfatah, A., Minhans, A.: Roundabout or traffic signal: a selection dilemma. J. Transp. Syst. Eng. **1**(1), 67–73 (2014)
48. Alozi, A.R., Hussein, M.: Multi-criteria comparative assessment of unconventional roundabout designs. Int. J. Transp. Sci. Technol. **11**(1), 158–173 (2022)
49. Çakici, Z., Murat, Y.S.: Sinyalize Dönel Kavşaklarda Diferansiyel Gelişim Algoritması ile Sinyal Süre Optimizasyonu. El Cezeri **8**(2), 635–651 (2021)
50. Ma, W., et al.: Integrated optimization of lane markings and timings for signalized roundabouts. Transp. Res. Part C Emerg. Technol. **36**, 307–323 (2013)
51. Ma, Q., Zeng, H., Wang, Q.: Traffic optimization methods of urban multi-leg intersections. Int. J. Intell. Transp. Syst. Res. **19**(2), 417–428 (2021)
52. Patnaik, A.K., et al.: Entry capacity modelling of signalized roundabouts under heterogeneous traffic conditions. Transp. Lett. **12**(2), 100–112 (2020)
53. Cheng, W., Zhu, X., Song, X.: Research on capacity model for large signalized roundabouts. Procedia Eng. **137**, 352–361 (2016)
54. Mohammed, H.H., Ismail, M.Q.: Efficiency assessment of a signalized roundabout and a traffic intersection in Baghdad City. Eng. Technol. Appl. Sci. Res. **11**(6), 7910–7916 (2021)
55. Fortuijn, L.G., Salomons, A.M.: Signalized Turbo Circle; Design and Performance (2015)
56. Jiang, Z.-h., et al.: Investigation on two-stop-line signalized roundabout: capacity and optimal cycle length. J. Adv. Transp. (2019)
57. Ghanim, M., et al.: Safety and operational performance of signalized roundabouts: a case study in doha. Procedia Comput. Sci. **170**, 427–433 (2020)

58. Panda, S.: Capacity Estimation at Signalized Roundabouts (2016)
59. Bie, Y., et al.: Stop-line setback at a signalized roundabout: a novel concept for traffic operations. J. Transp. Eng. **142**(3), 05016001 (2016)
60. Tracz, M., Chodur, J.: Performance and safety roundabouts with traffic signals. Procedia Soc. Behav. Sci. **53**, 788–799 (2012)
61. Bie, Y., Mao, C., Yang, M.: Development of vehicle delay and queue length models for adaptive traffic control at signalized roundabout. Procedia Eng. **137**, 141–150 (2016)
62. Kumar, R.: Capacity and Level of Service of Signalized Roundabouts in Urban Indian Context (2015)
63. Bawa, S., Damsere-Derry, J.: Assessment of the road safety performance of four roundabouts in Kumasi, Ghana. Int. J. Inj. Contr. Saf. Promot. **25**(3), 239–246 (2018)
64. Awadalla, D.M.M., de Albuquerque, F.D.B.: Identification of risk factors associated with fatal intersection crashes and assessment of the in-service safety performance of signalized intersections and roundabouts in Abu Dhabi. Safety **7**(4), 69 (2021)
65. Kabanga, D., Theuri, D.M., Kioi, D.: Evaluating unsignalized circular roundabout capacity using macroscopic traffic flow model. Glob. J. Pure Appl. Math. **18**(1), 205–219 (2022)
66. Hassouna, F.M.: Performance Analysis of Modern Roundabouts as an Alternative to Conventional Signalized Intersections: A Comprehensive Review
67. Alqahtani, A.A.F.: Comparative Analysis of Roundabout Capacity Analysis Methods. University of Dayton (2019)
68. Li, Z., et al.: Operational evaluation of two-lane roundabouts at freeway ramp terminals: comparison between roundabout and signalized interchanges. Transp. Res. Rec. **2637**(1), 99–113 (2017)
69. Ma, X., Yi, P., Bolla, V.: Priority metering control for an urban circular intersection. Civil Eng. J. **2** (2019)
70. Guo, R., Liu, L., Wang, W.: Review of roundabout capacity based on gap acceptance. J. Adv. Transp. (2019)
71. Saltık, E.C., Yaman, H.: Distribution of Time Headways on Approach Legs of an Urban Signalized Roundabout: A Case Study in Konya (2018)

Gap Acceptance Behaviour of Risky Right-Turning Motorcyclists from Minor Roads at Conventional and Unconventional T-Junctions

Lee Vien Leong, Hor Kuan Chan, and Shafida Azwina Mohd Shafie

Abstract Right turning from a minor road is the most critical movement at T-junction, and the risk of right-turning motorcyclists involved in an accident is extremely high because of their riding behaviour. This study was carried out at three types of T-junctions which are type-A (conventional T-junction), type-B (unconventional T-junction with short exit lane for right-turning minor road vehicles) and type-C (unconventional T-junction with short exit lane for through major road vehicles). Results indicated that type-A T-junction is the safest as it has a lower percentage of risky riding behaviour. However, motorcyclists at type-B T-junction behaved better as most of them only performed one risky behaviour (56%) or none (34%). Additionally, results showed that motorcyclists accepted a smaller gap of 1.639 s for forced entry than the usual gap of 7.262 s. For gap acceptance behaviour involving a motorcycle and car making a right turn together, the mean accepted gap for this behaviour was 9.318 s for type-A, 8.319 s for type-B and 8.758 s for type-C. In conclusion, the geometrical configuration of type-B T-junction provides the best performance as the motorcyclists are more well-behaved and record the smallest value of the mean accepted gap.

Keywords Gap acceptance behaviour · Motorcyclists' risky riding behaviour · T-junction · Traffic accident

1 Introduction

With the increase in vehicles on the roads, the motorcycle has become a popular choice for individual private transport, as it is convenient for short to medium-distance trips and affordability. Motorcyclists are a vulnerable group on the road

L. V. Leong (✉) · H. K. Chan · S. A. M. Shafie
School of Civil Engineering, Universiti Sains Malaysia, Engineering Campus, 14300 Nibong Tebal, Penang, Malaysia
e-mail: celeong@usm.my

N. Sabtu (ed.), *Proceedings of AWAM International Conference on Civil Engineering 2022 - Volume 3*, Lecture Notes in Civil Engineering 386,
https://doi.org/10.1007/978-981-99-6026-2_30

because motorcycles are comparatively small; therefore, other drivers tend to disregard them. In contrast to drivers, who are entirely covered by the vehicle's structure, motorcyclists are unprotected. Therefore, a motorcyclist is likelier to sustain severe injuries in an accident. Human (motorcyclist) behaviour is the leading cause of motorcycle fatalities [1]. Nevertheless, due to the small size, narrow width, short length and high power-to-weight ratio of motorcycles, motorcyclists are more adaptable on the road and tend to ignore traffic rules [2]. According to Ali Sahraei et al. [3], this characteristic enables the motorcyclist to get closer to the front stop line, resulting in a shorter delay for the motorcycle at a junction than other vehicles.

An intersection is where two or more roads cross or merge [3]. Drivers might change directions at an intersection to get to their destination. A priority junction, or unsignalized intersection, is an intersection without traffic signals that permit vehicles to change direction. Vehicle interactions at unsignalized intersections are particularly complex [4], as vehicles from different directions compete to occupy the same space simultaneously, which could lead to crashes [5]. Priority junctions are classified into two types: yield-controlled junctions and stop-controlled junctions. A T-junction, or three-legged junction, is a stop-controlled junction with three road segments (arms) and two arms for a straight road [6].

At a T-junction, the vehicle turning right from a minor approach must stop and waits for a safe gap before crossing to the major approach. The time interval between the first vehicle's rear bumper and the second vehicle's front bumper, travelling in the same direction and lane, is defined as the gap. Based on several literature studies, different authors give different interpretations of a gap, but they all express the same concept. In US HCM 2016 [7], the term "gap" refers to the time interval (time gap) and corresponding distance for a given speed (space gap) between the major-street vehicles entering an unsignalized intersection, measured from the back bumper to front bumper. Yan and Radwan [8] stated that a gap is defined as the difference in arrival time between two opposing through vehicles passing by the driver waiting to turn. Ali Sahraei et al. [3] explained that a gap is determined by evaluating vehicles' front or rear bumper passing a reference location. Based on another definition given by Advani et al. [5] explained that a gap is a time between the exit of a conflicting vehicle of higher priority and the arrival of a corresponding vehicle at the same reference line on the lane. Alexander et al. [9] defined a gap as the time interval between the arrivals of two consecutive vehicles. Campbell et al. [10] and Patil and Sangole [2] defined a gap as the time interval between two successive vehicles counted from the rear of the leading vehicle to the front of the following vehicle. However, this definition is only accurate if the major approaches have one lane in both directions and all vehicles move according to lane discipline. When several lanes are in each direction, the gap is measured between two consecutive cars regardless of their lateral location along the road's width.

A critical gap is a significant element in determining the delay and capacity of a minor road [4, 11–14]. The critical gap's accuracy highly influences capacity and delay accuracy [9]. According to the study of Mohan and Chandra [15], a minor road driver must determine if a gap in the opposing traffic stream is wide enough to complete the desired movement safely when entering the intersection. Generally, a driver accepts all gaps greater than their critical gap and ignores the rest. Thus, the critical gap determines the smallest value of the gap that a driver can consider. A critical gap is not a constant but varies from driver to driver and from time to time. The estimation of the critical distance is very intriguing since it cannot be deduced from field observations.

According to Harnen et al. [16], an intersection is high-risk because most recorded accidents have occurred there. The motorcycle–vehicle angle crash was rated as the most severe accident type at T-junctions, followed by the approach-turn accident. One of the risky entry behaviours of right-turning motorcyclists from the minor road is forceful turning which causes vehicles on the major road to slow down or stop. When there is a high volume of vehicles on the main road, motorcycles cannot turn right from the minor road. When they must wait an extended time, they become more aggressive and willing to tolerate short gaps to turn right [17–19]. They force the vehicles on the major road to slow down or stop as they cause a forced right turn. Consequently, this study will investigate the gap acceptance behaviour of motorcyclists that perform such forceful turning. Disregarding the principle of First-In-First-Out (FIFO) is another dangerous behaviour that right-turning motorcyclists always engage in [20]. Because motorcycles are small, they typically advance to the front of the queue and turn right together with other vehicles. Hence, this research will examine the entry behaviour of motorcyclists who turn right with a car with the same accepted gap.

A critical gap is essential in estimating the junction's capacity and level-of-service (LOS). Extensive research has been conducted on drivers' gap acceptance at the junction with cars; however, given the high incidence of motorcycle accidents, relatively few researchers have looked at the effect of risky riding behaviour on gap acceptance among motorcyclists. In addition, earlier studies solely focused on the time gap concept and rarely addressed the relationship between motorcyclist behaviour and the availability of gaps on the roadway [21]. Most research has also mainly focused on risky driving behaviour; however, no studies currently examine the effect of motorcyclists' risky riding behaviour on gap acceptance at T-junctions. Therefore, a comprehensive study of gap acceptance that emphasizes the behaviour of motorcyclists is required.

Unsignalized T-junctions in Malaysia can be broadly classified as either conventional or unconventional. For unconventional T-junctions, a short merging lane is added on the receiving approach of the far side of the major road to reduce conflicts between right-turning vehicles from the minor road and through-movement vehicles. It should be able to reduce the delay of right-turning vehicles as they are easier to merge. However, just a few studies look at the short lane's effectiveness and the influence of the junction's geometrical configuration on gap acceptance. Consequently, this research aimed to examine the effect of risky right-turning behaviour of motorcyclists on gap acceptance at conventional and unconventional T-junctions typically found in Malaysia.

2 Methodology

2.1 Site Selection

Three T-junctions (one conventional T-junction and two unconventional T-junctions) were selected for this study. They were categorized into type-A (typical two-lane two-way major road T-junction), type-B (unconventional T-junction with short entrance lane for right-turning major road vehicles and short exit lane for right-turning minor road vehicles) and type-C (T-junction with short entrance lane for right-turning major road vehicles and short exit lane for through major road vehicles). This study is conducted in the urban and suburban areas of Parit Buntar, Perak and Seberang Perai, Pulau Pinang.

First, several stop-controlled T-junctions that fulfil the required criteria were chosen and viewed in Google Maps. The satellite imagery and 360° interactive panoramic views of streets (Street View) in Google Maps provide a comprehensive picture of the junction configuration to ensure that the chosen junctions meet the basic requirements before the preliminary site survey. T-junctions with more than one lane at the near side of the major road are eliminated as this gap acceptance study is limited to only one row of vehicles from one direction. A staggered junction is avoided as the operation of a nearby junction will adversely affect the operation of the study junction.

2.2 Site Survey

During the site survey, the selected junctions were filtered. Two-lane two-way major road with wide lane width is eliminated if the width is large enough for two vehicles to move side by side. Besides, the road marking at the T-junction must be available and clear, especially the stop line on the minor road and the middle line on the major road. The "BERHENTI" sign must be installed on the minor road to alert drivers and motorcyclists to stop completely before turning. All sites should be located on level terrain, and adequate sight distances must be available for right-turning movement from the minor road. Another purpose of the site survey is to ensure the junctions have sufficient traffic volume to carry out the field study. Thus, to determine the site's suitability in traffic volume, the behaviour of on-site right-turning vehicles is recorded for 30 min at every selected junction during the site survey. The recorded videos were played back off-site to investigate the volume of vehicles, and the junctions with the insufficient volume of right-turning motorcycles were filtered. The location and space required to set up the video recording equipment were also reviewed to ensure that they can be positioned at a strategic location in order to avoid becoming a distraction to riders, as they might ride consciously and carefully in front of the camera. Finally, the three most suitable T-junctions that fulfil the criteria were selected.

Type-A is a typical or conventional T-junction with only one lane in each direction. The selected site is located at the intersection of Jalan Sekolah (major road) and Jalan Sentosa (minor road) with a coordinate of 5.122881, 100.486116 at Parit Buntar, Perak. The configuration of the selected site for type-A T-junction is illustrated in Fig. 1. The lane widths of the major road are 3.8 m on the near side and 4.0 m on the far side, while the road width of the minor road is 8.1 m. According to Arahan Teknik (Jalan) 8/86 [22], for conventional junctions located on R3 roads, each lane's width should be at least 3.0 m. Therefore, the configuration of the studied junction complies with the standard design.

(a)

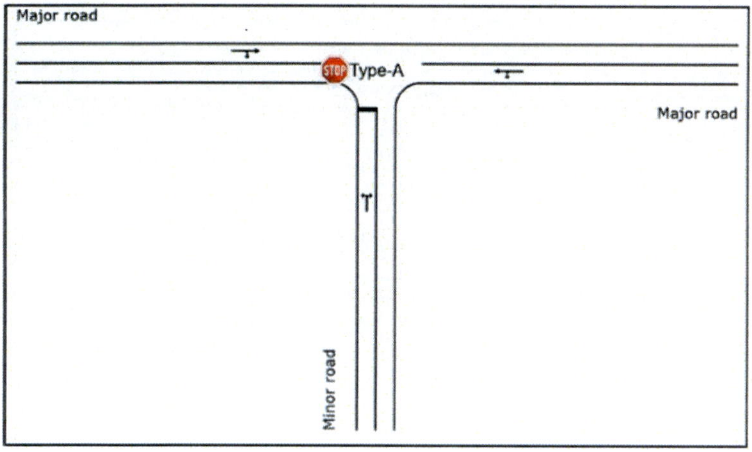

(b)

Fig. 1 Type-A T-junction showing **a** the satellite imagery layout and **b** the basic junction configuration

Type-B T-junction is an unconventional T-junction with an exclusive right-turning lane on a major road and a short exit lane on the receiving approach of a major road to accommodate right-turning vehicles from the minor road. An exclusive right-turning lane is provided so as not to disrupt the smooth flow of through traffic, while short exit lane facilitates the merging movement of right-turning vehicles from minor to major roads. The selected type-B T-junction is located at the intersection of Jalan Sekolah (major road) and Jalan Permai Utama (minor road) with a coordinate of 5.125122, 100.483028 at Parit Buntar, Perak. The geometrical configuration of the selected site is illustrated in Fig. 2.

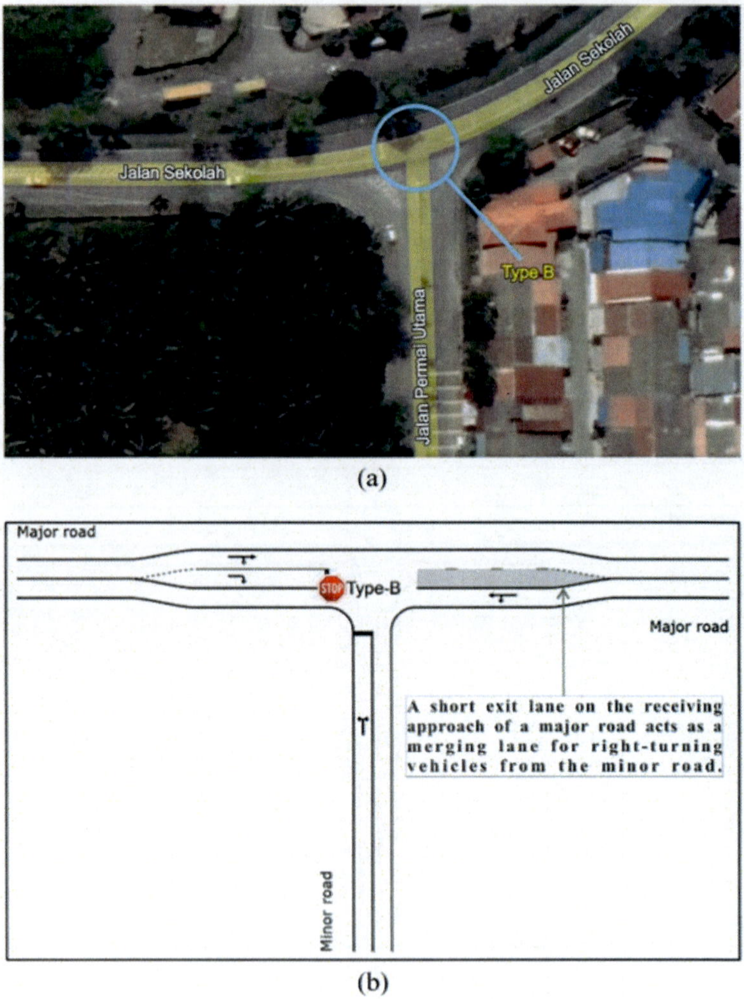

Fig. 2 Type-B T-junction showing **a** the satellite imagery layout and **b** the basic junction configuration

Lastly, the type-C T-junction is an unconventional T-junction similar to type-B, which also has an exclusive right-turning lane at the major road and a short exit lane on the receiving approach of the major road. The only difference is the short exit lane on the receiving approach on type-C major road acts as a merging lane for through movement on the receiving approach of the major road. In contrast, type-B acts as a merging lane for right-turning vehicles from the minor road. The selected type-C T-junction is located at the intersection of Jalan Permatang Tinggi (major road) and Jalan Markisah (minor road) with a coordinate of 5.316649, 100.461058 at Seberang Perai, Pulau Pinang. The dimension of the selected site is illustrated in Fig. 3.

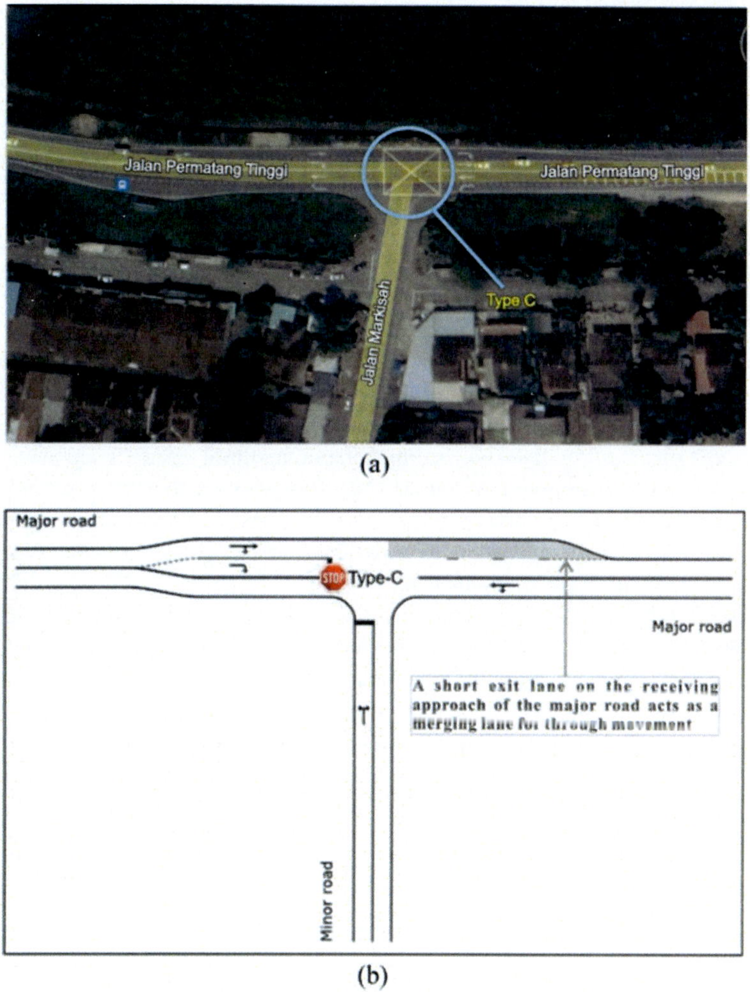

Fig. 3 Type-C T-junction showing **a** the satellite imagery layout and **b** the basic junction configuration

2.3 Data Collection

Data for this study were collected at the three selected stop-controlled T-junctions using the video recording technique. The video camera is set up strategically to capture the full movement of right-turning vehicles on the minor road and through movement vehicles on the near side of the major road in the intersection zone.

Data collection must be conducted during the peak hour of typical weekdays (Tuesday, Wednesday or Thursday) with the dry weather condition. Due to the unusual nature of the trip behaviour, Monday and Friday are considered weekdays with non-typical traffic flow. Public holidays and weekends were avoided as the traffic volumes are generally lower as most trips are optional and more likely for leisure purposes [23]. Besides, rainy weather will reduce the volume of motorcycles on the road as motorcyclists are not covered like other vehicles. Overall, data collection for field study should be made on typical working days to show the typical behaviour of road users. For this study, data were collected during the morning peak hours from 6:30 a.m. to 9:30 a.m. and evening peak hours from 4:30 p.m. to 7:30 p.m.

2.4 Data Extraction

All data are extracted from the recorded videos. The videos are watched off-site and replayed, if necessary, using Media Player Classic as it can extract gaps in second (s) up to three decimal places.

Through observation, five risky riding behaviours that can affect gap acceptance behaviour are picked, as given in Table 1. The frequency of each behaviour performed at all T-junctions and the number of risky riding behaviour performed by each motorcyclist was extracted. The frequency of each entry behaviour at the three T-junction types was recorded for further analysis.

Subsequently, the gap acceptance behaviour of right-turning motorcycles that enter forcefully into the major road is extracted. To do a comparative analysis between the mean accepted forceful gap of right-turning motorcycles and the mean accepted gap of right-turning motorcyclists that do not perform any risky riding behaviour, the accepted gaps of right-turning motorcyclists that do not perform any risky riding behaviour were extracted too. Consequently, the video footage also extracted the

Table 1 Risky riding behaviour of right-turning motorcyclists

Code	Risky riding behaviour of right-turning motorcyclists
R1	Not following the concept of First-In-First-Out (FIFO)
R2	Turning without fully stopping
R3	Not following the conventional right-turning path
R4	Not following priority rules
R5	Forceful turning into the major road

right-turning motorcycles that turn with a car together using the same accepted gap from minor road at each junction.

3 Results and Discussion

3.1 Frequency Distribution of Risky Riding Behaviour

283 cases were recorded at type-A T-junction, 114 cases from type-B T-junction and 445 cases from type-C T-junction. Based on the results shown in Fig. 4, motorcyclists are more likely to perform "R1—not following the concept of First-In-First-Out (FIFO)" with 51.75% and "R2—turning without fully stopping" with 30.70% at type-B T-junction compared with other T-junctions. This is possible because the major road's exclusive short exit lane makes motorcyclists feel safer when turning right. Hence, motorcyclists on the minor road are more likely to turn without much consideration of time.

Subsequently, motorcyclists at type-C T-junction show high aggressiveness as they are more likely to perform "R3—not following the conventional right-turning path" with 26.97%, "R4—not following priority rules" with 4.38% and "R5—forceful turning into major road" with 3.82% as compared with the other two T-junctions. It is believed to be caused by the geometrical configuration of a major road. Through observation on-site, the through-movement vehicles on the receiving approach of the major road will start to merge to the full lane from the short merging lane right after passing the exclusive right-turning lane. Hence, the exit lanes of the major road were

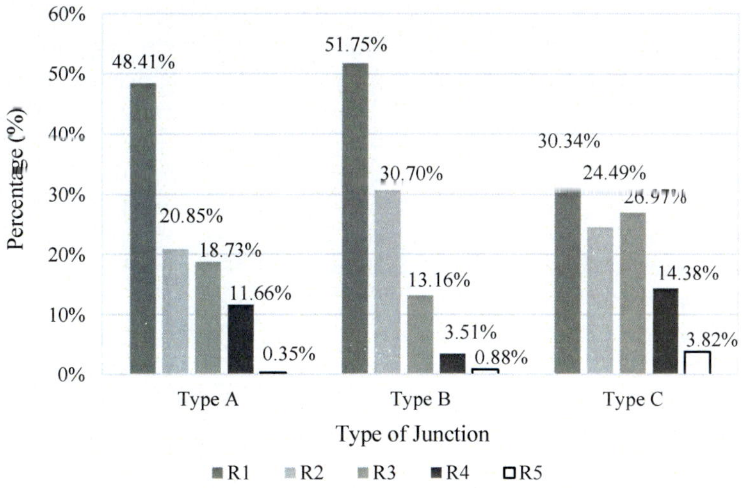

Fig. 4 Frequency (%) of risky riding behaviours based on their type

always occupied by through-movement vehicles. As a result, right-turning vehicles on the minor road, including motorcycles, cannot merge into the major road and are forced to perform such aggressive risky riding behaviours. Therefore, the results suggested that the geometrical configuration of type-A T-junction is the safest among the three T-junction types based on the frequency of right-turning motorcyclists performing risky riding behaviour.

Overall, "R1—not following the concept of First-In-First-Out (FIFO)" is observed to be the most performed risky behaviour at three T-junction types. This result indicates that most motorcyclists at T-junctions do not queue on the minor road but move to the stop line to reduce the waiting time. This behaviour impedes traffic flow on the minor road and causes disturbance and obstruction to the other vehicle's driver that stops at the stop line. However, it was believed that "R5—forceful turning into major road" is the most dangerous risky behaviour if compared with the other four risky behaviour. Nevertheless, based on the graph plotted in Fig. 4, its frequency percentage only ranged from 0.3% to 4% at three T-junction types. It showed that right-turning motorcyclists seldom make forceful turning at all types of T-junctions because they are aware of the high possibility of crashes and fatalities as the vehicles from major roads might be unable to stop and crash them at high speed. A side-impact collision between the vehicle on the major road with the departing vehicle from the minor road also caused a sudden slowing down or stopping on the major road and eventually induced rear-end crashes on the major road. Hence, forceful right turning into a major road from a minor is the most dangerous risky behaviour.

3.2 Number of Risky Riding Behaviour Performed by Each Motorcyclist

The number of risky riding behaviour performed by each motorcyclist is shown in Fig. 5. Type-A and type-C T-junctions have 1.36% and 0.67% of right-turning motorcyclists that performed four risky riding behaviours when turning right. However, none of the right-turning motorcyclists at type-B T-junction performed four risky riding behaviour, and only 1.33% of them performed three risky riding behaviour when turning right, which is still ranked at the lowest compared with type-A T-junction with 7.73% and type-C T-junction with 11.67%.

Figure 5 also shows signs that many motorcyclists performed only one risky riding behaviour or none at type-B T-junction compared to the other two T-junctions. This result indicates that right-turning motorcyclists behave better at type-B T-junction than type-A and type-C T-junctions. One possible explanation for this finding is the existence of a short exit lane on the major road which gives motorcyclists confidence and eventually reduces the need for them to take the risk.

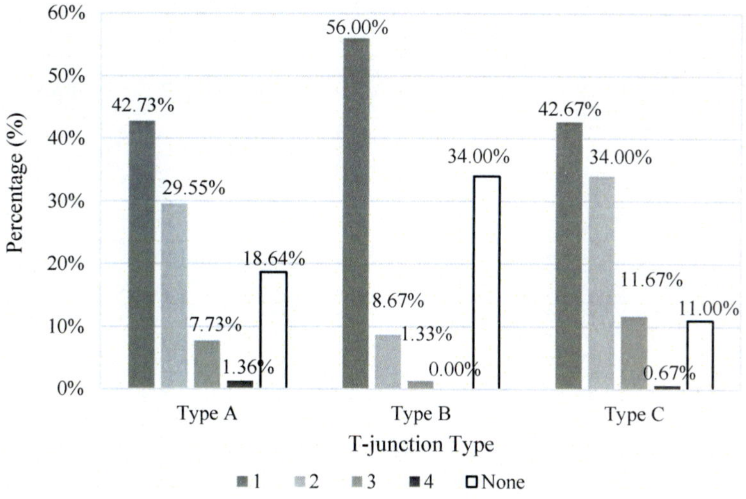

Fig. 5 Frequency (%) of risky riding behaviours based on the number of occurrences

3.3 *Comparative Analysis of Forceful Accepted Gap with Normal Accepted Gap*

Based on the recorded data, 0.35%, 0.88% and 3.82% cases of "R5—forceful turning into major road" were recorded at type-A, type-B and type-C T-junction. It was believed that forceful turning was induced by the high volume of straight-movement vehicles on the major road, which prolonged the waiting time of right-turning motorcyclists on the minor road. The volume of straight-movement vehicles on the major road was significantly higher at type-C T-junction, while a very low volume was observed at both type-A and type-B T-junctions. Hence, type-A and type-B T-junctions recorded a very low frequency of this risky riding behaviour, which was inadequate for an informative conclusion. Therefore, this study only focuses on type-C T-junction to determine the accepted gap of forceful right-turning motorcyclists.

An accepted forceful gap of a motorcyclist is defined as the gap used by a motorcyclist from a minor road to turn into a major road, which forces the vehicles on the major road at the near side or far side to reduce their speed or stop momentarily. To determine the gaps that would disqualify gap-acceptance behaviour, normal curves were drawn on the histograms to analyse the normality of the distribution using SPSS Statistics. Using the entire distribution of gap frequency, the normal curve is positively skewed, as shown in Fig. 6a, which suggests that the gaps fall towards the lower side of the scale and that there are only very few longer gaps. To obtain normally distributed gaps, accepted forceful gaps that are more than 3.0 s are not considered. The normal curve shown in Fig. 6b indicates that the accepted forceful gaps are normally distributed when the accepted forceful gaps are equal to or shorter

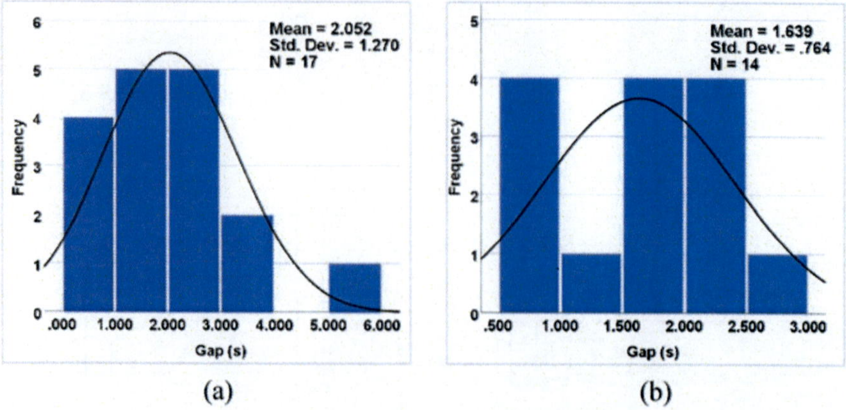

Fig. 6 Frequency distribution of **a** all accepted forceful gaps of right-turning motorcyclists and **b** accepted forceful gaps equal to or shorter than 3.0 s of right-turning motorcyclists

than 3.0 s. Inspection of the distribution of accepted forceful gaps provides meaningful information on the threshold gap that forceful right-turning motorcyclists will accept in which gap acceptance ceases for gaps longer than about 3.0 s.

The frequency distribution of accepted normal gap by right-turning motorcyclists that did not perform risky riding was studied and compared with the accepted forceful gap at the type-C T-junction. By using all accepted gaps, the normal curve is positively skewed, as shown in Fig. 7a, suggesting that the gaps fall towards the lower side of the scale and that there are few longer gaps. However, if we only considered accepted gaps less than or equal to 12.0 s, the curve was normally distributed, as shown in Fig. 7b. Hence, gaps larger than 12.0 s were not considered for accepted gaps by right-turning motorcyclists that did not perform any risky riding behaviour. In conclusion, motorcyclist gap acceptance behaviour was suspected of ceasing beyond the gap shorter than 12.0 s. However, these results only apply to the T-junction that is the same design as the type-C T-junction.

Therefore, a difference of 5.623 s was identified between the mean accepted forceful and normal gaps. The mean accepted gap for right-turning motorcyclists forced to turn into the major road is lower than the mean accepted gap for right-turning motorcyclists who do not perform risky riding behaviour, as given in Table 2. This result indicates that motorcyclists that turn forcefully into a major road will accept a gap much smaller than the gap they normally will accept to reach the far side of a major road. Although the result does not apply to type-A and type-B T-junctions, it represents a trend; forceful entries accept much smaller gap than usual.

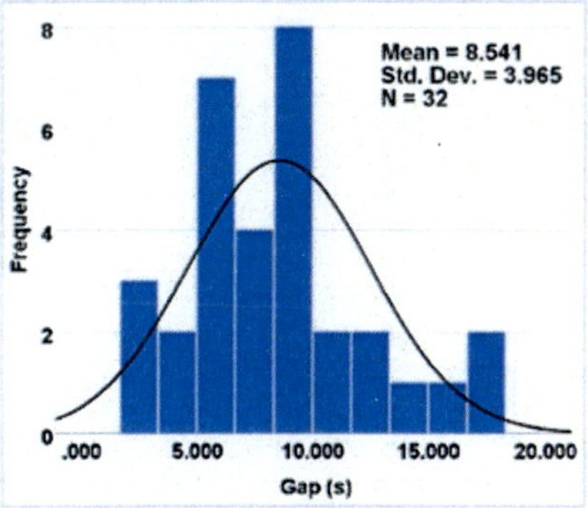

(a)

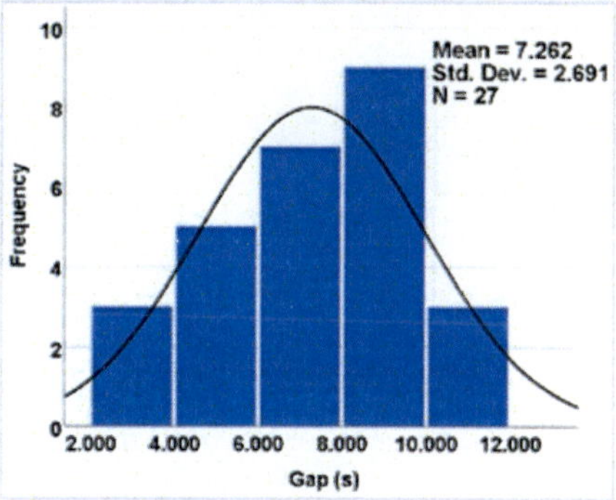

(b)

Fig. 7 Frequency distribution of **a** all accepted gaps of right-turning motorcyclists and **b** accepted gaps equal to or shorter than 12.0 s of right-turning motorcyclists

Table 2 Comparison of mean accepted forceful gap and mean accepted normal gap of right-turning motorcyclists

Right-turning motorcyclist with a forceful gap		Right-turning motorcyclist without any risky riding behaviour	
Frequency	Mean accepted forceful gap (s)	Frequency	Mean accepted normal gap (s)
14	1.639	27	7.262

3.4 Gap Acceptance Behaviour Involving Motorcycle and Car

According to Leong et al. [23] and Gattis et al. [24], when the accepted gap size exceeds 12.0 s, there is no significant information about driver gap acceptance behaviour. Including such data is likely to skew critical gap analysis results, causing the calculated critical gap size to be larger than it is. Hence, any accepted gaps larger than 12.0 s were eliminated in this study.

Subsequently, the frequency distribution histogram using an accepted gap of less than 12 s for a motorcycle turning right together with a car using the same gap from a minor road at type-A, type-B and type-C T-junctions were plotted in Fig. 8. All the data show normal curves indicating that the distributions of accepted gaps for three types of T-junctions are normally distributed when the accepted gaps are less than or equal to 12.0 s. This finding is similar to Leong et al.'s studies [23] and Gattis et al. [24].

According to results shown in Fig. 8 and summarized in Table 3, for type-A T-junction, the mean accepted gap is 9.318 s and standard deviation is 0.854; for type-B T-junction, the mean accepted gap is 8.319 s and standard deviation is 1.360; for type-C T-junction, the mean accepted gap is 8.758 s and standard deviation is 1.516. It shows that motorcyclists that turn right together with car at type-B T-junction have lower gap acceptance as compared with type-A and type-C T-junctions.

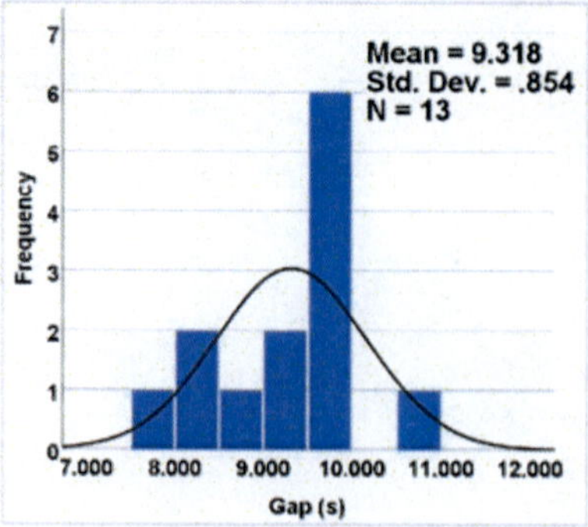

(a)

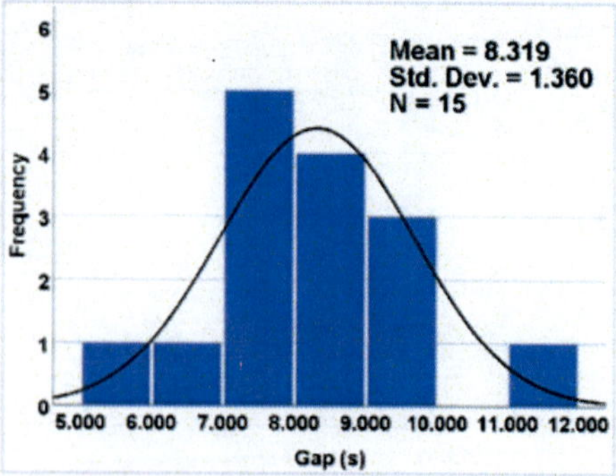

(b)

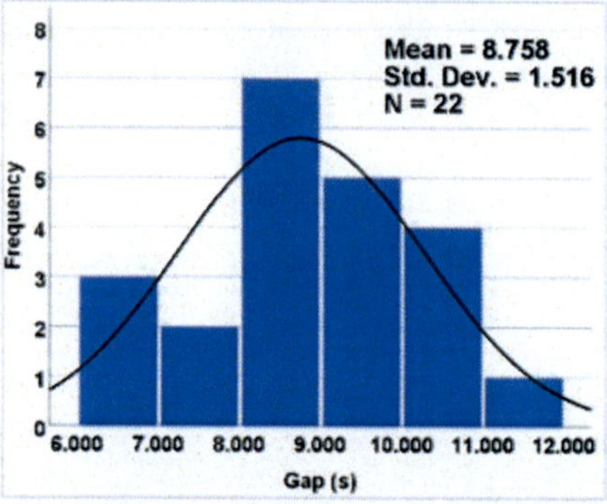

(c)

Fig. 8 Frequency distribution of accepted gaps for a motorcycle turning right together with a car from a minor road at the **a** type-A, **b** type-B and **c** type-C T-junctions

Table 3 Mean accepted gap for a motorcycle turning right together with a car from a minor road at the three types of T-junctions

T-junction type	Type-A	Type-B	Type-C
Mean accepted gap (s)	9.318	8.319	8.758
Standard deviation	0.854	1.360	1.516

4 Conclusion

Based on the frequency distribution of each behaviour at three T-junction types, it can be concluded that type-A T-junction is the safest among the three T-junction types regarding risky riding behaviour of right-turning motorcyclists as it has a lower percentage of risky riding behaviour. However, regarding operational performance, it can be concluded that the geometrical configuration of type-B T-junction provides the best performance for a motorcycle to turn right from a minor road as it has the smallest mean accepted gap of motorcycles turning right with cars from the minor road. Moreover, the motorcyclists at type-B T-junction were recorded to be more well-behaved than the motorcyclists at type-A and type-C T-junction by referring to the number of risky riding behaviour performed by each motorcyclist. The exclusive short exit lane for the minor-road right-turning vehicles on the junction's major road effectively made the drivers and motorcyclists feel safer and confident to turn. Drivers and motorcyclists only have to consider the gap between vehicles from the near side of the major road. Even so, having a short lane for through movement on the receiving approach of the major road at the T-junction was not recommended. As observed at type-C T-junction, the vehicles at receiving approach of the major road will merge into the full lane before they pass the junction, eventually blocking the way for the right-turning vehicles on the minor road to merge with the major road. The exclusive exit lane at type-C T-junction was ineffective in assisting the minor-road right-turning vehicles. Type-A T-junction is safer regarding risky riding behaviour, as motorcyclists are less likely to turn when no exclusive exit lane is provided. Still, it is also the T-junction with the lowest performance as the critical gap and follow-up time are the longest among the three T-junction types.

References

1. Sultan, Z., Ngadiman, N.I., Kadir, F.D., Roslan, N.F., Moeinaddini, M.: Factor analysis of motorcycle crashes in Malaysia. Plann. Malays. **4**(4), 135–146 (2016)
2. Patil, G.R., Sangole, J.P.: Behaviour of two-wheelers at limited priority uncontrolled T-intersections. IATSS Res. **40**(1), 7–18 (2016)
3. Ali Sahraei, M., Che Ouan, O., Parvin Hosseini, S.M., Hadi Almasi, M.: Establishing a new model for estimation of the control delay at priority junctions in Malaysia. Cogent Eng. **5**(1), 1–12 (2018)
4. Ashalatha, R., Chandra, S.: Critical gap through clearing behaviour of drivers at unsignalised intersections. KSCE J. Civ. Eng. **15**(8), 1427–1434 (2011)
5. Advani, M., Senathipathi, V., Gupta. N.J., Errampalli. M.: Defining and analyzing forceful gap behaviour at unsignalised intersections. Transp. Res. Rec. **2674**(8), 420–428 (2020)
6. Costa, M., Bichicchi, A., Nese, M., Lantieri, C., Vignali, V., Simone, A.: T-junction priority scheme and road user's yielding behaviour. Transp. Res. F Traffic Psychol. Behav. **60**, 770–782 (2019)
7. Transportation Research Board. HCM 2016, Highway capacity manual 6th edition a guide for multimodal mobility analysis. Volume 1: Concepts, National Academy of Sciences-Engineering-Medicine, Washington, DC, USA (2016)
8. Yan, X., Radwan, E.: Effect of restricted sight distances on driver behaviours during unprotected left-turn phase at signalized intersections. Transp. Res. F: Traffic Psychol. Behav. **10**(4), 330–344 (2007)
9. Alexander, J., Barham, P., Black, I.: Factors influencing the probability of an incident at a junction: results from an interactive driving simulator. Accid. Anal. Prev. **34**(6), 779–792 (2002)
10. Campbell, J.L., Lichty, M.G., Brown, J.L., Richard, C.M., Graving, J.S., Graham, J., Laughlin, M.O., Torbic, D., Harwood, D.: Human Factors Guidelines for Road Systems: Second Edition (2012)
11. Wu, N.: A new model for estimating critical gap and its distribution at unsignalised intersections based on the equilibrium of probabilities. In: Proceedings of the 5th International Symposium on Highway Capacity and Quality of Service, vol. 2, Technical Papers, pp. 517–526 (2006)

12. Karthika, P.T., Koshi, B.I.: Gap acceptance behaviour of drivers at T-intersections. Int. J. Eng. Res. Technol. **3**(11), 935–938 (2014)
13. Guo, R.J., Wang, X.J., Wang, W.X.: Estimation of critical gap based on Raff's definition. Comput. Intell. Neurosci. **236072**, 1–7 (2014)
14. Pawar, D.S., Patil, G.R.: Analyzing variations in spatial critical gaps at two-way stop-controlled intersections using parametric and non-parametric techniques. J. Traffic Transp. Eng. **8**(1), 129–138 (2021). (English Edition)
15. Mohan, M., Chandra, S.: Review and assessment of techniques for estimating critical gap at two-way stop-controlled intersections. Eur. Trans. Issue **61**, 1–18 (2016)
16. Harnen, S., et al.: Predictive model for motorcycle accidents at three-legged priority junctions. Traffic Inj. Prev. **4**(4), 363–369 (2003)
17. Kittelson, W.K., Vandeley, M.A.: Delay effects on driver gap acceptance characteristics at two-way stop-controlled intersections. Transp. Res. Rec. **1320**, 154–159 (1991)
18. Kyte, M., Zegeer, J., Lall, K.B.: Empirical models for estimating capacity and delay at stop-controlled intersections in the United States. In: Brilon, W. (ed.) intersection without traffic signals II: In Proceeding of an International Workshop, Bochum, Germany. Springer, New York, pp 335–361 (1991)
19. Wagner, F.A.: An evaluation of fundamental driver decision and reaction at an intersection. Highw. Res. Rec. **118**, 68–84 (1966)
20. Leong, L.V., Lee, J.H.: Impact of Impact of motorcyclists' travel behavior on delay and level-of-service at signalized intersections in Malaysia. Adv. Intell. Syst. Comput. **484**, 1165–1178 (2017)
21. Lord-Attivor, R., Jha, M.K.: Modeling gap acceptance and driver behaviour at stop controlled (priority) intersections in developing countries. In: Proceedings of the 6th WSEAS International Conference on Computer Engineering and Applications, and Proceedings of the 2012 American conference on Applied Mathematics, pp. 29–38. World Scientific and Engineering Academy and Society (WSEAS), Wisconsin (2012)
22. Public Works Department: A guide on geometric design of roads (Arahan Teknik Jalan 8/86). Road Branch, Public Works Department Malaysia, Kuala Lumpur (1986)
23. Leong, L.V., Lim, V.W.W., Goh, W.C.: Effects of short exit lane on gap-acceptance and merging behaviour of drivers turning right at unconventional T-junctions. Int. J. Civil Eng. **18**(1), 19–36 (2020)
24. Gattis, J.L., Low, S.T.: Gap acceptance at atypical stop-controlled intersections. J. Transp. Eng. **125**(3) (1999)

Intelligent Transport System and Image Processing: Developing Traffic Detection Program Based on Indonesian Highway Capacity Manual 1997 for Rural Road

Muchlisin Muchlisin and Nurtia Amandairst

Abstract It is important in this digitalization era to develop the intelligent transport system for the future transportation. This study aims to develop traffic detection program based on image processing and Indonesian Highway Capacity Manual 1997 for rural road. The methodology of this study uses Gaussian Mixture Model (GMM) method that base in image processing. This program will run of video input in the morning with 6 m height and 7 m, as well as video input in the afternoon with 6 m and 7 m height. The calculation results on the program will be compared with manual calculations to get the accuracy and percent error values. The highest accuracy results obtained in this program are 76.48% with an error value of 23.52% for video 6 m in the morning. For the lowest accuracy results found on the 7 m video in the afternoon with an accuracy of 43% with an error value of 57%, this caused by congested traffic during video capture.

Keywords Intelligent transport system · Image processing · Rural street · Traffic detection · Gaussian Mixture Model (GMM).

1 Background

In Indonesia, the population growth rate is increasingly high. This causes the level of transportation needs to be even greater which results in increased traffic volume. With the increase in traffic volume posing a big challenge in traffic research and planning [2]. In some conditions, traffic volume data is used in transportation planning. At present, to obtain traffic volume data, it still uses manual methods carried out by field observers. The manual method can be done, but with a high traffic volume it is considered less effective because it can cause human error. Problems in traffic

M. Muchlisin (✉) · N. Amandairst
Department of Civil Engineering, Faculty of Engineering, UMY, Kasihan, Indonesia
e-mail: muchlisin@umy.ac.id

N. Sabtu (ed.), *Proceedings of AWAM International Conference on Civil Engineering 2022 - Volume 3*, Lecture Notes in Civil Engineering 386,
https://doi.org/10.1007/978-981-99-6026-2_31

analysis require development to overcome existing problems so that they can be resolved easily [5].

According to Harwendhani et al. [3], one way to overcome traffic density and other problems can be to utilize intelligent transportation system technology; therefore, in this study modifications to the program were made that can detect the number of vehicles and classify vehicles according to their groups based on MKJI 1997 which automatically passes on road sections on the road outside the city. In this study, the Java programming language with the Gaussian Mixture Model (GMM) method that uses image processing was used.

The problem in this study is how the modification of making a vehicle detection program using the Java programming language and how much the accuracy and percent of errors (errors) obtained from the results of the program calculations with manual calculations. The purpose of this study is to make a vehicle detection program using the Gaussian Mixture Model (GMM) method based on vehicle classification from MKJI 1997, determine the conversion value of vehicle dimensions with pixel area, test the accuracy and number of errors on the program against manual calculations, and analyze factors that affect the accuracy of the program. The benefit of this research is to facilitate field observers in carrying out data collection on traffic flow volumes and become a reference for further research on vehicle detection programs.

1.1 Traffic Volume

Regarding traffic volume according to Government Regulation of the Republic of Indonesia Number 34 of 2006 traffic volume is the number of vehicles that pass certain road segments in a certain time unit. Traffic volume can also be interpreted as the number of vehicles passing a point on a road in a certain time interval, the unit of traffic volume is expressed in a vehicle unity time (pcu/hour).

The calculation of traffic volume usually still uses the manual method. The traffic volume on a road will vary depending on the daily, monthly and also annual volume in the composition of the vehicle. Traffic volume is also widely used in transportation planning.

1.2 Road Outside the City

According to the 1997 Indonesian Road Capacity Manual, out-of-town roads are roads that do not develop continuously. Types of out-of-town roads based on MKJI 1997, namely

(a) Undivided two-lane road.
(b) Undivided four-lane road (i.e., without median).

(c) A four-way two-lane road is divided (i.e., with a median).
(d) A divided two-way six-lane road.

In this study, data collection was carried out in the form of video traffic using a camera drone located on a road outside the city located on Jalan Brawijaya, Tamantirto, Kasihan, Bantul, and Yogyakarta which video traffic data will be input into the program.

1.3 Vehicle Classification

According to the 1997 Indonesian Road Capacity Manual which includes the types of vehicles out of town and the equivalent value of passenger cars (emp) are as follows:

(a) Motorcycle (MC), motorcycle with two or three wheels (motorbikes and three-wheeled vehicles). The equivalent of passenger cars on motorbike vehicles (MC) is 0.7.
(b) Light vehicles (LV), four-wheeled motorized vehicles with two axles that are 2–3 m away (passenger vehicles, three wheels, micro-buses, pick-ups and small trucks). The equivalent of passenger cars in light vehicles (LV) is 1.0.
(c) Medium-weight vehicles (MHV), motorized vehicles with two axles within 3.5–5 m (small buses, two axles with six wheels). The equivalent of passenger cars in medium heavy vehicles (MHV) is 2.00.
(d) Large bus (LB), a bus with two or three axles with an axle distance of 5–6 m. The equivalent of passenger cars in large bus vehicles (LB) is 2.5.
(e) Large trucks (LT), three-axle trucks and combination trucks at axle distance (first to second axle) < 3.5 m. The equivalent of passenger cars in large truck vehicles (LT) is 4.00.

The dimensions of vehicles that cross roads outside the city can be seen in Table 1.

Table 1 Dimension of rural vehicle types

No	Type of Vehicle	Length (m)	Wide (m)	High (m)	Volume (m^3)
1	MC	2.00	0.80	1.10	1.76
2	LV	4.19	1.66	1.70	11.82
3	MHV	7.50	2.20	3.00	49.50
4	LB	12.50	2.50	3.20	100
5	LT	12.94	2.60	3.30	111.03

1.4 Image Processing (Image Processing)

According to Sutoyo in Adistya and Muslim [1], image is an image, imitation, or similarity of an object. Image processing (image processing) is an attempt made to transform an image or image into another image by using certain techniques. Image processing based on image processing operations can be categorized as improving image quality and image recovery [12].

Background Subtraction

According to Augusta and Sari [11], Background Subtraction is widely used in an image-based application. Background Subtraction is often used in applications to count the number of vehicles passing a path in a traffic information system.

According to Putra and Afifah [7], Background Subtraction is a process to get a foreground object (an observed object) from a series of images on digital video. Background Subtraction is used to distinguish between the foreground (object) and the background. Background Subtraction uses the current image comparison method with background images to detect a movement. The process of working Background Subtraction can be seen in Fig. 1.

According to Andrew et al. Background Subtraction has a weakness in the video that conditions outside the room such as rain, wind that makes the video unstable and changes in lighting.

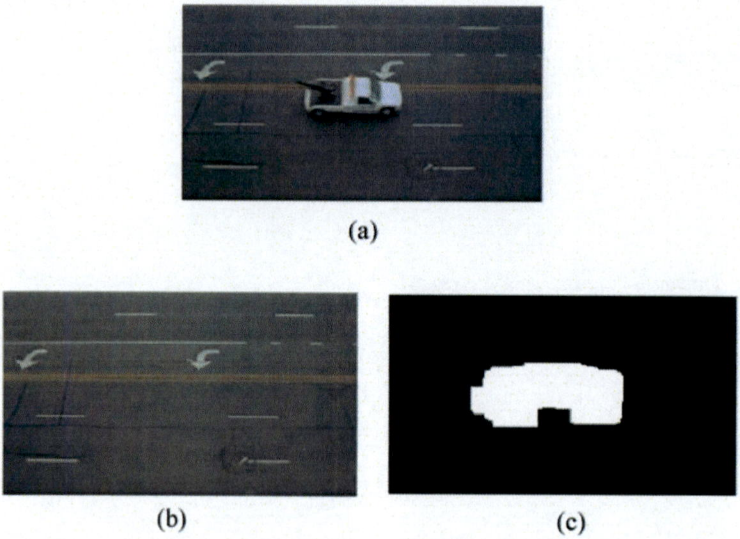

Fig. 1 **a** Object, **b** *background*, and **c** *foreground* object

1.5 Gaussian Mixture Model (GMM)

According to Indrabayu et al. [4], the Gaussian Mixture Model (GMM) is one of the popular methods of detecting and counting vehicles. GMM is one of the methods in background subtraction. According to Romadhony et al. [9], this method is used to describe the pixels of a background. Each pixel will be given the functions of a Gaussian component by GMM, with inputs in the form of pixel colors where GMM models are formed based on time. Two main components will be formed from the model, namely the background model with the foreground model. The background model is a model that reflects the background of an area being observed, while for the foreground model is a model that reflects observable objects.

Bounding Box Method

Bounding Box Method uses bounding boxes to mark each tracked object that limits the foreground with different colors to distinguish objects [8]. The dimensions of the object obtained from the results of the selection will be marked as a Bounding Box with the aim of knowing the pixel coordinates of the object that has been detected in the frame.

According to Sadewo et al. [10], the purpose of the bounding box is to find out the pixel coordinates of the objects detected in each frame. In this method, the image will be visualized through the dimensions of the object that the previous image has produced. This Bounding Box will determine the type of vehicle and the number of vehicles that cross the counter line will add the number of each Bounding Box that passes through it.

1.6 Java Programming Language

Java is an object-oriented programming language with elements such as C++ and other languages that have libraries that can be suitable for the internet environment [6]. As one of the libraries used in this study is the OpenCV libraries. This Java programming language can do many things in programming. To develop the Java programming language, an Integrated Development Environment (IDE) is required. In this study, the IDE used for development in Java is Intellij IDEA.

2 Research Methods

This study aims to create a vehicle detection program and also classify detected vehicles, this program is the result of a modification of the KS Traffic Analyzer program obtained on GitHub. The calculation results from the program will be compared with the results of manual calculations to get the accuracy and percent errors. The location

Fig. 2 Drone DJI Mavic Air

used as a place of research for taking traffic video data is on the road outside the city located on Jalan Brawijaya, Tamantirto, Kasihan, Bantul, and Yogyakarta.

In this study, the implementation of data collection in the form of traffic videos was carried out during the morning of April 4, 2019, at 06.30–09.00 WIB and in the afternoon of April 7, 2019, at 16.00–18.00 WIB. The program modification is carried out from November 2018 to April 2019. The research data consists of traffic videos on the streets outside the city and programs that will be modified which are obtained by downloading on an open source software project site, GitHub. The modified program is the name of the program display, modification of the classification based on MKJI 1997 on the program and modification of the pixel area dimensions of the vehicle in the program. The tools used in this study are camera drones as shown in Fig. 2 for retrieving traffic video data, Java Version 8 Update software, IntelliJ IDEA Community Edition 2018.3.2 software and computers.

2.1 Program Analysis Process

The first step taken in the process of analyzing this program is the input video file that will be used, then the program will read the frame on the video input to initialize the foreground and background using the Gaussian Mixture Model (GMM). The next step is the vehicle classification process using the bounding box method using the pixel area on the object. The results of this process are vehicle classification calculation data. The process of program analysis can be seen in Fig. 3.

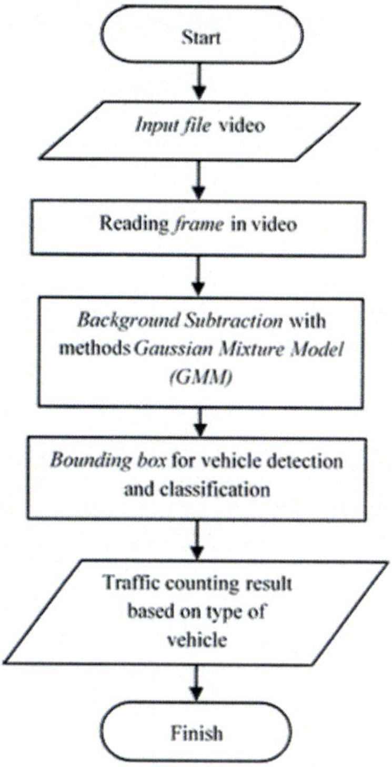

Fig. 3 Flowchart of process analysis program

3 Results and Discussion

3.1 Results of the Vehicle Detection and Classification Program

This program is a result of the modification of the KS Traffic Analyzer program obtained on GitHub. For the modified part, the classification of vehicles is based on MKJI 1997, dimensions of pixel area and several other names. In this study, a trial and error method was used to obtain accurate pixel area dimensions, pixel area dimensions, and conversion values from vehicle dimensions in this program can be seen in Table 2.

This study used four videos; there were two videos in the morning with a height of 6 m and 7 m and two videos in the afternoon with the same height of 6 m and 7 m. Videos are taken using a drone camera. Each video will be input into the program that has been made and then describe the line and speed of the video in the program. illustrating the video data collection at a height of 6 m can be seen in Fig. 4.

Each video was tested three times, and the results can be seen in Table 3 with videos of altitudes of 6 m and 7 m in the morning, Table 4 with videos of 6 m and 7 m in the afternoon.

Table 2 Dimensions of pixel area and value of conversion of rural street vehicles

Type of vehicle	Real volume	Dimension of vehicle (pixel area)		Conversion value	
		6 m	7 m	6 m	7 m
MC	1.76	0.2	0.15	0.113	0.085
LV	11.82	0.5	0.4	0.042	0.033
MHV	49.50	0.9	0.5	0.018	0.010
LB	100	1.4	0.6	0.014	0.006
LT	111.03	> 1.4	> 0.6	> 0.013	> 0.005

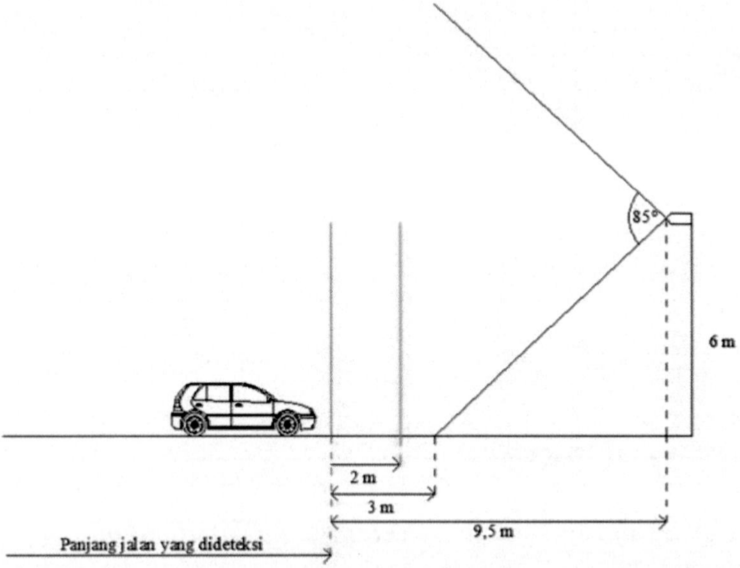

Fig. 4 Illustration of video data collection at 6 m height

Table 3 Vehicle calculation results in the program in the morning

Type of vehicle	Video height 6 m in the morning			Video height 7 m in the morning		
	Test 1	Test 2	Test 3	Test 1	Test 2	Test 3
MC	130	129	130	94	93	94
LV	16	16	16	19	19	19
MHV	4	4	4	2	2	2
LB	1	1	1	0	0	0
LT	0	0	0	5	5	5
Total	120	119	120	120	119	120

Table 4 Vehicle calculation results in the program in the evening

Type of vehicle	Video height 6 m in the evening			Video height 7 m in the evening		
	Test 1	Test 2	Test 3	Test 1	Test 2	Test 3
MC	71	71	71	19	20	21
LV	33	33	33	44	43	44
MHV	3	3	3	4	4	4
LB	4	4	4	2	2	2
LT	0	0	0	3	3	3
Total	111	111	111	72	72	74

Fig. 5 User interface of *U-COUNTER (rural area)*

The results of the research on making vehicle detection and classification programs are called U-COUNTER (Rural Area). The appearance of U-COUNTER (Rural Area) can be seen in Fig. 5.

3.2 *Manually Counted Results*

In this study, the results of calculations on the program will be compared with manual calculations which will then get the percent value of errors (errors) and the value of accuracy of the program. The results of manual vehicle calculations that have been done can be seen in Table 5.

Table 5 Manual vehicle counting results

Type of vehicle	Video in the morning		Video in the evening	
	6 m	7 m	6 m	7 m
MC	164	153	89	90
LV	28	31	54	62
MHV	4	5	5	4
LB	1	0	4	10
LT	0	5	0	3
Total	197	194	152	169

Table 6 Percent error value and accuracy

Video	Percent error (%)	Accuracy (%)
6 m in the morning	23.52	76.48
7 m in the morning	38.32	61.68
6 m in the evening	26.97	73.03
7 m in the evening	57.00	43.00

3.3 Discussion

Based on the results of the U-COUNTER program (Rural Area) and manual vehicle calculation, the percent value of the error is obtained, and the accuracy of the video is 6 m and 7 m in the morning and the video in the afternoon is 6 m and 7 m (Tables 1, 2, 3 and 4 in the Appendix). The reason for choosing a height of 6 and 7 m is because it adjusts to the height of the pole which is usually used for lighting and other utility functions. The percentage error value and accuracy can be seen in Table 6.

From the results of the percent error (error) and the value of accuracy, obtained the highest program accuracy is 76.48% in videos with a height of 6 m in the morning and for the lowest accuracy value found in the video height 7 m in the afternoon that is 43% The low value of accuracy in the 7 m video height in the afternoon is due to the congestion of traffic during video shooting and unstable weather. The density of this traffic resulted in the calculation of vehicles in the program becoming less effective, for example, what happened to the type of large bus vehicle (LB). As can be seen in Fig. 6, the density of traffic causes the vehicle that should be counted as a large bus (LB) to be a large truck (LT) because the calculated dimensions of pixels in the program are larger than large buses (LB). And less stable weather affects the calculation of the types of motorbike vehicles in this program, where this results in many undetected motorcycle vehicles.

Fig. 6 Error in program in detecting vehicles

4 Summary

The conclusions in this study are

(a) The vehicle detection program uses the Gaussian Mixture Model (GMM) method on out-of-town roads based on the 1997 MKJI classification named U-COUNTER (Rural Area).

(b) Conversion value from dimensions of vehicles with a video height of 6 m on motorbikes (MC) 0.113, light vehicles (LV) 0.042, medium heavy vehicles (MHV) 0.018, large bus vehicles (LB) 0.014, and large truck vehicles (LT) > 0.013. For videos with a height of 7 m the conversion value of a motorcycle (MC) 0.085, a light vehicle (LV) 0.033, a medium heavy vehicle (MHV) 0.010, a large bus vehicle (LB) 0.006 and a large truck vehicle (LT) > 0.005.

(c) The level of accuracy obtained in the video in the morning with a height of 6 m is 76.48% and an error of 23.52% and a video with a height of 7 m has an accuracy value of 61.68% with an error of 38.32%. For the video in the afternoon with a height of 6 m, the accuracy value is 73.03% and the error is 26.97% and the video with a height of 7 m is an accuracy of 43% with an error of 57%.

(d) The density of traffic and weather can affect the accuracy of this U-COUNTER (Rural Area) program.

References

1. Adistya, R., Muslim, M.A.: Deteksi dan Klasifikasi Kendaraan Menggunakan Algoritma Backpropagation dan Sobel. J. Mech. Eng. Mechatron. **1**(2) (2016)
2. Dewantoro, A.K., Tritoasmoro, I.I., Susatio, E.: Simulasi Dan Analisis Sistem Penghitung Kepadatan Lalu Lintas Dan Klasifikasi Kendaraan Berbasis Webcam Dengan Metode Background Subtraction. eProc. Eng. **2**(2) (2015)
3. Harwendhani, I.C., Ningrum, I.P., Sarita, M.I.: Sistem Pendeteksi Jumlah Mobil Dalam Intelligent Transport System (ITS) Menggunakan Metode Viola-Jones. semanTIK **2**(1) (2016)
4. Indrabayu, B., Achmad, A., Nurtanio, I., Mayasari, F.: Blob modification in counting vehicles using gaussian mixture models under heavy traffic. ARPN J. Eng. Appl. Sci. 7157–7163 (2015)
5. Memon, S., Bhatti, S., Thebo, L.A., Talpur, M.M.B., Memon, M.A.: A Video based vehicle detection, counting and classification system. Int. J. Image Graphics & Signal Processing **10**(9) (2018)
6. Nyura, Y.: Pembuatan Aplikasi Pembelajaran Bahasa Inggris Pada Handphone dengan J2ME. Informatika Mulawarman: Jurnal Ilmiah Ilmu Komputer **5**(3) (2016)
7. Putra, B.C., Afifah, Y.N.: Gaussian Mixture Model untuk Penghitungan Tingkat Kebersihan Sungai Berbasis Pengolahan Citra. Teknika: Engineering and Sains Journal **2**(1) (2018)
8. Rahmawati, L., Adi, K.: Rancang bangun penghitung dan pengidentifikasi kendaraan menggunakan Multiple Object Tracking. Youngster Phys. J. **6**(1) (2017)
9. Romadhony, A., Saidy, H.N., Sulistiyo, M.D.: Implementasi Metode Improved Adaptive Gaussian Mixture Model Background Subtraction dan Haar-Like Features untuk Menganalisis Status Kepadatan Kendaraan yang Melintas di Suatu Jalur pada Lampu Lalu Lintas. In: Indonesia Symposium on Computing 2015 (2015)
10. Sadewo, S.S., Sumiharto, R., Candradewi, I.: Sistem Pengukur Kecepatan Kendaraan Berbasis Pengolahan Video. Indonesian J. Electron. Instrum. Syst. **5**(2) (2015)
11. San Sayidul Akdam Augusta, Y.A., Sari, P.P.A.: Penentuan Jumlah Kendaraan Menggunakan Blob Detection dan Background Subtraction. Jurnal Pengembangan Teknologi Informasi dan Ilmu Komputer e-ISSN **3**(1) (2016)
12. Yuwono, M.J., Purnama, B., Sthevanie, F.: Penghitungan Kepadatan Kendaraan Di Jalan Tol Menggunakan Metode Gaussian Mixture Model Dan Kalman Filter. eProc. Eng. **2**(3) (2015)

Water Resources Engineering

Assessment of Water Security for a Sustainable Environment: An Indicator-Based Approach Applied in Penang, Malaysia

Rozi Abdullah, Rofiat Bunmi Mudashiru, and Nuridah Sabtu

Abstract When risks to water security increase it is necessary to enhance the management response. Water and climate change are closely related as water form a channel through which the effect of climate change can be experienced. This study has applied an indicator-based approach to establish a framework for assessing water security performance in Penang, Malaysia. The framework is targeted at supporting current contributions in attaining sustainable water security in the state. Analysis of the region's water security index was conducted and compared with international standards. The dimensions and indicators applied are based on the UN-water definition of water security, while the variables applied in the assessment were selected based on appropriate findings from the literature. A water security index ranging from a scale of 1–5 was obtained. The outcome of the assessment showed an overall good condition of water security in Penang with the potential to handle future changes for a sustainable and resilient environment. However, the findings have shown that there is room for improvement in water-related disasters, ecosystems preservation, and water governance to enhance the condition of water security in the state.

Keywords Sustainable · Penang · Water security · UN-water · Indicator-based · Dimensions

R. Abdullah · R. B. Mudashiru (✉) · N. Sabtu
School of Civil Engineering, Universiti Sains Malaysia, Gelugor, Malaysia
e-mail: bunmbello@gmail.com

R. Abdullah
e-mail: cerozi@usm.my

N. Sabtu
e-mail: nsabtu@usm.my

R. B. Mudashiru
Department of Civil Engineering, Federal Polytechnic Offa, Offa, Nigeria

N. Sabtu (ed.), *Proceedings of AWAM International Conference on Civil Engineering 2022 - Volume 3*, Lecture Notes in Civil Engineering 386,
https://doi.org/10.1007/978-981-99-6026-2_32

1 Introduction

Water security requires a dependable economic, environmental, and socially sustainable system while also tackling the diversification of associated risks to water [1]. It can be explained as a targeted aim for efficient water resources management by all related stakeholders [2]. Water security is required to be incorporated in the conceptualization of the Sustainable Development Goals (SDGs). The SDG system must address the goal and associated objectives of accomplishing water security as it will synthesize a variety of key developmental sectors been considered. These developmental sectors cover areas such as environmental sustainability, food and nutrition, health, energy, water, growth and employment, inequities, hunger, and conflict and vulnerability. Therefore, it is alright to point out that the contribution to water security is beneficial for visible immediate and sustainable impact on human development and economic growth [3]. The understanding of the word "nexus" in the context of Water, Energy, and Food (WEF) implies the inevitable interdependence between the three in such a way that the actions of the policies that make up the WEF sectors have an impact on each other and including on the ecosystems upon which all-natural and anthropogenic activities rely [4]. WEF is important for human wellness; it also represents an essential aspect in the reduction of poverty and the drive towards sustainable development. It also represents an integral part of the SDGs 2, 7, and 6. There are indications that in the next decades, there [5] will be an increase in the demand for WEF owing to an increase in population, urbanization, economic growth, change in diets, cultural diversity, technology development, international trade, and climate change [6]. Globally, 70–72% of water abstraction can be attributed to agricultural use, 16% to household use, and 12% to industrial use [7].

Water use includes agricultural production, food-processing supply chain, fishery processing, and forestry and as well as the production and transportation of energy in various forms [7, 8]. On the other hand, the food-processing supply chain takes up to 30% of total global energy usage [8]. The consumption of energy also cut across areas of food production, transportation, distribution, and in the water treatment, conveyance, and distribution. According to the report by [9], 90% of global power generation is obtained through the use of water. It was also noted in the world energy outlook report [10] that water abstraction for energy production will be increased by 20% and consumption by 85% mainly due to diversification towards higher efficiency in power production systems. Therefore, water security remains an integral park that links the web of the connection between food, energy, economic development, climate, human security problems that are projected to affect the world economy in the next two decades [2].

According to the report by the Department of Irrigation and Drainage (DID), Malaysia on WEF [11], the recent increase in per capita gross domestic product (GDP), the diversifying standard of living, population growth, and the total uncontrolled consumption in Malaysia is envisioned to increase demand of WEF nationally. If there are no relevant changes in the manner of production and consumption with population growth projection, there will be a need to increase agricultural and energy production globally by 70% and 50%, respectively. Such an increase will have a significant impact on water security [6, 12]. There are further threats associated with the projected climate change impact reaching 2050 that have indicated an adverse on the water supply security that is capable of reducing the quality of human wellness [13]. In the face of these challenges, there is a need to secure the WEF nexus through the approach of water security. Water security nexus aims at the adequate supply of the water resources through synergies empowerment and minimizing of trade-offs among the water sector [6].

The criteria that are appropriate for measuring water resource security in Malaysia include usage, availability, effectiveness, accessibility, reliability, and sustainability of the water resources [11]. There is no regional study on Penang's water security despite being under raw water pressure. Furthermore, there is also an increase in urbanization which adds more pressure on water security and linked to climate change projections. This current report aims to give a brief overview of these criteria as regards the current situation in Penang, Malaysia, and policies proposed in tackling the challenges of water security. In addition to this, the UN-water definition of water security is adopted as a framework applied in evaluating the water security to determine the performance of the water security index in Penang, Malaysia.

2 Water Security

SDG-6 addressed water security as access to potable water and sanitation. Water security is termed as accessibility to a sustainable and adequate amount of potable water for human wellness, livelihood sustenance, and socio-economic development to safeguard against water pollution, water-associated disasters, *and preserved ecosystem in a peaceful climatic and political environment* [14].

The major process in improving water security is to evaluate it. Several researchers and water organizations have made an effort to explain, design, and measure water security through different approaches as seen in works by [1, 2, 11, 15–21]. Earlier research in the area of water security was much focused on determining water security scope as applicable in *GWP* [22] as designed through accessible safe water supply at affordable cost, ecosystems protection, food security through water security, devising water wisdom, prioritizing water resources issues, and ensuring effective water governance. Recently, an improvement towards water security by *SWP* [23] focused more on the measurement of a context that was targeted at; water resources status, the development and achievements of water resources schemes and activities, and the functionality of water agencies in the management and operation of water resources. More recently, studies provide an assessment of water security through selected indicators as found in *Khan* et al. [16] where five dimensions and eleven water security were applied in the evaluation of water security of Islamabad. Similarly, *Babel* et al. [17] applied five dimensions and twelve water security indicators in analysing the water security index of Bangkok. Furthermore, a report by *SWP* [23] indicated that there have been attempts to synthesize water security indexes into individual scores with these trials conducted at a national scale through quantitative-based, qualitative-based measurements, or both. The national scale water security composite score rated between 0 and 100 was proposed by the Asian Water Development (ADB) with the consideration of five indicators which include resilience to water-related disasters, domestic, economic, urban, and environmental water security. Some past work on water security which describes the various approach in its implementation and assessment are summarized in Table 1.

Table 1 Past studies assessment approach for evaluating water security

Source	Scale	Issues	Framework for assessment	Findings
[21]	Basin	Water shortage and environmental issues	Integrated catchment management	Studies of catchment-scale hydrological processes, water use efficiency, integrated catchment management, governance, and market-based methods for water resources management
[20]	National	Complex trade-offs that characterize water management processes	Three hydrology types applying water and growth "S-curve"	Contributing to the development of both complementary institutions and infrastructure in a balanced approach
[5]	Basin	Water quality quantity in relation to aquatic ecosystems and human health	Water security status indicators	Incorporating research output into community water governance
[18]	Local	Water pollution and environmental deterioration due to extensive urbanization recently	Potable water security, flood mitigation security, and water environment security	Proposed a decision-making support model for water resource management
[19]	Local	Water pollution, flood disaster, and environmental deterioration	Flood mitigation security, economic development, water supply, and water environment security	The water security model based on policymakers and public input provides a platform for policies for improving issues of water security

2.1 Indicator-Based Approach

The building and application of an indicator-based system require the opinion of different people [24] or through benchmarks from research findings. This current study applies a set of indicators as a scoring tool for evaluating the performance of water security against a sustainable index. To effectively measure the progress that results in sustainable development, it is necessary to apply some principles such as stated by Pintér et al. [25]: (i) adequate motivating purpose, (ii) consideration of vital factors, (iii) appropriate scope, (iv) suitable framework and indicators, (v) transparence, (vi) adequate communications, (vii) extensive involvement, and (viii) adequate sustainability capacity. By adopting these principles, the level of uncertainty associated with the assessment can be decreased to a reasonable extent.

3 Methodology

Applying the UN-water definition of water security, the report of water security in Penang is given from four dimensions namely potable water supply and sanitation for all, water-related disasters and its resilience, ecosystems preservation and economic development, and water governance and policy responses for water security issues as illustrated in Fig. 1.

The methodological approach applied in this study is described in Fig. 2. The first step entails setting up the objectives required to obtain water security index (WSI). This involves determining the relevant factors that determines water security. The second step involves the identification of the dimensions required for the evaluation of the WSI and the possible indicators possible for the measurement of the selected dimensions. The third step involved obtaining the data related to the selected indicators to enable the WSI evaluation. The fourth step involved applying the WSI measurement reference with respect to each indicator on a scale of 1–5, while the last step involved analysing the data of the study area with the reference indicators benchmark scores from past findings.

3.1 Study Area

The study area is bounded by the longitudes of 100°12′45″ E to 101°07′50″ E and latitudes of 4°58′45″ N to 6°6′20″ N. It is situated in the north-western part of Peninsular Malaysia. Penang has its capital in George Town which is a modern city characterized by various business hubs, shopping malls, and skyscraper buildings among others. It covers an area of 1048 km^2 with an estimated population of 1,774,600 from the 2010 population census as of 2019 (Department of Statistics, Malaysia). The preliminary GDP per capita current prices for 2019 stands at 52,937(RM) and an

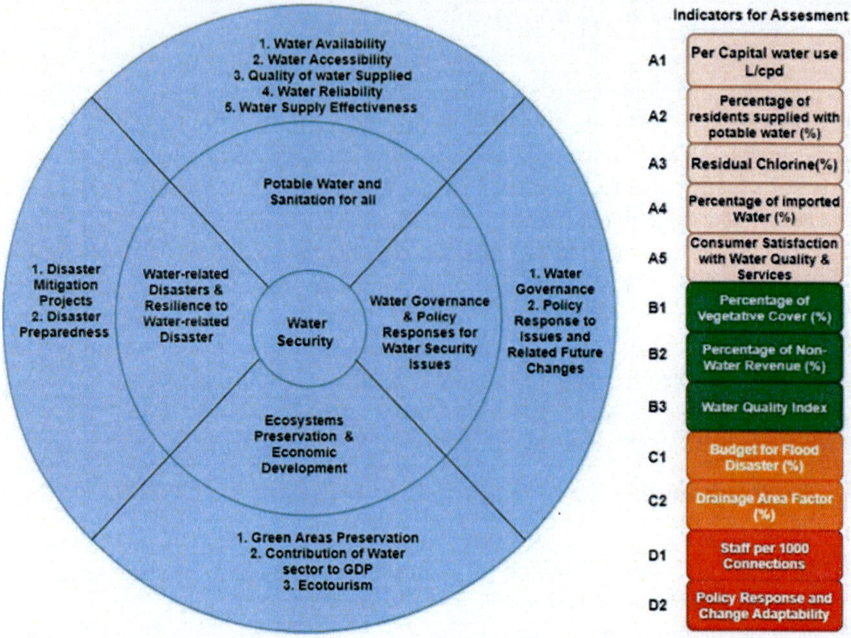

Fig. 1 Framework for assessment of water security in Penang, Malaysia

Fig. 2 Framework for assessment of water security in Penang, Malaysia

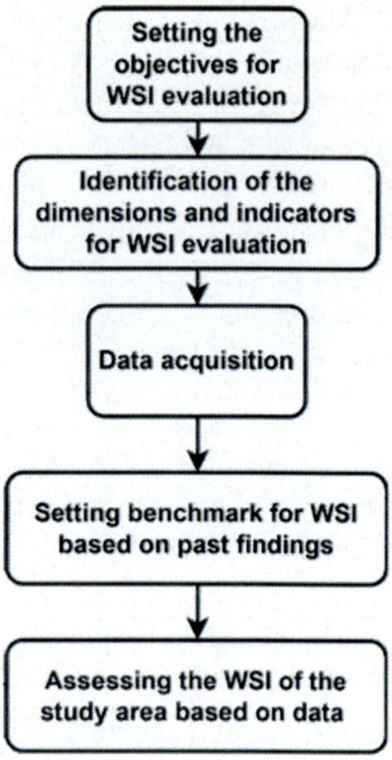

estimated GDP of 51,115(RM) for 2018. The entire water supply chain is governed by the Perbadanan Bekalan Air (PBA) Pulau Pinang (PBAPP) which gained the sole subsidiary ownership of PBA Holdings Berhad (PBAHB) in 2002 [26, 27]. The PBAPP handles raw water from point of extraction to distribution and billings.

The number of registered domestic and commercial water consumers in Penang in 2019 is 635,250. It has a total water catchment area spanning over 62.9km^2 with four available dams. The total storage water capacity reaches a volume of 107,106 million litres. There are currently nine water treatment plants in Penang at a design capacity of 1599 million litres/day each. Additionally, Penang has 61, 46, and 96 numbers of treated water reservoirs, treated water towers, and booster pump stations, respectively. Figure 3 shows the distribution of treatment plants and current dams' capacity in Penang.

3.2 *Framework for Assessing Water Security*

In this section, the dimensions that form the framework are discussed, and possible variables for measuring the respective dimension's indicator are discussed. The corresponding data for each indicator are obtained through secondary data of water resources and the economic development of Penang.

Potable water supply and sanitation for all

This dimension indicates the capability of the government to provide potable water for both consumptive and non-consumptive use while ensuring the hygiene of all its citizens. The indicators for this dimension are explained in dimensions B1–B3.

(a) **Water availability (A1)**

Water availability considers all types of water available for the daily activities of the community. The Perbadanan Bekalan Air (PBA) Penang is a private corporation responsible for water supply in Penang since 2002 [26]. According to its report (on water security towards 2050), the state of Penang accommodates an estimated population of 1.78 million people. The report stated that the state contributed a total of 96.9 billion in RM value to the GDP of Malaysia. The report indicates that there has been no case of water rationing in Penang between 1999 and 2019. The domestic per capita consumption is an indicator that can evaluate the availability of water. The domestic per capita between 2015 and 2019 is given in Table 2.

(b) **Accessibility (A2)**

This indicates the number of consumers that have access to potable water supply within the community. An indicator that is capable of measuring water accessibility can be explained through the percentage of the population served or through service area coverage through potable water supplied [16, 17]. The data extracted from the report of the road to privatization by the PBAPP [26, 29] show that in 2012 100% of urban and 99.7% of rural consumers were served with potable water. The

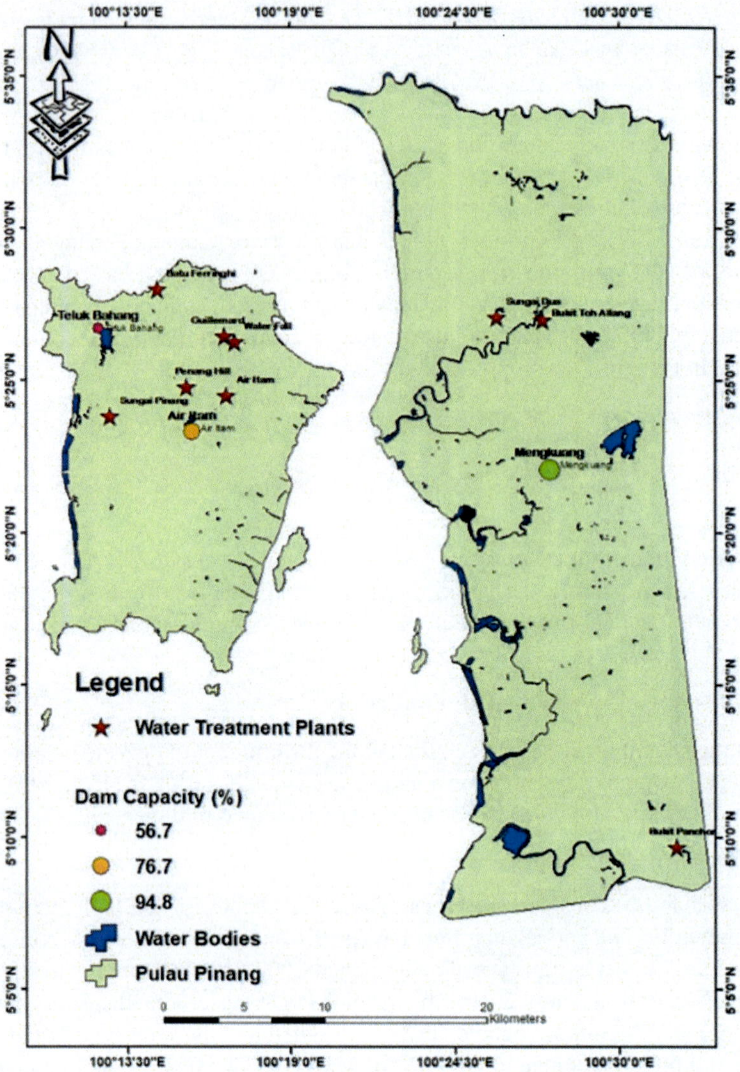

Fig. 3 Study area map showing some water supply system

Table 2 Per capita domestic consumption

1. Year	2. L/c/d
3. 2010	4. 255
5. 2016	6. 286
7. 2017	8. 276
9. 2018	10. 278
10. 2019	12. 281

Source * (PBA) Penang [28]

Table 3 Quality of treated water for 2017 (357 samples)

Parameters	Result	MOH standard
E-coli+ residual chlorine	0.00	0.05
Turbidity (NTU)	0.00	0.15
E-coliform bacteria	0.00	2.00
Residual chlorine	0.03	1.85
Aluminium	2.27	10.20

Source * [31]

consumers in Penang have water at a lower tariff in comparison to the average tariff of other member states in Malaysia according to [11, 28]. This has encouraged more consumption than any other state in Malaysia for 2018.

(c) **Quality of water supplied (A3)**

This is to ensure the standard of water supplied is up to the required standard by the World Health Organization. Water treatment plants usually perform quality tests on water supplied such as PH level, coliform count, turbidity, total chlorine, and residual chlorine among others. Although many indicators can be used to evaluate water quality, residual chlorine is a suitable indicator that measures the quality of water at the point of delivery to consumers [17]. Table 3 shows the result of selected parameters of water quality for 2017 from 567 samples. Also, a press statement released by the PBAPP in 2018 mentioned that all 3054 samples of treated water tested exceeded the Ministry of Health's (MOH) quality assurance parameter for E. Coli+ residual chlorine, E-coliform bacteria, turbidity (NTU), free residual chlorine, and aluminium [30].

(d) **Water reliability (A4)**

The PBA Penang report indicates that the state is under raw water pressure because of its small geographical presence. Although potable water supply has not been an issue in Penang since 1999, however, the emphasis has been made on the fact that about 80% of the raw water withdrawal is made from a single source which is the Sungai Muda. A summary of water production and consumption between 2010 and 2016 is given in Table 4. The average increment in the water production shows the reliability of the system in meeting the demand for water supply (average increment = 1.97) in the state. Additionally, in 2018, the state's average water consumption was reported to be 840 ML/day. The projection for consumption is expected to reach 1483 ML/day, 1696 Ml/day, and 1884 ML/day by 2030, 2040, and 2050, respectively. It however predicted that the single source of raw water cannot meet the demand for water beyond 2025 in the state. The PBAPP implemented some programmes as part of its obligation for continuous improvement of water supply service to its customers which include continuous maintenance and replacement of obsolete pipes, quick emergency repairs, rapid and completion of repair works. Currently, Penang does not rely on any external source for water supply. The percentage of water imported for use is an indicator that can be applied in measuring how reliable the water supply system of a state is [17].

(e) **Water supply effectiveness (A5)**

Water supply can be termed as effective/successful if it has availed more than 95% of its population with potable water, affordable water tariff, provision of water for the poor, and a daily supply of water [33]. The study by [34] **indicated a high efficiency** in the PBA Penang water supply system. The authors applied operating expenditure, the volume of non-revenue water, and length of the network as input and volume delivered, service area, and the number of connections as output for the determination **of the water supply efficiency. Additionally, the key performance indicators for the PBA Penang for 2012 have shown high efficiency in its water supply services as** indicated in Table 5. Additionally, Marina [35] stated that as custodians of the state's water resources, PBAPP's key performance indicators (KPIs) include 100% water supply coverage to urban areas and 99.8% supply coverage in rural areas, supply of water on a round-the-clock basis with no incidence of water rationing or disruption, and treated water quality to be one of the best in Malaysia based on the test results for 3150 treated water samples taken in 2019. An important indicator applied in measuring water supply effectiveness is through the consumers' satisfaction. Table 6 also shows the rating of water quality and the PBAPP's service by 1650 domestic and 130 industrial consumers for 2012.

Ecosystems Preservation and Ecotourism for Economic Activity Development

This dimension explains the ability of the state to preserve and maintain its water bodies, resources, and ecosystem. The indicators for this dimension are explained in dimensions B1–B3.

(a) **Green Area Preservation (B1)**

The PBAPP in its effort to enhance water security through the Penang state government has implemented the Water Supply (Catchment Area) Order 2004 which

Table 4 Annual variations in water production and consumption between 2010 and 2016

Year	Production (Million Litres)	Increase in Production (%)	Consumption (Million Litres)	Increase in consumption (%)
2010	349,324	4.8	287,584	5.9
2011	347,123	– 0.6	283,159	– 0.8
2012	352,762	1.6	290,625	2.6
2013	360,709	2.3	295,109	1.5
2014	363,198	0.7	296,898	0.6
2015	370,286	2.0	296,662	2.0
2016	385,241	4.1	302,627	2.0
Mean	361,235	2.13		1.97

Source * (PBA) Penang, adapted from [32]

Table 5 Key performance indicator of PBA Penang (2012 and 2013)

Key performance indicator	Penang 2012	Penang 2013	Average in Malaysia
Urban population served (%)	100	100	96.9
Rural population served (%)	99.7	99.7	90.7
Pipeline network density (km/km^2)	3.93	3.93	1.65
Domestic population served (km/pipeline)	391	–	206
Average connections by an employee (number)	445	451	354
Non-revenue water (%)	17.6	18.2	36.4
Average domestic water tariff (RM/m^3)—initial use 35m^3	0.31	0.31	0.66
Average industry water tariff (RM/m3)—initial use 500m3	1.19	1.19	1.45

Source * [26]

Table 6 Water quality and service rating for 2012

Consumer type	Tap water quality					
Rating	Excellent	Good	Average	Poor	Very Poor	Total
Domestic	10	59	21	9	1	100
Industrial	11	69	11	7	2	100
	Service					
Domestic	16	65	16	2	1	100
Industrial	11	77	12	–	–	100

Source * [26]

prohibits possible destructive commercial activities within an area (up to 62.9km^2 proximity) of green and forest areas that have been identified as catchment areas. These green and preserved areas are now set aside as eco-friendly zones for activities such as jungle trekking, hiking, camping, sightseeing, and the creation of minimal-impact ecotourism initiatives that enable favourable human interaction in a natural environment. The order was further improved in 2009 to allow eco-friendly and ecotourism schemes in the downstream areas. The preservation of the green areas helps keep the environment in its natural state for different species of plants and animals which in turn promotes the sustainability of potable water supply and ecotourism [26]. The percentage of vegetative cover is an appropriate indicator in evaluating the preservation of the ecosystem. The land use in Penang as provided by Penang Geographic Information System Centre (PeGIS) for 2019 indicates 51.23% are covered by forest area, 42.80% by urbanized area, 3.99% by bareland, 1.03% by cultivated area, and 0.94% by waterbodies. This indicates that the area consists of 30.5% non-vegetative cover and 69.5% vegetative cover [36].

(b) **Contribution of the water sector to Penang's GDP (B2)**

The development revenue report extracted from the Penang State Financial Statement, 2016, and Penang State Budgets, 2017 and 2018, indicated an estimated RM 20.5 and 20.0 million respectively as revenue collection for development projects such as the sale of raw water. Furthermore, the GDP growth rate for the water sector showed an increase from 5.8% in 2015 to 6.7% in 2016 and 6.0% in 2017 [37]. Non-revenue water is an indicator capable of measuring the economic value of water [16] as effective management of NWR is very important in any water-stressed state. PBAPP is targeting a low percentage of NRW in all its sectors of the water supply chain [26]. In 1999, the level of NRW in Penang was as much as 23.9% of the total water produced. After the introduction of the NRW reduction programme by the PBAPP, the NRW levels were maintained at values less than 20% for up to six years. Strategies for NRW reduction management such as active leakage control, water pressure management, pipe repair and asset management helped the PBAPP to reduce the level of NRW to 18.4% in 2011 [38]. The value of NRW for 2012 is presented in Table 5, and according to Jones et al. [39], NRW was 18.2%, 18.3%, 19.9%, 21.5%, and 21.9% in 2013–2017, respectively, for Penang.

(c) **Ecotourism (B3)**

Ecotourism is an area of tourism that has potential economic opportunities such as employment opportunities for boatmen, nature guides, food chain supply operators, homestay, hotel facility operators, and most importantly a source of income for residents. Even though ecotourism comes with the above-mentioned opportunities, it is important to bear in mind that a balance must be struck between economic benefits and preservation of nature to promote ecosystem sustainability [40]. Penang houses various nature locations which are operational such as the Penang Hill, Habitat, Botanical Gardens, Entopia, Penang National Park, Escape, and the bird watching at Teluk Air Tawar-Kuala Muda [37]. According to Vaghefi [41], the total income for homestays of 4512 tourists amounted to RM401, 117. Also, approximately 78% of green areas are preserved as Permanent Reserved Forests which are managed sustainably for a potential source of income. Various indicators have been applied in measuring ecotourism environmental protection [42] such as ecological environmental quality, planning among others, but water quality has proven to be an important indicator [43]. This is because activities related to ecotourism at any location require a measure of anthropogenic activity. According to the Department of Environment, Penang, the overall river basin water quality can be categorized as moderately polluted with the water quality index slightly increasing by 0.8% between 2015 and 2016 [37]. The water quality index (WQI) of 11 rivers in Penang Island indicated a mean value of 54.67 (Class III) [44, 45] for 2012–2013.

Water-related Disasters and Resilience

Flood and drought are extreme events and are increasingly related to climate change and global warming. Drought is not as an issue in Penang as the average annual rainfall ranges in 2670 mm, but on the other hand, the flood is a major recurring problem in Penang with the worst ever occurrence in November 2017 [46]. The resilience of a community to flood disasters can be measured through its disaster mitigation projects and preparedness.The indicators for this dimension are explained in dimensions C1–C2.

(a) **Disaster Mitigation Projects (C1)**

The government of Penang has implemented several projects to address flooding through the DID and the Majlis Bandaraya Pulau Pinang (MBPP) following the flood mitigation report from the Japan International Cooperation Agency (JICA) between 1995 and 2010. Some of the details of such projects are listed in Table 7, while an extended report can be found in the masterplan report of the DID Penang for the Timur Laut for 2018 [46].

The budget stipulated for flood mitigation indicates how much the state government is investing to control and minimize the impact of flooding. According to the data extracted from the Penang State budget for development expenditure by four topmost selected development projects between 2016 and 2018, there was a 25.9% increase between the budget for flood mitigation of 2016 and 2017. Table 8 shows the budget allocation between 2016 and 2018 for flood mitigation.

(b) **Disaster Preparedness (C2)**

Some of the key aspects in flood preparedness are the development of flood hazard or risk map, prediction and early warning system awareness, public awareness, disaster relief planning and management, guidelines, and standard operating procedures [17]. The disaster and risk reduction management of Penang has proposed four ways to enhance disaster management, and they include public awareness and education, an enhanced early warning system, creating of a disaster management department at the state level, and a better land-use planning system [47]. The number of flood risk maps for the entire Penang is not ascertained, although flood hazard, damage, and risk maps are available for the Timur Laut area of Penang [46].

Table 7 Some selected flood mitigation projects within Penang

Flood mitigation project	Catchment area
Pond construction with a complete pump and curb drainage system	S10 and S18
Construction of river diversion system	Sungai Jelutong
Drainage construction	Jalan Gurdwara linking S18, Jalan Perak, and Jalan Peel
Pond construction	Sungai Dondang
Preparations of diversion drawings	Sungai Air Terjun Diversion to Sungai Babi
Improvement works	Sungai Gelugor, Dua, Air Itam, Dondang, and Pinang

Source * [46]

Table 8 Budget allocation for flood mitigation for 2016–2018

Development Projects	Budget (RM million)			Percentage Change (%)		
	2016	2017*	2018*	2016	2017*	2018*
DID	13.61	14.46	12.35	2.6	6.2	−14.6
Deepening and repair of rivers	3.13	4.55	4.35	12.8	45.5	−4.4
Flood mitigation	2.84	3.57	1.81	−42.2	25.9	−49.4
Total budget for development expenditure	307.46	887.44	684.97			

* = Estimated Budget
Source * Penang State Budget 2018 [37]

Water governance and current policies for addressing water security challenges

This dimension explains how the state can manage its water sector and handle projected changes. The indicators for this dimension are explained in dimensions D1–D2

(a) **Water Governance (D1)**

Water governance plays a key role in solving water management issues. The PBAPP has been rated as one of the most successful private-owned water supplying bodies in Malaysia, and it is worthy to note that it is involved in the entire process of the water supply chain business [33]. According to the water security assessment report by Babel et al., Khan et al. [16, 17], water governance can be measured by indicators such as institutional factors, awareness and support for water security, staff productivity, and adaptability factors. Indicators that are capable of measuring water governance and its sustainability include NRW, staff per 1000 connection, and proportion of metered consumers [48]. In the case of this study, staff productivity (staff/1000 m connections) is applied as the indicator in measuring water governance as applied in the study by [16].

(b) **Policy Response to Issues and projected Future Changes (D2)**

The PBA Penang has listed five objectives to enable sustainable water security for the state of Penang with the mindset that demand will increase in correlation with the projected population increase and the immediate stress on the raw water source. The five objectives include

(i) Implementing holistic non-revenue water management
(ii) Raising/reviewing the value of water (tariffs and water conservation surcharge—WCS)
(iii) Increasing public awareness and participation
(iv) Promoting water saving devices (WSDs)
(v) Developing an additional water resource.

In 2009, the (PBA) Penang authorized the master plan study for potable water supply in the state till 2050. One of the key reports of the master plan study was that **the Sungai Muda will not be sufficient to supply the required water demand beyond** 2025. In May 2019, a press statement was released by the CEO of the PBAPP on how to tackle water key supply issues in the state [28]. The three major issues listed were high per capita domestic consumption rate, high subsidized rate of domestic water supply, and ways to get an additional raw water source. The proposed policies to address these key issues stated were

(i) The increase in per capita consumption has to be reduced as Penang's per capita consumption was the highest in Malaysia: The Federal Government is working on various policies to meet the targeted average consumption of 180L/c/day by 2025.
(ii) High subsidized rate of domestic water supply: water tariff review is in progress to encourage water preservation. The tariff setting procedure is awaiting the government's approval.
(iii) Ways to get an additional raw water source: there is a plan to obtain the solution from the Sungai Perak Raw Water Transfer Scheme (SPRWTS) which commissioning is planned for 2025. If successful, this will ensure raw water supply security till 2050.

In addition to this, the government of Penang is proposing a nature-based climate adaptation programme for the urban areas of Penang Island to improve urban resilience, health, and ecosystem sustainability towards impacts from climate change such as extreme weather events. This adaption programme is targeted at enhancing **storm water management to decrease flooding and improving the regulation of other** climatic conditions and was endorsed in January 2020 by the Ministry of Energy, Science, Technology, Environment and Climate Change (MESTECC) [49, 50].

4 Results and Discussion

The performance of the four dimensions and the twelve indicators used in explaining water security in Penang is conducted by adopting the water security index scale in a similar study by *Babel* et al. [17]. The study created reference normalized values for ranking each indicator as presented in Tables 9 and 10.

The respective performance rating was conducted using available data and reference values for each indicator. The performance of water supply and sanitation for all indicates an overall high score of 5, while some of the indicators for water-related disaster, water governance, and ecosystem preservation had values below 5 as shown in Fig. 4. This might be due to the effort put into these indicators by the government as extracted from the data applied in this study. This is indicative of a need to improve in these sectors especially water-related disasters to enhance the water security of the state. The overall findings show a good performance for all the dimensions applied in the rating water security in the state.

Table 9 Explanation of water security index (WSI)

Normalized performance score	Water security condition	Linguistic description of water security condition	Reference
< 1.5	Poor	It lacks an appropriate water governance system, has no plans for future changes, and has various water security issues	[17]
1.5–2.4	Moderate	It is characterized by water some water security challenges, requires improvement in its water governance system and how to plan for future changes	
2.5–3.4	Good	It is relatively water-secured concerning most dimensions, has a fair water governance system that can plan for some future changes	
3.5–4.4	Very Good	It is quite water-secured concerning most dimensions, has a good water governance system that can plan for some future changes	
≥ 4.5	Excellent	It is highly water-secured concerning most dimensions, has no water security issues, and has a very good water governance system that is fully capable of tackling issues and future changes	

5 Conclusion

This study aims to give an overview of water security in Penang by applying the UN-water definition for building the framework to access its performance. The framework applied was developed to support practical contributions towards achieving sustainable water security in Penang but not as a tool for setting standards. The assessment carried out is relevant in the sense that great effort is required at the state level to enhance water security at the national level. The findings are targeted at describing the situation of water security in Penang, and this has shown that there is room for improvement in tackling water-related disasters. The findings from the study show that water security is gaining necessary attention and sustainable measures are

Table 10 Performance of water security in Penang

Dimension	Indicator	Variable for assessment	Available data						Reference values					WSI performance	Source
			2012	2015	2016	2017	2018	2019	1	2	3	4	5		
Potable water supply and sanitation for all	A1	Per capita water use (L/cpd)	–		286	276	278	281	< 20	21–50	51–90	91–100	> 100	Excellent—5	[17, 51]
	A2	Percentage of population with access to potable water (%)	99.85		–	–		99.80	< 60	60–70	70–80	80–90	> 90	Excellent—5	[15, 17]
	A3	Residual chlorine %	–		–	100			≤ 60	61–70	71–80	81–90	91–100	Excellent—5	[16, 52]
	A4	Percentage of imported water (%)	0	0	0	0	0	0	> 50	30–50	10–30	0–10	0	Excellent—5	[53]
	A5	Customer satisfaction with water quality and services (%)	1% rated very poorly	–	–	–			Applied Likert Scale form survey feedback					Excellent— ≥ 4.5	[17]
Ecosystems preservation and economic development	B1	Percentage of vegetative cover (%)						51.23	< 10	10–20	20–35	35–50	> 50	Excellent—5	[17, 41]

(continued)

Table 10 (continued)

Dimension	Indicator	Variable for assessment	Available data						Reference values					WSI performance	Source
			2012	2015	2016	2017	2018	2019	1	2	3	4	5		
	B2	Percentage of non-revenue water (%)	17.6	19.9	21.5	21.9		22.9	> 25	25–20	20–15	15–10	< 10	Good	[16]
	B3	Water quality index	54.67	54.67					< 60	60–70	70–90	90–100	100	Good	[17, 42–44]
Water-related disasters and resilience	C1	Budget for **flood** disaster (%)			1.94	0.91	0.9		0	0–1	1–5	5–10	> 10	Moderate, Good	[17]
	C2	Drainage factor						51.23	< 10	10–20	20–35	35–50	> 50	Excellent—5	“
Water governance and current policies for addressing water security challenges	D1	(Staff/1000 connections)	2.24, 2.21						< 3	3–4	4–5	5–6	7–8	Poor—1.5	[16, 48]
	D2	Ability to adapt to future changes	Nature-based climate adaptation programme, Penang Green Agenda 2030, and sustainable water supply 2050											Excellent— ≥ 4.5	Expert opinion

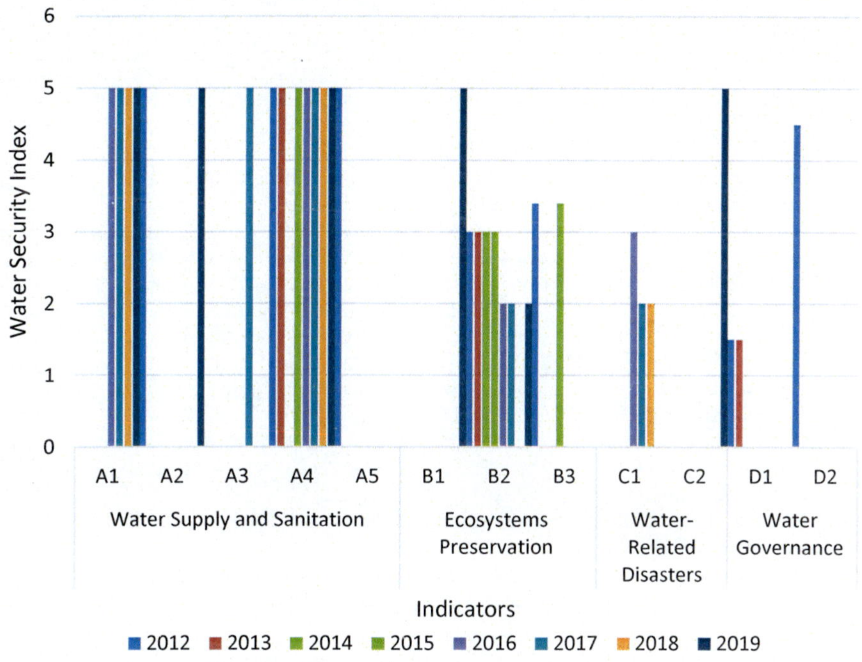

Fig. 4 Performance of water security for Penang based on available data

being addressed to tackle future scenarios. This is quite commendable because water security is a bedrock for a thriving and sustainable city.

One of the significant urban area weather-related issues in Penang, Malaysia, is flood-related disasters resulting from progression in urbanization over recent years in Penang, Malaysia. This challenge has been addressed in the water security analysis in this study. The result from the study has additionally shown a high index in water supply and sanitation and above-average performance in ecosystem preservation. The effective implementation of the framework in this study makes a need to study for other regional parts of Malaysia to give a robust framework and report of the national water security. It will also create a medium for understanding the water security challenges to support practical contributions at the state and national level for a sustainable and resilient environment. A proper qualitative investigation on water governance is a recommendation for future studies to present a robust situation of water security in the region.

References

1. Marttunen, M., Mustajoki, J., Sojamo, S., Ahopelto, L., Keskinen, M.: A framework for assessing water security and the water-energy-food nexus-the case of Finland. Sustain **11**(10), 1–24 (2019). https://doi.org/10.3390/su11102900
2. WEF: Water Security: the Water-Food-Energy-Climate Nexus: the World Economic Forum Water Initiative (2011). https://doi.org/10.5860/choice.49-1001
3. UN-Water: Water Security & the Global Water Agenda : A UN-Water Analytical Brief (2013)
4. Strasser, L., Stec, S.: Transboundary Diagnostic Analysis. Thematic Report on the Resource Nexus (Phase I of the Water-Food-Energy-Ecosystems Nexus Assessment of the Drin Basin). GEF Project "Enabling Transboundary Cooperation and Integrated Water Resources Management in the Ext," 12, Kyrristou str., 10556 Athens, Greece (2020)
5. Norman, E.S., Dunn, G., Bakker, K., Allen, D.M., de Albuquerque, R.C.: Water security assessment: integrating governance and freshwater indicators. Water Resour. Manag. **27**(2), 535–551 (2013). https://doi.org/10.1007/s11269-012-0200-4
6. Hoff, H.: Understanding the Nexus. Background paper for the Bonn2011 Nexus Conference. In: Bonn2011 Conference The Water, Energy and Food Security Nexus Solutions for the Green Economy, no. November, pp. 1–52 (2011)
7. UN-Water: Summary Progress Update 2021 : SDG 6—Water and Sanitation for All, Geneva, Switzerland (2021)
8. FAO: The state of the World's land and water resources for food and agriculture (SOLAW)—managing systems at risk, Earthscan. Food and Agriculture Organization of the United Nations, Rome and Earthscan, London (2011)
9. UN-Water: Water and Energy, Paris (2014). https://doi.org/10.1016/S0007-1536(78)80057-6
10. IEA: World Energy Outlook 2012. International Energy Agency (2012)
11. Hezri, A.: An Overview Study of Water-Energy-Food Nexus in Malaysia (2018)
12. Pluschke, L., Duboi, O., Flammini, A., Puri, M.: Environment and Natural Resources Management Working Paper in the Context of the Sustainable Energy for All Initiative Walking the Nexus Talk : Assessing the Water-Energy-Food Nexus (2013) [Online]. http://www.fao.org/3/a-i3959e.pdf
13. FAO: Energy-smart food for people and climate. Issue Paper. Food Agric. Organ. United Nations 66 (2011)
14. UN-Water: What is water security? What is Water Secur. (2013) [Online]. http://www.unwater.org/publications/water-security-infographic/
15. ADB: Asian Water Development Outlook 2013: Measuring Water Security in Asia and Pacific (2013). [Online]. https://www.adb.org/sites/default/files/.../asian-water-development-outlook-2013.pdf
16. Khan, S., Guan, Y., Khan, F., Khan, Z.: A comprehensive index for measuringwater security in an UrbanizingWorld: The case of Pakistan's capital. Water (Switzerland) **12**(1) (2020). https://doi.org/10.3390/w12010166
17. Babel, M.S., Shinde, V.R., Sharma, D., Dang, N.M.: Measuring water security: A vital step for climate change adaptation. Environ. Res. **185**(November 2019), 109400 (2020). https://doi.org/10.1016/j.envres.2020.109400
18. Yang, F., Shao, D., Xiao, C., Tan, X.: Assessment of urban water security based on catastrophe theory. Water Sci. Technol. **66**(3), 487–493 (2012). https://doi.org/10.2166/wst.2012.182
19. Shao, D., Yang, F., Xiao, C., Tan, X.: Evaluation of water security: an integrated approach applied in Wuhan urban agglomeration, China. Water Sci. Technol. **66**(1), 79–87 (2012). https://doi.org/10.2166/wst.2012.147
20. Grey, D., Sadoff, C.W.: Sink or swim? Water security for growth and development. Water Policy **9**(6), 545–571 (2007). https://doi.org/10.2166/wp.2007.021
21. Xia, J., Zhang, L., Liu, C., Yu, J.: Towards better water security in North China. Water Resour. Manag. **21**(1), 233–247 (2007). https://doi.org/10.1007/s11269-006-9051-1
22. GWP: Towards water security: a framework for action. Glob. Water Partnersh. (2000)

23. SWP: Monitoring the improvement of water. Sustain. Water Partnersh. (2018) [Online]. https://www.globalwaters.org/resources/assets/swp/monitoring-improvement-water-security-swp-toolkit-6
24. Zhou, Y., Yi, P., Li, W., Gong, C.: Assessment of city sustainability from the perspective of multi-source. Sustain. Cities Soc. **70**(March), 102918 (2021). https://doi.org/10.1016/j.scs.2021.102918
25. Pintér, L., Hardi, P., Martinuzzi, A., Hall, J.: Bellagio STAMP: principles for sustainability assessment and measurement. Ecol. Indic. **17**, 20–28 (2012). https://doi.org/10.1016/j.ecolind.2011.07.001
26. Maidinsa, J.: Perbadanan Bekalan Air Pulau Pinang public listing corporatisation, the road to & beyond. Ingenieur **3**(March), 54–65 (2014). https://doi.org/10.1525/9780520941229
27. Weng, C.N.: The Perbadanan Bekalan Air Pulau Pinang Sdn Bhd (PBAPP): a good example of corporate social responsibility of a private water company in Malaysia Chan Ngai Weng. Seminar (2002)
28. Barakbah, S.N.b.S.F.: Penang Has To Tackle 3 Key Water Supply Issues, pp. 4–7 (2019) [Online]. http://www.pba.com.my/pdf/news/2019/30052019_PBAHB_AGM_PR3A.pdf
29. Academy of Science Malaysia: Study on the Current Issues and Needs for Water Supply and Wastewater Management in Malaysia (2015) [Online]. http://www.akademisains.gov.my/download/StudyontheCurrentIssuesandNeedsforWaterSupplyandWastewaterManagementinMalaysia-volume2.pdf
30. Maidinsa, J.: PBAPP Receives IKM 'Laboratory Excellence Award' for the 2nd Consecutive Year (2018)
31. PBAPP: Annual Report 2017 (2017)
32. Penang Institute: Penang Statistics Quarter2, 2017, 10 Brown Road, 10350 Penang, Malaysia (2017)
33. Chan, N.W.: Issues and challenges in water governance in Malaysia. Iran. J. Environ. Heal. Sci. Eng. **6**(3), 143–152 (2009)
34. Munisamy, S.: Efficiency and ownership in water supply: evidence from Malaysia. Int. Rev. Bus. Res. Pap. **5**(6), 148–260 (2009)
35. Marina, E.: Penang a fine example of how to manage water resources well. New Straits Times 8–9 (2020). https://www.nst.com.my/opinion/columnists/2020/09/622859/penang-fine-example-how-manage-water-resources-well
36. Raja, R.B.R.S.: Preliminary Environmental Impact Assessment for Proposed Residential Development at Solok Tanjung Bunga on Lots 4181, 4182 and 4183, Bandar Tanjung Bungah, Daerah Timur Laut, Pulau Pinang (2015)
37. Penang Institute: Penang Economic and Development Report (2019)
38. Lai, C.H., Chan, N.W.: A Study of Non-Revenue Water Management in Penang as an Example of Good Water Governance, no. March 2014 (2013)
39. Jones, L.J.N., Kong, D., Tan, B.T., Rassiah, P.: Non-revenue water in Malaysia: Influence of water distribution pipe types. Sustain **13**(4), 1–16 (2021). https://doi.org/10.3390/su13042310
40. Pociovalişteanu, D.M., Niculescu, G.: Sustainable development through eco-cultural tourism. Eur. Res. Stud. J. **13**(2), 149–160 (2010). https://doi.org/10.35808/ersj/280
41. Vaghefi, N.: Ecotourism: A Sector where Sustainability is Everything (2019)
42. Wang, L.E., Zhong, L., Zhang, Y., Zhou, B.: Ecotourism environmental protection measures and their effects on protected areas in China. Sustain **6**(10), 6781–6798 (2014). https://doi.org/10.3390/su6106781
43. Ken Darrie, T.: Ecotourism and Water Quality: Linking Management, Activities and Sustainability Indicators in the Caribbean, vol. 3424496, p. 393 (2010) [Online]. http://search.proquest.com/docview/755720586?accountid=178282
44. Nurul-Ruhayu, M.-R., An, Y.J., Khairun, Y.: Detection of river pollution using water quality index: a case study of tropical rivers in Penang Island, Malaysia. OALib **02**(03), 1–8 (2015). https://doi.org/10.4236/oalib.1101209
45. Andrews, K.: The water quality index (Wqi) in Penang Island. J. Water Resour. (JWR) **2**(2), 17–20 (2019)

46. DID: Kajian Pelan Induk Sistem Saliran Mesra Alam Bagi Daerah Timur Laut, Pulau Pinang (Review of Eco-Friendly Drainage System Master Plan for Timur Laut area, Penang). Jabatan Pengairan dan Saliran Negeri Pulau Pinang—Department of Irrigation and Drainage, Penang (2018)
47. Penang Green Council: Penang Green Agenda 2030 Title: Green Economy, no. June (2020)
48. ESAWAS, "Regional Benchmarking of Water Supply and Sanitation Utilities 2016/2017 Report," 2018.
49. S. and C. C. (MESTECC) Ministry of Energy, Science, Technology, M. B. P. P. (MBPP), J. P. D. S. (JPS), and T. City: Proposal To The Adaptation Fund Part I : Nature-based climate adaptation programme for the urban areas of Penang Island (2020)
50. Mutiara, B.: Penang Climate Action Week kicks off | Buletin Mutiara (2019). https://www.buletinmutiara.com/. Accessed 09 Mar 2021 [Online]. https://www.buletinmutiara.com/penang-climate-action-week-kicks-off/
51. Howard, G., Bartram, J.: Domestic Water Quantity, Service Level and Health (2003). https://doi.org/10.1016/S0009-9260(98)80189-X
52. EPA: Parameters of Water Quality Interpretation and Standards (2001). https://doi.org/10.4315/0022-2747-35.12.715
53. UN-Water: Water, a shared responsibility: the United Nations world water development report 2, vol. 44, no. 05. United Nations Educational, Scientific and Cultural Organization (UNESCO), France (2006). https://doi.org/10.5860/choice.44-2732

Land Cover Mapping Based on Open-Source Data and Software: Kelantan Area Case Study

Raidan Maqtan, Faridah Othman, Wan Zurina Wan Jaafar, and Ahmed Elshafie

Abstract Land cover map preparation is essential for catchment management, hydrological analysis, and flood mapping. The process of deriving accurate land cover maps is challenging and requires a certain level of skill and experience. Cloud cover and the heterogeneity of the area under consideration increase the challenge of classification, especially in tropical areas. Recently, a wide range of free digital data and open-source software have been developed and become easily accessible. Using these data and software will help overcome many of the obstacles related to data availability and analysis in the fields of environment and water resources. In this study, the construction of the land cover map of Kelantan State in Malaysia and the accuracy assessment were performed using open-source data, and software has been demonstrated. Open satellite maps and the QGIS software were used for analysis with an overall accuracy of 91.46%.

Keywords Land cover · Open-source data · QGIS · Kelantan

1 Introduction

Natural land cover has changed substantially in the past centuries due to human activities but has increased significantly in the past two decades and is anticipated to increase continuously [1]. The human-caused changes in land cover and land

R. Maqtan (✉) · F. Othman · W. Z. W. Jaafar · A. Elshafie
Department of Civil Engineering, Universiti Malaya, 50603 Kuala Lumpur, Malaysia
e-mail: ridan76@yahoo.com

F. Othman
e-mail: faridahothman@um.edu.my

A. Elshafie
e-mail: elshafie@um.edu.my

R. Maqtan
Department of Civil Engineering, University of Aden, Aden, Yemen

N. Sabtu (ed.), *Proceedings of AWAM International Conference on Civil Engineering 2022 - Volume 3*, Lecture Notes in Civil Engineering 386,
https://doi.org/10.1007/978-981-99-6026-2_33

use are the major causes of degradation in the global environment. Furthermore, the main human-related sources of greenhouse gases in the atmosphere are deforestation and irrigation until the start of the industrial era of fossil fuel consumption [2]. The loss in natural forest cover proportionally increases with population growth and agriculture products [3]. Land cover and climate changes may lead to extreme weather-related disasters, such as flooding. Therefore, identifying and quantifying land cover is crucial to understanding the land surface characteristics and the consequences of its changes and performing hydrological and hydraulic analyses on a catchment or a floodplain area.

Peninsular Malaysia was totally forested in the late 1800s. Three decades ago, the forest cover became less than half of the original spread. Agriculture use, especially tree crops, has basically replaced part of the natural forests in Peninsular Malaysia. In 1988, approximately 83% of the area devoted to agricultural use was for oil palm, rubber, cacao, and coconut [4]. According to the Food and Agriculture Organization [5], the highest forest land cover changes from 1990 to 2015 were in the tropics. Particularly, Malaysia has the largest rate of deforestation among the tropical countries worldwide [6]. From 1990 to 2010, approximately 8.6% of Malaysia's forest land cover changed, with a deforestation rate of 96,000 ha/year [7].

Several studies discussed issues related to flooding and land cover mapping in different parts of Peninsular Malaysia, including Kelantan State. Pradhan et al. [8] recognized that the susceptibility of Kota Bharu (Kelantan State's capital) to flooding is medium–high through a logistic regression model and GIS. Pradhan and Youssef [9] developed a flood susceptibility map of Kelantan by using hydrologic and hydraulic models with GIS for different return periods. After the 2014 flood disaster, several studies conducted for Kelantan State focused on this severe flood event [10–15]. Land cover data and mapping are among the main pillars of nearly all these studies. However, the details provided on how the land cover map is constructed and assessed are insufficient. Additionally, studies using open-source data and software are very limited, especially for Malaysia. Therefore, this study aims to demonstrate in detail a simple method of land cover classification using open-source GIS software and the free remote sensing data on Kelantan State.

2 Study Area

Kelantan is a state in Peninsular Malaysia, located in the northeast between latitudes 4°30′–6°30′ north and longitudes (101°10′–102°40′) east, Fig. 1. Kelantan State occupies an area of 15,060 km^2, , which is approximately 4.45% of the total land area of Malaysia, with a population of 1.539 million. Kelantan State has ten districts, namely Kota Bharu, Pasir Mas, Tumpat, Bachok, Tanah Merah, Pasir Puteh, Kuala Krai, Machang, Gua Musang, and Jeli. The climate of Kelantan is tropical and rainy throughout the year, with an average annual rainfall of approximately 2500 mm and an average temperature of approximately 28 °C. The highest rainfall is usually recorded during northeast monsoon, which starts from late October until early January. Some

Fig. 1 Study area

studies reported that the average combined annual evaporation and interception loss is approximately 1200 mm, and the average annual runoff is approximately 1300 mm. The Kelantan basin is characterized by dense vegetation and a general soil thickness of 1 m (some areas may have a depth of 18 m) [9].

The northward downstream part of Kelantan, including the capital city Kota Bharu, is vulnerable to flooding due to several factors, including urbanization, geographical characteristics, and proximity to the South China Sea [8]. Floods occur nearly every year in this part of Kelantan during the northeast monsoon, which hits the eastern coast of Peninsular Malaysia. Among the recorded floods events in Kelantan are the flooding in 1993 (3500 evacuees and 2.2 million US$ in damages), 2003 (2 deaths, 2200 evacuees, and 1,460,000 US$ in damages), and 2004 (12 deaths, 10,400 evacuees, and 370,000 US$ in damages) [16]. The most recent severe flood event that hit the east coast of Peninsular Malaysia was in December 2014, with more than 230,000 evacuees from the Kelantan, Terengganu, and Pahang states in addition to those from other states that were slightly affected. This event caused 21 deaths and approximately 0.7 billion to 1 billion RM in damages, and the most affected region was Kuala Krai in Kelantan [13, 17].

3 Methodology and Data

One of the challenges of land cover mapping in the tropics using optical remote sensing images is the existence of dense clouds in most days of the year [18]. Synthetic aperture radar (SAR) remote sensing is an alternative for solving this problem. However, SAR images are limited in availability and coverage compared with optical remote sensing satellite images. In the present study, we used Landsat-8-Operational Land Imager (OLI)-Thermal Infrared Sensor (TIRS) for the land cover classification of the study area. Landsat-8-OLI-TIRS was launched in February of 2013, collecting "image data for nine visible, near-infrared, shortwave infrared bands, and two longwave thermal bands" [18].

The images are available at USGS Earth Explorer website (https://earthexplorer.usgs.gov/). The available Landsat-8 images for the last three years were examined, and the most recent images with the lowest cloud cover percentage were selected. The image acquired on March 8, 2020 over the study area has the lowest land cloud cover of only 3.22%. Different blocks of multiband Landsat-8 images were downloaded and preprocessed for analysis using the QGIS software [19]. QGIS is an open-source and free software, which is very useful and powerful in analyzing spatial data and geographical maps [20–26]. The Semi-Automatic Classification Plugin (SCP) [27] was used for pre- and post-processing and the supervised classification of the Landsat-8 band set.

Landsat-8-OLI-TIRS comes with 11 bands (1–11). Bands 1, 8, 9, 10, and 11 were excluded from processing and the calculation process [27]. The other bands were converted to reflectance and used as a band set for the classification process. The types of land covers of the area basically include permanent water bodies, vegetation, built-up areas, and bare soil. The study area was intensively investigated using existing land cover maps, the information in previous studies [8, 10, 14, 18, 28], satellite images, and the standard OpenStreetMap (OSM). The vegetation types include natural forest, tree agriculture fields (most of which are rubber and palm oil), and crops fields (paddy and others). Accordingly, the adopted land cover classes include water bodies (rivers and lakes), trees fields, crops fields, forests, built-up areas, bare soils, and croplands. The cropland class include some fields of crops that appear non-vegetated during the time of classification.

The classification process then starts by selecting a suitable number of regions of interest (ROI) for each land cover class. The selection process for cover classes were enhanced by google satellite and OSM in addition to using different RGB colors compensation. A plot of spectral signatures was used also to assess the accuracy during and after the classification process. An example of spectral signature plot is shown in Fig. 2. Given the small percentage of cloud cover in this image, it was considered as an unclassified class with a zero value. Additionally, part of the cloud cover was masked by image editing at locations with a known land cover class from other data sources.

The Normalized Difference Vegetation Index (NDVI) map for the study area was constructed using Eq. (1) [21, 29] as shown in Fig. 3. NIR and RED represent the

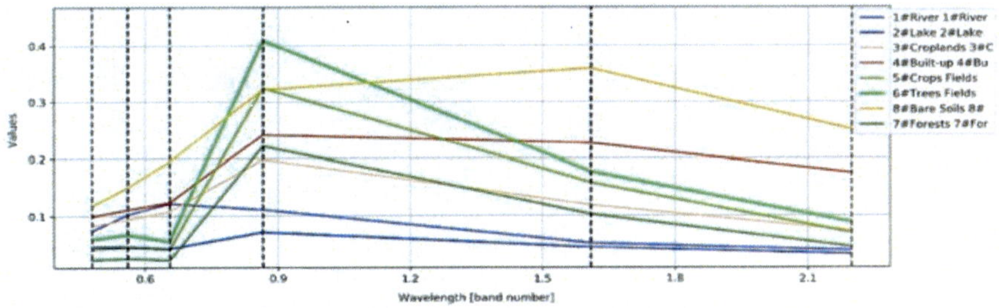

Fig. 2 Spectral signature plot of classified land cover classes

respective near-infrared and red (visible) lights reflected by vegetation. The NDVI map indicates that the forest land cover class has a NDVI value equal or greater than 0.8. This information is very useful in enhancing the classification process, especially in differentiating between forest and other vegetations.

$$\text{NDVI} = \frac{\text{NIR} - \text{RED}}{\text{NIR} + \text{RED}} \tag{1}$$

4 Accuracy Assessment

Accuracy assessment must be performed on a land cover map after the classification process to evaluate the overall accuracy and reliability of the map generated. Generally, the accuracy of a land cover classified map is assessed using different data sources as references for the same classified area. Satellite images of higher resolution than the classified image can be used as reference data in accuracy assessment, and open satellite imagery is a cost-free source that can enhance image interpretation [30]. In this study, Sentinel-2 satellite image, Google satellite, and OSM were used as reference data for the accuracy assessment process. High-resolution Sentinal 2 images are freely available at the Copernicus website. The overall methodology flowchart of the classification process is summarized in Fig. 4.

Determining the sample size for the accuracy assessment is important for reliable and accurate classification. The process described by [30, 31] for accuracy assessment was followed in this study. The sample size for accuracy assessment was determined through stratified random sampling, and the number of samples was approximated using Eq. (2). In the stratification process [32], a simple random sample for each class (i.e., stratum) is determined from the total sample.

$$N = \left(\frac{\sum_i^C W_i * S_i}{S_o} \right)^2, \tag{2}$$

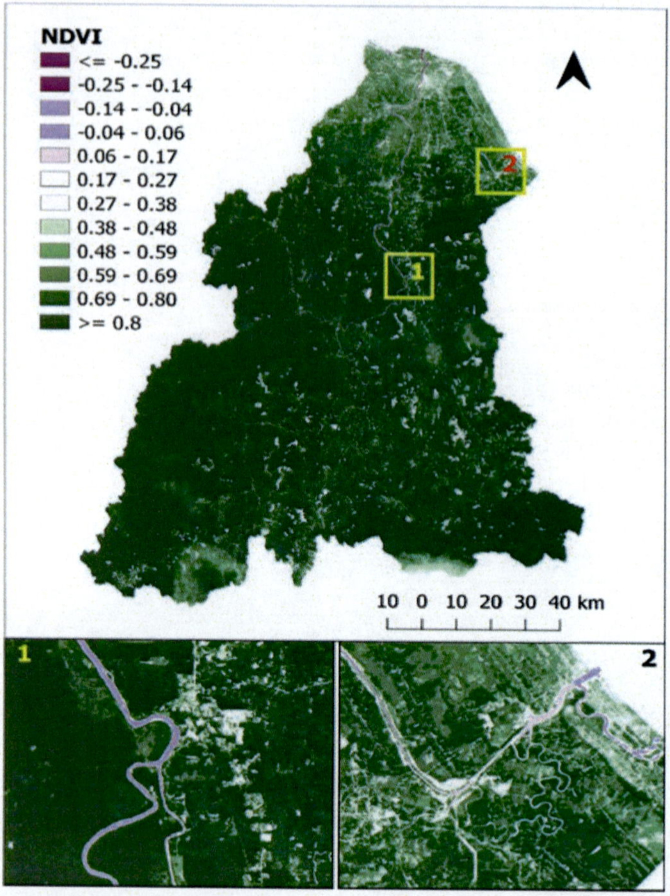

Fig. 3 NDVI map of Kelantan

where N is the designed sample number (total number of ROIs), W_i is the area proportion mapped of class i, S_i is the stratum's standard division calculated by Eq. (3). In Eq. (3), U_i is the user's accuracy conjectured values, S_o is the expected standard error, and C is the total number of land cover classes.

$$S_i = \sqrt{U_i(1 - U_i)} \tag{3}$$

The error matrix was then constructed based on the area proportions to quantify the user's accuracy (UA), the producer accuracy (PA), the standard error (SE), the 95% confidence interval (CI) area, the Kappa coefficient, and the overall accuracy of classification. Equation (4) was used to calculate the overall accuracy (O). P_{jj} in this equation are the values in the diagonal of the error matrix, representing the correctly classified area proportion of the strata (land cover classes).

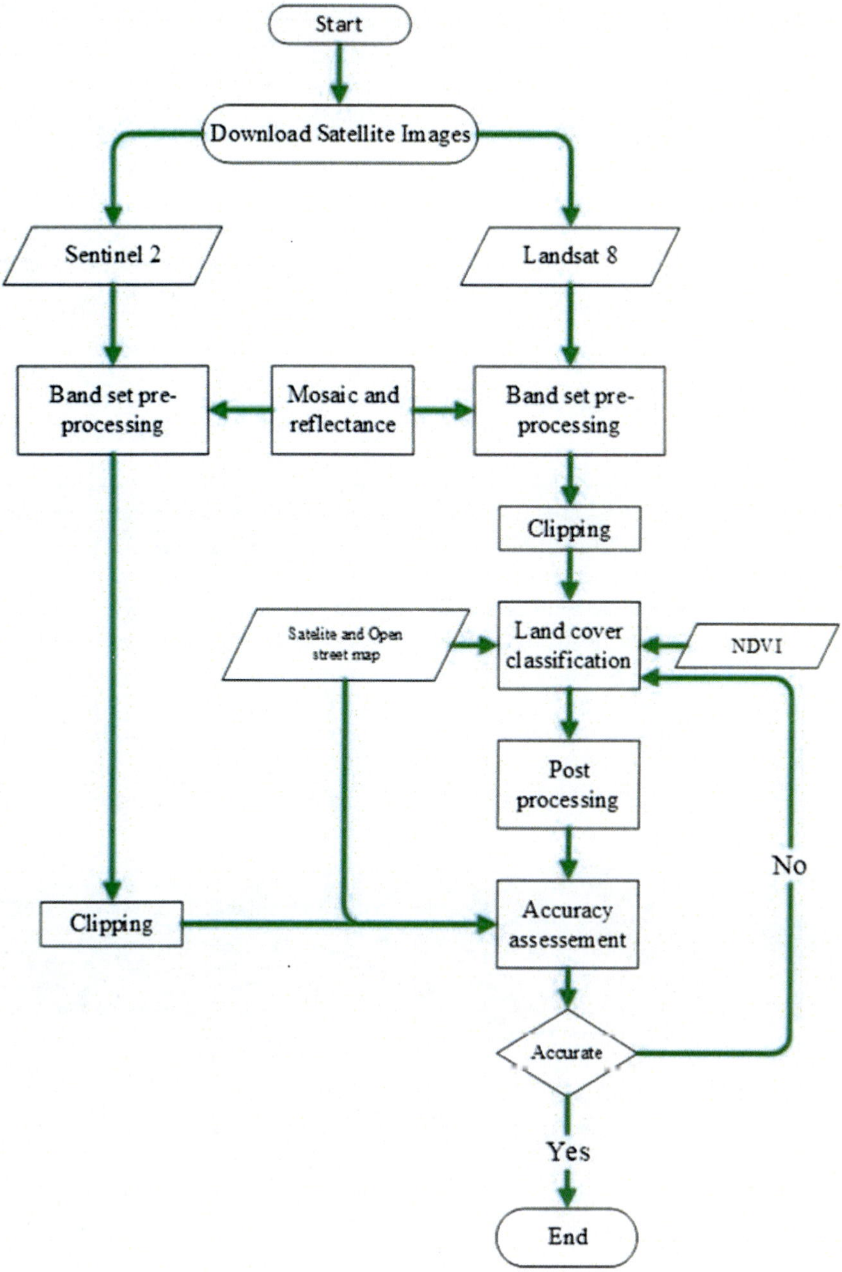

Fig. 4 Methodology flowchart

$$O = \sum_{j=1}^{C} P_{jj} \quad (4)$$

The final accuracy agreement of land cover classification depends on the goals of classification process. In this context, for instance, some researchers proposed that Kappa hat values more than 0.75 indicate strong agreement whereas values less than 0.4 indicate poor agreement, and the ideal accuracy for standard LULC maps is at least 90% [25]. However, numerous researchers have recommended that adopting of error matrix to express accuracy should become standard reporting practice [33].

5 Results and Discussion

Supervised classification was used to produce the land cover map of the study area. The maps generated are shown in Figs. 5, 6, and 7, and the summary of the classification report is given in Table 1. The error matrix based on the classified area proportion is illustrated in Table 2. The classification of different land cover classes was based on the remote sensing spectral signature of each class in the Landsat-8 satellite image. The calculated accuracy assessment report is summarized in Table 3.

The largest part of Kelantan is covered by forests followed by tree agricultures of rubber, palm oil, and other types. Differentiating between different types of vegetation was challenging, especially between the forest and tree fields and the built-up and bare lands, because the pixels of the image classifying according to their spectral signatures, which are similar in some cases. Using band combination with true and

Fig. 5 Part of original Landsat-8 image with false color composite at left, and the same part in the classified land cover map at right for comparison. The legends of classes are same as in Fig. 7

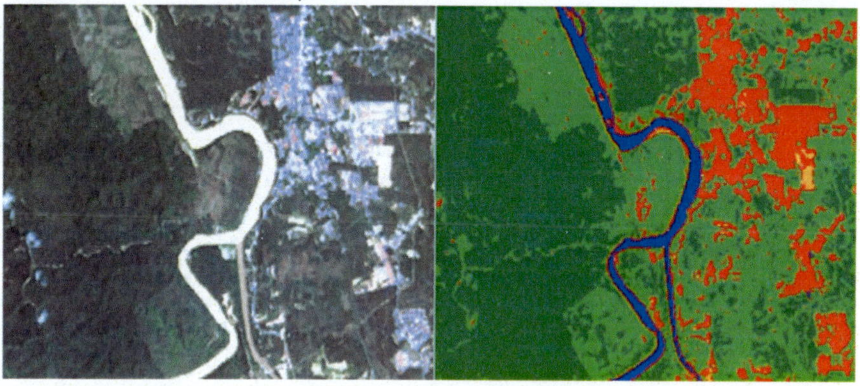

Fig. 6 Landsat-8 image of true color composite at left, and the same portion in the classified map at right for comparison. The legends of classes are same as in Fig. 7

false colors, the NDVI, and Goggle satellite imagery was effective in image interpretation and the recognition of different types of vegetation. However, the built-up and bare soil classes were not well-separated. A large proportion of bare soil was classified as built-up area as shown in the error matrix of classification in Table 2 and the very low PA in Table 3 (and high omission error), probably because the built-up and bare land classes have a similar spectral signature and occupy small areas adjacent to each other, where a 30-m resolution may be insufficient. In this case, a more accurate ground truth reference data is recommended. Nevertheless, the overall accuracy of classification of 91.46% for entire area is very good.

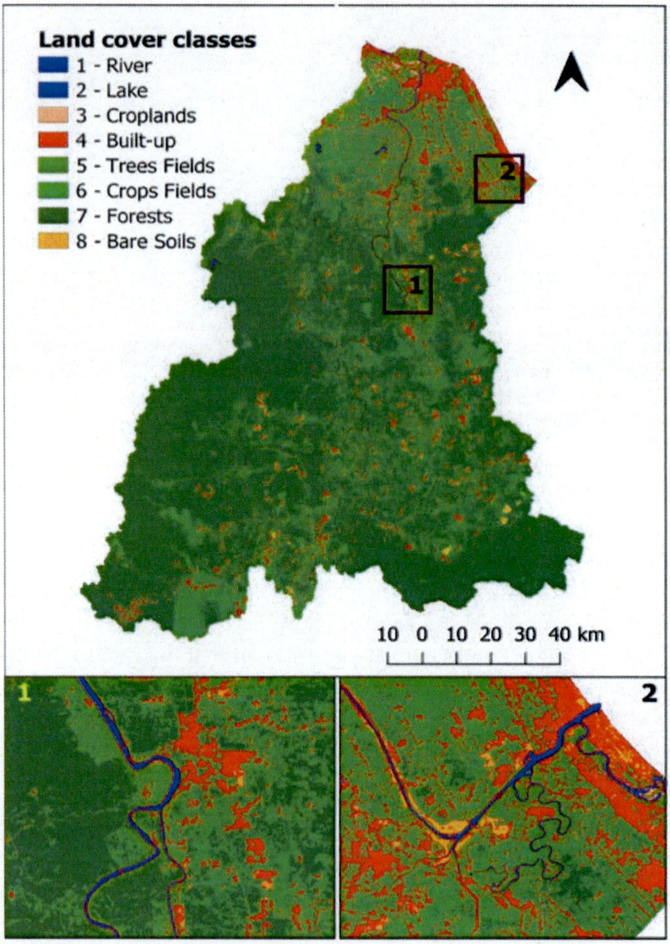

Fig. 7 Land cover classification map of Kelantan

Table 1 Classification report

Class order	Land cover	Pixel sum	Percentage (%)	Area [km^2]
1	Rivers	73,125.0	0.44	65.8
2	Lakes	8830.0	0.05	7.9
3	Agriculture fields	69,837.0	0.42	62.9
4	Built-up areas	1,143,861.0	6.84	1029.5
5	Trees fields	5,284,500.0	31.58	4756.1
6	Crops fields	218,771.0	1.31	196.9
7	Forest	9,830,430.0	58.75	8847.4
8	Bare soil	104,001.0	0.62	93.6
Total		6,733,355.0	100.00	15,060.0

Table 2 Area proportions error matrix of classified strata

Classified classes order	1	2	3	4	5	6	7	8
1	0.0042	0.0002	0	0	0	0	0	0
2	0.0001	0.0005	0	0	0	0	0	0
3	0	0	0.0031	0.0002	0.0004	0	0.0004	0
4	0	0	0	0.0319	0.0114	0	0	0.0251
5	0	0	0	0	0.3021	0	0.0137	0
6	0	0	0	0	0.0019	0.0112	0	0
7	0	0	0	0	0.0318	0	0.5557	0
8	0	0	0	0	0.0003	0	0	0.0059
Total	0.0042	0.0007	0.0031	0.0321	0.3478	0.0112	0.5699	0.031

Table 3 Accuracy assessment report of classification

Classified classes order	1	2	3	4	5	6	7	8
Estimated Area (km^2)	63.32	10.44	47.14	483.56	5238.17	168.77	8582.22	466.39
Standard Error, SE	0.0002	0.0002	0.0004	0.0063	0.0157	0.0010	0.0149	0.0061
SE area	3.34	3.34	6.24	95.42	235.74	15.41	224.14	92.24
95% CI area	6.54	6.54	12.24	187.03	462.05	30.20	439.31	180.79
PA [%]	98.74	68.49	100.00	99.35	86.85	100.00	97.52	19.07
UA [%]	95.00	90.00	75.00	46.67	95.65	85.71	94.59	95.00
Kappa hat	0.95	0.90	0.75	0.45	0.93	0.86	0.87	0.95
Overall accuracy [%] =	91.46							
Kappa hat classification =	0.85							

6 Conclusion

The land cover classification map of Kelantan was generated using open-source remote sensing data and the open-source software QGIS. Landsat-8 and Sentinel-2 satellite images were utilized to generate and assess the land cover map, respectively. Additionally, the classification and assessment processes were enhanced by using the NDVI and other standard satellite imagery. Information from previous reports and maps of the area were also of great value. Kelantan has a heterogeneous vegetation cover of different agricultures and forests, requiring advance techniques and data to accurately determine the types of crops. Even the different water bodies (i.e., rivers, lakes, and ponds) in the area appear in different colors in the satellite images, which may be due the different degrees of water pollution. Nevertheless, the use of QGIS and the SCP plugin for classification in this study contributed to the accuracy of classification. Simplicity and data and software availability are the main advantages of this classification process.

Acknowledgements The authors would like to thank the QGIS Development Team, USGS, and Copernicus developers for providing open-source remote sensing data and analysis to public. We extend our thanks to Universiti Malaya Research Grant (GPF070A-2018) for providing the financial support for this research.

References

1. Joshi, N., Baumann, M., Ehammer, A., Fensholt, R., Grogan, K., Hostert, P., et al.: A review of the application of optical and radar remote sensing data fusion to land use mapping and monitoring. Remote Sens. **8**, 1–23 (2016)
2. Turner, B.L., Lambin, E.F., Reenberg, A.: The emergence of land change science for global environmental change and sustainability. Proc. Natl. Acad. Sci. U S A. 20666–20671 (2007)
3. Defries, R.S., Rudel, T., Uriarte, M., Hansen, M.: Deforestation driven by urban population growth and agricultural trade in the twenty-first century. Nat Geosci. Nat. Publishing Group **3**, 178–181 (2010)
4. National Research Council: Sustainable Agriculture and the Environment in the Humid Tropics. National Academies Press, Washington (1993)
5. FAO: Global Forest Resources Assessment 2015: How are the world's forests changing? Food Agric. Organ. United Nations (2015)
6. Malaymail. Malaysia in Bad Light over Rapid Deforestation [Internet] (2015). Malaymailonline.com. Accessed 16 Nov 2019. https://www.malaymail.com/news/malaysia/2015/08/09/malaysia-in-bad-light-over-rapid-deforestation/948481
7. Masum, K.M., Mansor, A., Sah, S.A.M., Lim, H.S.: Effect of differential forest management on land-use change (LUC) in a tropical hill forest of Malaysia. J Environ Manage. **200**, 468–474 (2017)
8. Pradhan, B., Shafiee, M., Pirasteh, S.: Maximum flood prone area mapping using RADARSAT images and GIS: Kelantan River basin. Int. J. Geoinform. **5**, 11–23 (2009)
9. Pradhan, B., Youssef, A.M.: A100-year maximum £ ood susceptibility mapping using integrated hydrological and hydrodynamic models: Kelantan River Corridor, Malaysia. J. Flood Risk Manag. **4**, 189–202 (2011)

10. Asmat, A., Mansor, S., Saadatkhah, N., Adnan, N.A.: Land use change effects on extreme flood in the Kelantan basin using hydrological model. Isfram **2015**, 221–236 (2016)
11. Azad, W.H., Sidek, L.M., Basri, H., Fai, C.M., Saidin, S., Hassan, A.J.: 2 dimensional hydrodynamic flood routing analysis on flood forecasting modelling for Kelantan River Basin. MATEC Web Conf (2016)
12. Jaafar, A.S., Sidek, L.M., Basri, H.: An overview: flood Catastrophe of Kelantan watershed in 2014. Isfram **2015**, 17–29 (2016)
13. Shamshuddin, J., Panhwar, Q.A., Othman, R., Ismail, R., Jol, H., Yusoff, M.A.: Effects of December 2014 great flood on the physico-chemical properties of the Soils in the Kelantan Plains, Malaysia. J. Water Resour. Prot. **08**, 263–276 (2016)
14. Wan, K.M., Billa, L.: Post-flood land use damage estimation using improved Normalized Difference Flood Index (NDFI3) on Landsat 8 datasets: December 2014 floods, Kelantan, Malaysia. Arab J. Geosci. **11** (2018)
15. Alias, N.E., Mohamad, H., Chin, W.K., Yusop, Z.: Rainfall analysis of the Kelantan Big yellow flood 2014. J. Teknol. **4**, 83–90 (2016)
16. Duminda, E, Perera, P, Lahat, L. Fuzzy logic based flood forecasting model for the Kelantan River basin. J. Hydro-Environ. Res. Elsevier B.V, 1–12 (2014)
17. Buslima, F.S., Omar, R.C., Jamaluddin, T.A., Taha, H.: Flood and flash flood geo-hazards in Malaysia. Int. J. Eng. Technol. **7**, 760–764 (2018)
18. Pour, A.B., Hashim, M.: Application of Landsat-8 and ALOS-2 data for structural and landslide hazard mapping in Kelantan, Malaysia. Nat. Hazards Earth Syst. Sci. **17**, 1285–1303 (2017)
19. QGIS Development Team: Quantum GIS Geographic Information System [Internet]. Open Source Geospatial Foundation Project 2015. Accessed 13 Nov 2020. https://qgis.org/en/site/
20. Ruwaimana, M., Satyanarayana, B., Otero, V., Muslim, A.M., Muhammad Syafiq, A., Ibrahim, S., et al.: The advantages of using drones over space-borne imagery in the mapping of mangrove forests. PLoS ONE **13**, 1–22 (2018)
21. Arekhi, M., Goksel, C., Sanli, F.B., Senel, G.: Comparative evaluation of the spectral and spatial consistency of Sentinel-2 and Landsat-8 OLI data for Igneada longos forest. ISPRS Int. J. Geo-Inform. **8** (2019)
22. Priyadi, P.: Analysis of spatio temporal change of land use of chrysanthemum farm in semarang regency using landsat image 8 OLI. Indonesia J. Comput. Model **1**, 49–54 (2018)
23. Correia, R., Duarte, L., Teodoro, A.C., Monteiro, A.: Processing image to geographical information systems (PI2GIS)—A learning tool for QGIS. Educ. Sci. **8** (2018)
24. Hamad, S.: Updating topographic maps at scale 1:250000 for Libyan territory using quantum GIS (QGIS) and open geospatial data: Libya Topo-Project. J. Geogr. Stud. **4**, 22–34 (2020)
25. Khan, A.S., Khan, A.B.: Assessment of land use and land cover dynamics in shingla river basin using multi temporal satellite imageries. Int. J. Emerg. Technol. **11**, 263–269 (2020)
26. Congedo, L., Sallustio, L., Munafò, M., Ottaviano, M., Tonti, D., Marchetti, M.: Copernicus high-resolution layers for land cover classification in Italy. J. Maps. Taylor Francis **12**, 1195–1205 (2016)
27. Congedo, L.: Semi-Automatic Classification Plugin Documentation (2016)
28. Samsurijan, M.S., Abd Rahman, N.N., Syakir Ishak, M.I., Masron, T.A., Kadir, O.: Land use change in Kelantan: review of the environmental impact assessment (EIA) reports. Malaysian J. Soc. Sp. **14**, 322–331 (2018)
29. Rouse Jr., J.W., Haas, R.H., Schell, J.A., Deering, D.W.: Monitoring vegetation systems in the great plains with erts. NASA SP-351, 3rd ERTS-1 Symposium (1974)
30. Olofsson, P., Foody, G.M., Herold, M., Stehman, S.V., Woodcock, C.E., Wulder, M.A.: Good practices for estimating area and assessing accuracy of land change. Remote Sens. Environ. **148**, 42–57 (2014)

31. Congedo, L.: From GIS to Remote Sensing: Accuracy Assessment of a Land Cover Classification [Internet] (2013). Web page (From GIS to Remote Sensing). Accessed 16 Nov 2020. https://fromgistors.blogspot.com/2019/09/Accuracy-Assessment-of-Land-Cover-Classification.html
32. Cochran, W.G.: Sampling Techniques, 3rd edn. John Wiley & Sons, New York (1977)
33. Congalton, R.G.: A review of assessing the accuracy of classifications of remotely sensed data. Remote Sens. Environ. **37**, 35–46 (1991)

Energy Consumption of Variable Speed Pumps Under Transient-State Condition in Water Distribution Networks

Rehan Jamil, Hamidi Abdul Aziz, and Mohamad Fared Murshed

Abstract This article aims at drawing a comparison between the power consumption and cost incurred for variable speed pumps (VSPs) working under steady-state and transient-state conditions in a water distribution network. A 24-h variable water demand pattern was attached to a residential community water distribution network in EPANet 2.2 for simulation along with introducing a pipe leak at a critical node to assimilate the transient-state condition. It was observed that there was a significant difference in the energy consumption of VSPs working at lower and high speeds when compared with their nominal speed. The results show an energy cost cut of 40% at 15% lower speed and an energy cost hike of 50% at a 15% higher speed of VSP than its nominal speed. A mathematical relation is developed to estimate the energy cost difference at any speed of VSP working under the transient-state condition in a water distribution network.

Keywords Variable speed pump · Energy consumption · Water distribution network · Transient flow · Pipe leak · EPANet

1 Introduction

According to a study, the energy consumption in the world for pumping systems is estimated to be more than 45% of the total amount [1]. Whether it is water supply, air conditioning, sewage transfer, or any industrial use, pumps consume a huge amount of energy. The above figure expands to 25–50% of the total energy consumption of the whole facility when this parameter is filtered to certain industries. Researchers

R. Jamil · H. A. Aziz (✉) · M. F. Murshed
School of Civil Engineering, Engineering Campus, Universiti Sains Malaysia, 14300 Nibong Tebal, Pulau Pinang, Malaysia
e-mail: cehamidi@usm.my

H. A. Aziz
Solid Waste Management Cluster, Engineering Campus, Universiti Sains Malaysia, 14300 Nibong Tebal, Pulau Pinang, Malaysia

N. Sabtu (ed.), *Proceedings of AWAM International Conference on Civil Engineering 2022 - Volume 3*, Lecture Notes in Civil Engineering 386,
https://doi.org/10.1007/978-981-99-6026-2_34

have always been keen on developing such methods in the form of retrofitting, load management, and other fabrication techniques to lower the energy consumed by the electric motors of the pumping systems.

It is a well-known fact that when a fixed-speed pump is operated at a lower speed, it consumes less energy but at the same time, the efficiency of the pump is compromised as well. When only fixed-speed pumps (FSP) were in operation, the objective of energy management was achieved by altering the speed of the pump to the water demand by using flow control devices. However, it was not a simple task to do the speed adjustment of FSP as the available solution at that time was not appropriate for all pumping systems. Such complex solutions, in some cases, proved to be costlier than the energy cost intended to be saved. With the advent of technology, the idea of variable speed pumps (VSP) came into existence somewhere around the early 2000s. As the name suggests, the VSP can work at different speeds automatically while delivering the same amount of water at the required pressure, fulfilling the objective of lowering energy consumption in case of low water demand conditions. In addition to the low energy consumption, the VSPs offer a few other benefits as well including the ability to adapt to the variable water demand by itself, reduction in frictional losses in the piping system, simple mechanics creating less noise, and increased durability. Since its advent, due to these benefits, VSPs have become the most commonly used pumps for water distribution networks [2].

One of the major problems found in water distribution networks is the pipe leak [3–5]. A leak in a pipe may occur due to various reasons. A pipe may burst at a weaker location due to excessive water pressure or it may get damaged by construction or earthworks resulting in a pipe hole. A leak in a water distribution network results in a huge amount of water loss as well as pressure. In addition to this when a water network is being fed by a pumping system, it causes excessive energy loss due to the additional operation of pumps to overcome the lost amount of water [6].

An extensive literature review was performed to pick the latest research trends related to the use and application of VSPs for water distribution networks. Bortoni et al. [7] tried to optimize the use of variable speed pumps by connecting them in parallel in a network. Wu et al. [8] used hydraulic models and various operational scenarios for variable speed pumps to assess the hydraulic characteristics of the water networks. Marchi et al. [9] assessed the efficiency of variable speed pumps for water networks by considering the mechanical components of the pumps including their electric motor, rotational shaft, and controllers. Georgescu et al. [10] performed an experimental study for the estimation of the efficiency of variable speed pumps by using EPANet and compared it with the actual model. In a similar kind of study, they used EPANet for the simulation of various control methods for centrifugal pumps [11]. Dunca et al. [12] did experimental work for testing the efficiency of EPANet regarding variable speed pumps. They compared the results obtained by experiments and by simulation for a water network. A perfect match was obtained in the final parametric values. Page et al. [13] proposed new controllers for the pressure management of water networks by using a remote control system for variable speed pumps. They found the new controllers efficient. Lydon et al. [14] also performed a pressure management-related study by using three pumps. Nowak et al. [15] developed a decision support system for the operation of variable speed pumps working in a water network. They found that a large number of options are created by using this method. Abdallah and Kapelan [16] devised a pump scheduling method for optimizing energy costs in water networks by using fixed and variable speed pumps. Luna et al. [17] and Dini et al. [18] also performed a similar study by using pump operational scheduling as a method to optimize the efficiency of pumps and water networks. They found that by using this method 15% of energy can be saved on average. Briceno-Leon et al. [2] used different control strategies for fixed and variable speed pumps in water networks to obtain the number of pumps and their configuration required for a specific case. The effects of VSPs on water quality have also been studied by researchers [16, 19].

After going through the available literature, it was found that a sufficient amount of study has been done on the application of VSPs for water distribution networks. However, there is no literature available that discusses the effect on the energy consumption of a VSP in the case of a transient flow condition or a pipe leak in a water distribution network having a variable demand pattern. As pipe leaks contribute a lot towards the loss of water and pressure in a water distribution network and when coupled with the variable speed pumping system, huge differences regarding the energy consumption of the pump were expected. Hence, this task was taken up to overcome the research gap and various hydraulic simulations were performed to study this aspect and highly significant results were obtained.

2 Methodology and Analysis Options

The water supply pipe network used for the study consists of 82 nodes, pipe joints, and 122 pipe sections as shown in Fig. 1a. The area selected for the research is an underdeveloped housing community in the vicinity of Islamabad, Pakistan located at 33°30′N, 73°5′E. The housing community is designed for 1800 residential units along with commercial, educational, and recreational buildings and green areas. The expected population of the community is at least 10,000.

The pipe material used for the water supply network is high-density polyethylene (HDPE). The diameter of pipes ranges from 65 to 125 mm. The water supply network is fed with a set of centrifugal VSPs connected in parallel. The network is spread over an area of 85 hectares (0.85 km^2) and the general topography of the study area is mildly sloping at the rate of 5–8%. The water distribution network along with the elevation contour map is shown in Fig. 1b.

The methodology used to achieve the objectives of the research is shown as a flow diagram in Fig. 2. The hydraulic analysis of the water distribution network consists of two types of simulation. The first is steady-state and the second is the transient-state analysis of the water distribution network. A steady-state condition reflects the balanced hydraulic conditions of the network which usually lasts for longer periods. On the other hand, a slight disturbance caused due to sudden closure of the valve,

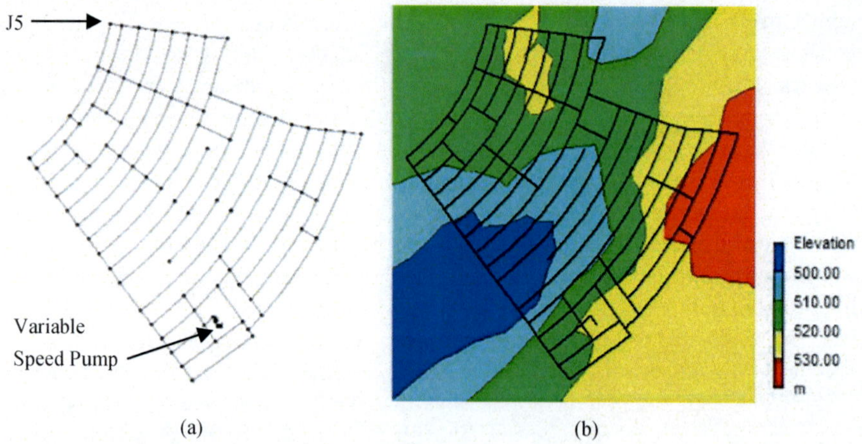

Fig. 1 **a** Water distribution network with location of critical node and VSP, **b** Water distribution network superimposed on contour plan

pump throttle, or a pipe leak initiates a transient-state condition in the system which may last for a short time up to a few seconds or minutes [20].

All hydraulic simulations were run by using the Hazen-Williams equation in EPANet 2.2. A roughness coefficient of 150 was used for plastic pipes [21]. A residential water demand pattern of 24 h was used for the consumption of water. The water demand at each node was calculated and assigned based on the design guidelines provided by the Capital Development Authority, Islamabad, Pakistan [22]. Typically, residential water demand is maximum in the morning and it reduces during the latter part of the day. A small peak is observed late in the evening as well. The demand pattern used for the simulation is shown in Fig. 3 which shows the demand multipliers based on the actual water demand.

The rotational speed of a centrifugal pump controls its energy efficiency parameters. The equations that involve pump speed, flow, pressure, and power consumption are known as affinity laws [23]. One of the equations relating the power consumption of VSP to its rotational speed is given below [21].

$$\frac{P_1}{P_2} = \left(\frac{n_1}{n_2}\right)^3 \left(\frac{d_1}{d_2}\right)^3 \tag{1}$$

where P_1 and P_2 are the shaft power in kW at nominal rotational speed n_1 and variable rotational speed n_2 respectively. d_1 and d_2 are the external diameters of the rotor. For a pump working at a specific speed, the external diameter of the rotor undergoes negligible change.

After obtaining the simulation results from EPANet 2.2 regarding the energy consumption and cost of the energy, regression analysis is performed to observe the relation between the parameters and to create a mathematical relation for the difference in cost of energy and pump rotational speed. A typical linear regression

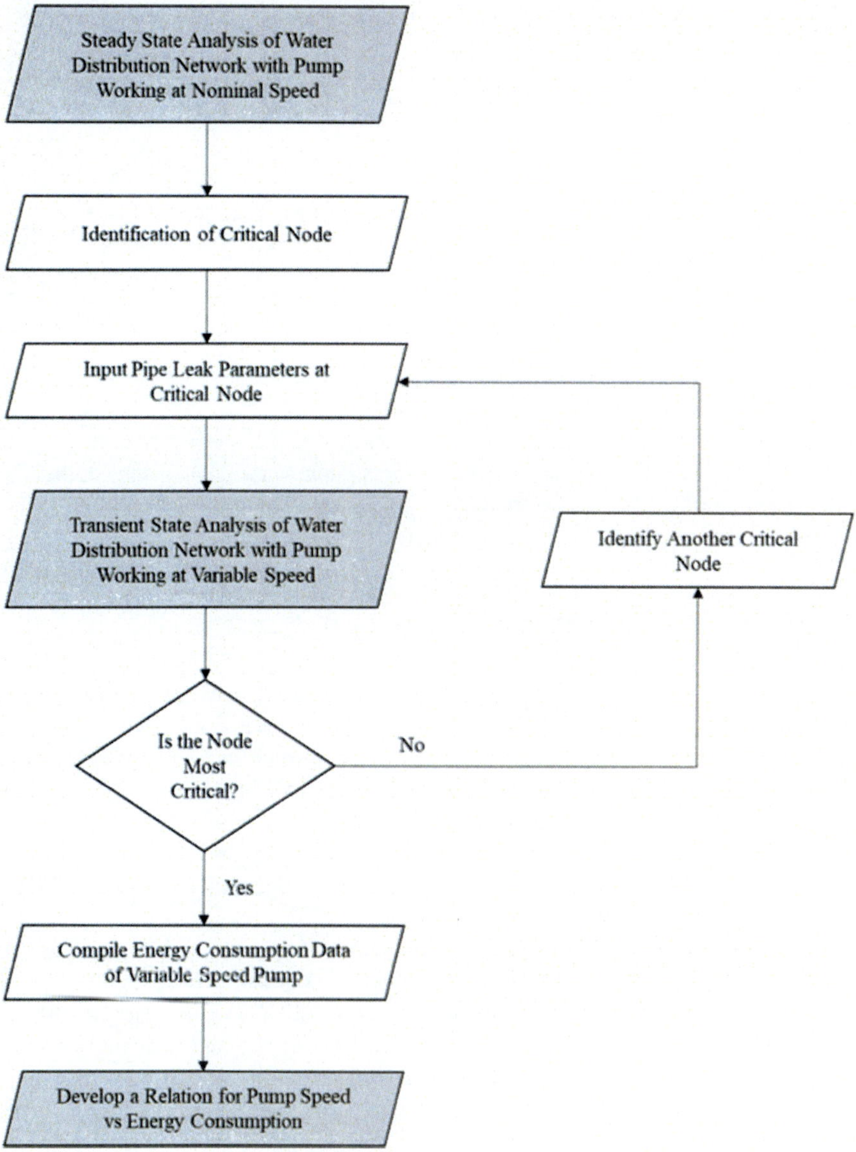

Fig. 2 Flow diagram of research work methodology

analysis equation is of the following form [24].

$$y = mx + c \tag{2}$$

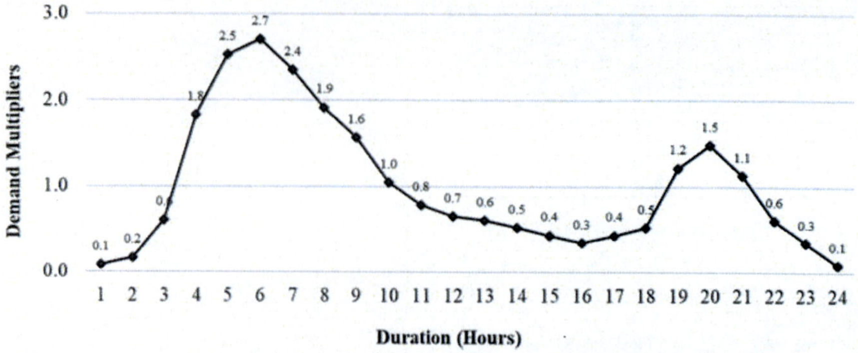

Fig. 3 24-h typical residential water demand pattern

where y is the dependent variable, x is the independent variable, m is the coefficient, and c shows the value of the intercept.

3 Results and Discussions

The initial steady-state analysis of the water distribution network was performed to make sure that the diameter of all the pipes was correct and various hydraulic components were working well. The initial pump setting was kept in such a way that the minimum pressure at any pipe node remained above 1 bar (10 m) to ensure sufficient water pressure at all locations of the network. At nominal speed, the pump setting was found to be delivering 50 lps at 40 m of the head. The contour profile showing the pressure of water in the network at peak demand hours under normal and steady-state conditions is shown in Fig. 4.

After the initial simulation, various critical nodes in the network were identified based on maximum and minimum values of water demand, pressure, elevation, or distance. It was found that the farthest node is the most significant node creating the highest amount of disturbance in the system. The farthest node was identified as J5 as shown in Fig. 1a. To induce the transient-state condition in the network a pipe leak was introduced at J5. Pipe leak parameters were entered in the EPANet 2.2 at the node properties box and the pump was allowed to run at variable speeds. Pump curves shown in Fig. 5 were entered into the system by considering the pump works at six different speeds other than the nominal speed value of 1. The variable pump speed values were taken as 0.85, 0.9, 0.95, 1.05, 1.1, and 1.15 of the nominal speed of the pump.

Hydraulic variables such as pressure, head loss, and velocity were measured at the critical node J5 during the simulation for every 5% increase or decrease in the nominal speed of the pump. Variation in pressure was observed at the node with variation in the speed of the pump. However, it was significant to note that the pattern of the

Fig. 4 Pressure distribution in water distribution network for steady-state condition at peak demand

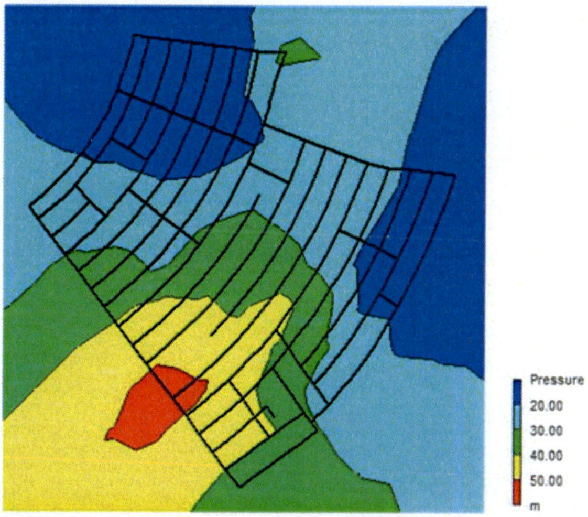

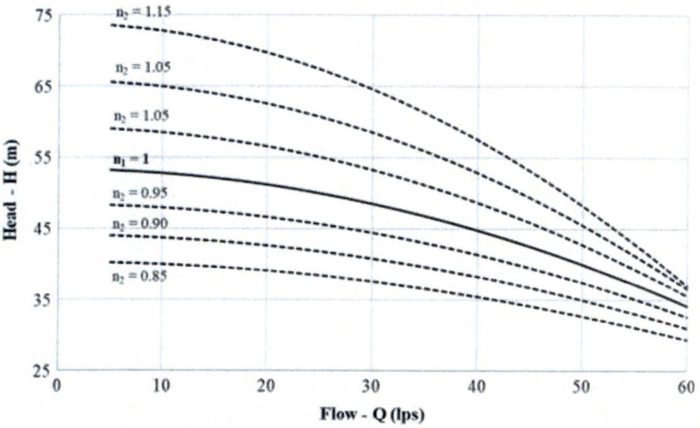

Fig. 5 Variable speed pump characteristic curves at various speeds

variation remained the same. With higher pump speeds, the pressure at J5 was higher and when the pump was working at low speed due to low demand, the pressure at the node also dropped. Figure 6 shows the pressure profile drawn for the pipe network from VSP to J5 which is the longest path from the source to any node in the network. It illustrates three pressure profile curves for three different pump speeds valued at 0.85 of the nominal speed, 1 of the nominal speed itself, and 1.15 of the nominal speed of the pump.

Another comparison for pressure was drawn at J5 for both steady-state and transient-state conditions of the network as shown in Fig. 7. It was observed that during the transient-state condition when a leak was introduced at J5, the pressure at peak load was found to be less than without the leak. It is also noticeable that the loss

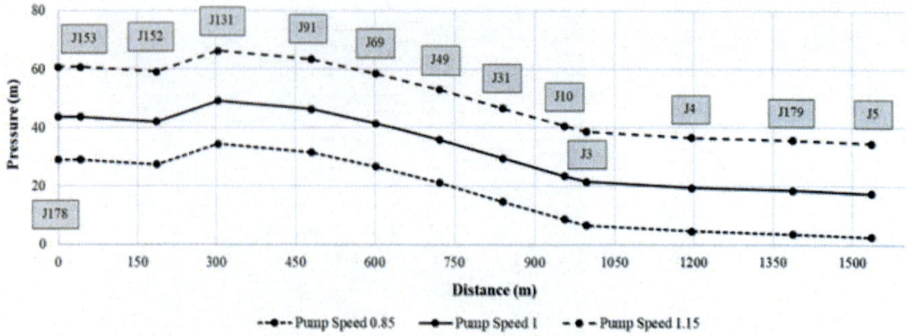

Fig. 6 Pressure profile from VSP to critical node J5 following the longest path

Fig. 7 Pressure at J5 for steady-state and transient-state conditions at different speed of pumps

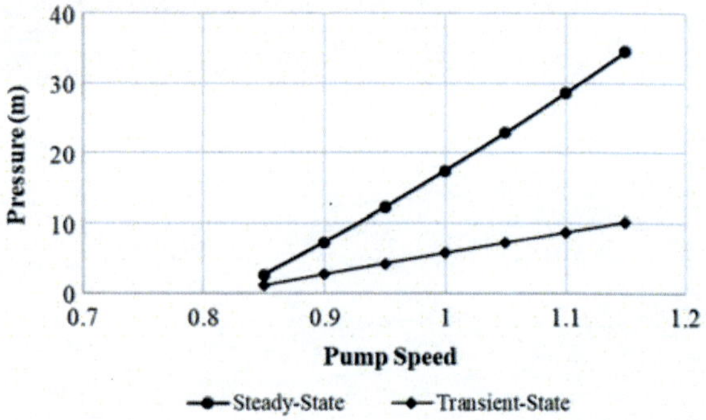

of pressure is higher at higher pump speeds and it becomes lower up to negligible as soon as the pump speed drops. This result obtained from the simulation validated the fact that the pump had to work at higher speeds to cater to the loss of water during a transient-state condition.

This increase in the working of the pump resulted in overall higher energy consumption and energy costs. The simulation results show that the installed pump consumes more power at the peak water demand working at higher speeds when compared with its operation at the nominal speed at the same peak water demand load. The statistics given in Table 1 can be taken into consideration.

The cost of energy incurred by the variable speed pump is USD 76.45/day when it is working at 15% lower than its nominal speed resulting in a decrease of 40% in cost, whereas this value increases to USD 191.94/day when the pump is working at a speed 15% higher than its nominal value creating a difference of above 50% in cost. It is evident that the pump consumes more power and energy during a transient-state condition and the amount of energy consumption increases with the increase in pump speed and vice versa.

The results obtained in this research are similar to the ones reported by various researchers. For an instance, Darweesh concluded that energy costs can be reduced by up to 20% by managing the operation of VSPs efficiently [5], whereas Abdallah and

Table 1 Energy consumption of VSP and cost difference results

Pump speed	Power consumption (kWh)		Cost of energy consumed (USD/day)	% Difference in cost with reference to nominal speed
	Average	Peak		
0.85	15.93	22.75	76.45	−40.06
0.9	19.19	26.76	92.12	−27.78
0.95	22.74	31.46	109.15	−14.42
1	26.57	36.59	127.55	–
1.05	30.72	42.18	147.45	15.60
1.1	35.17	48.28	168.84	32.37
1.15	39.99	55.01	191.94	50.48

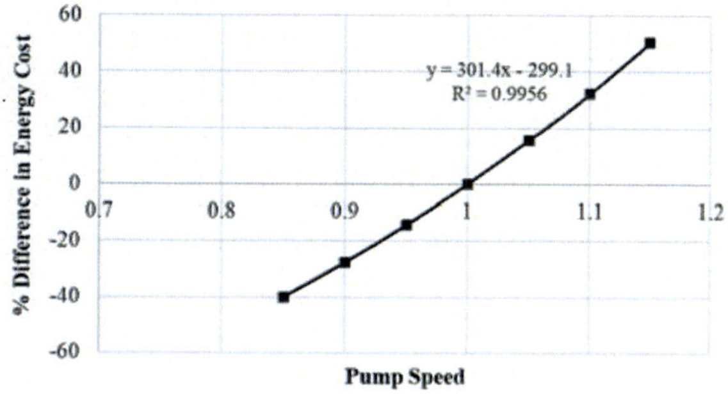

Fig. 8 Trend line for difference in energy cost at various speeds of pump during transient-state condition

Kapelan provided a scheduling model for VSPs and obtained a significant amount of cut in energy costs [16].

The trend of cost difference based on pump speed, calculated in Table 1, is plotted on a graph as shown in Fig. 8 and a mathematical relation is obtained between the two parameters.

By using statistical analysis a correlation coefficient (r) was calculated for the two parameters. It was found that the pump speed and the cost difference occurred due to the pump speed having a strong positive relationship with each other as shown by the correlation coefficient value of 0.99. The linear regression analysis produced the following equation for a VSP working at a specific speed.

$$y = 301.4x - 299.1 \tag{3}$$

In Eq. (3), y represents the independent variable which is the difference in energy consumption (ΔE) and x is the dependent variable which is the pump speed (n) in this case. Hence we can reformat Eq. (3) and the newly derived equation can be written as follows.

$$\Delta E = 301.4n - 299.1 \tag{4}$$

The above-mentioned Eq. (4) can be utilized to forecast and calculate the difference in the cost of energy that shall occur at any pumping speed with respect to its nominal speed for a VSP connected to a water distribution network under transient-state conditions.

4 Conclusions

Considering the huge amount of energy consumption and the cost incurred by variable speed pumping systems, a comparative study was performed on a water distribution network under transient-state conditions. The simulation results on EPANet 2.2 show a clear difference in the energy consumption of the pumps when they were allowed to run at variable speeds when compared with when they were run at a nominal speed. A decrease in energy consumption of pumps and cost was observed with a decrease in the pump speed while delivering the same amount of water. On the other hand, the energy consumption and cost were increased when the pump was run at higher nominal speeds. After compiling the data obtained from the software for all pump speeds, a statistical analysis was performed to find a relation between the energy consumption and the speed of VSP. The proposed mathematical relation is found to be reliable for estimating the difference in energy consumption a VSP would cause while working at a specific speed for a water distribution network under transient-state conditions. The overall results show that they are consistent with the theory of the variable speed pumps but as there was no such mathematical relation for energy consumption of VSP working under transient-state conditions, this article provided one to fill the research gap.

References

1. Waide, P., Brunner, C.U.: Energy-Efficiency Policy Opportunities for Electric Motor-Driven Systems. Working Paper—Energy Efficiency Series, International Energy Agency—IEA, Organization of Economic Co-Operation and Development—OECD (2011)
2. Briceno-Leon, C.X., Iglesias-Rey, P.L., Martinez-Solano, F.J., Mora-Melia, D., Fuertes-Miquel, V.S.: Use of fixed and variable speed pumps in water distribution networks with different control strategies. Water **13**, 479 (2021)
3. Qasem, A., Jamil, R.: GIS-based financial analysis model for integrated maintenance and rehabilitation of underground pipe networks. J. Perform. Constr. Facil. **35**(5), 04021046 (2021)
4. Kim, S.H.: Multiple leak detection algorithm for pipe network. Mech. Syst. Signal Process. **139**, 106645 (2020)
5. Darweesh, M.: Assessment of variable speed pumps in water distribution systems considering water leakage and transient operations. J. Water Supply Res. Technol. AQUA **67**(1), 99–108 (2018)
6. Jamil, R.: Performance of a new pipe material UHMWPE against disinfectant decay in water distribution networks. Clean Technol. Environ. Policy **20**(6), 1287–1296 (2018)
7. Bortoni, E.C., Almeida, R.A., Viana, A.N.C.: Optimization of parallel variable-speed-driven centrifugal pumps operation. Energ. Effi. **1**, 167–173 (2008)

8. Wu, Z.Y., Tryby, M., Todini, E., Walski, T.M.: Modeling variable-speed pump operations for target hydraulic characteristics. J. Am. Water Works Assoc. **101**(1), 54–64 (2009)
9. Marchi, A., Simpson, A.R., Ertugrul, N.: Assessing variable speed pump efficiency in water distribution systems. Drinking Water Eng. Sci. **5**, 15–21 (2012)
10. Georgescu, A.M., Cosoiu, C.I., Perju, S., Georgescu, S.C., Hasegan, L., Anton, A.: Estimation of the efficiency for variable speed pumps in EPANET compared with experimental data. In: 16th Conference on Water Distribution System Analysis, WDSA 2014. Procedia Engineering, vol. 89, pp. 1404–1411 (2014)
11. Georgescu, A.M., Georgescu, S.C., Cosoiu, C.I., Hasegan, L., Anton, A., Bucur, D.M.: EPANET simulation of control methods for centrifugal pumps operating under variable system demand. In: 13th Computer Control for Water Industry Conference, CCWI 2015. Procedia Engineering, vol. 119, pp. 1012–1019 (2015)
12. Dunca, G., Piraianu, V.F., Roman, R., Ciuc, P.O., Georgescu, S.C.: Experimental versus EPANET Simulation of Variable Speed Driven Pumps Operation. Sustainable Solutions for Energy and Environment, EENVIRO 2016. Energy Procedia **112**, 100–107 (2017)
13. Page, P.R., Abu-Mahfouz, A.M., Mothetha, M.L.: Pressure management of water distribution systems via the remote real-time control of variable speed pumps. J. Water Resour. Plan. Manag. **143**(8), 04017045 (2017)
14. Lydon, T., Coughlan, P., McNabola, A.: Pressure management and energy recovery in water distribution networks: development of design and selection methodologies using three pump-as-turbine case studies. Renew. Energy **114**, 1038–1050 (2017)
15. Nowak, D., Krieg, H., Bortz, M., Geil, C., Knapp, A., Roclawski, H., Bohle, M.: Decision support for the design and operation of variable speed pumps in water supply systems. Water **10**, 734 (2018)
16. Abdallah, M., Kapelan, Z.: Fast pump scheduling method for optimum energy cost and water quality in water distribution networks with fixed and variable speed pumps. J. Water Resour. Plan. Manag. **145**(12), 04019055 (2019)
17. Luna, T., Ribau, J., Figueiredo, D., Alves, R.: Improving energy efficiency in water supply systems with pump scheduling optimization. J. Clean. Prod. **213**, 342–356 (2019)
18. Dini, M., Hemmati, M., Hashemi, S.: Optimal operational scheduling of pumps to improve the performance of water distribution networks. Water Resour. Manage **36**, 417–432 (2022)
19. Darweesh, M.: Impact of variable speed pumps on water quality in distribution systems. Water SA **44**(3), 419–427 (2018)
20. Sela, L., Xing, L.: An overview of the transient simulation in water distribution networks (TSNet). In: World Environmental and Water Resources Congress 2020, Henderson, Nevada (2020)
21. Rossman, L.A., Woo, H., Tryby, M., Shang, F., Janke, R., Haxton, T.: EPANet 2.2 User Manual. Water Infrastructure Division, Centre for Environmental Solutions and Emergency Response, U.S. Environmental Protection Agency, Cincinnati, OH. EPA/600/R-20/133 (2020)
22. CDA: Modalities and Procedures for Development of Private Housing/Farm Housing Schemes. The Gazette of Pakistan, Capital Development Authority, Islamabad, Pakistan, Doc. No. 5104(20)/Ex. Gaz. (2020)
23. Simpson, A.R., Marchi, A.: Evaluating the approximation of the affinity laws and improving the efficiency estimate for variable speed pumps. J. Hydraul. Eng. **139**(12), 1314–1317 (2013)
24. Bluman, A.G.: Elementary Statistics—A Step by Step Approach. McGraw-Hill, New York. ISBN 9780073534978 (2009)

Reliability of Rainwater Harvesting System Using Yield-After-Spillage Algorithm

Nur Shazwani Muhammad and Mee Wai Lin

Abstract Malaysia receives a lot of rain yearly, making rainwater harvesting system a high potential for usage in commercial buildings, industry and residential areas. The rainwater collected can be used for irrigation systems and as a source of indoor non-potable use. This study is conducted to test the reliability of rainwater harvesting system at Sultan Abdul Aziz Shah Airport in Subang, Selangor. The yield-after-spillage algorithm is used to perform the rain data analysis and calculation of reliability of rainwater harvesting system. The parameters analyzed in this study are annual rainfall pattern, roof surface area, storage tank size and rainwater demand. Generally, the result of the analysis shows that the reliability of rainwater harvesting system increases as the parameters of roof catchment area and tank capacity increase, but decreases when water demand increases. The maximum reliability is 51.1% for moderate water demand and 30.6% for high water demand during the year with average yearly rainfall. High reliability can be achieved with a minimum roof surface area value of 5000 m^2 and tank size of 140 m^3 for an average rainfall year. For moderate water demand, the increase in the reliability is less than 5% after exceeding 100 m^3 tank size with roof surface area of 20,000 m^2. For these specifications and by using the rainfall data from 1994 to 2014, at least 29.3% of the water can be supplied from the rainwater storage tank.

Keywords Rainwater harvesting system · Reliability · Yield-after-spillage algorithm · Sultan Abdul Aziz Shah Airport (SZB)

N. S. Muhammad (✉) · M. W. Lin
Department of Civil Engineering, Faculty of Engineering and Built Environment, Universiti Kebangsaan Malaysia, 43600 Bangi, Selangor, Malaysia
e-mail: shazwani.muhammad@ukm.edu.my

M. W. Lin
Sepakat Setia Perunding Sdn. Bhd, Wisma SSP, Taman Serdang Raya, 43300 Seri Kembangan, Selangor, Malaysia

N. Sabtu (ed.), *Proceedings of AWAM International Conference on Civil Engineering 2022 - Volume 3*, Lecture Notes in Civil Engineering 386,
https://doi.org/10.1007/978-981-99-6026-2_35

1 Introduction

Rainwater harvesting system is a technology that collects and stores rainwater for daily use. There are various types of rainwater harvesting systems, e.g., a simple system which consists of a rainwater storage tank and a more complex structure that contains pumps, tanks and purification system. In Malaysia, the problem of water cut still occurs in some areas such as Kuala Lumpur, Petaling, Klang, Shah Alam, Kuala Selangor, Hulu Selangor, Gombak and Kuala Langat [1], although the country has abundant water resources compared with other countries [2–4]. Hence, Malaysia has a high potential to use this system effectively. Although guidelines for rainwater harvesting were published since 1999, the utilization of this system is still low due to several factors such as long investment returns as well as low community acceptance [2]. Therefore, the government should encourage and recommend the widespread use of rainwater harvesting systems as one of the alternatives to address this critical problem, especially in high-density urban areas that have reached the maximum use of surface water resources. Previous researchers have studied extensively on the design, advantages, performance, feasibility and reliability of water harvesting systems around the world, e.g., in Pakistan [5], Southern Italy [6] and the United States [7]. However, the number of studies on rainwater harvesting systems in Malaysia is limited [2].

Lani et al. [2] have listed the prospects and challenges of implementing rainwater harvesting systems in Malaysia. Among the challenges encountered are the initial and maintenance costs as well as the reliability of the system. Studies on the reliability of rainwater harvesting systems are rarely conducted in commercial buildings and public airport facilities with high potential that has large rain catchment roof areas. Areas such as airports need abundant water supply to manage routine operations and infrastructure demands such as toilet flushing, lawn and crop irrigation, outdoor pool water supply and general cleaning activities. A consistent supply of water is essential to ensure that airport operations run smoothly.

In Malaysia, more than 98% of water supply is from surface water resources such as rivers and dams, while groundwater only contributes less than 2% [8]. Therefore, the quality of river water becomes one of the factors that affects the distribution of clean water resources [9]. Malaysia has experienced rapid development and urbanization and this to some extent has had a detrimental effect on river basins in Malaysia [9, 10]. Severe river water pollution has caused water supply disruptions in some areas recently, especially in the Klang Valley [11]. This problem should be resolved as it provides negative impacts such as health issues [12], stunted economic development in various sectors [13] and socioeconomic impacts [14]. Due to population growth [15] and the development of the industrial, tourism as well as agricultural sectors, the demand for water supply has increased drastically. Problems will arise when the water supply is insufficient to meet the water demand or when pollution of water supply sources cause disruption to the supply of treated water.

Therefore, this study is conducted to evaluate the reliability of rainwater harvesting system in a well-known public transport facility building in Malaysia, Sultan Abdul

Aziz Shah Airport in Subang to reduce the dependence on the supply of treated water from Pengurusan Air Selangor Sdn. Bhd. The term reliability is defined as a measure of the system's ability to provide the water supply needed by users for crop irrigation and non-potable water demand such as toilet flushing. This information will determine the suitability of the implementation of the rainwater harvesting system as an alternative water supply other than the existing water supply.

2 Methodology

2.1 Study Area and Data

Sultan Abdul Aziz Shah Airport (SZB) also known as Subang Skypark is an airport located in Subang, Petaling District, Selangor, Malaysia with latitude 03° 07′ 54″ N and longitude 101° 32′ 54″ E shown in Fig. 1. The airport opened in 1965 was once known as Subang International Airport where at that time SZB was the main flight gateway to Malaysia before the opening of Kuala Lumpur International Airport in Sepang in 1998. Currently, only turboprop passenger aircraft operates from SZB as well as domestic destinations flights in Malaysia and several regional destinations in Indonesia, Thailand and Singapore. The location is also one of the largest base operators in the region, with more than 100,000 ft^2 of indoor hangar space and SZB is a private aviation hub in Malaysia [16].

Figure 2 shows the total annual rainfall at SZB from 1994 to 2014. Rainfall data was obtained from the Malaysian Meteorological Department for Subang station. Total annual rainfall ranges from 2292.4 mm (2005) to 3455 mm (2006). Rainfall data is a fixed parameter in this study.

Daily rainfall amounts and changes in rainfall patterns are categorized for three rainfall conditions, namely dry year, average year and wet year. The rainfall condition categories are based on the 10th, 50th and 90th percentiles for annual rainfall at Subang station. The 10th and 90th percentiles were selected for the driest year and the wet year because the amount of rainfall in those percentiles had a higher probability of occurring when compared with the driest or wettest years on record. Extreme dry or wet conditions are rare under normal conditions of annual rainfall distribution. The mean year for rainfall data is a qualitative term and is represented by the median rather than the arithmetic mean [17].

2.2 Water Balance Model

The tank of rainwater harvesting system is filled with rainwater collected from the roof of building. The volume of rainwater is calculated by assuming that the rainfall is constant throughout the day using the following equation:

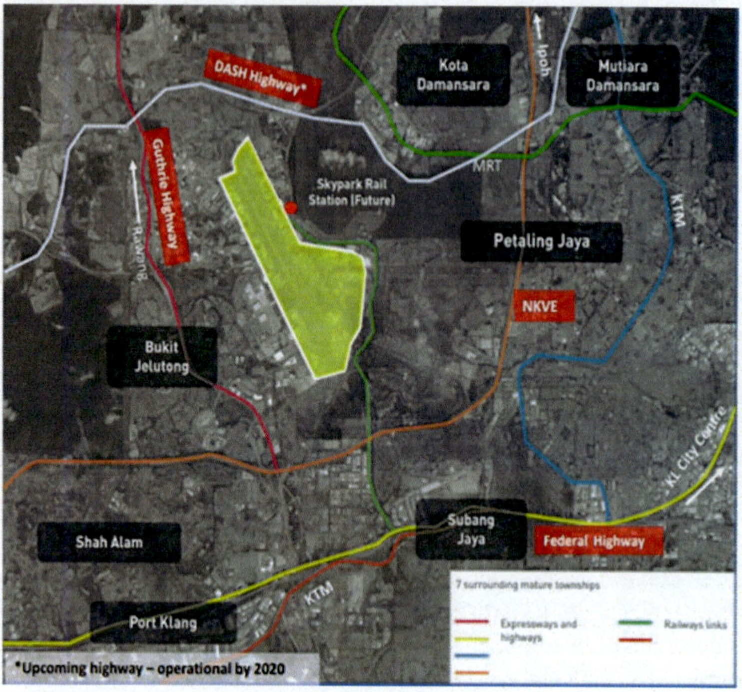

Fig. 1 Location of Sultan Abdul Aziz Shah Airport

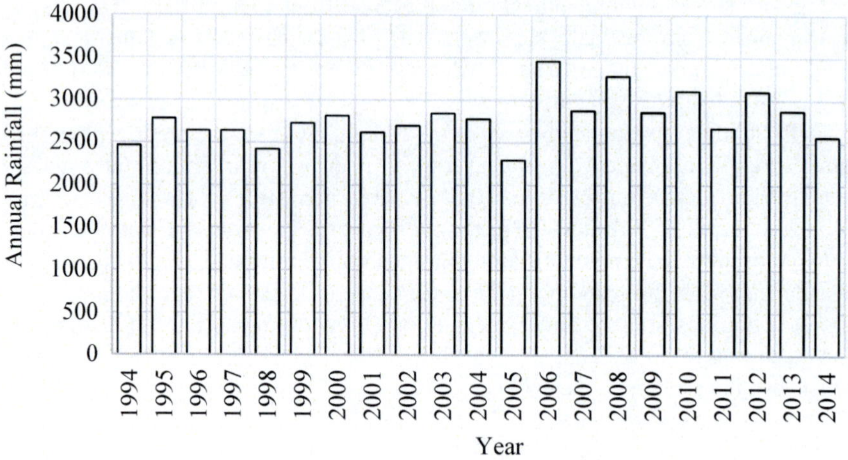

Fig. 2 Annual rainfall at SZB

$$Q_t = \Phi.A_s.R_t = A.R_t \tag{1}$$

where Q_t is the volume of inflow into the tank at time t (m^3), Φ is the water runoff coefficient for water loss, R_t is the amount of rainfall at time t (m), A_s is the total surface area (m^2) and A is the effective surface area (m^2). The loss of water due to evaporation from the storage tank is negligible. In this study, Φ is set with a value of 0.9. Surface area of roof is measured using Google Earth.

First flush is also modeled for this study. The first flushing is defined as the initial period of rainwater runoff in which the concentration of contaminants is much higher than that of subsequent rainwater. Roof runoff quality is influenced by rainwater quality and roof type such as material, slope and roof length [18]. A reduction of 0.33 mm of the first rainwater was from the daily rainfall amount as the first rinse can significantly improve the quality of water collected from the roof [19]. For this study, all daily water balance simulations were performed by subtracting 0.33 mm from the daily rainfall data.

Water demand for this study considered only toilet flushing and garden irrigation. Water demand for toilet flushing is estimated using the value estimated by Suruhanjaya Perkhidmatan Air Negara (SPAN) [20] which is 25 L per passenger. Passenger traffic data is obtained from Malaysia Airports Holdings Berhad's annual reports. For garden irrigation, surface area of garden is measured using Google Earth. In this study, turf grass is assumed to cover the entire area measured. CROPWAT 8.0 software is used for the calculation of water demand for garden irrigation based on data such as soil, climate and crops. The software was developed by the Land and Water Development Division under the United Nations Food and Agriculture Organization (FAO). CROPWAT 8.0 also generates irrigation schedules for garden irrigation. The generation of the irrigation schedule is based on the daily soil–water balance. All computational procedures used in this software are based on FAO publications [21]. Weather data such as monthly average minimum and maximum temperatures, average monthly humidity, average monthly wind speeds as well as average monthly sunlight are required and obtained from historical climate data [22].

Analysis of water demand is conducted as there was a very significant difference of water demand between the period of 1994–1998 and the period of 1999–2014, hence making the reliability analysis of rain air harvesting system to be unsuitable for future use. Therefore, the annual water demand is reanalyzed and divided into three categories of high, medium and low water demand rate using the total annual water demand from the period 1999 to 2014 only. The 90th, 50th and 10th percentiles represents the high, medium and low water demand, respectively. Daily water demand is calculated by dividing the total water demand per year with the number of days for each year, assuming that the daily water demand is the same for each day for both toilet flushing and garden irrigation. The high, medium and low daily water demand rates are given in Table 1. These values are used to calculate the reliability of rainwater harvesting system for wet, average and dry years.

To study the reliability of rainwater harvesting system, the yield-after-spillage (YAS) algorithm has been widely used [23]. The YAS algorithm is shown by the following equations:

Table 1 Water demand rate

Water demand rate	Daily water demand (m^3)
High	141.1
Medium	77.5
Low	6.0

$$Q_{D_t} = \max\begin{cases} V_{t-1} + Q_t - S \\ 0 \end{cases} \quad (2)$$

$$P_t = \min\begin{cases} D_t \\ V_{t-1} \end{cases} \quad (3)$$

$$V_t = \min\begin{cases} V_{t-1} + Q_t - P_t \\ S - P_t \end{cases} \quad (4)$$

Q_{D_t} is the volume removed as overflow from the storage tank at time t (m^3), V_t is the volume stored in the tank at time t (m^3), P_t is the supply of rainwater from the storage tank at time t (m^3), D_t is the water demand at time t (m^3) and S is the storage capacity of the tank. The reliability or performance of rainwater harvesting systems can generally be measured in terms of water saving, E_T as shown in the following equation:

$$E_T = \frac{\sum_{t=1}^{T} P_t}{\sum_{t=1}^{T} D_t}.100 \quad (5)$$

where T is the period of analysis considered and E_T is the amount of water supplied relative to total demand. If the storage tank is empty, the remaining water supply demand will be supplied by the municipal water source. The frequency of full, partially filled and full tank is also analyzed. A full tank is defined as a tank fully filled with water until it reaches maximum tank capacity at the beginning of day t. A partially filled tank means the tank is not empty and filled with rainwater, but does not reach the tank capacity at the beginning of day t, whereas an empty tank is defined as a tank having no water at the beginning of day t.

In this study, the tank size variable parameter ranged from 0 to 200 m^3, with an increase interval of 20 m^3. The tank size is set as 20 m^3 per tank, e.g., if 100 m^3 is the required tank size, five tanks are considered necessary to be implemented in the study area.

3 Results and Discussion

3.1 Reliability

Analysis as shown in Fig. 3a–c is done for medium water demand for wet, average and dry year by using tank size of 200 m^3 and roof area of 20,000 m^2. Among the three years with different rainfall patterns, the wet year has the highest total annual full tank frequency of 114 days and the lowest frequency during a dry year with 96 days of full tank. For empty tanks, the maximum annual frequency is 126 days during the average year and the minimum value is 95 days during the wet year, with a difference of 8.5%. This indicates that the rainfall pattern has a higher effect on the frequency of empty tanks than the frequency of full tanks. Therefore, the average rainfall year will be analyzed using high and low water demand rates as it has the highest frequency of empty tanks.

Figure 4a, b show the frequency of empty tank, partially filled tank and full tank for high and low water demand rates, respectively, for an average rainfall year in 2004 using a tank size of 200 m^3 and a roof area of 20,000 m^2. The total annual full tank frequency is the lowest for high demand rate which is 95 days, while the highest is estimated for the low water demand rate with 163 days. The total annual empty tank frequency has a maximum value for high water demand which is 162 days and the minimum value is zero for low water demand. The empty tank frequency still shows a higher difference compared with the full tank frequency for different water demands. Moreover, when compared with the annual rainfall pattern, water demand rate shows a more significant effect.

A reliability analysis is carried out as shown in Fig. 5a–d for average rainfall year for different tank sizes and surface areas. In general, the increase in reliability of rainwater harvesting system is highly dependent on both factors, namely the roof surface area and the size of the tank. High reliability value can be achieved with the use of a minimum tank size of 140 m^3 for different surface areas for low water demand. However, high reliability is not achievable when the tank size is increased to 200 m^3 and the roof surface area is maximized to 20,000 m^2 for high and medium water demand. The reliability of the rainwater harvesting system has the highest increase for the medium water demand rate, with a maximum value of 51.5% when the tank size is increased from 0 to 200 m^3, compared with high and low water demands. This proves that the increase in tank size has the greatest impact on the reliability of medium water demand. The maximum reliability increase rate was 4.1% for low water demand and 7.3% for increment of roof surface area. The highest increase in reliability can also be seen for the medium water demand for the variable roof surface area parameter, with the highest increase value of 10.9% when the surface area of 5000 m^2 is increased to 10,000 m^2. Surface area difference also has the highest impact on the reliability of medium water demand. The rate of increase in reliability has a decrease with increasing roof surface area and tank size. For medium water demand, the increase in reliability is less than 5% when exceeding 100 m^3 with a roof surface

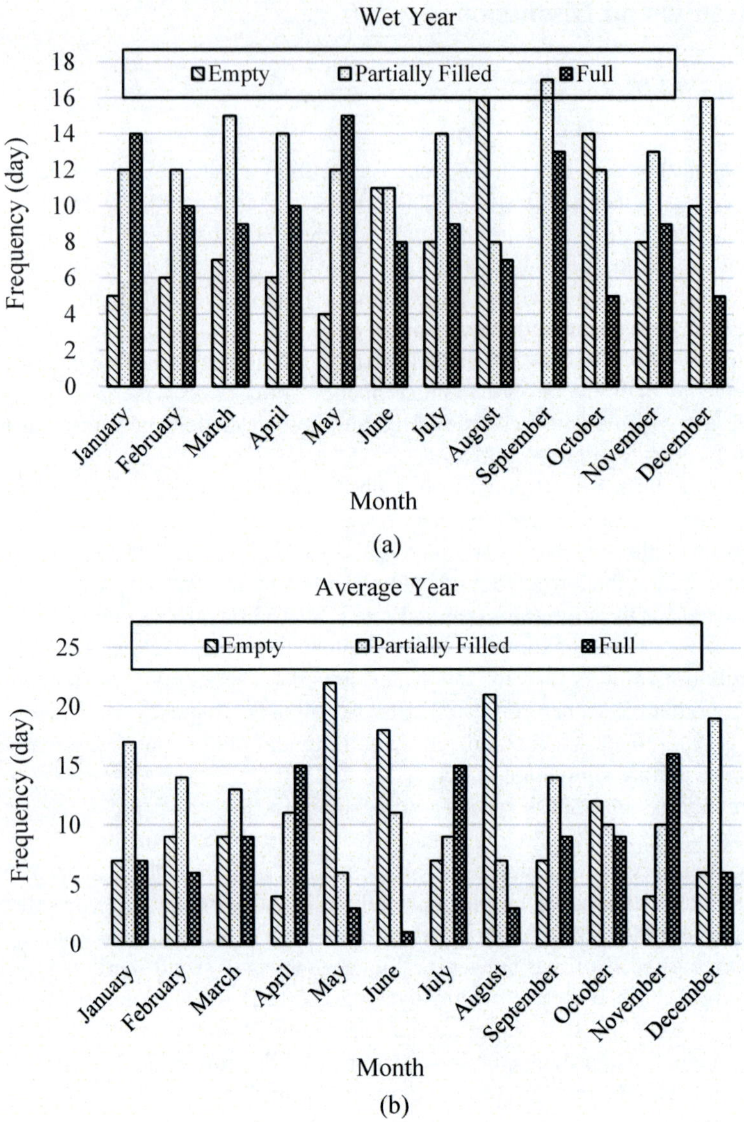

Fig. 3 Frequency of full tank, partially filled tank and empty tank for medium water demand

area of 20,000 m^2. Additionally, the increase in reliability is less than 5% when the roof surface area exceeds 10,000 m^2 with a minimum tank size of 120 m^3.

Figure 6a, b show the reliability of rain harvesting system for wet and dry years, respectively, for a surface roof area of 20,000 m^2 using high, medium and low water demand rates. The graphs of wet and dry years are almost the same and the difference

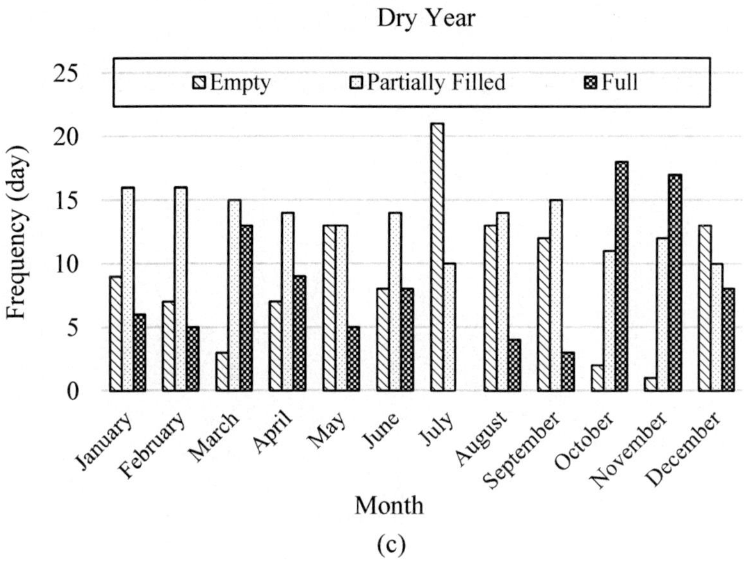

(c)

Fig. 3 (continued)

is only 3.4% for medium water demand for 140 m^3 tank size. The increase in reliability during wet year for medium water demand is more than 5% with a minimum of 140 m^3 tank size when compared with the reliability of average rainfall year. The reliability difference between the three different rainfall patterns ranged from 0 to 7.1% only. This proves that the rainfall pattern does not have a significant impact on the reliability of the rainwater harvesting system compared with the rainwater demand pattern. The maximum value for low demand during the dry year is only 99.5% with tank size of 100–200 m^3 because the dry year of 1994 is the first year the analysis started. The assumption has been made that at the beginning of the analysis the water supply in the storage tank starts with a value of zero. Rain only occurred on the second day of the start of the analysis. Therefore, high reliability of the rainwater harvesting system in that year could not be reached despite having low rainwater demand.

4 Conclusion

This study evaluates the reliability performance of rainwater harvesting system in SZB located in Subang. For the analysis of reliability with different water demand, a high reliability cannot be achieved for high and medium water demand despite using the maximum roof surface area and tank size, as well as different annual rainfall patterns. The maximum reliability is 51.5% for medium water demand and 30.6% for high water demand during an average rainfall year. High reliability can

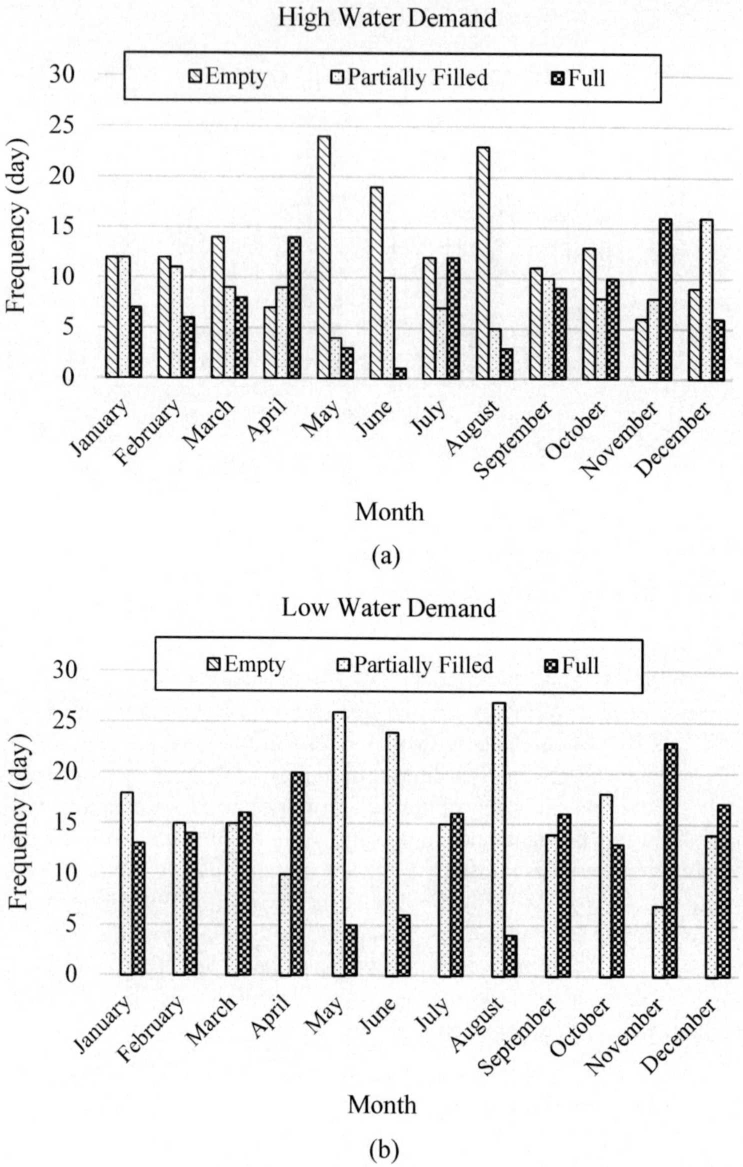

Fig. 4 Frequency of full tank, partially filled tank and empty tank for high and low water demand

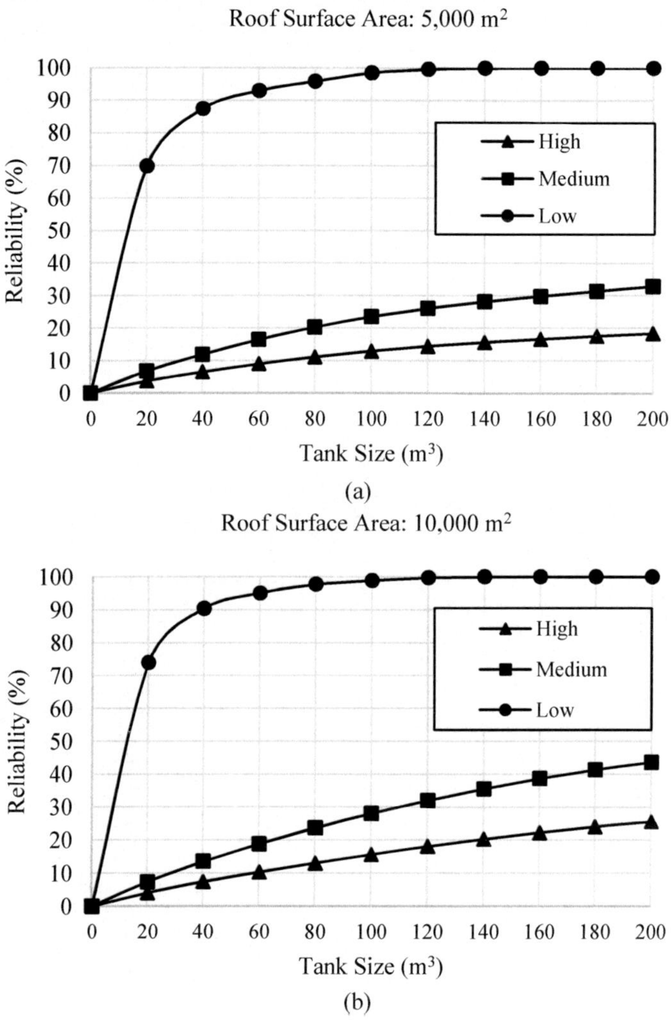

Fig. 5 Reliability for average rainfall year

be achieved with a minimum value of 5000 m^2 roof surface area and 140 m^3 tank size during an average rainfall year. Optimum tank size can be determined through tank size, roof surface area and annual rainfall pattern. For medium water demand, the increase in reliability is less than 5% when exceeding 100 m^3 tank size with a roof surface area of 20,000 m^2. Additionally, using similar specifications and rainfall data from 1994 to 2014, at least 29.3% of the water supply can be supplied from rainwater storage tanks.

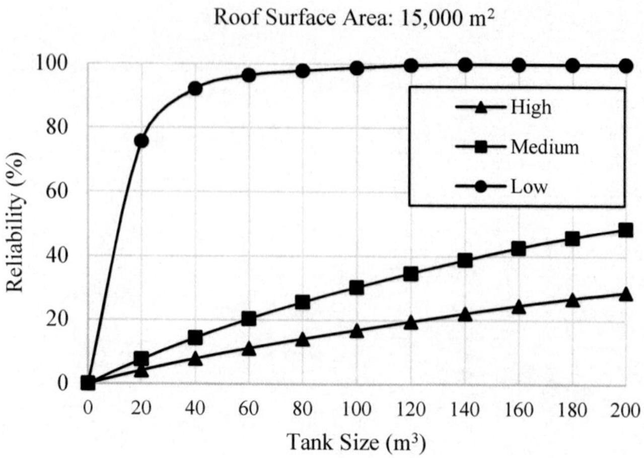

(c)

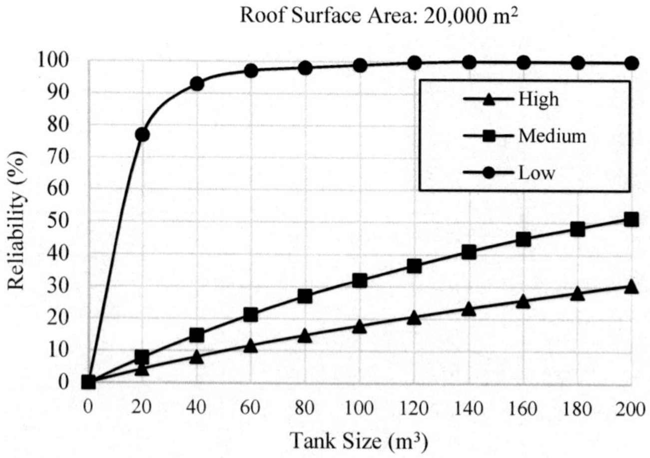

(d)

Fig. 5 (continued)

Rainwater harvesting system is potentially one of the alternatives to solve water scarcity issue. Further studies need to be conducted to identify the suitability to implement a rainwater harvesting system in Malaysia and various factors that contribute to the increase in reliability of such system in the country.

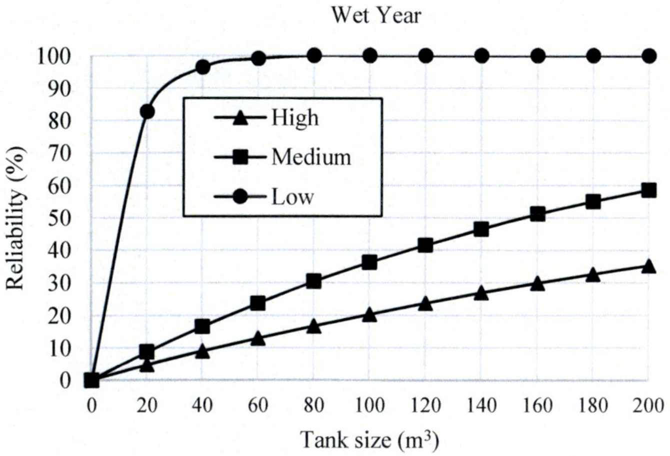

(a)

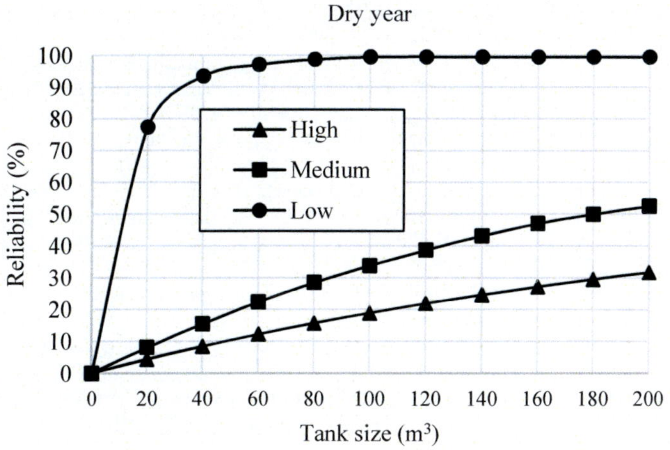

(b)

Fig. 6 Reliability for wet and dry years

Acknowledgements The authors are grateful for the funding provided by Universiti Kebangsaan Malaysia through Geran Universiti Penyelidikan (Grant No. GUP-2021-012).

References

1. Aminnuraliff, M.: Tujuh kawasan alami gangguan bekalan air. Sinar Harian (2020)
2. Lani, N.H.M., Yusop, Z., Syafiuddin, A.: A review of rainwater harvesting in Malaysia: prospects and challenges. Water (Switzerland) **10**, 1–21 (2018). https://doi.org/10.3390/w10040506

3. Muhammad, N.S., Abdullah, J., Julien, P.Y.: Characteristics of rainfall in Peninsular Malaysia. J. Phys.: Conf. Ser. **1529** (2020).https://doi.org/10.1088/1742-6596/1529/5/052014
4. Muhammad, N.S., Julien, P.Y., Salas, J.D.: Probability structure and return period of multiday monsoon rainfall. J. Hydrol. Eng. **21**, 04015048 (2015). https://doi.org/10.1061/(ASCE)HE.1943-5584.0001253
5. Ali, S., Zhang, S., Yue, T.: Environmental and economic assessment of rainwater harvesting systems under five climatic conditions of Pakistan. J. Clean. Prod. **259**, 120829 (2020). https://doi.org/10.1016/j.jclepro.2020.120829
6. Liuzzo, L., Notaro, V., Freni, G.: A reliability analysis of a rainfall harvesting system in Southern Italy. Water (Switzerland) **8**, 1–20 (2016). https://doi.org/10.3390/w8010018
7. Jones, M.P., Hunt, W.F.: Performance of rainwater harvesting systems in the southeastern United States. Resour. Conserv. Recycl. **54**, 623–629 (2010). https://doi.org/10.1016/j.resconrec.2009.11.002
8. The Malaysian Water Industry Status and Outlook Report 2020–2021 Edition. The Malaysian Water Association. (2020)
9. Wan Mohtar, W.H.M., Abdul Maulud, K.N., Muhammad, N.S., et al.: Spatial and temporal risk quotient based river assessment for water resources management. Environ. Pollut. **248**, 133–144 (2019). https://doi.org/10.1016/J.ENVPOL.2019.02.011
10. Ahmad Kamal, N., Muhammad, N.S., Abdullah, J.: Scenario-based pollution discharge simulations and mapping using integrated QUAL2K-GIS. Environ. Pollut. **259**, 113909 (2020). https://doi.org/10.1016/J.ENVPOL.2020.113909
11. FMT Reporters.: Water cuts in Selangor again after river pollution. Free Malaysia Today (2020)
12. Utsev, J.T., Aho, M.I.: Water shortage and health problems in Benue State-Nigeria: impacts and prospects for solutions. Impacts Prospects Solutions **1**, 79–85 (2012)
13. Nobre, C.A., Marengo, J.A., Seluchi, M.E., et al.: Some characteristics and impacts of the drought and water crisis in Southeastern Brazil during 2014 and 2015. J. Water Resour. Prot. **08**, 252–262 (2016). https://doi.org/10.4236/jwarp.2016.82022
14. Mohamed, N.N.: Egyptian food insecurity under water shortage and its socioeconomic impacts. In: Handbook of Environmental Chemistry, vol. 74, pp. 245–273. (2019).https://doi.org/10.1007/698_2018_239
15. Masron, T., Yaakob, U., Mohd Ayob, N., Mokhtar, A.S.: Population and spatial distribution of urbanisation in Peninsular. Malaysia J. Soc. Space **2**, 20–29 (2012)
16. Malaysia Airports Holdings Berhad. Sultan Abdul Aziz Shah Airport (2020). https://www.malaysiaairports.com.my/. Accessed 1 Jan 2021
17. Lawrence, D., Lopes, V.L.: Reliability analysis of urban rainwater harvesting for three Texas cities. J. Urban Environ. Eng. **10**, 124–134 (2016). https://doi.org/10.4090/juee.2016.v10n1.124134
18. Gikas, G.D., Tsihrintzis, V.A.: Effect of first-flush device, roofing material, and antecedent dry days on water quality of harvested rainwater. Environ. Sci. Pollut. Res. **24**, 21997–22006 (2017). https://doi.org/10.1007/s11356-017-9868-6
19. Freni, G., Liuzzo, L.: Effectiveness of rainwater harvesting systems for flood reduction in residential urban areas. Water (Switzerland) **11** (2019).https://doi.org/10.3390/w11071389
20. Uniform Technical Guidelines Water Reticulation and Plumbing. Suruhanjaya Perkhidmatan Air Negara (SPAN), Selangor, Malaysia. (2018)
21. Allen, R.G., Luis, S.P., Raes, D., Smith, M.: Crop Evapotranspiration—Guidelines for Computing Crop Water Requirements—FAO Irrigation and Drainage Paper 56. FAO—Food and Agriculture Organization of the United Nations, Rome (1998)
22. Weather Underground.: Data cuaca di Lapangan Sultan Abdul Aziz Shah (2021). https://www.wunderground.com/
23. Zhang, S., Jing, X., Yue, T., Wang, J.: Performance assessment of rainwater harvesting systems: Influence of operating algorithm, length and temporal scale of rainfall time series. J. Clean. Prod. **253**, 120044 (2020). https://doi.org/10.1016/J.JCLEPRO.2020.120044

Multivariate Flood Damage Model: A Case Study of 2013 Kuantan Flood

Sumiliana Sulong, Noor Suraya Romali, and AbdullahMukmin Ahmad

Abstract Flood damage assessment is a crucial element to be considered in the implementation of efficient flood risk management. Hence, with the aim to examine the significant correlation between the impact and resistance parameters with the degree of flood damages, this study utilized the multiple linear regression (MLR) technique to develop a multivariate flood damage model of 2013 Kuantan flood. This study focuses on estimating flood damage in urban areas for residential and commercial buildings. The damage type assessed in this study is direct tangible damage which includes damage to buildings structure and its contents. The study's findings revealed that socio-economy and property characteristics have also shown a significant correlation to the degree of flood damage. Flood depth was found to have a significant effect on all damage categories as in previous studies. However, in addition to flood depth, flood damage is also influenced by the value of properties, income, building material, building ownership status and type of properties. The MLR models validations are showing reasonable results compared with other studies with mean absolute error (MAE), root mean square error (RMSE), mean bias error (MBE), coefficient of variation (CV) which are closer to 0 and HR is close to 0.9.

Keywords Flood risk · Flood damage assessment · Flood damage influencing factors · Multivariate analysis

1 Introduction

Flood damage assessment is an essential element in flood risk management, whereas it is a significant evolution in the transformation from flood hazard to flood risk. Methods of flood damage assessment are well developed worldwide where the majority of methods for determining flood damage assessment are through the

S. Sulong · N. S. Romali (✉) · A. Ahmad
Faculty of Civil Engineering Technology, Universiti Malaysia Pahang Al-Sultan Abdullah, Gambang, Kuantan, Malaysia
e-mail: suraya@ump.edu.my

N. Sabtu (ed.), *Proceedings of AWAM International Conference on Civil Engineering 2022 - Volume 3*, Lecture Notes in Civil Engineering 386,
https://doi.org/10.1007/978-981-99-6026-2_36

unit loss [1–3], and the damage function curve method [4–7]. Recently authors are developing the multivariate flood estimation model using more than one influencing variables [3, 8–11].

The unit loss or unit cost estimation approach refers to the direct calculation of loss to individual properties, which are added together to give a total loss figure for the event in the study area [1]. The unit loss function is derived from the empirical data of each property (or unit) through the questionnaire survey of the affected population and properties to estimate the incurred loss, or the use of average figures and an adapted damage function [1, 2]. Other than that, the total damage cost can be computed by finding the number of properties being flooded and multiplying it with the individual unit cost estimation [12]. The curve method is more widely used approach for flood loss assessment due to its simplicity [13]. White [14] was the first to create the flood depth-damage curve to represent the relationship between damage and flood depth in 1960s. Flood damage curves relate the damages in terms of monetary (absolute damage curve) or damage factor in percentage of the estimated total replacement value of the property (relative damage curve) to given flood parameters (also known as flood damage influencing factors) [7–9, 15]. However, it does not represent the distribution of the losses over flooded area due to the neglect of variation in hydrological impacts and building vulnerability parameters in developing the flood-depth damage function.

In recent studies, multivariate flood damage model is an increasingly popular method in modelling flood damage assessment. The researchers developed a multivariate flood damage assessment model representing the relationship between impact parameters (hazard), resistance parameters (vulnerability) and the extent of flood damage [8–11, 16–18]. The most widely used techniques in producing the multivariate flood damage model are multiple linear regression (MLR) [8, 17], Bayesian network (BN) [17, 19], artificial neural network (ANN) [16], regression tree (RT) [10] and random forest (RF) [9, 11, 17]. The process of performing a regression needs empirical flood damage data of the study area. The outcome allows confidently determining which factors are significant most and how they influence each other. Apart from modelling, this regression technique has been applied to generate appropriate new data points to counter the data scarcity problems and deal with incomplete data [11]. However, multivariate flood damage models that consider various damage influencing parameters rather than the ones considered in flood depth damage functions can only be effectively developed if a sufficient and detailed empirical dataset is available [20].

This study aims to develop a multivariate flood damage model to residential and commercial sectors and includes structural and contents damage that correlates flood impacts and resistance parameters to the magnitude of flood damage. This newly derived multivariate flood damage model applies the multiple linear regression (MLR) technique. The inputs to the model are the empirical data from the 2013 flood event in Kuantan River Basin. The model encapsulates more influencing parameters and is divided into three categories, i.e. the hydrological characteristics, socio-economic situation and property characteristics.

2 Methodology

2.1 Study Area

In recent years, Kuantan city has experienced some extreme flood events because it is located near the Kuantan River Basin (KRB) waterway. The most recent flood was in December 2001, December 2012 and November 2020. All of these flood events were as a result of the northeast monsoon. The flood event from 3rd to 5th December 2013 was caused by the prolonged heavy rainfall, high tides and rapid urbanization activity, which is considered the worst flood experienced in Kuantan's recorded history. The return periods of this flood events have been estimated to be approximately 100 years, based on the flood frequency analyses. Furthermore, the water and power supplies were disrupted, agricultural activity was impacted, roads, buildings and personal belongings suffered extensive damage, and insurance claim dramatically increased [21–23]. In addition, the Kuantan River Basin (Fig. 1) is suitable to be selected as the main study area because the KRB is a vital watershed and the key of economic activity at Kuantan city has a high population density.

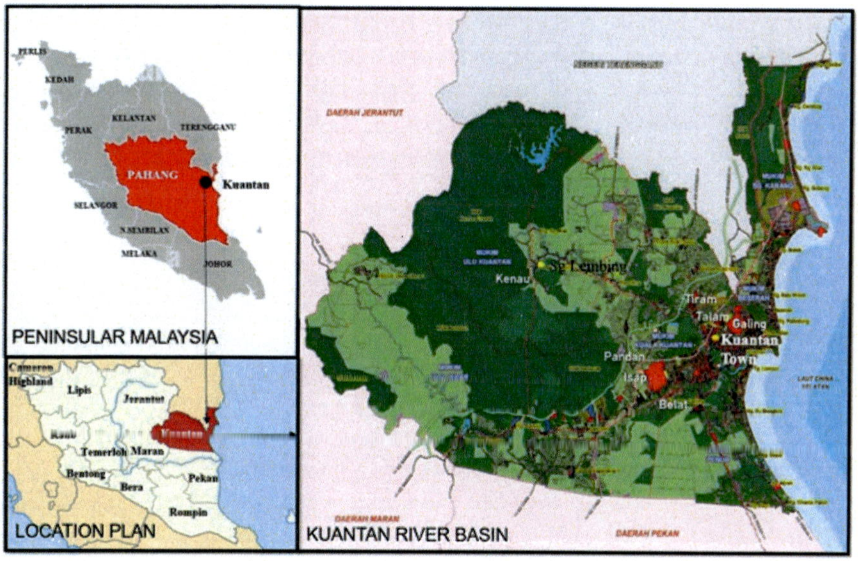

Fig. 1 Kuantan River Basin

2.2 Data

The land-use map was provided by the Department of Town and Country Planning Malaysia, PLAN Malaysia, Pahang and the estimated number of people and households affected by a flood was obtained from the population, and housing census for Malaysia provided by the Department of Statistics, Malaysia (DOS) [21]. Other than that, potential sampling areas were selected based on the previous massive flood event through the information provided by DID Pahang (flood reports), newspaper, Internet sources, government intensive study reports, satellite images (google maps and google street view) and field visit. The target sample size of the population in the affected area was determined based on Krejcie and Morgan required sample table [24, 25] and sampling equation. The respondents for each category were randomly selected according to the target sample size.

A total of 454 respondents from residential sector and 287 from the commercial sector was gathered using face-to-face, telephone interviews and online surveys from 1st April to 15th June, 2021. The questionnaire was designed as a closed-end type where the respondent can respond in a single word, in a short phrase or multiple choices answer. Two separate sets of questionnaires were prepared for residential and commercial sectors. Each set consists of four main sections according to the questionnaire information theme; demographic data, residential or commercial property information, flood data and flood damage data. Following the approach by Refs. [11, 16], the size of the dataset was separated into two partitions; 75% of the respective data was used to train the model and another 25% was used to validate the predicted relative damage.

2.3 Flood Damage Influencing Parameters

The important step in developing the multivariate flood damage model is selecting the flood damage influencing parameters. The parameters that have been selected for residential and commercial categories are displayed in Table 1. Accordingly, based on the average value of damaged items relative to the price of the property or cost of affected contents, the description of damages were published in a percentage of damages for the building structure and contents damage [10, 11]. The value of damaged items obtained from the respondent is either replacement, reconstructing or insurance compensation. The property's price is the current value of the property or affected contents in the year of the flood. Meanwhile, simultaneously, the independent variables were divided into three categories, i.e. the impact characteristics, property characteristics and socio-economic situation [8, 10, 11, 20]. For the residential and commercial sector, the suggested impact characteristics are the observed flood depth and flood duration, and the properties variables are the price of properties, type of property, number of storeys and building material. While the socio-economic

Table 1 Selection of flood damage influencing factors

Dependent variables	Independent variables
Building damage Content damage	*Impact variables* Flood depth and flood duration *Property variables* Price of property, type of property, number of storeys and building material *Socio-economic variables* Household/business income, occupational background, ownership status of property, business income, business type and business size

variables in the residential damage model are household income, occupational background and ownership status of the property, in contrast, for the commercial sector, the socio-economic variables include the information about business income, business type, business size and the information of the ownership status.

Following the approach by Refs. [26, 27], the collected data were statistically studied considering the possibility of potential relations between the resistance parameters and the magnitude of damage data using Pearson's correlation analysis. In the Pearson correlation coefficient analysis, the positive coefficient indicates a positive relationship between variables and negative coefficient means inverse correlation [17].

For impact variables, the range of floodwater depth was classified into 0.5 m range intervals, while the flood duration range into 1–3 days, 4–7 days and more than 7 days. The records from the database concern the observed above floor flood depth. For the price of the property variables, the price of property for residential sector was ranging from low price houses, LPH (RM80,000 and below), medium price houses, MPH (more than RM80,000 and RM200,000), and high price houses, HPH (more than RM200,000). While for the commercial sector, the price of the property was ranged according to the price increase (lower than RM100,000 between RM100,000 and RM200,000, and more than RM200,000). The household income range was classified according to the percentage of total income that is less than RM5000, more than RM5000 to RM10,000, and more than RM10,000. Similarly, to business income, the prescribed range were RM10,000 and below, above RM10,000 to RM20,000, above RM20,000 to RM50,000 and more than RM50,000. The classification and range for other variables are as in Table 2.

2.4 *Multivariate Flood Damage Assessment Model*

After the selection of flood damage influencing parameters was made, the multivariate regression analysis was performed using multiple linear regression (MLR). This probabilistic approach was used to assess flood damage with a linear functional form equation based on the coefficient value of each input variable from the regression

Table 2 Description of selected flood damage influencing parameters

Categories		Variables	Variables character; variation	
			Residential sector	Commercial sector
Independent variables				
Flood impact	FD	Flood depth	C; 0–3.5 m above floor	C; 0–2.5 m above floor
	FDur	Flood duration	C; 1–10 days	C; 2–10 days
Socio-economic characteristics	OWN	Ownerships status	N; own (o) and rent (r)	N; own (o) and rent (r)
	INC	Income	C; RM350–RM15,000	C; RM1500–RM85,000
	OCC	Occupation	N; salaried employees (se) and business (b)	–
	TCom	Business type	–	N; retail (r) and services (s)
	SCom	Business size	–	N; 2–10 employees (mac), 11–20 employees (s) and > 20 employees (m)
Property characteristics	PProp	Price of property	C; RM35,000–RM410,000	C; RM37,000–RM680,000
	TProp	Type of property	N; terrace (t), bungalow (b) and traditional house (th)	N;terrace (t), single unit (su) and warehouse (wh)
	MProp	Building material	N; concrete (c), timber (t) and concrete and timber (ct)	N; concrete (c), timber (t) and concrete and timber (ct)
	SProp	Number of storeys	N; 1-storey, 2-storey, house-on-stilt (hos)	N; 1-storey, 2-storey and 3-storey
Dependent variables				
Damage	DP_{rs}	Damage percentage for residential structural damage		C; 0–100
	DP_{rc}	Damage percentage for residential content damage		C; 0–100
	DP_{cs}	Damage percentage for commercial structural damage		C; 0–100
	DP_{cc}	Damage percentage for commercial content damage		C; 0–100

result. The probabilistic machine learning used in this study was IBM SPSS Statistic Version 28.0.

In constructing the prediction models, the MLR models show the linear relationship between independent and dependent variables. The coefficient of determination value, R^2 value obtained from regression analysis was used to check regression performance [8, 18]. The estimated variable regression coefficient (β_1 to β_i) and significance level (at p-value less than 0.05) was used to measure the most influencing factors to the damages for the residential and commercial sectors [8, 18]. The correlation pattern whether positively or negatively correlated for each variable is determined by the value of the variable's regression coefficient (β_i).

In essence, this model applied ordinary least-squares (OLS) regression technique to include multiple damage influencing parameters by simply adding significant variables to the regression equation [8, 18]. This technique determines the coefficient by representing the relationship of dependent variables (y) and the predicted multiple independent variables (x_1 to x_i). The significance of the relationship between each independent variable and dependent variable consumption was determined by the 5% significance level (p-value is less than 0.05). The functional form of the MLR with OLS technique used is shown in Eq. 1.

$$y = \beta_0 + \beta_1 x_1 + \beta_2 x_2 + \cdots \beta_i x_i \quad (1)$$

When interpreting the results of MLR, beta coefficients are valid while holding all other variables constant, where i is the number of independent variables with the 5% significance level, y is the dependent variable, x is the independent variable, β_0 is the constant (at y-intercept) and β_i is the estimated variable regression coefficient.

Model performance and validation is a significant step in the flood damage model development [4, 9, 26]. The robustness performance of the model ensures the completeness and correctness of the estimation. At the same time, validation establishes its credibility to determine the degree to which it can reasonably represent the assessment [26]. The validations are crucial for testing the reliability of the models' predictions for evaluating damage and ensuring the predictive outcome's precision. Following the approach of [11, 17], the validation of the flood damage model for this study was tested with 25% of the respective empirical dataset which is not used for the training of the model. Accordingly, five statistical measures were used to evaluate the performance of the models; mean absolute error (MAE) and mean bias error (MBE) to test the precision of the model and root mean square error (RMSE) and coefficient of variation (CV) to check the residual variations of the model and the Hit Ratio (HR) to examine the model's reliability [4, 9, 26].

3 Result and Discussion

3.1 Multivariate Regression Analysis

The descriptive analysis of the data obtained from the field survey was presented using frequency distribution charts as shown in Fig. 2. Descriptive analysis was carried out after the smoothing process by discarding the unreliable characteristics data. The result of the field survey found that 173 respondents from 443 residential records were comprised of 39.1% experienced flood with a depth of 1.1–1.5 m, which is the majority of recorded flood depths for the residential sector in the 2013 floods, followed by 19.2% experienced flood with a depth of 0.6–1.0 m, 12.2% with a flood depth below 0.5 m, 11.5% for 1.6–2.0 m, 8.8% for 3.1–3.5 m, 5.4% for 2.1–2.5 m and 3.8% for 2.6–3.0 m. Most of respondents experienced 3–7 days of flood duration which accounted for 48.8%, followed by 47.0% of respondents experienced with flood duration between 1 and 3 days and 4.3% above 7 days. The majority of respondents were salaried employees, which was 89.4% compared with 10.6% of respondents who were businessmen. The average household income was between RM1000 and RM5000, which was 67.9%. Among the 443 respondents, 30.7% earned RM5000–RM10,000, while only 1.4% earned more than RM10,000. 91.2% of the respondents were the owner of the house, while the rest were tenants. The properties were divided into three classes according to the price; LPH, MPH and HPH. Majority of the houses experienced in 2013 flood are MPH, which was 48.8%, 35.2% were LPH and 16% were HPH. Most of the surveyed houses are terrace house (77%) with concrete as the building materials. Meanwhile, 16.5% were traditional houses and 6.5% were bungalows. Majority of the residential buildings were built from the concrete materials and single-storey type.

In terms of business profile, from the 274 databases, majority of the business premises experienced flood with a depth of 1.0–1.5 m, which was 46.7 and 54.4% experienced with flood duration 3–7 days. From the records, 72.6% traded goods and 27.4% ran services type of business. 31.8% of the premises were owned by the respondents, while 68.2% were rented. The distribution of the data shows the majority of the business premises value over RM300,000, which was 89.4%. Most of the premises were built from concrete materials and two-storey buildings.

3.2 *Important Damage Influencing Parameters*

Table 3 gives the Pearson correlation coefficient of the selected flood damage influencing parameters and the damage percentage. The Pearson correlation coefficient test results revealed there is a relationship between impact, resistance parameters (consist of socio-economic and property characteristics) and the percentage of damage. The results show that the selected parameters are suitable to be considered in the multivariate regression analysis.

MLR results for residential and commercial categories are given in Table 4. The variable in which the value of coefficients is zero was excluded. This means that there is no linear relationship between the independent variable (parameter that affects flood damage) and the dependent variable (relative damage) [17]. The positive coefficient means that an increase of independent variables value is associated with an increased independent value, and a negative coefficient means that independent variables and dependent variables change in opposite directions [18].

Based on the results of the MLR analysis, the residential structural damage shown has a negative correlation to property price and single unit building types. However, it is positively correlated with flood depth and double storey buildings. This situation

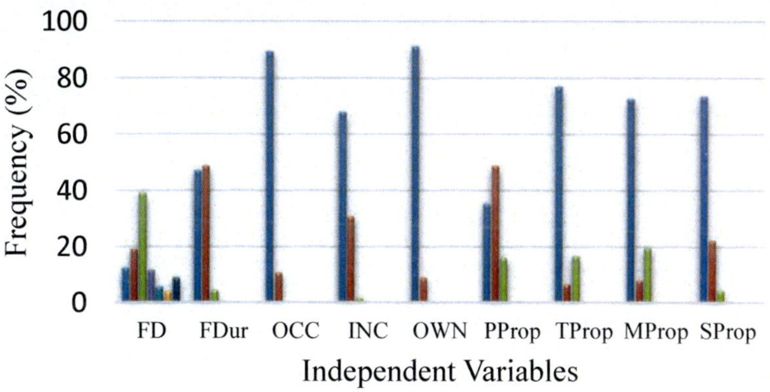

(a)

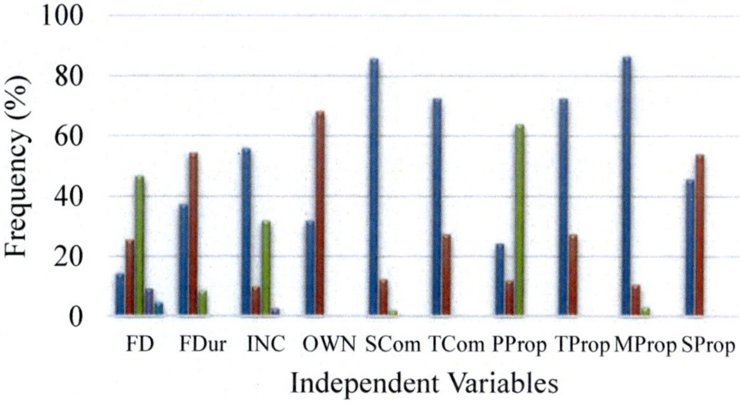

(b)

Fig. 2 Descriptive analysis for **a** residential sector **b** commercial sector

Table 3 Pearson correlation test result for residential and commercial category

Residential sector

	FD	FDur	INC	OCC	OWN	PProp	TProp	SProp	MProp
DP_{rs}	**0.301**	**0.202**	0.03	0.089	0.074	**− 0.147**	0.063	**− 0.099**	0.034
DP_{rc}	**0.282**	**0.116**	**0.177**	0.066	− 0.058	**0.193**	0.043	− 0.025	− 0.083

Commercial sector

	FD	FDur	INC	OWN	SCom	TCom	PProp	TProp	SProp	MProp
DP_{cs}	**0.307**	**0.176**	− 1.115	**− 0.372**	0.014	0.099	**− 0.551**	**0.583**	**− 0.324**	**0.449**
DP_{cc}	**0.275**	**0.198**	**0.264**	**− 0.154**	**0.254**	**0.364**	**− 0.122**	**0.266**	− 0.002	**0.120**

Note Significant correlations (5% significance level) are marked bold

Table 4 Multiple linear regression results for residential and commercial sector

Model	DF_{rs}		DF_{rc}	
	β	Sig	β	Sig
Constant, β_0	1.769		38.715	
FD	**0.95**	**0.00**	**4.65**	**0.00**
FDur	0.12	0.12	0.61	0.07
INC	0.00	0.02	0.00	0.07
OWN_o	–	–	–	–
OWN_r	– 0.25	0.65	0.34	0.88
OCC_se	–	–	–	–
OCC_b	0.71	0.13	0.97	0.63
PProp	**-9.10^{-6}**	**0.00**	**3.0^{-5}**	**0.00**
TProp_t	–	–	–	–
TProp_b	0.24	0.70	– 1.28	0.60
TProp_th	0.89	0.13	– 0.55	0.82
MProp_c	–	–	–	–
MProp_t	**– 1.60**	**0.00**	– 1.23	0.64
MProp_ct	– 0.35	0.45	– 1.23	0.54
SProp_1	–	–	–	–
SProp_2	0.78	0.04	– 1.17	0.51
SProp_hos	– 0.87	0.28	– 1.95	0.58

Model	DF_{cs}		DF_{cc}	
	β	Sig	β	Sig
Constant, β_0	3.464		36.349	
FD	0.10	0.62	**5.33**	**0.01**
FDur	**– 0.16**	**0.01**	**1.44**	**0.01**
INC	-6.92^{-7}	0.94	0.00	0.00
OWN_o	**0.55**	**0.01**	–	–
OWN_r	–	–	– 1.72	0.38
SCom_mac	0.23	0.48	–	–
SCom_s	– 0.52	0.57	1.15	0.72
SCom_m	–	–	– 6.12	0.50
TCom_r	–	–	–	–
TCom_s	0.22	0.36	**7.64**	**0.00**
PProp	**-3.97^{-6}**	**0.00**	6.32^{-7}	0.93
TProp_t	–	–	–	–
TProp_su	**1.51**	**0.00**	5.10	0.05
TProp_wh	**2.68**	**0.04**	12.08	0.35
MProp_c	–	–	–	–

(continued)

Table 4 (continued)

Model	DF_{cs}		DF_{cc}	
	β	Sig	β	Sig
MProp_t	**1.06**	**0.00**	1.86	0.56
MProp_ct	0.80	0.13	− 1.84	0.72
SProp_1	− 0.39	0.12	0.42	0.86
SProp_2	–	–	–	–
SProp_3	0.26	0.86	− 21.91	0.12

Note Significant correlations, Sig. (5% significance level) are marked bold

shows that the level of flood damage to building structures is higher for low price building. It is possible that low price buildings use poor quality materials or do not have flood-resistant features. In regard to timber type house that correlates negatively to structural damage, this result is contrasted with the finding of the study by Zin et al. [8] in Myanmar. Through observation in KRB, most of the houses built from timber are of the stilt house type or traditional house. This type of house is elevated from the ground (with stilts) to avoid flooding. Therefore, the risk of damage is less for this factor. The residential content damage is positively correlated to flood depth and price of property. This expected result shows that the magnitude of damage increases with the increase in flood depth and accords with previous study [2, 28]. In relation to the property's price, this result shows the possibility that more expensive houses have more valuable and luxurious furniture. Therefore, the risk of damage is higher.

For the commercial category, the damage was mostly influenced by flood depth, flood duration, building ownership status, type of property (single unit building and warehouse type), timber building and type of business (service type). Among the possible reasons for the building ownership factor is that homeowners are more willing to use high quality and durable materials to their homes to reduce the risk of damage in the event of a flood. Meanwhile, single unit buildings and warehouses have been identified as having a high risk of structural damage. The possible reason is that single unit buildings and warehouses normally are large and have many built-in structures to accommodate trade goods. Apart from that, single unit buildings are usually built on private land and some of them are built in low-lying areas. Flood duration was found to be negatively correlated with structural damage and inversely correlated with content damage. From the data observed in the study area, building in the area that is often flooded for more than seven days experiences flooding almost every year and does not contain high-value finishes. Therefore, the effect of the flood is less even if the duration of the recorded flood exceeds seven days. In addition, the service business also has the most influential effect on commercial content damage. In the study area, these types of service-based businesses are clinics, banks and motor vehicle service centres. This type of business has high-value equipment that is difficult to relocate in the event of a flood. Therefore, the risk of damage for this type of business is high.

3.3 Model Performance and Validation

The performance of regression analysis for residential and commercial damage has shown a satisfactory R^2 value when compared with previous studies [8, 18]. For the residential category, the R^2 value obtained is 0.61 for the analysis of residential sector, and it is equivalent to the results of a study conducted by [18] by obtaining an R^2 value of 0.61 and [8] which obtained 0.35. While for the commercial category, the R^2 value is also satisfactory which is 0.59 compared with the study by Yu et al. [18] who obtained 0.57.

The validation of flood damage estimations by MLR model shows reasonable results compared with other recently published models [11, 26]. The results show an average MAE of 0.17 and an average RMSE of 0.65. The mean of MBE of around 0 with negative values due to a slight underestimation of the predicted values. These errors are larger than those reported for the validation of the multivariate model using other regression analysis techniques [11, 26]. Meanwhile, the means CV and HR values, respectively, 0.32 and 0.76 are around the values from the study conducted by Nafari [26]. For the commercial sector, the mean of MAE is approximately 0.06. These are lower errors compared with the validation results of the multivariate models by Sieg et al. [11], who observed an MAE of 0.18. The RMSE has a mean value of approximately 0.49 which is lower than residential sector result. The mean value of the MBE is around 0, which was also observed by Refs. [20, 26]. The variation of the errors was checked based on the distribution of the residuals (CV) that shows a little variation in the result, and the value is lower than the result reported in the study by Nafari [26]. The performance of the model is more reliable for commercial sector, since the mean of HR indicator is 0.79 (which is close to perfect prediction values). Overall, the models perform well, however the prediction errors are still relatively large, due to the high variation of damage values especially for the damage estimation to content damage for both residential and commercial sectors (Table 5).

Table 5 Validation of flood damage to residential and commercial sector

Sector	Model performance indicator				
	MAE	RMSE	MBE	CV	HR
DF_{rs} (LPH category)	0.025	0.168	− 0.005	0.481	0.83
DF_{rs} (MPH category)	0.022	0.161	0.007	0.577	0.74
DF_{rs} (HPH category)	0.060	0.256	0.002	0.640	0.75
DF_{rc} (LPH category)	0.120	0.687	0.046	0.091	0.76
DF_{rc} (MPH category)	0.115	0.828	0.081	0.120	0.75
DF_{rc} (HPH category)	0.677	1.792	− 0.095	0.113	0.71
DF_{cs}	0.010	0.090	0.001	0.535	0.78
DF_{cc}	0.114	0.900	− 0.026	0.159	0.79

4 Conclusion

In summary, the development of multivariate flood damage models required large amounts of observed data, and this model is an empirical and site-specific approach [8–10, 17]. Thus, transferring the damage models to another urban area and a flood event, might be does not result in an accurate relationship between the extent of damages and the impacts of flood. Therefore, to allow the transferability of the model to different locations and times, this proposed multivariate flood damage assessment framework can be tailored using empirical dataset collected from the new case study.

In this study, the MLR technique was applied to establish the relationship between flood damages against the hydrological, socio-economic, property characteristics and magnitude of flood damage. In addition, multivariate flood damage model also identifies the most significant flood damage influencing parameters. The regression analysis conducted in this study confirmed that flood damage is not only influenced by the flood depth, but also by socio-economic and property characteristics. The study's finding revealed that flood depth, flood duration, price of property, ownership status, building material, type of building and type of business are the parameters that most significantly affect the magnitude of flood damage.

The validation of the new multivariate flood damage model was performed in comparison with the empirical data obtained from the field survey and shows reasonable results compared with other recently published models. Based on the findings of this study, we recommend that other regression techniques can be applied to identify the approach that produces the best results. Future research is encouraged to calibrate the model with damage data from other flood events. Compilation of results from the study is useful in model transferability and creating flood damage estimation guidelines for Malaysia.

Acknowledgements The authors would like to thank Ministry of Education of Malaysia (MOE) for providing financial support under Fundamental Research Grant Scheme no. FRGS/1/2019/TK01/UMP/02/2 (University reference RDU1901155) and Universiti Malaysia Pahang Al-Sultan Abdullah for technical support and encouragement.

References

1. Mohd Mushar, S.H., Kasmin, F., Syed Ahmad, S.S., Kasmuri, E.: Flood damage assessment: a preliminary study. Environ. Res. Engineyono **75**(3), 55–70 (2019)
2. Win, S., Kawasaki, A., Zin, W.W.: Flood damage assessment for flood prone agricultural areas in the Bago River Basin. ICSE Conf. Pap. (2018)
3. Sulong, S., Romali, N.S.: Flood damage assessment: a review of multivariate flood damage models. Int J GEOMATE **22**(93), 106–113. (2015)
4. Dottori, F., Figueiredo, R., Martina, M.L.V., Molinari, D., Scorzini, A.R.: INSYDE: a synthetic, probabilistic flood damage model based on explicit cost analysis. Nat. Hazards Earth Syst. Sci. **16**(12), 2577–2591 (2016)

5. Velasco, M., Cabello, À., Russo, B.: Flood damage assessment in urban areas. Application to the Raval district of Barcelona using synthetic depth damage curves. Urban Water J. **13**(4), 426–440 (2016)
6. Scorzini, A.R., Frank, E.: Flood damage curves: new insights from the 2010 flood in Veneto, Italy. J. Flood Risk Manag. **10**(3), 381–392 (2017)
7. Frongia, S., Ruiu, A., Sechi, G.M.: Water depth-damage functions for flood direct tangible damage evaluation in built-up areas in Sardinia (Italy). Eur. Water **57**, 3–9 (2017)
8. Zin, W.W., Kawasaki, A., Hörmann, G., Acierto, R.A., San, Z.M.L.T., Thu, A.M.: Multivariate flood loss estimation of the 2018 Bago flood in Myanmar. J. Disaster Res. **15**(3), 300–311 (2020)
9. Malgwi, M.B., Schlögl, M., Keiler, M.: Expert-based versus data-driven flood damage models: a comparative evaluation for data-scarce regions. Int. J. Disaster Risk Reduct. **57** (2021)
10. Nafari, R.H., Ngo, T., Mendis, P.: An assessment of the effectiveness of tree-based models for multi-variate flood damage assessment in Australia. Water (Switzerland) **8**(7) (2016)
11. Sieg, T., Vogel, K., Merz, B., Kreibich, H.: Tree-based flood damage modeling of companies: damage processes and model performance. Water Resour. Res. **53**, 6050–6068 (2017)
12. Olesen, L., Löwe, R., Arnbjerg-Nielsen, K.: Flood damage Assessment—Literature Review and Application to the Elster Creek Catchment. Cooperative Research Centre for Water Sensitive Cities (2016)
13. Amirebrahimi, S., Rajabifard, A., Mendis, P., Ngo, T.: A framework for a microscale flood damage assessment and visualization for a building using BIM–GIS integration. 1–461 (2016)
14. White, G.F.: Choice of adjustment to floods. Research paper No. 93. Department of Geography, University of Chicago, Illinois, United States (1964)
15. Martínez-Gomariz, E., Forero-Ortiz, E., Guerrero-Hidalga, M., Castán, S., Gómez, M.: Flood depth-damage curves for Spanish urban areas. Sustainability **12**(7) (2020)
16. Amadio, M., Scorzini, A.R., Carisi, F., Essenfelder, A.H., Domeneghetti, A., Mysiak, J., Castellarin, A.: Testing empirical and synthetic flood damage models: the case of Italy. Nat. Hazards Earth Syst. Sci. **19**(3), 661–678 (2019)
17. Wagenaar, D., De Jong, J., Bouwer, L.M.: Multi-variable flood damage modelling with limited data using supervised learning approaches. Nat. Hazard. **17**(9), 1683–1696 (2017)
18. Yu, I., Necesito, I.V., Kim, H., Cheong, T.S., Jeong, S.: Development of multivariate flood damage function for flood damage assessment in Gunsan City, Korea. Korean Soc. Hazard Mitig. **17**(2), 247–258 (2017)
19. Paprotny, D., Kreibich, H., Morales-Nápoles, O., Wagenaar, D., Castellarin, A., Carisi, F., Bertin, X., Terefenko, P., Merz, B., Schröter, K.: Flood exposure and vulnerability estimation methods for residential and commercial assets in Europe. In: EGU General Assembly (2020)
20. Carisi, F., Schröter, K., Domeneghetti, A., Kreibich, H., Castellarin, A.: Development and assessment of uni- and multivariable flood loss models for Emilia-Romagna (Italy). Nat. Hazards Earth Syst. Sci. **18**(7), 2057–2079 (2018)
21. DID: The establishment of Sungai Kuantan integrated river basin management plan. Final Report, Dep. Irrig. Drain **1937**, 162–173 (2018)
22. Safiah Yusmah, M.Y., Bracken, L.J., Sahdan, Z., Norhaslina, H., Melasutra, M.D., Ghaffarianhoseini, A., Sumiliana, S., Shereen Farisha, A.S.: Understanding urban flood vulnerability and resilience: a case study of Kuantan, Pahang. Malaysia. Nat. Hazards **101**(2), 551–571 (2020)
23. DID: Updating of condition of flooding and flood damage assessment in Malaysia. Final Report, Dep. Irrig. Drain. **2**(May), 1–74 (2012)
24. Krejcie, R.V., Morgan, D.W.: Determining sample size for research activities. Educ. Psychol. Measur. **30**(3), 607–610 (1970)
25. Ahmad, H., Halim, H.: Determining sample size for research activities: the case of organizational research. Bus. Rev. **2**(1), 20–34 (2017)
26. Nafari, R.H.: Flood damage assessment in urban areas. Doctoral dissertation, University of Melbourne, Australia (2018)

27. Komolafe, A.A., Herath, S., Avtar, R.: Establishment of detailed loss functions for the urban flood risk assessment in Chao Phraya River basin, Thailand. Geomat. Nat. Haz. Risk **10**(1), 633–650 (2019)
28. Romali, N.S.: Flood damage and risk assessment for urban area in Segamat. Doctoral dissertation, University of Technology, Malaysia (2018)

Performance Quantification of Flood Hydrograph Models Based on Catchment Morphometry Derived from 30 m SRTM DEM

I. Gede Tunas, Yassir Arafat, and Rudi Herman

Abstract Synthetic hydrograph unit (SUH) is one type of hydrological models for predicting floods at a watershed outlet. The accuracy of this model, apart from being dependent on the distribution and intensity of the rainfall, is also largely determined by the characteristics of the watershed, especially the area of the watershed and the configuration of the drainage network system. The characteristics of the watershed that can be described by morphometric and fractal properties, where the determination of which is very dependent on the resolution of the digital elevation model (DEM) data as the basis for derivation. This paper aims to evaluate the performance of SUH models based on catchment morphometry derived from Shuttle Radar Topography Mission (SRTM) DEM with a resolution of 30 m. This topographic elevation data is one of the free medium-resolution DEMs produced by the US National Geospatial-Intelligence Agency (NGA) and National Aeronautics and Space Administration (NASA). Two flood hydrograph models based on catchment morphometry—GAMA I and ITS-2—were evaluated by quantifying their performance using root mean squared error (RMSE) criteria. This performance indicator was measured based on predicted and measured unit hydrograph in one of sub-watersheds of Palu, Sulawesi, Indonesia, namely Bangga Catchment. Predicted unit hydrograph is interpreted from morphometric and fractal catchment parameters, while the observed unit hydrograph is derived from several flood hydrograph data. The evaluation results showed that the two hydrograph models that were examined showed very good performance with an RMSE below 1. This indicates that the 30 m SRTM DEM used to derive catchment parameters is still relatively good in defining the characteristics of the catchment. However, for applications to other catchments using the same DEM type, the rainfall characteristic must be carefully identified in order to represent the rainfall distribution pattern across the catchment.

Keywords Synthetic unit hydrograph · ITS-2 model · Digital elevation model

---

I. G. Tunas (✉) · Y. Arafat · R. Herman
Department of Civil Engineering, Universitas Tadulako, Jalan Soekarno-Hatta Km. 9, Palu, Central Sulawesi 94117, Indonesia
e-mail: tunasw@yahoo.com

N. Sabtu (ed.), *Proceedings of AWAM International Conference on Civil Engineering 2022 - Volume 3*, Lecture Notes in Civil Engineering 386,
https://doi.org/10.1007/978-981-99-6026-2_37

1 Introduction

The use of digital elevation model (DEM) data for hydrological modeling is currently very intense. The fundamental advantage of using this data is that identification of the morphometric parameters of the catchment such as area, river network system, surface slope and several other parameters can be easily and quickly carried out [1, 2]. However, the accuracy of the derivation of these parameters depends on the resolution of the data used. The interpretation accuracy is proportional to the resolution of the DEM data.

In hydrological analysis, especially those related to rainfall-runoff transformations, the use of DEM data can be applied at the catchment model development stage. Catchment is one of the most important elements in the transformation process. Apart from being influenced by the nature of the rainfall as an input, the discharge characteristics of the transformation are also very much determined by the characteristics of the catchment as a transformation medium [3, 4]. The quality of DEM, which is expressed by resolution, greatly affects the quality of the resulting catchment model.

Several researchers have conducted studies on the application of DEM for watershed modeling, especially for spatial-based hydrological analysis such as the Soil and Water Assessment Tool (SWAT), HEC-HMS, Coupled Routing and Excess Storage (CREST), Non-point Area Source Watershed Environment Response Simulation (ANSWERS), TOPMODEL and other hydrological models [5–10]. Azizian and Shokoohi [11], Nagaveni et al. [12] and Zhang et al. [13], conducted a study on the use of SRTM DEM data as a basis for catchment delineation. The results of delineation and stream reconstruction are used as the basis for flow modeling. Several other researchers have also conducted the same study, using different types of DEM [14–16].

The content of the study in this paper is considered very important because its application will be tested with hydrological models developed in Indonesia, such as: GAMA I and ITS-2. Both of these hydrograph models function to estimate the flood peak with catchment morphometry as the main parameter. The study of DEM applications in these two models is expected to provide information about the opportunities for using the 30 m SRTM DEM data as a basis for compiling a catchment model. The objective of the research is to evaluate the performance of the SUH Models based on catchment morphometry derived from the SRTM DEM.

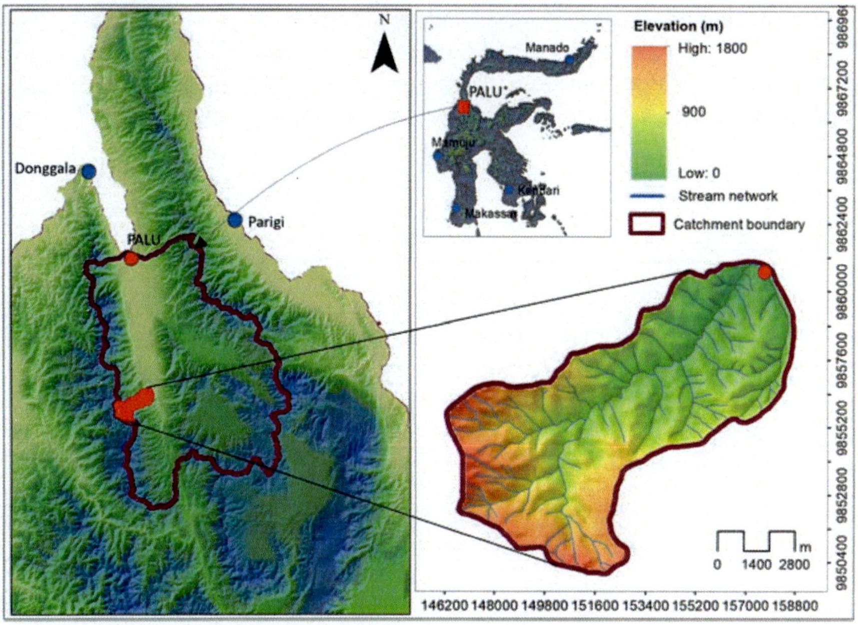

Fig. 1 Detailed location of study site

2 Material and Methods

2.1 Site

This research site is located in the sub-watersheds of Palu, in this case is Bangga catchment. This catchment occupies the western side of Palu watershed with an area of approximately 68.28 km^2 (Fig. 1). This catchment is one of the important sub-catchments in the study area, because apart from being a source of water supply, this catchment is a flood-prone area that contributes greatly to the stability of flow and sediment transport in the Palu River.

2.2 Data

The main data for this study were DEM data recorded by the Shuttle Radar Topography Mission (SRTM). This data can be obtained from https://earthexplorer.usgs.gov/ [17]. DEM with a resolution of 30 m is available not only for Indonesia area but also throughout the world. Apart from the official website, this DEM can also be accessed for free on several other sites.

Other data used for analysis was the observed unit hydrograph data. This data is a type of secondary data obtained from the results of research and previous publications. This data is used to measure the performance of the hydrograph model based on the catchment morphometric parameters derived from the DEM data.

2.3 Methods

This research is based on the results of unit hydrograph derivation using GAMA I and ITS-2 models. The catchment morphometric parameters used to determine the unit hydrograph parameters were interpreted from the SRTM DEM data with a resolution of 30 m using the geographic information system tool. Unit hydrograph parameters are calculated using the basic formula for each model. For the GAMA I model, the basic equation is stated by the following formula [18, 19]:

$$T_p = 0.43\left(\frac{L}{100\ \mathrm{SF}}\right)^3 + 1.0665\ \mathrm{SIM} + 1.2775 \tag{1}$$

$$Q_p = 0.1836 A^{0.5886} \mathrm{JN}^{0.2381} T_p^{-0.4008} \tag{2}$$

$$T_b = 27.4132 T_p^{0.1457} S^{-0.0986} \mathrm{SN}^{0.7344} \mathrm{RUA}^{0.2574} \tag{3}$$

where T_p = time of rise (hour), L = main stream length (km), Q_p = peak flow (m^3/s), T_b = time of base (hour), and S = average main stream slope. Other supporting parameters such as SF refer to the source factor, SIM = symmetry factor, JN = total number of stream junctions, SN = source frequency, and RUA = relative upstream area of catchment.

Similar equation for ITS-2 model, contains three main hydrograph parameters and one hydrograph shape equation as follows [20–22]:

$$T_p = C_1(0.102\ L - 0.162\ D - 0.524\ R_L + 1.24) \tag{4}$$

$$T_b = C_2(0.136\ A - 43.0\ S + 11.5) \tag{5}$$

$$Q_p = \frac{R}{3.6T_p}\frac{A}{A_{\mathrm{SUH}}} \tag{6}$$

where $q(t)$ is unit discharge (dimensionless). C_1, C_2, and C_3 are coefficients of hydrograph main parameters. D is density of stream network (km/km^2) and R_L is ratio of stream length [14]. A is area of catchment (km^2), R is unit rainfall (mm), A_{SUH} is area under hydrograph shape, and T = time (hour).

The performance of the unit hydrograph model is measured by comparing the discharge prediction results with the measured unit hydrograph using the following root mean square error (RMSE) approach:

$$\mathrm{RMSE} = \sqrt{\frac{1}{N}\sum_{i=1}^{N}\left(Q_{o_i} - Q_{s_i}\right)^2} \tag{7}$$

where N = the numbers of data and Q_{o_i} and Q_{o_i} are measured flow (m^3/s) and predicted flow (m^3/s). The best indication of RMSE is exactly equal to zero.

3 Result and Discussion

3.1 Morphometric Parameters

GeoTIFF is a DEM data format that can be accessed on its official page. This type of data format is very compatible with various types of GIS applications: ArcGIS, QGIS, Map Info, and other applications. The DEM data selected is the DEM which covers the entire study area (Fig. 2a). The area studied is located in the elevation range between 150 and 1800 m above sea level (Fig. 2b). In this type of DEM, the appearance of objects that illustrate the topography is relatively clear, especially the terrain ridges, valleys, slopes, and plains. The stream network configuration is also quite clear.

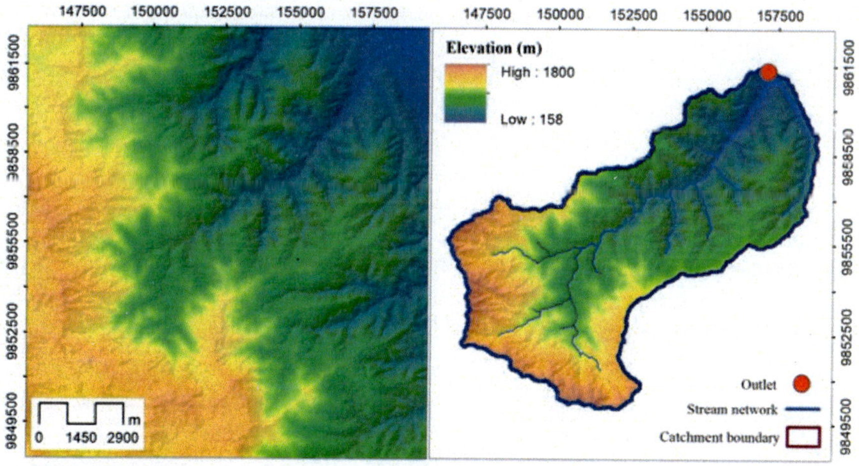

(a) 30m SRTM DEM (b) Basin boundary

Fig. 2 Boundary and stream network of the Bangga catchment derived from 30 m SRTM DEM

Table 1 Morphometric parameters of the Bangga catchment

No	Morphometric parameters	Symbol	Unit	Dimension
1	Catchment area	A	km^2	68.28
2	Area of the catchment at the upstream of gravity center	AU	km^2	35.26
3	Width of the catchment at the point 0.75 L from the gauging site	WU	km	7.99
4	Width of the catchment at the point 0.25 L from the gauging site	WL	km	5.69
5	Average main stream slope	S	–	0.0859
6	Main stream length	L	km	16.48
7	Total length of the first order streams	$\Sigma L1$	–	68.90
8	Total length of streams of all order	ΣLu	–	116.96
9	Total number of segments of the first order streams	$\Sigma N1$	–	85
10	Total number of stream segments	ΣNu	–	120
11	Total number of stream junctions	JN	–	84
12	Drainage density	D	km/km^2	1.88
13	Ratio of stream length	RL	–	0.90

Delineation and generation of a stream network based on GIS produces a catchment model as shown in Fig. 2. The outlet point is fixed on the hydrometry gauge, located upstream of Bangga Weir. The delineated catchment shows an elongated shape with a ratio of 3:1 to the width. Drainage density decreases with reduction of elevation. In detail, the morphometric parameters associated with this delineation process can be seen in Table 1 with reference to Fig. 3. These morphometric parameters serve as the basis for determining hydrograph parameters.

3.2 *Unit Hydrograph*

The hydrograph parameters of the two synthetic unit hydrograph models were computed from the catchment morphometric parameters generated from the DEM data. The three hydrograph parameters of the two models are T_p, T_b and Q_p using the corresponding formula according to Eqs. 1–3 for GAMA I and Eqs. 4–6 for ITS-2. The values of the three main parameters are then formulated to define a hydrograph shape in the form of a curve including a rising limb and a recession limb. The form of the two synthetic unit hydrographs is compared with the measured unit hydrograph obtained from the results of research and previous publications.

Observation in Fig. 4, the two predictive unit hydrographs show the conformity of the shape with the measured unit hydrograph. On the rising and falling sides, the proximity of the GAMA I curve is better than that of the ITS-2 curve. However,

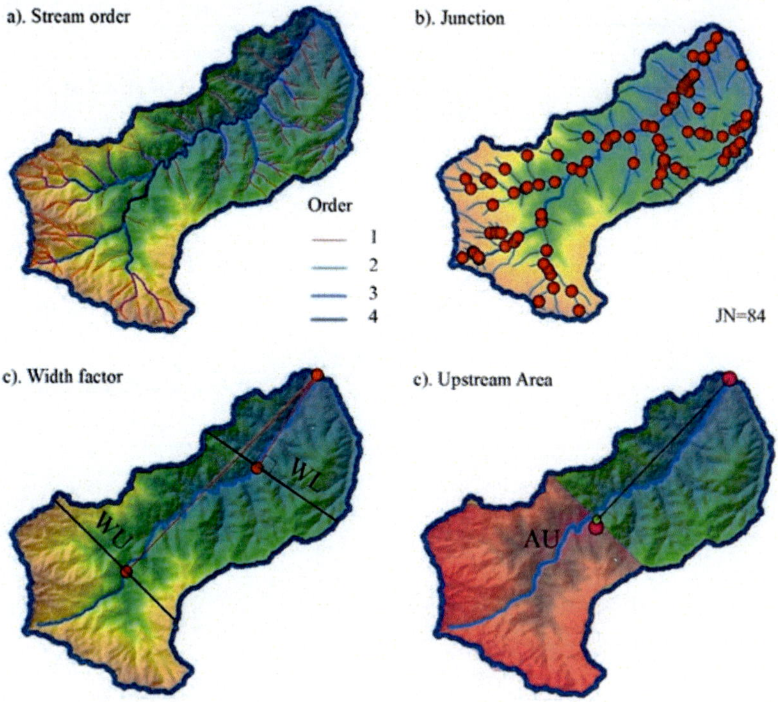

Fig. 3 Morphometric parameters derived from 30 m SRTM DEM

at the hydrograph peak, the ITS-2 discharge deviation was lower than the GAMA I discharge deviation. Even the ITS-2 peak discharge was almost the same as the observed hydrograph peak discharge. Both hydrographs show the advantages of each hydrograph parameter.

Quantitatively, the performance of the two hydrograph models can be measured by various types of indicators, one of which is RMSE. All performance measuring indicators compare all predicted and observed flows at the same time step. The results of the analysis show that the performance of both models is very good with RMSE < 1 (Table 2). This indicator shows that the 30 m SRTM DEM is sufficient to derive the morphometric parameters of the catchment.

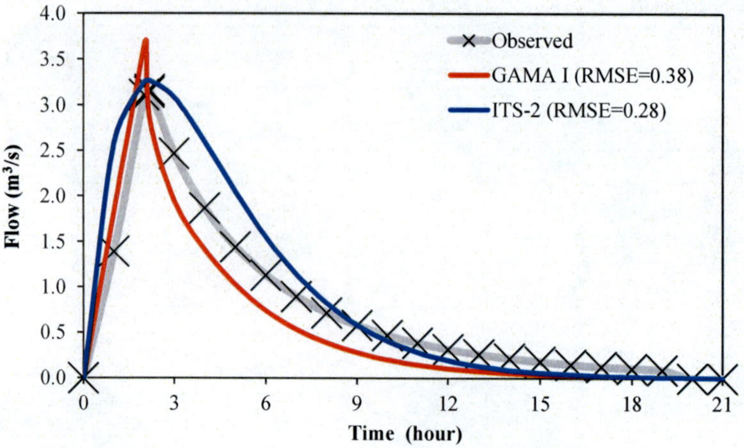

Fig. 4 Observed and simulated hydrographs using GAMA I and ITS-2

Table 2 Measurement of simulation performance using RMSE

Observed UH		GAMA I		ITS-2	
Time (h)	Hydrograph ordinate (m^3/s)	Hydrograph ordinate (m^3/s)	$(Q_{sim} - Q_{obs})^2$	Hydrograph ordinate (m^3/s)	$(Q_{sim} - Q_{obs})^2$
0	0.00	0.00	0.00	0.00	0.00
1	1.40	1.97	0.33	1.72	0.11
2	3.11	3.76	0.92	3.23	0.01
2.18	3.17	3.65	0.29	3.39	0.15
2.79	2.60	2.78	0.15	3.55	0.14
3	2.47	1.92	0.31	3.53	0.40
4	1.87	1.39	0.23	3.10	0.29
5	1.44	1.01	0.19	2.40	0.08
6	1.13	0.73	0.16	1.73	0.00
7	0.90	0.53	0.14	1.18	0.02
8	0.72	0.38	0.11	0.77	0.05
9	0.58	0.28	0.09	0.49	0.08
10	0.47	0.20	0.07	0.31	0.08
11	0.39	0.15	0.06	0.19	0.08
12	0.32	0.11	0.05	0.11	0.06
13	0.26	0.08	0.03	0.07	0.05
14	0.22	0.06	0.03	0.04	0.04
15	0.18	0.04	0.02	0.02	0.03
16	0.15	0.03	0.02	0.01	0.02
17	0.13	0.02	0.01	0.01	0.02
18	0.11	0.02	0.01	0.00	0.01
19	0.09	0.01	0.01	0.00	0.01
20	0.00	0.01	0.00	0.00	0.00
21	0.00	0.01	0.00	0.00	0.00
Total			3.22		1.71
RMSE			0.38		0.28

4 Conclusion

The performance of the two flood hydrograph models has been evaluated based on morphometric parameters derived from the 30 m SRTM DEM data. The two hydrological models: GAMA I and ITS-2 are rainfall-runoff transformation models based on synthetic unit hydrographs. The morphometric parameters generated from the DEM data illustrate the catchment configuration especially with regard to the catchment boundary for determining its area, network system stream, surface slope, and special parameters related to the fractal catchment characteristics. This fractal parameter is specifically used in the ITS-2 model.

The results of geographic information system analysis inform that the morphometric parameters derived from the DEM data show high detail of the object features and that it is very adequate for determining the synthetic unit hydrograph parameter. The use of DEM data as the basis for catchment modeling was then evaluated on the level of accuracy of the applied hydrograph model. Based on the evaluation results, both models show very good performance with an RMSE below 1. This indicates that the 30 m SRTM DEM data is very adequate for catchment modeling with regard to applications for hydrograph modeling.

Acknowledgements The authors acknowledge the Ministry of Research and Technology/National Research and Innovation Agency for its support through the Fundamental Research Program, Grant No: 155/E4.1/AK.04.PT/2021.

References

1. da Cunha, E.R., Bacani, V.M.: Morphometric characterization of a watershed through SRTM data and geoprocessing technique. J. Geogr. Inf. Syst. **8**, 238–247 (2016)
2. Vieceli, N., Bortolin, A.B., Mendes, L.A., Bacarim, G., Cemin, G., Schneider, V.E.: Morphometric evaluation of watersheds in Caxias do Sul City, Brazil, using SRTM (DEM) data and GIS. Environ. Earth Sci. **73**, 5677–5685 (2015)
3. Walega, A., Ksiazek, L.: Influence of rainfall data on the uncertainty of flood simulation. Soil Water Res. **11**, 277–284 (2016)
4. Shrestha, R.R., Dibike, Y.B., Prowse, T.D.: Modelling of climate-induced hydrologic changes in the Lake Winnipeg watershed. J. Great Lakes Res. **38**(3), 83–94 (2012)
5. Yan, B., Fang, N.F., Zhang, P.C., Shi, Z.H.: Impacts of land use change on watershed streamflow and sediment yield: an assessment using hydrologic modelling and partial least squares regression. J. Hydrol. **484**(25), 26–37 (2013)
6. Jodar-Abellan, A., Valdes-Abellan, J., Pla, C., Gomariz-Castillo, F.: Impact of land use changes on flash flood prediction using a sub-daily SWAT model in five Mediterranean ungauged watersheds (SE Spain). Sci. Total Environ. **657**(20), 1578–1591 (2019)
7. Alaghmand, S., Abdullah, R., Abustan, I., Said, M.A.M., Vosoogh, B.: Gis-based river basin flood modelling using HEC-HMS and MIKE11-Kayu Ara river basin, Malaysia. J. Environ. Hydrol. **20**(16), 1–16 (2012)
8. Daffi, R.E., Ahuchaogu, I.I.: Delineation of river watershed and stream network using ILWIS 3.7. 1 Academic. Asian J. Environ. Ecol. **4**(4), 1–8 (2017)
9. Emam, A.R., Kappas, M., Linh, N.H.K., Renchin, T.: Modeling and runoff mitigation in an ungauged basin of Central Vietnam using SWAT model. Hydrology **4**(1), 1–17 (2017)
10. Sahour, H., Mokhtari, A., Ghahfarokh, S.S.: Rainfall-runoff modeling using remotely sensed data and the hydrologic modeling system (HEC-HMS). Ecol. Environ. Conserv. **22**(4), 163–173 (2016)
11. Azizian, A., Shokoohi, A.: Effects of data resolution and stream delineation threshold area on the results of a kinematic wave based GIUH model. Water SA **41**(1), 61–70 (2015)
12. Nagaveni, C., Kumar, K.P., Ravibabu, M.V.: Evaluation of TanDEMx and SRTM DEM on watershed simulated runoff estimation. J. Earth Syst. Sci. **128**, 1–10 (2019)
13. Zhang, H., Li, Z., Saifullah, M., Li, Q., Li, X.: Impact of DEM resolution and spatial scale: analysis of influence factors and parameters on physically based distributed model. Adv. Meteorol. **8582041**, 1–10 (2016)

14. Rokni, K., Ahmad, A., Hazini, S.: Comparative analysis of ASTER DEM, ASTER GDEM, and SRTM DEM based on ground-truth GPS data. J. Teknologi **76**(1), 97–102 (2015)
15. Buakhao, W., Kangrang, A.: DEM resolution impact on the estimation of the physical characteristics of watersheds by using SWAT. Adv. Civ. Eng. **8180158**, 1–9 (2016)
16. Saksena, S., Merwade, V.: Incorporating the effect of DEM resolution and accuracy for improved flood inundation mapping. J. Hydrol. **530**, 180–194 (2015)
17. USGS Homepage. https://earthexplorer.usgs.gov/. Last accessed on 28 Apr 2021
18. Brotowiryatmo, S.H.: Hydrology: Theory, Problem and Solving. Nafiri Offset, Yogyakarta (2000) [in Indonesian]
19. Huyskens, R.B.H.: Accuracy of flood estimation with the Gama I unit hydrograph. MSc thesis. Faculty of Civil Engineering Delft University of Technology, Delf (1991)
20. Tunas, I.G., Tanga, A., Oktavia. S.R.: Impact of landslides induced by the 2018 Palu Earthquake on flash flood in Bangga River Basin, Sulawesi, Indonesia. J. Ecol. Eng. **21**(2), 190–200 (2020)
21. Tunas, I.G.: The application of ITS-2 model for flood hydrograph simulation in large-size rainforest watershed, Indonesia. J. Ecol. Eng. **20**(7), 112–125 (2019)
22. Tunas, I.G., Anwar, N.: A flood forecasting model based on synthetic unit hydrograph of ITS-2. In: IEEE Xplore Conferences, vol. 9084614, pp. 42–46 (2020)

Determination of Downstream Hydraulics Geometry Parameters

Zuliziana Suif, Mira Azmiza Azmi, Nordila Ahmad, Othman, and Jestin Jelani

Abstract The hydraulic geometry term represents the relation between the dynamic characteristics of a river stream and its discharge. The river characteristics quantitatively obtained by describing these parameters as power functions of discharge employing one coefficient and exponent. The aim of the study is determining the downstream hydraulic geometry parameters and to obtain a relation between flow discharge and sediment discharge with hydraulic geometry parameters of the channel. The depth, velocity, width, surface roughness and suspended sediment concentration data are collected at three selected stations along the channel. The results show that the downstream hydraulic geometry parameters for flow discharge relations b, f, m, p, a, c, k and N are discovered as 0.0417, 0.8889, 0.0789, 0.4580, 1.9842, 1.9710, 0.3693 and 1.8056, respectively. For sediment discharge relations, the hydraulic geometry parameters b, f, m, p, a, c, k and N are shown as − 0.0454, 0.9054, 0.3090, 0.1739, 1.1712, 2.1230, 0.3348 and 0.9325, respectively. The values of exponent and coefficient, (b, f, m, p, a, c, k and N) varies within some limits imposed more by ways the adjustment of stream power is dispersed across variables. This is because the scales parameters are varying and depends on the hydraulics parameters of channel. In addition, according to the results, the sediment and flow discharge relations at Station 1 were determined being more relevant than those at a different station when it comes to the correlation coefficient with average co coefficient and exponent of 0.75 and 0.25, respectively. Generally, the findings from this study are intended to be beneficial for manager and planner of water resources initiatives in the basin.

Keywords Hydraulic geometry · Power functions · Flow discharge · Sediment discharge

Z. Suif (✉) · M. A. Azmi · N. Ahmad · Othman · J. Jelani
Department of Civil Engineering, Faculty of Engineering, National Defence University of Malaysia, Kuala Lumpur, Malaysia
e-mail: zuliziana@upnm.edu.my

N. Sabtu (ed.), *Proceedings of AWAM International Conference on Civil Engineering 2022 - Volume 3*, Lecture Notes in Civil Engineering 386,
https://doi.org/10.1007/978-981-99-6026-2_38

1 Introduction

The depth, velocity, breadth, and surface roughness of rivers, as well as their hydraulic geometry characteristics, are related to their discharge as simple power functions at a particular cross-section. Hydraulic geometry of a river does not remain exactly the same. It shows variations over a time period as a function of discharge. Hydraulic geometry parameters are important in representing the form of a stream. The characteristics of a river can be determined by using the hydraulic geometry parameters. The measurable hydraulic characteristic such as depth, velocity, width, and surface roughness represents the form of a river. These measurable hydraulic characteristics will be used in the power function analysis. Depth, velocity, width, and surface roughness indicate the flow discharge as a power function.

Most commonly used and very effective hydraulic geometry estimation method was introduced by Leopold and Maddock [1]. They discovered that several quantitatively measurable hydraulic properties of stream channels, such as width, depth, velocity, and suspended sediment load are fluctuating with discharge as simple power functions for a particular river cross-section. The functions derived from various cross-sections along the river differ only in the numerical values of the coefficients and exponents. The functions derived from numerous cross-sections alongside river differs simply in regards of coefficients and exponents numerical values [1, 2]. Leopold and Maddock [1] introduced hydraulic geometry theory by analysing the cross-sectional data of 20 rivers and proposed a power law relation between the flow discharge (Q), average water depth (H), average flow velocity (V), B is width water surface and surface roughness (n) as follows;

$$B = aQ^b \tag{1}$$

$$H = cQ^f \tag{2}$$

$$V = kQ^m \tag{3}$$

$$n = NQ^p \tag{4}$$

where a, b, c, f, k, m, N and p are numerical constants in which b, f, m and p are its exponents; meanwhile a, c, k and N are its coefficients.

Equations (5) and (6) indicate that the depth, velocity, width, and surface roughness of rectangular channels fulfil the continuity equation, which is to mean that the summation of exponents and multiplication of coefficients are equal to 1. The relation of hydraulic geometry assumes that the flow of water is uniform and steady. The exponents and coefficients are calculated using the power function approach.

$$a \times c \times k \times n = 1 \tag{5}$$

$$b + f + m + p = 1 \tag{6}$$

The continuity equation which means summation of the exponents and multiplication of coefficients are equal to 1. The continuity equation requires that $a \times c \times k \times N$ and $b + f + m + p$ both equal unity at a channel cross-section, but not when derived from multiple transects with different shapes. These continuity equations were only useful on rectangular channels.

Watershed management is critical due to increasing population and human activities, particularly in UPNM area. Therefore, an appropriate catchment area management is required to fulfil water resource demand. For this purpose, the understanding of the concepts and behaviours of river channel hydraulic geometry is necessary. Hydraulic geometry implies to the exponential correlations with hydraulic geometry parameters and flow discharge. The aims of this study are to determine the downstream hydraulic geometry parameters and to obtain relations between flow discharge and sediment discharge with hydraulic geometry parameters of the Universiti Pertahanan Nasional Malaysia (UPNM) channel.

2 Study Area

The study area in determining the hydraulic geometry parameter located in Universiti Pertahanan Nasional Malaysia (UPNM) encompasses an area of roughly 1.51 km^2. Three (3) stations were utilized in UPNM area. The first (S1) and second station (S2) are trapezoidal channels that are man-made, while the Earth channel is the third station (S3). All the locations of the station are shown in Fig. 1.

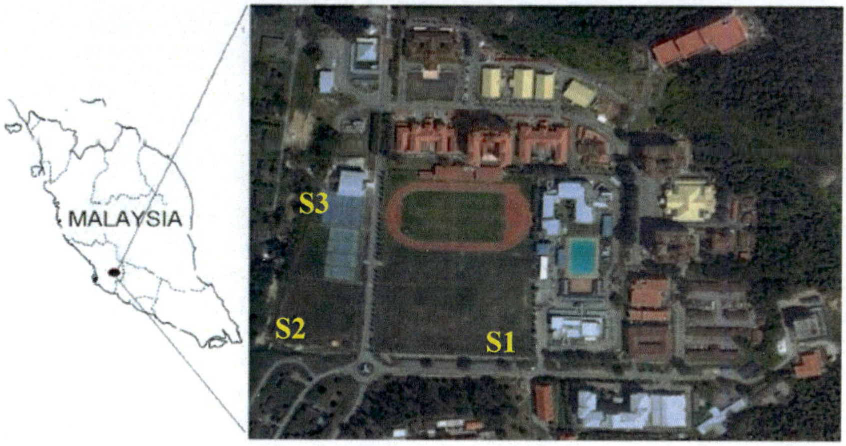

Fig. 1 Location of utilized stations in UPNM area

3 Methodology

For this study, the purpose of installation of the equipment is to determine the value of hydraulics geometry parameters in selected stations around UPNM area. The parameters such as velocity (V), depth (D), discharge (Q), surface width (B) and surface roughness (n) were determined using the Flow Tracker2 as shown in Fig. 2. Then, the downstream hydraulic geometry with depth, width, velocity and surface roughness were analyzed using power function analysis. Moreover, the field water sample was collected and tested for suspended sediment concentration. The total suspended solid (TSS) test was performed on the field water sample in the laboratory.

Fig. 2 Flow Tracker2

Fig. 3 Measurement of hydraulic geometry parameters

3.1 *Measuring of Hydraulic Geometry Parameters*

Flow Tracker2 is an equipment that measures hydraulic geometry parameters. The installation of equipment purpose to be estimate the hydraulic geometry of the specified channel [2]. The equipment was setup on the field, as well as all data will be obtained there. The hydraulic geometry parameters obtained from the selected stations were recorded. The Flow Tracker2 is easy to operate and can measure water discharge, velocity, depth, temperature, coordinate of the location, and a variety of other functions. The data was measured like depth, velocity and flow discharge of channel at various stations by utilizing the equipment (Fig. 3).

3.2 *Collecting of Water Sample*

Three (3) water samples were obtained in total with one taken each week to test the suspended sediment concentration. Moreover, the water sample data was obtained at depth of 0.3 m from the surface at the centre in the main channel. Since the sample point is shallow, the technique for obtaining the sample does not pose many difficulties. The sample are transferred to laboratory and tested by TSS testing to determine the concentration of suspended sediment.

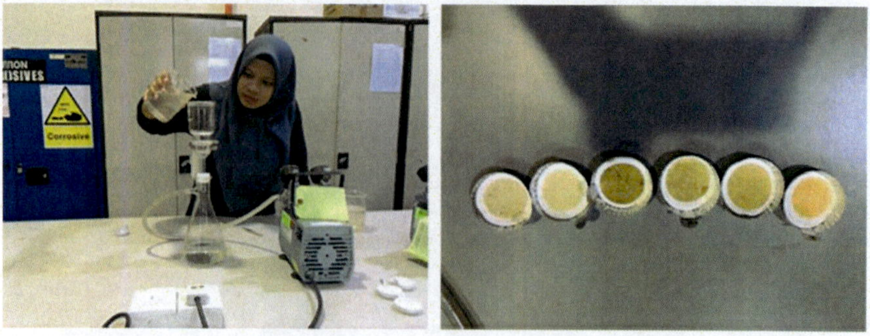

Fig. 4 Total suspended solid (TSS) test

3.3 Suspended Sediment Concentration

A 1.5 L container was used to collect water samples from all stations once a week. The sample of water was sent to laboratory for analysis. Firstly, weigh the filter disc and paper, and record the data. Afterwards, select a sample volume that will result of total suspended solids not greater than 200 mg. Next, apply vacuum while placing the filter on the base and clamping the funnel. To attach the filter against the base, wet it with a tiny amount of distilled water. Mix the sample by shaking vigorously and quantitatively transfer 100 ml of sample to the filter using a large orifice and volumetric pipette. Continue to apply vacuum after the sample has gone through filter to remove all traces of water. Next, rinse the pipette and funnel onto the filter with small volume of distilled water. Lastly, carefully remove the disc filter from the base. Dry for at least one hour at 103–105 °C. Later, cool in an oven and weigh. Aluminum weighing pan that has weighted with glass filter paper first before the glass filter paper used to filter the water sample (Fig. 4).

4 Results and Analysis

4.1 Relations of Flow Discharge and Hydraulic Geometry

Graph in Fig. 5 indicates a relation between the data on depth (H), mean width (B), mean flow velocity (V), roughness (n) and flow discharge (Q) for S1. The plotted data has shown increase in trends with increasing flow discharge (Q) for S1 which is a man-made channel and located in the upstream part in this study. The exponent average values of b, f, m and p are 0.1009, 0.7946, 0.1181, 0.3503, respectively. While, the empirical equation by employing a power function analysis is derived using statistical tool and shown in Fig. 5. Meanwhile, the coefficient average values of a, c, k and N are 1.4527, 1.3173, 0.5077 and 0.2977. From power function, graph

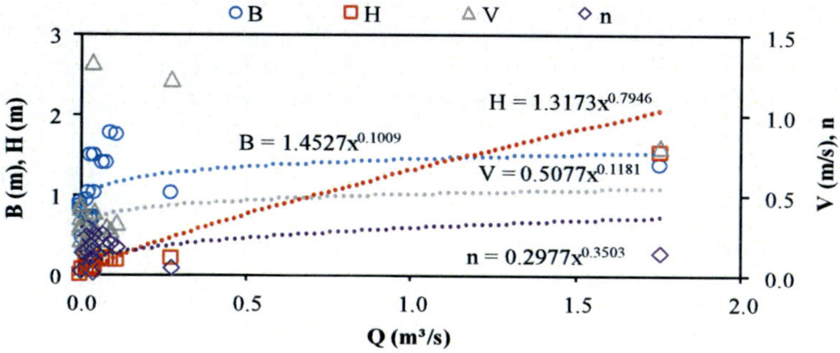

Fig. 5 Relations of flow discharge with hydraulic geometry parameters at S1

of H shows the highest value of data. It indicates the rates at which depth increased as a function of discharge was the highest [3]. It also shows that the highest value of flow discharge (Q) is 1.758 m^3/s.

The station S2 is a man-made channel. The graph in Fig. 6 shows data of width (B), mean depth (H) and roughness (n) were noticed increasing trend with increasing flow discharge (Q) [4]. The value of the exponent (b, f, m and p) are − 0.0017, 1.2811, − 0.2680, 1.1126, while its coefficient (a, c, k and N) are 2.4167, 3.6823, 0.1085 and 4.9764. Meanwhile velocity (V) graph is the only graph that decreases with increasing discharge and which signifies slow flow. Huang and Nanson [3] examined bank vegetation and discovered that dense bank vegetation creates narrower channels, whereas bed vegetation increases the roughness and caused wider channels, lower in flow velocity and therefore no significant difference in depth [4]. Thus, the depth (H) graph only slightly increases against flow discharge (Q) that indicates shallow. In addition, it also shows that low rates of velocity (V) decrease with low rates of width (B) increase when against flow discharge.

Figure 7 illustrates how the depth (H), mean width (B) and mean flow velocity (V) of S3 increased as the flow discharge (Q) increased. S3 is a natural channel and located at downstream. For S3, the exponent values b, f, m and p are 0.0259, 0.5911,

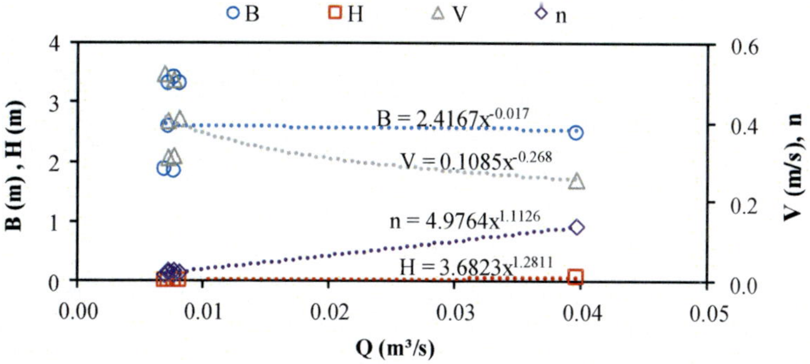

Fig. 6 Relations of flow discharge with hydraulic geometry parameters at S2

0.3866 and − 0.0890. Likewise, the coefficient values *a*, *c*, *k* and *N* value for S3 are 2.0832, 0.9133, 0.4918 and 0.1426. The roughness (*n*) power graph was observed and is the only graph that shows decreasing with the increasing flow discharge (*Q*). This is because erosion decreases the size of bed load material and reduces the surface roughness of channel. Kolberg and Howard [5] ascertained hydraulic geometry parameters demonstrate variations, relying on the bed materials of alluvial channel. Velocity of this channel rises downstream when more water comes towards the channel and reduces the size of bed load. Huang and Warner [6] indicated that the coefficient of hydraulic geometry parameters is correlated to hydraulic roughness (Manning's *n*), bank strength and slope [4]. Thus, the highest value of flow discharge (*Q*) on this station is 1.608 m^3/s which is the value of fifth week and it shows the highest value because of rain.

Table 1 gives the summary of exponent and coefficient values at each station. Exponent *b* and *m* shows the lowest value on S2 among all stations. This is due to the presence of vegetation, grass and sand along the channel. As a result, the channel is becoming narrower and the water flow is becoming unstable. This is because the natural channel is not vulnerable and always affected by water flow. Moreover, during heavy rain, it will either become narrower or enlarge the channel because the wall of the channel having corrosion. In this study, it shows the channel becomes narrower during heavy rain.

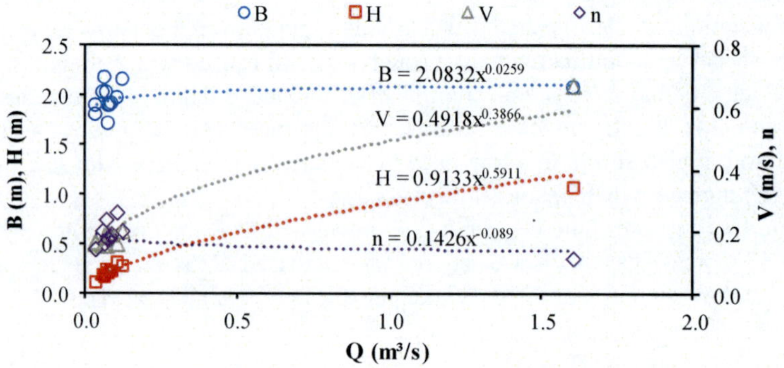

Fig. 7 Relations of flow discharge with hydraulic geometry parameters at S3

Table 1 Exponents and coefficients of flow discharge and hydraulic geometry parameters

Station	Exponent				Coefficient			
	b	*f*	*m*	*p*	*a*	*c*	*k*	*N*
S1	0.1009	0.7946	0.1181	0.3503	1.4527	1.3173	0.5077	0.2977
S2	− 0.0017	1.2811	− 0.2680	1.1126	2.4167	3.6823	0.1085	4.9764
S3	0.0259	0.5911	0.3866	− 0.0890	2.0832	0.9133	0.4918	0.1426
Ave	0.0417	0.8889	0.0789	0.4580	1.9842	1.9710	0.3693	1.8056

The exponent values show that exponent of f at S2 shows the highest value from the other exponent from all stations. This is because S2 has a high depth and it has slower flow of water. Conversely, S3 has low value of exponent f. In general, the higher the exponent value of f, the lower the value of exponent m. Rhodes [7] indicated that the exponent values for high flow rates can be considerably different from those for low flow rates [8]. This explained why their values of exponent m are higher than S2.

Lastly, the value of exponent p shows the highest on S2 due to the lowest rates of erosion because of low flow. Erosion reduces the size of bed load material and decreases the roughness of the channel. It rises downstream as tributaries and groundwater flow supply more water to the channel, and the channel becomes more efficient with distance downstream. S2 shows rougher surface is having a higher roughness exponent and less retarding effect on the water flow, lower flow rate is produced [9]. As a conclusion, flow rate and roughness exponent were influenced by bed roughness.

The above discussion shows that the values of exponents, b, f, m and p do not possess fixed values; rather they vary over certain ranges dictated by the way the adjustment of stream power is distributed among variables. Furthermore, the scale parameters are variable, depending on the channel hydraulics. This observation should be helpful with regionalization of scale factors.

4.2 Relations of Sediment Discharge and Hydraulic Geometry

Figure 8 shows data on width (B), mean depth (H), mean flow velocity (V) and roughness (n) that were observed to increase with increasing sediment discharge (Q_s) for S1 which is a man-made channel. The exponent value of b, f, m and p for Station 1 are 0.0126, 0.6662, 0.2262 and 0.1180, while coefficient value of a, c, k and N are 1.0453, 0.9913, 0.8167 and 0.1285, respectively. From the power function, graph of B shows the highest value of data signifies rates of width (B) increased as a function of discharge was the highest. It also shows the highest value of sediment discharge (Q_s) is 0.241 kg/m^3, which is the most inconsistent and far from the other data. This is because the weather for that week is always cloudy and usually raining heavily in the evening. Climate change, such as more frequent and heavy rain events, can worsen erosion and cause more silt to flow into rivers, lakes and streams. Sediment loading from storm-water runoff can be increased by more frequent and heavy rain events.

While, the S2 is a natural channel and the value of the exponent (b, f, m and p) are − 0.1980, 0.5153, − 0.3540, 0.7027 as shown in Fig. 9, whereas the coefficient value on a, c, k and N are 0.5800, 0.1092, 0.0703 and 0.7862. The plotted graph shows data of mean depth (H) and roughness (n) were observed to increase with increasing sediment discharge (Q_s). Meanwhile velocity (V) graph and width (B) graph is the graph that shows decreasing with the increasing sediment discharge (Q_s) [4]. This is

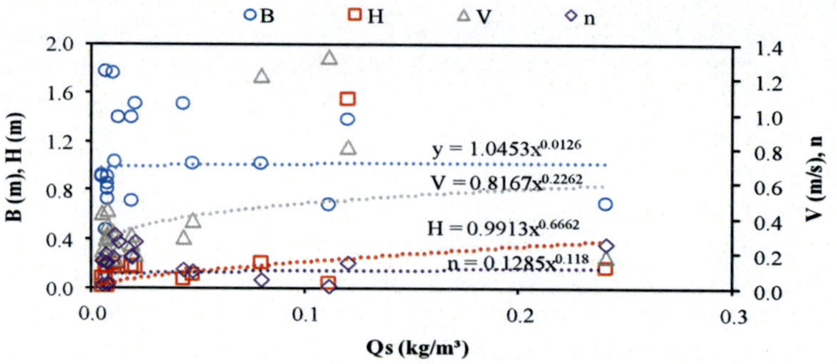

Fig. 8 Relations of sediment discharge and hydraulic geometry parameters at S1

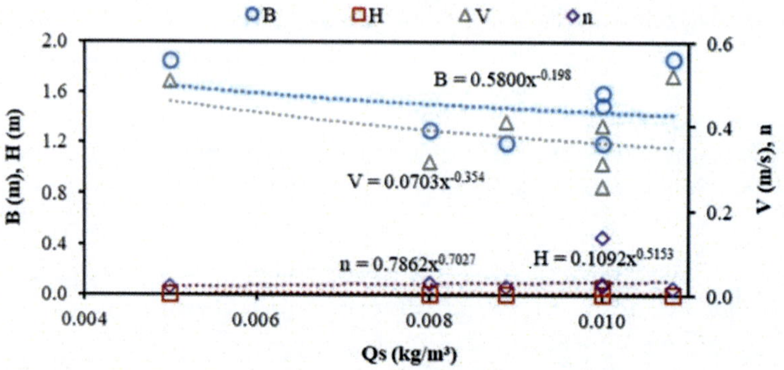

Fig. 9 Relations of sediment discharge and hydraulic geometry parameters at S2

because the data of width (B) affect data of velocity of this channel [9]. The erosion on this channel is the highest since the width (B) always becomes narrower when raining. When the width (B) is narrower, the slower the velocity (V) of the water.

S3 station is downstream natural channel. Figure 10 shows the exponent values b, f, m and p are -0.0060, 0.6459, -0.0750 and 0.4513, respectively. Likewise, the coefficient, a, c, k and N value for S3 are 1.8897, 5.2685, 0.1175 and 1.8827. The graph in Fig. 10 presents data of width (B), mean depth (H) and roughness (n) that were observed to increase with increasing sediment discharge (Q_S). However, the velocity (V) power graph was observed as the only graph that shows slightly decreasing with the increasing sediment discharge (Q_S). Huang and Nanson [3] examined bed vegetation increases the roughness and causes wider channels, reduction in flow velocity and no significant change in depth [4]. As shown, there is significant change in depth (H) against sediment discharge (Q_S) but roughness (n) linearly increases. It indicates the higher the roughness (n), the lower the velocity (V) when against sediment discharge (Q_S).

Table 2 gives the summary of the exponent and coefficient values of hydraulic geometry parameters for all selected stations. The exponent values show that f at S3

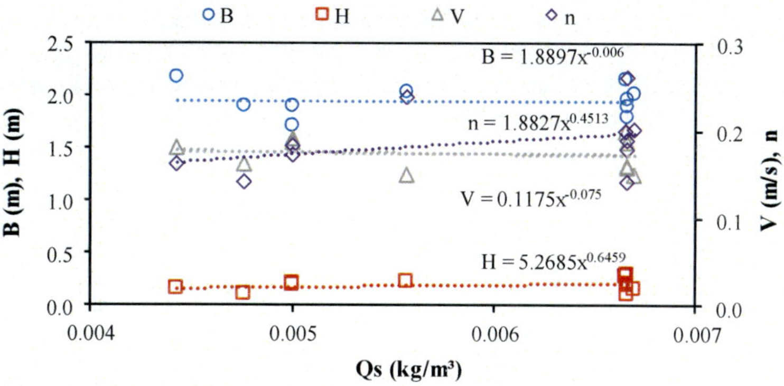

Fig. 10 Relations of sediment discharge and hydraulic geometry parameters at S3

has the highest value from the other exponent from all stations. This is because S3 is located at the downstream station where the location is steep and has the higher velocity. S3 has a higher depth and it can carry higher value of sediment discharge. That explains why S3 has high value of exponent *f*. The higher the depth, the more the water contains in a channel, the higher sediment that can be carried.

The exponent value of *m* at S2 shows the lowest value because along the channel there is some plant, grass and sand along the channel. It is observed to be making the channel become narrower and make the flow of water unstable. This is because the natural channel is not vulnerable and always affected by water flow. During heavy rain, it will either become narrower or widen the channel because the wall of the channel having corrosion [10]. In this study, it shows the channel becomes narrower when there is heavy rain. The smaller the sediment size and load and the lower the flow velocity and stream power, the more stable the stream.

Thus, the value of exponent *p* shows the highest on S2 because S2 shows the lowest rates of erosion because of low flow [9]. Erosion reduces the size of bed load material and decreases the roughness of the channel. It increases downstream as more water is added to the channel and because the channel becomes more efficient with distance downstream [10]. S2 shows rougher surface is having a higher roughness exponent and less retarding effect on the sediment flow, lower sediment rate is produced. As a conclusion, flow rate and roughness exponent were influenced by bed roughness.

Table 2 Exponents and coefficients of sediment discharge and hydraulic geometry parameters

Station	Exponent				Coefficient			
	b	*f*	*m*	*p*	*a*	*c*	*k*	*N*
S1	0.0126	0.6620	0.2262	0.1180	1.0453	0.9913	0.8167	0.1285
S2	− 0.1980	0.5153	− 0.3540	0.7027	0.5800	0.1092	0.0703	0.7862
S3	0.0491	1.5390	1.0547	− 0.2990	1.8897	5.2686	0.1175	1.8827
Ave	− 0.0454	0.9054	0.3090	0.1739	1.1712	2.1230	0.3348	0.9325

The above discussion shows that exponents, b, f, m, and p, do not possess fixed values; rather they vary over certain ranges, depending on the way the adjustment of stream power is distributed among variables [11]. Furthermore, the scale parameters are variant, depending on channel hydraulics, and this observation should help with regionalization of scale parameters.

4.3 Sediment Rating Curve ($Q - Q_s$)

Rivers have a capacity to transport large volume of sediment while conveying water. Hence, the term sediment discharge (Q_s) becomes an important factor for water resources management. The sediment volume transported by streams is critical for all hydraulic engineering projects. Power function analysis was utilized to determine the relations between the channel flow discharge and the sediment discharge. In the analysis, it was observed that the sediment discharge (Q_s) is increasing with increasing flow discharge (Q) in all stations. This sediment rating curve is expected to be useful for the study on the sediment in the future.

Figure 11 shows the relations of sediment discharge (Q_s) against flow discharge (Q) at S1. The exponent value of S1 is 0.3665 and its coefficient is 0.0680. The data were observed and show that the sediment discharge (Q_s) was increasing against increasing flow discharge (Q). The highest value of sediment discharge (Q_s) is at S1. This is because S1 has fast flow. In general, the greater the flow, the more sediment that will be conveyed [11]. Water flow can be strong enough to suspend particles in the water column as they move downstream, or simply push them along the bottom of a waterway. Transported sediment may include mineral matter, chemicals and pollutants and organic material.

Figure 12 shows the relations of sediment discharge (Q_s) against flow discharge (Q) at S2. The exponent value of S2 is 0.0789 and its coefficient is 0.0126. The data were observed and show that the sediment discharge (Q_s) was increasing against

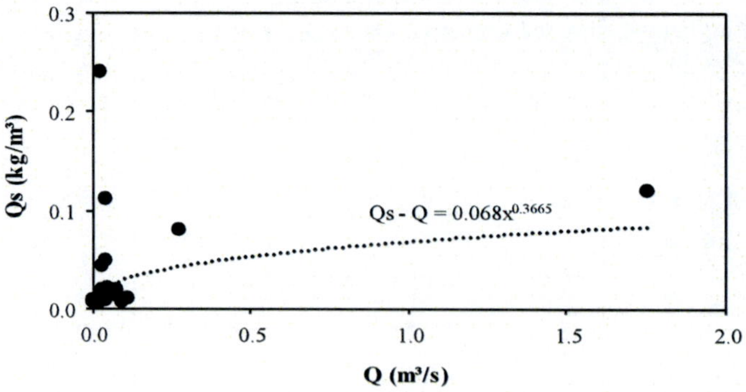

Fig. 11 Sediment rating curves ($Q_s - Q$) for S1

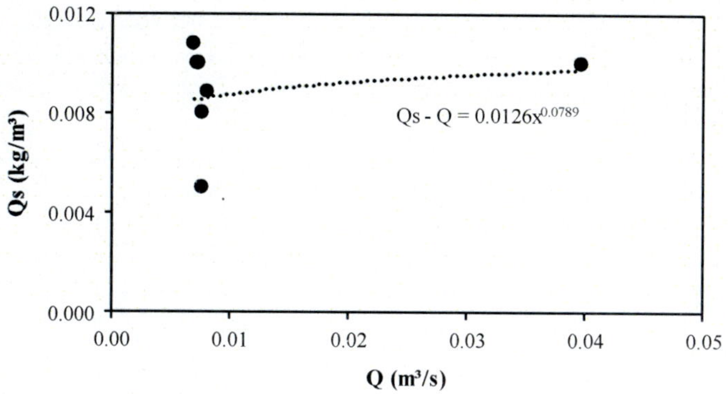

Fig. 12 Sediment rating curves ($Q_s - Q$) for S2

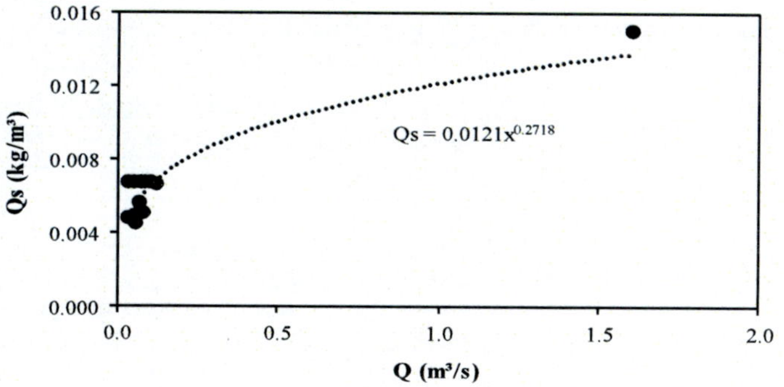

Fig. 13 Sediment rating curves ($Q_s - Q$) for S3

increasing flow discharge (Q). The lowest value of sediment discharge (Q_s) is at S2. This is because the flow of water at S2 is slow. The slower the flow, the lesser sediment that will be conveyed.

Furthermore, Fig. 13 shows the relations of sediment discharge (Q_s) against flow discharge (Q) at S3. The exponent value of S3 is 0.2718 and its coefficient is 0.0121. The data were observed and show that the sediment discharge (Q_s) was increasing against increasing flow discharge (Q).

5 Conclusion

Sediment transport and hydraulic geometry parameters in a river are important for water resources management, as well as planning and mitigating extreme hydrological events such as floods and droughts. As in the study, downstream hydraulic geometry analyses were performed at three selected stations on UPNM channel once a week for seven weeks. In order to evaluate the relation between flow discharge and river flow characteristics, the studies were carried out in the form of a power function. The UPNM channels may adapt a consistent pattern calculated using the hydraulic geometry concept.

The presented results of exponents and coefficients given in for downstream hydraulic geometry accurately describe the general hydraulic geometry relations for UPNM channel. These relations can be inferred for other basins when they have similar characteristics. Hydraulic geometry parameters for UPNM were compared well with previously reported theoretical values [9]. Nonetheless, diverse geologic, climatic, physiographic and morphologic contexts apply the fundamental hydraulic geometry method, resulting in a wide variety of exponents and coefficients in downstream hydraulic geometry.

In conclusion, for downstream hydraulic geometry, mean depth, width, mean flow velocity and surface roughness were shown to rise as the flow discharge increased [9]. Additionally, to hydraulic geometry analysis, the relations between sediment discharge and flow discharge has also been explored. According to the findings, the relations in S1 were discovered to be more significant than other stations in regard with correlation coefficient. Depending on this relation, the volume of the sediment transported in UPNM can be estimated. The output from this study is expected to shed light for hydraulic works in UPNM channel.

Acknowledgements A part of this research was funded by short-term grant of Universiti Pertahanan Nasional Malaysia under grant UPNM/2019/GPJP/TK/9.

References

1. Leopold, L.B., Maddock, T.: The Hydraulic Geometry of Stream Channels and Some Physiographic Implications pp. 252. U.S. Geological Survey Professional Paper (1953)
2. Fan, X., Giovanni, C., Zeng, Z., Ian, T., Leicheng, G., Qing, H.: A universal form of power law relationships for river and stream channels. Geophys. Res. Centre **47**, e2020GL090493 (2020). https://doi.org/10.1029/2020GL090493
3. Huang, H.Q., Nanson, G.C.: Vegetation and channel variation; a case study of four small streams in Southeastern Australia. Geomorphology **18**, 237–249 (1997)
4. Yuce, M.I., Esit, M., Karatas, M.C.: Hydraulic geometry analysis of Ceyhan River, Turkey. SN Appl. Sci. **1**(763), 1–11 (2019)
5. Kolberg, F.J., Howard, A.D.: Active channel geometry and discharge relations of U.S. Piedmont and Midwestern streams: the variable exponent model revisited. Water Resour. Res. **31**(9), 2353–2365 (1995)
6. Huang, H.Q., Warner, R.F.: The multivariate controls of hydraulic geometry: a casual investigation in terms of boundary shear distribution. Earth Surf. Proc. Landf. **20**, 115–130 (1995)
7. Rhoads, B.L.: A continuously varying parameter of downstream hydraulic geometry. Water Resour. Res. **27**(8), 1865–1872 (1991)
8. Singh, V.P.: On the theories of hydraulic geometry. Int. J. Sedim. Res. **18**, 196–218 (2003)
9. Suif, Z., Wahi Anuar, S.S., Ahmad, N., Othman, M., Che Osmi, S.K.: Hydraulics geometry analysis of UPNM channel. Environ. Sci. Eng. 17–26 (2021)

10. Francalanci, S., Lanzoni, S., Solari, L., Papanicolaou, A.N.: Equilibrium cross section of river channels with cohesive erodible banks. J. Geophys. Res.: Earth Surf. **125**, e2019JF005286 (2020). https://doi.org/10.1029/2019JF005286
11. Gleason, C.J.: Hydraulic geometry of natural rivers: a review and future directions. Prog. Phys. Geogr.: Earth Environ. **39**(3), 337–360 (2015). https://doi.org/10.1177/0309133314567584

Computational Fluid Dynamic Analysis at Dam Spillway Due to Different Upstream Water Levels

N. H. Hassan, S. Z. A. S. Kamarulbahrin, M. H. Zawawi, A. Z. A. Mazlan, M. A. Abas, M. R. R. M. A. Zainol, and M. A. Kamaruddin

Abstract This paper analyses the fluid dynamic on the dam spillway structure when the water spilling event occurred. The parameters determined in this study are velocity and pressure. A 3D Computational Fluid Dynamics (CFD) model of the dam spillway structure and the fluid boundary condition was developed using FLUENT. The result of the CFD model shows that hydrostatic pressure for all water levels was observed at the upstream section. The highest velocity and pressure occur at the highest water level which are 3.2 m/s and 1.526 MPa, respectively, at the spillway wall. High water velocity may contribute to critical stress due to the high pressure. All of the cases from 146 to 153 m water level show the fluid flows which are at reasonable velocity and pressure at the spillway. However, the continuous interaction with water towards the granite may affect the scouring effect of the granite in the long-term run.

Keywords Spillway · Velocity · Pressure · Computational Fluid Dynamic (CFD)

N. H. Hassan · M. H. Zawawi (✉)
Department of Civil Engineering, College of Engineering, Universiti Tenaga Nasional, 43000 Kajang, Selangor, Malaysia
e-mail: MHafiz@uniten.edu.my

S. Z. A. S. Kamarulbahrin
Department of Electrical and Electronic Engineering, College of Engineering, Universiti Tenaga Nasional, 43000 Kajang, Selangor, Malaysia

A. Z. A. Mazlan · M. A. Abas
School of Mechanical Engineering, Universiti Sains Malaysia, 14300 Nibong Tebal, Penang, Malaysia

M. R. R. M. A. Zainol
School of Civil Engineering, Universiti Sains Malaysia, 14300 Nibong Tebal, Penang, Malaysia

M. A. Kamaruddin
Environmental Technology Division, School of Industrial Technology, Universiti Sains Malaysia, 11800 George Town, Penang, Malaysia

N. Sabtu (ed.), *Proceedings of AWAM International Conference on Civil Engineering 2022 - Volume 3*, Lecture Notes in Civil Engineering 386,
https://doi.org/10.1007/978-981-99-6026-2_39

1 Introduction

A spillway is a structure built to provide a safe release of flood water from the dam downstream [1]. If the reservoir reaches the full supply level (FSL) and the inflow of water keeps increasing, the water will be drawn from the top of the reservoir through the spillway [2]. The present spillways exist in two types which can be controlled and uncontrolled [2, 3]. Kenyir spillway is an uncontrolled spillway that consists of no gates to control the water level in the reservoir.

The spillway section of Kenyir Dam regulates the water flow from the upstream to the downstream section based on the water level of the lake. Spilling event would happen when the water level at the reservoir exceeds 145 m. Normally, the water level is 135 m throughout the year, but in 2014, the highest water level ever achieved was at 148 m. This paper is presenting the effect of different water levels to velocity and pressure over time at spillway part using Computational Fluid Dynamic (CFD). The simulation was calculated using the scenario of water level at 146, 148, 150 and 153 m.

CFD is based on the Navier–Stokes equations [4]. These equations describe how the hydraulic parameters such as velocity and pressure of a moving fluid are related [5]. These equations can be applied to both compressible and incompressible fluids [6]. The ultimate objective of CFD is to understand the physical events that occur in the flow of fluids around and within the designated object or structure [7]. Navier–Stokes equations are the governing equations of CFD [7]. By applying the mass, momentum and energy conservation, the mass continuity equation and momentum equation can be derived [7–9].

2 Methodology

2.1 Development of Model and Boundary Condition

For geometry setup, a 3D model drawing of the spillway was developed according to as-built drawing that was prepared for designing the spillway. The spillway consists of two different materials which are concrete and natural rock (granite). Figure 1 shows the detail view of the spillway parts.

The boundary condition is an important procedure that needs to be done for the finite element (FE) model before the simulation is operated. The boundary condition of the FE model has to be set accordingly to the real behaviour of the dam structure. Fluid domain is modelled to include from 146 to 153 m upstream from the crest level as the water starts to spill after exceeding 145 m height.

Fluid domain and the schematic diagram of the boundary conditions (BC) associated with the fluid domain are illustrated in Fig. 2. Periodic boundary condition (continuous water flow) will be set at the cut-off sections to allow continuous flow

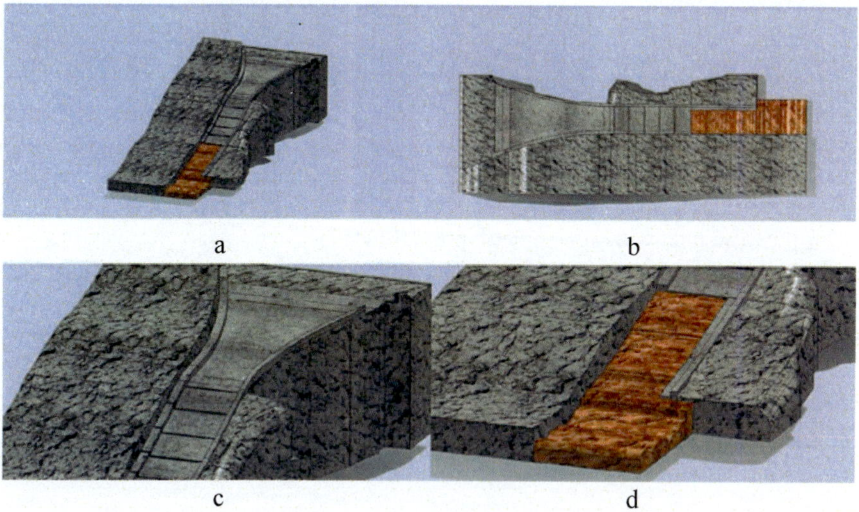

Fig. 1 Detail views of spillway; **a** Isometric view of the spillway, **b** Top view of the spillway, **c** Detail view of spillway concrete part and **d** Detail view of spillway natural rock part

Fig. 2 Diagram of fluid domain

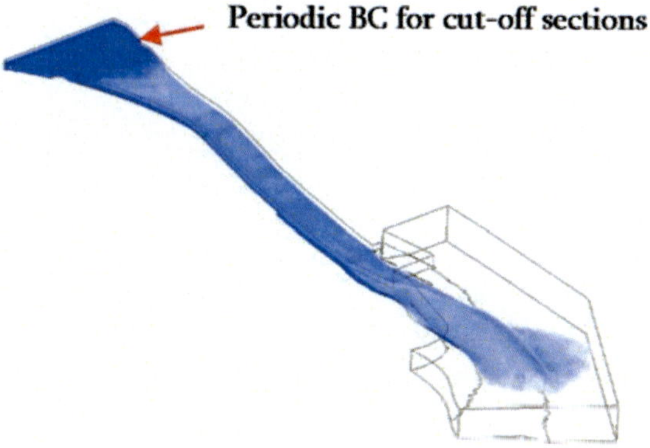

of water that is influenced by the water level, contact with structure and gravity on mimic the real flow profile from the upstream level.

2.2 Numerical Simulation

Water inlet was set at the bottom surface upstream of the gate to resemble a rise of water level and filling over the gateways. In order to maintain the intended water heights, the boundary type of water inlet was pressure inlet, P_{inlet}, with calculated values based on hydrostatic pressure;

$$P_{\text{inlet}} = \rho g h, \quad (1)$$

where ρ is water density, g is Earth's gravitational field strength and h is the water depth from surface level. The water outlet was set as periodic pressure outlet with zero-gauge pressure (relative to atmospheric pressure) as water is flowing out in contact with the atmosphere. This will ensure continuous flow of water thereby allowing similar upstream water level as observed on-site. The top boundary was set as air pressure inlet with atmospheric pressure value.

Since the fluid of interest is incompressible water in an open-channel turbulent flow, the pressure-based solver with standard k-ε turbulence and volume of fluid (VOF) model was chosen in FLUENT. The time discretization employed in this study was implicit. The spillway section of Kenyir Dam regulates the water flow from the upstream to the downstream section based on the water level of the lake. The level of water will be operated based on 146, 148, 150 and 153 m in the simulation.

3 Result and Discussion

3.1 Computational Fluid Dynamic Analysis

After simulation results have been completed, Kenyir spillway was assessed quantitatively to ensure that the spillway is able to withstand the fluid pressures due to continuous flow without losing its integrity. The presented results were based on each individual operating conditions. Figures 3, 4, 5 and 6 show the results of velocity and pressure at spillway during water spilling.

Volume flow rate at spillway inlet is 3311 m^3/s, 4851 m^3/s, 8655 m^3/s and 13,678 m^3/s, while the velocity that recorded at spillway inlet is 1.05 m/s, 1.52 m/s, 2.2 m/s and 3.3 m/s, respectively. Hydrostatic pressure for all water levels was observed at the upstream section. The flow of water is diverged once exits the spillway. This situation could make the water accumulate in the river and create back flow (vortex). The highest velocity and pressure occur at the highest water level which are 3.2 m/

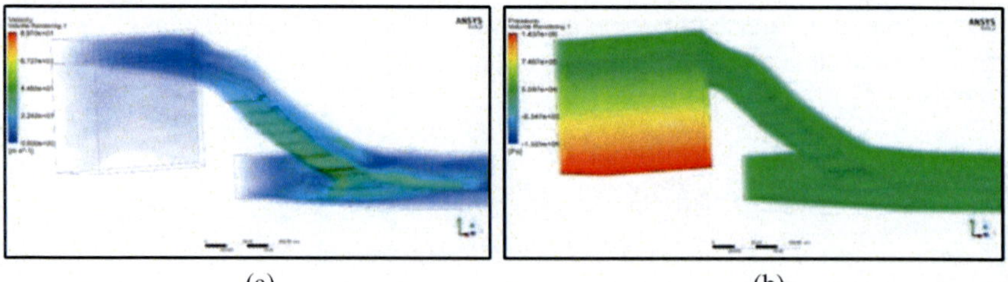

(a) (b)

Fig. 3 Results of **a** velocity and **b** pressure of water level at 146 m

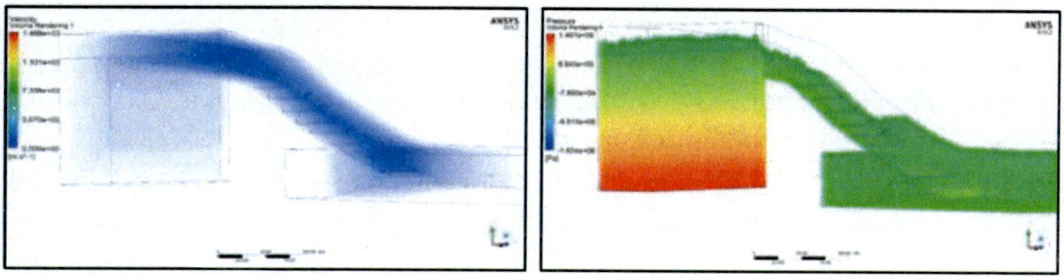

(a) (b)

Fig. 4 Results of **a** velocity and **b** pressure of water level at 148 m

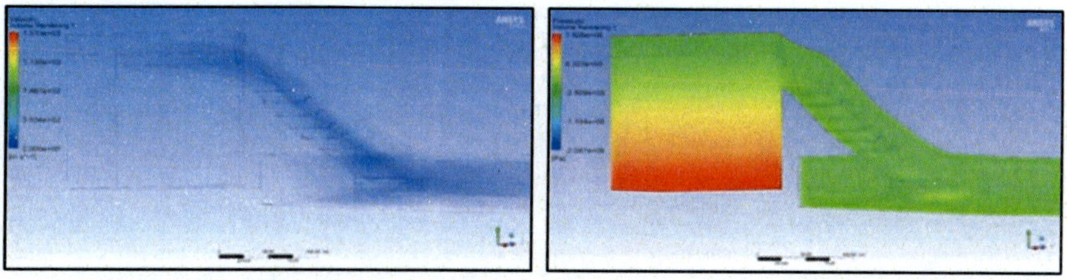

(a) (b)

Fig. 5 Results of **a** velocity and **b** pressure of water level at 150 m

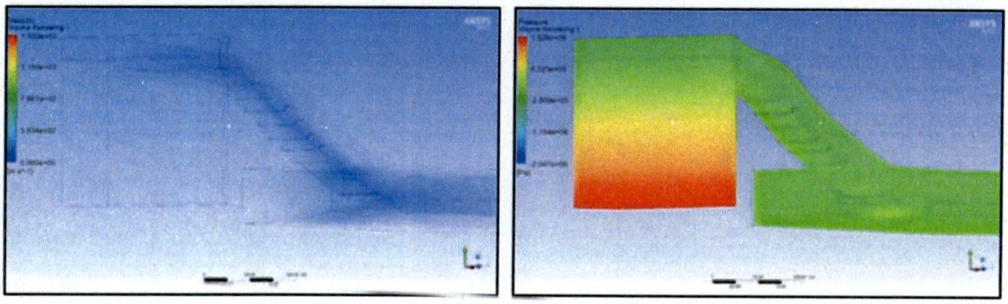

(a) (b)

Fig. 6 Results of **a** velocity and **b** pressure of water level at 153 m

s and 1.526 MPa, respectively, at spillway wall. High water velocity may contribute to critical stress due to the high pressure.

4 Conclusion

Based on fluid flow analysis, it can be concluded that water level at reservoir that exceeds spillway height would affect the velocity of water release at the spillway. All of the cases from 146 to 153 m water level show the fluid flows which are at the reasonable velocity and pressure [7]. However, the continuous interaction with water towards the granite may affect the scouring effect of the granite in long-term run. Other than that, vortex at downstream area may occur and water level will rise at the river. It may cause flood problem and other structural failure. Therefore, further study on impact of water towards structure and vortex analysis should be done. For mitigation works, another design on enhancing the efficiency of the spillway needs to be proposed [9].

References

1. Kamaruddin, M.A.: Fluid-Structure Interactions Study on Hydraulic Structures : A Review, vol. 020244 (2018)
2. Date, V., Dey, T., Joshi, S.: Numerical modeling of flow over an ogee crested spillway under radial gate: VOF and MMF model. J. Appl. Mech. Eng. **6**(5) (2017)
3. Chanel, P.G.: An Evaluation of Computational Fluid Dynamics for Spillway Modeling, p. 84 (2008)
4. Kim, D.G., Park, J.H.: Analysis of flow structure over ogee-spillway in consideration of scale and roughness effects by using CFD model. J. Civ. Eng. KSCE **9**(2), 161–169 (2005)
5. Ho, D.K.H., Boyes, K.M., Donohoo, S.M.: Investigation of spillway behaviour under increased maximum flood by computational fluid dynamics technique. In: 14th Australasian Fluid Mechanics Conference, no. December, pp. 577–580 (2001)
6. Pedley, T.J.: Introduction to fluid dynamics. Sci. Mar. **61**(Supl.1), 7–24 (1997)
7. Zawawi, M.H., Saleha, A., Salwa, A., Hassan, N.H., Zahari, N.M., Ramli, M.Z.: A Review : Fundamentals of Computational Fluid Dynamics, vol. 020252 (2018)
8. Teng, P.: CFD Modelling of Two-Phase Flows At Spillway a Erators, no. February (2017)
9. Muda, Z.C.: Computational Fluid Dynamic Analysis at Dam Spillway due to Different Gate Openings, vol. 020245 (2018)

Vibration Effect to Kenyir Dam Intake Section During a Spill Event

N. H. Hassan, M. H. Zawawi, A. Z. A. Mazlan, A. Arbain, and M. A. Abas

Abstract Kenyir Dam located in Terengganu, Malaysia is a hydropower dam which served the purpose of generating electricity to the surrounding areas. Water is basically flown from the upstream to downstream area through the intake and pressure tunnel sections. In some cases, due to large volume of water, the dam structure might experience significant vibration effect from water spill event. In this study, the flow-induced vibration during critical water spilling of 148 m water level at intake section of Kenyir Dam is investigated using ANSYS software. The results from both frequency domain and operational deflection shapes (ODS) are compared with the natural frequencies and mode shapes of the dam, in order to determine the safety precautions during dam operating condition. From the results of water level 148 m, the transient vibration response from the water flow is occurred at operating frequency of 2.745 Hz. There is no resonance phenomenon during this water spilling condition since the natural frequency of the intake section occurred at 3.5 Hz. However, some precautions should be taken since the deflection shapes of the dam structure might be higher above this water level.

Keywords Vibration · Water spilling · Intake

1 Introduction

Flow-induced vibration is a vortex shedding phenomenon due to high velocity of water flow and can significantly risk the structure. Other than that, this vibration can affect the mechanical equipment as well as reduce the performance of the structure.

N. H. Hassan · M. H. Zawawi (✉)
Department of Civil Engineering, College of Engineering, Universiti Tenaga Nasional, 43000 Kajang, Selangor, Malaysia
e-mail: MHafiz@uniten.edu.my

A. Z. A. Mazlan · A. Arbain · M. A. Abas
School of Mechanical Engineering, Universiti Sains Malaysia, 14300 Nibong Tebal, Penang, Malaysia

N. Sabtu (ed.), *Proceedings of AWAM International Conference on Civil Engineering 2022 - Volume 3*, Lecture Notes in Civil Engineering 386,
https://doi.org/10.1007/978-981-99-6026-2_40

Dam structure might fail when resonance phenomenon happens, whereby the natural frequencies coincided with the operating frequencies of the water flow. However, it is very challenging to diagnose this fault before it becomes critical [1]. According to [2], a forced vibration procedure can be used to estimate the fundamental and vibration frequency that associated with the mode shapes of the structure.

Besides, modal analysis (MA) is an alternative way to study structural parameters under certain conditions [3]. Through this analysis, the dynamic characteristics of structure can be determined, improved and optimized [4]. Dynamic characteristics such as natural frequencies and mode shapes are important to be studied in order to determine structural behaviour [5]. Other quantities such as the displacement and stress are also important to be calculated to identify and evaluate the vibration performances [4]. In general, modal analysis can be utilized in solving natural tendencies of the structure and understand the causes of vibration [3].

Over the years, numerous studies have been conducted on flow-induced vibration with various samples. Study on the effect of reservoir water level fluctuation to the nonlinear behaviour of concrete dam has been conducted in [5]. Another study has been conducted by [6] which aimed to study the pressure distribution and velocity profile of discharge water flow from the gate opening. Recently, many researchers used fluid–structure-interaction (FSI) method to solve flow-induced vibration studies [7]. FSI analysis can be done using either monolithic or partitioned methods, whereby partitioned method can be divided into one or two-ways interaction [8]. Benra and team [9] have studied both methods and found that two-way FSI is more accurate and reliable. In order to examine the safety of the dam structure, the finite element (FE) method is often utilized to estimate the dynamic response of the dam. This method is a mathematical tool that can be used to solve various engineering problems [10]. The mathematical model of the geometry is expressed in terms of finite number degree of freedom and then been connected for interpretation of the FEM [11]. Even though FE method is a powerful tool to estimate the dynamic characteristics of the structure, Sato and Obuchi [12] are still investigating the result accuracy level from this analysis. Chaphalkar and team also added that it is necessary to access the FE solution technique and if the criteria do not meet, the numerical solution must be repeated with refined solution parameters [4].

Another method that can be used to study the vibrations of the dam structure is through operational deflection shape (ODS). By performing this method, a full dynamic response of the structure could be obtained as well as the vibration pattern of the structure with influence by known or unknown operating frequencies [13]. In addition, the deformation characteristics of vibrating structure can also be animated [5]. There is difference between MA and ODS, whereby MA can only be used for linear and stationary motion, while ODS can be used for nonlinear and nonstationary structural motion [1]. Katunin [7] stated that ODS can provide more reliable dynamic information than MA since ODS can produce information at any single harmonic frequency and not restricted to the natural frequencies.

In this study, the vibration of Kenyir Dam intake structure is investigated in terms of MA and ODS using numerical method. FE model of the intake section is modelled,

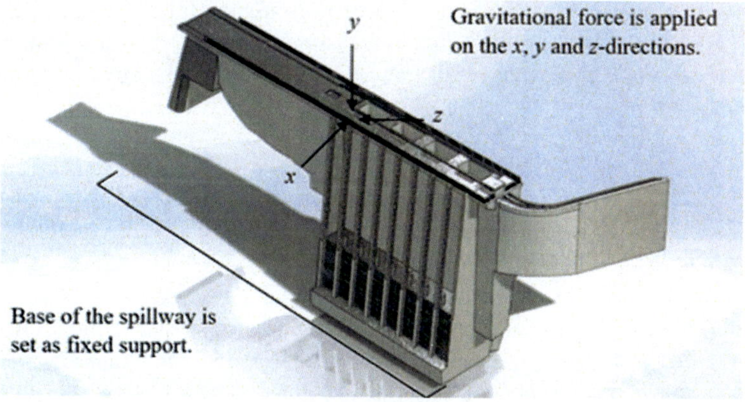

Fig. 1 Real-scale 3D model and structural boundary condition of the intake section

and simulation study is performed using ANSYS software. Results of both MA and ODS analyses are then compared to find out the formation of resonance phenomenon.

2 Methodology

2.1 3D Modelling and Structural Boundary Condition

The 3D model of the intake section is constructed using SolidWorks software, as shown in Fig. 1. Intake section is located at the upstream area of the dam and it is the first structure that water will be impacted before it starts flowing from upstream to downstream areas. In this study, the critical water level of 148 m will be studied in detail to determine the intake section level of safety during operating condition.

2.2 Modal and Harmonic Response Analysis

Both modal and harmonic response analyses will be conducted using ANSYS software. These analyses are performed to determine the vibration characteristics of the structure such as the mode shapes and natural frequencies. The boundary condition is set according to the real case on the site, whereby all the bottom faces are set as fixed support as shown in Fig. 1. Each part of the structure is applied with the real material properties as in real dam and the type of meshing used is tetrahedral. MA is specifically been carried out to determine the dynamic properties of the intake structure. First 30 mode shapes are determined to evaluate the deflection of the structure and the most significant one is selected for further discussion. Input forces with magnitude of 9.81 N are applied in all three directions (-*x*, -*y* and -*z*). These forces acted as gravity forces that reacted in all three directions since this structure simulated independently.

2.3 Fluid Structure Interaction (FSI) Analysis

This analysis is conducted to determine the interaction between the intake structure and fluid flow. The boundary condition of the fluid at inlet and outlet sections are set as velocity and pressure, respectively, and the velocity magnitude of water is set as 1 m/s. This FSI analysis is made up of two analysis systems and one integrated coupling system in ANSYS software. The analysis systems are transient structural and fluent whereas the integrated system is system coupling. The transient structural is used to study the deflection of the structure during fluid flow, while fluent used to study the flow and type of fluids that are going to act on the dam structure. The system coupling element is the combination of transient structural and fluent. All these three elements are linked together for data sharing. Through this analysis, the result that will be obtained is more accurate and stable.

For the transient structural system, the type of mesh used is tetrahedral due to its faster simulation time and much more convenient to be applied. In the analysis, the step end-time is set to 5 s and the minimum and maximum sub-step is set as 1 and 100, respectively. In order to maintain the intended water heights, the boundary type of water inlet was set as pressure's inlet with calculated values based on hydrostatic pressure. The water outlet was set as periodic pressure outlet with zero-gauge pressure relative to atmospheric pressure since the water is flowing out in contact with the atmosphere. This is also to ensure the continuous flow of water similar to upstream water level as observed on-site. The boundary conditions for the intake structure was set as no-slip wall, while the top boundary was set as air pressure inlet with atmospheric pressure value, as shown in Fig. 2. The data obtained from this analysis is then converted to frequency domain data through fast Fourier transform (FFT) equation.

3 Results and Discussion

3.1 Natural Frequencies and Mode Shapes

Figure 3 shows the mode shapes of the intake section determined from MA. From the figure, the mode shapes of the intake are corresponded to the flow of water towards the downstream area with an amplitude value of 8.093×10^{-3} m and the natural frequency occurred at 3.5 Hz. Any vibration within this frequency value must be

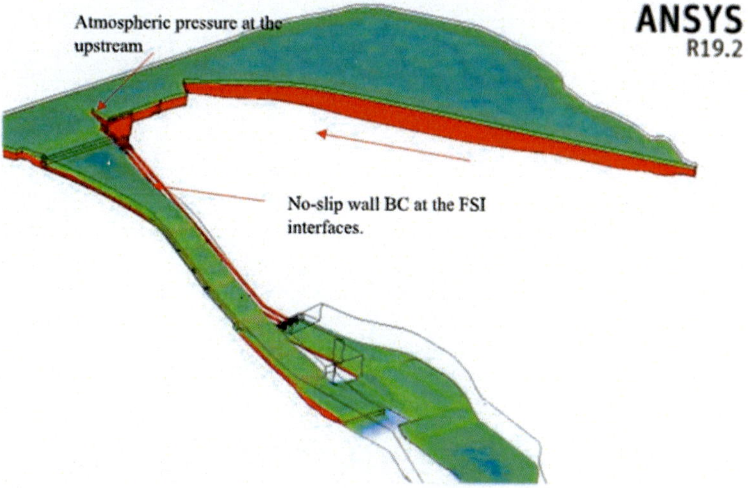

Fig. 2 Boundary condition and fluid domain of dam structure

avoided to prevent resonance phenomenon from happening. From the harmonic response analysis, the highest deflection occurred in x-direction with deflection amplitude of 2.283×10^{-7} m, as shown in Fig. 4. The highest peak of the graph represents the natural frequency value of 3.5 Hz.

Fig. 3 Mode shape of intake section

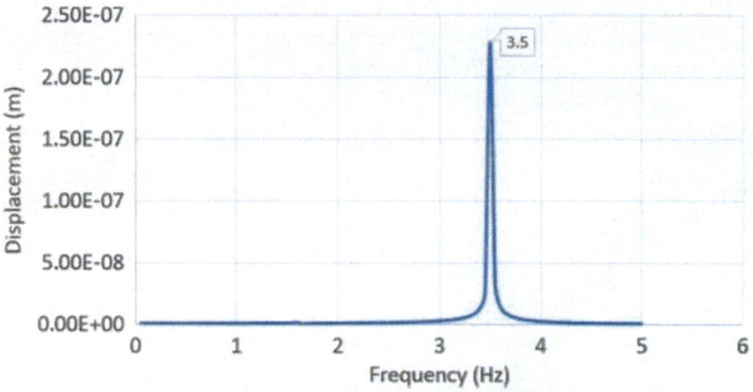

Fig. 4 Deflection amplitude at x-direction

3.2 *Operational Deflection Shapes (ODS)*

Figure 5 shows the ODS result of the Kenyir Dam intake section. From the figure, the maximum displacement mainly occurred at the side wall of trash rack and there is no significant effect at the penstock location where the water is flown.

The FSI simulation of the intake section is conducted as in real operating condition, whereby the base of the structure is set as fixed support. Figure 6 shows the ODS displacement frequency response resulting from the FSI study. During the water spilling at critical water level of 148 m, the transient vibration effect with amplitude value of 0.0033 m at operating frequency of 2.745 Hz is not coincided with the natural frequency of the intake structure (3.5 Hz). Thus, there is no resonance phenomenon that can be expected and the intake section is considered safe during this water spilling condition.

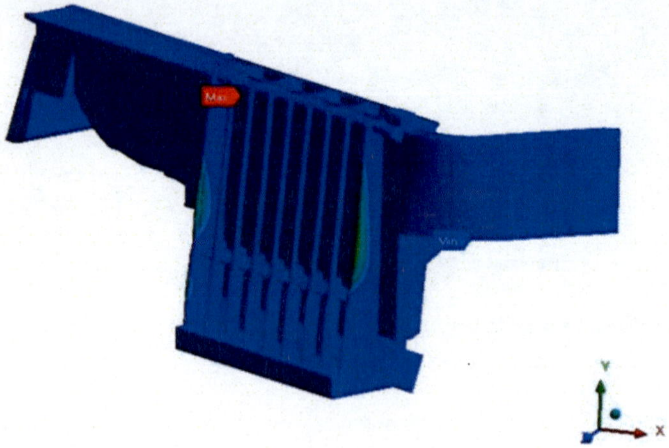

Fig. 5 Maximum displacement at intake section

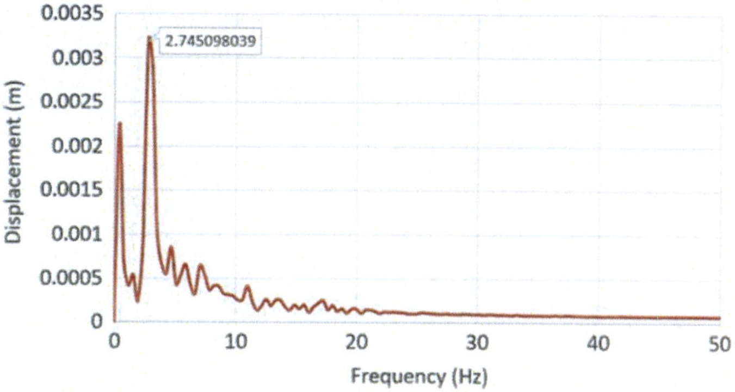

Fig. 6 ODS displacement frequency response

4 Conclusion

From this study, the following conclusions can be made:

- The ODS and MA results of the intake section have been obtained in terms of deflection shapes and frequency domain graphs. From MA result, the highest deflection occurred at one of trash rack area with natural frequency of 3.5 Hz and exhibits displacement of 8.093×10^{-3} m, while for ODS, the highest deflection occurred at sidewall of trash rack area with natural frequency of 2.745 Hz and displacement of 0.0033 m, during critical spilling at water level of 148 m.
- From both the results, there is no significant coincidence between natural frequencies and operating frequency at 148 m water level. However, precautions should be taken during water spilling above this water level since the deformation magnitude can be increased gradually.

References

1. Saravanan, K., Sekhar, A.: Crack detection in a rotor by operational deflection shape and kurtosis using laser vibrometer measurements. J. Vib. Control **19**(8), 1227–1239 (2012)
2. Chugh, A.K.: Natural vibration characteristics of gravity structures. Int. J. Numer. Anal. Methods Geomech **31**, 607–648 (2007)
3. Ramli, M.I., Nuawi, M.Z., Abdullah, S., Salleh, M.S., Teknikal, U., Jaya, H.T.: The study of EMA effect on modal identification: a review. J. Mech. Eng. Technol **9**(1), 103–121 (2017)
4. Chaphalkar, S.P., Khetre, S.N., Meshram, A.M.: Modal analysis of cantilever beam structure using finite element analysis and experimental analysis. Am. J. Eng. Res. **4**(10), 178–185 (2015)
5. Parsa Mahmoudi, P., Mirzabozorg, H., Varmazyari, M., Aghajanzadeh, S.M.: Effect of foundation nonlinearity on seismic response of an existing arch dam. Civ. Eng. J. **2**(5), 197–207 (2016)
6. Sanchez, D., Salazar, J.E.: On the effects of water discharge through radial gates at the Caruachi dam. In: Proceedings of ASME 2010 10th Biennial Conference on Engineering Systems Design and Analysis, pp. 1–5 (2010)

7. Katunin, A.: Damage assessment in composite structures using modal analysis and 2d undecimated fractional wavelet transform. J. Vibroengineering **16**(8), 3939–3950 (2014)
8. Ahamed, M., Atique, S., Munshi, M., Koiranen, T.: Open access a concise description of one way and two way coupling methods for fluid-structure interaction problems. American Journal of Engineering Research (AJER) **3**, 86–89 (2017)
9. Benra, F., Dohmen, H.J., Pei, J., Schuster, S., Wan, B.: A comparison of one-way and two-way coupling methods for numerical analysis of fluid-structure interactions. J. Appl. Math. **853560**, 1–16 (2011)
10. Erhunmwun, I., Ikponmwosa, U.: Review on finite element method. J. Appl. Sci. Environ. Manag. **21**(5), 999–1002 (2017)
11. Zeidan, B.A., Seleemah, A.A., Rashed, A.A.: Static and dynamic analysis of concrete gravity dams. Thesis (2015)
12. Sato, H., Obuchi, Y.: A study on earthquake responses of actual rock-fill dam and numerical analyses. In: 12th World Conference on Earthquake Engineering (2000)
13. Wang, S.S., Zhang, Y.F., Cao, M.S., Xu, W.: Effects of contraction joints on vibrational characteristics of arch dams: experimental study. Shock. Vib. **327362**, 1–8 (2015)

Numerical Study of Incompressible Flow Past a Circular Cylinder at Low Reynolds Number Using COMSOL Multiphysics

Mohamad Faizal Ahmad, Mohd Ridza Mohd Haniffah, Ahmad Kueh, and Erwan Hafizi Kasiman

Abstract Incompressible turbulent flow past a stationary circular cylinder at Reynolds number of 10,000 was investigated using COMSOL Multiphysics in a two-dimensional simulation. At this Reynolds number, the flow pattern was in a transition from periodic to chaotic behaviour due to small eddies developing at the wall of the circular cylinder. The aim is to investigate a couple of turbulence models to predict the flow behaviour by comparing the result of vorticity patterns, drag and lift coefficients with the numerical result of previous researchers. This study employed two turbulence models, Spalart Allmaras and Menter's Shear Stress Transport. Wall treatment and wall distance initialization options were also investigated. The SA model was able to give better result as compared to SST model when compared to standard drag coefficients values for circular object. The wall treatment was able to perform very well in determining how wall treatment should be treated, either using wall function or resolve the boundary layer with RANS depending on the maximum $y+$ that correlates with the mesh size on the wall of circular cylinder. The simulation with wall distance initialization turned off successfully imitates the result of previous research in terms of CD and CL curves pattern and wake pattern for the specific case of incompressible flow past a stationary circular cylinder.

Keywords Turbulence Model · Spalart Allmaras · COMSOL Multiphysics

M. F. Ahmad (✉) · M. R. M. Haniffah · E. H. Kasiman
Faculty of Civil Engineering, Universiti Teknologi Malaysia (UTM), 81310 Johor Bahru, Malaysia
e-mail: faizal_9273@yahoo.com

M. R. M. Haniffah
e-mail: mridza@utm.my

A. Kueh
Department of Civil Engineering, Faculty of Engineering, Universiti Malaysia Sarawak (UNIMAS), 94300 Kota Samarahan, Sarawak, Malaysia

N. Sabtu (ed.), *Proceedings of AWAM International Conference on Civil Engineering 2022 - Volume 3*, Lecture Notes in Civil Engineering 386,
https://doi.org/10.1007/978-981-99-6026-2_41

1 Introduction

In the past few decades, incompressible fluid flow past a circular cylinder has been studied due to its importance in many engineering applications such as circular piers in a river, marine piling, marine riser, pipeline, and cable on seabed. Most turbulence models and computational fluid dynamics (CFD) platforms have difficulties in modelling the vortex dynamics at the leeward of the circular cylinder eventhough the circular shape is simple geometry. They are due to the flow instabilities in the wake region, in the boundary layer and the separating shear layer. Direct numerical simulation (DNS) and large eddy simulation (LES) is the best method to move forward but the computational cost is very expensive. The computational cost and reliability of a CFD simulation should be balanced in order to achieve fast and reliable results [1].

Monk and Chadwick [11] investigated the performance of three turbulence models, which were SA, *k*-omega, and Transition SST turbulence models by simulating airflow past a delta wing at low Reynolds number and then comparing it with wind tunnel data. All the turbulence models gave similar results that failed to validate the experimental data at high angle of attack of 10° and above. But Spalart Allmaras model slightly outperforms the other two turbulence models. Water flow past a circular cylinder at different Reynolds numbers ranging from subcritical, critical and supercritical regimes were investigated by Ahmad et al. [2] and Qiu et al. [13] with different turbulence models and platforms including LES and detached eddy simulation (DES). Numerical prediction by LES had better agreement with experimental data than the other simulation since LES has the highest computational cost.

A new method was invented by Durante et al. [6] through modification of the Vortex Particle Method by meshless Lagrangian approach, called Diffused Vortex Hydrodynamics (DVH). The method was tested with water flow past a circular cylinder at different Reynolds numbers ranging from 100 to 500,000. The results were comparable to the other numerical methods but overpredicted when compared to the experimental data due to two-dimensional limitation. The authors also mentioned that the computational cost was comparable with the Finite Volume Method (FVM) with the same performance. The same case also was simulated via two dimensional LES by Singh and Mittal [16]. Then also overpredicted the results and the drag crisis was earlier than the experimental data. The authors also tested the simulation with and without subgrid scale model that led to very similar results.

This study aims to investigate the capability of COMSOL Multiphysics turbulence models, a Computational Fluid Dynamic (CFD) transient solver through quantitative and qualitative comparisons against existing numerical models which are the DVH and Large Eddy Simulation (LES). The Menter's Shear Stress Transport (SST) and Spalart Allmaras turbulence model were employed in two-dimensional RANS simulation. All simulations were simulated at Reynolds number of 10,000 and the vorticity field, drag, and lift coefficients were compared with existing data. Two additional set-ups of wall treatment and wall distance initialization options in COMSOL

Multiphysics were tested to validate the present simulation with the previous work's results.

2 Turbulence Modelling

Turbulence modelling is a tool in CFD that predicts turbulent flow behaviour by using mathematical models. Most fluid flows in the world are turbulent, such as airflow over an aircraft wing, river flow past piers, marine riser, cable on seabed, etc. The challenges in development of turbulent model there is no analytical theory to predict the exact behaviour of turbulent flow. These turbulence models are simplified constitutive equations that predict the statistical evolution of turbulent flows. There are several most common turbulence models that are used in engineering and science fields. The most common turbulence model is the *k*-epsilon turbulence model. It was invented in 1973 by Launder [8] and Spalding that can simulate fast and reliable prediction in inviscid flow, but the drawback of this turbulence model is it does not resolve the boundary layer and models it with wall treatment. Thus, it cannot be used in drag calculation problems. Then, Wilcox [18] invented *k*-omega in 1988 that resolves the boundary layer at the wall via RANS. The *k*-omega gives better performance than the *k*-epsilon model, but it has instability at free-flow region and quite sensitive to the initial guess of the solution. The other two turbulence models are the Spalart Allmaras and Menter's Shear Stress Transport turbulence model, which were invented in the same year, 1994 [9].

2.1 *Spalart–Allmaras Turbulence Model (SA)*

Spalart and Allmaras [17] invented a one-equation turbulence model and introduced a new variable for an undamped kinematic eddy viscosity, ν^{t} and can resolve the flow field all the way down to the wall. SA model is a low Reynolds number model that was originally developed for aerodynamics application and it is fast and relatively robust. Based on NASA [15], the SA model provided by COMSOL Multiphysics is SA model with Rotation Correction (SA-noft2-R), modified by Dacles-Mariani et al. [4]. This correction to the SA model reduces the eddy viscosity in regions where vorticity exceeds strain rate, such as in vortex core regions where pure rotation should not produce turbulence. The modification is passive in thin shear layers where vorticity and strain are very close. The mathematics of SA Model in COMSOL Multiphysics is presented below:

$$\rho\frac{\partial \mathbf{u}}{\partial t} + \rho(\mathbf{u}\cdot\nabla)\mathbf{u} = \nabla\cdot\left[-p\mathbf{I} + (\mu + \mu_{\mathrm{T}})\left(\nabla\mathbf{u} + (\nabla\mathbf{u})^{\mathrm{T}}\right)\right] + \mathbf{F} \tag{1}$$

$$\rho\nabla\cdot(\mathbf{u}) = 0 \tag{2}$$

$$\frac{\partial \nu^{\mathrm{t}}}{\partial t} + (\mathbf{u} \cdot \nabla)\nu^{\mathrm{t}} = C_{\mathrm{b1}} S^{\mathrm{t}} \nu^{\mathrm{t}} - C_{\mathrm{w1}} f_{\mathrm{w}} \left(\frac{\nu^{\mathrm{t}}}{\ell_{\mathrm{w}}} \right)^{2} + \frac{1}{\sigma_{\overline{\nu}}} \nabla \cdot \left(\left(\nu + \nu^{\mathrm{t}} \right) \nabla \nu^{\mathrm{t}} \right) + \frac{C_{\mathrm{b2}}}{\sigma_{\overline{\nu}}} \nabla \nu^{\mathrm{t}} \cdot \nabla \nu^{\mathrm{t}} \tag{3}$$

$$C_{\mathrm{w1}} = \frac{C_{\mathrm{b1}}}{\kappa_{\mathrm{v}}^{2}} + \frac{1 + C_{\mathrm{b2}}}{\sigma_{\overline{\nu}}} \tag{4}$$

$$f_{w} = g \left(\frac{1 + C_{w3}^{6}}{g^{6} + C_{w3}^{6}} \right)^{\frac{1}{6}} \tag{5}$$

$$g = r + C_{w2}\left(r^{6} - r\right), r = \min\left(\frac{\nu^{t}}{S^{t} \kappa_{v}^{2} \updownarrow_{w}^{2}}, 10 \right) \tag{6}$$

where Eqs. (1) and (2) are the RANS equations while Eq. (3) is the SA turbulence model with the undamped kinematic eddy viscosity term. The parameter μ_{T} is the turbulent viscosity, S^{t} is the rotation correction, $\updownarrow_{\mathrm{w}}$ is the distance to the closest wall and ν is the kinematic viscosity. The constant turbulence model parameters are $C_{\mathrm{b1}} = 0.1355$, $C_{\mathrm{b2}} = 0.622$, $\sigma_{\overline{\nu}} = \frac{2}{3}$, $C_{\mathrm{w2}} = 0.3$, $C_{\mathrm{w3}} = 2$, and $k_{\upsilon} = 0.41$.

2.2 *Menter's Shear Stress Transport Turbulence Model (SST)*

The strengths and weaknesses of the *k*-epsilon and *k*-omega turbulence model gives an idea for Menter to combine both models in order to eliminate the drawback of both models. Menter had the idea in 1994 to combine their strength in simulating turbulent flow in which the *k*-omega is applied to resolve the inner region of the boundary layer and applies *k*-epsilon for the free stream flow. The combination reduces the computational load for the original *k*-omega model and consequently reduces the computational time for external flow cases. The model is known as a low Reynolds number model because it resolves the boundary layer with RANS. Based on NASA [15], the SST model provided by COMSOL Multiphysics is SST-2003, modified by Menter et al. [10] after 10 years of Standard SST. The mathematics of SST Model in COMSOL Multiphysics is presented below:

$$\rho \frac{\partial \mathbf{u}}{\partial t} + \rho(\mathbf{u} \cdot \nabla)\mathbf{u} = \nabla \cdot \left[-p\mathbf{I} + (\mu + \mu_{\mathrm{T}})\left(\nabla \mathbf{u} + (\nabla \mathbf{u})^{\mathrm{T}} \right) \right] + \mathbf{F} \tag{7}$$

$$\rho \nabla \cdot (\mathbf{u}) = 0 \tag{8}$$

$$\rho \frac{\partial k}{\partial t} + \rho(\mathbf{u} \cdot \nabla)k = \nabla \cdot [(\mu + \mu_{\mathrm{T}} \sigma_{k}) \nabla k] + P - \beta_{0}^{*} \rho \omega k \tag{9}$$

$$\rho\frac{\partial\omega}{\partial t}+\rho(\mathbf{u}\cdot\nabla)\omega=\nabla\cdot[(\mu+\mu_{\mathrm{T}}\sigma_{\omega})\nabla\omega]+\frac{\gamma}{\mu_{\mathrm{T}}}\rho P-\rho\beta_0\omega^2$$
$$+2(1-f_{v1})\frac{\sigma_{\omega 2}\rho}{\omega}\nabla_k\cdot\nabla_{\omega} \tag{10}$$

$$\mu_T=\rho\frac{a_1k}{max(a_1\omega,Sf_{v2})} \tag{11}$$

$$P=\min(P_k,10\rho\beta_0\omega k),\ P_k=\mu_{\mathrm{T}}\left[\nabla\mathbf{u}:\left(\nabla\mathbf{u}+(\nabla\mathbf{u})^T\right)\right] \tag{12}$$

where Eqs. (7) and (8) are exactly the same as Eqs. (1) and (2), and Eqs. (9) and (10) is the SST turbulence model. The blending functions f_{v1} and f_{v2} are calculated with several parameters, S is the characteristic magnitude of the mean velocity gradients, μ_{T} is the turbulent viscosity, k is the turbulence kinetic energy and ω is the specific rate of dissipation. The constant turbulence model parameters are $a_1=0.31$, $\beta_0^*=0.09$ and $\sigma_{\omega 2}=0.856$.

2.3 Wall Distance

The wall distance in turbulence modelling is reciprocal distance to the selected walls. The value will be small when the object is far from the respective walls and more significant when closer [14]. This function affects the calculation of RANS to differentiate between free-stream and wall-bounded flow regions. In COMSOL Multiphysics, the wall distance is calculated in stationary study method before solving the selected turbulence model. The mathematics of Wall Distance Initialization (WDI) is presented below:

$$\nabla G\cdot\nabla G+\sigma_w G(\nabla\cdot\nabla G)=(1+2\sigma_w)G^4 \tag{13}$$

where $G=2/l_{\mathrm{ref}}$ on the selected wall, l_{ref} is reference normal length to the selected wall and σ_w is the smoothing parameter.

3 Numerical Set-Ups

A fixed circular cylinder with a diameter of $D=0.1$ m was constructed in COMSOL Multiphysics, which solves the Navier–Stokes equations via the finite-element method. The fluid domain was designed at 1.1 m high and 2.0 m length to reduce the simulation time. The domain used by Durante et al. [6] was up to 30D high and 100D in length. The centre of the circular cylinder was located horizontally at the middle and is 0.55 m from the water inlet of the domain as shown in Fig. 1. The fluid's density and dynamic viscosity are 1000 kg/m^3 and 1×10^{-3} kg/m s, respectively.

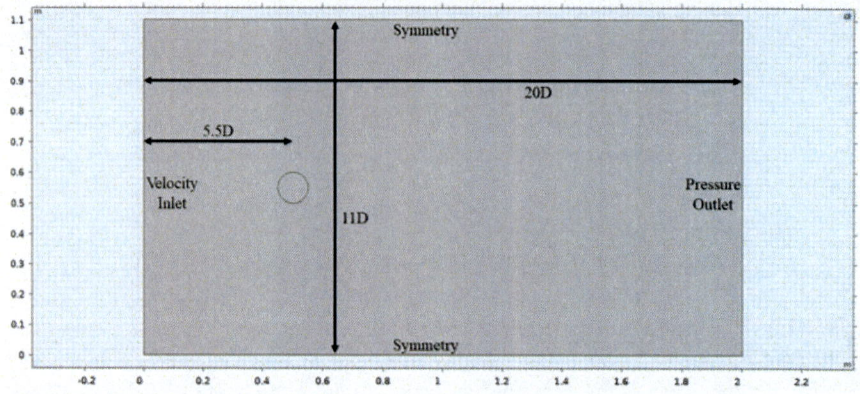

Fig. 1 Computational domain dimensions with boundary conditions

The velocity boundary condition, a uniform normal inflow was specified of the inlet. The pressure boundary condition was applied on the outlet boundary which is sufficiently far from the cylinder to allow the wake to dissipate naturally. Symmetric boundary conditions were set on the two side boundaries to avoid flow disturbance from the side boundaries. No-slip wall boundary condition was specified on the lines of the circular cylinder.

Triangular unstructured grids were generated with refinement close to the circumference of the circle. Note that very fine rectangular structured grids were used around the circle in order to capture the boundary layers. The example of the rectangular structured grids around the circle and refined triangular unstructured grids are presented in Fig. 2. For the given domain, the total number of mesh is 105,106, with the initial smallest mesh size around the circular cylinder of 0.007 m. The mesh size was further refined, and the effect of wall distance initialization was investigated.

All simulations were simulated with a normal inflow velocity of 0.1 m/s corresponding to Reynolds number of 10,000. The hydrodynamic drag and lift coefficients were calculated using the Morison equation [12],

$$C_d = \frac{2F_d}{\rho u^2 A} \tag{14}$$

where ρ is the fluid density, A is the projected frontal area or line, u is the free-stream velocity, and F_d is the drag force computed by the numerical simulation.

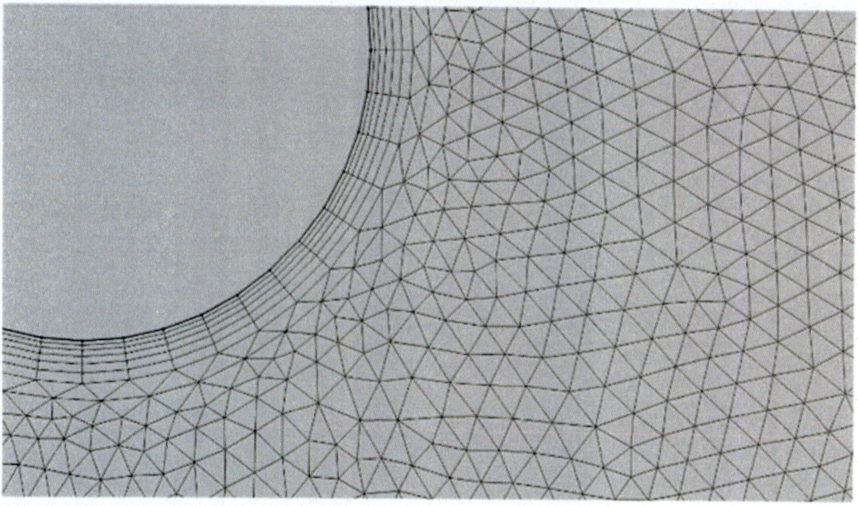

Fig. 2 Meshing in COMSOL Multiphysics

4 Result and Discussion

4.1 Turbulence Model

Figure 3 shows simulation results of drag and lift coefficient (CD and CL) via SA and SST turbulence model in COMSOL Multiphysics with wall distance initialization. The results between the turbulence models are different eventhough the simulation was carried out with an identical mesh and set-up. But on average for CL, both turbulent models give almost the same value of zero. Both turbulence models have similar $y+$ which is around 6 but the computational time was different where it takes 84 and 37 min for SST and SA turbulence models, respectively. SA turbulence model has an advantage due to its formulation of the one-equation model that resolves all fluid behaviour in both the boundary layer and free-stream flow. Meanwhile, SST turbulence model has two main equations that calculate turbulence kinetic energy and dissipation rate. Hence, SST turbulence model has a longer computational time than SA turbulence model and is in line with the results.

All graphs give a complete sinusoidal pattern after the simulations are stable at around 30s, but SST turbulence model has higher amplitude for both drag and lift coefficients. The calculation of the energy dissipation rate, which takes into account the high adverse pressure gradient is a dominant factor in the SST model. The frequency of the alternating vortex generation is almost the same at 0.213 and 0.238 Hz for CL of SA and SST respectively and 0.417 and 0.5 Hz for CD of SA and SST respectively. Interestingly the result shows that both models gave the same behaviour at the beginning until the occurrence of vortex shedding at 20s. According to Cengel and Cimbala [3], the mean CD of 2D circular cylinder simulation is 1.2 for

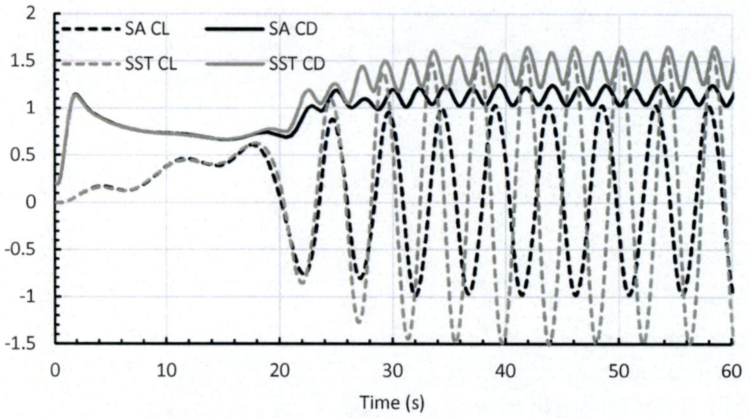

Fig. 3 Drag and lift coefficients of SA and SST turbulence model in time series

Reynolds numbers ranging from 1×10^4 to 2×10^5 which is laminar boundary layer or subcritical regime. After averaging the stable sinusoidal curve of drag coefficient for both models, mean CD of SA and SST turbulence models is 1.14 and 1.44, respectively. Since the result of mean SA CD is closer to 1.2, it proves that SA turbulence model has a better prediction of fluid flow past a circular cylinder than SST turbulence model.

Figure 4 shows vorticity field in z-direction where blue indicates vorticity in clockwise direction and red indicates vorticity in anticlockwise direction. Alternating vortex formation at the leeward of circular cylinder is due to instability of boundary layer at the wall where in which the separation points changes in location over the length of the circular cylinder. Both turbulence models results were selected by the peak amplitude of lift coefficient for comparison purposes. Since both turbulence models have different frequencies, thus comparing at the same instantaneous time is not practical. As can be seen, the SA turbulence model dispersed the energy of vortices faster than SST turbulence model, as has been discussed for Fig. 3. Consequently, SST turbulence model has higher drag due to the magnitude of vortices at the leeward of the circular cylinder is higher and it has greater drop in pressure.

The arc length against pressure coefficient curve was plotted in Fig. 5. Arc length zero corresponds to the windward of the circular cylinder while arc length 0.16 m corresponds to the leeward part. Negative pressure coefficient indicates suction pressure. Most of the suction pressure is at the middle of the arc length to the end, contributing to the drag force. The curves correlate to the drag and lift of the circular cylinder where SST turbulence model has higher negative pressure at the middle of the bottom arc length, indicating higher amplitude of lift as in Fig. 3. SST turbulence model also has higher negative pressure at the end of arc length of the circular cylinder that correlates to higher drag coefficient as in Fig. 3. The pressure coefficient at stagnation point should be equal to 1 as the velocity at the front of circular cylinder is zero.

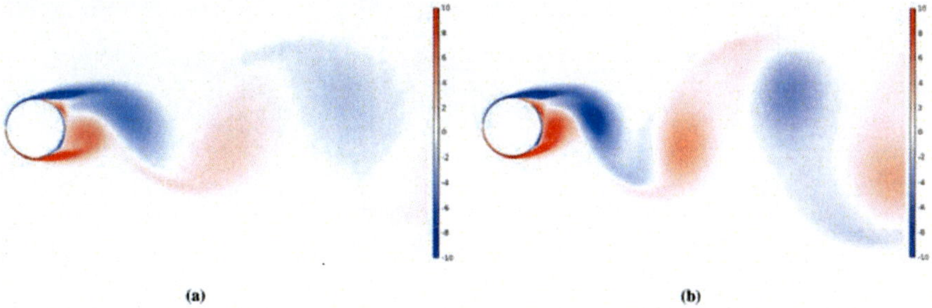

Fig. 4 Vorticity field in z-direction, **a** SA at 58 s (left), **b** SST at 58.3 s (right)

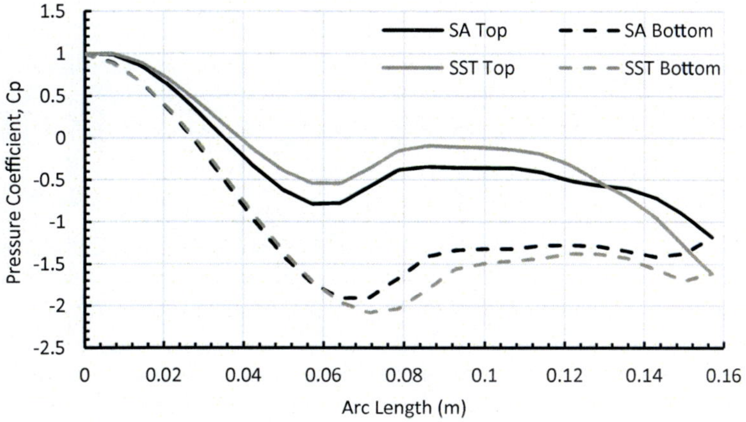

Fig. 5 Pressure coefficient on the wall of the circular cylinder

4.2 *Wall Treatment*

Figure 6 shows drag coefficient of SA turbulence model with three sizes of mesh. The maximum $y+$ for the simulation result are 0.3, 0.2 and 0.04 corresponding to SA fine, SA finer and SA extra-fine, respectively. All meshes with auto wall treatment converge to the same amplitude of the sinusoidal curve. Time taken for the simulations to stabilize is correlated to the size of mesh, in which the finer the meshing, the longer the time taken for the simulation to be stable. SA fine was also simulated with a different wall treatment which is Low Reynolds wall treatment. The result also converges to the other three simulations. This is because the auto wall treatment will shift to Low Reynolds wall treatment as the maximum $y+$ is lower than 1 [7].

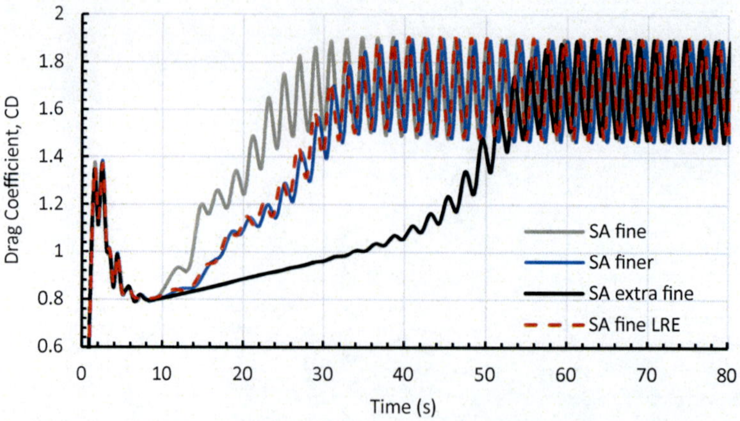

Fig. 6 Simulations with different mesh sizes with automatic versus LRE wall treatment for fine mesh

4.3 Wall Distance Initialization

Eventhough the simulations converged for all mesh sizes, the results of CD and CL curves pattern still do not imitate the result of previous researchers [6]. The simulation with Wall Distance Initialization (WDI) was then turned off to produce obviously different behaviour from the simulation with WDI turned on, as default option in COMSOL Multiphysics. This is as shown in Fig. 7. The CD and CL curve patterns changed from perfect sinusoidal curve patterns into chaotic curve patterns that are quite similar to the results of Durante et al. [6]. They simulated using meshless Lagrangian approach called diffused vortex hydrodynamics. The CD curve pattern at the beginning of the simulation during the development of the first set of vortices at near wake was similar, and at this period the wake is symmetrical. The mean CD of the present study is slightly higher than mean CD of Durante et al. [6] which is 1.8 and 1.7, respectively. When compared to result of LES model by [16], the mean SA CD COMSOL Multiphysics is similar at 1.8. Thus, the simulation with WDI turned off is better than the simulation with WDI turned on for this specific case of water flow past a circular cylinder at Reynolds number of 10,000.

Figure 8 shows vorticity field of SA extra-fine simulation with WDI both turned on and off. The behaviour near and far wake of the simulation with WDI turned on is similar to the result of Durante et al. [6], but with Reynolds number of 1000, which shows an incorrect behaviour. On the other hand, the behaviour of wake with existence of vortex dipole with WDI turned off is exactly similar to Durante et al. [6] results for the same Reynolds number. The wake patterns correlate to the CD and CL curve pattern where simulation with WDI turned on shows a periodic pattern alternating positive and negative vortex along the vortex street that produces periodic sinusoidal patterns. Figure 8 (right) shows there is no vortex street at the leeward of the circular cylinder simulation with WDI turned off which correlates with the

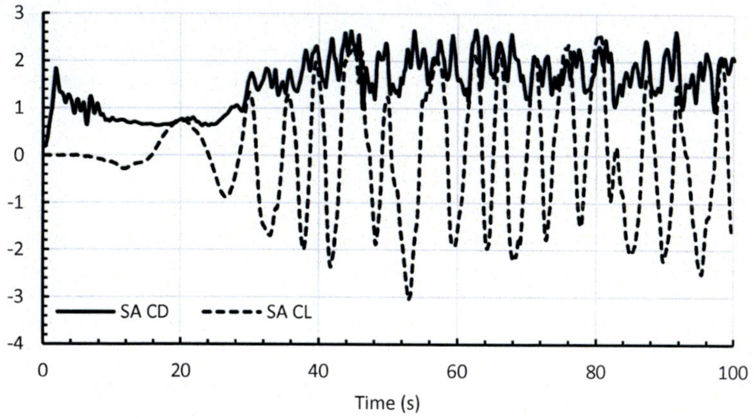

Fig. 7 Drag and lift coefficient of SA extra fine with WDI turned off

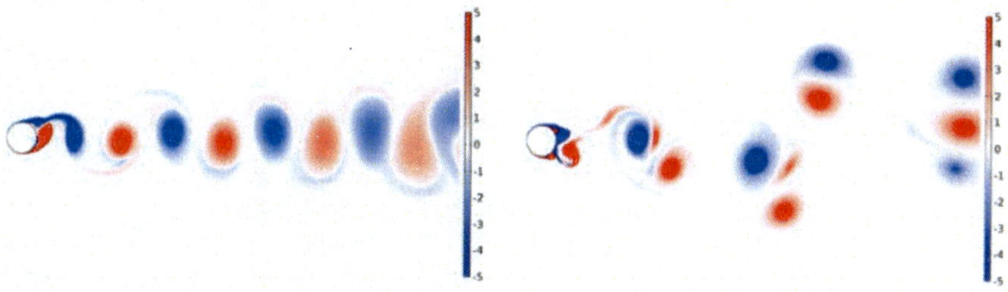

Fig. 8 Vorticity field of SA extra fine with WDI turned on (left) and off (right)

chaotic curve pattern in Fig. 7. Besides, Durante et al. [5] also simulated with elliptic cylinder shape at Reynolds number of 10,000 and produced the same behaviour of chaotic wake with the existence of vortex dipole.

5 Conclusion

In this chapter, a stationary circular cylinder was simulated with incompressible flow at Reynolds number of 10,000 using COMSOL Multiphysics, that solves Reynolds Average Navier–Stokes (RANS) via finite element method (FEM) approach. The focus is to investigate the capability of available turbulence models of SA model and SST model. The effects of wall treatment and wall distance initialization options were also investigated in order to validate the present simulation results with previous researchers [6, 16]. The simulation by Spalart Allmaras turbulence model is better than Menter's Shear Stress Transport turbulence model due to its computational time is nearly two times faster and the difference is minimal. The wall treatment was able to perform very well in determining how wall treatment should be treated, either using wall function or resolve the boundary layer with RANS depending on the maximum $y+$ that correlates to the mesh size on the wall of circular cylinder. The simulation with wall distance initialization (WDI) turned off successfully imitates the result of previous research in terms of CD and CL curves pattern and wake pattern for the specific case of incompressible flow past a stationary circular cylinder. Hence, the correct set-up of Spalart Allmaras turbulence model is capable of giving comparable results to the other models such as diffused vortex hydrodynamics (DVH) and Large Eddy Simulation (LES) with certain limitations.

References

1. Abu-Zidan, Y., Mendis, P., Gunawardena, T.: Optimising the computational domain size in CFD simulations of tall buildings. Heliyon **7**(4), e06723 (2021)
2. Ahmad, M.F., Haniffah, M.R.M., Kueh, A., Kasiman, E.H.: Numerical study on drag and lift coefficients of a marine riser at high Reynolds number using COMSOL Multiphysics. In: IOP Conference Series: Earth and Environmental Science, vol. 476, No. 1, p. 012075 (2020). IOP Publishing
3. Cengel, Y.A., Cimbala, J.M.: Fluid Mechanics Fundamentals and Applications. The McGrawHill Companies, Inc. (2013)
4. Dacles-Mariani, J., Kwak, D., Zilliac, G.: On numerical errors and turbulence modeling in tip vortex flow prediction. Int. J. Numer. Meth. Fluids **30**(1), 65–82 (1999)
5. Durante, D., Giannopoulou, O., Colagrossi, A.: Regimes identification of the viscous flow past an elliptic cylinder for Reynolds number up to 10000. Communications in Nonlinear Science and Numerical Simulation 105902 (2021)
6. Durante, D., Rossi, E., Colagrossi, A., Graziani, G.: Numerical simulations of the transition from laminar to chaotic behaviour of the planar vortex flow past a circular cylinder. Commun. Nonlinear Sci. Numer. Simul. **48**, 18–38 (2017)
7. Fontes, E.: COMSOL Multiphysics Blog (2017). https://www.comsol.com/blogs/why-should-i-use-automatic-wall-treatment-for-my-cfd-modeling. Last accessed 9/12/2021
8. Launder, B.E., Spalding, D.B.: The numerical computation of turbulent flows. In: Numerical prediction of flow, heat transfer, turbulence and combustion, pp. 96–116. Pergamon (1973)
9. Menter, F.R.: Two-equation eddy-viscosity turbulence models for engineering applications. AIAA J. **32**(8), 1598–1605 (1994)
10. Menter, F.R., Kuntz, M., Langtry, R.: Ten years of industrial experience with the SST turbulence model. Turbulence, Heat and Mass Transfer **4**(1), 625–632 (2003)
11. Monk, D., Chadwick, E.A.: Comparison of turbulence models effectiveness for a delta wing at low Reynolds numbers. In: 7th European conference for aeronautics and space sciences (EUCASS) **10**, (2017)
12. Morison, J.R., Johnson, J.W., Schaaf, S.A.: The force exerted by surface waves on piles. J. Petrol. Technol. **2**(05), 149–154 (1950)
13. Qiu, W., Lee, D.Y., Lie, H., Rousset, J.M., Mikami, T., Sphaier, S., Tao, L., Wang, X., Magarovskii, V.: Numerical benchmark studies on drag and lift coefficients of a marine riser at high Reynolds numbers. Appl. Ocean Res. **69**, 245–251 (2017)
14. Ruhl, C.: COMSOL Multiphysics Blog (2014). https://www.comsol.com/blogs/tips-using-wall-distance-interface. Last accessed 9/12/2021

15. Rumsey, C.: NASA Langley Research Center Turbulence Modeling Resource (2021). https://turbmodels.larc.nasa.gov/sst.html. Last accessed 9/12/2021
16. Singh, S.P., Mittal, S.: Flow past a cylinder: shear layer instability and drag crisis. Int. J. Numer. Meth. Fluids **47**(1), 75–98 (2005)
17. Spalart, P., Allmaras, S.: A one-equation turbulence model for aerodynamic flows. In: 30th Aerospace Sciences Meeting and Exhibit, p. 439 (1994)
18. Wilcox, D.C.: Reassessment of the scale-determining equation for advanced turbulence models. AIAA J. **26**(11), 1299–1310 (1988)

Comparison of Water Velocity of Physical and Numerical Modelling Analysis on a Downscale Spillway

F. Nurhikmah, M. H. Zawawi, N. H. Hassan, and N. M. Zahari

Abstract The study on the effect of velocity on the spillway was done conducted using physical and numerical model analysis. The 1:40 scale physical model was developed, where the velocity was measured in fourteen measurements points for three different velocity flows. Numerical modelling was analysed using Computational Fluid Dynamic (CFD) to compare the accuracy of the physical model analysis. Results from the study indicate that the maximum percentage difference between the physical model and numerical analysis in velocity values do not exceed 10% for all three cases. Point Q3 shows the highest value of velocity with 0.036 m/s for the physical model and 0.039 m/s for numerical modelling. The percentage of difference at point Q3 for numerical modelling and the physical model result is 7.69%. Meanwhile, points P5 and S5 indicate the lowest value of velocity, where the physical model and numerical modelling shows the same which is 0.006 m/s. The percentage of difference at points P9 and S5 for numerical modelling and the physical model result is 0.00%. It can be concluded that the velocity flow is higher at the energy dissipater parts compared to the stilling basin area due to the dispersion of water flow along the spillway.

Keywords Velocity flow · Physical model · Computational Fluid Dynamic

F. Nurhikmah (✉) · M. H. Zawawi · N. H. Hassan · N. M. Zahari
Department of Civil Engineering, College of Civil Engineering, Universiti Tenaga Nasional (UNITEN), Jalan IKRAM-Uniten, 43000 Kajang, Selangor, Malaysia
e-mail: nurhikmah.ghazali@uniten.edu.my

M. H. Zawawi
e-mail: MHafiz@uniten.edu.my

N. Sabtu (ed.), *Proceedings of AWAM International Conference on Civil Engineering 2022 - Volume 3*, Lecture Notes in Civil Engineering 386,
https://doi.org/10.1007/978-981-99-6026-2_42

1 Introduction

1.1 Physical Model of Spillway Structure

For nearly a century, physical modelling has been employed in the design and analysis of hydraulic structures [1]. Most often, the design process has included the preliminary design development based on empirical and theoretical approaches. The construction of 2D or 3D physical scale model arrangement or numerous scenarios would be conducted to validate whether that the hydraulic performance was justifiable and to extract the data for the design input [2].

The previous study shows that many research was conducted on the scaled-down physical model to analyse the hydraulic parameters on spillways. A 1:30 scale model of the Susu spillway was constructed at the Utah Water Research Laboratory at Utah State University (USA) to study the pressure flow behaviour at the stair-stepped spillway chute with 2.4 m high steps, in particular the piezometric pressures and transient pressures on the tread and riser of the steps [3]. Another study on the Batu dam spillway using a 1:25 scale physical model was conducted to investigate the hydraulic behaviour of the spillway and its ancillary structures under a range of design discharges, the discharge capacity of the prototype and the measures the hydraulic parameters [4]. A physical model with a distorted small 1:80 vertical scale for Gaara dam was constructed to study the effects of scales on the flow pattern over the crest of a standard ogee spillway of small dams [5].

1.2 Numerical Modelling for Verification Purpose

Since in middle of the last century, a wide range of applications of Computational Fluid Dynamic (CFD) has been commonly used in both industry and research thoroughly [6, 7]. Navier–Stokes equations are the essential foundation for solving the CFD problem especially involving fluid flow [8].

CFD methods are constantly being developed and together with improvements in computational resource capabilities, CFD has reached the stage where analysis of complex phenomena such as multiphase, free surface, highly turbulent flows is possible for a large number of design variations in a reasonable timescale [9]. Recent developments in CFD codes have meant that simulations of this nature show good agreement with experimental data [10, 11] because of the reliability of the numerical results produced when simulating complex phenomena. Hydraulic engineers are enthusiastic about validating the capacity of CFD software since it can give a quicker and more affordable solution than physical modelling [12].

Table 1 The scale downsize of the model from 1:1 to 1:40

Parameter	1:1 scale value	1:40 scale value
Length (y)	39 m	0.975 m

2 Methodology

2.1 *Physical Model Construction and Analysis*

A comprehensive and geometrically similar model of the dam spillway was built to a scale of 1:40, invoking the Froude number similarity for studying the velocity, pressure and specific energy. The proposed physical model scale is 1:40 scale, extending from the upstream and downstream portion of the dam. The sizes of the proposed dam physical model are 0.975 m (wide), 1.6 m for upstream and downstream lengths (including water tank) and 1.5 m (height). As the model factor is 1:40, the scale downsized for the spillway length was shown as in Table 1. The velocity measurements were taken at fourteen different points with three different velocities; Case 1 is 0.025 m/s, 0.045 m/s for Case 2 and 0.060 m/s for Case 3. Figure 1 shows the points taken along the spillway.

Figure 2 shows the equipment used to determine flow velocity is the Velocity Meter. A flow meter is the instrument used to evaluate, track and monitor extremely low water and other conductive fluids. It is designed essentially for laboratory and specialised industrial use, where the miniature head of the flow sensing probe can be inserted into small ducts and channels [6]. It is suitable to measure velocities accurately in hydraulic models of river estuaries, dams harbours and many more in addition to field measurements of clean water flows.

2.2 *Comparison of Physical Model Using Numerical Modelling*

To justify the physical model study, the velocity for numerical modelling was compared with velocity obtained from the physical model analysis. The physical model analysis was carried out and the result obtained was then compared with the Computational Fluid Dynamics (CFD) simulation. The numerical analysis using Ansys was conducted to validate the experimental data to ensure the reliability and precision results.

The position of Plane 1 is at − 0.36 m to the left part of the spillway model and where points P1, P2, P3, P4 and P5 were located as shown in Fig. 3. Plane 2 was located at 0 m within the centre of the spillway model as shown in Fig. 4. This is where points Q1, Q2, Q3 and Q4 are situated at. Whereas, Fig. 5 illustrate the points S1, S2, S3, S4 and S5 were pinpoint on the Plane 3 which located at 0.36 m on the right part of the spillway physical model.

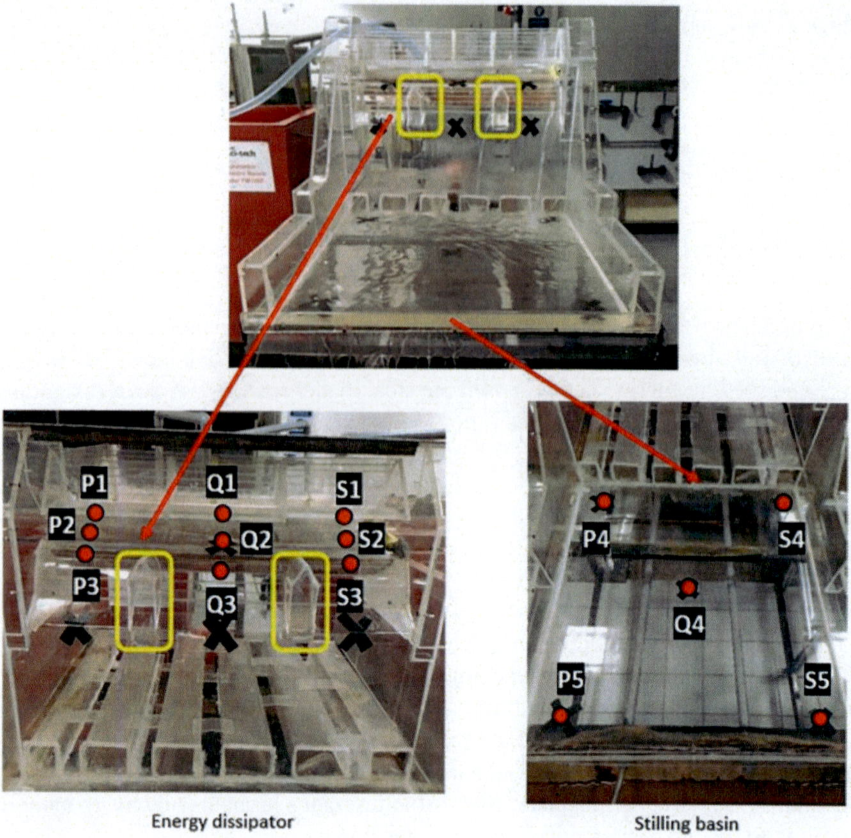

Fig. 1 The area for velocity measurement

3 Results and Discussion

3.1 Physical Model Analysis

The velocity distribution along the spillway for different discharges is shown in Table 2. The velocity flow for Case 1 is 0.025 m/s, 0.045 m/s for Case 2 and 0.060 m/s for Case 3. For all three discharges, it can be seen that the minimum and maximum velocities at the flow surface on the spillway structure.

For Case 1, the velocity at point P1 is 0.020 m/s, point P2 is 0.027 m/s, point P3 is 0.030 m/s, point P4 is 0.008 m/s and point P5 is 0.005 m/s. Points P1 until P5 was located along the Plane 1. The velocity measurement for point Q1 is 0.021 m/s, 0.027 m/s for point Q2, 0.031 m/s for point Q3, and lastly 0.007 m/s for point Q4. Whereas, Plane 3 is where points S1 until S5 are located. The velocity measurement

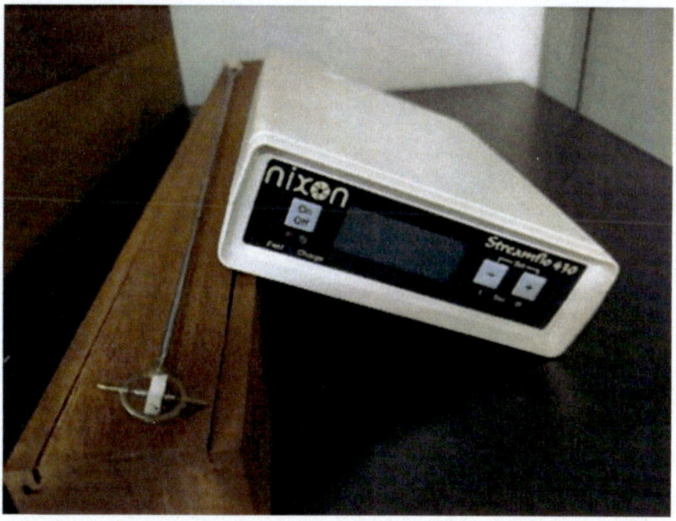

Fig. 2 Velocity meter

Fig. 3 Plane 1

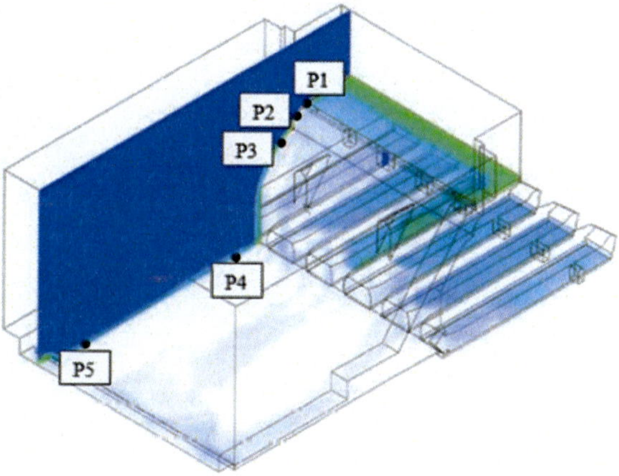

for point S1 is 0.020 m/s, point S2 is 0.027 m/s, point S3 is 0.030 m/s, point S4 is 0.008 m/s and point S5 is 0.005 m/s.

For Case 2, the velocity at point P1 is 0.023 m/s, point P2 is 0.029 m/s, point P3 is 0.035 m/s, point P4 is 0.009 m/s and point P5 is 0.006 m/s, which located on Plane 1. Plane 2 is where points Q1 until Q4 are located on the centre of the spillway model. The velocity measurement for point Q1 is 0.024 m/s, 0.031 m/s for point Q2, 0.036 m/s for point Q3, and lastly 0.010 m/s for point Q4. Whereas, Plane 3 is where points S1 until S5 are located. The velocity measurement for point S1 is 0.023 m/s, point S2 is 0.029 m/s, point S3 is 0.035 m/s, point S4 is 0.009 m/s and point S5 is 0.006 m/s.

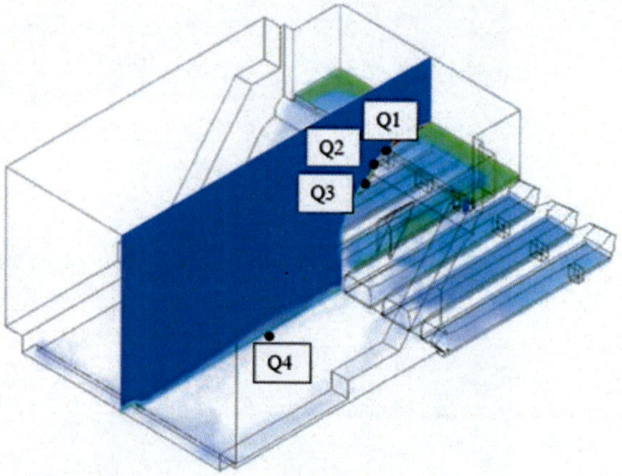

Fig. 4 Plane 2

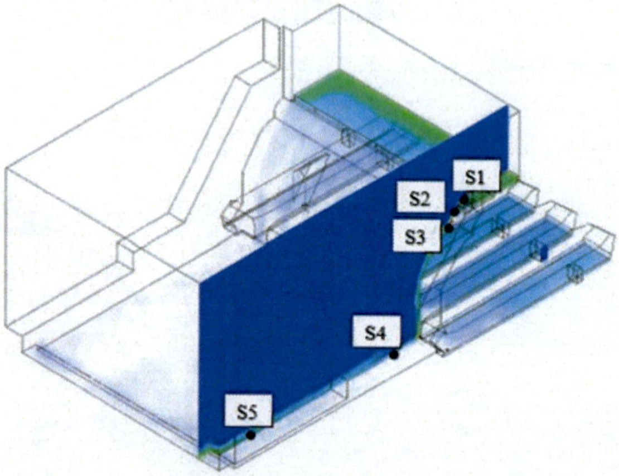

Fig. 5 Plane 3

For Case 3, the velocity at point P1 is 0.0283 m/s, point P2 is 0.034 m/s, point P3 is 0.039 m/s, point P4 is 0.013 m/s and point P5 is 0.009 m/s, which located on Plane 1. Plane 2 is where points Q1 until Q4 are located on the centre of the spillway model. The velocity measurement for point Q1 is 0.029 m/s, 0.035 m/s for point Q2, 0.040 m/s for point Q3, and lastly 0.014 m/s for point Q4. Whereas, Plane 3 is where points S1 until S5 are located. The velocity measurement for point S1 is 0.028 m/s, point S2 is 0.034 m/s, point S3 is 0.039 m/s, point S4 is 0.014 m/s and point S5 is 0.009 m/s.

Based on the result, all three cases shows that the highest velocity is at Case 3 point Q3 (0.040 m/s) which located near the energy dissipator area. Due to the conversion of all potential energy to kinetic energy, water rushing over a spillway has an extremely high kinetic energy cause the increasing of the fluid velocity. The high amount of energy dissipate contributes in high stress distribution through collision of water with the surface of dam stilling basin. Concurrently, the lowest velocity measurement for

Table 2 Velocity measurement using velocity meter

Plane	Point	Case 1 (m/s)	Case 2 (m/s)	Case 3 (m/s)
Plane 1	P1	0.020	0.023	0.028
	P2	0.027	0.029	0.034
	P3	0.030	0.035	0.039
	P4	0.008	0.009	0.013
	P5	0.005	0.006	0.009
Plane 2	Q1	0.021	0.024	0.029
	Q2	0.027	0.031	0.035
	Q3	0.031	0.036	0.040
	Q4	0.007	0.010	0.014
Plane 3	S1	0.020	0.023	0.028
	S2	0.027	0.029	0.034
	S3	0.030	0.035	0.039
	S4	0.008	0.009	0.014
	S5	0.005	0.006	0.009

all three cases is at point P5 and S5 with value of 0.005 which at Case 1. The area above the bed floor where the flow is mainly stable on the stilling basin towards the outlet.

3.2 *Numerical Model Analysis*

The input velocity from upstream at the top of spillway is 0.045 m/s, which is same as Case 2 in physical model analysis. The points P1, Q1 and S1 are located at the spillway crest area where the upstream area before the weir whereas points P2, Q2 and S2 are pointed on the nappe area where the water flows over the dam. Points P3, S3 and Q3 are pinpoint along the energy dissipator area. Meanwhile the position for points P4, P5, Q4, S4 and S5 are fixed at the stilling basin area of the spillway.

Points P1, P2, P3, P4 and P5 are located at Plane 1 as shown in Fig. 3. The velocity measurement using numerical analysis at point P1 is 0.025 m/s, P2 is 0.031 m/s, P3 is 0.037 m/s, P4 is 0.010 m/s and P5 is 0.006 m/s.

Points Q1, Q2, Q3 and Q4 are located at Plane 2 as shown in Fig. 4. The velocity measurement using numerical analysis at point Q1 is 0.026 m/s, Q2 is 0.035 m/s, Q3 is 0.039 m/s and Q4 is 0.011 m/s.

Next, points S1, S2, S3, S4 and S5 are located at Plane 3 as shown in Fig. 5. The velocity measurement using numerical analysis at point S1 is 0.026 m/s, S2 is 0.031 m/s, S3 is 0.037 m/s, S4 is 0.010 m/s and S5 is 0.006 m/s. The velocity measurements are decreasing as the water flow is moving towards the downstream.

3.3 Validation of Physical Model and Numerical Analysis

As shown in Table 3, physical model and numerical simulation analysis result were compared in term of velocity at the points along the spillway with the highest and lowest velocity. At point Q3 (along the energy dissipater area) shows the highest value of velocity with 0.036 m/s for physical model and 0.039 m/s for numerical modelling. The percentage of difference at point Q3 for numerical modelling and physical model result is 7.69%. At point P5 and S5 (end of stilling basin) indicate the lowest value of velocity, where the physical model and numerical modelling shows the same which is 0.006 m/s. The percentage of difference at points P9 and S5 for numerical modelling and physical model result is 0.00%. The result shows that both of the difference percentage at energy dissipater and stilling basin for velocity are less than 10%.

It was found that the flow velocity is higher near the energy dissipator compared at the stilling basin. There is a reduction of 90% of flow velocity as the flow reaches the stilling basin. This revealed the efficacy of energy dissipators in absorbing the preponderance of the kinetic energy contained in water flow before it reaches the downstream stilling basin. In addition to the energy dissipators, other places of high velocity flow are directly beneath the energy dissipators, where the water stream vertically falls on from the energy dissipators as a result of gravity. The water flows disperse across the stilling basin eventually, and the velocity is decreased.

Table 3 Comparison between physical model and numerical analysis

Plane	Point	Velocity (m/s)		Percentage difference (%)
		Physical model (m/s)	Numerical simulation (m/s)	
Plane 1	P1	0.024	0.025	4.00
	P2	0.030	0.031	3.23
	P3	0.035	0.037	5.41
	P4	0.010	0.010	0.00
	P5	0.006	0.006	0.00
Plane 2	Q1	0.025	0.026	3.85
	Q2	0.032	0.035	8.57
	Q3	0.036	0.039	7.69
	Q4	0.010	0.011	9.09
Plane 3	S1	0.024	0.026	7.69
	S2	0.030	0.031	3.23
	S3	0.035	0.037	2.70
	S4	0.010	0.010	0.00
	S5	0.006	0.006	0.00

4 Conclusion

From the study, a 1:40 scale physical model was constructed to study the velocity flow along the spillway. The, the results from the physical model were compared to the numerical simulation (CFD) to assure the accuracy of the results. The percentage difference between physical modelling and numerical simulation (CFD) resulting in velocity flow are compatible to each other which is less than 10%. In hydraulic properties analysis, it was demonstrated that there is a reasonable degree of agreement between physical and numerical models.

References

1. Preece, D., Williams, S.B., Lam, R., Weller, R.: 'Let's get physical': advantages of a physical model over 3D computer models and textbooks in learning imaging anatomy. Anat. Sci. Educ. **6**(4), 216–224 (2013). https://doi.org/10.1002/ase.1345
2. Gurav, N.V.: Physical and numerical modeling of an orifice spillway. Int. J. Mech. Prod. Eng. **3**(10), 71–75 (2015)
3. Lesleighter, E.J., Tullis, B.P., Andrews, D.: Stepped Spillway Model Pressures Characteristics, Susu Dam Malaysia, pp. 25–27, June 2014. https://doi.org/10.14264/uql.2014.20
4. Hamzah, S.B., Kamarul, M., Samion, H., Mohamad, M.F.: Utilising physical model for design assessment: proposed alterations of Batu Dam. Malaysia **8**(3), 202–205 (2016). https://doi.org/10.7763/IJET.2016.V8.885
5. Kamel, A.H.: Simulation of Flow Patterns Over Spillway of Small Dams Using Models with Distorted Small Simulation of Flow Patterns Over Spillway of Small Dams Using Models, November 2015
6. Ringgau, D., Anyi, M.: The Implementation and Calibration a Low Cost Propeller Type Current Meter in the Laboratory, vol. 1, no. 1, pp. 128–135 (2020)
7. Yang, Y., Zhou, B., Post, J., Scheepers, E., Reuter, M., Boom, R.: Computational fluid dynamics simulation of pyrometallurgical processes. In: Fifth International Conference on CFD Process Industries (2006)
8. Zawawi, M.H., Swee, M.G., Zainal, N.S., Zahari, N.M., Kamarudin, M.A., Ramli, M.Z.: Computational fluid dynamic analysis for independent floating water treatment device. AIP Conf. Proc. **1885** (2017). https://doi.org/10.1063/1.5002316
9. Zawawi, M.H. et al.: A review: fundamentals of computational fluid dynamics (CFD). AIP Conf. Proc. **2030** (2018). https://doi.org/10.1063/1.5066893
10. Perrig, A.: Hydrodynamics of the Free Surface Flow in Pelton Turbine Buckets, vol. 3715 (2007)
11. Panthee, A., Neopane, H.P., Thapa, B.: CFD Analysis of Pelton Runner, vol. 4, no. 8, pp. 4–9 (2014)
12. Faridmehr, I., Nejad, A.F., Baghban, M.H.: Numerical and Physical Analysis on the Response of a Dam's Radial Gate to Extreme Loading Performance (2020)

Water Flow Velocity Analysis on Small-Scale Spillway Using Particle Image Velocimetry (PIV)

F. Nurhikmah, M. H. Zawawi, N. H. Hassan, and N. M. Zahari

Abstract The spillway design for a dam project needs to deal with many hydraulics issues based on the study of the physical model. This study aims to investigate the flow characteristic and the effect of velocity along the spillway structure. The durability of the spillway structure was identifying by the area that experienced high velocity. In this study, twelve measurements points with three different velocity flow (0.025, 0.045 and 0.060 m/s) along the spillway were conducted. The scale-down physical model for the spillway was conducted for particle image velocimetry (PIV) analysis. Result from the study shows that the highest velocity value is at point Q3 on Case 3 which is 0.040 m/s and the lowest is at points P5 and S5 on Case 1 with 0.005 m/s. It can be concluded that the velocity flow is higher near the energy dissipater parts compared to the stilling basin area due to the water force acting on the spillway structure. To ensure the structural integrity sustainability, further study on the future mitigation measures can be taken.

Keywords Particle Image Velocimetry · Velocity flow · Physical model

1 Introduction

A spillway is an essential hydraulic structure in the construction of a dam. It is built to withstand any forces and flow conditions from stagnant or flowing water loading [1]. They must be so designed as to last over their design life span and be resistant to deterioration, ageing, and forces due to weather extremes and vibrations [2]. The integrity of a dam spillway structure study is very vital as it is one of the main parts of a dam. It is the part where water will be spilt at a certain water level.

F. Nurhikmah (✉) · M. H. Zawawi · N. H. Hassan · N. M. Zahari
Department of Civil Engineering, College of Civil Engineering, Universiti Tenaga Nasional (UNITEN), Jalan IKRAM-Uniten, 43000 Kajang, Selangor, Malaysia
e-mail: nurhikmah.ghazali@uniten.edu.my

M. H. Zawawi
e-mail: MHafiz@uniten.edu.my

N. Sabtu (ed.), *Proceedings of AWAM International Conference on Civil Engineering 2022 - Volume 3*, Lecture Notes in Civil Engineering 386,
https://doi.org/10.1007/978-981-99-6026-2_43

Consequently, the dam integrity study was crucial for the structural integrity safeguard and operation of the dam and hydroelectric generation [3, 4]. In the past works of literature, the structural reliability of the dam was mainly investigated from the perspective of the interaction with fluid flow. The particle image velocimetry (PIV) was adopted to analyse the dynamic nature of water velocity flow.

PIV techniques have been commonly utilized in numerous engineering studies and research works involving measurements of flow because it can calculate the whole sur- face flow velocity with minimum disturbance of the instrumentation [5]. PIV is an Image-processing technique to track water movement from one point to another to measure the surface flow velocity. Researches using PIV is continuously raising occasionally even though it is a new measuring technique [6, 7]. The development of non-intrusive flow measurement techniques for measuring scalar quantities and vectors in- flow is the most significant in the latest times. Thus, this study used PIV as a device to measure the surface flow velocity along the proposed small-scale spillway.

2 Methodology

2.1 Construction of 1:40 Scale Physical Model

Model size should preferably be sized appropriately for the purpose of assuring accuracy of similitude and measurements. A 1:40 physical model was used to construct the experiment as illustrated in Fig. 1. The sizes of proposed physical model are 1.6 m (length) $\times$ 0.975 m (width) $\times$ 1.5 m (height). The scaling down was executed by scaling down the model geometry and velocity flow on the basis of Froude number similarity. This results in the following model and prototype relationships in Table 1.

2.2 Particle Image Velocimetry (PIV)

Particle Image Velocimetry (PIV) is a non-intrusive, whole-filed optical method that provides instantaneous velocity information in the fluid [8]. The flow is seeded with illuminate tracer particles irradiate the target area with a light sheet at least twice within a short time interim. The camera captures the target area and images each light pulse in different image frames. The particle displacement between the light pulses can be utilized to determine the velocity vectors.

As illustrated in Fig. 2, the entry side of the fibre bundle is round while the fibres at the distal end area are arranged along a straight line. A light sheet can then be formed by projecting this line into the area under investigation using a short focal length cylindrical lens. The illumination setup of the system is shown in Fig. 3.

Fig. 1 Physical model of 1:40 scale

Table 1 Table captions should be placed above the tables

Scale 1:40	
Length	1:40
Velocity (m/s)	1:6.324
Discharge (m^3/s)	1:10,119

3 Result and Discussion

From the previous study by Hassan [9] at the Chenderoh Dam spillway using numerical simulation (1:1 scale), it shows various velocity profiles according to the operating conditions of the spillway and the highest value recorded is 0.38 m/s for operating condition 4.9 m sector gate opening [9]. One of the cases in this study was conducted similar to the velocity value which is Case 3 as shown in Table 4.2. Thus, this study matches the velocity profile of the previous study to verify the real scale.

PIV is used to measure the surface flow velocity. Figure 4 shows points taken to validate the surface flow velocity on the small-scale spillway. 14 different points along the spillway with three different cases was tested. The input velocity flow for Case 1 is 0.025 m/s, 0.045 m/s for Case 2 and 0.060 m/s for Case 3. The points P1,

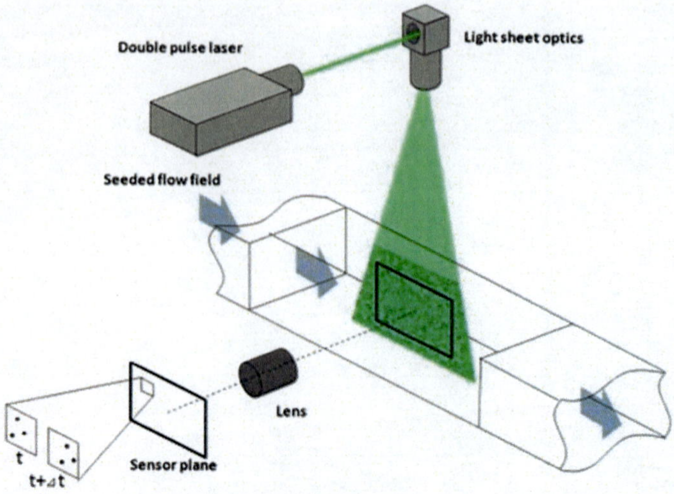

Fig. 2 Schematic of HP-LED based on light sheet illumination

Q1 and S1 are located at the spillway crest area where the up- stream area before the weir, whereas, points P2, Q2, and S2 are pointed on the nappe area where the water flows over the dam. Points P3, Q3 and S3 are pinpoint along the energy dissipator area. Meanwhile the position for points P4, P5, Q4, S4 and S5 are fixed at the stilling basin area of the spillway. Figure 4 shows the points along the spillway physical model for PIV experiment.

3.1 Case 1

The input velocity flow for Case 1 is 0.025 m/s. Using Particle Image Velocimetry (PIV) was recorded accordingly. The velocity value at point P1 is 0.021 m/s, point P2 is 0.028 m/s, point P3 is 0.032 m/s, point P4 is 0.008 m/s, point P5 is 0.005 m/s, which located on Plane 1. Plane 2 is where points Q1 until Q4 are located on the centre of the spillway model. The velocity value for point Q1 is 0.022 m/s, 0.029 m/s for point Q2, 0.033 m/s for point Q3, and lastly 0.007 m/s for point Q4. Whereas, Plane 3 is where points S1 until S5 are located. The velocity value for point S1 is 0.021 m/s, point S2 is 0.028 m/s, point S3 is 0.032 m/s, point S4 is 0.008 m/s and point S5 is 0.005 m/s. The following Table 2 shows the velocity value for Case 1.

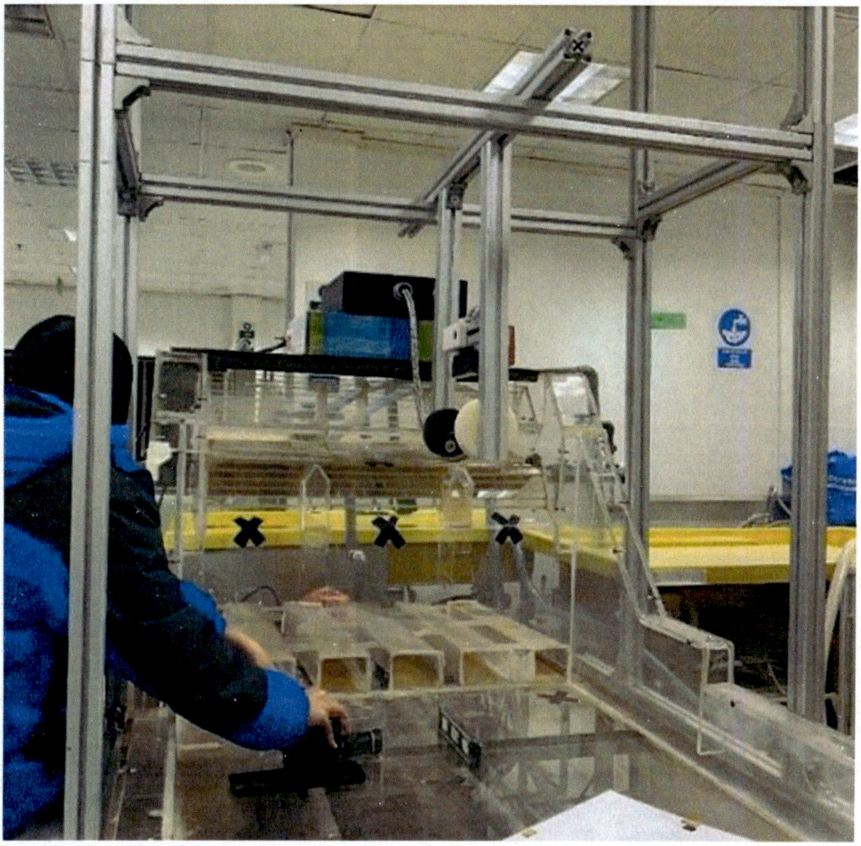

Fig. 3 Illumination setup of the system

3.2 *Case 2*

The initial velocity flow for case 1 is 0.045 m/s. Using Particle Image Velocimetry (PIV) was recorded accordingly. The measurement at point P1 is 0.024 m/s, point P2 is 0.030 m/s, point P3 is 0.035 m/s, point P4 is 0.010 m/s, point P5 is 0.006 m/s, which located on Plane 1. Plane 2 is where points Q1 until Q4 is located in the centre of the spillway model. The velocity measurement for point Q1 is 0.025 m/s, 0.032 m/s for point Q2, 0.036 m/s for point Q3, and lastly 0.010 m/s for point Q4. Whereas, Plane 3 is where points S1 until S5 is located. The velocity measurement for point S1 is 0.024 m/s, point S2 is 0.030 m/s, point S3 is 0.035 m/s, point S4 is 0.010 m/s and point S5 is 0.006 m/s. The following Table 3 shows the velocity value for Case 2.

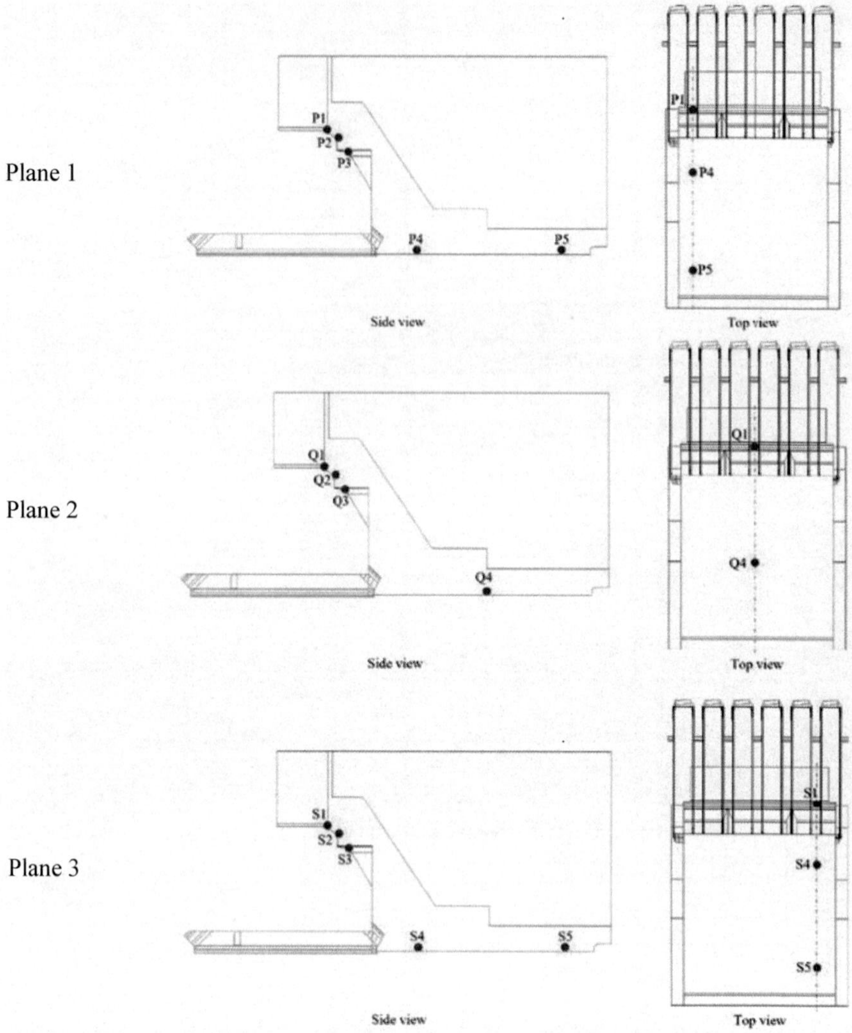

Fig. 4 Points along the spillway for the PIV experiment

3.3 *Case 3*

The initial velocity flow for case 1 is 0.060 m/s. Using Particle Image Velocimetry (PIV) was recorded accordingly. The measurement at point P1 is 0.029 m/s, point P2 is 0.035 m/s, point P3 is 0.039 m/s, point P4 is 0.014 m/s, point P5 is 0.010 m/s, which located on Plane 1. Plane 2 is where points Q1 until Q4 are located in the centre of the spillway model. The velocity measurement for point Q1 is 0.029 m/s, 0.036 m/s for point Q2, 0.040 m/s for point Q3, and lastly 0.015 m/s for point Q4.

Table 2 Velocity value for Case 1

Plane	Point	Velocity
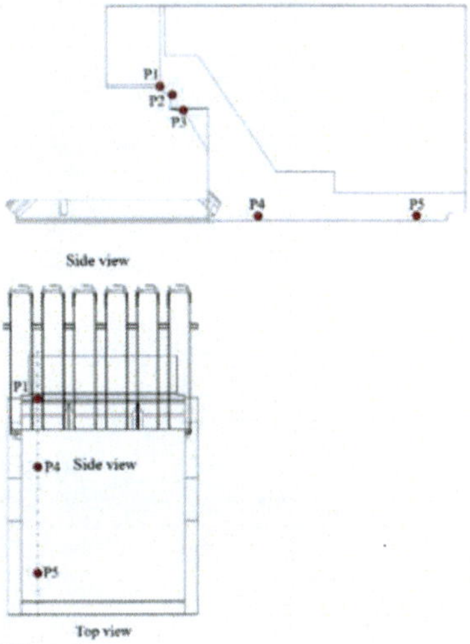 Plane 1	P1 = 0.021 m/s	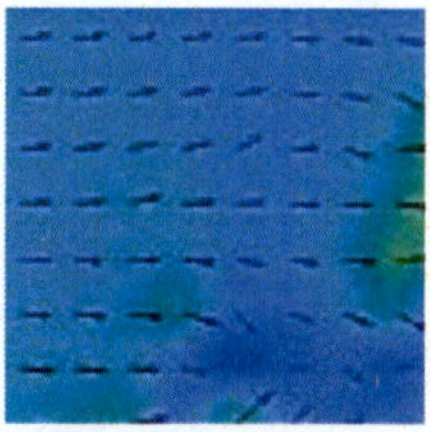
	P2 = 0.028 m/s	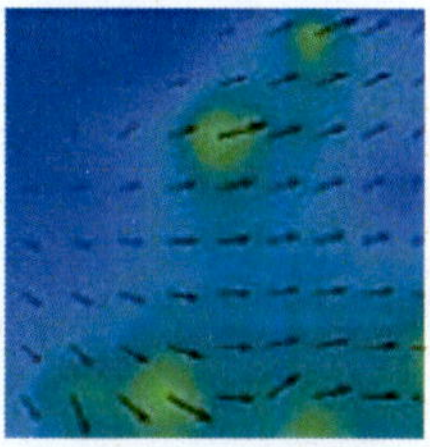
	P3 = 0.032 m/s	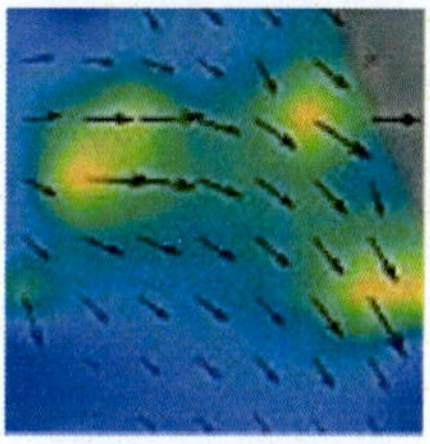
	P4 = 0.008 m/s	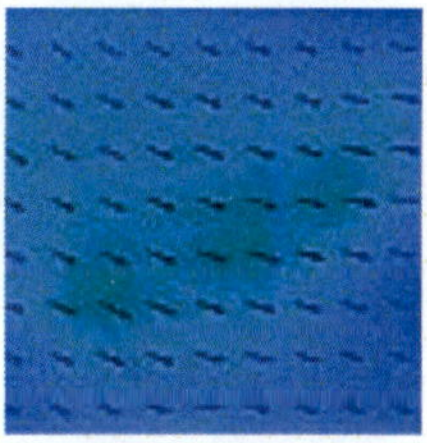
	P5 = 0.005 m/s	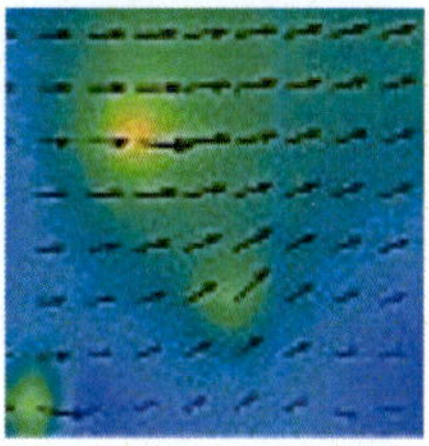

(continued)

Table 2 (continued)

Plane	Point	Velocity
	Q1 = 0.022 m/s	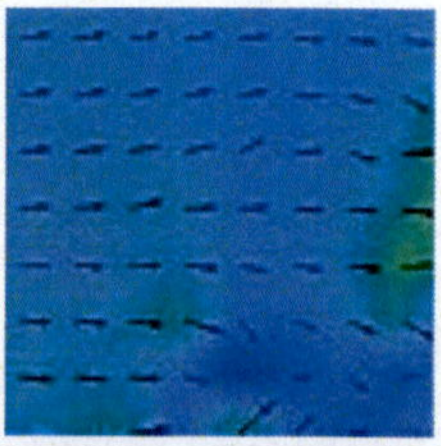
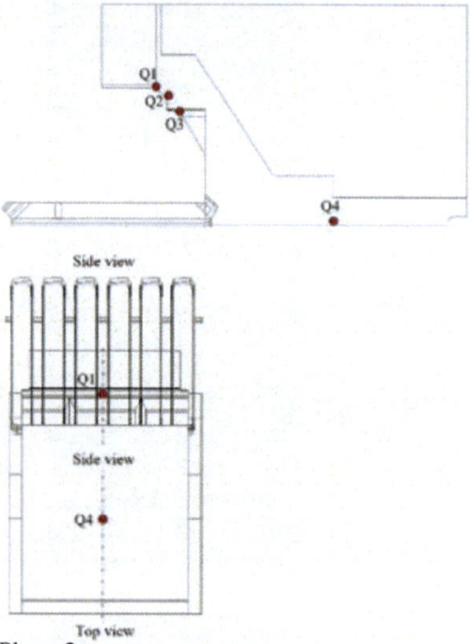 Plane 2	Q2 = 0.029 m/s	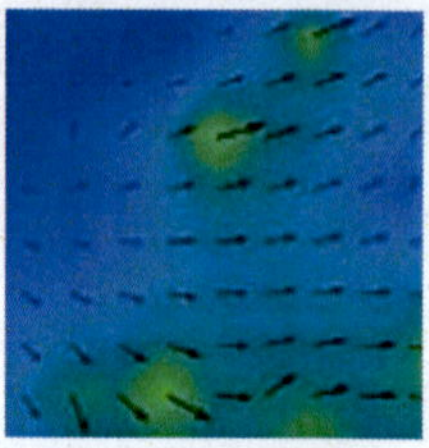
	Q3 = 0.033 m/s	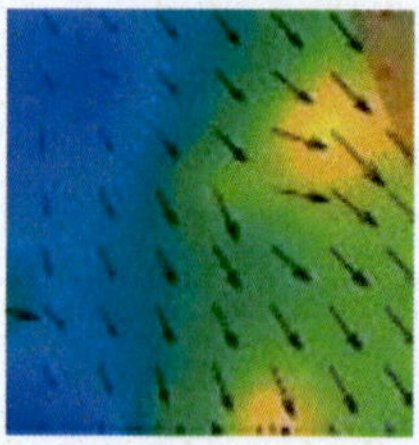
	Q4 = 0.007 m/s	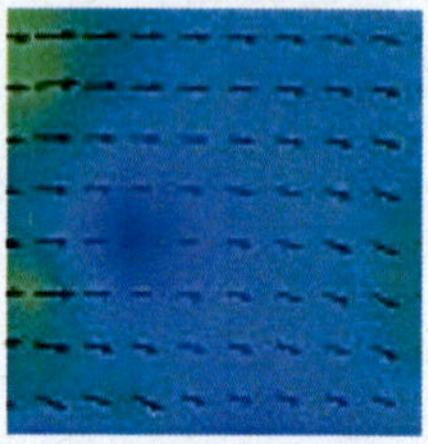
	S1 = 0.021 m/s	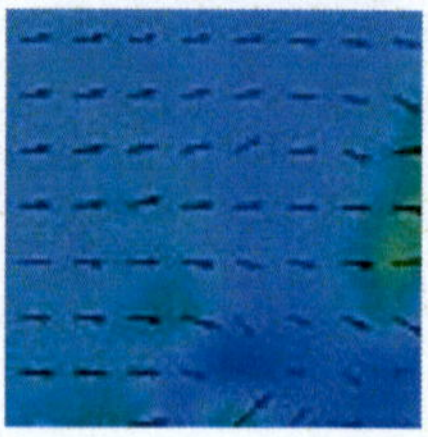

(continued)

Table 2 (continued)

Plane	Point	Velocity
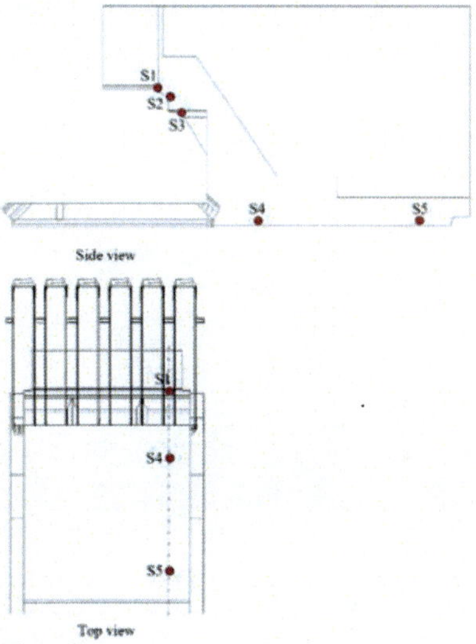 Plane 3	S2 = 0.028 m/s	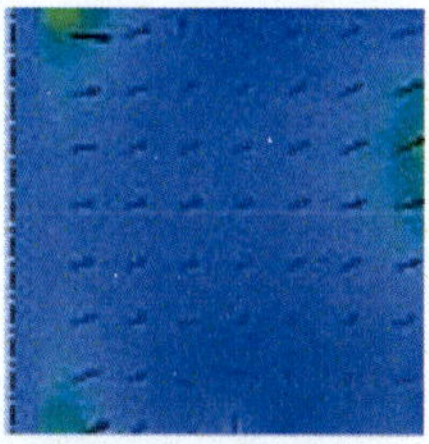
	S3 = 0.032 m/s	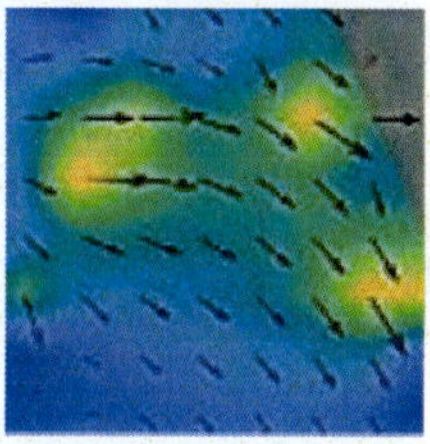
	S4 = 0.008 m/s	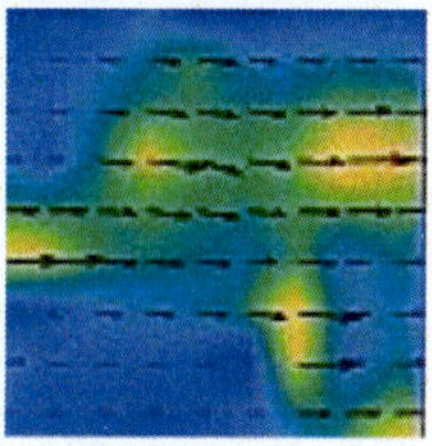
	S5 = 0.005 m/s	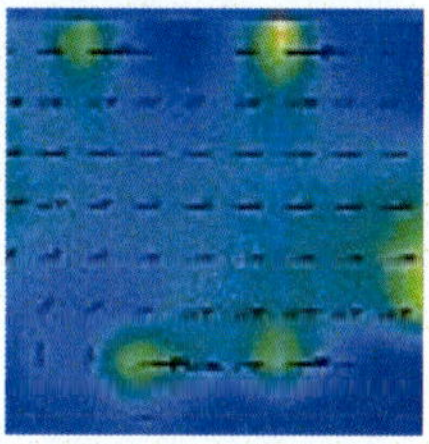

Whereas, Plane 3 is where points S1 until S5 are located. The velocity measurement for point S1 is 0.029 m/s, point S2 is 0.036 m/s, point S3 is 0.039 m/s, point S4 is 0.014 m/s and point S5 is 0.010 m/s. The following Table 4 shows the velocity value for Case 3.

From the results, all three cases shows that point Q3 pinpoint near the energy dissipator area has the highest reading with 0.033 m/s, 0.036 m/s and 0.040 m/s for Case1, Case 2 and Case 3 respectively. Meanwhile, the lowest reading also indicates at the same points for all three cases which is at P5 and S5 where the points located at

Table 3 Velocity value for Case 2

Plane	Point	Velocity
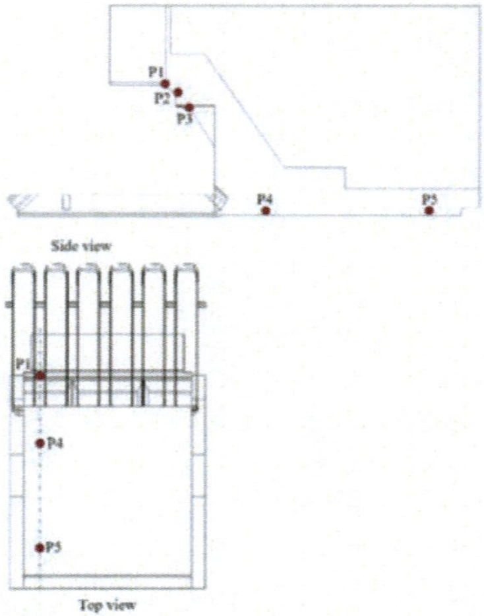 Plane 1	P1 = 0.024 m/s	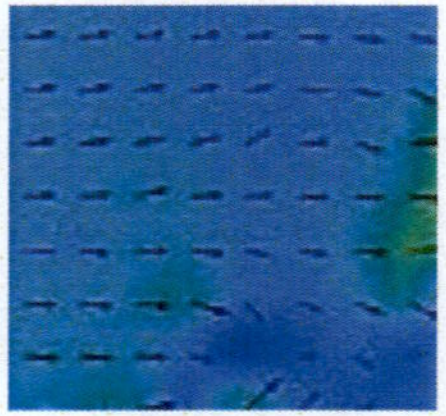
	P2 = 0.030 m/s	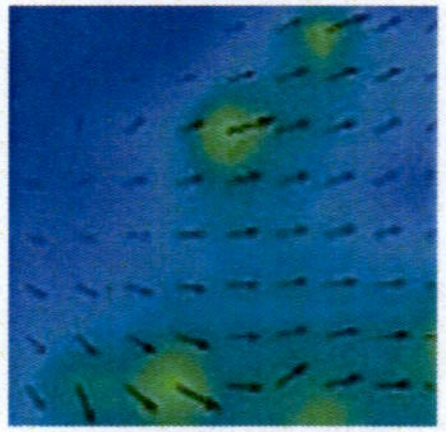
	P3 = 0.035 m/s	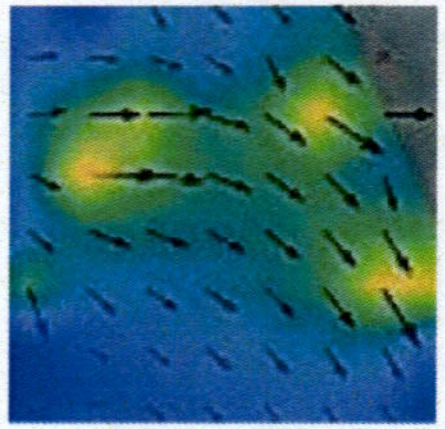

(continued)

Table 3 (continued)

Plane	Point	Velocity
	P4 = 0.010 m/s	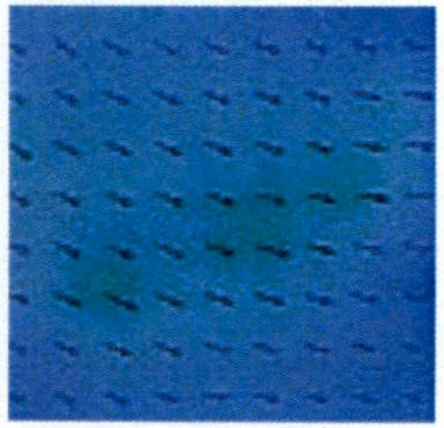
	P5 = 0.006 m/s	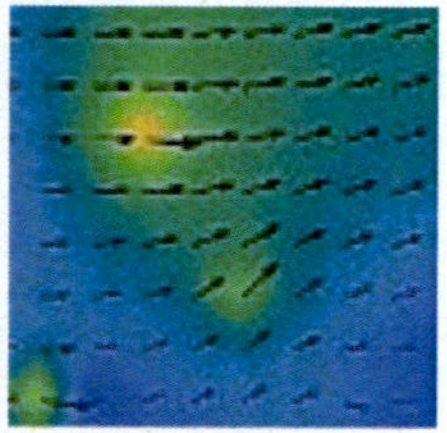

(continued)

Table 3 (continued)

Plane	Point	Velocity
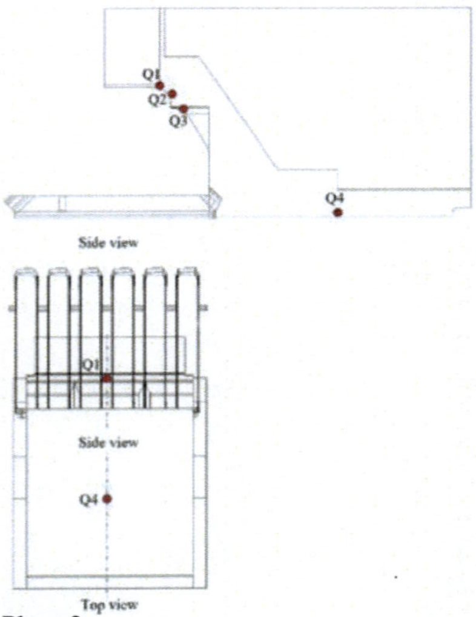 Plane 2	Q1 = 0.025 m/s	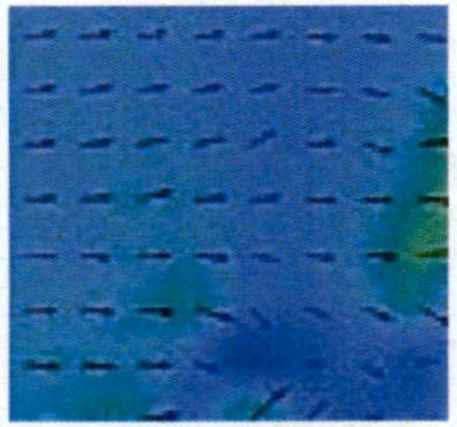
	Q2 = 0.032 m/s	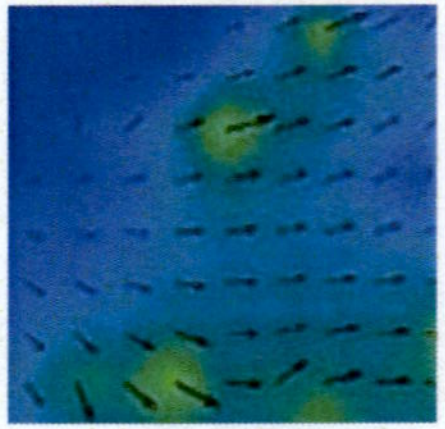
	Q3 = 0.036 m/s	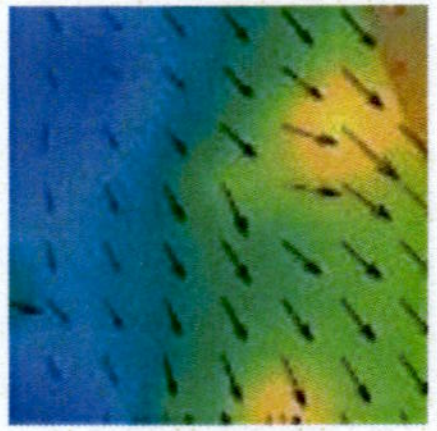

(continued)

Table 3 (continued)

Plane	Point	Velocity
	Q4 = 0.010 m/s	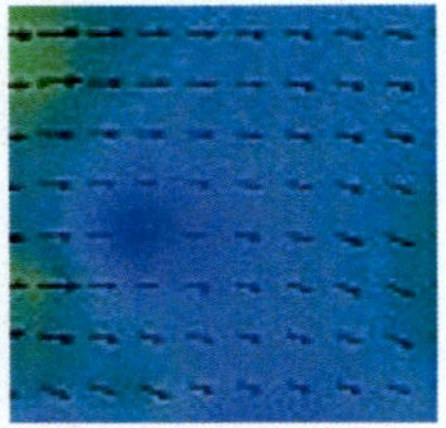
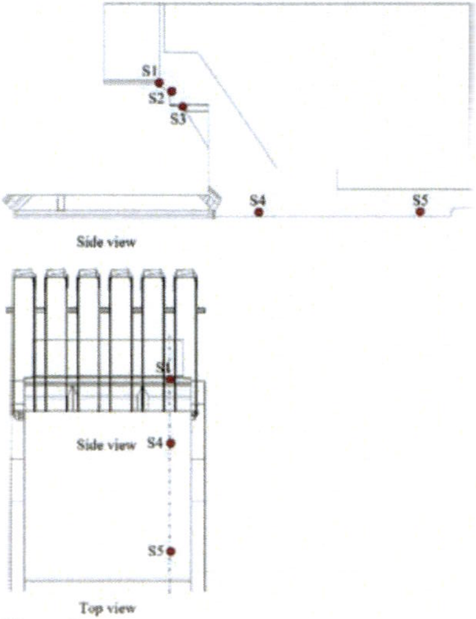 Plane 3	S1 = 0.024 m/s	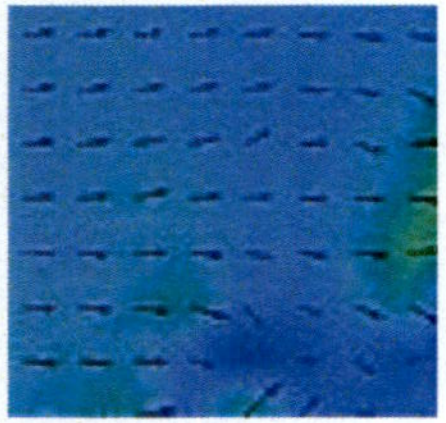
	S2 = 0.030 m/s	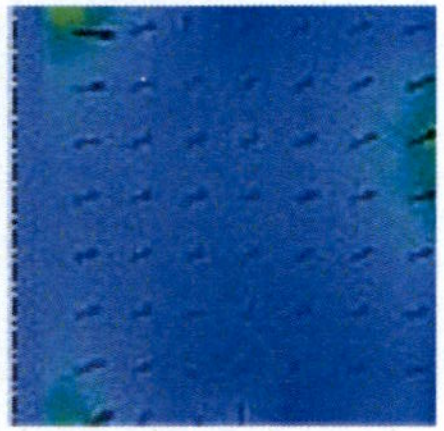

(continued)

Table 3 (continued)

Plane	Point	Velocity
	$S3 = 0.035$ m/s	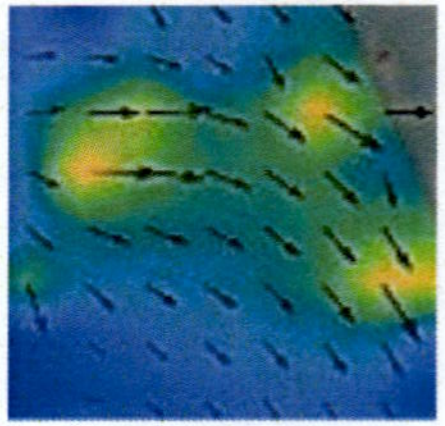
	$S4 = 0.010$ m/s	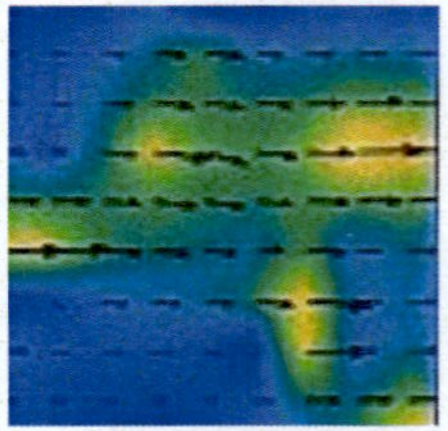
	$S5 = 0.006$ m/s	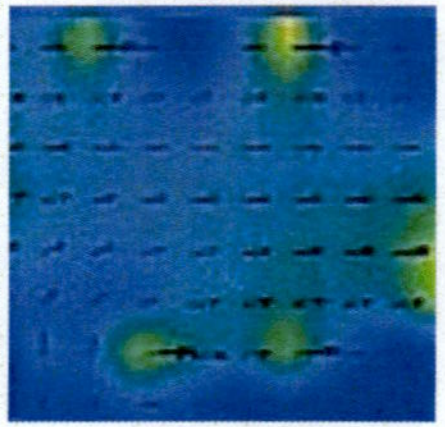

Table 4 Velocity value for Case 3

Plane	Point	Velocity
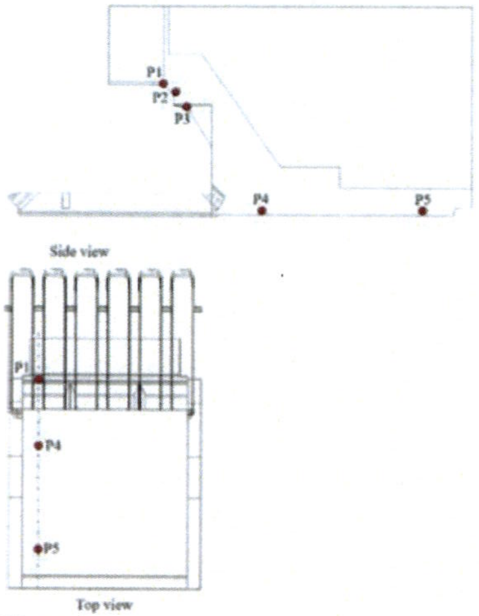 Plane 1	P1 = 0.029 m/s	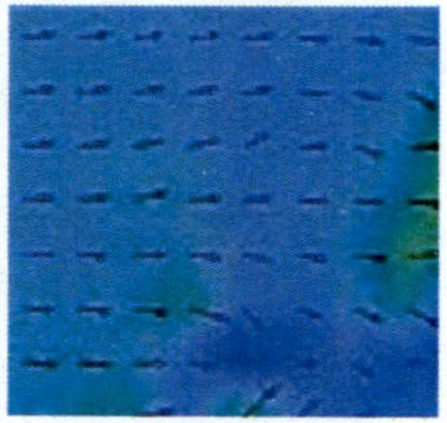
	P2 = 0.035 m/s	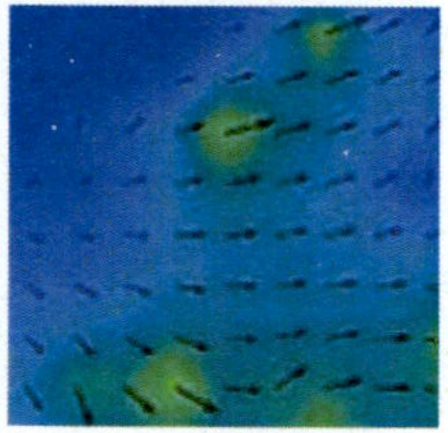

(continued)

Table 4 (continued)

Plane	Point	Velocity
	P3 = 0.039 m/s	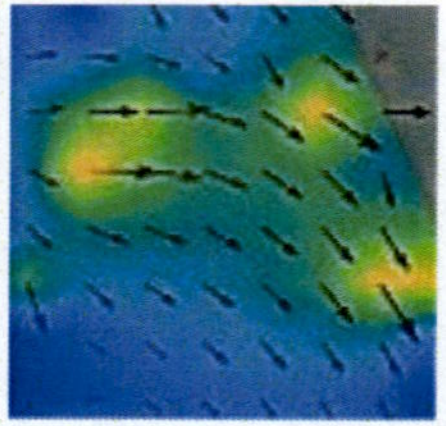
	P4 = 0.014 m/s	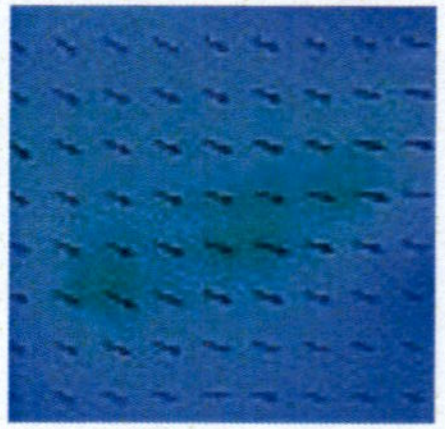
	P5 = 0.010 m/s	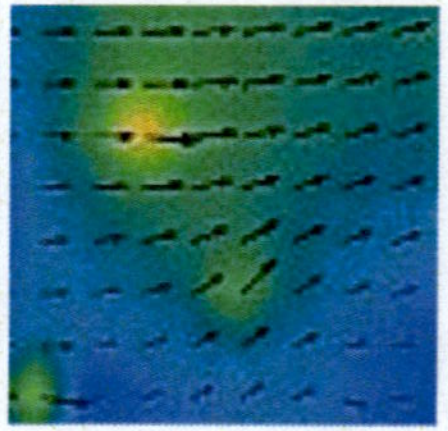

(continued)

Table 4 (continued)

Plane	Point	Velocity
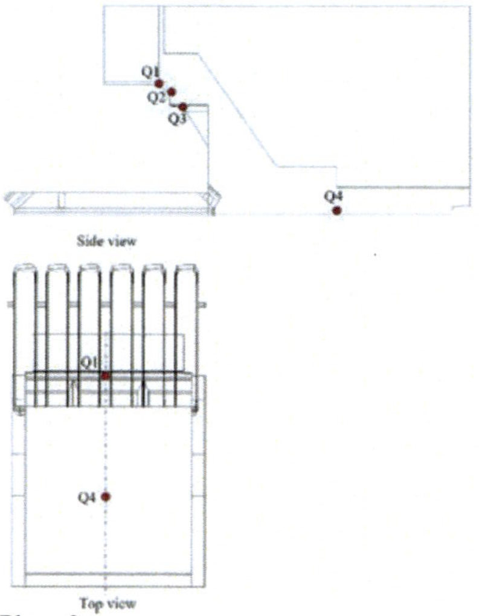 Plane 2	Q1 = 0.029 m/s	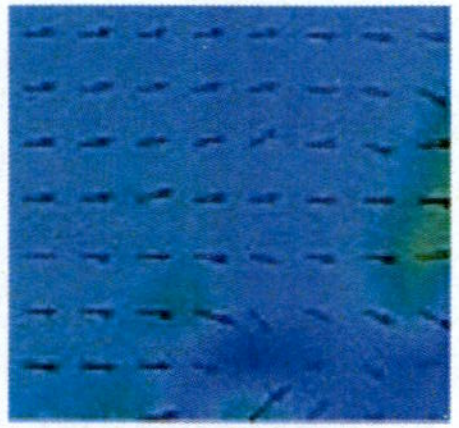
	Q2 = 0.036 m/s	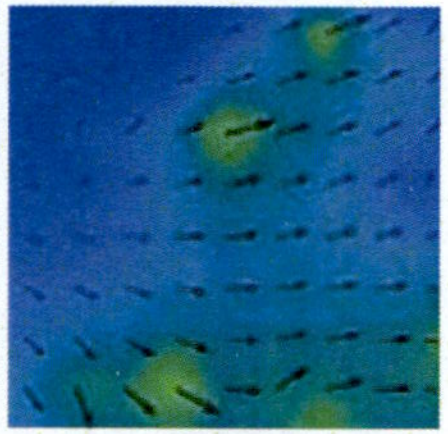
	Q3 = 0.040 m/s	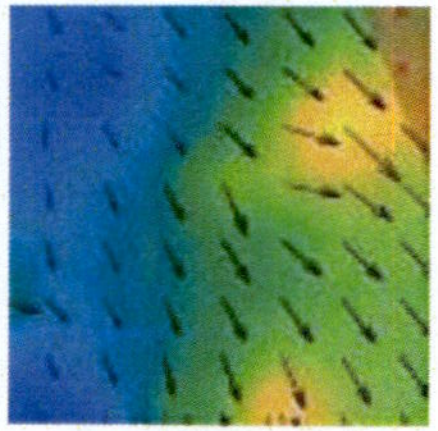

(continued)

Table 4 (continued)

Plane	Point	Velocity
	Q4 = 0.015 m/s	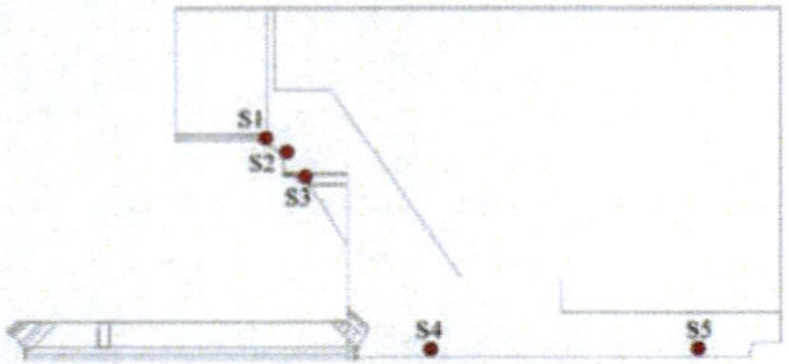
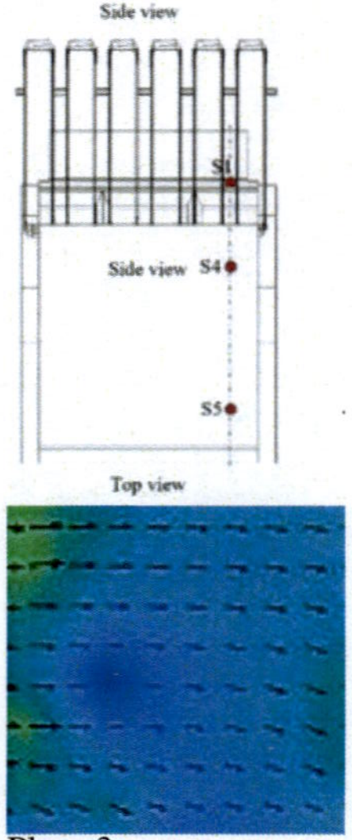 Plane 3	S1 = 0.029 m/s	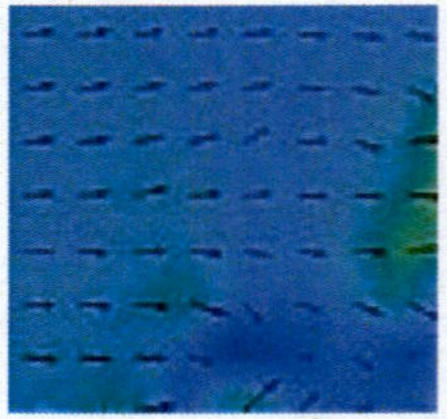
	S2 = 0.036 m/s	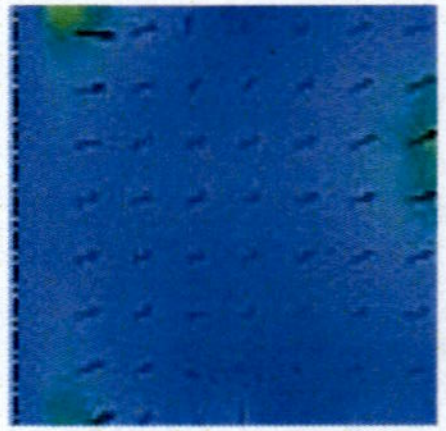

(continued)

Table 4 (continued)

Plane	Point	Velocity
	S3 = 0.039 m/s	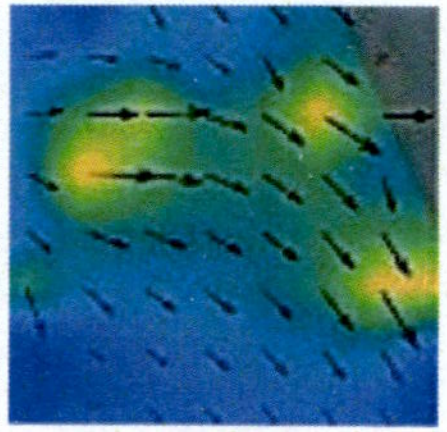
	S4 = 0.014 m/s	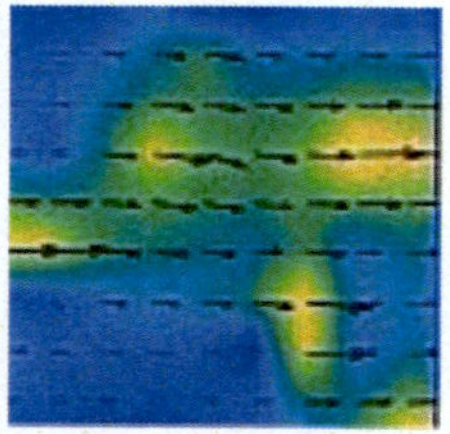
	S5 = 0.010 m/s	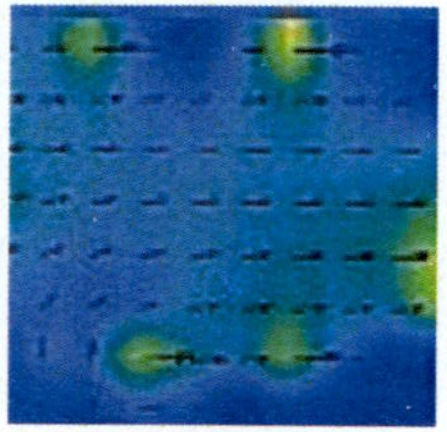

the end of the spillway stilling basin. Case 1 shows the value of 0.005 m/s, 0.006 m/s for Case 2 and 0.010 m/s for Case 3.

The images have vague surface flow visualization; hence the vector of the surface flow profile will be accurately plotted. Based on the visual inspection, all the flow direction in this study is flowing from left to right. Basically, the flow direction from left to right will generate positive vectors values in the software (MatPIV) and vice versa [7].

4 Conclusion

In conclusion, the application of Particle Image Velocimetry (PIV) was used to study the surface water velocity on the dam spillway analysis. A spillway physical model of 1:40 small scale was fabricated to conduct the experiment. From the results, the highest velocity value is at point Q3 on Case 3 which is 0.040 m/s and the lowest is at points P5 and S5 on Case 1 with 0.005 m/s. This demonstrated the effectiveness of energy dissipators in absorbing most of the kinetic energy of water flow before it reaches the downstream stilling basin. Other regions of high velocity flow are directly downward the energy dissipators where the water stream vertically fall from the energy dissipators, due to the gravity. Eventually, the water flows redistribute over the stilling basin and reduced the velocity. In addition, the PIV results visualized that the velocities flow magnitudes are consistent for the data analysed. This confirms the vectors produced from the experiment are able to represent the surface flow clearly.

References

1. Kumcu, S.Y.: Investigation of flow over spillway modeling and comparison between experimental data and CFD analysis. KSCE J. Civ. Eng. **21**(3), 994–1003 (2017). https://doi.org/10.1007/s12205-016-1257-z
2. Zawawi, M.H. et al.: Fluid-structure interactions study on hydraulic structures: A review. AIP Conf. Proc. **2030** (2018). https://doi.org/10.1063/1.5066885
3. Siacara, A.T., Napa-García, G.F., Beck, A.T., Futai, M.M.: Reliability analysis of earth dams using direct coupling. Journal of Rock Mechanics and Geotechnical Engineering **12**(2), 366–380 (2020). https://doi.org/10.1016/j.jrmge.2019.07.012
4. Yi, P., Liu, J., Xu, C.: Reliability analysis of high Rockfill dam stability. Math. Probl. Eng. **2015**, 512648 (2015). https://doi.org/10.1155/2015/512648
5. Hassan, N.H. et al.: A review on applications of particle image velocimetry a review on applications of particle image velocimetry. In: 2nd Joint Conference on Green Engineering Technology & Applied Computing 2020 (2020). https://doi.org/10.1088/1757-899X/864/1/012149
6. Abas, M., Jamil, R., Rozainy, M., Zainol, M., Adlan, M., Keong, C.: PIV study of aeration efficient of stepped spillway system PIV study of aeration efficient of stepped spillway system. Int. Conf. Innov. Res. (2017). https://doi.org/10.1088/1757-899X/209/1/012090
7. Kuok, K.K., Chiu, P.C.: Particle Image Velocimetry for Measuring Water Flow Velocity, no. January 2013 (2018)

8. Ahmed, R., Rasheed, J.: A comparison study between CFD analysis and PIV technique for velocity distribution over the Standard Ogee crested spillways. Heliyon **6**, no. October 2019, p. e05165 (2020). https://doi.org/10.1016/j.heliyon.2020.e05165
9. Hassan, N. et al.: Water Spilling Fluid Dynamic Analysis on Sector Gate Opening, vol. 32, no. 3, pp. 261–265 (2016)

Sustainable Infiltrated Drainage System with Porous Sublayer Modules

Zulkiflee Ibrahim, Mazlin Jumain, Fenny Anak Baseng, Abdul Hadi Mohd Subki, Nor Zehan Hamzah, Ezat Faiz, Md Ridzuan Makhtar, Roslee Ishak, and Nurfarhain Mohamed Rusli

Abstract The main problem in the current urban drainage system is the inefficiency of the drainage system to accommodate high stormwater discharge in the system. This experimental study was aimed to investigate the effectiveness of porous sublayer modules in an infiltrated drainage system. Primarily, this chapter focused on the influence of different lengths of porous sublayer module on the infiltration, flow depth and velocity distributions in the drainage system. Laboratory experiments were carried out for three porous sublayer module lengths to drain diameter ratio (Ls/d) of 0.7, 1.0 and 1.4 for two drain bed slopes and four drain section spacing. The study found that the optimum (Ls/d) for 0.7 and 1.4 which allowed more than 75% of surface runoff to infiltrate into underground storage, as stated in MSMA for On-Site Detention. The flow velocity in the drain reduced by more than 80% with the application of porous sublayer modules. It was also found that the spacing between drain sections was a significant factor to lower the drain velocity. Its velocity increased as the water travelled further downstream and reduced as the drain spacing became larger.

Keywords Urban Drainage · Infiltration · Porous Sublayer Modules · Underground Storage · Laboratory Experiment

Z. Ibrahim (✉) · M. Jumain · F. A. Baseng · A. H. M. Subki · M. R. Makhtar · R. Ishak
School of Civil Engineering, Faculty of Engineering, Universiti Teknologi Malaysia, UTM Johor Bahru, 81310 Johor, Malaysia
e-mail: zulkfe@utm.my

N. Z. Hamzah · E. Faiz
Ranhill Syarikat Air Johor, Johor Bahru, Johor, Malaysia

Z. Ibrahim · M. Jumain
Centre for River and Coastal Engineering, Universiti Teknologi Malaysia, UTM Johor Bahru, 81310 Johor, Malaysia

N. M. Rusli
Research Institute for Sustainable Environment, Universiti Teknologi Malaysia, UTM Johor Bahru, 81310 Johor, Malaysia

N. Sabtu (ed.), *Proceedings of AWAM International Conference on Civil Engineering 2022 - Volume 3*, Lecture Notes in Civil Engineering 386,
https://doi.org/10.1007/978-981-99-6026-2_44

1 Introduction

Flash floods which are natural disasters may occur within an extremely short period of time and without any warning. Occurrence of flash floods can be considered increasing due to several contributing factors such as the inefficient existing drainage system coupled with inadequate provision of maintenance of the constructed infrastructures [1–3]. The impact of the urban flood is severe and can be life-treating. A study in Dehradun, India found that urban flood promoted negative impacts to the pollution to the riverbed, vulnerable urban services, sensitive building materials and construction, socio-economic vulnerability, old buildings and older building construction, damage to infrastructure, life and property loss, severe traffic jam and health issue due to spilling garbage [4]. The causes of urban flood depend on the location of the urban area. For instance, an urban area situated in flood plain area is more vulnerable to flood risk than a higher elevation area.

In Malaysian urban areas such as Kuala Lumpur and Penang, inefficient urban stormwater drainage has been highlighted as one of the causes of flash floods. The drainage systems in some housing areas that have been in place for 10–20 years with no improvement and maintenance will increase the possibility of more severe flooding which could lead to more property damage and loss of lives [5, 6]. Additionally, disturbance to the road traffic flows due to flash flood water inundation has become major problem to the urban communities particularly during peak working hours [7, 8]. Due to increase of urbanisation activities, the impervious surface such as parking lots, roads, buildings, and compacted soil do not allow rain to infiltrate into the ground and more runoff is generated [9].

Meanwhile, post-storm flash floods are becoming more frequent events in the urbanised areas in Malaysia which are probably due to the global warming effect especially climate change. The Best Management Practices (BMP) for Sustainable Urban Drainage System (SUDS) such as On-Site Detention (OSD), permeable pavements, biofiltration swales, bioretention basins, and ponds for stormwater quantity and quality control were introduced in the Urban Stormwater Management Manual (MSMA) for Malaysia since 2001 [10]. The Sustainable Urban Drainage Systems (SUDS) are recommended techniques towards solving flash flood, water scarcity and water pollution problems. The SUDS have been adopted as a component of integrated urban water management which to ensure that future developments are implemented using an integrated urban water management approach. Nevertheless, the conventional precast concrete drains are still favourable in construction industry due to financial causes in Malaysia. The concept of infiltration or porous surface is one of the approaches being recognised to reduce surface runoff volume.

The interaction between infiltration concept and porous sublayer modules for reducing surface runoff in a drainage system remains unexplored. Various characteristics of porous sublayer materials and its influence are not yet fully known. Thus, an experimental study on the effectiveness of integrating the infiltration concept into a drainage system was carried out in the School of Civil Engineering, Universiti Teknologi Malaysia (UTM). It is expected that an effective infiltrated precast concrete drainage system which can minimise flooding problem and be accepted as a practical approach in local construction industry. This research is a continuation to works discussed in [11].

The aim of the study was to analyse the experiments conducted on the effectiveness of introducing the infiltration concept into the drainage system to reduce runoff in a drainage system. This study focussed on the application of aggregate or porous sublayer module length to the efficiency of the drainage system. Thus, the objectives of research were to investigate the effectiveness of different porous sublayer module lengths on the infiltration rate, flow depth and velocity profiles in the drainage system.

2 Methodology

The first stage of this research was the planning and preparation of experimental setup which consisted of (i) inlet or inflow water tank, (ii) adjustable slope steel supporting structure (or testing rig) for precast concrete drain testing, and (iii) water flow collection systems located at downstream and underneath the testing rig. The experiments on capability of open drainage systems to convey stormwater were conducted in a 4 m long and 0.2 m wide drainage model constructed in the Hydraulic Laboratory in Universiti Teknologi Malaysia (UTM). Figure 1 illustrates the laboratory experimental research setup with testing rig, drain section and porous sublayer modules.

The performance of the drains with four different gaps between drain sections were evaluated based on percentage flow distribution. Selected drain longitudinal or

Fig. 1 Laboratory experimental research setup with testing rig, drain section and porous sublayer module

bed slopes (So) of 1/300 and 1/600 were studied which represented low and high drainage bed gradients based on actual condition in construction site practiced in civil engineering. This experimental study covered the impact of different porous sublayer module lengths to drain diameter ratio, (Ls/d) on the infiltration rate, drain flow depth and velocity in a half-circular drain system with 6-in. (15.2 cm), 9-in. (22.9 cm) and 12-in. (30.5 cm) long porous sublayer modules. The thickness of the porous modular was fixed at 3 in. (7.6 cm).

Main parameters measured were inflow and outflow rates in drains, infiltration flow rate in underneath flow collector, water surface profiles (or hydraulic gradients) and streamwise velocity along the drains. Among the equipment deployed in the experimental work were land survey total station, digital point gauge, and miniature current meter. All the flow rates were determined using volumetric approach. This paper focused on the influence of various porous sublayer module lengths to drain diameter ratio, (Ls/d) on the infiltration rate, flow depth as well as velocity in the drainage system.

3 Results and Discussion

3.1 Flowrates Distribution

The main parameter studied in this research was the infiltration capability with the presence porous sublayer modules in the drainage system. Figures 2 and 3 present the ratio of infiltrated flow from drain to the total flow in the drainage system, ($Q_{\text{inf}}/Q_{\text{total}}$) for drain bed slopes of 1/300 and 1/600, respectively. In general, the infiltration rates increased as the spacing between drains (s/d) increased. Detailed discussion on this result can be found in [11].

The study found that the highest infiltration rate occurred at (Ls/d) of 0.7. Besides, the highest infiltration rates for *Ls/d* of 0.7, 1.0 and 1.4 at the largest spacing (s/d) of 11.1% were up to 98 to 99% of the total flow rate in the drainage system. Meanwhile, the optimum (Ls/d) for 0.7 and 1.4 were at (s/d) of 4.4% since they allowed more than 75% of surface runoff to infiltrate into the lower layer for storage, as illustrated in MSMA [12] for On-Site Detention (OSD). The optimum drain spacing (s/d) for (Ls/d) of 1.0 was at 6.7%. The lowest infiltration rate for (Ls/d) of 1.0 was at (s/d) of 4.4%. It was observed that the results were almost similar for both drain bed slopes of 1/300 and 1/600 (see Figs. 2 and 3).

3.2 Drain Flow Depth

The flow depth in the drain was also studied. The study on flow depth profiles on both drain bed slopes showed that the lowest flow depth distribution. Figure 4 displays

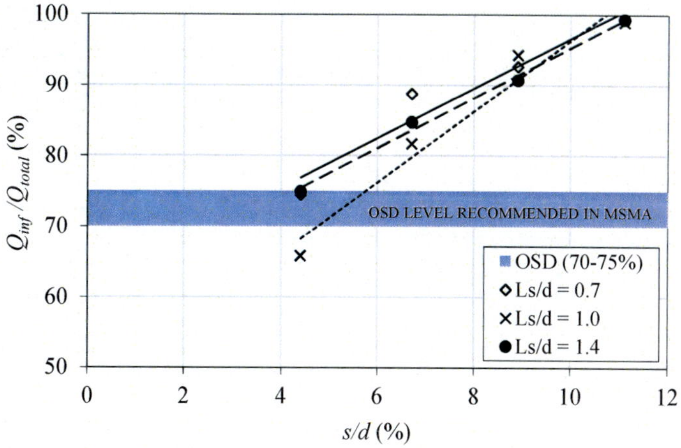

Fig. 2 ($Q_{\text{inf}}/Q_{\text{total}}$) for different drain spacing (s/d) and sublayer module lengths (Ls/d) on drain bed slope of 1/300

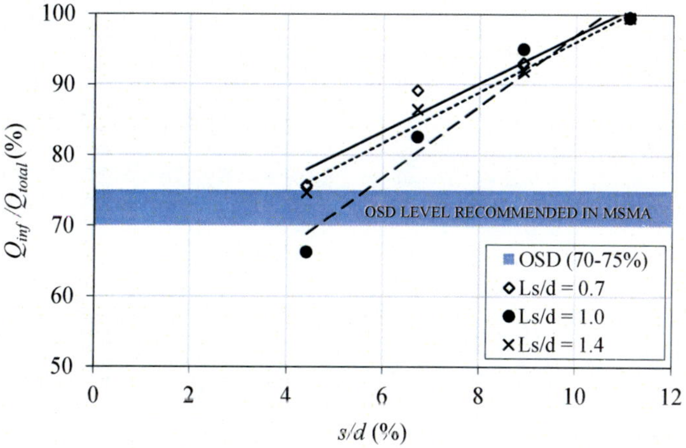

Fig. 3 ($Q_{\text{inf}}/Q_{\text{total}}$) for different drain spacing (s/d) and sublayer module lengths (Ls/d) on drain bed slope of 1/600

the flow depth (H/d) for various (s/d) on bed slope of 1/300. The lowest flow depth distribution resulted at (Ls/d) *of* 0.7, while (Ls/d) of 1.4 exhibited the highest drain flow depth distribution. The lowest flow depth in the drainage system was identified at largest drain spacing (s/d) of 11.1% took place at the most downstream point, (x/L) of 0.83 in which the resulted flow depth (H/d) was about 0.30.

The sublayers in the drain were able of reducing the flow depth along the drain. The quantitative measure indicated that the flow depth along the drain reduced up to 67, 72 and 60% of the flow depth at upstream section of the drain for (Ls/d) of 0.7, 1.0 and 1.4, respectively. The trendline for (Ls/d) of 1.0 showed a steep profiles

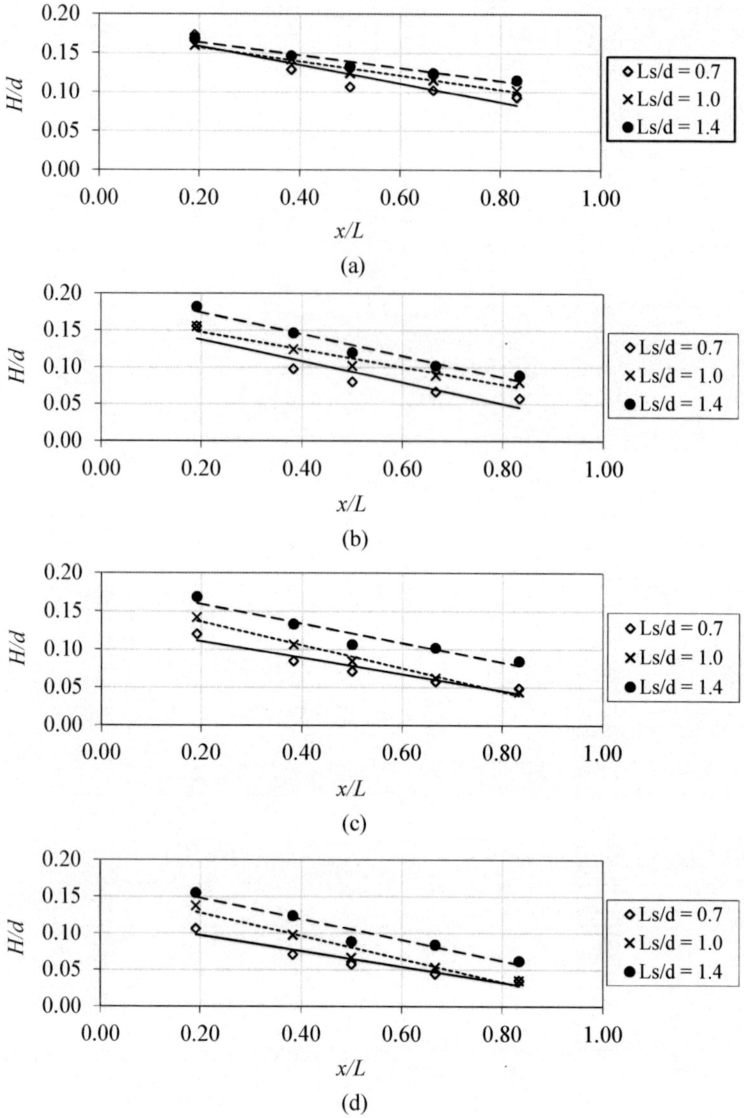

Fig. 4 Flow depth (H/d) profiles for various (s/d) on bed slope of 1/300 at s/d of **a** 4.4%, **b** 6.7%, **c** 8.9% and **d** 11.1%

of flow depth for large (s/d) in Fig. 4c and d compared the other sublayer lengths (Ls/d). The results for all drain spacing (s/d) indicated that (Ls/d) of 0.7 was the most effective sublayer length in reducing the flow depth along the drainage system.

3.3 Drain Velocity

The velocity distributions along the drainage system were analysed and plotted. The measured velocity in drain section with spacing (U) was normalised by the measured initial flow velocity in drain (U_o). Figure 5 illustrates the profiles of normalised (U/U_o) along the drainage system with bed slope of 1/300 and various drain spacing ratios. In general, the velocity in the drain reduced as the drain spacing (s/d) became larger. It was noticed that the (U/U_o) increased from upstream to downstream of the drainage system due to the lowering of flow depths at downstream sections, as illustrated in the Sect. 3.2 earlier. For (s/d) of 4.4, 6.7 and 8.9%, velocity profile at largest sublayer length (Ls/d) of 1.4 was the lowest compared to other (Ls/d). In contrast, the velocity distribution for (Ls/d) of 1.4 in Fig. 4d became higher than the other two (Ls/d) at the largest drain spacing (s/d) of 11.1%. However, it was noticed that only a small velocity difference between difference (Ls/d) for this case.

4 Conclusion

The application of aggregate or porous sublayer modules in a sustainable drainage system was experimentally investigated by comparing the effectiveness of infiltration approach in the drainage system. The findings of the study are (i) the optimum (Ls/d) for 0.7 and 1.4 were at (s/d) of 4.4% since they allowed more than 75% of surface runoff to infiltrate into the lower layer for storage, as illustrated in in MSMA (DID, 2012) for On-Site Detention (OSD). (ii) The sublayers in the drain were able of reducing the flow depth along the drain. The results for all drain spacing (s/d) indicated that (Ls/d) of 0.7 was the most effective sublayer length in reducing the flow depth along the drainage system. (iii) The flow velocity in the drain reduced by more than 80% with the application of porous sublayers. It was found that the spacing between drain sections (s/d) was important factor to reduce the velocity in the drainage system. The drain velocity increases as the flows further downstream and reduces as the drain spacing becomes larger.

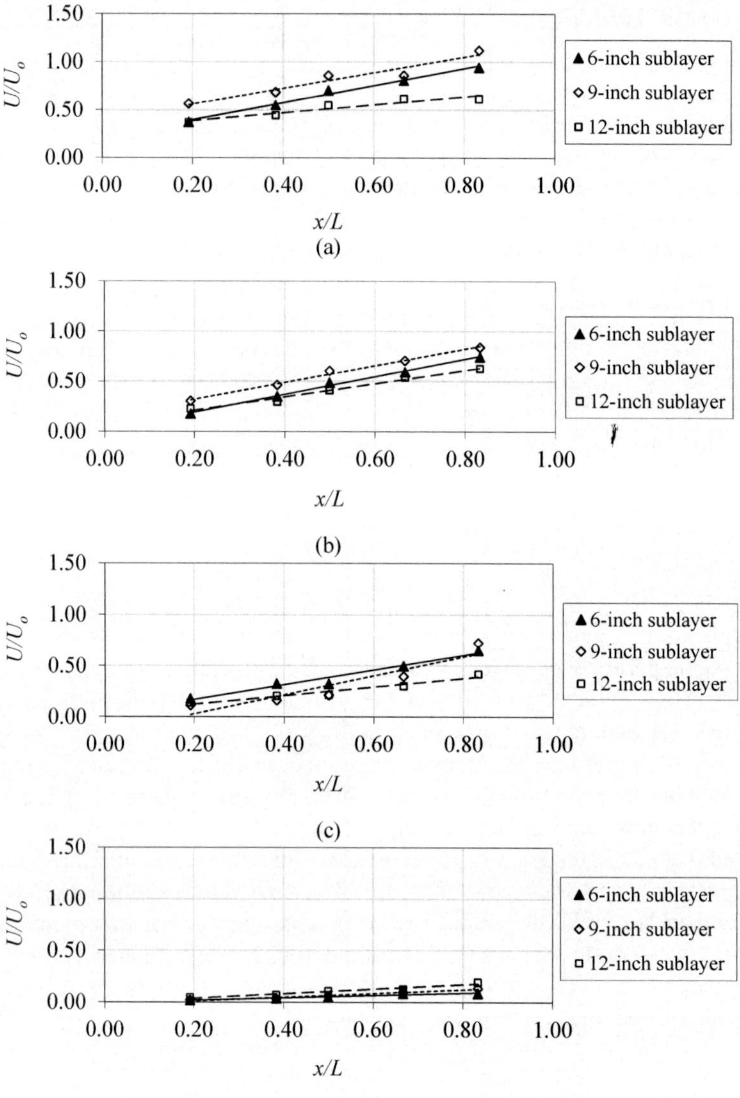

Fig. 5 (U/Uo) profiles in the drainage system for various (s/d) on bed slope of 1/300 at s/d of **a** 4.4%, **b** 6.7%, **c** 8.9% and **d** 11.1%

Acknowledgements The financial of this research were supported by Universiti Teknologi Malaysia (UTM) and Ministry of Higher Education (MOHE) under vote numbers 14J57 and 15J61. The authors wish to express a sincere appreciation to others who were involved either directly or indirectly during the implementation of this experimental research.

References

1. Desa, M.N.M., Lariyah, M.S., Ahmad Husaini, S., Leong, T.M., Anita, A.: Current trend in integrated urban stormwater management practice in the tropical region Malaysia. In: 12th International Conference on Urban Drainage, Porto Alegre/Brazil, 11–16 Sept 2011
2. Thorndahl, S., Willems, P.: Probabilistic modeling of overflow, surcharge and flooding in urban drainage using the 1st order reliability method and parameterization of local rain series. Water Res. **42**(1–2), 455–466 (2008)
3. Sani, G.D., Barzani, G.M., Ekhwan, T.M., Musa, G.A.: Floods in Malaysia historical reviews, causes, effects and mitigations approach. International Journal of Interdisciplinary Research and Innovations **2**(4), 59–65 (2014)
4. Bansal, N., Mukherjee, M., Gairola, A.: Causes and impact of urban flooding in Dehradun. Rorkee: International Journal of Current Research (2015)
5. Liew, Y.S., Selamat, Z., Ghani, A.A., Zakaria, N.A.: Performance of a dry detention pond: case study of Kota Damansara. Selangor, Malaysia, Urban Water Journal **9**(2), 129–136 (2012)
6. Doocy, S., Daniels, A., Murray, S., Kirsch, T.D.: The human impact of floods: a historical review of events 1980–2009 and systematic literature review. April 16. Review PLOS Currents: Disasters (2013)
7. Abdullah, N.I.: Clogged drains to blame for flash floods, 8 Oct 2018. https://m.malaysiakini.com
8. Muthiah, W.: Morning downpour causes flash floods, traffic jams in Klang and Shah Alam, 7 Nov 2017. https://www.thestar.com.my/news
9. Hoyer, J., Dickhaut, W., Kronawitter, L., Weber, B.: Water sensitive urban design principles and inspiration for sustainable stormwater management in the city of the future. Manual, Integrated Project Global Change and Ecosystems (2011)
10. Beecham, S.: Water sensitive urban design (Chap. 23). In: Daniels, C.B. (ed.) Adelaide: Water of a City. Wakefield Press (2010)
11. Ibrahim, Z., Fadzil, A.B., Demun, A.S., Jumain, M., Makhtar, M.R., Alias, N., Rusli, N.M., Baseng, F.: Infiltrated sustainable urban drainage system for storm water control. Water Management and Sustainability in Asia **23**, 53–64 (2021)
12. Department of Irrigation and Drainage, Malaysia. Urban Stormwater Management Manual for Malaysia (MSMA) (2nd ed.) (2012)

Design of Individual On-Site Storage Pond for Supplementary Irrigation During Dry Period in Malaysia

Nadiatul Amira Ab Ghani, Puay How Tion, Nor Azazi Zakaria, and Nasehir Khan E. M. Yahaya

Abstract Paddy and water are strongly connected to each other. In recent years, climate change has increased drought frequency, resulting in water scarcity in paddy sector in Malaysia. Therefore, in this study, a methodology was developed to design an individual on-site storage pond for supplementary irrigation purpose. The storage pond collects return flow from the paddy field for reuse purpose during dry season. First, the inflow discharge into the paddy field was estimated from field data. Field data was collected in both the main and off seasons. A simplified water balance model was then developed to estimate the amount of return flow from the paddy field, by utilizing the inflow discharge estimated from the field data. The return flow was used to design the individual on-site storage pond and both the inlet and outlet flow control structures. Results from this study showed that a simple methodology can be applied to design an individual on-site storage pond for water reuse purpose. This simple methodology can serve as a design guideline.

Keywords Individual on-site Storage Pond · Supplementary Irrigation · Paddy Field

1 Introduction

Paddy cultivation is highly dependent on water. Too much or too little water will affect the paddy production. In recent years, climate change has caused water scarcity. Water scarcity poses a serious problem to the agricultural industry, especially to

N. A. A. Ghani · P. H. Tion (✉) · N. A. Zakaria
River Engineering and Urban Drainage Research Centre (REDAC), Universiti Sains Malaysia, Nibong Tebal, Pulau Pinang, Malaysia
e-mail: redac_puay@usm.my

N. K. E. M. Yahaya
National Hydraulic Research Institute of Malaysia (NAHRIM), 43300 Seri Kembangan, Selangor, Malaysia

N. Sabtu (ed.), *Proceedings of AWAM International Conference on Civil Engineering 2022 - Volume 3*, Lecture Notes in Civil Engineering 386,
https://doi.org/10.1007/978-981-99-6026-2_45

paddy cultivation, which uses a large amount of water compared to other crops [1]. It is estimated that 1 kg of paddy requires 2–3 m^3 of water [2].

Climate change causes severe drought which directly reduces the amount of available water in the water catchment area, and directly affects water supply. At the same time, the increase in population will push the domestic water demand upward. With competition from the domestic demand, the water security in the agricultural sector is affected. The paddy cultivation is especially affected by the competition from domestic water demand and vulnerable to the climate change effect. This is because data in the 2000s shows that 90% out of the 70% of water withdrawal for the agricultural industry is used for paddy cultivation [3].

In general, there are two methods of paddy planting, that is, the transplanting method and direct seeding method [4]. In transplanting method, seedlings are first raised in seedbeds before they are transferred into the field. In the direct seeding, seeds are broadcasted directly in the field by hand. If the seeds are broadcasted on dry soil, the process is referred to as the dry direct seeding. On the contrary, if the seeds are broadcasted onto shallow standing water, the process is referred to as the wet direct seeding. In the wet direct seeding method, field preparation starts with first flooding to level the soil. Second flooding is carried out to saturate the soil for seed broadcasting [5]. The volume of water necessary for this stage is assumed to be 200 mm [5]. Third flooding is carried out to keep the water level in the field around 20–50 mm during the germination and tillering period [6].

In Malaysia, the gravity-based irrigation system is widely practiced in paddy granary [7]. Water is transferred from the main reservoir to the paddy granary through a series of irrigation network by potential energy. Since the irrigation system operates based on gravity flow, the water level in the irrigation channel must be maintained above the minimum operating level to ensure optimum performance, regardless of the amount of water demand from the paddy field in the granary. Thus, such condition reduces the efficiency of the irrigation system [8]. Azmi et al., [9] showed that by the 2000s, 90% of paddy cultivation area in Malaysia is using the direct-seeding method for planting. Although direct seeding reduces the laborious work in transplanting process and uses less water in the field, it sometimes burdens the irrigation system. In direct seeding method, after seeds are broadcasted, farmers must pay more attention to water level in each plot to prevent flood or dry soil condition [8]. This leads to non-uniformity in irrigation activity. Therefore, the water level in the irrigation channel must be kept at operational level for a longer time and this condition adds burden on the irrigation system.

Peninsular Malaysia is likely to experience higher risk of very dry condition in the future [10]. Therefore, farmers need to have irrigation alternatives to reduce total paddy loss during dry period. Individual on-site storage pond can be a viable alternative to provide supplementary irrigation and help to reduce loss of paddy harvest during dry period. Compared to the centralized storage pond, the individual on-site storage pond allows farmers to micro-manage supplementary irrigation during dry periods. Therefore, in this study, a methodology to design an individual on-site storage pond for supplementary irrigation purpose will be presented.

2 Methodology

An individual on-site storage pond to gather the excess water from paddy irrigation is proposed for reuse purpose. This irrigation excess water is also referred to as the return flow. The proposed storage pond is individual type as the storage pond is constructed inside the paddy field plot. Individual storage pond will allow farmer or owner of the field plot to micro-manage supplementary irrigation during dry period.

During paddy field preparation, germination and tillering stages, the field will be flooded several times. Field flooding is accomplished by discharging water into the field through an inlet opening in the subsidiary channel (point 1 in Fig. 1). The purpose of the first flooding is to allow farmer to perform soil leveling. This is followed by the second flooding to saturate the soil and prepare the field for seeding. Wet direct seeding method is practiced at the study site. The paddy field is subsequently subjected to a third flooding to raise the water level during the paddy germination and tillering stages.

First, utilizing the water depth at the subsidiary irrigation channel, this study empirically estimates the input discharge into the paddy field (Fig. 1). Throughout the entire planting season, the water depth is manually measured on the spot. A simplified water balance model is then developed to estimate the return flow from the paddy field. It is assumed that the return flow is maximum when the soil has achieved saturated condition after the second flooding. Therefore, only the return flow from the third flooding is used to design the on-site storage pond. Additionally, it is assumed that the irrigation system is the only source of water supply for estimating the return flow. The return flow discharge is then used to design an on-site storage pond and its inlet and outlet control structures. Finally, flow routing is performed to check the overall design.

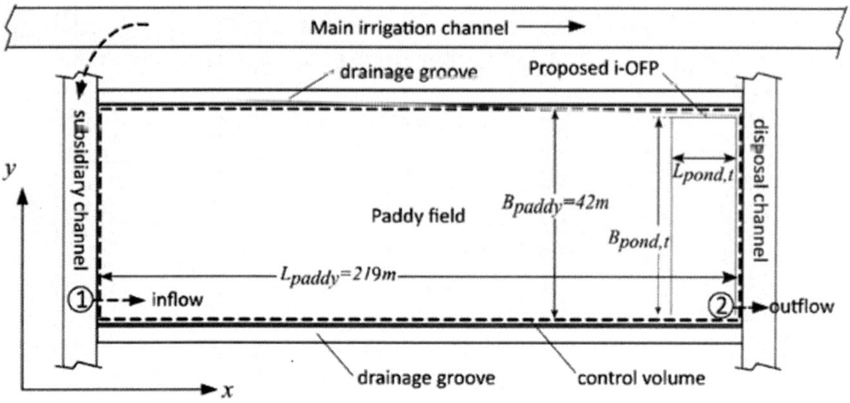

Fig. 1 Layout of paddy field plot at the study site

2.1 Study Site

The location chosen for this study is located at Kampung Sungai Pinang, Simpang Empat, Semanggol, Perak. The GPS coordinate for this site is 4°56′23.1″N, 100°35′49.6″E. This area is under the Kawasan Pembangunan Pertanian Bersepadu Kerian [Integrated Agriculture Area Kerian (IADA)] Sungai Manik granary. The IADA Kerian granary is established for more than 30 years. The study site is irrigated by the water from the Bukit Merah reservoir, which is situated about 10 km away from the study site. Paddy cultivation is carried out twice a year. The main season starts in September and ends in February, and the off season starts in March and ends in August. The area of the paddy field is roughly 9198 m^2. The total paddy field length, L_{paddy} is 219 m and its width, B_{paddy} is 42 m. The layout of the paddy field plot is shown in Fig. 1. The paddy field is irrigated by a gravity-based irrigation system. Water is supplied from the subsidiary channel to the paddy field through a rectangular opening at the subsidiary channel (point 1 in Fig. 1). Flow discharge into the paddy field is controlled by a sluice gate, which is installed at the rectangular opening. To drain the water out from the paddy field, an overflow pipe is installed at the other end of the paddy field (point 2 in Fig. 1). The overflow pipe level can be adjusted to control the water depth in the paddy field.

2.2 Water Balance Model

Inflow water into the paddy field comes from two sources: the irrigation water from the subsidiary channel and the rainfall. The outflow from the paddy field is consisted of three components: outflow or return flow into the disposal channel, water infiltrated through the soil surface and percolated into ground, and water evaporated into the atmosphere through the evapotranspiration process. The evapotranspiration process is a combination of the evaporation from the soil and vegetation surface and transpiration from the paddy plant.

In this study, the rainfall contribution is not considered for the design of the storage pond. This is because during dry period, the amount of rainfall is low and negligible, and the paddy depends entirely on the irrigation system. As a result, it is essential to store the paddy field's return flow. Therefore, the estimation of return flow is required to design the on-site storage pond.

The water balance model for the steady-state flow condition in the paddy field can be represented by Eq. (1) by applying the law of conservation of mass, and assuming the paddy field plot as a control volume.

$$Q_{\text{inflow}} = fA + E_T A + Q_{\text{return}} \tag{1}$$

The term on the left-hand side of Eq. (1) consists of the inflow component, Q_{inflow}, which is the inflow discharge into the paddy field. The term on the right-hand side are the outflow components: Q_{return} is the return flow from the paddy field, f is the infiltration rate through the soil surface and E_T is the field evapotranspiration rate. A is the arable surface area. f and E_T have the same dimension of $[L/T]$ and both Q_{inflow} and Q_{return} have the dimension of $[L^3/T]$. Under the assumption of steady flow condition, Q_{inflow}, Q_{return}, E_T and f are the averaged value over the irrigation period.

2.3 *Inflow Discharge into the Paddy Field*

The sluice gate in the subsidiary channel (Fig. 2) is pulled up to allow water to discharge into the paddy field. The inflow discharge through the sluice gate $Q_{\text{sluice}}(t)$ can be estimated by using the sluice gate equation (Eq. 2) [11].

$$Q_{\text{sluice}}(t) = \frac{C_c}{\sqrt{k}} W B \sqrt{2g(h(t) - h_1(t))} = \frac{C_c}{\sqrt{k}} W B \sqrt{2g(h - C_c W)} \tag{2}$$

Here, $h(t)$ is the water depth behind the sluice gate in the subsidiary channel, which is manually measured during the main and off seasons. W and B are the height and width of the sluice opening, respectively. W, B and $h(t)$ have the same dimension of $[L]$. C_c is the contraction coefficient, and its value depends on $W/h(t)$ ratio. According to the Clemmens et al. [12], the value of k is around 1.01 – 1.02 for

Fig. 2 Sluice gate at the subsidiary channel

field scale gate. For this study, the value of $k = 1.01$ is used. The value of $C_c = 0.62$ is taken from Belaud et al. [11] using $a = W/h = 2.5\text{cm}/58.9\text{cm} = 0.042$. Since the downstream of the gate on site is not flooded, $h_1(t) = 0$. Therefore, Eq. (2) can be simplified to the form in Eq. (3).

$$Q_{\text{sluice}}(t) = \frac{0.62}{\sqrt{1.01}} WB\sqrt{2gh(t)} \approx 0.62WB\sqrt{2gh(t)} \tag{3}$$

The time-averaged of $Q_{\text{sluice}}(t)$ over the entire flooding period is used to approximate Q_{inflow} as follows,

$$Q_{\text{inflow}} = \frac{1}{\Delta t}\int Q_{\text{sluice}}(t)\text{d}t \tag{4}$$

In this study, two flooding periods are being considered, that is: the third flooding period in both the main and off seasons. The main season is from the 1st until 20th December 2019 and the off season is from the 13th until 27th June 2020.

2.4 Maximum Return Flow Volume

The maximum return flow discharge occurs when water loss through evapotranspiration and infiltration process is minimum during the irrigation period. The average maximum return flow rate $Q_{\text{return,max}}$ can be calculated by using Q_{inflow}, as shown in Eq. (5).

$$Q_{\text{return,max}} = Q_{\text{inflow}} - f_{\min}A - E_{\text{Tmin}}A \tag{5}$$

Here, $f_{\min}$ and $ET_{\min}$ are the minimum soil infiltration rate and minimum evapotranspiration rate, respectively. $f_{\min}$ and $ET_{\min}$ are assumed to be constant during the irrigation period. The annual crop evapotranspiration rate at the IADA Kerian granary is in range of 3.5–4.76 mm/day [13]. Therefore, the value of $ET_{\min} = 3.5$ mm/day is used. Based on the information obtained from the Kerian IADA management office, the soil is clayey type at the site. In saturated condition, the minimum infiltration rate for clayey soil is 1 mm/h [14]. Therefore, $f_{\min} = 1$ mm/h is used in this study.

The average maximum return flow volume $\overline{V}_{\text{return,max}}$ is evaluated by integrating $Q_{return,max}$ with time as follows,

$$\overline{V}_{\text{return,max}} = Q_{\text{return,max}}\Delta t \tag{6}$$

Consequently, the amount of water used per unit area v_{paddy} during the flooding period can be estimated as follows,

$$v_{\text{paddy}} = \frac{1}{A}\left(Q_{\text{inflow}} - Q_{\text{return,max}}\right)\Delta t \tag{7}$$

The dimension for v_{paddy} is $[L]$.

3 Pond Sizing and Flow Control Structure

3.1 Pond Sizing

The individual on-site storage pond is designed with two control structures, that is, the inlet and outlet structures. The storage pond is constructed at the downstream side of the paddy field (Fig. 1) to collect the return flow from the paddy field. The excess flow from the storage pond is discharged into the disposal channel. The water in the storage pond will be used for supplementary irrigation during the dry period.

For first trial design, the length of the storage pond at the top surface $L_{\text{pond},t}$ is set to 8.76 m. This length is 4% of L_{paddy}. The width of the storage pond at the top surface $B_{\text{pond},t}$ is set to 33.7 m. This length is slightly smaller than B_{paddy}. Figure 3 shows the longitudinal cross section of the proposed storage pond. The top surface area of the storage pond $A_{\text{pond},t}$ is about 295 m^2 which is equivalent to approximately 3.2% of the total area paddy field. The slope of the storage pond side wall is designed as 1:2 (V:H) with fencing. The slope and fencing requirement are based on the recommendation by Urban Stormwater Management Manual for Malaysia (MSMA) [15] which requires that for any pond having a side wall steepness of more than 1:4, a fence or rail is required for safety purpose. Since the pond is fenced, the permanent storage pond depth can be equal or more than 1.5 m. Therefore, in this study, the permanent water depth h_{pond} is designed as 1.5 m and the maximum water depth $h_{\text{pond,max}}$ is designed as 1.9 m. The maximum pond depth is designed to cater extreme rainstorm event. Based on this dimension, the permanent pond storage volume v_{pond} is 385 m^3.

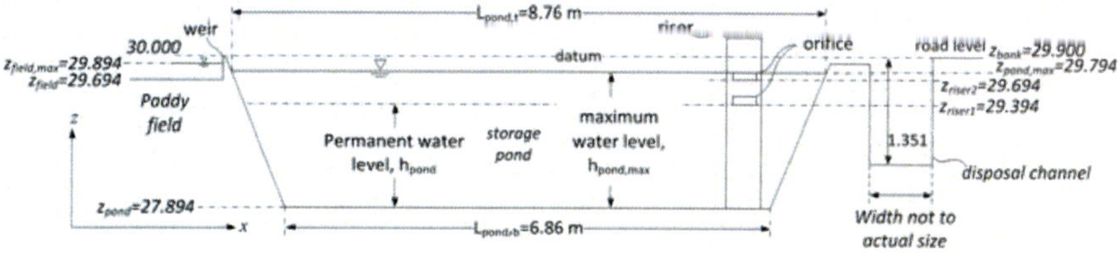

Fig. 3 Longitudinal cross section of the storage pond with design water and orifice levels

3.2 Flow Control Structure

A weir with a sharp crested edge is designed to alter and control the return flow into the storage pond. The weir can be adjusted to control the ponding depth in the paddy field. The maximum ponding depth of 20 cm is used, based on the average reading taken from the study site during the third flooding period. From Fig. 3, the return flow passes through the weir and discharges into the storage pond. Once the water in the storage pond reaches the maximum depth, it will be discharged into the disposal channel through the orifice placed on a riser structure, as shown in Fig. 3. The riser consists of two square orifices (0.3 m width by 0.1 m height). The orifices are placed 0.2 m vertically apart. The lower orifice is designed to discharge excess flow from the storage pond to maintain the water level in the storage pond at the design permanent water level. The upper orifice is designed to increase the disposal rate when combined with the lower orifice. Higher disposal rate is required to discharge flow during extreme storm event and prevent overflow from the storage pond. It is important to discharge the excess flow from the storage pond into the receiving water body to recharge the ground water. The designed water level in the paddy field and the storage ponds are shown in Fig. 3 with land survey datum (LSD) assumed as 30.000 m.

3.3 Flow Routing

Flow routing is performed for the paddy field and storage pond, based on indication method [16], to check the overall design and regulate the outflow discharge from the storage pond. The routing process is performed by using the inflow discharge during the third flooding period in the off season (from 13th June 2020 until 27th June 2020). The third flooding period in the off season is chosen for flow routing because the inflow discharge during this period is higher than in the main season, as shown in Table 1. In this study, routing cycle starts when water level in the paddy field reaches the design ponding depth of 20 cm. The inflow discharge for storage pond routing $Q_{\text{routing}}(t)$ is generated by using $Q_{\text{sluice}}(t)$ under the condition of minimum infiltration and evapotranspiration rate, as shown in Eq. (8).

$$Q_{\text{routing}}(t) = Q_{\text{sluice}}(t) - f_{\min} A - ET_{\min} A \tag{8}$$

Therefore, $Q_{\text{routing}}(t)$ also represents the return flow from the paddy field.

Table 1 Calculation of inflow discharge and maximum return flow discharge

Description	Third flooding in the main season	Third flooding in the off season
Irrigation period	1st–20th Dec 2019 (20 days)	13th–27th June 2020 (15 days)
Paddy field width, B_{paddy}	42 m	
Paddy field length, L_{paddy}	219 m	
Area, $A = L_{paddy} B_{paddy}$	9198 m^2	
Minimum infiltration rate, f_{min}	1 mm/h	
Minimum evaporation rate, E_{Tmin}	3.5 mm/day	
Average inflow discharge, Q_{inflow} (Eq. 4)	0.00370 m^3/s	0.00496 m^3/s
Average maximum return flow $Q_{return,max}$ (Eq. 5)	0.0007883 m^3/s	0.00205 m^3/s
Average total return flow volume $\overline{V}_{return,max}$ (Eq. 6)	1362 m^3	2650 m^3
Total volume of water used in paddy field per area v_{paddy} (Eq. 7)	0.547 m	m

4 Result

4.1 Flow Depth Behind the Sluice Gate

The flow depth behind the sluice gate h in the subsidiary channel observed during the third flooding period in both the main and off seasons are shown in Fig. 4. Zero depth means no reading was taken at site. The h value is used to calculate Q_{sluice} and subsequently Q_{inflow}. The result for Q_{sluice} is shown in Fig. 5.

4.2 Result of Analysis

During the third flooding period in the main and off season, the flow depth behind the gate was observed and used to calculate Q_{sluice}. The value of Q_{sluice} is calculated based on Eq. (3) and used to calculate Q_{inflow}. The Q_{inflow} for the third flooding in the main and off season is 0.00370 m^3/s and 0.00496 m^3/s respectively as shown in Table 1. $Q_{return,max}$ for the third flooding period in the main and off season are 0.0007883 m^3/s and 0.00205 m^3/s which is about 21% and 41% of Q_{inflow} respectively. This figure reveals that return flow can be collected and can be restored for reuse purpose. Thus, wastage of irrigation water can be reduced. $\overline{V}_{return,max}$ during the third flooding in the main season (20 days duration) is 1362 m^3, which is calculated using the Eq. (6). For the third flooding in the off season (15 days duration) $\overline{V}_{return,max}$ is 2650 m^3. The

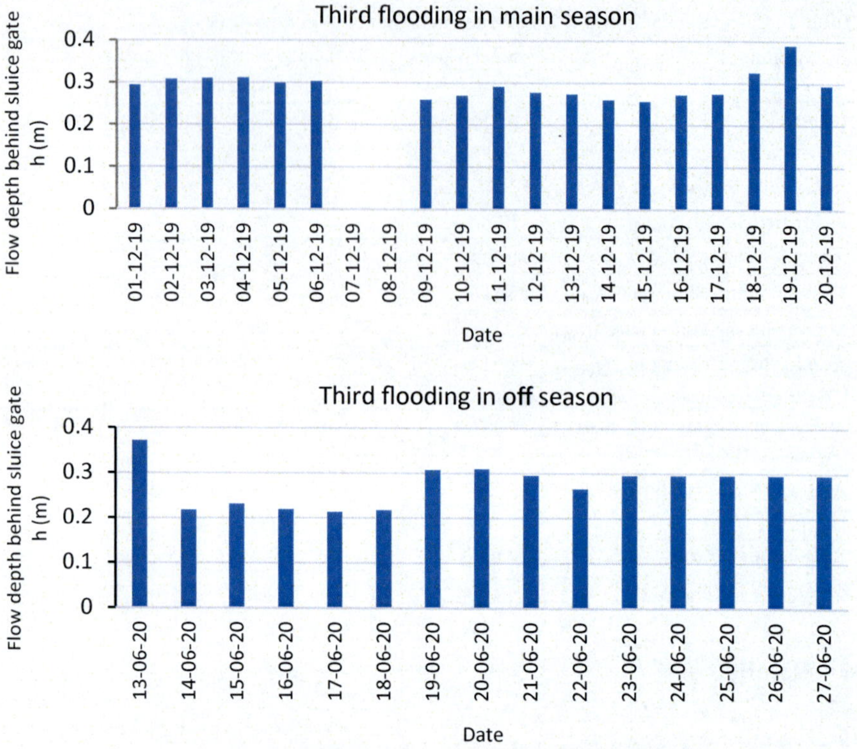

Fig. 4 Flow depth behind the sluice gate in the subsidiary channel during the main and off season

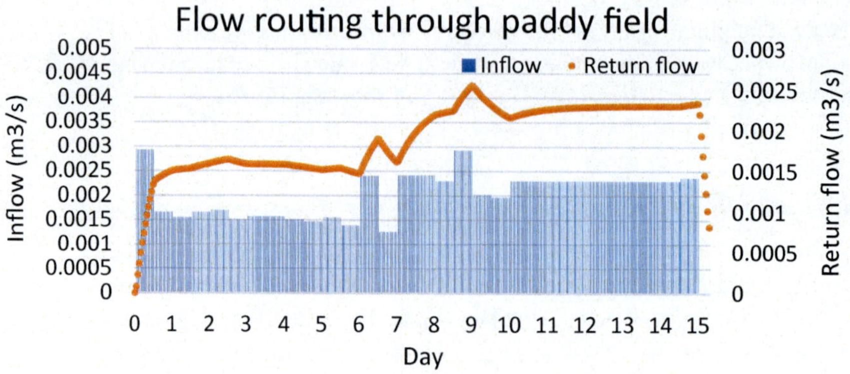

Fig. 5 Inflow and outflow hydrograph in the paddy field

designed storage pond has a permanent storage capacity v_{pond} of 385 m^3. Therefore, about 27% and 14% of $\overline{V}_{\text{return,max}}$ is captured and stored in the designed storage pond during the third flooding in the main and off season, respectively. In addition, by using Eq. (7), v_{paddy} for the third flooding period both seasons is calculated as 0.547 m and 0.410 m, respectively.

4.3 *Result of Flow Routing*

The return flow hydrograph generated from the flow routing through the paddy field is shown in Fig. 5. It is evident that the variability of the inflow discharge reflects the variability of the return flow. In the beginning, under steady inflow discharge, the return flow rises gradually. However, the return flow dropped suddenly when the inflow discharge abruptly drops at day 15. The hydrograph of the return flow is used as the inflow for storage pond flow routing.

Figure 6 shows the hydrographs of the return flow from the paddy field (as inflow to the storage pond) and outflow from storage pond. The outflow hydrograph shows that it takes about 2.4 days for the storage pond to be filled up to the design permanent water level h_{pond}. The storage pond filling process is represented by the flat section at the initial stage of the outflow hydrograph. Excess flow from the storage pond occurs when the water level in the storage pond is higher than h_{pond} and this is represented by the rising limb of the outflow hydrograph. This occurs at around $t = 2.4$ days.

By integrating the outflow hydrograph using trapezoidal method [17] in Fig. 6 with time, the total excess flow from the storage pond is estimated to be 2124 m^3. The total excess flow volume can also be estimated by subtracting v_{pond} from $\overline{V}_{\text{return,max}}$, which yield 2262 m^3 (2650 m^3–385 m^3). This value is 6.5% more than the value estimated by integration method (2124 m^3). The small difference in the estimation

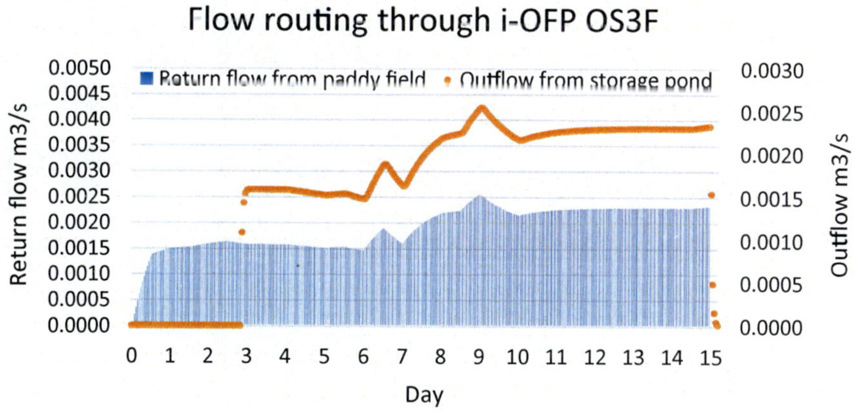

Fig. 6 Inflow and outflow hydrograph in the storage pond

of the total excess flow volume, thus, validates the flow routing process in paddy field and storage pond.

5 Conclusion

This research presents a simple methodology to design an individual on-site storage pond for reuse purpose during dry period. The water depth behind the sluice gate in the subsidiary channel was measured for two successive planting seasons, and the results were utilized to estimate the inflow discharge into the paddy field. By using the water balance model the return flow for both the main and off seasons were estimated. An individual on-site storage pond was suggested at the downstream of the paddy field to capture and store the return flow. Following that, flow routing through the paddy field and the storage pond was performed to validate the design and estimate the outflow discharge. Even though the study is site specific, the simple methodology is useful to design an on-site storage pond.

References

1. Haruna, S.N., Hanafiah, M.M.: Consumptive use of water by selected cash crops in Malaysia. Malaysian J. Sustain. Agric. **1**(2), 06–08 (2017)
2. Sharma, B.B., Kumar, N.: Iot-based intelligent irrigation system for paddy crop using an internet-controlled water pump. Int. J Agric. Environ. Inf. Syst **12**(1), 21–36 (2021)
3. Ahmed, F., Siwar, C., Begum, R.A.: Water resources in Malaysia: issues and challenges. J. Food, Agric. Environ **12**(2), 1100–1104 (2014)
4. Y. Singh et al.: Direct seeding of rice and weed management in the irrigated rice-wheat cropping system of the Indo-Gangetic Plains (2008)
5. Brouwer, C., Heibloem, M.: Irrigation water management: irrigation water needs in: water harvesting. Train Man No. 3 (1986)
6. JICA, DID: The Study on Modernization if Irrigation Water Management System in the Granary areas of Peninsular Malaysia, Draft Final Report, Volume-II, Annexes (1998).
7. Bockari-Gevao, S.M., Wan Ishak, W.I., Azmi, Y., Chan, C.W.: Analysis of energy consumption in lowland rice-based cropping system of Malaysia. Songklanakarin J. Sci. Technol. **27**(4), 819–826 (2005)
8. Toriman, M.E., Mokhtar, M.: Irrigation: types, sources and problems in Malaysia. In: Teang, S.L. (ed.) Irrigation Systems and Practices in Challenging Environments. InTech, Rijeka, Crotia (2012)
9. Azmi, M., Chin, D. V., Vongsaroj, P., Johnson, D.E.: Emerging issues in weed management of direct-seeded rice in Malaysia, Vietnam, and Thailand. In: Toriyama, K., Heong, K.L., Hardy, B. (eds.) Rice is Life: Scientific Perpectives for the 21st Century, pp. 196–198. Internal Rice Research Institute, Tsukuba, Japan (2005)
10. Wan Zin, W.Z., Jemain, A.A., Ibrahim, K.: Analysis of drought condition and risk in Peninsular Malaysia using standardised precipitation index. Theor. Appl. Climatol. **111**(3–4), 559–568 (2013)
11. Belaud, G., Cassan, L., Baume, J.: Calculation of contraction coefficient under sluice gates and application to discharge measurement. J. Hydraul. Eng. **135**(12), 1086–1091 (2009)

12. Clemmens, A.J., Strelkoff, T.S., Playán, E.: Field verification of two-dimensional surface irrigation model. J. Irrig. Drain. Eng. **129**(6), 402–411 (2003)
13. Hamidon, N., Harun, S., Malek, M.A., Ismail, T., Alias, A.: Prediction of paddy irrigation requirements by using statistical downscaling and CROPWAT models: a case study from the Kerian Irrigation Scheme in Malaysia. J. Teknol. **76**(1), 281–288 (2015)
14. Pitt, R., Lantrip, J.: Infiltration through disturbed urban soils. J. Water. Manag. Model. **104**, 1–10 (2000)
15. DID.: Urban Stormwater Management Manual for Malaysia, MSMA, 2nd edn. DID, Kuala Lumpur (2012)
16. Gordon, N.D., McMahon, T.A., Finlayson, B.L., Gippel, C.J., Nathan, R.J.: Stream Hydrology: An Introduction for Ecologists, 2nd edn. Wiley, England (2004)
17. Moin, P.: Fundamentals of Engineering Numerical Analysis. Cambrigde University Press, New York (2010)

Application of GETFLOWS and HEC-RAS in Assessing Sediment Balance Within River Estuary

Siti Nurwajihah Abu Bakar, Abdul Hakim Salleh, Mu'izzah Mansor, Mohd Aminur Rashid Mohd Amiruddin Arumugam, Mohd Remy Rozainy Mohd Arif Zainol, Safari Mat Desa, Nasehir Khan E.M Yahaya, and Fatehah Mohd Omar

Abstract River plays an important role in the human need as it provides water for human usage, irrigation, agriculture and industry as well as a range of other ecosystem services other than intrinsic and biodiversity values. Managing the river can lead to many benefits and convenience. However, due to lack of proper management, rivers can be easily polluted due to human activities. Sediment is one of the components that can damage the ecosystem and diversity of the river especially in local spots which involves soil erosion. Heavy rainstorms can cause an excessive erosion event,

S. N. Abu Bakar · A. H. Salleh · M. Mansor · M. R. R. Mohd Arif Zainol · F. Mohd Omar (✉)
School of Civil Engineering, Universiti Sains Malaysia, Engineering Campus, 14300 Nibong Tebal, Penang, Malaysia
e-mail: cefatehah@usm.my

S. N. Abu Bakar
e-mail: snurwajihah@gmail.com

A. H. Salleh
e-mail: abd.hakim@usm.my

M. Mansor
e-mail: muizzahmansor0@gmail.com

M. R. R. Mohd Arif Zainol
e-mail: ceremy@usm.my

M. A. R. Mohd Amiruddin Arumugam · S. Mat Desa
River Basin Research Centre, National Water Research Institute of Malaysia (NAHRIM), Lot 5377,Jalan Putra Permai, 43300 Seri Kembangan, Selangor, Malaysia
e-mail: rashid@nahrim.gov.my

M. R. R. Mohd Arif Zainol
River Engineering and Urban Drainage Research Centre (REDAC), Universiti Sains Malaysia, Engineering Campus, 14300 Nibong Tebal, Penang, Malaysia

N. K. E.M Yahaya
Water Quality Laboratory, National Water Research Institute of Malaysia (NAHRIM), Lot 5377,Jalan Putra Permai,, 43300 Seri Kembangan, Selangor, Malaysia
e-mail: nasehir@nahrim.gov.my

N. Sabtu (ed.), *Proceedings of AWAM International Conference on Civil Engineering 2022 - Volume 3*, Lecture Notes in Civil Engineering 386,
https://doi.org/10.1007/978-981-99-6026-2_46

however, most soil erosion happens gradually over time and is very hard to notice without constant monitoring. Furthermore, the sediment will be mobilized and transported along the river and eventually stored in the bottom of the river, but usually it will deposit near the estuary. A sediment modeling is needed to carter this problem as to predict the behavior of the sediment based on the hydrological components. The comparison between the 1D (HEC-RAS) and 3D (GETFLOWS) will be discussed in this paper to check the suitability and the validity of the model in sediment studies.

Keywords Sediment balance · GETFLOWS · HEC-RAS · Estuary · Sediment transport · Soil erosion · Clean water and sanitation

1 Introduction

The rapid urbanization is one of the factors has an adverse impact on the future land use and soil sustainability. Rapid development occurs when the number of populations increases along with agricultural production, construction as well as other anthropogenic activities [1, 2]. In rivers, erosion takes place as a natural process as the soil erosion typically caused by water which initiates the detachment process of soil particles via raindrops and flowing water and soil particles move downslope [3].

Soil erosion occurs because of the weathering of soil via water, especially when the soil is exposed in rainstorms events for a long time which will then carry sediment along the river [4]. Flooding will result from the transportation of the sediment from the surrounding activities along the river will be deposited at the bottom of the river where the depth of the river will be affected. Practically, forests result in less soil erosion than agricultural land use, whereas agricultural plantation area tends to hasten erosion, which worsens river water quality due to the high transportation of suspended sediment loads [5].

The GETFLOWS simulation model allows to analyze and estimate the river's current condition in three-dimensional (3D) models. River simulation in 3D models required input details of meteorological records, land cover, geology, and soil characteristic from secondary data collection [6]. The outcome of the simulation can predict and estimate the water cycle (surface water, groundwater and seawater) for the future by comparing the field monitoring data based on concentration of the suspended sediment, river erosion and deposition and also water flow rates.

The widely used software application to simulate sediment transportation is HEC-RAS. The computation of this software utilizes one dimensional (hydraulic property, cross section averaged from RAS's hydraulic engines to analyze sediment transport rates and revolution of channel geometry based on sediment continuity calculations) [7, 8]. This enables the calculation for river aggregation or erosion, temporal entrainment, transport and deposition of sediment and alteration of the cross sections.

2 Source, Transport and Storage of Sediment

Sediments originates from anthropogenic activities and natural processes. These two factors that are human conditioning or nature can contribute to soil erosion. Centuries ago, soil erosion triggered by the improper land use practices has been regarded as a global threat to soil sustainability and food security [38, 39] and major environmental and agricultural sector [40]. Ouillon [41] discussed the sediment mobility process includes the process of erosion and the role of water to transport the sediment. Unconsolidated sediments are transported by erosion match the energetic forces that able to drive the sediment downstream of the river. Physical weathering and chemical weathering induce the degradation of rocks and soil, after which particles are eroded, removed and carried into the waterbody. Rainfall and surface runoff act as the driving force behind soil erosion. Meanwhile, the bottom shear and turbulence level acting as the primary transport agent for all collected material.

The deposition of sediment into a waterway can significantly degrade the water quality and aquatic habitat. Accumulation of sediments in a waterway has high level of suspended solid concentration and less light penetration. Elevated water temperature causing the dissolved oxygen levels to drop [9].

2.1 *Estuarine Sediment*

Terrestrial matter is transported to the sea by the estuarine confluence between the two water bodies that are the freshwater and seawater. The coastal region has been severely affected by the surrounding's rapid economic development. The problems that arise at the coastal area due to the significant anthropogenic inputs (industrialization and urbanization activities, and ecological issues) transported by estuaries. Consequently, a few researchers agreed that the ecosystem at the estuary can be count as one of the most exploited and universally endangering environment systems [10–13].

The geology and morphology of the estuaries depend on the landscape setting. The primary origin is the crucial component to the classification of the estuary itself because it varies substantially [14, 15]. Townend et al. [15] asserts that the origin is influenced by the antecedent landform such as surface deformation of the hard geology, marine derived embayment and fluvial/glacial river valleys. Accordingly, there were evidence confirmed by researchers, namely Townend et al. [15], Dalrymple [16] and Rees et al. [17] that the marine transgression of estuaries produced from marine and fluvial sediment is transported toward river valley under the effect of sea-level rise.

The floodplain is formed alongside the estuarine based on the volumetric difference between freshwater and seawater. The formation of the estuaries natural landscape setting is depicted in Fig. 1. The alluvial estuaries in river valleys respond to marine tide as to maintain its position in the tidal frame where the estuary moves

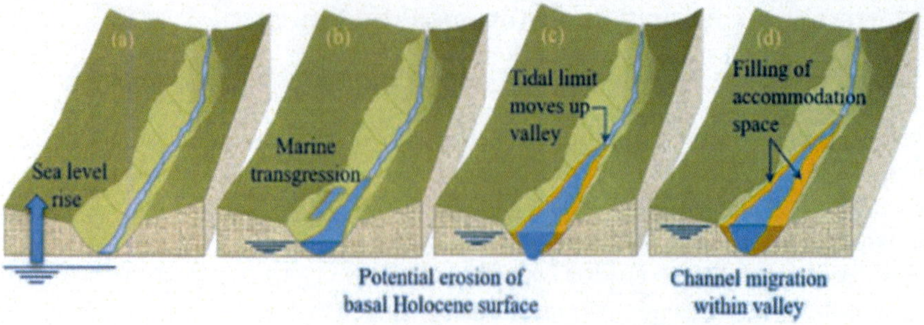

Fig. 1 Progression of river valley due to effect of sea level rise (*Source* [15])

landwards indicating the implication of marine transgression. The kinematic movement along the estuary morphology can potentially cause erosion. It enables the relative significance of the space open to sediment deposition from river and marine. The system is more susceptible to changes in sediment supply or the rate of sea-level rise when the size of the floodplain is smaller because the transgression distance is reduced. The changes in the estuary form with the greater landward movement will manipulate the sediment demand. Conversely, more space is available when the floodplain is separated from the estuary. Therefore, restoration of the estuarine landscape will rely on the availability of sediment and the rate of sea-level rise.

2.2 *Sediment Balance*

In river system, the concept of sediment balance in rivers describes the equilibrium between the amount of sediment supplied to a water channel and capacity of the flow to transport that sediment. Long-term sediment supply to rivers, sediment transport via rivers and sediment storage in watersheds collectively known as the sediment regime, generally achieve a state of dynamic equilibrium resulting in distinct channel morphologies [18, 19]. Wohl et al. [20] mentioned that the natural sediment regime is hardly observable, given the degree of human alteration to land cover (inputs) and instream modification (storage and movement). As a result, researchers differentiate between natural and balanced sediment regimes, where balanced sediment regime happens when the water flow have enough energy to carry sediment is proportional to availability of sediment over a specified period and the river shape is stable in equilibrium. He further stated that when the whole river system such as water and sediment is altered where the ecosystem and biota are attuned.

2.3 *Equation Involved in Sediment Balance*

2.3.1 Fringe's Equation

A study made by Frings et al. [21], an analysis to quantify the downstream fluxes of different sediment particles size through the Rhine River for the period 1990–2010 and identifying sediment source in the upstream and sediment deposition within the channel. As illustrated in Fig. 2 shows the sediment balance components for both sediment input, output and storage. The sediment inputs consist of sediments carried from the upstream area, sediments carried from the tributaries, riverbank erosion, artificial sediments fed into the river channel. While the sediment outputs and storages occur when the sediments are exit from the targeted channel, riverbed sediments removed by river mining, sediments in the floodplains and/or groin fields, riverbed material abrasion. The parameters used in this study are sediment transport analysis, bed-level analysis, sediment budget analysis and grain size analysis.

Frings mentioned that sediment budget is the balance of whole sediment process ($I - O = \Delta S$) between the mass of sediment entering the targeted river (I), the mass of sediment exiting the targeted area (O) and the differences in sediment mass stored inside the targeted river area (ΔS) as shown in Eq. (1). However, a mathematical sediment transport model is required to analyze the stored sediment in the channel where which benefits from a precise spatial distribution and more reliable prediction [22]. The sediment balance equations are extracted as shown below (tonnes/area):

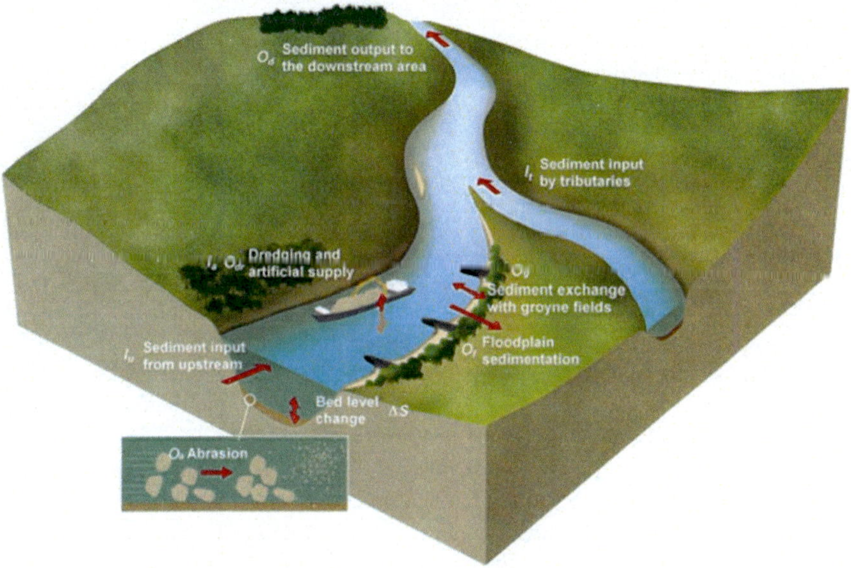

Fig. 2 Scheme of sediment balance components (*Source* [21])

$$I_{\text{tributary}} + I_{\text{Upstream}} + I_{\text{sediment feeding}} + I_{\text{bank erosion}} - O_{\text{dredging}} - O_{\text{downstream}} - O_{\text{abrasion}} - O_{\text{floodplains/groynefields}} = \Delta S \quad (1)$$

where

ΔS—change in the storage of sediments.

(a) Sediment Input (I)

I_{Upstream}—sediments carried from upper part of the river.

$I_{\text{tributary}}$—sediments carried from the river tributaries.

$I_{\text{bank erosion}}$—sediments generated from bank erosion.

$I_{\text{sediment feeding}}$—sediments artificially dump into the river.

(b) Sediment Output (O)

$O_{\text{downstream}}$—sediment mobilized to the end of the river.

$O_{\text{floodplains/groinfields}}$—sediments retained in the floodplains and/or groin fields.

O_{abrasion}—river-bed material abrasion.

Finally, the changes in the net morphological sediment stored in the study area (ΔS) are shown in Eq. (2):

$$\Delta S = ((\Delta z - \Delta zt)/\Delta t)) \cdot W \cdot L \cdot \Delta \mathrm{P}s(1 - p) \quad (2)$$

where

ΔS—net changes in sediment.

$\Delta z - \Delta zt$—river-bed change in time interval Δt.

W—river width.

L—river segment's length.

$\mathrm{P}s$—specific weight of sediments.

p—porosity of the bed material.

The results demonstrate that the suspended solid, which included clay and silt as characteristics and was also known as a wash load, was transported more frequently than the bed load. According to Fig. 3, the transported sand dominated over gravel in the sense of morphologically sediment cycle. While the mobility of coarse gravel (including cobbles) remained small toward downstream and once more the fine gravel is increasing.

Fring et al. [21] provide further evidence that according to budget analysis of sand and gravel being supplied to the targeted river segment is limited. There were 3 major sediment source which occurs in the study area that are 1/3 total of sediment flux originated from the upstream, 1/3 was supplied by the bed degradation and 1/3

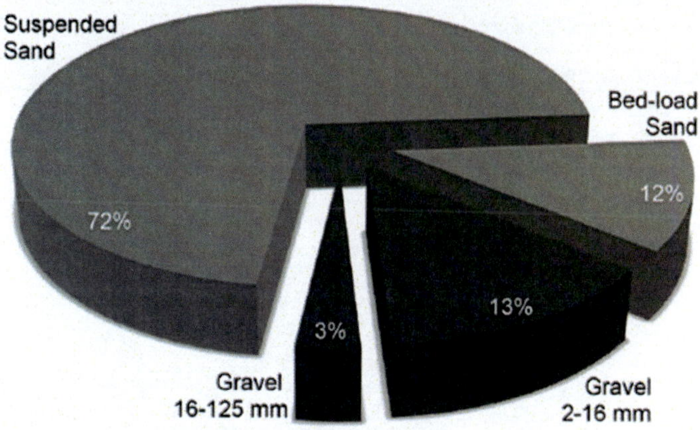

Fig. 3 Composition of the sediment load at the study area

was added artificially by humans to stabilize the bed as demonstrated in Fig. 4. The transition zone between gravel and sand, mining-induced subsidence and places with tertiary sand close to the bed surface is where bed degradation is the most severe. The study confirms that the erosion occurs on the riverbed will generate high sand and fine gravel loads. While the bed slope and flow velocities decrease in further downstream, the coarse gravel and cobble loads will decrease due to a reduced sediment mobility. Just a little amount of sediment was lost to abrasion can be found throughout the study.

3 GETFLOWS

The General-purpose Terrestrial Fluid Flow simulator or GETFLOW is a simulation code for numerical modeling of multiple flow analyzes code to minimize the discrepancy between numerical simulation and observation. The surface flow data is in two dimension, while the subsurface data is in three dimension. A study by Hazart et al. [23] stated that the GETFLOWS are used to compute the water balance component that will be used to train the surrogate model to obtain the global water balance indicator. This is also to estimate the movement of water flow in surface and subsurface sections of a basin watershed which requires a thorough analysis of precipitation patterns, land use, soil properties, and hydrological modeling. This modeling ensures the data to be obtained and estimation of water balance without having additional steps to execute an expensive hydrological model. It is also said suitable to be the new support-decision strategy for regional watershed stakeholders lacking numerical modeling knowledge [24]. In other instances, the GETFLOWS are used to simulate the hydromechanical behavior during gas migration and consider the mechanical stability of engineered barrier system [42].

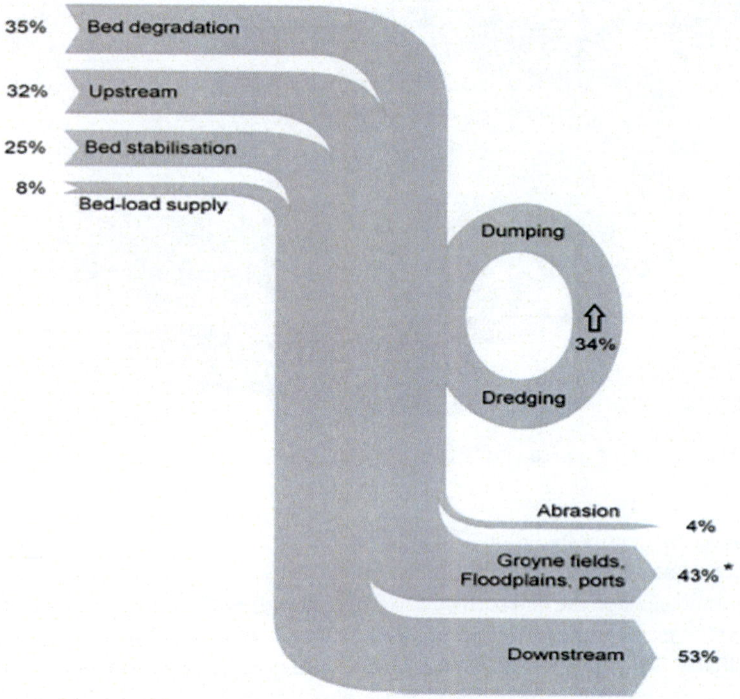

Fig. 4 Sediment budget (*Source* [21])

3.1 Application of GETFLOWS in Environmental Monitoring (Sediments)

According to study by Mori et al. [25], a fully integrated watershed modeling simulator was developed to simulate the mobility of radionuclides. The GETFLOWS also be utilized in the study of the radiocesium (^{137}Cs) fate and transport process from Fukushima Dai-ichi earthquake and the subsequent tsunami to reproduce the redistribution of ^{137}Cs in an actual watershed. The study objectives were achieved through few key assumptions in the modeling. The initial assumption by using the diffusive wave approximation of the shallow water equations is that surface water dynamics, including that of rivers, streams, and hillslopes, will be addressed. Second, a two-phase isothermal compressible air or water flow model of the fluid system is used. Third, the concentration level of radionuclide and suspended sediment does not impact fluid properties such as compressibility, viscosity and fluid density.

Fourth, the appearance of suspended sediment can only be seen on the surface water, while colloid transport in groundwater is ignored. Fifth, the surface soil composed of different particles grain sizes and can easily be detached by the water flow. Sixth, the fate and transport of radionuclides is influenced by suspended sediment, surface and subsurface water. All the assumptions above are used to understand the conceptual model of the watershed system. Data needed in mathematical part of the governing equations are the fluid flow, radionuclide transport-coupled processes and sediment. Fluid flow is represented by coupled surface and subsurface fluid flows. From the generalized Darcy Law, the continuity and the shallow water equations for surface flow serve as the governing equation. The model can generate an accurate estimation of the water saturation, air pressure and temperature for the entire watershed in both surface and subsurface. The required major parameters to conduct the modeling of the radionuclide transport model are listed in Table 1.

The phases in modeling surface/subsurface water flows, sediment and radionuclide transport coupled are summarized into a flowchart (Fig. 5).

Figure 6 shows the schematic diagram of fallout radionuclide redistribution in the watershed system, where the radionuclides deposited on the land surface can be transported by sediment, surface water flow and subsurface water flow in the watershed system. Aqueous phase and solid phase are the two primary transport media for radionuclides. Both in surface and subsurface water flow, the radionuclide redistribution is entirely interconnected with each other [25]. Whereas contaminated sediment particles contained radionuclides element can be mobilized in surface water flows, but groundwater was assumed otherwise. Both surface water and groundwater become the media to transport radionuclide species into surface water bodies.

Another work by Mori et al. [25] simulated the fate and mobility of nitrogen coupled with biogeochemical kinetics reaction. In this study, the kinetic reaction between several chemical elements (i.e., ammonium nitrogen, nitrate nitrogen, etc.) and microbial activities was taken into account. The exchange of polluted water in surface and subsurface can be calculated though the interaction on land surface, where nitrogen loads from point and nonpoint source can be identified.

To grasp better understanding in the conceptual model, this study considers the generalized fluid flow as a compressible, isothermal. (multiphase and multicomponent fluid approach.) They considered the diverse distribution of meteorological conditions, hydrologic processes, land use/land cover (LULC), topography, soil surface and water. Surface water (streams, hill slope and reservoir flows) is portrayed as a depth-averaged, diffusive was approximation including. The concentration of nitrogen levels was mostly regulated by continuous groundwater discharge for a long period where it can be predicted through the discharge of nitrogen from subsurface water to the rivers and the lake water [25]. It further stated that the coding determines water flows, surface water flows, subsurface air and sediment transport by soil erosion, suspension within surface water flows and re-deposition. Cesium-137 transport was estimated in both forms; particulate and dissolved [25, 27].

According to Kitamura et al. [6], a study was made by using GETFLOWS to simulate the sediment migration within five basins focusing around Fukushima Daichi Nuclear Power Plant (FDNPP). The model is used to design and develop as to treat soil erosion from rain splash erosion and hydraulic erosion/deposition. In order to achieve that, the model used to simulate surface and subsurface flows in a fully coupled way. As stated by Sakuma et al. [28], the code is applied for sediment transport, where it simulates raindrop-induced soil detachment, including the impacts of interception by forest canopies, and direct erosion by surface water flows. The model

Table 1 List of major parameters required for the radionuclide transport assessment

Categories	Required specification	Unit	Notation	References
1. Meteorology	Rate of precipitation	(mm/d)	H	[6, 25, 26]
	Air temperature	(C)	T_a	[25, 26]
	Wind velocity	(m/s)	U	[26]
	Daytime length	(s)	H	[25, 26]
	Solar radiation	W/m^2	S_a	[26]
	Saturation vapor pressure	(Pa)	E	[25]
2. Land surface processes	Elevation	(m)	Ξ	[6, 25, 26]
	Manning's roughness parameter	(m − 1/3 s)	n	[6, 25, 26]
	Canopy cover density	(–)	Cc	[6, 25]
	Albedo	(–)	A	[26]
	Bulk transfer coefficient	(–)	C_H	[26]
	Vegetation cover density	(–)	C_L	[6]
	Canopy height	(m)	PH	[6, 25]
3. Subsurface fluid flow	Intrinsic permeability	(m/s)	K	[6, 25, 26]
	Effective porosity	(–)	ϕ	[25, 26]
	Void ration	(–)	e	[6]
	Relative permeability	(–)	kr	[25, 26]
	Capillary pressure	(Pa)	P_c	[25, 26]
4. Sediment transport	Particle density	(kg/m^3)	Pss	[6, 25]
	Grain size composition	(–)	H	[6, 25]
	Soil detachability index	(g/j)	ke	[6, 25]
	Sediment settling velocity	(m/s)	V_{ss}	[6]
	Cohesive strength	(Pa)	J	[25]
	Transport capacity derived from experiment		c, n	[6]
	Bed material cohesiveness (dimensionless)	(–)	β	[6]
5. Radionuclide transport	Decay constant	(1/s)	λ	[25, 26]
	Soil/rock density	(kg/m^3)	Ps	[25, 26]
	Molecular diffusion coefficient	(m^2/s)	Dcw	[25, 26]
	Distribution coefficient	(L/kg)	Kd	[25]
	Tortuosity factor	(–)	t	[25, 26]
	Dispersion length	(m)	aL, aT	[25, 26]
	Decay constant	l/s	λ	[26]
	Heat capacity	J/kg/K	cr	[26]
6. Heat transport	Thermal conductivity	W/m/K	κ	[26]

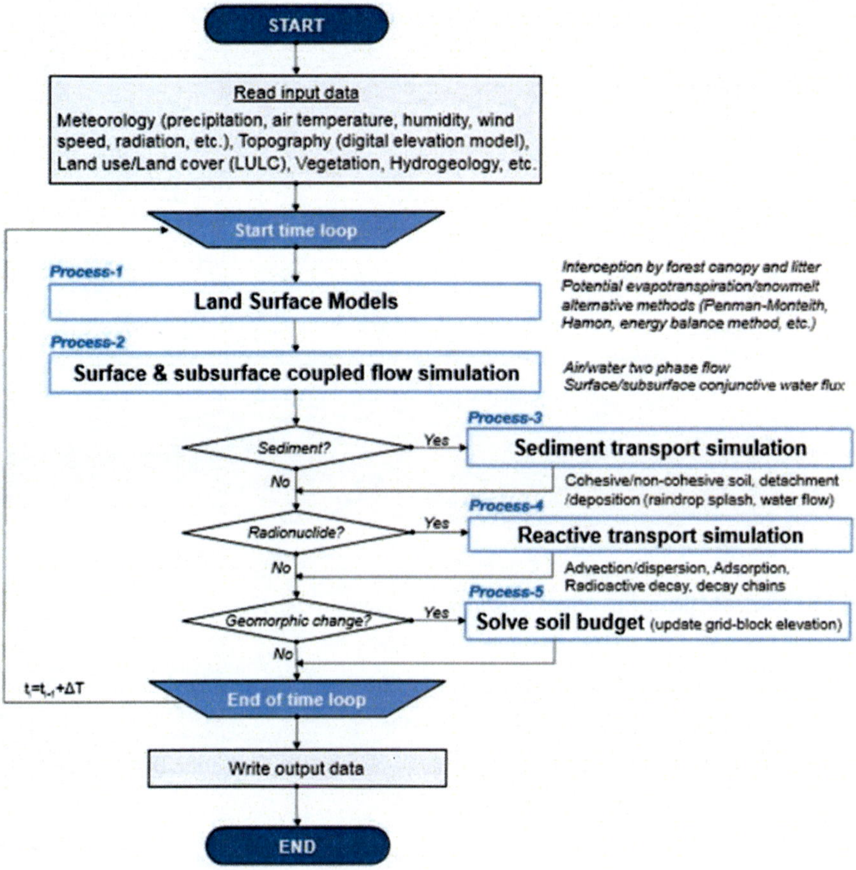

Fig. 5 Coupled simulations flowchart of surface and subsurface water flows, sediment and radionuclide transport (*Source* [25])

input parameters relating to rain splash erosion, such as land use, canopy height and coverage and vegetation type. In this case, to study the radiocesium transport and discharge between basins near the FDNPP following heavy rainfall events.

Sakuma et al. [27] conducted a study to assess the amount of ^{137}Cs redistribution that occurred in the Oginosawa River catchment over a certain period. The goal of study was to grasp the knowledge regarding the difference on the relative contributions between adjacent land to channels and forested areas far from channels to ^{137}Cs input to the watercourses, respectively. This study also emphasized the need to recognize the effect of decontamination work on soil erosion and movability of sediment rates within the catchment area.

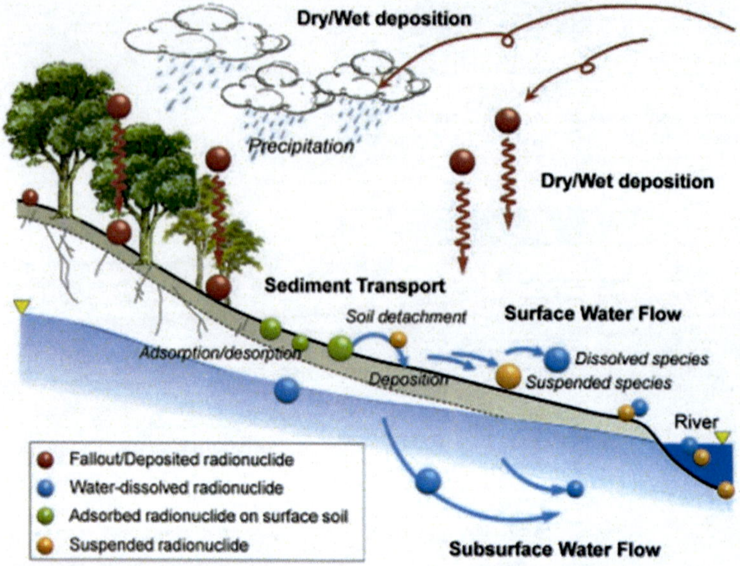

Fig. 6 Schematic illustration of fallout radionuclide redistribution [25]

3.2 Benefits and Drawbacks of GETFLOWS Application

Table 2 gives the benefits and the drawbacks of GETFLOWS modeling application in monitoring as well as other applications.

Table 2 Strength and weakness of GETFLOWS simulator [23, 24, 25, 27]

Advantages	Limitation
• No need extra steps to execute a costly hydrological model in water balance estimation • Can construct the conjunctive water flow through watershed components such as rivers, hillslopes and unsaturated and saturated aquifers without requiring any assumption • The simulation automatically covers the whole river basin in terms of surface water flows, therefore there will be no problems with disconnection and reconnection of river segment • The code can be transferred globally to the numerous different aquifer systems near convergent margins	• Inaccuracy—the variability of data caused by the heterogeneity of raindrop size, soil moisture, land cover density and other factors • The need of field monitoring in some cases—the complexity of modeling sediment and pollutant discharge from paddy fields and land adjacent

4 HEC-RAS

Hydrologic Engineering Center's-River Analysis System (HEC-RAS) is a modeling application that simulate the water flow through rivers and other channels. It was developed by the US Department of Defense, Army Corps of Engineers in order to manage the rivers [43]. This software includes four types of one-dimensional hydraulic components for steady flow, unsteady flow simulation, movable boundary sediment transport analysis and water quality analysis, all of which use a common representation of geometric data and hydraulic computation [29, 30, 44]. HEC-RAS is also an effective tool to simulate the runoff relying on the channel morphology [31]. That said, the HEC-RAS 1D model is often applied to analyze sediment transport. HEC-RAS' version 4.0 was the first to incorporate calculations for 1D sediment transport [7]. The sediment transport model from HEC-RAS requires hydraulic variables (velocity, flow depth and shear stress) and sediment properties to evaluate the transport capacity for cohesive and non-cohesive soils. The user can compute transport potential. (temporal entrainment, deposition and alteration of the cross sections to reflect aggregation or erosion via the HEC-RAS software.)

4.1 Application of HEC-RAS in Sediment Modeling

HEC-RAS was applied by Joshi et al. [8] to develop a sediment transport model analysis in the river channel. The first data used in this study is geometric data that was generated from a digital elevation model with a resolution of 10 m by 10 m using ArcGIS and the HEC-GeoRAS extension This model has to be calibrated and validated to guarantee measurement accuracy and that it fulfill the specified functional goal. Making use of different sediment transport functions and Manning's roughness coefficient, calibration and validation were completed. By assuming the sediment transport equation from the main hydraulic variables, the sediment transport rate can be determined in sediment studies. The predicted and measured sediment transport rates usually differ from each other. According to this study, Meyer-Peter and Mullet (MPM) had the best fit to the field data in contrast to the other six (6) equations. The sediment transport analysis indicated variations in riverbed pattern as well as the areas of river that are vulnerable to erosion or deposition. Consequently, by combining the model, output and local knowledge may assist to mitigate the problem drove by sediment. It shows that most sediments accumulate in the upper stream of the river. Yet, downstream area displays different outcomes. Additionally, the pattern sediment distribution is also not uniform. In short, having more cross section area, bed load and gradation of sediment load data can help to develop a more reliable model for predicting the future sediment behavior.

Research by Foti et al. [32], this study to determine the impact of river removal by analyzing the entire basin to determine the implications of sediment balance. Additionally, it is possible for this model to locate the feasible area and also estimate the potential evolution of river morphology caused by withdrawals. The process of sediment withdrawals will be removed far from the riverbank to avoid affecting them and without eroding the riverbed. Sediment removal will extract from the riverbed that has equal to or higher than that of conjoining lands to minimize the flooding risk (Fig. 7). A total of 1 m in height and 100 m in width removal of accumulated sediments at the middle part of the targeted river section.

In order to carry out the research successfully, there are three (3) main phases involved throughout the process. The early phase involved the development using GIS software to perimeter and morphometrically characterize the river basin and its hydraulically and sedimentological homogeneous sub-basins. Utilizing the HEC-HMS software, the next phase was created to assess the hydrological balance of the basin and its sub-basins. In the final phase, by combining two software of HEC-RAS

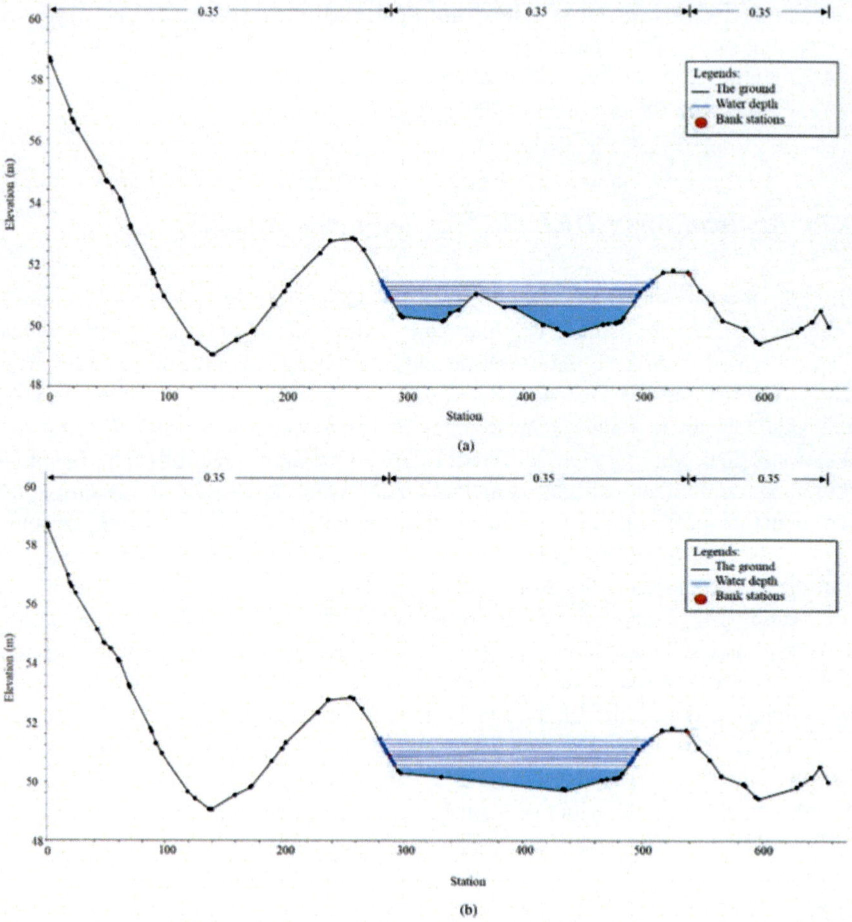

Fig. 7 **a** Before sediment withdrawal, **b** after sediment withdrawal

Table 3 Advantages and disadvantages of HEC-RAS modeling [8, 33–35]

Advantages	Disadvantages
• 1D models can still be used to lengthy river reached because of their straightforward computation and calibration • Need smaller in amount of hydrologic data for calibration and validation • Fast computation	• Capable of developing 1D modeling in sediment studies • Does not consider salinity or how organics can influence fine sediment

software and SIAM model to discover the regions that were experiencing erosion, deposition and equilibrium.

4.2 *Benefits and Drawbacks of HEC-RAS Application*

Table 3 gives the benefits and the drawbacks of HEC-RAS modeling application in monitoring as well as other applications.

5 Conclusion

This research review's purposes are to help the reader to grasp the concept of sediment transport in the river system by understanding the process of the sediment starting from the source, transportation until the deposition of the sediment and also to determine the benefits of using HEC-RAS and GETFLOWS. This is significant because rivers have different types of situations and management according to the current event that can affect the river system in the specific time interval. There has been much research and discussion conducted on the behavior and the mobility of the sediment. Most of the research found was on the composition of the sediment and sediment load. Apart from that, this review showed that the benefit of using GETFLOW, it can develop a 3D modeling and it can illustrate the whole hydrological cycle based on the sediment mobility. As for the HEC-RAS, it has been widely used by many researchers in this field to identify the mobility of the sediment and it is in 1D modeling and easy to use. More research and testing are required to gain a better prediction in identifying the behavior of the sediment by using sediment modeling.

Acknowledgements This study is in collaboration and financially supported by National Water Research Institute of Malaysia (NAHRIM) (Grant no. 304.PAWAM.6050432.I136).

References

1. Ding, L., Chen, K.L., Cheng, S.G., Wang, X.: Water ecological carrying capacity of urban lakes in the context of rapid urbanization: a case study of East Lake in Wuhan. Physics and Chemistry of the Earth, Parts A/B/C **89–90**, 104–113 (2015)
2. Reitsma, K.D., Dunn, B.H., Mishra, U., Clay, S.A., DeSuttere, T., Clay, D.E.: Land-use change impact on soil sustainability in a climate and vegetation transition zone. Agronomy Journal. **10**(6) (2015). https://doi.org/10.2134/agronj15.0152
3. Yusof, M.F., Jamil, N.R., Leaw, C.N.I., Aini, N., Manaf, A.L.: Land use change and soil loss risk assessment by using geographical information system (GIS): a case study of lower part of Perak River. IOP Conf. Ser.: Earth Environ. Sci. **37**, 012065 (2016). https://doi.org/10.1088/1755-1315/37/1/012065
4. Bagarello, V., Di Stefano, C., Ferro, V., Pampalone, V.: Predicting maximum annual values of event soil loss by USLE-type models. CATENA **155**, 10–19 (2017)
5. Ouyang, W., Wu, Y., Hao, Z., Zhang, Q., Bu, Q., Gao, X.: Combined impacts of land use and soil property changes on soil erosion in a mollisol area under long-term agricultural development. Sci. Total Environ. **613–614**, 798–809 (2018)
6. Kitamura, A., Kurikami, H., Sakuma, K., Malins, A., Okumura, M., Machida, M., Mori, K., Tada, K., Tawara, Y., Kobayashi, T., Yoshida, T., Tosaka, H.: Redistribution and export of contaminated sediment within eastern Fukushima Prefecture due to typhoon flooding. In: Earth Surface Processes and Landforms. Wiley Online Library (2016)
7. Brunner, G.W., Gibson, S.: Sediment transport modelling in HEC RAS. Impacts of Global Climate Change (2005)
8. Joshi, N., Lamichhane, G.J., Rahaman, M.M., Kalra, A., Ahmad, S.: Application of HEC-RAS to study the sediment transport characteristics of Maumee River in Ohio. World Environmental and Water Resources Congress (2019). https://doi.org/10.1061/978084482353.024
9. Osuagwu, J.C., Nwachukwu, A.N., Nwoke, H.U., Agbo, K.C.: Effect of soil erosion and sediment deposition on surface water quality: case study of Otamiri River. Asian Journal Engineering and Technology (2014)
10. Li, H., Lin, L., Ye, S., Li, H., Fan, J.: Assessment of nutrient and heavy metal contamination in the sea water and sediment of Yalujiang Estuary. Mar. Pollut. Bull. (2017). https://doi.org/10.1016/j.marpolbul.2017.01.069
11. Liu, J.Q., Yin, P., Chen, B., Gao, F., Song, H.Y., Li, M.N.: Distribution and contamination assessment of heavy metals in surface sediments of the Luanhe River Estuary, northwest of the Bohai Sea. Mar. Pollut. Bull. **109**, 633–639 (2016)
12. Lotze, H.K., Lenihan, H.S., Bourque, B.J., Bradbury, R.H., Cooke, R.G., Kay, M.C., et al.: Depletion, degradation, and recovery potential of estuaries and coastal seas. Science **312**, 1806–1809 (2006)
13. Liu, R.M., Men, C., Liu, Y.Y., Yu, W.W., Xu, F., Shen, Z.Y.: Spatial distribution and pollution evaluation of heavy metals in Yangtze estuary sediment. Mar. Pollut. Bull. **110**, 564–571 (2016)
14. Hume, A.D., Herdendorf, C.E.: A geomorphic classification of estuaries and its application to coastal resource management. Journal of Ocean and Shoreline Management. **11**, 249–274 (1988)
15. Townend, I., Zhou, Z., Guo, L., Coco, G.: A morphological investigation of marine transgression in Estuaries. Earth. Surf. Processes. Land. **46**(5) (2020)
16. Dalrymple, R.W.: Incised valleys in time and space: an introduction to the volume and an examination of the controls on valley formation. Society for Sedimentary Geology 5–14 (2006)
17. Rees, J.G., Ridgway, J., Ellis, S., Knox, R.W.O., Newsham, R., Parkes, A.: Holocene sediment storage in the Humber Estuary. Geological Society, London, Special Publications **166**, 119–143 (2000)
18. Church, M.: Channel morphology and typology. In: The River Handbook, pp. 126–143. Blackwell Scientific Publishers, Oxford (1992)

19. Gilbert, J.T., Wilcox, A.C.: Sediment routing and floodplain exchange (SeRFE): A spatially explicit model of sediment balance and connectivity through river networks. Journal of Advance in Modelling Earth Systems (2020). https://doi.org/10.1029/2020MS002048
20. Wohl, E., Bledsoe, B.P., Jacobson, R.B., Poff, N.L., Rathburn, S.L., Walters, D.M., Wilcox, A.C.: The natural sediment regime in rivers: broadening the foundation for ecosystem management. Bioscience **65**(4) (2015). https://doi.org/10.1093/biosci/biv002
21. Frings, R.M., Doring, R., Beckhausen, C., Schuutrumpf, H., Vollmer, S.: Fluvial sediment budget of a modern, restrained river: the lower reach of the Rhine in Germany (2014)
22. Maldegem, D.C., Mulder, H.P.J., Langerak, A.: A cohesive sediment balance for the Scheldt estuary. Netherlands Journal of Aquatic Ecology **27**(2–4), 247–256 (1993)
23. Hazart, A.,Mori, K., Tada, K.,Tosaka, H.: Using surrogate modelling for fast estimation of water budget component in a regional watershed. In: International Congress on Environmental Modelling and Software. 7th International Congress on Environmental Modelling and Software, San Diego, USA (2014)
24. Hosono T., Yamada C., Shibata T.: Coseismic groundwater drawdown along crustal ruptures during the 2016 Mw 7.0 Kumamoto Earthquake. Lawrence Berkeley National Laboratory (2019)
25. Mori, K., Tada, K., Tawara, Y., Ohno, K., Asami, M., Kosaka, K., Tosaka, H.: Integrated watershed modelling for simulation of spatiotemporal redistribution of post-fallout radionuclides: application in radiocesium fate and transport processes derived from the Fukushima accidents. Environ. Model. Softw. **72**, 126–146 (2015)
26. Rahman, S.A.T.M., Hosono, T., Tawara, Y., Fukuoka, Y., Hazart, A., Shimada, J.: Multiple-tracers-aided surface-subsurface hydrological modeling for detailed characterization of regional catchment water dynamics in Kumamoto area, southern Japan. Hydrogeol. J. **29**, 1885–1904 (2021)
27. Sakuma K., Malins A., Funaki H., Kurikami H., Niizato T., Nakanishi T., Mori K., Tada K., Kobayashi T., Kitamura A., Hosomi M.: Evaluation of Sediment and 137Cs Redistribution in the Oginosawa River Catchment near the Fukushima Dai-ichi Nuclear Power Plant Using Integrated Watershed Modelling (2018)
28. Sakuma K., Kitamura A., Malins A., Kurikami H., Machida M., Mori K., Tada K., Kobayashi T., Tawara Y., Tosaka H.: Characteristics of radio-cesium transport and discharge between different basins near to the Fukushima Dai-ichi Nuclear Power Plant after heavy rainfall events (2017)
29. Amir, H.H., Ehsan, Z.: Evaluation of HEC-RAS ability in erosion and sediment transport forecasting. World Appl. Sci. J. **17**(11), 1490–1497 (2012)
30. Hasani, H.: Determination of flood plain zoning in Zarigol river using the hydraulic model of HEC-RAS. International Research Journal of Applied and Basic Sciences (2013)
31. Thakur, B., Parajuli, R., Kalra, A., Ahmad, S., Gupta, R.: Coupling HEC RAS and HECHMS in precipitation runoff modelling and evaluating flood plain inundation map, World Environmental and Water Resources Congress **2017**, 240–251 (2017)
32. Foti, G., Barbaro, G., Manti, A., Foti, P., Torre, A.L., Geria, P.F., Puntorieri, P., Tramontana, N.: A methodology to evaluate the effects of river sediment withdrawal: the case study of the Amendolea River in southern Italy. Aquat. Ecosyst. Health Manage. **23**(4), 465–473 (2020). https://doi.org/10.1080/14634988.2020.1807248
33. Horritt, M., Bates, P.: Evaluation of 1D and 2D numerical models for predicting river flood inundation. J. Hydrol. **268**(1–4), 87–99 (2002)
34. Panin, N., Jipa, D.: Danube River sediment input and its interaction with the northwestern Black Sea. Estuar. Coast. Shelf Sci. **54**(3), 551–562 (2019)
35. Werner, M.: Impact of grid size in GIS based flood extent mapping using a 1D flow model. Phys. Chem. Earth Part B **26**(7–8), 517–522 (2001)
36. Kitamura, A., Kurikami, H., Sakuma, K., Malins, A., Okumura, M., Machida, M., Mori, K., Tada, K., Tawara, Y., Kobayashi, T., Yoshida, T., Tosaka, H.: Redistribution and export of contaminated sediment within eastern Fukushima prefecture due to typhoon flooding. Earth Surf. Proc. Land. (2016). https://doi.org/10.1002/esp.3944

37. Mori, K., Tawara, Y., Hazart, A., Tada, K., Tosaka, H.: Simulating Nitrogen Long-term fate and transport processes at a regional scale with a surface and subsurface fully-coupled watershed model. In: 21st International Congress on Modelling and Simulation, Gold Coast, Australia (2015)
38. Jacks, G.V., Whyte, R.O.: The rape of the earth: a world survey of soil erosion. Rape. Earth. World. Survey. Soil. Erosion. (1939)
39. Hassan, M.A., Roberge, L., Church, M., More, M., Donner, S.D., Leach, J., Ali, K.F.: What are the contemporary sources of sediment in the Mississippi River? Geophys. Res. Lett. **44**, 8919–8924 (2017)
40. Borelli, P., Robinson, D.A., Panagos, P., Lugato, E., Yang, J.E., Alewell, C., Wuepper, D., Montanarella, L., Ballabio, C.: Land use and climate change impacts on global soil erosion by water (2015–2070). Proc. Nat. Acad. Sci. **117**(34), 1–8 (2020)
41. Ouillon, S.: Why and how do we study sediment transport? Focus on coastal zones and ongoing methods. Water **10**(4), 390 (2018)
42. Yamamoto, S., Kumagai, M., Koga, K., Sato, S.: Mechanical stability of engineered barriers in a subsurface disposal facility during gas migration based on coupled hydromechanical modelling. Geol. Soc. Lon. Spec. Publ. **415**(1) (2015)
43. Robert, W.C.J., Karen, F., William, J.: Auto-integrating multiple HEC-RAS flood-line models into catchment-wide SWMM flood forecasting models. AWRA Hydrol. Watershed. Manage. Tech. Committee **10**(1), 1–15 (2012)
44. Traore, V.B., Bop, M., Faye, M., Giovani, M.: Using of HEC-RAS model for hydraulic analysis of a river with agriculture vocation: a case study of the Kayanga River Basin, Senegal. Am. J. Water. Res. **3**(5), 147–154 (2015)

Application of Computational Fluid Dynamics in an Outer Channel of Orbal Biological System

N. M. Kamal, N. A. Saad, N. A. Zakaria, J. Abdullah, and N. A. Azizan

Abstract Orbal Biological System (OBS) is one of the modified oxidation ditches claimed as the most energy efficient, safest, and has the lowest maintenance. The OBS can be modified to meet a wide assortment of influence conditions and effluent requirements. There are three channels in OBS, outer channel, middle channel, and inner channel. The outer channel of OBS has the largest volume among the three channels, which is about 50% of the system's total volume. It is crucial to study the operating system of the outer channel of OBS as it contributes to the highest energy consumption in the system. The computational fluid dynamics (CFD) model is developed to have a better picture of the OBS. The simulation in CFD was used to study the behaviour of the influence in the outer channel of OBS and the velocity distribution for different operating conditions in the outer channel of OBS. The result of the CFD model was used to study the relationship between the velocity and the number of aeration discs being used. This study shows that the volume fraction of water is higher at the bottom of the ditch, while the volume fraction of air is higher at the top surface of the ditch. This shows that the CFD model can represent the OBS in reality. From the model created, the velocity inside the ditch decreases when the water is deeper. The research provides a better understanding of the model flow behaviour of a biological treatment system using the simulation tool and can contribute to a better design of wastewater treatment process models in future.

Keywords Biological treatment · Computational fluid dynamics · Flow behaviour · Orbal Biological System · Oxidation ditch · Wastewater treatment

---

N. M. Kamal · N. A. Saad (✉) · N. A. Zakaria · J. Abdullah · N. A. Azizan
River Engineering and Urban Drainage Research Centre (REDAC), Engineering Campus, Universiti Sains Malaysia, 14300 Nibong Tebal, Penang, Malaysia
e-mail: aidasaad@usm.my

N. Sabtu (ed.), *Proceedings of AWAM International Conference on Civil Engineering 2022 - Volume 3*, Lecture Notes in Civil Engineering 386,
https://doi.org/10.1007/978-981-99-6026-2_47

1 Introduction

Worldwide, a lot of wastewater is produced every day from home and industrial sources, creating problems like a water shortages and deteriorating the environment. Municipal energy consumption in emerging nations is largely attributed to the water sector [8]. Wastewater treatment facilities represent a sizable portion of energy-intensive water-related operations. When oxygen is given for a biological system through aeration processes, a substantial quantity of energy is used [6]. The Orbal Biological System (OBS), which was created to provide an optimum approach to the biological treatment process, is one of the modified oxidation ditches that is employed. To accommodate a variety of influent conditions and effluent needs, the OBS is easily modifiable. OBS typically had three concentric elliptical or round channels: an inner channel, a middle channel, and an outer channel. OBS's outer channel has 50–55% of the system's overall volume. Controlling oxygen supply allowed for the creation of alternating oxic-anoxic environments in a single channel, and simultaneous nitrification and denitrification in the OBS allowed for the removal of nitrogen [3, 13]. The study of the mechanics of fluids, including liquids, plasmas, gases, and the forces acting on them, is the focus of the field of physics known as computational fluid dynamics (CFD). CFD is a well-liked modelling method that is frequently used to simulate and evaluate the performance of diverse processes [2]. CFD has emerged as a crucial instrument for developing answers for fluid flows that interact with or without solids. CFD is a method for mathematically simulating a fluid flow-related physical event and using computational power to simulate, analyse, and solve it numerically [11]. Due to the culture of relying on tried-and-true approaches in a traditional industry like water and wastewater treatment, CFD models are also frequently utilised as an add-on to conventional design processes. Aside from that, it gained popularity over conventional approaches to wastewater treatment modelling because it is a very accurate process that enables evaluation of engineering systems, which are expensive, difficult, and risky to be replicated in laboratory, pilot, or field circumstances [4, 5]. The most effective application of computational fluid dynamics is to simulate integrated physical, chemical, and/or biological processes involved in the design and operation of wastewater treatment (WWT), however up to this point, it should be understood for unit process design and troubleshooting [10].

This section explains the numerical procedure to perform the CFD analysis in ANSYS FLUENT. In order to get a clearer picture of the real conditions inside the outer channel of OBS, a three-dimensional CFD model is developed. The geometry model of OBS was created according to the design layout and dimensions that provided by IWK. *X*-axis corresponded to the fluid flow direction, *Y*-axis indicated the width of the outer channel, while *Z*-axis represents the depth of the fluid inside the OBS.

2 Methodology

This section provides an explanation of the numerical technique that must be followed in order to carry out the CFD analysis using ANSYS FLUENT. A three-dimensional computational fluid dynamics (CFD) model is being built in order to obtain a more accurate representation of the conditions that exist within the outer channel of OBS. IWK was responsible for providing the design layout as well as the dimensions, which were used in the creation of the geometry model of OBS. The X-axis represented the direction that the fluid was moving in, the Y-axis denoted the width of the outside channel, and the Z-axis displayed the depth of the fluid that was contained within the OBS.

2.1 Geometry Creation

ANSYS FKUENT's Design Modeller was utilised throughout the process of developing the geometric model. In order to get the geometry ready for analysis in ANSYS Workbench, the geometry must first be created in Design Modeller and then modified. As can be seen in Fig. 1, the entire length of the model is 91.90 m, while its overall width is 54.56 m. These results are derived from the actual readings taken at the OBS location. The geometrical structure that was constructed included three channels: the outside channel, the middle channel, and the inner channel. The CFD model that was constructed, on the other hand, is solely for the outer channel of OBS. The greatest channel is the outer channel, which is also where the influent is found. The length of the straight section along the long axis is 37.34 m, while the width of the outside channel is 12.50 m and the thickness of the wall is 0.36 m. The water in this system is a total of 4.4 m deep. The water that comes from the distribution enters the system via the outside channel, travels along the outer channel, and eventually enters the centre channel via the effluent penstock.

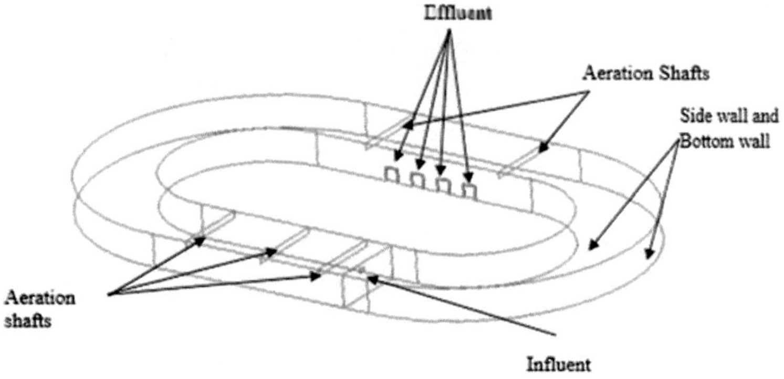

Fig. 1 Geometrical layout of 3D model

2.2 Boundary Condition

The boundary conditions include internal face borders, wall, repeating, and pole barriers, as well as flow inlets and exit boundaries. There are inlet, outlet, fan, and wall in this model. The 3D CFD model's inlet's mass flow boundary condition is applied. It does so because it enables the overall pressure to change in reaction to the inside solution. The penstock between the outer and middle channels of the system is the effluent of the outer channel of OBS. Since the exit flow velocity and pressure are unknown because of the intricate nature of the flow inside the ditch, the effluent of the outer channel of OBS was termed as outflow. The fan model represented the system aeration shafts. OBS's bottom and side walls were considered non-slip walls. Table 1 displays the boundary conditions of the 3D model. Figure 2 displays the boundary conditions of the model and Fig. 3 depicts the plan view sketch of the OBS and the location of the sample point.

Table 1 3D model boundary condition

Name	Level
Influent	Mass flow inlet (Mass flow rate is 580 kg/s)
Materials	Fluid (Water-liquid, air)
Aeration shafts	Fan
Effluent	Outflow

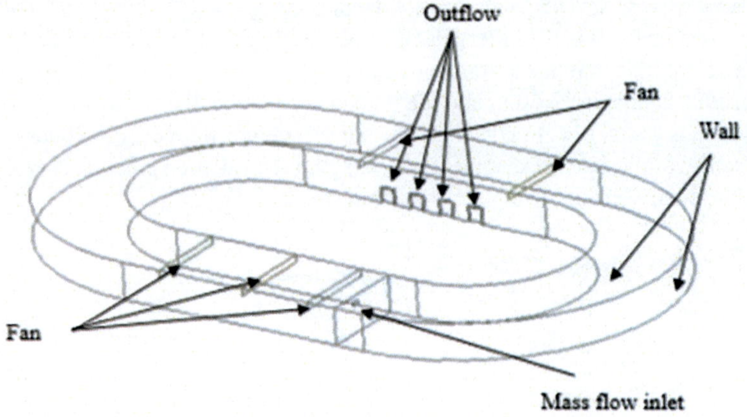

Fig. 2 Boundary conditions of CFD model

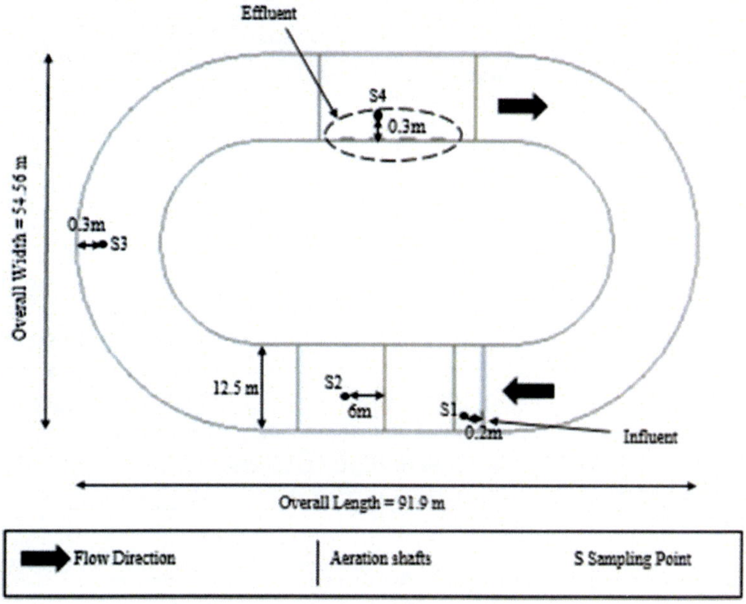

Fig. 3 Plan view sketch of the OBS and the location of sampling points

Table 2 Material properties

Water-liquid (fluid)	
Density	998.2 kg/m^3
Viscosity	0.001003 kg/ms
Molecular weight	18.0152 kg/kg mol
Air (fluid)	
Density	1.225 kg/m^3
Viscosity	1.7894e−05 kg/ms
Molecular weight	28.966 kg/kg mol

2.3 Material Properties

In the CFD model, there are two phases of fluid were involved. Water and air properties have been defined as summarised in Table 2.

2.4 Solver Setting

The default value of standard k-epsilon turbulence model is used for the simulation. The *k*-epsilon model is the method of calculating turbulent viscosity. The

Table 3 CFD model case set-up

Parameter	Detail
Space	3D
Time	Unsteady
Multiphase	Volume of fluid (VOF)
Viscous	Standard *k*-epsilon turbulence model
Wall treatment	Standard wall functions

k-epsilon model is the simplest 'complete models' of turbulence which are two-equation models by the FLUENT's user guide, in which the solution of two separate transport equations allows the turbulent velocity and length scales to be independently determined. The standard *k*-epsilon model is a semi-empirical model based on model transport equations for the turbulence kinetic energy (k) and its dissipation rate (ε). Table 3 gives the case set-up of the CFD model.

3 Results and Discussions

The flow behaviour and velocity throughout the outer channel of OBS were investigated by developing a 3D CFD model. The outer channel of OBS was designed with 5 aeration shafts. To demonstrate the real condition of outer channel of OBS as it was designed, the CFD model was developed with 5 aeration shafts attached to it. The transient model was chosen as the time variations inside the ditch, as the flow characteristics in the domain may change with time [1]. Two-phase model (liquid–gas) and an open-channel system was applied. Material properties for both water and air were defined and included in the system. Solid was not included in the simulation due to the small value of the suspended solid in the wastewater which is only 0.01–0.1% [7]. The actual flow process can be captured by defining the model as open-channel flow. The reason is the surface of the tank is exposed to the atmosphere. The atmospheric conditions act on the tank model, therefore, the mixing process facilitated by the mixers can be visualised well.

3.1 Volume Fraction of Air and Fluid (Water) Across the Ditch

The results of volume fractions of fluids (water) and air can be obtained as the model is defined as open-channel flow. It is due to the influence of atmospheric air. Figure 4 represents the distribution of two phases, which is air and water that involved in this study. Variation of volume fractions shows that anoxic and aerobic conditions are really occurring throughout the OBS [9]. The volume fractions represent the real conditions in OBS as an open-channel system.

The volume fraction of water was described in Fig. 4a. The blue colour represents less or zero volume of water, while red represents the maximum volume of water. As expected, the volume of water is maximum at the bottom of the tank, and lower at the top of the tank. It is due to the exposure of atmospheric air in an open-channel flow modelling. Then, Fig. 4b shows the volume fraction of air. The composition of air is obviously higher at the top of the tank when compared to the bottom of the tank. This shows that oxygen mixing that facilitated by the mixers is occurring.

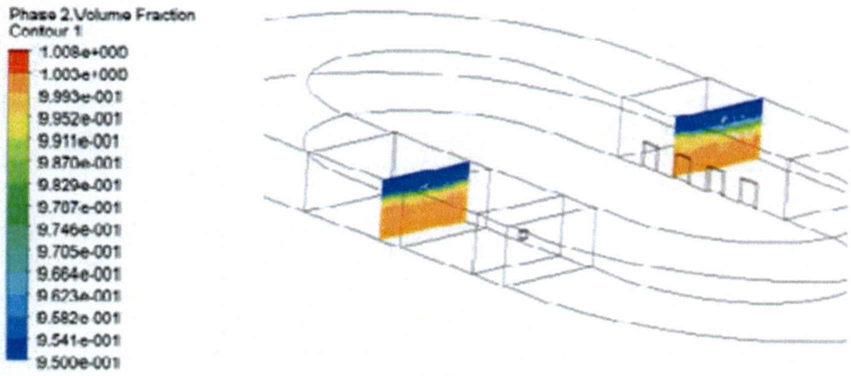

(a) Water

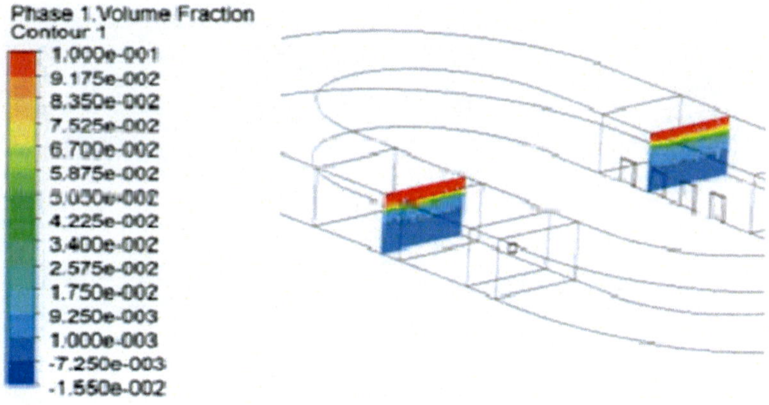

(b) Air

Fig. 4 Contours of volume fraction of **a** water **b** air

3.2 Flow Analysis of the OBS Model

The contours of the velocity magnitude at the top surface of the OBS model are shown in Fig. 5. The maximum velocity at the top surface is at the aeration disc. This shown that the aeration disc functioned well to maintain the flow in the system. This figure shows the flow distribution at different depths inside the system. Figure 6 shows the cross-sectional views of velocity distribution, which the variation of velocity inside the ditch can be seen. The maximum velocity can be obtained at the top of the ditch, and the velocity reduced at the bottom of the ditch. Besides, the vectors of the velocity magnitude can be obtained from CFD. The vectors show the movement of the water inside the ditch. It shows that the direction of water inside the ditch is on the right way, which is in clockwise direction. These vectors plot is shown in Fig. 7. The flow's vectors near the aeration shafts is shown in this figure, which shows that the model is successfully represented the system, with the aeration discs that functioning well as the water flow in one direction as can be seen during site observation.

Each sampling point's velocity magnitudes has been extracted and put into the line graph in Fig. 8's representation. The average flow velocity when all aeration shafts are opened is 0.71 m/s, according to the simulation's results. The OBS has a maximum velocity of 1.20 m/s and a low velocity of 0.19 m/s. At sampling point 1, which is close to the influent, the line graph displays the greatest value. At sample point 2, the velocity then begins to decline, and sampling point 3, the lowest value, follows. This is as a result of sample point 3's distance from the aeration discs. Due to the surrounding aeration discs, the velocity increased at sample point 4 [12]. The

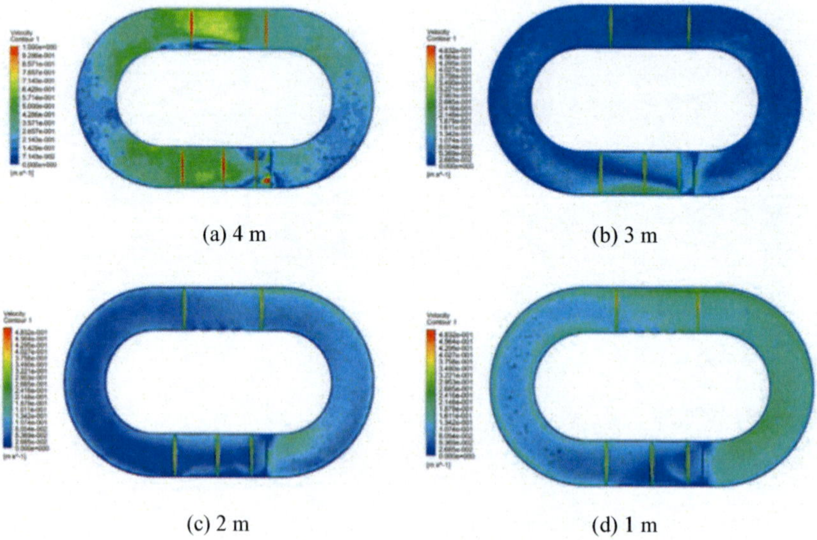

(a) 4 m (b) 3 m

(c) 2 m (d) 1 m

Fig. 5 Contours of velocity magnitude at **a** 4 m depth, **b** 3 m, **c** 2 m, **d** 1 m

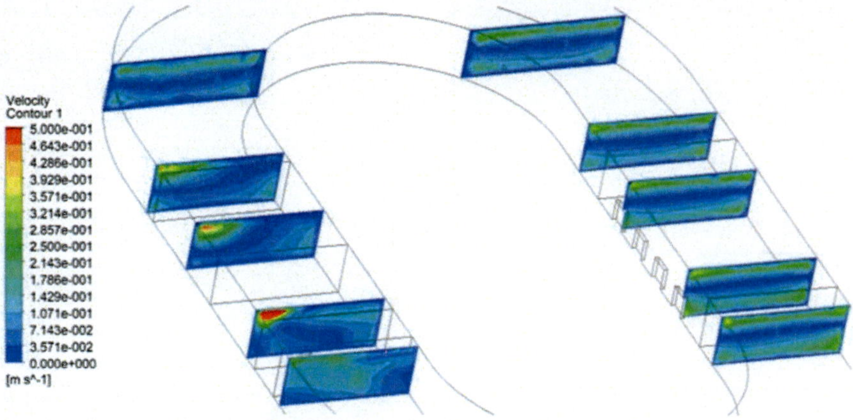

Fig. 6 Contour of velocity magnitude at various cross section

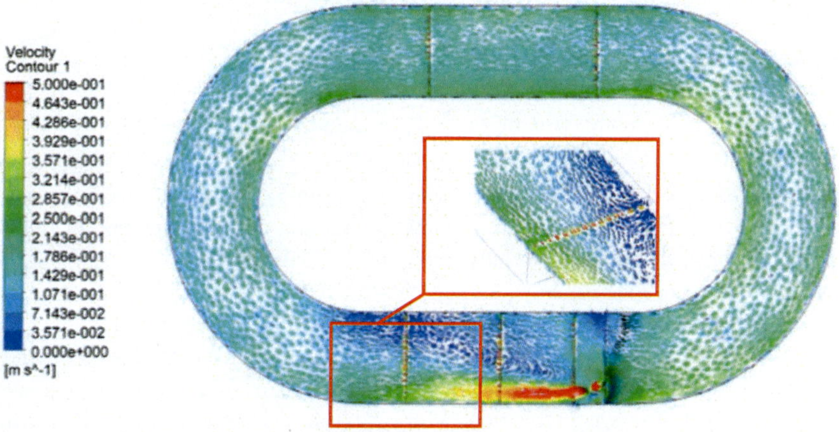

Fig. 7 Vectors coloured by velocity magnitude at 4 m depth (near the aeration shafts)

3D CFD model was validated by the overall results, which showed a respectable agreement with the real onsite measurements. To make sure the CFD model can be utilised to examine the hydraulic behaviour of OBS, the average deviation is computed. With the real data, the average variance for this CFD model is 23%. As a result, the model may be utilised to examine the flow behaviour in OBS in depth.

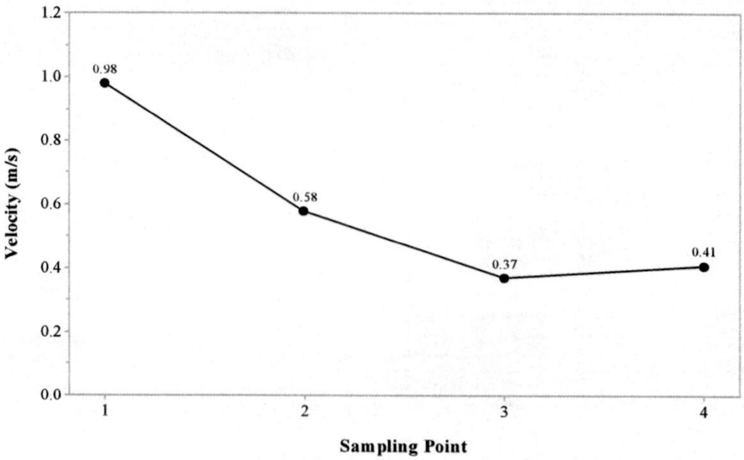

Fig. 8 Velocity magnitudes at every sampling point given by CFD model

4 Conclusions

As a means of simulating the flow behaviour within the outer channel of OBS, a computational fluid dynamics (CFD) model was constructed. Additionally, a computational fluid dynamics (CFD) model can be utilised to investigate the OBS outer channel's various operational states. Using a computational fluid dynamics (CFD) model, one may calculate the relative volumes of water and air. It reveals that air occupies a larger volume than water near the ditch's top surface, while water occupies a larger volume than air near the ditch's bottom. This demonstrates that the CFD model accurately mimics the OBS. According to the simulation, the speed of water inside the ditch decreases as the water level rises. Outside of that, the outer channel's velocity is still within acceptable ranges for operating the OBS's systems there. This is due to the fact that an increase in velocity can lead to a rise in the DO content of the water, and that an excessive amount of oxygen can induce corrosion or air binding at the shafts.

Acknowledgements A special thanks to the Ministry of Higher Education (MOHE) for funding this publication under the programme of the Higher Institution Centre of Excellence (HICoE) (311.PREDAC.4403901).

References

1. Akan, A.O.: Open Channel Hydraulics, 1st edn. Elsevier (2006)
2. Bitog, J.P.P., et al.: Optimised hydrodynamic parameters for the design of photobioreactors using computational fluid dynamics and experimental validation. Biosys. Eng. **122**, 42–61 (2014). https://doi.org/10.1016/j.biosystemseng.2014.03.006
3. Daigger, G.T., Littleton, H.X.: Simultaneous biological nutrient removal: a state-of-the-art review. Water Environ. Res. **86**(3), 245–257 (2014). https://doi.org/10.2175/106143013X13736496908555
4. Karpinska, A.M., Bridgeman, J.: CFD-aided modelling of activated sludge systems—a critical review. Water Res. **88**, 861–879 (2016). https://doi.org/10.1016/J.WATRES.2015.11.008
5. Liu, X., Zhang, J.: Computational fluid dynamics: applications in water, wastewater, and stormwater treatment: EWRI computational fluid dynamics task committee. Computational Fluid Dynamics: Applications in Water, Wastewater and Stormwater Treatment, 1–224 (2019). https://doi.org/10.1061/9780784415313
6. Odlare, M.: Introductory chapter for water resources. In: Reference Module in Earth Systems and Environmental Sciences. Elsevier (2014). https://doi.org/10.1016/b978-0-12-409548-9.09035-7
7. Rahbar, M.S., et al.: Color removal from industrial wastewater with a novel coagulant flocculant formulation. Int. J. Environ. Sci. Technol. **3**(1), 79–88 (2006). https://doi.org/10.1007/BF03325910
8. Rathour, R., Kalola, V., Johnson, J., Jain, K., Madamwar, D., Desai, C.: Treatment of various types of wastewaters using microbial fuel cell systems. In: Biomass, Biofuels, Biochemicals: Microbial Electrochemical Technology: Sustainable Platform for Fuels, Chemicals and Remediation, pp. 665–692 (2019). https://doi.org/10.1016/B978-0-444-64052-9.00027-3
9. Saad, N.A., et al.: Study of velocity profiles of orbal biological system (OBS) using computational fluid dynamics (CFD). Journal of Computational Methods in Sciences and Engineering **19**(2), 447–454 (2019). https://doi.org/10.3233/JCM-181145
10. Samstag, R.W., et al.: CFD for wastewater treatment: an overview. Iwaponline. Com. (2016). https://doi.org/10.2166/wst.2016.249
11. Simscale: What is Computational Fluid Dynamics (CFD)? (2021). https://www.simscale.com/docs/simwiki/cfd-computational-fluid-dynamics/what-is-cfd-computational-fluid-dynamics/
12. SUEZ Homepage: Water Handbook—Aeration (2019). https://www.suezwatertechnologies.com/handbook/chapter-04-aeration, last accessed 2021/12/20.
13. Wang, X., Gao, C., Jin, P., Zhang, Y., Xie, Y., Chen, T., Zhang, A.: Nitrogen removal performance and bacterial community in a full-scale modified Orbal oxidation ditch with internal nitrate recycle and biocarriers. Journal of Water Process Engineering **40**, 101791 (2021). https://doi.org/10.1016/J.JWPE.2020.101791

Experimental Modal Analysis and Operational Deflection Shape Analysis for Kenyir Dam Spillway Physical Model

N. H. Hassan, M. H. Zawawi, A. Z. A. Mazlan, M. A. Abas, M. R. R. M. A. Zainol, M. R. M. Radzi, and M. A. Kamaruddin

Abstract Kenyir Dam spillway physical model was developed to determine the vibration characteristics of the dam spillway due to different spilling events. This study focused on determining the natural frequency of the dam spillway in order to compare it with the mode shape due to spilling events. The physical model scale was 1:50 and similitude analysis using Froude law was used for scaling down purposes. There were 2 different water levels simulated for the experiment which were 2.92 and 2.96 m, and for each water level spilling, operational deflection shape (ODS) analysis was done for the spillway body. The most dominant mode shape was observed at the wall of the spillway at the downstream area. This mode shape corresponded to the natural frequency of 2801.1 Hz and cause an overall deflection amplitude of 0.353 m for a water level of 2.96 m. The numerical analysis was completed in order to compare the experimental result and it showed that the percentage differences for all parts were below 10%, which is generally below an allowable error percentage. The conclusion from this study was the experimental natural frequencies and mode

N. H. Hassan · M. H. Zawawi (✉) · A. Z. A. Mazlan
Department of Civil Engineering, College of Engineering, Universiti Tenaga Nasional, 43000 Kajang, Selangor, Malaysia
e-mail: MHafiz@uniten.edu.my

M. A. Abas
School of Mechanical Engineering, Universiti Sains Malaysia, 14300 Nibong Tebal, Penang, Malaysia

M. R. R. M. A. Zainol
School of Civil Engineering, Universiti Sains Malaysia, 14300 Nibong Tebal, Penang, Malaysia

M. R. M. Radzi
Hydro Life Extention Program (HELP), Business Development (Asset) Unit, TNB Power Generation Division, PJX HM Shah Tower, Jalan Persiaran Barat, 46050 Petaling Jaya, Selangor, Malaysia

M. A. Kamaruddin
Environmental Technology Division, School of Industrial Technology, Universiti Sains Malaysia, 11800 Penang, Malaysia

N. Sabtu (ed.), *Proceedings of AWAM International Conference on Civil Engineering 2022 - Volume 3*, Lecture Notes in Civil Engineering 386,
https://doi.org/10.1007/978-981-99-6026-2_48

shapes for all the parts show good agreement with the numerical analysis of vibration characteristics of Kenyir Dam spillway physical model.

Keywords Spillway · Experimental modal analysis (EMA) · Operational deflection shape (ODS) · Natural frequency · Deflection

1 Introduction

Uncertainties of vibration induced from the internal and external sources spillway such as the flow of water, vibration of the hydropower mechanical machine and other related sources have increased the spillway insecurity [1]. Vibration analysis such as experimental model analysis (EMA) and operational deflection shapes (ODS) of dam structure specifically on a spillway has to be investigated in order to have a reliable dam structure that can avoid any obliteration of the structure in the future [2].

Numerical modeling techniques have been increasingly applied in numerous engineering applications nowadays [3]. Information regarding the dynamic characteristics of the dam structure in terms of natural frequencies, mode shapes and frequency response function (FRF) over spillways has historically been obtained through the use of physical model experiments [4].

A physical model represents the actual prototype to be built in real life and the model is used as a tool to find the most technical and economical solution [5]. There are some numerical ratios and scales that must be taken into account when constructing the physical model [6]. The scale ratio between the physical model and prototype is important for examining and measuring processes that are difficult to observe in reality [7].

As the scale ratio is reduced, an error caused by the scale effect is increased and the results obtained from the model and prototype are getting away [8]. Similitude analysis was used to perform experiments on the scaled physical model version of the spillway structure. It was operated on the basis of Froude number similarity because of the minimal and negligible viscosity effects [9].

Numerical simulation is performed to examine the operational deflection shape (ODS) of the spillway, while a physical model experiment is conducted to ensure the reliability of numerical simulation by experimental modal analysis (EMA). In this study, results obtained using numerical simulation and physical model experiments were compared.

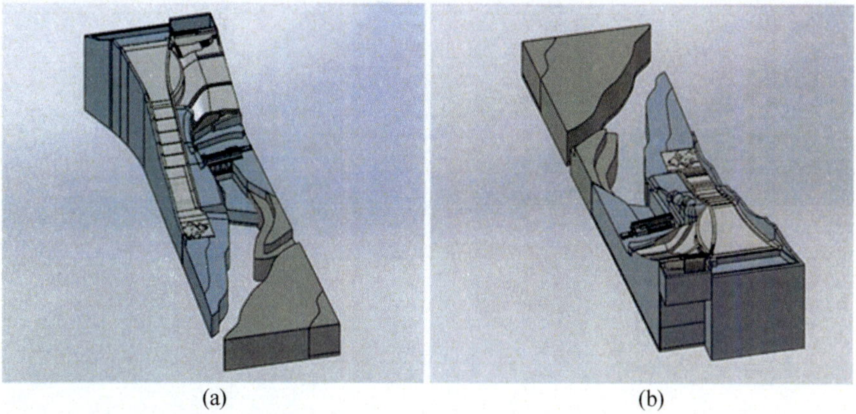

Fig. 1 **a** Downstream and **b** upstream 3D views of Kenyir Dam spillway

2 Methodology

2.1 Development of Simulation and Physical Models

The model should preferably be sized appropriately for the purpose of assuring accuracy of similitude and measurements. A 3D model drawing was developed using computer-aided design (CAD) software with the proposed scale as shown in Fig. 1. Operational deflection shape (ODS) was conducted using the 3D model.

The physical model comprising of the immediate vicinity upstream and downstream of the Kenyir Dam spillway was constructed in cement brick and granite with the scale ratio of 1:50 as shown in Fig. 2. The experimental study was focused on the spillway structure.

2.2 Operational Deflection Shape

Deflection shapes of a structure under different operating conditions can be determined by the analysis of ODS. It contains both forced and resonant vibration components. It also could detect the cause of excessive levels of vibration and/or structural stress. This study is important in order to verify the simulation result of the real spillway. Since the actual experimental is difficult to be conducted at the real site, a downscaled physical model is constructed and both experimental and simulation are compared and verified.

Fig. 2 Physical model of Kenyir Dam spillway

2.3 Experimental Modal Analysis

EMA is an experimental method to determine the dynamic characteristics of the dam structure in terms of natural frequencies, mode shapes and FRF. It determines the natural frequencies and mode shapes of a structure and verify accuracy and calibrate a finite element model (FE). The instrumentation used for the EMA includes the impact hammer, accelerometer, calibrator, LMS software and LMS SCADAS as illustrated in Fig. 3. The roving hammer method was used in this measurement, where the accelerometer is fixed at one point and the impact hammer was knocked from one point to another to complete the measurement. There were total of 9 points of each location for the measurement. This analysis was conducted to determine the mode shapes, natural frequencies and FRF graph.

3 Result and Discussion

3.1 Operational Deflection Shape Analysis

Figure 4 shows the meshed FE model of the 1:50 scale spillway structure. From the modal analysis result, the most dominant mode shape was observed at the wall of the spillway at the downstream area as illustrated in Fig. 5. This mode shape

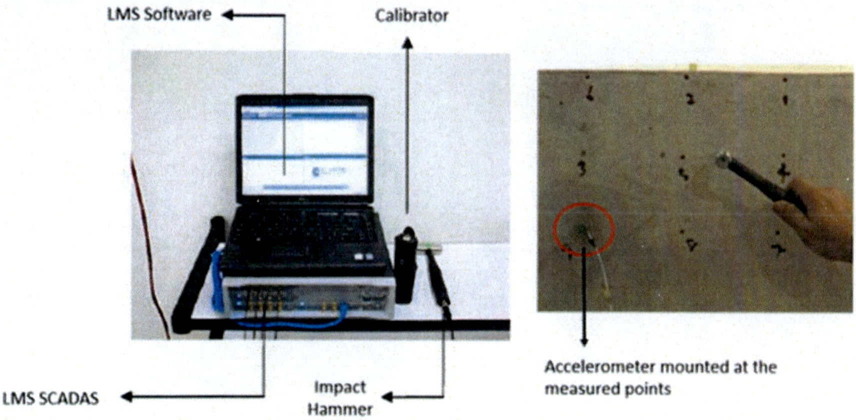

Fig. 3 Instrumentation for EMA and location of measurement

corresponded to the natural frequency of 2801.1 Hz and cause an overall deflection amplitude of 0.353 m. In term of FRF graph as shown in Fig. 6, it is clear that the highest deflection occurred at *z*-axis direction with amplitude value of 9.9187×10^{-15} m.

Figure 7 shows the relationship between the mode shape of modal analysis and ODS of FSI analysis for the water spilling of 2.92 m (the equivalent water level of 146 m in the real spillway). During this water spilling, the transient vibration effect with deflection amplitude value of 5.5×10^{-8} m at the operating frequency of 40.39 Hz did not coincided with the natural frequency of the spillway at 2801.1 Hz; hence, there was no resonance phenomenon existed.

Fig. 4 Meshed FE model of the spillway

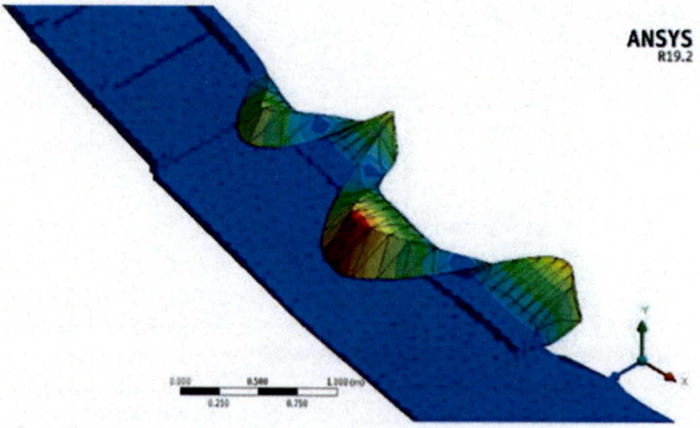

Fig. 5 Mode shape at natural frequency of 2801.1 Hz

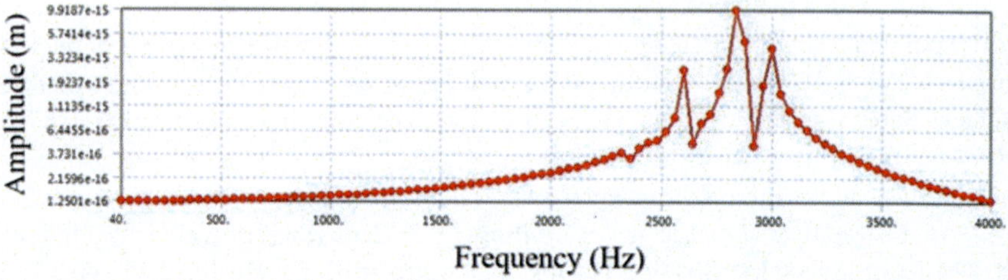

Fig. 6 FRF graph for spillway section in z-axis direction

Figure 8 shows the relationship between the mode shape of modal analysis and ODS of FSI analysis for the water spilling of 2.96 m (the equivalent water level of 148 m in the real spillway). During this water spilling, the transient vibration effect with deflection amplitude value of 3.39×10^{-9} m at the operating frequency of 32.94 Hz did not coincided with the natural frequency of the spillway at 2801.1 Hz; hence, there was no resonance phenomenon existed.

3.2 *Experimental Modal Analysis*

The location for EMA measurement was at downstream and upstream dividers as shown in Fig. 9. The results of the EMA for each of the physical model dam locations are shown in Figs. 10 and 11.

Based on the results obtained, the following summary can be made as given in Table 1. The locations were determined identically as in the simulation for validation purpose. The percentage difference for experiment and simulation results for EMA at downstream part is 1.35%, while at upstream part is 1.76%. It shows that the

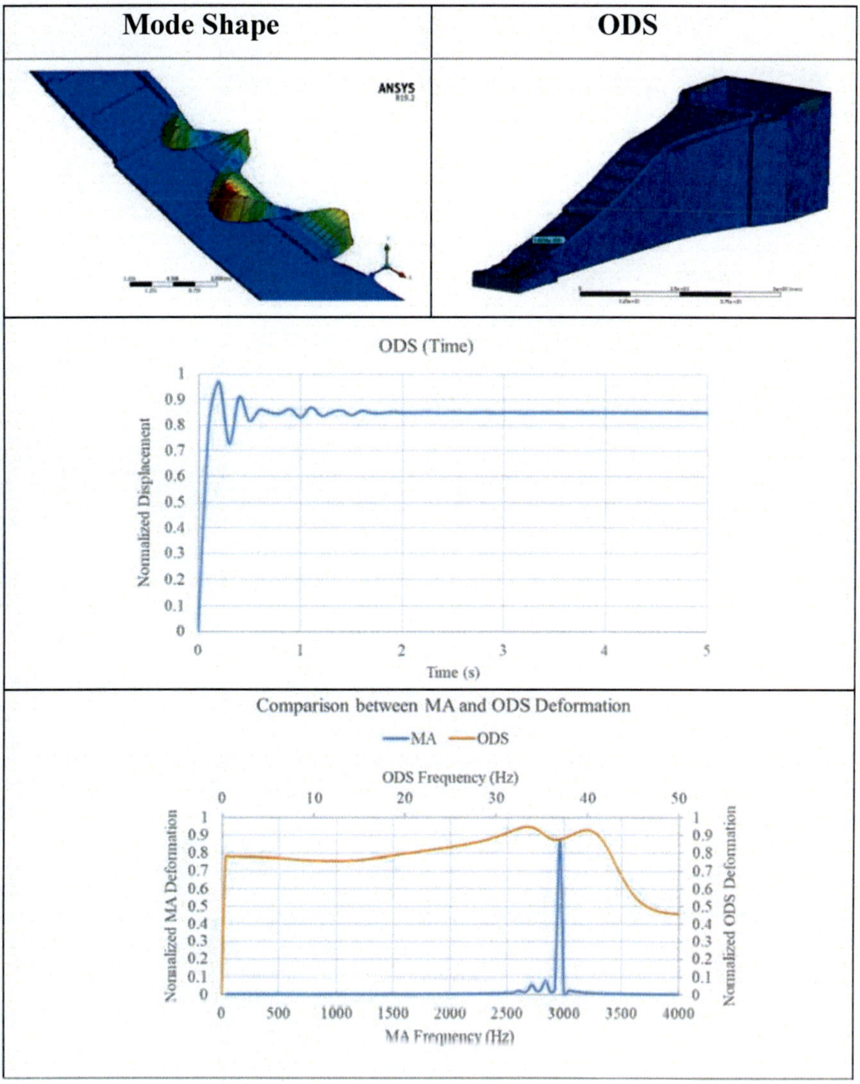

Fig. 7 Relationship between mode shape and ODS of spillway (2.92 m)

percentage differences for all parts are below 10%, which is generally below an allowable error percentage for experimental and simulation studies comparison.

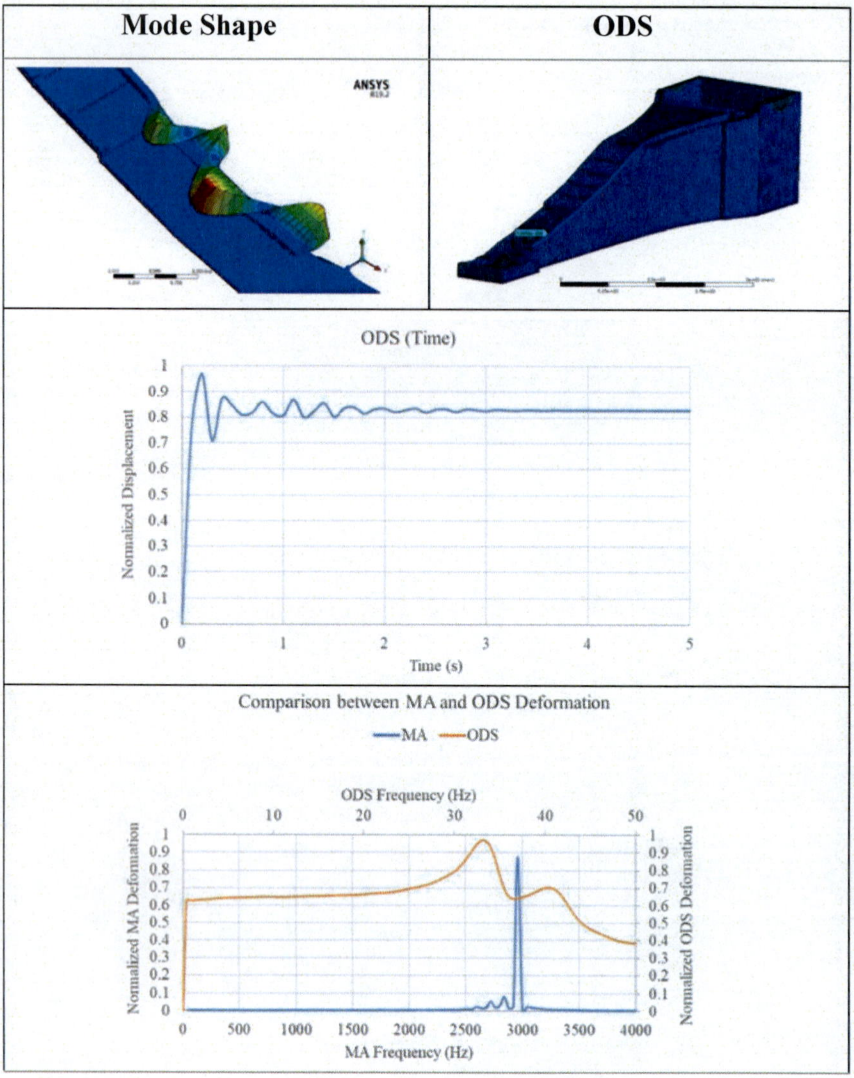

Fig. 8 Relationship between mode shape and ODS of spillway (2.96 m)

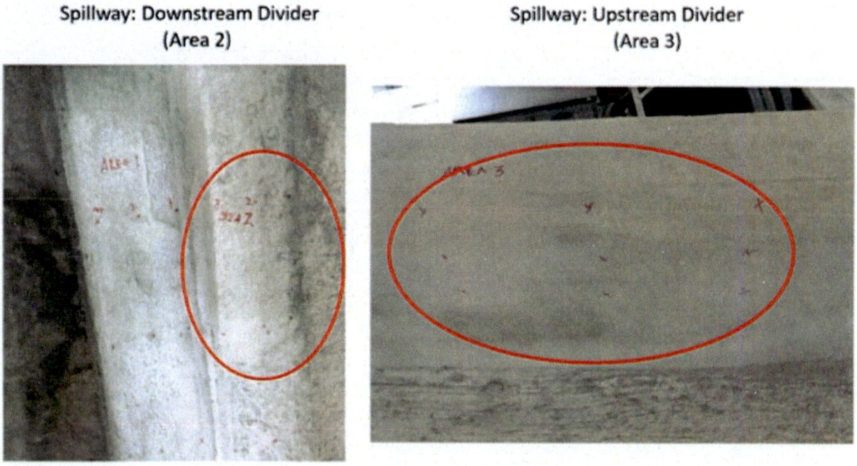

Fig. 9 Location of EMA measurement at downstream and upstream divider of the spillway

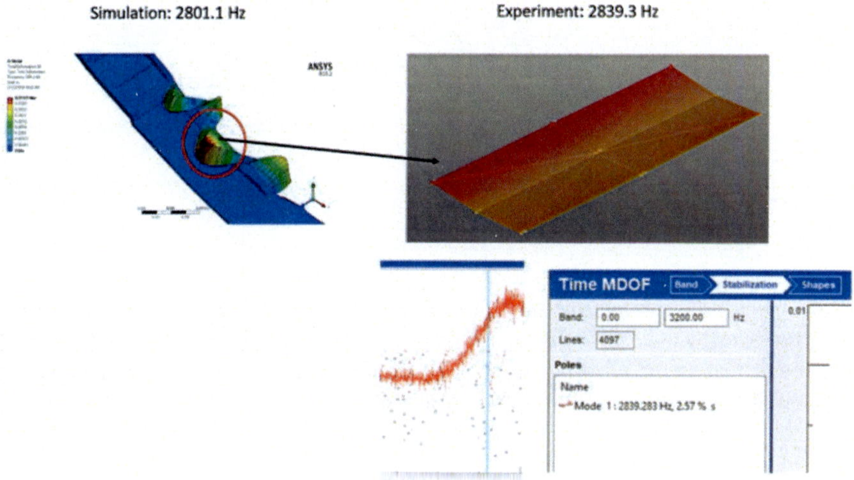

Fig. 10 EMA results comparison between simulation and experiment at downstream divider

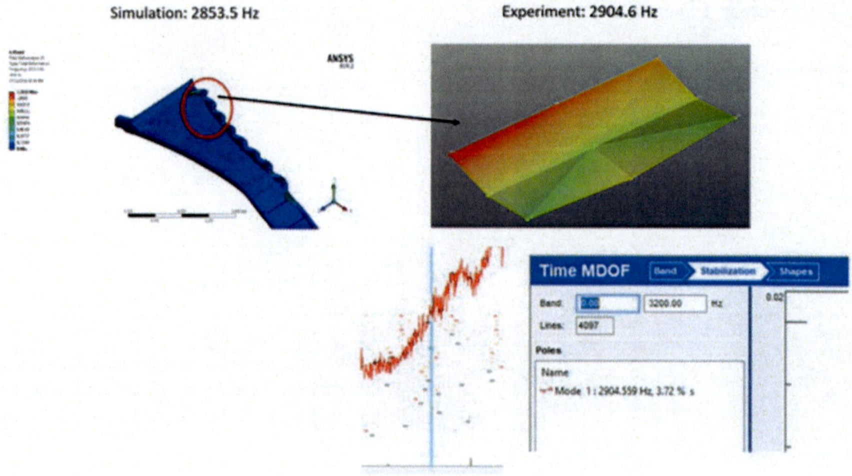

Fig. 11 EMA results comparison between simulation and experiment at upstream divider

Table 1 EMA result summary between experiment and simulation

Locations	Experiment (Hz)	Simulation (Hz)	Percentage differences (%)
Downstream	2839.3	2801.1	1.35
Upstream	2904.6	2853.5	1.76

4 Conclusion

From the results, the natural frequencies and mode shapes for all the parts show an agreement to the build physical model of the Kenyir Dam spillway. It can be concluded that the result obtained from physical model experiment validates the simulation values.

References

1. Lewin, J.: Spillway gate design features which can cause vibration. In: Ensuring Reservoir Safety into the Future: Proceedings of the 15th Conference of the British Dam Society, pp. 145–156 (2008)
2. Lian, J., Zheng, Y., Liang, C., Ma, B.: Analysis for the vibration mechanism of the spillway guidewall considering the associated-forced coupled vibration. Appl. Sci. **9**(12) (2019)
3. Mirabi, M.H., Akbari, H., Alembagheri, M.: Detailed vibrational analysis of unbalanced morning glory spillways using coupled finite volume-finite element method. SN Appl. Sci. **3**(1), 1–16 (2021)
4. Crookston, B.M., Anderson, A., Shearin-Feimster, L., Tullis, B.P.: Mitigation investigation of flow-induced vibrations at a rehabilitated spillway. In: ISHS 2014-Hydraulic Structures and

Society-Engineering Challenges and Extremes: Proceedings of the 5th IAHR International Symposium on Hydraulic Structures, pp. 1–8 (2014)
5. Kiricci, V., Celik, A.O.: Modeling hydraulics structures with Computational Fluid Dynamics (CFD). vol. 2014, pp. 585–591 (2014)
6. Arias, I., Knap, J., Chalivendra, V.B., Hong, S., Ortiz, M., Rosakis, A.J.: Numerical modelling and experimental validation of dynamic fracture events along weak planes. Comput. Methods Appl. Mech. Eng. **196**(37–40), 3833–3840 (2007)
7. Heiner, B.J., Shupe, C.C.: El Vado Dam—service spillway modification—physical model study bureau of reclamation. (2017)
8. Kumcu, S.Y.: Investigation of flow over spillway modeling and comparison between experimental data and CFD analysis. KSCE J. Civ. Eng. **21**(3), 994–1003 (2017)
9. Zaki, S.A.A., et al.: Numerical and physical model analysis comparison for velocity of water at spillway. Water Resour. Dev. Manag. 408–416 (2020)

Implication of Uncertainty in River Gauged Data and the Rating Curve Representations to Flood Quantiles: Case Studies from Stations Across Peninsular Malaysia

Ammar Ulwan Mohd Alayudin and Balqis Mohamed Rehan

Abstract River discharge data are important in hydrological modeling and flood frequency analysis (FFA), e.g., in model tuning and determination of flood quantiles. A time series of river discharge is commonly generated from the application of a rating curve of gauged water level and discharge at a specific river cross-section. The curve fit using the data points inevitably comes with errors because of the imperfect knowledge on the stage and velocity measurements during individual gauge, for example, due to scour/fill, vegetation, and backwater effects. The errors challenge the reliability of a single rating curve over the governing dataset. This study aims to analyze how uncertainty in the rating curve generated from a complete record of gauged data can provide additional information in flood frequency studies when consider error bounds. A methodology was developed to tailor the aim of the study to the available data from stations across the Peninsular Malaysia. All river gauging stations in Peninsular Malaysia under the Department of Irrigation and Drainage (DID) databank were first examined and finally selected based on a set criteria (e.g., length of data availability and missing data). Uncertainty in the rating curve was represented by error distribution with upper and lower limit curves. A time series of discharges was then generated using the rating curves. Results show that there are 15 gauging stations from the overall 103 across Peninsular Malaysia complies with the conditions set for this study. The findings suggest that explicit consideration of rating curve uncertainty in streamflow data is potentially useful in hydrological modeling and flood frequency studies to manage long-term water security.

Keywords Uncertainty · Time-variant rating curve · Single rating curve

A. U. Mohd Alayudin · B. Mohamed Rehan (✉)
Department of Civil Engineering, Universiti Putra Malaysia, Serdang, Malaysia
e-mail: balqis@upm.edu.my

N. Sabtu (ed.), *Proceedings of AWAM International Conference on Civil Engineering 2022 - Volume 3*, Lecture Notes in Civil Engineering 386,
https://doi.org/10.1007/978-981-99-6026-2_49

1 Introduction

Accurate estimation of discharge in rivers is essential for hydrological and hydraulic analyses. This includes in flood forecasting, where well-defined stage-discharge curves based on a large range of flow observations are required [1], and in designs, where highly uncertain rating curves could cause costly over-designed or dangerously under-designed infrastructures or flood defenses [2]. Furthermore, the generated flows from rating curves can also affect results of climate change impact studies, where the flow time series serves as a basis of hydrological model calibration and validation [3]. How it influences the statistical characteristics of streamflow could cause a decrease/increase in projected extreme events [4]. Therefore, understanding the hydrologic data uncertainty is necessary to be able to not under or overestimate predictions.

The stage and discharge measurements used to drive rating curve's parameters are subjected to epistemic error [5, 6]. Epistemic error in rating curve is the imperfect knowledge on the stage and velocity measurements during individual gaugings that can be due to several factors, such as weed growth, scour and redeposition of the bed gravels in a braided river, and unrestricted high flows [7]. The factors lead to dispersion in stage-discharge gauging data and an unstable relationship between stage and discharge, which exacerbate over a long period [8–10]. The assumptions made in the development of a rating curve, such as the type of model used to fit the observed data can also contribute to uncertainty. There are different methods to analyse the errors caused by uncertainty in a stage-discharge relationship. An extensive literature review of the methods for estimating uncertainty in the rating curves can be found in [11, 12]. In principle, a good representation of a stage-discharge curve would need to be constructed from a large range of measurement data [13]. These errors could affect the derived streamflow time series [14].

Majority of river basins in Malaysia has lack of sufficient data to develop a reliable rating curve, hence the inevitable uncertainty in the developed rating curve. Flood management and implementation of a control measure is under the jurisdiction of the DID. The current practice by DID to construct a daily mean streamflow time series is by using periodic data points to establish a rating curve that result in time invariant rating curve for the entire historical years. Most of the time it is based on yearly data points that result in yearly rating curve (YRC) across the recorded years. It is not known specifically what parameter used in the YRC model fit in the system, but the general equations used is known to be either power or polynomial function depending on the best fitted model. Although this method captures the variability of rating curves over time, the downfalls of having smaller data points could lead to unrealistic representation of rating curves. This raises a question whether a generated streamflow time series could be improved with a single rating curve accompanied by an error range (SRCU) that leverage upon a greater number of data points.

The aim of this study is to analyze how uncertainty in rating curves generated from a full recorded gauged data can better represent extreme flow data. This was achieved by first investigating the characteristics of hydrometric data across the Peninsular

Malaysia and analyzing the statistical difference of flow time series derived from a changing rating curve over the records, YRC, and a single rating curve over the records, SRCU. Finally, flood quantiles of flood frequency analysis derived from the generated flow time series of YRC were compared against the SRCU.

2 Study Area

Malaysia covers an area of about 329,758 km^2 consists of 11 states in Peninsular Malaysia and two states in Borneo (Sabah and Sarawak). Peninsular Malaysia spans approximately 500 miles from north to south and around 200 miles from east to west with a tropical climate. More than one-half of the total area is covered by mountains with the tallest mountain over 500 feet above the sea level [15]. The temperature is moderated by the presence of the surrounding oceans [15], while humidity is usually high, and the average annual rainfall is 2500 mm (98 in) [16].

Floods in Malaysia are one of the most regular natural disasters affecting the country, which occurs nearly every year especially during the monsoon season. Widespread flooding seen in many states in Malaysia in the last year, 2021 was caused by heavy rainfall which creates extremely high river flow. Massive flood losses across nearly every state in Peninsular Malaysia due to flood inundation impact lives, properties, buildings, infrastructures, and others.

There are 103 hydrometric stations across Peninsular Malaysia with different period of records. Some of the hydrometric stations has historical records of below 20 years length. Figure 1 shows the location of 78 active hydrometric stations with above 20 years historical records of water level in Peninsular Malaysia. There are 14 hydrometric stations located in urban areas, while the other 64 are located in rural areas.

3 Methodology

The methodology in Fig. 2 was designed in such a way that the derived streamflow time series using the yearly rating curve (YRC) and single rating curve with uncertainty bands (SRCU) can be compared. Exploratory data analysis (EDA) is the first step in the methodology to understand the raw data characteristics before analyzing them further. The EDA includes gathering and eliminating invalid information, organizing data for processing, and selecting the stations for analysis. The second step is to develop rating curves using readily available rated tables from DID (i.e., data points of measured water level and flow using undefined power or polynomial functions) and gauged data (i.e., data points of water level and computed flow using velocity-area method) with consideration of uncertainty bounds. Historical water level and developed rating curve were then used to generate streamflow time series and next to estimate flood quantiles.

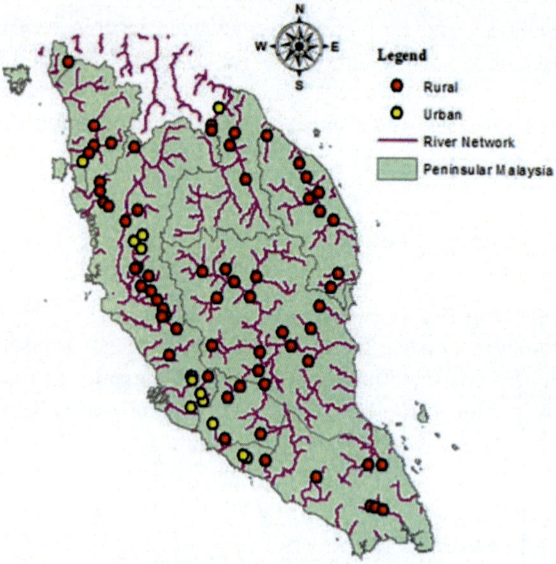

Fig. 1 78 active hydrometric stations in Peninsular Malaysia

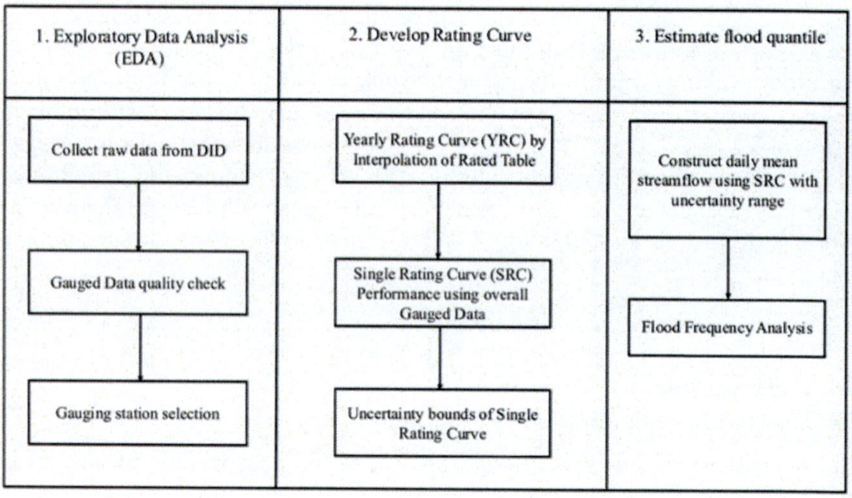

Fig. 2 Overall research methodology

3.1 Exploratory Data Analysis

Collect raw data from Department of Irrigation and Drainage (DID)

Several data were collected from the DID which are gauged data (individual gauging at site), daily mean water level, daily mean streamflow, rated table (represents YRC), and riverbed level above datum. The data were retrieved from the Tideda system, or a website managed by DID called Hydronet H_2O, whichever available. This study targets the annual maximum series of historical streamflow data to determine flood quantiles, which requires long records of hydrological information.

Gauged data quality check

In preparing for the SRCU rating curve model fit, gauged data from each DID's station was investigated. One example of the time series is shown in Fig. 3. Invalid data was observed through eyeball test and realistic justification. For example, those below the riverbed level and with recorded water level as zero when gauged discharge is non-zero, and vice versa was removed. The remaining gauged data was then used for rating curve model fitting using the power function.

Gauging station selection

Each hydrometric station comes with daily mean streamflow data, daily mean water level, gauged data, rated table, and riverbed level. Hydrometric stations that have insufficient information on gauged data and rated table were not included in this study. There are various factors that lead to insufficient information, such as poor maintenance and measurement errors when setting up equipment. In this case, we are focusing on the automated daily mean water level data that were collected at the hydrometric stations to identify insufficient information. The percentage of missing data set in this study is 1–10% of daily mean water level time series, as this is considered as minor or negligible information loss [17]. Programming code was developed in R Software to analyse the missing data. The minimum record for water levels was also set to be 20 years for each station and stations with historical records less than this was discarded from further analysis. Numerous studies have used limited historical data in their studies, usually between 10 and 20 years [18]. Finally, the water level station with less than 10% of missing data and more than 20 years of available record was brought forward for rating curve model fits. Rating curves of power regression were fitted onto the gauged data of each station and followed by further screening using the regression coefficient. Only water level stations with

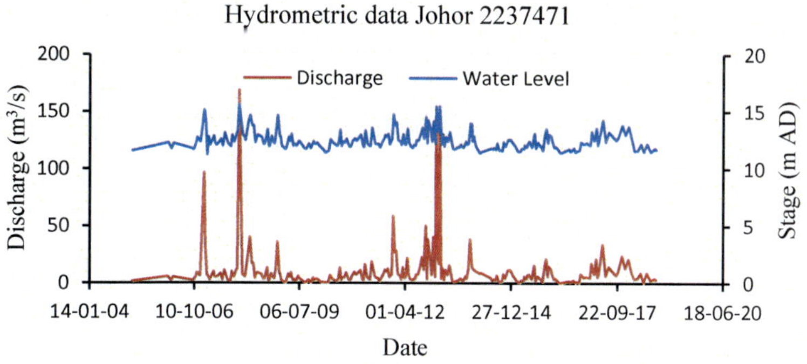

Fig. 3 Example of gauged data plot of water level and discharge

regression coefficient, R^2, more than 0.7 were brought forward. In summary, the stations that were selected are those with missing daily water level data less than 10%, historical records of more than 20 years and R^2 of rating curve above 0.7.

3.2 *Development of Single Rating Curve with Uncertainty (SRCU)*

Single Rating Curve (SRC) performance

Power function is suitable to provide an accurate representation of rating curve [19]. It is the type of algebraic equations that are commonly fitted to stage-discharge data [20] and has the advantage of being a monomial function (i.e., only one parameter needs to be estimated). Another suitable function is through polynomial model fit, but it requires estimation of several parameters that increase the variance and complexity. Once the model fitting on stage-discharge datapoints and determination of *R*-squared are completed, a smaller group of hydrological stations were selected based on model performance ratings, as given in Table 1. In this study, hydrological stations with regression coefficient, R^2, below 0.7 were eliminated. Figure 4 shows examples of the single fitted rating curve for stations in Johor, where Fig. 4a with a regression coefficient $R^2 > 0.7$, and Fig. 4b with $R^2 < 0.7$.

Table 1 Model performance ratings [21]

Value range	Performance classification
$0.7 < R^2 < 1$	Very good
$0.6 < R^2 < 0.7$	Good
$0.5 < R^2 < 0.6$	Satisfactory
$R^2 < 0.5$	Unsatisfactory

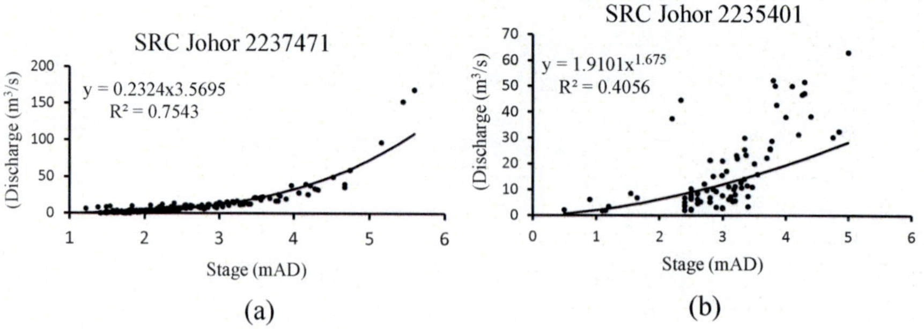

Fig. 4 Single rating curve plot

Uncertainty bounds

In order to quantify discharge uncertainty stems from the uncertain single rating curve, a standard deviation of assumed error distribution is set as in Eq. (1). The variance of the distribution was set to yield a confidence interval three times of the error distribution from the 'true' discharge [22]. The error distribution is truncated, as in Eq. (2). Equations (3), (4), and (5) are the power function that includes the error distribution for the uncertainty bound. An example of developed uncertainty boundaries are shown in Fig. 5.

$$\sigma = 0.04Q \tag{1}$$

$$3\sigma = 0.12Q \tag{2}$$

$$\text{Single Rating Curve} = kx^{P} \tag{3}$$

$$\text{Upper Quantile} = 0.12kx^{P} + kx^{P} \tag{4}$$

$$\text{Lower Quantile} = kx^{P} - 0.12kx^{P} \tag{5}$$

where k is constant of variation and P is power of function.

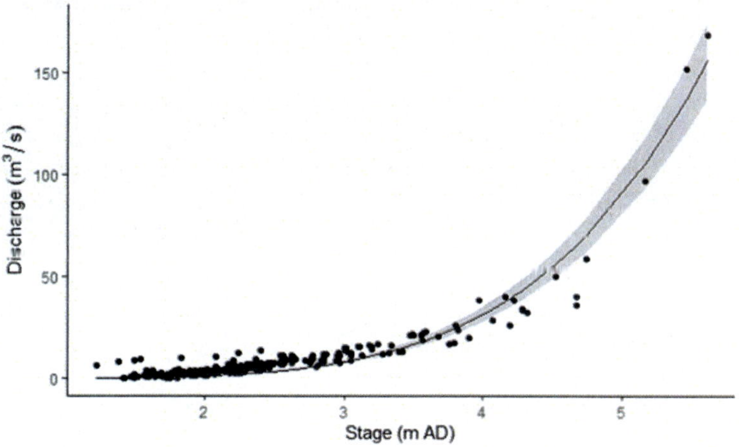

Fig. 5 Single rating curve plot with uncertainty bands (SRCU)

3.3 Estimation of Flood Quantile

Annual Maxima Streamflows using SRCU

R is a highly convenient tool for implementing various statistical operations. It is attractive for various data science applications because it provides aesthetic visualization tools. New daily mean flow data of each selected gauged station was derived from the SRCU priorly developed (Eqs. (3), (4), and (5)) and the station's daily mean water level. From the daily mean flow data, annual maxima streamflow series were then extracted using *R* and compared to the annual maxima flow series given by DID, which were derived from YRC.

Flood Frequency Analysis

Frequency analysis plays an important role in hydrological modeling and has been applied in calculation of design values for dams and many other hydraulic constructions. GEV distribution was used to represent the underlying probability of the annual maxima flows in this study. The GEV distribution parameters, i.e., location, scale, and shape, specify the distribution center, the mean deviation, and the distribution tail behavior, respectively. GEV distribution is the limiting distribution for block maxima dataset. In this study, it was parameterized by using L-moments. L-moments are a conventional way to estimating parameters in probability distribution, and the parameters are more accurate for small samples and subject to less bias in estimation. L-moments, being linear functions of the data, are less sensitive than the classical moments to sampling variability or measurement errors in the extreme data values. Figure 6 gives the mathematical equation behind FFA using GEV distribution. Flood quantiles were then estimated for a range of return period.

Where *u* is the location parameter, ∝ is the scale parameter, and *k* is the shape parameter.

Cumulative density function (CDF)	$F(x) = exp\left[-\left(1-\kappa\frac{(x-u)}{\alpha}\right)^{\frac{1}{\kappa}}\right]$
Quantile estimation	$x(F) = \begin{cases} u + \frac{\alpha}{\kappa}(1-(-\log(1-F))^{\kappa}), & \kappa \neq 0 \\ u - \alpha\log(-\log(1-F)), & \kappa = 0 \end{cases}$
Probability density function (PDF)	$y = \begin{cases} (1-\frac{\kappa(x-u)}{\alpha})^{\frac{1}{\kappa}}, & \kappa \neq 0 \\ e^{-\frac{x-u}{\alpha}}, & \kappa = 0 \end{cases}$ $f(x\|u,\alpha,\kappa) = \frac{1}{\alpha}y^{1-\kappa}e^{-y}$

Fig. 6 General form of GEV for stationary models [23]

4 Results and Discussion

4.1 Rating Curve

Exploratory data analysis using R software shows that only 25 out of 103 hydrometric stations in Peninsular Malaysia have more than 20 years of historical records and at most 10% missing data. Gauged data for the 25 stations were fitted with a power function returned only 15 stations with regression coefficient, R^2, higher than 0.7. Chosen hydrometric stations mapped in Geographic Information System (GIS) is shown in Fig. 7. Two stations located at an urban area.

The performance of SRCU and YRC is given in Table 2. Majority of the hydrometric stations show that the YRC propagates steeper than the SRCU over the water level, except for one station in Pulau Pinang. The reason of the steeper curve is because the number of gauged data is less at the upper water level range as compared to the lower water level range. The small number of samples agrevate the performance of YRC. This is, however, exceptional for the one station in Pulau Pinang. The station has a historical daily mean water level of 60 years, but only 62 datapoints of gauged data. DID practice is that, any year without a rating curve will use the previous year's rating curve [24]. The total number of developed YRC was not corresponding to the total number of gauged data, which indicates that each of the developed YRC was not using the same amount of gauged data.

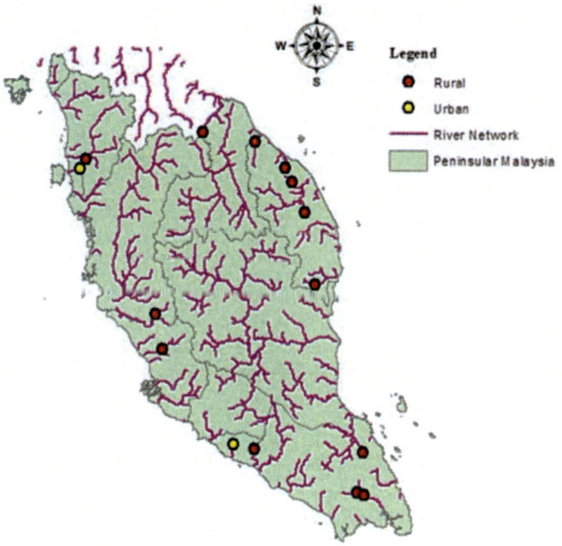

Fig. 7 15 hydrometric stations with < 10% missing data, > 20 years historical daily mean water level and $R^2 > 0.7$ for single rating curve

Table 2 Characteristics of input data and performance of single rating curve

Region	Station ID & Year Start-End	Rural/Urban	%NA Daily	Total Year	River Bed (m)	No. of Gaug-ed Data	Gauge Data		No. of Yearly Rating Curve availa-ble	Single Rating Curve R^2	Remark
							Start	End			
Johor	2237471 (1970-2019)	Rural	6.68	50	10	250	2005	2018	19	0.9438	SRC within YRC
	1836402 (1977-2019)	Rural	0	43	6	236	2005	2018	19	0.7551	YRC steeper
	1737451 (1970-2019)	Rural	8.2	50	0.8	268	2005	2018	26	0.7566	YRC steeper
Kelan-tan	5818401 (1991-2019)	Rural	9.2	29	17.7	67	2004	2017	17	0.8316	YRC steeper
Melaka	2322413 (1960-2019)	Urban	10	60	2	237	2002	2018	29	0.7262	YRC steeper
	2224432 (1960-2019)	Rural	9	60	3.6	225	2002	2018	32	0.8933	SRC within YRC
Pulau Pinang	5505412 (1961-2018)	Rural	5.1	57	0.8	104	2004	2019	31	0.8825	SRC within YRC
	5405421 (1960-2019)	Urban	6.9	60	4.5	62	2002	2015	30	0.9217	SRC Steeper
Selan-gor	3813411 (1961-2019)	Rural	5.8	59	14.5	326	2004	2019	24	0.9143	SRC within YRC
	3414421 (1960-2019)	Rural	6.7	60	3	337	2001	2019	23	0.8722	SRC within YRC
Ter-engga-nu	5724411 (1992-2019)	Rural	7.2	28	2	68	2009	2017	8	0.9266	SRC within YRC
	5428401 (1979-2019)	Rural	7.7	41	5	190	2004	2017	21	0.7265	YRC steeper
	5229436 (1982-2019)	Rural	3.7	38	6	165	2004	2013	21	0.8471	YRC steeper
	4930401 (1991-2019)	Rural	5.4	29	19	166	2004	2017	9	0.9237	YRC steeper
	4131453 (1997-2019)	Rural	9.6	23	9	88	2004	2017	12	0.8577	YRC steeper

Type I: SRC within YRC
Type II: YRC steeper
Type III: SRC steeper

4.2 Derived Annual Maxima Time Series

Type I: Single Rating Curve with Uncertainty bands within Yearly Rating Curve.

There are six hydrometric stations that are in the category of SRCU within the YRC. Figure 8 shows an example of such occurences. The stations under this criteria found to be governed by a large number of gauged data points. The overlapped of YRC and SRCU increases the reliability of the fitted curve. The variations of high and low flows bring a positive impact to the generated rating curve(s) and create a more accurate relation across the whole range. In Fig. 9, an example of the time series of reconstructed and original annual maxima streamflow (AMSF) is shown. It can be seen that the time series of AMSF generated from the two different source of rating curves show similar patterns and small magnitude difference.

Type II: Yearly Rating Curve steeper than Single Rating Curve.

Most hydrometric stations are in the category where YRC is steeper than SRCU. The mathematical structure of the polynomial model has causes the generated flow to dramatically increases as the water stage increases. Meanwhile, a power model fit causes a steadier flow increment along the higher water level. Due to this behavior, the SRCU yields a lower slope than the other rated YRCs. As an example, Figure 11 shows that the reconstructed AMSF from SRCU has significantly low magnitude of flows as compared to the AMSF from YRC, except for the period of 2019 due to its specific developed rating curve.

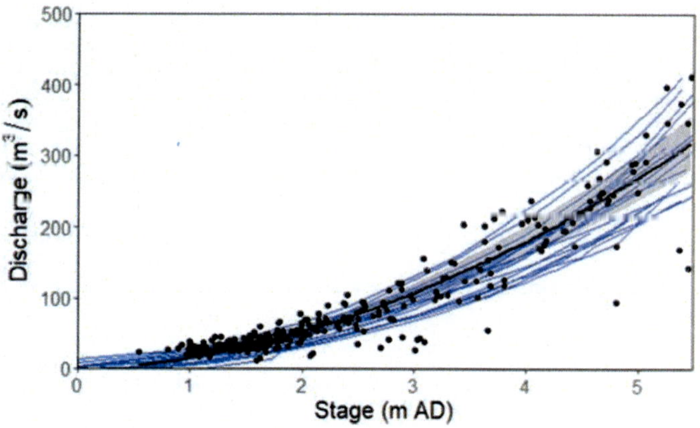

Fig. 8 Rating curve plot for Selangor (ID: 3414421). The blue lines are YRCs, their generated AMS series and the estimated flood quantile curve. The black lines are SRCU, with the gray-shaded envelope is the uncertainty bands

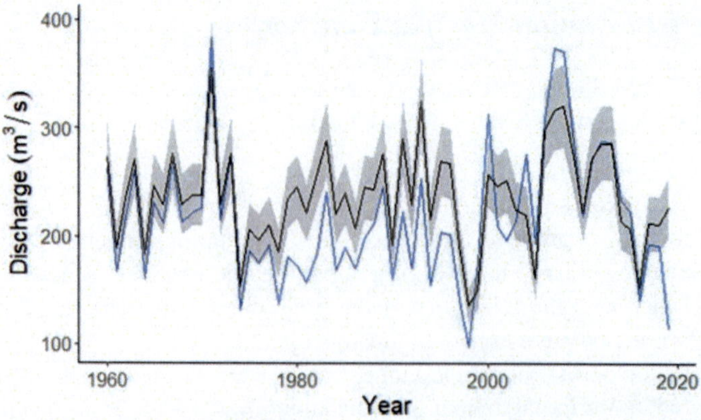

Fig. 9 Annual maxima streamflow (AMSF) time series plot for Selangor (ID: 3414421). The blue lines are the AMSF generated using YRCs. The black line is the AMSF generated using SRC. The gray-shaded envelope is the AMSF generated using the uncertainty bands of SRC

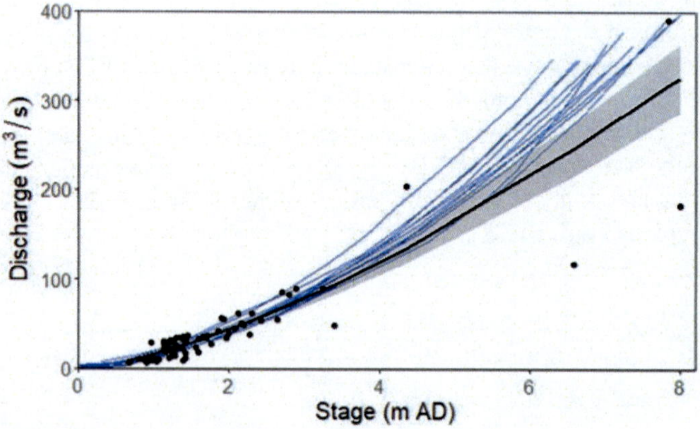

Fig.10 Rating curve plot for Terengganu (ID: 4131453). The blue lines are YRCs and their generated AMS series. The black lines are SRCU with gray-shaded envelope of the uncertainty bands

Type III: Single Rating Curve with Uncertainty bands steeper than Yearly Rating Curve.

Only one hydrometric station is in the category where SRCU is much steeper that the YRC. An example is given in Figure 12. It illustrates that the lower gauged data points increase the variation in the YRC. In another example shown in Fig. 13, the reconstructed AMSF shows significantly higher flows than the original AMSF for the years 1960–2006. Due to the high variation of YRC developed between the year 2006 and 2019, the AMSF shows relatively high magnitude of flows.

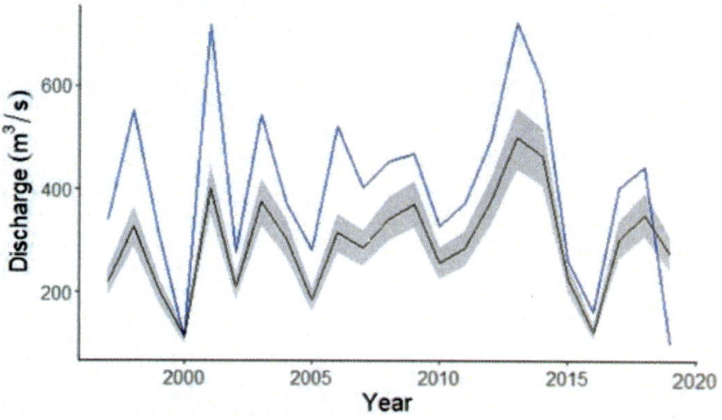

Fig.11 Annual maxima streamflow (AMSF) time series plot for Terengganu (ID: 4131453). The black line are the AMSF generated using SRCU with the grey-shaded envelope is the AMSF generated using the uncertainty bands

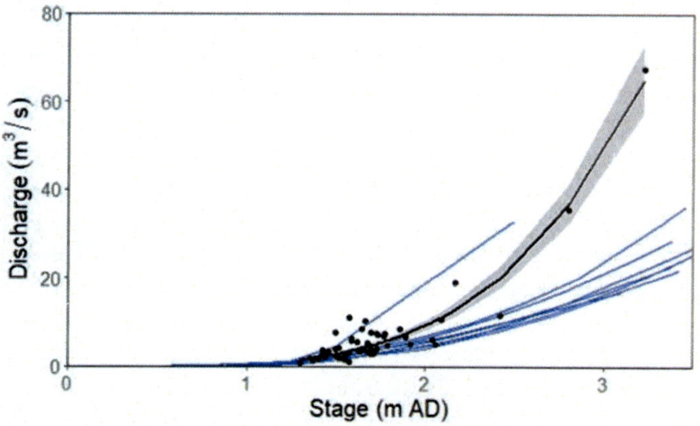

Fig. 12 Rating curve plot for Pulau Pinang (ID: 5405421). The blue lines are YRCs and the black lines are SRC. The gray shaded envelope is the uncertainty bands of SRCU

4.3 Flood Quantiles

4.3.1 Annual Maxima Streamflow

Table 3 gives statistical comparisons of all hydrometric stations that were analyzed. All 15 stations shows decrement in coefficient of variation for the derived annual maxima streamflow (AMSF) using SRCU compared to the AMSF derived using YRC. The range between minimum and maximum of AMSF is also decreasing when using SRCU. Extreme difference between minimum and maximum magnitude of AMSF from YRC is generally questionable and some seems to be illogical and

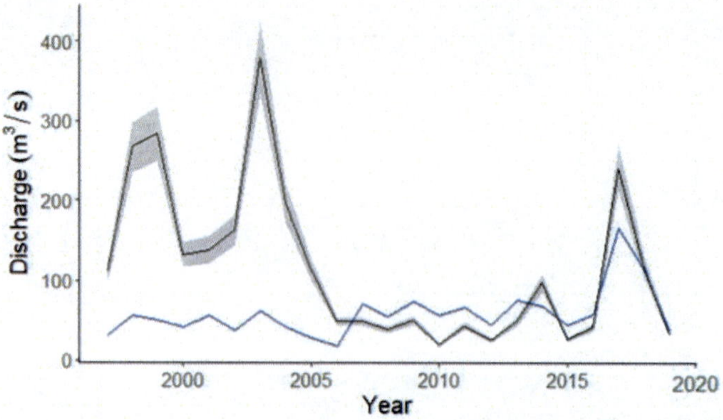

Fig. 13 Annual maxima streamflow (AMSF) time series plot for Pulau Pinang (ID: 5405421). The black line are the AMSF generated using SRC. The gray-shaded envelope is the AMSF generated using the uncertainty bands of SRC.

impossible to be justified as a result of physical alterations. As an example, some with magnitude difference above 2000 m^3/s between the minimum and maximum AMSF (station 2237471). Yet, the same station shows reduced magnitude difference from SRCU of about 250 m^3/s between the minimum and maximum.

The results suggests that SRCU can provide more reliable and realistic AMSF compared to the one using YRC. YRC may have limited high values due to the small samples governing the curves. Furthermore, the limited samples inevitably include 'outliers' from measurement errors and further extrapolation can lead to 'too high' or 'too low' extreme flows for a given stage. Meanwhile, SRCU is driven by a large number of datapoints, that can even out the effects of measurement errors from the larger samples despite it downfall of not able to capture time-changing rating curve. When essential information on what causes the measurement discrepancies and measurements are surrounded with uncertainty, consideration of deviation/uncertainty range of errors using SRCU can improve representation of streamflow time series.

4.3.2 Flood Frequency Analysis

Figures 14, 15, and 16 are examples of the flood quantile estimates for stations classified in each I, II, and III category, respectively. Flood quantiles estimated from the GEV distribution of AMSF generated using YRC and SRCU show significant variations between type II and type III. This is due to the high variations of YRCs, which influence the development of streamflow and affecting the flood quantile estimates. On the other hand, the performance of SRCU is influenced by the good model fit (R-squared is higher than 0.7). Moreover, the results indicate that the uncertainty of streamflow affects the flood quantile estimates, hence the danger of over or under

Table 3 Statistical output of annual maxima streamflow time series derived from YRC and SRCU

Region and station ID	Hydrological time series	Annual maxima					
		Mean	SD	CV	Min	Max	Max–Min
Johor	Water level	15.4	0.7	4.6	13.8	17.2	
2,237,471	Streamflow (YRC)	181.0	383.9	212.1	8.5	2188.9	2180.4
	Streamflow (SRCU)	105.6	51.5	48.8	28.1	269.6	241.5
Kelantan	Water level	22.9	1.3	5.7	19.3	24.7	
5,818,401	Streamflow (YRC)	236.4	111.8	47.3	13.8	480.0	466.2
	Streamflow (SRCU)	56.1	23.1	41.2	5.9	93.7	87.8
Melaka	Water level	5.6	0.8	13.7	3.1	7.4	
2,322,413	Streamflow (YRC)	50.7	28.4	56.0	1.2	142.6	141.4
	Streamflow (SRCU)	30.8	10.9	35.3	3.5	62.5	59.0
P. Pinang	Water level	7.5	1.3	16.8	4.6	9.6	
5,505,412	Streamflow (YRC)	534.0	238.2	44.6	132.1	1317.3	1185.2
	Streamflow (SRCU)	635.8	231.5	36.4	178.8	1070.6	891.8
Selangor	Water level	19.0	0.6	3.1	17.5	21.0	3.5
3,813,411	Streamflow (YRC)	188.5	76.2	40.4	28.8	568.2	539.4
	Streamflow (SRCU)	202.4	58.2	28.7	77.2	422.2	345.0
Terengganu	Water level	8.1	2.2	27.4	2.7	12.9	
5,724,411	Streamflow (YRC)	1132.3	749.0	66.2	87.5	2627.0	2539.5
	Streamflow (SRCU)	865.5	441.5	51.0	75.6	2112.7	2037.1

estimate the flows using YRC. YRC could over or underestimate the extreme flows because of the high variation of AM series. Meanwhile, stations classified as type I show similar magnitude of estimated flood quantiles using either YRC or SRC. Although both results in a comparable flood quantiles, SRCU is arguably preferable than YRC as it is governed by larger sample sizes and accounts for uncertainty.

The extrapolation of high flows when using YRC is subjected to serious errors if not controlled. The over or under estimates resulting from it can have significant implications on flood management and water resources planning. Under conditions of limited stage-discharge information and pooling fewer flow measurements for the rating curve, the YRC unable to produce realistic variability of extreme flows leading to the erroneous estimates of flood quantiles [25]. This highlights that continuous monitoring of data quality is essential. When flow data are imprecise or inaccurate, unnecessary costs or poor ecological or societal outcomes follow [26]. In cases where

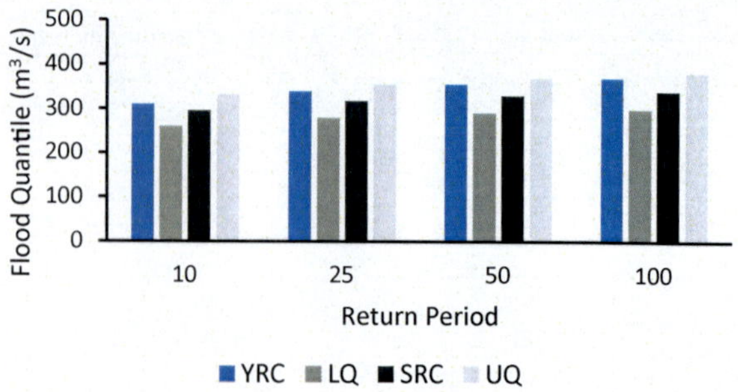

Fig. 14 Flood quantiles in bar chart for a range of return period (Selangor, station ID: 3414421). LQ and UQ stand for lower and upper quantile

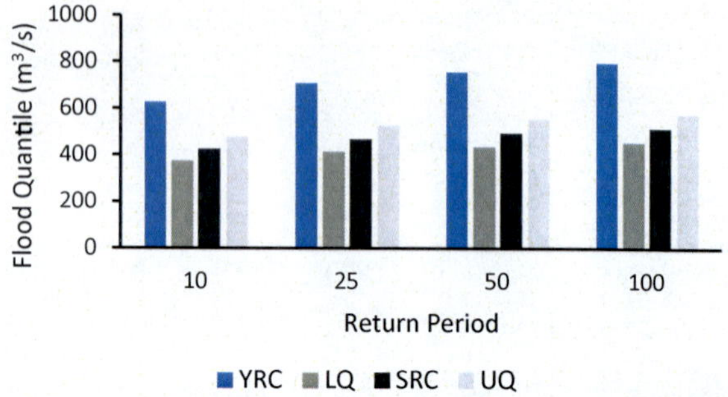

Fig. 15 Flood quantile in bar chart for a range of return period (Terengganu, station ID: 4131453). LQ and UQ refer to lower and upper quantile

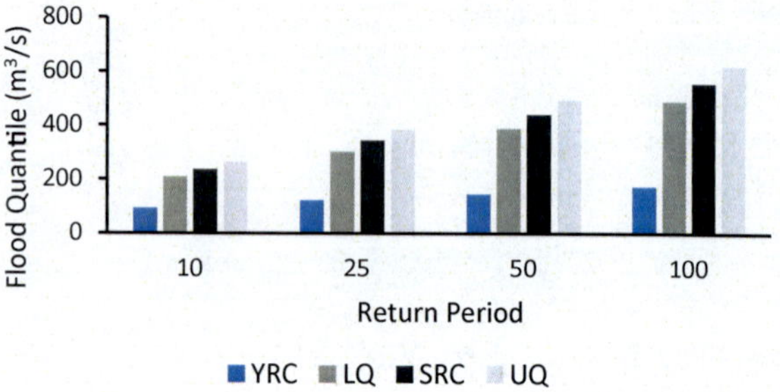

Fig. 16 Flood quantile in bar chart for a range of return period (Pulau Pinang, station ID: 5405421)

historical information is absents, reanalysis of rating curves can provide better representation of long-term flow series and consequently better flood quantile estimates useful for research or application purposes. Overestimating a design flood could lead to higher bridge building costs, for example. Meanwhile, underestimation could lead to a higher failure risk [27]. Uncertainty in rating curves need to be recognized as a way to avoid such over or under estimations, because if not it can lead to erroneous outcome.

5 Conclusion

This study presents the implication of rating curve uncertainty by applying two distinct approaches for rating curve development. The effect of the distinct approaches were carried forward into the associated streamflow time series of gauging stations in Peninsular Malaysia. It was found that the developed single rating curve with uncertainty range (SRCU) enhance the overall reliability of developed rating curve when tested on the basis of the flood discharge. The total number of collected gauged datapoints and the variability across the water level greatly influence the better performance of SRCU as compared to the yearly rating curve (YRC). As for the methods to quantify uncertainty in river discharge measurements that arises from the rating curve's imprecision, the designed methodological framework in this study successfully assessed the uncertainty. The resulting SRCU with consideration of uncertainty range improves implicit uncertainty resulting from rating curves that pushes the way forward for better decisions in flood management-related decision.

References

1. Ocio, D., Le Vine, N., Westerberg, I., Pappenberger, F., Buytaert, W.: The role of rating curve uncertainty in real-time flood forecasting. Water Resour. Res. **53**, 4197–4213 (2017). https://doi.org/10.1002/2016WR020225
2. Jiang, S., Kang, L.: Flood frequency analysis for annual maximum streamflow using a non-stationary GEV model. E3S Web Conf. **79**, 03022 (2019)
3. Harmel, R.D., Smith, P.K.: Consideration of measurement uncertainty in the evaluation of goodness-of-fit in hydrologic and water quality modeling. J. Hydrol. **337**, 326–336 (2007)
4. Serago, J.M., Vogel, R.M.: Parsimonious nonstationary flood frequency analysis. Adv. Water Res. **112**, 1–16 (2018)
5. Coxon, G., Freer, J., Westerberg, I.K., Wagener, T., Woods, R., Smith, P.J.: A novel framework for discharge uncertainty quantification applied to 500 UK gauging stations. Water Resour. Res. **51**, 5531–5546 (2015). https://doi.org/10.1002/2014WR016532
6. Schmidt, A.R.: Analysis of stage discharge relations for open channel flow and their associated uncertainties. Ph.D. Thesis, University of Illionois, Champaign, IL, USA, 2004, p. 328 (2002)
7. McMillan, H.K., Westerberg, I.K.: Rating curve estimation under epistemic uncertainty. Hydrol. Processes **29**, 1873–1882 (2015). https://doi.org/10.1002/hyp.10419
8. Jalbert, J., Mathevet, T., Favre, A.-C.: Temporal uncertainty estimation of discharges from rating curves using a variographic analysis. J. Hydrol. **397**, 83–92 (2011). https://doi.org/10.1016/j.jhydrol.2010.11.031
9. Guerrero, J.-L., Westerberg, I.K., Halldin, S., Xu, C.-Y., Lundin, L.-C.: Temporal variability in stage–discharge relationships. J. Hydrol. **446–447**, 90–102 (2012). https://doi.org/10.1016/j.jhydrol.2012.04.031
10. Fenton, J.D.: On the generation of stream rating curves. J. Hydrol. **564**, 748–757 (2018). https://doi.org/10.1016/j.jhydrol.2018.07.025
11. Le Coz, J.A.: Literature Review of Methods for Estimating the Uncertainty Associated with Stage-Discharge Relations. WMO, Lyon, France (2012)
12. Kiang, J., Gazoorian, C., McMillan, H., Coxon, G., Le Coz, J., Westerberg, I., Belleville, A., Sevrez, D., Sikorska, A., Petersen-Øverleir, A., Reitan, T., Freer, J., Renard, B., Mansanarez, V., Mason, R.: A comparison of methods for streamflow uncertainty estimation. Water Resour. Res. **54**(10), 7149–7176 (2018)

13. Kim, S., Shin, J., Seo, W., Lyu, S.: Development of stage-discharge rating curve using hydraulic performance graph model. In: 12th International Conference on Hydroinformatics, HIC. (2016)
14. McMillan, H., Westerberg, I., Krueger, T.: Hydrological data uncertainty and its implications (2018). https://doi.org/10.1002/wat2.1319
15. Barr, M., Bean, R., Berger, M.T., Yi, C.C., Chong, T., et al.: World and Its Peoples: Malaysia, Philippines, Singapore, and Brunei. Marshall Cavendish Corporation, New York (2008)
16. Saw, S.-H.: The population of Peninsular Malaysia. In: Institute of Southeast Asian Studies, 2nd ed., Institute of Southeast Asian Studies, Singapore (2007)
17. Widaman, K.F.: Missing data: What to do with or without them. **71**(1), 210–211 (2006)
18. Keast, D., Ellison, J.: Magnitude frequency analysis of small floods using the annual and partial series. Water **5**(4), 1816–1829 (2013)
19. De Lorenzo, G., Macchione, F.: The power function for representing the reservoir rating curve: morphological meaning and suitability for dam breach modelling. Water Resour. Manage **30**, 4861–4881 (2016)
20. Stajnko, J., Jecl, R., Perc, M.: Measurement and modeling of suspended sediment dynamics in watercourses (2020). ISSN 1743-3533
21. Moriasi, D.N., Arnold, J.G., Van Liew, M.W., Bingner, R.L., Harmel, R.D., Veith, T.L.: Model evaluation guidelines for systematic quantification of accuracy in watershed simulations. ASABE **50**, 885–900 (2007)
22. McMillan, H., Freer, J., Pappenberger, F., Krueger, T., Clark, M.: Impacts of uncertain river flow data on rainfall-runoff model calibration and discharge pre-dictions. Hydrol. Process. **24**(10), 1270–1284 (2010)
23. Rehan, B.M.: Risk-based flood protection decisions in the context of climatic variability and change [PhD thesis]. University of Oxford (2016)
24. DID. Stage-Discharge Curves. Hydrological Procedure No.10 (2019)
25. Vieira, L.M.D.S., Sampaio, J.C.L., Costa, V.A.F., Eleutério, J.C.: Assessing the effects of rating curve uncertainty in flood frequency analysis. RBRH **27** (2022). https://doi.org/10.1590/2318-0331.272220220012
26. Fatdillah, E., Rehan, B.M., Rameshwaran, P., Bell, V.A., Zulkafli, Z., Yusuf, B., Sayers, P.: Spatial estimates of flood damage and risk are influenced by the underpinning DEM resolution: a case study in Kuala Lumpur, Malaysia. Water **14**, 2208 (2022). https://doi.org/10.3390/w14142208
27. Rehan, B.M., Hall, J.W.: Uncertainty and sensitivity analysis of flood risk management decisions based on stationary and nonstationary model choices. In: 3rd European Conference on Flood Risk Management, FLOODrisk, Lyon, vol. 7, pp. 20003–20012 (2016)

Hydraulic Properties of Flow Over Different Types of Spillways: A Review

Nur Azwa Muhamad Bashar, Mohd Remy Rozainy Mohd Arif Zainol, Mohd Sharizal Abdul Aziz, Mohd Hafiz Zawawi, and Ahmad Zhafran Ahmad Mazlan

Abstract A review of hydraulic properties of flow over the spillway via physical experiment and computational approaches is presented in this paper. The hydrodynamic analysis of flow via the computational fluid dynamics (CFD) approach is vital in solving complex flow problems where the field and experimental assessments are limited. The most selected and efficient turbulence model is the RNG k-ε model. Free water surface profile, Froude number, flow regime, velocity, and pressure distribution were identified as the most studied parameters of flow over the spillway structure. These hydraulic characteristics of flow were affected by the geometrical configuration such as the type and arrangement of the spillway. The skimming flow was recognized as the most frequent type of flow in the actual field. The stepped and hybrid types were identified as the most studied concerning functionality in controlling the velocity and dealing with rapidly varied flow types as compared to the ogee-type. Understanding the hydraulic properties of flow is mandatory and

N. A. M. Bashar
School of Civil Engineering, Engineering Campus, Universiti Sains Malaysia, Seri Ampangan, 14300 Nibong Tebal, Pulau Pinang, Malaysia

School of Civil Engineering, College of Engineering, Cawangan Pulau Pinang, Universiti Teknologi MARA, Kampus Permatang Pauh, 13500 Pulau Pinang, Malaysia

N. A. M. Bashar
e-mail: nurazwa.bashar@student.usm.my; nurazwa.bashar@uitm.edu.my

M. R. R. M. A. Zainol (✉)
River Engineering and Urban Drainage Research Centre (REDAC), Engineering Campus, Universiti Sains Malaysia, Seri Ampangan, 14300 Nibong Tebal, Pulau Pinang, Malaysia
e-mail: ceremy@usm.my

M. S. A. Aziz · A. Z. A. Mazlan
School of Mechanical Engineering, Engineering Campus, Universiti Sains Malaysia, Seri Ampangan, 14300 Nibong Tebal, Pulau Pinang, Malaysia
e-mail: msharizal@usm.my

M. H. Zawawi
Department of Civil Engineering, College of Engineering, Universiti Tenaga Nasional, 43000 Kajang, Selangor, Malaysia
e-mail: Mhafiz@uniten.edu.my

N. Sabtu (ed.), *Proceedings of AWAM International Conference on Civil Engineering 2022 - Volume 3*, Lecture Notes in Civil Engineering 386,
https://doi.org/10.1007/978-981-99-6026-2_50

important to ensure the serviceability of the spillway structure to safely transfer the water to the downstream section and to minimize the massive impacts of hydraulic jump together with the presence of air–water interface, cavitation, and downstream scour problem. The potential future research work includes an in-depth relationship between the hydraulic characteristic of flow between the efficient spillway structure and the baffle block structure in controlling the hydraulic jump in the stilling basin area.

Keywords Spillway · Hydraulic characteristics · Clean water and sanitation · RNG k-ε model · Velocity · Pressure · Skimming flow · Energy dissipation

1 Introduction

Over a few decades, a dam has been identified as one of the main hydroelectric power generation, flood control facilities, and water storage infrastructures to cater to a continuous water supply demand. In addition, the dam itself should be properly built (new dam infrastructure) with effective and efficient management for a sustainable operation (available constructed dam). This promising effort is in line with the Sustainable Development Goals (SDGs) No. 6 and No. 7, namely clean water and sanitation, and clean energy, respectively [1]. This effort was frequently and critically considered due to the current and future sustainability concerns and operability of the infrastructure and its resources. In-depth, a proper policy together with the list of scheduling efforts should be in line for both short- and long-term planning. Thus, this effort requires continuous support from both top to bottom and vice versa including policy implementation (technical design code and guidelines), pre-construction (planning and site survey), during construction (construction method and materials as well as identified problem and limitation), and post-construction (maintenance, operationality, and monitoring aspects including preparedness towards hazard and related system for vibration and cracking detection due internal and external factors).

The functionality and safety related to the structural stability of a dam mainly in an aging condition should be a priority due to its current and future positive and negative impacts on human and environmental factors. Historical and current rainfall input data, actual and predicted flow determination and simulation, and reliable technical design on dam structures (reservoir/water intake, radial gate, spillway, stilling basin, orifice, tailrace, etc.) should be considered in ensuring the functionality and safety of the infrastructure. For instance, the validated data on the hydrostatic force on the dam structure should be identified and estimated correctly by conducting the stability analysis at the technical design stage [2]. Other important factors including maintenance and monitoring (seismic and structural monitoring) efforts [3, 4] are vital with regards to hazard preparedness due to unpredictable global situations concerning climate changes impacts (rainfall patterns imbalanced and dam overtopping due to severe flooding effect) [5–9] and natural disaster problems (earthquake-induced strong ground vibration or shaking) [10, 11].

The spillway is known as one of the dam structural components to provide excess water transfer from the reservoir via inlet to the stilling basin structure and downstream section [12]. Thus, understanding the hydraulic characteristic of flow behavior is crucial to avoid any shortcomings in either short-term or long-term periods. Furthermore, the characteristic of flow result from different spillway types and the installation of an energy dissipation structure at the end of the spillway toe are added to these. Several spillway structures are considered and studied by previous researchers in terms of traditional and innovative design types. The type of traditional spillway structure was ogee-crested, stepped, and siphon spillways [2, 13–17] while the innovative spillway structure includes flat, pooled type, hydraulic-jump-step-spillway, and quarter-circular crested stepped spillway [13, 18–21].

Conventionally, flow characteristics on the hydraulic structure can be measured via field measurement. However, this option is time-consuming and provides limited site access. Therefore, a hydraulic physical model study based on laboratory pilot testing has been introduced to cope with this limitation. The physical experimental model is beneficial as it was studied at an early stage even before the development of the numerical model to understand the actual causal impact situation and be as an actual reference together with actual data acquisition [14, 19–21]. However, this selection is costly and required a longer time due to structural design and construction as well as regular system maintenance in terms of pumps and water tanks together with other related systems to ensure efficient water flow. Thus, numerical simulation provides inexpensive, accurate, and shorter time as it considers computer simulation and offers validation of flow conditions observed during field or physical model experiments [2, 22]. Recently, most researchers consider numerical simulation in dealing with the more complex fluid flow analysis on the spillway [2, 13, 15–18, 20, 22–25].

This paper aims to provide a short review of the hydraulic characteristics of flow (such as flow pattern, Froude number, velocity, and pressure distribution) over a spillway via both physical model experiments and numerical simulation works. The paper is organized as follows: First, an overview of the literature on traditional and modified spillway structures is presented. Second, the hydraulic characteristic of flow is studied based on the experimental and numerical set-up afterward with a conclusion of the reviewed study. The article ends with a future direction of research.

2 Traditional and Modified Spillway Structures

The spillway structure is divided into traditional and hybrid types. Table 1 shows summaries of the different types of spillway design based on the specific purpose and contribution.

Practically, the spillway is designed and constructed either freely or controlled by gated- and ungated-type structures. These gated or ungated types may function

Table 1 Overview of the different types of spillway design, purposes, and contributions

Author(s)	Categories/Type of spillway	Purpose	Contribution
Ghaderi et. al. [18]	Pooled stepped spillways • Flat steps, reference pooled steps (RPS), and notch pooled steps (NFS) • Fully pooled steps (FPS) • Zig-zag pooled steps (ZPS) • Central pooled steps (CPS) • Two-sided pooled steps (TPS)	Impacts of the geometrical characteristics of the flat and pooled step on the: • Energy dissipation • Hydraulic characteristics of flow (flow patterns' properties, velocity rates, and pressure distribution), • Inception point of air entrainment • Turbulent kinetic energy	• Best energy dissipation performance (flat step configuration) • Improved 5.8% of the efficiency performance (notch pooled step configuration) • Smaller interfacial velocities (flat step configuration) • The pressure rate is higher at the entrance of pooled configuration compared to the flat step configuration • The maximum pressure in the notch pooled configuration is decreasing near the step pooled configuration
Zhou et. al. [13]	Hydraulic-jump-stepped spillway (HJSS) Traditional stepped spillway	Performance of the HJSS on the: • Flow pattern • Energy dissipation • Air entrainment characteristics • Time-averaged pressure (horizontal and vertical condition)	• Reduction of cavitation damage • Higher energy dissipation • Better air entrainment performance • Reasonable pressure distribution • 69% increment of unit discharge capacity (traditional step spillway)
Karim and Mohammed [14]	Ogee-crested spillway (OS)	Performance of the OS on the: • Mean velocity • Vertical velocity distribution • Locations of the maximum velocity, d_m	• Understanding of flow over the ogee-crested spillway in a controlled environment for heads higher than the design head and selection of the vertical location for maximum velocity

(continued)

Table 1 (continued)

Author(s)	Categories/Type of spillway	Purpose	Contribution
Imanian and Mohammadian [15]	Ogee-crested spillway (OS)	Performance of the OS on the: • Flow separation • Discharge coefficient	• Understanding of the hydrodynamic field (flow separation or significant streamline curvature) • Flow separation zone grows linearly (increasing hydraulic head) • Discharge coefficient is increased due to decreasing pressure on the ogee crest (increasing of head ratio)
Li and Zhang, 2018 [22]	Pooled stepped spillway	• Understanding	• Understanding of the hydrodynamic field (flow regime, pressure distribution, and vortex detection)
Daneshfaraz and Ghaderi [25]	Inverse curvature ogee spillway	Performance of the inverse curvature OS on the: • Pressure profile	• Understanding the flow behavior on the ogee spillway • Understanding the relationships between pressure with inverse curvature • Maximum pressure on the inverse curve is reduced by increasing the radius of curvature (economical cost related to radius curvature construction)
Felder and Chanson 2016 [20]	Stepped spillway • Flat uniform and non-uniform configuration • Flat, pooled, and porous pooled steps • Combination of flat and pooled steps	Impacts of the geometrical characteristics of the flat, pooled, porous pooled, and combination of flat and pooled step on the: • Energy dissipation • Hydraulic characteristics of flow (flow patterns properties, velocity rates) • Inception point of air entrainment	• Understanding the flow behavior on the different configurations of flat uniform and non-uniform as well as flat, pooled, porous pooled, and combination of flat and pooled steps

(continued)

Table 1 (continued)

Author(s)	Categories/Type of spillway	Purpose	Contribution
Krisnayanti et al. 2016 [21]	Stepped spillway • Flat uniform • Pooled steps	Impacts of the geometrical characteristics of the flat and pooled step on the: • Energy dissipation • Hydraulic characteristics of flow (flow patterns properties, velocity rates)	• Understanding the flow regime on the different configurations of flat and pooled steps as well as energy dissipation
Johnson and Savage 2006 [2]	Ogee spillway	Performance of the OS (with the presence of tailwater) on the: • Discharge • Pressure	• Understanding the flow behavior, velocity, and pressure distribution on the ogee spillway with the submergence spillway
Savage and Johnson 2001 [17]	Ogee spillway	Performance of the OS (without the presence of tailwater) on the: • Discharge • Pressure	• Understanding the flow behavior, velocity, and pressure distribution on the ogee spillway without the submergence of the spillway

as an overflow structure mainly during the full capacity of the reservoir. The difference between these two types is that the ungated type may allow water to freely overflow from the fully occupied reservoir, while the gated one is operated based on the full capacity level and follow the water release schedule [12]. Conventionally, the ogee-crested spillway is the most frequently selected and designed type chosen by the engineer due to its broad functionality (higher amount of water being released), serviceability, and economical factors [2, 13, 15–19, 23–25]. Most of the conducted studies were done experimentally and numerically by referring to the standard guidelines and previous literature for results comparison and validation. The problem with ogee-type spillways is both uncontrolled and controlled gate structures may contribute to the energy dissipation problem and scour effect at the downstream section of the spillway. This problem is getting worsened with the presence of fluid jet impacts from upstream to the transfer section or downstream section due to the sloping condition of the ogee-crest spillway [12]. Thus, a focused study related to this matter is done by Daneshfaraz and Ghaderi [25] on the inverse curvature curve ogee spillway to study the relationship between the effects of pressure distribution with the modification of the radius of curvature concerning the economic aspects.

The introduction of a steps spillway provides an improvement on the flow characteristic distribution (three-dimensional) and is suitable for the skimming flow regime [18, 20]. In addition, an effort has been conducted to improve the energy dissipation capacity by designing a more reliable and efficient stepped spillway either single or hybrid concepts. Thus, this innovative spillway design provides a higher ability to convey the massive and high-speed flood flow from the upper section to the end of the spillway section by protecting the respected section from scour and cavitation problems as well as reducing the hydraulic jumps or fluid turbulent impacts [18, 22]. Interestingly, the cavitation problem is eliminated or reduced concerning the hydraulic jump reduction (length).

The innovative stepped spillway, for instance, pooled steps with several configurations (fully, zig-zag, double, etc.), and hydraulic-jump-stepped spillway (HJSS) are designed with the novel steps, and the system has given further attention to global researchers based on the resulting energy dissipation improvement. Thus, the focus on the traditional and innovative step spillway via experimental and numerical study is significantly increased due to the complicated fluid flow problem such as the air–water interface, the flow regime, and the energy loss process due to the hydraulic jump problem [20, 22]. The design of stepped spillways of both traditional and innovative has stressed the geometrical step and configuration. This is the main key factor for the high energy dissipation rate. The energy dissipation may be improved by the improvement of the amount of entrained air. The higher energy dissipation leads to a reasonable and appropriate downstream stilling structure concerning the construction cost.

In conclusion to this point, the most effective and gain interesting spillway structure among the researcher for energy dissipation improvement is the stepped spillway. This interest has been driven by the higher energy dissipation rate condition with the presence of several steps on the structure. The stepped spillway is more robust as compared to the ogee-type (free fall flow) where it could control the velocity jet and deal with the rapidly varied flow type. In addition, the stepped type is better in handling skimming flow problems, which is the actual condition of flow on-site. Furthermore, the innovation of single and hybrid stepped may enhance the better understanding of flow especially related to the three-dimensional complex hydrodynamic problem. In addition, the problem related to an air–water interface may be reduced or solved depending on the situation and condition. Thus, a proper investigation of the geometrical configuration of the stepped spillway should be conducted to differentiate the effect between them. The focus also may be extended to connect the ogee type with the stepped type (hybrid type), and the relationship between these combinations may be studied to evaluate the performance of these hybrid spillway types in terms of energy dissipation rate and air–water interface problem.

3 Remarked Hydraulic Properties of Flow Over Spillway Structures

The hydraulic characteristics of flow over a spillway structure were summarized in Table 2.

Table 2 Hydraulic properties of flow studied by different researchers

Author(s)	Type of study/Model	Q (m^3/s)/ Unit Q (m^2/s)	V (m/s)	Fr. No	Re. No	Flow regime	Free water surface profile and streamlines	P (KPa)	Turbulent kinetic energy, (J/kg)	C_d
Ghaderi et. al [18]	Numerical models—CFD: FLOW-3D® model—volume of fluid (VOF) Hydrodynamic model—three-dimensional Navier–stokes equations [Reynolds-averaged Navier–Stokes (RANS)] and the continuity equation Turbulence model—RNG k-ε	√	√	√	√	√	√	√	√	NA
Zhou et. al. [13]	Hydraulic physical model (inflow channel, weir, aeration basic, stepped spillway, outflow channel) Scale: 1:40	√	√	√	NA	√	√	√	NA	NA
Karim and Mohammed [14]	Physical model and numerical model	√	√	NA	NA	√	√	√	NA	√
Imanian and Mohammadian [15]	Numerical models—CFD: OpenFOAM—volume of fluid (VOF) Turbulence model—standard k-ε, realizable k-ε, k-ω SST and LRR (Launder–Reece–Rodi) RNG k-ε	√	√	NA	NA	√	√	√	NA	√

(continued)

Table 2 (continued)

Author(s)	Type of study/Model	Q (m^3/s)/ Unit Q (m^2/s)	V (m/s)	Fr. No	Re. No	Flow regime	Free water surface profile and streamlines	P (KPa)	Turbulent kinetic energy, (J/kg)	C_d
Li and Zhang [22]	Numerical models—CFD: ANSYS 16.0—volume of fluid (VOF) Turbulence model—RNG k-ε	√	√	√	NA	√	√	√	NA	NA
Daneshfaraz and Ghaderi [25]	Numerical models—CFD: fluent—volume of fluid (VOF) Turbulence model—RNG k-ε	√	√	NA	NA	NA	√	√	NA	NA
Krisnayanti et al. [21]	Physical experimental model	√	√	√	NA	√	√	NA	NA	NA
Felder and Chanson [20]	Physical experimental model	√	√	√	√	√	√	NA	NA	NA
Johnson and Savage [2]	Numerical models—CFD: FLOW-3D® model—volume of fluid (VOF) Existing literature	√	√	NA	NA	√	√	√	NA	NA
Savage and Johnson [17]	Physical model, numerical model, and existing literature	√	√	NA	NA	√	√	√	NA	NA

The understanding of the hydraulic characteristic of flow is vital, especially for design purposes. The previous design of hydraulic structures especially spillways only depended on the technical guidelines and design handbook. However, the sustainability of the hydraulic structure (ability to function and serve within its operational and design life) provides an interest among the researchers to focus on the area. In addition, natural disasters including climate change impacts show an alert to the post-construction (maintenance and monitoring purposes) activity. The presence of hydraulic jump phenomena may contribute to the scour and cavitation problem in the downstream section [18, 21].

Conventionally, the actual site assessment is conducted together with the experimental work. However, in dealing with the complex flow problem and on-site limitation (safety purposes) plus limited experimental work related to complex flow problems, a numerical simulation may be a good and alternative option. Numerical analysis is a robust technique with a complex flow solution (hydrodynamic analysis especially for 3D flow). An arrangement or configuration has been done numerically without any doubt via the specific model. Based on the review data, most of the researchers used several types of computational fluid dynamics models either 1D, 2D, or 3D flow models based on the basic equation of three-dimensional Navier–Stokes equations: Reynolds-averaged Navier–Stokes (RANS) and continuity equation for the hydrodynamic model analysis. The type of most applicable turbulence model is the RNG k-ε model [1, 13, 25]. Furthermore, another researcher also applies other types of models, namely standard k-ε, realizable k-ε, k-ω SST, and Launder–Reece–Rodi (LRR) in addition to the RNG k-ε [15].

The most studied hydraulics parameters are velocity and pressure distribution as shown in Table 2. Based on these parameters, it can be said that there is the main impact of velocity and pressure in the formation of several types of hydraulic jumps. In addition, the most influential point for hydraulic failure is located at the stilling basin section where the water is being transferred from the inlet via spillway structure to the stilling basin. Here the formation of a hydraulic jump occurs (the intensity is dependent on the velocity, pressure, and sloping condition). In this section, it is vital to provide additional structures for energy dissipating purposes. Normally, the type of flow over the spillway can be divided into nappe, transition, and skimming flow. Interestingly, the actual flow condition on the field is the skimming-type flow which is affected by the site condition.

Daneshfaraz and Ghaderi [25] conducted a study related to the impact of inverse curvature on the pressure rate and found that the maximum pressure is achieved by increasing the inverse linear curvature. This study was less investigated by the researcher and provided a long-term value as it would be an economical option. Other studies are related to the hydraulic characteristic (discharge characteristics and pressure distribution) of flow over submerged ogee with the effect of tailwater, via physical mean and numerical simulation in comparison with the previously established standard guidelines and literature. In addition, the usage of a numerical model for hydrodynamics analysis accurately simulates the flow and pressure distribution and gives significant findings especially related to the understanding of pressure distribution related to a submergence case study [2]. The pressure distribution and discharge rate are affected by the crest and spillway geometry, upstream flow depth, together with the downstream depth. The analysis of pressure on the submerged spillway is important due to the resisting force to overturn or slide at the end of the spillway section [2].

Zhou et al. [13] study on the effect of time-averaged pressure on the face of a stepped spillway: horizontal and vertical conditions via physical model experimental. The obtained result shows the pressure distributions on the horizontal step surfaces exhibited S-shaped variations with the maximum and minimum values occurring at the downstream end and upstream start, respectively. The study on this variation is good since the whole direction of pressure on the spillway was analyzed and identified. Some studies and experiments were also performed on flat and pooled stepped spillways, providing information on flow patterns and dissipation performance [18, 20, 21].

Ghaderi et al. [18] investigated the simulated velocity, pressure, and total kinetic energy distributions agreed with the previous literature and within the consistency of the experimental data. The free water surface of flow was identified horizontally on flat and FSP steps while unstable in other configurations. The appearance of streamlines is decreasing on notch pooled steps due to the presence of vortex conditions. The maximum velocity was detected on the surface of the pool steps, while the minimum velocity value occurred in the bottom section of each of the steps. Maximum pressure was found to be higher in the beginning section of the pooled step compared to the flat step, while no negative pressure was found throughout the horizontal step section of flat and pooled types. The lowest pressure is also improved on the horizontal and vertical sections of the notched pool step surface compared to the flat and simple pooled. Thus, the cavitation risk due to the water pressure is mitigated. The excellent energy dissipation was founded on notched pool steps as compared to others. Thus, making this type of geometry can be one of the great options for energy dissipating structures.

In conclusion, most of the changes in the hydraulics characteristics of flow were contributed by the different types of the spillway, geometrical configuration, and flow characteristics including discharge and Froude number, length of the steps (for the stepped spillway). The skimming flow type is said to be the real type of flow regime on the actual engineering site compared to other types of flow (nappe and transition). Furthermore, the rapidly varied flow was found to be experienced in the flow over the spillway due to the sloping condition from the inlet to the end of the spillway connected to the stilling basin section. The understanding of the hydraulic characteristic of flow over the spillway is crucial mainly related to the complex flow problem. Furthermore, the application of numerical analysis for hydrodynamic flow provides an alternative to solving fluid flow problems (from simple to complicated case studies) other than experimental and actual site determination. Furthermore, it is of utmost importance in choosing the energy dissipation structure for the industrial application within the scope (problem-solving strategies), time, and cost. The stepped spillway and hybrid spillway is found to be the best option in terms of energy dissipation structure as compared to the ogee type. This is said to be the best option for the industrial practitioner if there are no issues related to the budget. However, the appropriate option such as the beneficial contribution of the inverse curvature ogee-type study is found to be economically feasible and reasonable to be applied as the industrial application or dam authority is facing constraints with the budget.

4 Conclusion

In conclusion, the understanding of the characteristic of flow over the spillway is a crucial part before the design (new dam infrastructure), repairing of the available structure, and implementation stage (construction and post-construction). The structural functionality for a maximum and optimum safe water transfer in a hydraulic structure (from the reservoir via spillway to the downstream section) is of utmost importance due to the risky condition. The connection between the fluid and structural aspects is crucial in dealing with the cavitation and scour effect due to the hydraulic jumps problem at the end of the spillway toe. This knowledge is essential when planning and dealing with natural disasters and hazard preparedness mitigation strategies (short- and long-term periods). Furthermore, the hydraulic characteristic of flow is useful as a guideline to choose the most applicable energy dissipation structure at the optimum function within the quality, scope, and budget. The arrangement of geometrical configuration is important to achieve the best option for energy dissipation structure. The numerical modeling of flow mainly the hydrodynamic simulation and analysis is vital before the current and future practices to overcome the field and hydraulic model limitation. In addition, the frequent changes in flow conditions due to environmental changes may affect the important parameters including Froude number, velocity, and pressure distribution.

5 Future Direction of Research

The future direction of the study is listed and explained based on the review conclusion and potential research gap based on the influence factor, variables as well as case study, and method. The potential research work includes an in-depth relationship between the hydraulic characteristic of flow between the efficient spillway structure and the baffle block structure in the stilling basin area. Other is the critical review of the hydraulic jump types and efficient energy dissipating structure.

References

1. UN.: (2015). https://www.un.org/development/desa/disabilities/envision2030.html. Last accessed 20 Nov 2021

2. Johnson, M.C., Savage, B.M.: Physical and numerical comparison of flow over ogee spillway in the presence of tailwater. J. Hydraul. Eng. **132**(12) (2006). https://doi.org/10.1061/(ASCE)0733-9429(2006)132:12(1353)
3. Oliveira, S., Alegre, A.: Seismic and structural health monitoring of Cabril dam. Software development for informed management. J. Civ. Struct. Health Monit. **10**(3) (2020). https://doi.org/10.1007/s13349-020-00425-0
4. Wei, P., Lin, P., Peng, H., Yang, Z., Qiao, Y.: Analysis of cracking mechanism of concrete galleries in a super high arch dam. Eng. Struct. **248**, 113227 (2021). https://doi.org/10.1016/j.engstruct.2021.113227
5. Kostecki, S., Banasiak, R.: The catastrophic of the Niedow dam—the causes of the dam's breach, its development, and consequences. Water **13**, 2524 (2021)
6. Ahn, J.M., Lee, S., Kang, T.: Evaluation of dams and weirs operating for water resource management of the Geum River. Sci. Total Environ. **478**, 103–115 (2014). https://doi.org/10.1016/j.scitotenv.2014.01.038
7. Daneshi, A., Brouwerb, R., Najafinejad, A., Panahi, M., Zarandian, A., Maghsood, F.F.: Modelling the impacts of climate and land use change on water security in a semi-arid forested watershed using InVEST. J. Hydrol. **593**, 125621 (2021). https://doi.org/10.1016/j.jhydrol.2020.125621
8. Li, W., Zhu, J., Fu, L., Zhu, Q., Guo, Y., Gong, Y.: A rapid 3D reproduction system of dam-break floods constrained by post-disaster information. Environ. Model. Software 104994 (2021). https://doi.org/10.1016/j.oceaneng.2020.108545
9. Devkota, B., Imberger, J.: Upper and middle Tiete river basin dam-hydraulic system, travel time and temperature modeling. J. Hydrol. **475**, 12–25 (2012). https://doi.org/10.1016/j.jhydrol.2012.07.025
10. Chai, S., Tian, W., Yu, L., Wang, H.: Numerical study of ground vibrations caused by cylindrical wave propagation in a rock mass with a structural plane. Shock Vibr. 1–9 (2020). https://doi.org/10.1155/2020/4681932
11. Rutqvist, J., Cappa, F., Rinaldi, A.P., Godano, M.: Modeling of induced seismicity and ground vibrations associated with geologic CO_2 storage, and assessing their effects on surface structures and human perception. Int. J. Greenhouse Gas Control **24**, 64–77 (2014). https://doi.org/10.1016/j.ijggc.2014.02.017
12. Chanson, H.: Hydraulics of Open Channel Flow 2nd edn. Butterworth-Heinemann, Great Britain (2004). ISBN No.: 9780750659789
13. Zhou, Y., Wu, J., Ma, F., Qianc, S.: Experimental investigation of the hydraulic performance of a hydraulic-jump-stepped spillway. KSCE J. Civ. Eng. 1–7 (2021). https://doi.org/10.1007/s12205-021-1709-y
14. Karim, R.A., Mohammed, J.R.: A comparison study between CFD analysis and PIV technique for velocity distribution over the standard ogee crested spillways. Heliyon **6**(10), e05165 (2020)
15. Imanian, H., Mohammadian, A.: Numerical simulation of flow over ogee crested spillways under high hydraulic head ratio. Eng. Appl. Comput. Fluid Mech. **13**(1), 983–1000. https://doi.org/10.1080/19942060.2019.1661014
16. Ghafourian, A., Jahromi, H.M., Bajestan, M.S.: Hydraulic of siphon spillway by physical and computational fluid dynamics. World Appl. Sci. J. **14**(8), 1240–1245 (2011)
17. Savage, B.M., Johnson, M.C.: Flow over ogee spillway physical and numerical model case study. J. Hydraul. Eng. ASCE **127**(8), 640–649 (2001). https://doi.org/10.1061/(ASCE)0733-9429(2001)127:8(640)
18. Ghaderi, A., Abbasi, S., Di Francesco, S.: Numerical study on the hydraulic properties of flow over different pooled stepped spillways. Water **13**, 710 (2021). https://doi.org/10.3390/w13050710
19. Parsaie, A., Haghiabi, A.H.: The hydraulic investigation of circular crested stepped spillway. Flow Meas. Instrum. **70**, 101624 (2019). https://doi.org/10.1016/j.flowmeasinst.2019.101624
20. Felder, S., Chanson, H.: Simple design criterion for residual energy on embankment dam stepped spillways. J. Hydraul. Eng. ASCE **142**(4), 04015062, 1–11 (2016). https://doi.org/10.1061/(ASCE)HY.1943-7900.0001107. ISSN 0733–9429

21. Krisnayanti, D, Suhardjono, S., Dermawan, V, Sholichin, M.: Energy dissipation of skimming flow on flat and pooled stepped spillways. J. Teknologi **78**(8) (2016). https://doi.org/10.11113/jt.v78.5973
22. Li, S., Zhang, J.: Numerical investigation on the hydraulic properties of the skimming flow over pooled stepped spillway. Water **10**, 1478 (2018)
23. Ghaderi, A., Abbasi, S., Abraham, J., Azamathulla, H.M.: Efficiency of trapezoidal labyrinth shaped stepped spillways. Flow Meas. Instrum. **72**, 101711 (2020). https://doi.org/10.1016/j.flowmeasinst.2020.101711
24. Saidin, M.S I., Abdul-Aziz, M.S., Zainol, M.R.R.M.A.: Numerical investigation of hydraulic characteristics on 3D unsymmetrical smooth spillway. J. Adv. Res. Fluid Mech. Thermal Sci. **64**(1), 107–116 (2019)
25. Daneshfaraz, R., Ghaderi, A.: Numerical investigation of inverse curvature ogee spillway. Civ. Eng. J. **3**(11), 1146 (2017). https://doi.org/10.28991/cej-030944

Comparison of Multi-satellite Rainfall Data in Runoff Model

Puji Harsanto, Dian Kusumaningrum, Djoko Legono, Adam Pamuji Rahardjo, and Rachmad Jayadi

Abstract This research was performed to analyze the Global Precipitation Measurement (GPM) 3IMERGHHE and the Tropical Rainfall Measuring Mission (TRMM) 3B42RT in a direct runoff model. The area studied was in Gendol River watershed in Indonesia between February 19th, 2017, and February 26th, 2017. The observed rainfall based on Automatic Rainfall Recorder (ARR) and observed water level based on Automatic Water Level Recorder (AWLR) are used to calibrate the excess rainfall by using the phi index method. The results showed that satellite data could describe rainfall spatially. The TRMM 3B42RT data distinctly propagated to streamflow simulations with a significant magnitude of peak flow. The correction of the satellite rainfall data effectively improved the direct runoff model. In general, the GPM 3IMERGHHE is more feasible compared to TRMM 3B42RT.

Keywords Satellite rainfall data · GPM 3IMERGHHE · TRMM 3B42RT · Direct runoff · Gendol watershed

P. Harsanto (✉) · D. Kusumaningrum
Universitas Muhammadiyah Yogyakarta, Yogyakarta, Indonesia
e-mail: puji_hr@umy.ac.id

D. Kusumaningrum
e-mail: dian.kusumaningrum.2015@ft.umy.ac.id

D. Legono · A. P. Rahardjo · R. Jayadi
Universitas Gadjah Mada, Yogyakarta, Indonesia
e-mail: djokolegono@ugm.ac.id

A. P. Rahardjo
e-mail: adam.pamudji.rahardjo@ugm.ac.id

R. Jayadi
e-mail: rjayadi@ugm.ac.id

N. Sabtu (ed.), *Proceedings of AWAM International Conference on Civil Engineering 2022 - Volume 3*, Lecture Notes in Civil Engineering 386,
https://doi.org/10.1007/978-981-99-6026-2_51

1 Introduction

Predicting the flow design, or direct runoff is one of the important steps in many rivers' practical works. Unfortunately, some of the rivers do not have a series of recorded flows. As a substitute, the rainfall-runoff flow model is applied to convert excess rainfall into direct runoff. However, it has a problem due to the availability of rainfall data in short intervals (hourly), Tan et al. [1]. Commonly, in practical work, the analysis uses the rainfall gauge as a representative of the rainfall in a point and it produces a problem because the rainfall has spatial variability, Hou et al. [2].

Recently, satellite remote sensing technology has developed rapidly. Global and regional rainfall products from various satellite missions and projects ranging from a half-hour to monthly resolutions are available at Giovanni, Hou et al. [2]. Furthermore, it provides opportunities for scientists to study and evaluate satellite-based rainfall data, Feyera and Gebremichael [3] and Moazami et al. [4].

Yogyakarta is a city located in Java, Indonesia which has an active volcano, called Mt. Merapi. It is one of the most active volcanoes in the world. Yogyakarta lies at the southern part of the mountain. Every watershed located under the active volcano has sediment hazard and risk for people living along the river, Bélizal et al. [5]. Consequently, estimating the streamflow in the Yogyakarta area is an important effort to prevent damage or losses due to the flood. This study presents the performance of satellite rainfall data to estimate the hydrograph flow in the river. In the future, the results of the research can be used to assist in risk management in Yogyakarta.

2 Study Area

The upstream Gendol River watershed is in the southern slope of Mt. Merapi and at the coordinates 110.4105 W, − 7.7798 N, 110.5015 E, and − 7.5086 S (Fig. 1). The area of the watershed is 62.74 km^2 and the river has a 25 km length.

3 Data and Methods

3.1 Rainfall Satellite Data

The satellite rainfall data that were used in this research are TRMM 3B42RT three-hourly and the GPM 3IMERGHHE half-hourly. TRMM is a collaboration between the United States NASA and NASDA-Japan. The activity is to measure rainfall in the tropical area with a wide range of sensors from orbit and low inclination. GPM is a constellation-based satellite mission specifically designed to provide observations of rainfall data from space to improve the understanding of the hydrology cycle. Satellite-based rainfall data are an average rainfall prediction above an area. This

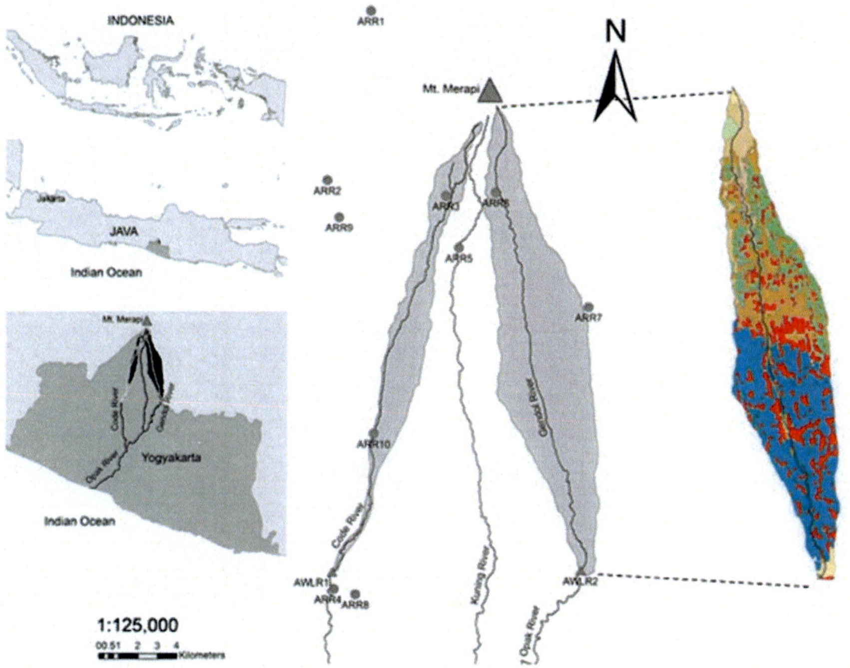

Fig. 1 Watershed and land use in Gendol River

makes the data a prediction that does not necessarily reflect true rainfall. The estimation is based on weather conditions that include clouds, temperature, humidity, and wind. The area is a grid with a rectangular shape. The shape has a grid of 11 km × 11 km. Therefore, conducting the selection of data pairs for both ground-based data and satellite rainfall data with river flow data is important in this research.

3.2 Ground Base Rainfall and Streamflow Data

Ground-based recording rainfall data is a point recording data. Because rainfall is spatially distributed, the rainfall that occurs at the recording station does not reflect the overall rainfall event in the watershed. The selection of data pairs is done by making a graph of flow and rainfall. The pair data is chosen visually to describe a rainfall event and it is followed by a hydrograph with a rising limb and falling limb.

Figures 2, 3, and 4 are rainfall data taken from satellites and Automatic Rainfall Recorder (ARR) stations in the period of the 19th to 29th of February 2017. Data are available every hour. Flow data are collected from the Automatic Water Level Recorder (AWLR). These data come from the Civil Engineering and the Department

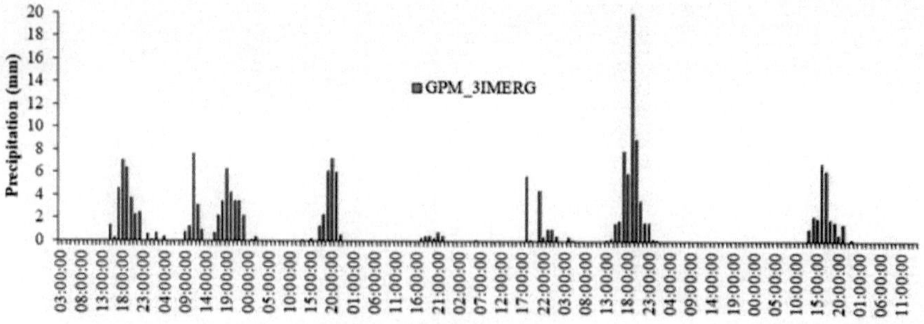

Fig. 2 GPM 3IMERGHHE satellite rainfall data at Gendol River Watershed Yogyakarta within 19–26 February 2017

of Environment at Gadjah Mada University. The selection of rainfall and flow data is based on rainfall events that produce flood hydrographs.

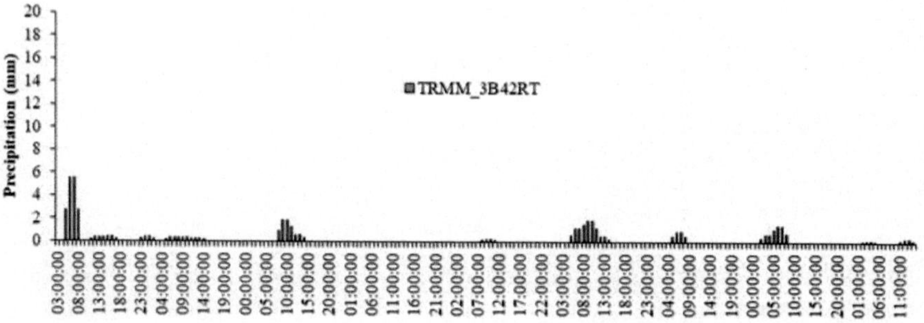

Fig. 3 TRMM 3B42RT satellite rainfall data at Gendol River Watershed Yogyakarta within 19–26 February 2017

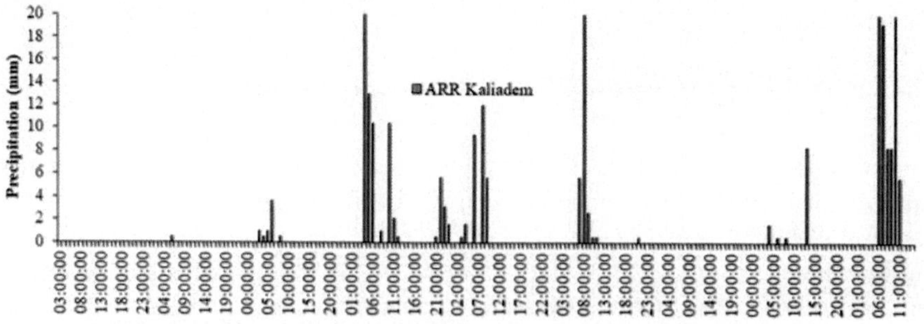

Fig. 4 Ground-based rainfall data at ARR Kaliadem Yogyakarta within 19–26 February 2017

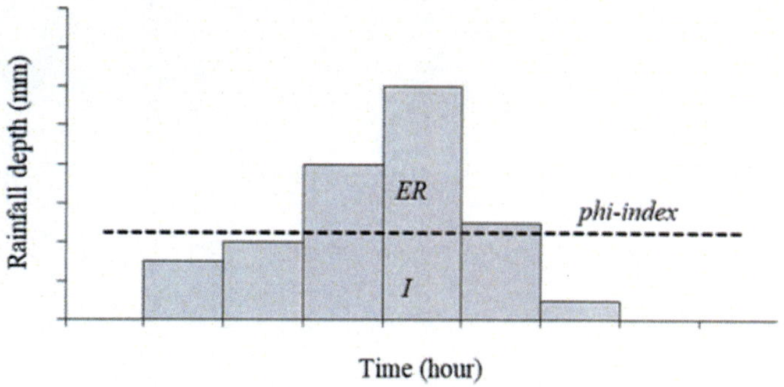

Fig. 5 Phi-index method for calculating effective rainfall

3.3 *Effective Rainfall and Direct Runoff*

Streamflow hydrograph consists of direct runoff (DRH) and base flow (BF), that is, DRH = Streamflow − BF. Furthermore, the effective rainfall (ER) is the rainfall that becomes DRH at the outlet of the watershed. In other words, ER could be defined as the rainfall that is neither retained on the land surface nor infiltrated into the soil and evaporated into the air. DRH and ER represent the same total quantity but in different units. The ER is usually in mm plotted against time (t) and ER multiplied by watershed area (A) gives the total volume of direct runoff which is the same as total volume DRH flow the outlet. The ER is obtained by using the Phi index method (constant loss) to consider the infiltration's effect. The one parameter model relating to runoff is the phi-index model. The Phi index (Fig. 5) of the catchment is defined as capacity of infiltration to produce the actual total runoff for a given amount of rainfall.

The conversion of ER to DRH in this research is based on the instantaneous unit hydrograph (IUH). The concept of the IUH is specifically used for the small duration of the effective rainfall. The unit hydrograph (UH) analysis was later generalized into the instantaneous unit hydrograph; it is used for lumped direct runoff simulation. A unit hydrograph gives the temporal variation of direct surface runoff resulting at a given location in a river from a unit rainfall excess for an effective duration uniformly in the watershed. When this effective duration is infinitely small, the UH corresponds to an IUH, Bhunya, et al. [6]. The modified calculation for the IUH used the series of n linear reservoirs, Nash [7]. The recent result shows that the model has a good correlation between the observed and simulated hydrographs, Ghumman et al. [8].

The concept of IUH model explains that the watershed is represented as a cascade of linear reservoirs. The rainfall is assumed to be uniform along the entire watershed. In each reservoir, the continuity equation and a storage function are applied. A formulation to obtain the IUH and DRH is shown in Eqs. 1 and 2, respectively.

$$q(t) = \frac{1}{(n-1)!K}\left(\frac{t}{K}\right)^{n-1} e^{-t/K} \tag{1}$$

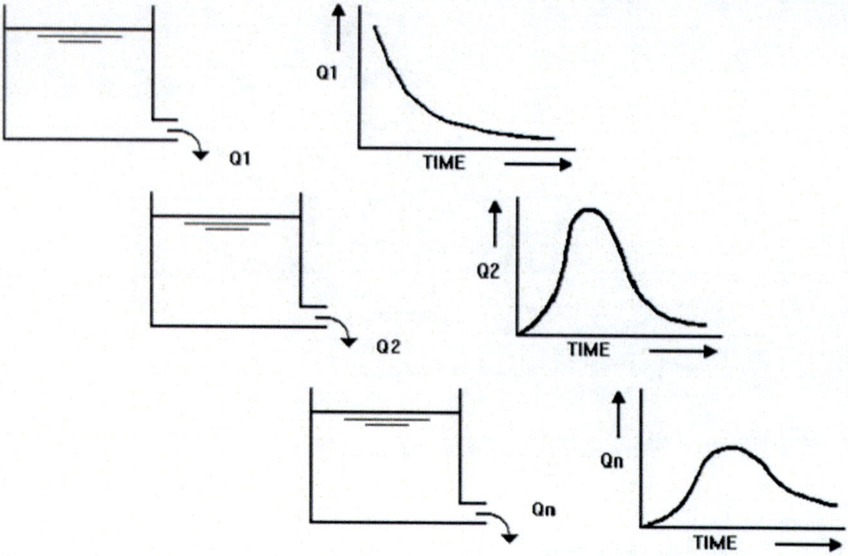

Fig. 6 IUHs of the three reservoirs by Nash model parameters

$$Q(t) = 2.778\frac{1}{\Delta t}q(t)A \tag{2}$$

where $q(t)$ is ordinate of IUH, n is a few cascades reservoir, K is the storage coefficient, A is area of watershed (km^2), e is 2.718, Δt is time interval (hour), $Q(t)$ is DRH (m^3/s) and 2.778 is a metric constant. The superposition procedure is used for another duration of effective rainfall. Figure 6 is the IUH model by NASH using three reservoirs [9]. In this research, after the calibration, the IUH model uses 2 (two) reservoirs.

The separation BF form $Q(t)$ is an important task of the unit hydrograph theory model application. Unfortunately, there is no satisfactory technique for extracting the DRH from the $Q(t)$, Beven [10] and Bedient et al. [11]. The methods of base flow separation in this research are the straight-line method. The separation of BF is achieved by joining with a straight line from the beginning of the rising limb to a point on the recession limb representing the end of DRH.

4 Results and Discussions

The depth and duration magnitude of rainfall data from satellites have different values than measurement data. There is a tendency that in duration, satellite rainfall data are longer than the measurement data. The average rainfall duration for ARR, GPM, and TRMM data are 4.3 h, 11.3 h, and 9 h, respectively, while the depth of rainfall from satellite data tends to be smaller than the measurement data. The average rainfall intensity for ARR, GPM, and TRMM data are 58.0 mm/h, 4.2 mm/h, and 1.9 mm/h, respectively. The direct runoff flow model was calculated. First is streamflow hydrograph analysis from the AWLR station. The base flow is predicted from the average data during the dry season and the day with no rain. The direct runoff is obtained from subtracting observed flow data with the predicted baseflow. To obtain data pairs, it is necessary to plot the rainfall and flow data in one graph. Comparison of hydrographs by using observed values and model values obtained using IUH methods are given for TRMM, GPM and ARR data studied in this work. Figures 7, 8, and 9 are the comparison of DRH from the 19th to 26th of February 2017.

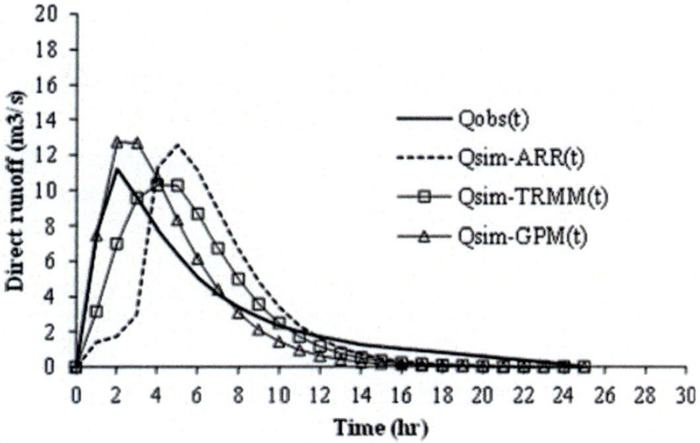

Fig. 7 Comparison of model and observed hydrograph on 19–20 February 2017

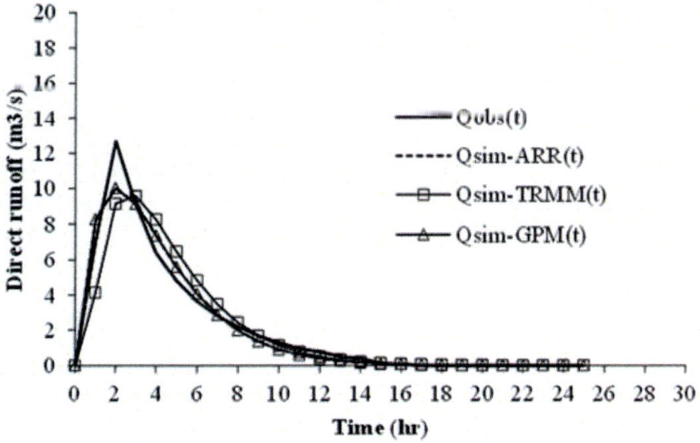

Fig. 8 Comparison of model and observed hydrograph on 20–21 February 2017

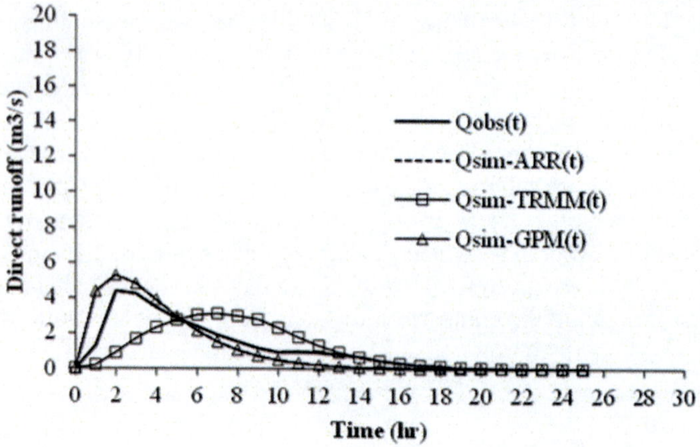

Fig. 9 Comparison of model and observed hydrograph on 25–26 February 2017

The model using TRMM data showed that the peak flow is slightly underestimated compared to observed data flow. Time of concentration in the model with input the TRMM rainfall data tends to be longer than observed data. The satellite data have more long duration in one rainfall event and have a small quantity. The model using GPM data showed that the peak flow is slightly overestimated compared to observed data flow. Time of concentration in the model with input the GPM rainfall data tends to be the same as observed data. Time parameters such as the time to peak, time of concentration, and lag time serve as a fundamental input to all methods of estimating flood hydrograph in ungauged areas. Therefore, appropriate watershed response time estimation directly impacts estimated peak discharges. The catchment response time is inconsistent, and it varies widely as applied in modern flood hydrology practice. An underestimated time to peak results in an over-designed hydraulic structure, on the other hand, would result in under design, Gericke and Smithers [12].

5 Conclusions

An IUH concept is conducted to compute DRH, in which the input is the ground base rainfall and satellite data in this research. GPM 3IMERGHHE and TRMM 3B42RT represent a similar spatial pattern in February 2017. The results obtained using GPM are better than TRMM.

References

1. Tan, M.L., Ibrahim, L., Duan, Z., Arthur, P., Cracknell, A.P., Chaplot, V.: Evaluation of six high-resolution satellite and ground-based precipitation products over Malaysia. Remote Sens. **7**(2), 1504–1528 (2015)
2. Hou, A.Y., Kakar, R.K., Neeck, S., Azarbarzin, A.A., Kummerow, C.D., Kojima, M., Oki, R., Nakamura, K., Iguchi, T.: The global precipitation measurement mission. Bull. Am. Meteorol. Soc. **95**, 701–722 (2014)
3. Feyera, A.H., Gebremichael, M.: Evaluation of high-resolution satellite precipitation products over very complex terrain in Ethiopia. J. Appl. Meteorol. Climatol. **49**, 1044–1051 (2010)
4. Moazami, S., Golian, S., Kavianpour, M.R., Hong, Y.: Comparison of PERSIANN and V7 TRMM multi-satellite precipitation analysis (TMPA) products with rain gauge data over Iran. Int. J. Remote Sens. **34**(22), 8156–8171 (2013)
5. Bélizal, E., Lavigne, F., Hadmoko, D.S., Degeai, J.P., Dipayana, G.A., Mutaqin, B.W., Marfai, M.A., Coquet, M., Mauff, B.L., Robin, A.-K., Vidal, C., Cholik, N., Aisyah, N.: Rain-triggered lahars following the 2010 eruption of Merapi volcano, Indonesia: a major risk. J. Volcanol. Geotherm. Res. **261**, 330–347 (2013)
6. Bhunya, P.K., Berndtsson, R., Singh, P.K., Hubert, P.: Comparison between Weibull and gamma distributions to derive synthetic unit hydrograph using Horton ratios. Water Resour. Res. **44**, 1–17 (2008)
7. Nash, J.E.: The form of the instantaneous unit hydrograph. Hydrol. Sci. Bull. **3**, 114–121 (1957)
8. Ghummana, A.R., Ghazawa, Y., Maguida, A.R.H., Zafar, A.: Investigating parameters of geomorphic direct runoff hydrograph models. Water Resour. **46**(1), 19–28 (2017)
9. US Army Corps of Engineer.: Engineering and Design: Flood-Runoff Analysis (Engineer Manual 1110-2-1417) (1994)
10. Beven, K.J.: Rainfall–Runoff Modelling, The Primer, 360 pp. Wiley, New York (2001)
11. Bedient, P.B., Hoblit, B.C., Gladwell, D.C., Vieux, B.E.: NEXRAD radar for flood prediction in Houston. J. Hydrol. Eng. **5**, 269–277 (2000)
12. Gericke, O.J., Smithers, J.C.: Review of methods used to estimate catchment response time for the purpose of peak discharge estimation. Hydrol. Sci. J. **59**(11), 1935–1971 (2014)

Comparative Study of Radiation-Based Reference Evapotranspiration Models for Tropical Urban Region

Muhamad Askari, Umar Faruk Lawan, Jazaul Ikhsan, Mohd Remy Rozainy Mohd Arif Zainol, and Nor Azazi Zakaria

Abstract Estimation of reference evapotranspiration (ETo) using the FAO Penman–Monteith model become more popular in developing countries. However, it involves complete weather variables data. In order to provide an alternative solution, we evaluated twenty-one radiation-based ETo models against 20-years observed pan evaporation data from Subang meteorological station, Malaysia. To fill the missing data was based on Julian day-based mean imputation. Outlier was tested using Tukey's detection method prior to estimating the ETo. The percentage of error (% error) in total values, % error in maximum values, % error in minimum values, % error in average values, root mean squared error (RMSE), and the paired *t*-test of the prediction accuracy were employed to evaluate the performance of the models. The results showed that the Conchrane-Orcutt 2014 model was performed better than the other radiation-based ETo models. The present study recommended the Conchrane-Orcutt 2014 model as the possible alternative ETo model for practical hydrological application according to the smallest difference of RMSE values as compared to the ETpan and due to the less number of required weather variables.

Keywords Urban region · Radiation-based evapotranspiration model · FAO Penman-Monteith · Tukey's outlier detection · Prediction accuracy

M. Askari (✉)
Crop Production Programme, Faculty of Sustainable Agriculture, Universiti Malaysia Sabah, 90000 Sandakan, Sabah, Malaysia
e-mail: muhamad.askari@ums.edu.my

M. Askari · J. Ikhsan
Department of Civil Engineering, Universitas Muhammadiyah Yogyakarta, Kasihan, Bantul, Yogyakarta, Indonesia

U. F. Lawan
Nigerian College of Aviation Technology, Zaria, Kaduna State, Nigeria

J. Ikhsan · M. R. R. M. A. Zainol · N. A. Zakaria
River Engineering and Urban Drainage Research Centre (REDAC), Universiti Sains Malaysia, Nibong Tebal, Malaysia

N. Sabtu (ed.), *Proceedings of AWAM International Conference on Civil Engineering 2022 - Volume 3*, Lecture Notes in Civil Engineering 386,
https://doi.org/10.1007/978-981-99-6026-2_52

1 Introduction

Evapotranspiration (ET) has vital agricultural, ecological, and hydrological roles. It is an essential factor of hydrological budgets at various spatial dimensions and is a crucial variable for understanding regional biological processes [1–3]. It is used in estimating actual evapotranspiration (AET) in rainfall-runoff and ecosystem modeling [4–7]. It is also used in estimating the amount of water evaporating into the atmosphere [8, 9]. The level of ET from the surface is mainly dependent on the availability of moisture on the surface and the amount of energy available to evaporate that moisture.

Evapotranspiration estimation nowadays has been given more considerable attention by many researchers due to it being important in determining the actual water loss. Previous researches have been conducted to estimate ET in Malaysia including ET estimation due to changes in climate [7, 10], rate of ET in oil palm catchment areas [11, 12], and ET model for Muda irrigation project [13–15]. Most of these studies were attributed to agricultural functional zones. Noticeably, ET estimation in the urban functional zones is still lacking.

Malaysia is a developing country with a hot and humid tropical climate. The development of this country especially in the area of the capital city such as Kuala Lumpur, Petaling Jaya, and Subang Jaya grows rapidly. One of the factors is expected due to the increase in population growth. When the population is increasing, the land use tends to change to an urbanized area such as residential, industrial, and commercial purposes. Another concern of urban development is its effects on the soils system by changing vegetation clearing, topsoil removal, grading, and compaction [16]. As evapotranspiration has a site-specific characteristic, the ET estimation for one place cannot be directly applied to other places having different climatic conditions and land use [14].

The aim of the present study is to evaluate the performance of radiation-based ETo models for the tropical urban region where complete climate data are still lacking. It is expected that the study will support water resource managers and the development/planning authorities to better manage the water toward sustainable water resources management in the region.

2 Material and Methods

The study area is represented by Subang (Sbg) meteorological station (ID: 48647). It is one of the main meteorological stations which is located at Subang Jaya district in the state of Selangor, Malaysia (see Fig. 1). It is located about 30 km south of Kuala Lumpur City. Geographically, it is located at 3° 07′ N and 101° 33′ E, and the elevation of 16.5 m above mean sea level. The temperature remains constant with a maximum value between 30 and 32 °C, while a minimum value between

23 and 24.5 °C. The relative humidity is around 78–85% and the annual rainfall of approximately 3300 mm [17].

The available meteorological data comprised of maximum (T_{max}), average daily wind speed at 2 m height ($u2$), rainfall, solar radiation (R_s), and pan evaporation

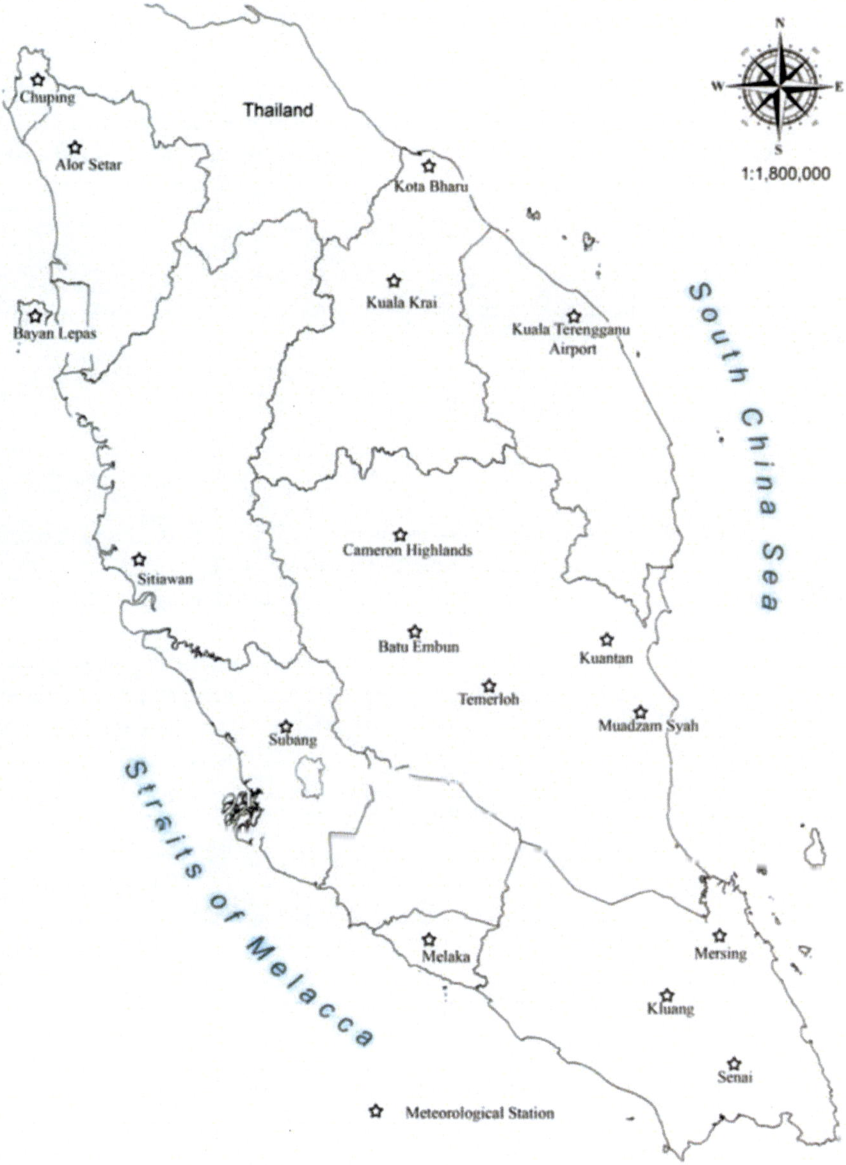

Fig. 1 Location of Subang meteorological station in Peninsular Malaysia [19]

(Epan). All the available data are presented in Table 1. The selected input variables for estimating ET_o models are T_{max}, T_{min}, T_{avg}, RH, $u2$, and R_s. In the present study, we analyzed radiation-based ET_o models as it is presented in Table 2 and FAO Penman–Monteith (FPM) model as presented below [18]:

$$\mathrm{FPM} = \frac{0.408\Delta(Rn - G) + \gamma \frac{900}{T+273} u_2 (e_s - e_a)}{\Delta + \gamma(1 + 0.34u_2)}$$

Prior to further calculating ET_o, Julian day-based mean imputation [20] was used to fill the missing meteorological data. Eventually, Tukey's outlier filter was used to detect the outlier data, which are the values in a quartile range 1.5 below the lower quartile or 1.5 above the quartile [21]:

$$Y < (Q1 - 1.51\mathrm{QR}) \quad \text{or} \quad Y > (Q3 + 1.51\mathrm{QR})$$

where $Q1$ = lower quartile, $Q3$ = upper quartile, and IQR = inter quartile range = $Q3 - Q1$.

The pan evaporation (E_{pan}) was converted into pan evapotranspiration (ET_{pan}) according to the following equation [18]:

$$\mathrm{ET}_{\mathrm{pan}} = E_{\mathrm{pan}} K_{\mathrm{p}}$$

where K_p is the pan coefficient. In the present study, K_p is set to 0.75 which refers to Table 5 of the FAO Irrigation and Drainage Paper No. 56 [18].

Performance evaluation of each ET_o model was conducted using statistical error indexes such as percentage of error in total values, percentage of error in maximum values, percentage of error in minimum values, percentage of error in mean values, and root mean squared error (RMSE). The performance indicators for the present study are summarized in Table 3. Based on the prediction accuracy (error) of the

Table 1 Climate data at Subang meteorological station during 1968–2009

Period	Tmax	Tmin (°C)	Tavg	RH (%)	$u2$ (m s^{-1})	Rainfall (mm)	R_s (MJ m^{-2})	Epan (mm)
Jan 1968–Dec 1974	Available	Available	Available	Available	Available	Available	Not available	Not available
Jan 1975–Dec 1984	Available	Available	Available	Available	Available	Available	Available	Not available
Jan 1985–Jun 2004	Available	Available	Available	Available	Available	Available	Available	Available
Jun 2004–Dec 2009	Available	Available	Available	Available	Not available	Available	Available	Available

Table 2 The selected radiation-based daily potential evapotranspiration methods

No	Models	Equation	Reference
1	Makkink 1957	$ET_o = 0.61\left(\frac{\Delta}{\Delta+\gamma}\right)\left(\frac{R_s}{58.5}\right) - 0.12$	[2]
2	Turc 1961a	$ET_o = \frac{0.013T}{T+15}(R_s + 50)$	[2]
3	Turc 1961b	$ET_o = 0.013\left(\frac{T}{T+15}\right)(R_g + 50)\left[1 + \frac{50-RH}{70}\right]$	[24]
4	Priestly Taylor 1972a	$ET_o = \frac{\alpha\Delta R_n}{\lambda(\Delta+\gamma)}$	[24]
5	Ritchie 1972	$ET_o = \frac{\Delta R_n}{\lambda(\Delta+\gamma)}$	[24]
6	Jensen-Haise 1963a	$ET_o = \frac{C_t(T-T_x)R_s}{\lambda}$	[25]
7	Hargreaves 1975	$ET_o = \frac{0.0135(T+17.8)R_s}{\lambda}$	[25]
8	Jensen-Haise 1963b	$ET_o = \frac{R_s}{\lambda}(0.025T_{mean} + 0.08)$	[26]
9	Caprio 1974	$ET_o = 6.1 \times 10^{-6} R_s(1.8T + 1)$	[26]
10	Jensen-Haise 1963c	$ET_o = \frac{R_a T_{mean}}{40\lambda\rho}$	[27]
11	Priestly Taylor 1972b	$ET_o = \alpha\frac{\Delta}{(\Delta+\gamma)} \times \frac{R_n}{\rho\lambda}$	[27]
12	McGuiness 1972	$ET_o = 2.54(0.0082T - 0.19)\frac{R_s}{1500}$	[22]
13	Ritchie 1990	$ET_o = \alpha(3.87 \times 10^{-3} R_s)(0.6T\ \min_{max})$	[22]
14	Irmak 2003a	$ET_o = -0.642 + 0.1749R_s + 0.0353T_a$	[22]
15	Irmak 2003b	$ET_o = -0.478 + 0.156R_s - 0.0112T\ \min_{max}$	[22]
16	Abtew 2003	$ET_o = \frac{1}{56}\left(\frac{R_s T_{max}}{\lambda}\right)$	[28]
17	Rs based 2003	$ET_o = -0.611 + 0.149R_s + 0.079T$	[29]
18	Rn based 2003	$ET_o = 0.489 + 0.289R_n + 0.023T$	[29]
19	Copais 2006	$ET_o = m_1 + m_2c_2 + m_3c_1 + m_4c_1$	[30]
20	Valiantzas 2013	$ET_o = 0.039R_s\sqrt{T+9.5} - (0.19R_s^{0.6}\varphi^{0.15}) + 0.048(T+20)\left(1 - \frac{RH}{100}\right)U^{0.15}$	[30]
21	Conchrane-Orcutt 2014	$PET = -0.755 + (0.257 \times R_n) + (0.062 \times T_{max})$	[31]

selected model, the possible alternative ET_o model was then selected using the paired student *t*-test.

3 Results and Discussion

Table 4 shows the summary of the meteorological data that are used as the input variables for the ET_o model, and the measured pan evaporation (E_{pan}). The range of values for the variables of T_{max}, T_{min}, T_{avg}, R_s, $u2$, and RH are 21.1 °C–26.5 °C,

Table 3 Performance indicators [32] for evaluating ETo models

Error index	Equation
% the absolute error in total value estimate	$\frac{\lvert P_{total} - O_{total} \rvert}{O_{total}} \times 100$
% the absolute error in maximum value estimate	$\frac{\lvert P_{max} - O_{max} \rvert}{O_{max}} \times 100$
% the absolute error in minimum value estimate	$\frac{\lvert P_{min} - O_{min} \rvert}{O_{min}} \times 100$
% the absolute error in mean value estimate	$\frac{\lvert P_{mean} - O_{mean} \rvert}{O_{mean}} \times 100$
Root mean squared error (RMSE)	$\sqrt{\frac{\sum (P-O)^2}{n}}$

The estimated ET_o represents predicted values (P), while the estimated pan evapotranspiration (ET_{pan}) represents observed values (O)

29.2 °C–36.4 °C, 24.3 °C–30.5 °C, 1.7 MJ m^{-2}–26.5 MJ m^{-2}, 0.1 m s^{-1}–3 m s^{-1}, 65.2%–95.4%, respectively. The pan evaporation data range from 0.1 mm to 10.8 mm.

Figure 2 describes that the Jensen-Haise 1963c model provides the highest estimated ET_o values among others. On the contrary, Turc 1961b model attributed the least estimated ET_o values among others. Furthermore, the Conchrane-Orcutt 2014, Valiantzas 2013, Ritchie 1972, and Copais 2006 models show a better performance for ET_o estimation (see Fig. 3). The results are in line with the statistical analysis which is presented in Table 5. According to Table 5, Conchrane-Orcutt 2014, Valiantzas 2013, and Copais 2006 are slightly underestimated values as compared to the ET_{pan} by 0.11, 0.16, and 0.27 mm day^{-1} with the RMSE values of 0.96, 0.98, and 0.99, and % error of mean values of 3.6, 4.9, and 8.4%, respectively. Noticeably, the Ritchie 1972 model was noted to slightly overestimate as compared to the ET_{pan} by 0.03 mm day^{-1} with an RMSE value of 0.94 and % error of mean value of 0.7%. The Jensen-Haise 1963b, Priestly Taylor 1972b, McGuiness 1972, and Jensen-Haise 1963c models demonstrate the poor performances among others with the RMSE values of 2.13, 2.41, 3.40, and 7.00, respectively. The poor performance of the Jensen-Haise model obtained in the present study is similar to the results found in [22, 23]. The model selection was decided based on the criteria attributed

Table 4 Statistical description of climate data at Subang meteorological station

Parameter	T_{max} (°C)	T_{min} (°C)	T_{avg} (°C)	R_s (MJ m^{-2})	RH (%)	$u2$ (m s^{-1})	$Epan$ (mm)
Max	36.4	26.5	30.5	26.5	95.4	3	10.8
Min	29.2	21.1	24.3	1.7	65.2	0.1	0.1
Mean	32.8	23.8	27.4	15.7	80.4	1.5	4.3
Median	32.9	23.8	27.4	15.9	80.6	1.5	4.2
St Dev	1.3	0.9	1.1	3.0	5.3	0.5	1.5
Skewness	– 0.24	– 0.01	– 0.01	– 0.3	– 0.2	0.4	0.3
CV	0.04	0.04	0.04	0.19	0.07	0.33	0.35

Table 5 Statistical performance of the radiation-based ET_0 models versus ET_{pan}

Models	% error in total estimate	% error in max estimate	% error in min estimate	% error in mean estimate	RMSE
Conchrane-Orcutt 2014	3.6	33.4	313.3	3.6	0.96
Valiantzas 2013	4.9	29.5	222.2	4.9	0.99
Ritchie 1972	0.7	34.1	870.8	0.7	0.94
Copais 2006	8.4	32.8	395.2	8.4	1.00
Abtew 2003	18.8	12.9	442.2	18.8	1.19
Irmak 2003a	4.5	38.1	715.2	4.5	0.95
Makkink 1957	11.4	39.4	172.3	11.4	1.01
Ritchie 1990	21.8	15.4	462.9	21.8	1.23
Turc 1961a	11.8	27.1	916.9	11.8	1.03
Hargreaves 1975	23.4	13.5	471.6	23.4	1.27
Rs based 2003	21.6	29.7	2151.6	21.6	1.16
Caprio 1974	51.6	9.2	596.8	51.6	2.03
Priestly Taylor 1972a	27.1	16.9	1125.4	27.1	1.31
Jensen-Haise 1963a	54.1	10.5	608.7	54.1	2.10
Turc 1961b	35.6	48	294.3	35.6	1.48
Irmak 2003b	27.4	27.1	2330.1	27.4	1.28
Rn based 2003	28.2	26.2	2286.2	28.2	1.30
Jensen-Haise 1963b	55.1	11.2	613.5	55.1	2.13
Priestly Taylor 1972b	67.0	6.4	1521.5	67.0	2.41
McGuiness 1972	99.3	99.3	98.4	99.3	3.40
Jensen-Haise 1963c	216.0	48.1	11,421.5	216.0	7.00

by the least value of the arithmetic mean of each performance indicators. Therefore, Conchrane-Orcutt 2014 is the best model among others. Moreover, Figs. 4 and 5 show the graphical comparison between the FAO Penman–Monteith (FPM) and the ETpan. Noticeably, the estimated FPM indicates a slightly overestimated value of 0.36 mm day^{-1} which is attributed to the RMSE value of 1.03 and % error of mean values of 11.3% (Table 6).

By comparing the statistical performance of the Conchrane-Orcutt 2014 and FAO Penman–Monteith (FPM) model, it is clearly understood that % of the absolute error in total, maximum, minimum, and mean value estimates of Conchrane-Orcutt 2014 model are always lower than those indicators of FPM model. The RMSE of Conchrane- Orcutt 2014 model is also lower than the RMSE of FPM model. In the present study, we can evaluate that the additional meteorological variables do not always contribute a positive effect to the ability of the model to better fit the ET_{pan}.

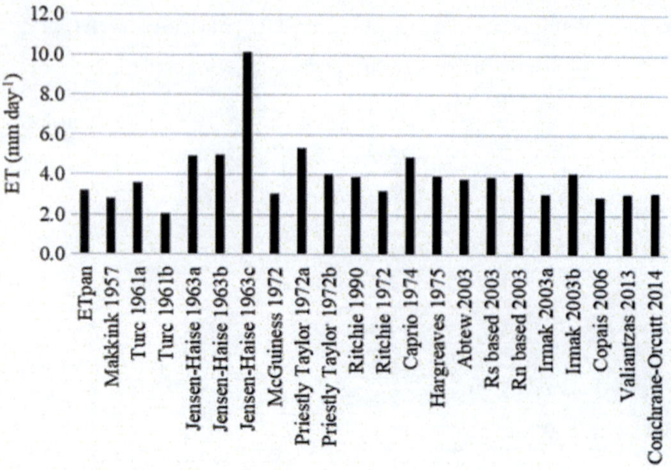

Fig. 2 Yearly average graphical evaluation of the radiation-based ET_o models and ET_{pan}

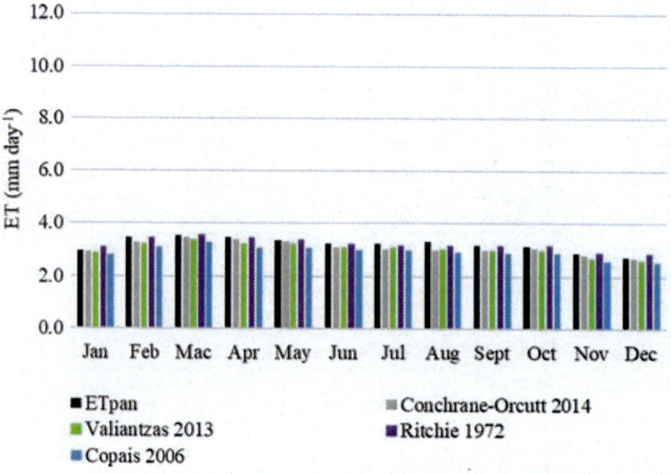

Fig. 3 Comparison of 20-year mean monthly radiation-based selected ET_o models and ET_{pan}

With the hypothesized mean difference of the prediction accuracy (error) between the models are equals to zero, and the alpha level being equal to 0.05, it was found that probability ($T \leq t$) for two-tail is always less than the alpha level. It means that we rejected the hypothesized mean difference of the prediction accuracy (error) between the models are equals to zero. In another word, there was a statistically significant difference between the Conchrane-Orcutt 2014 and FPM models. In the present study, by considering the number of meteorological variable data involved in the model as well as the small difference of RMSE, Conchrane-Orcutt 2014 model is regarded as one of the best PET models among others.

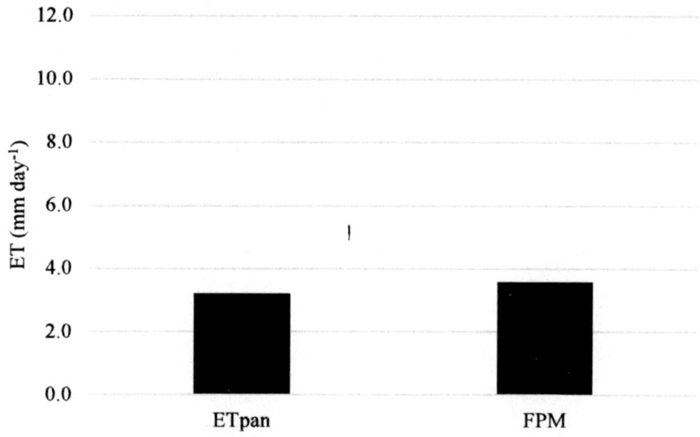

Fig. 4 Yearly average graphical evaluation of the FAO Penman–Monteith (FPM) and ET_{pan}

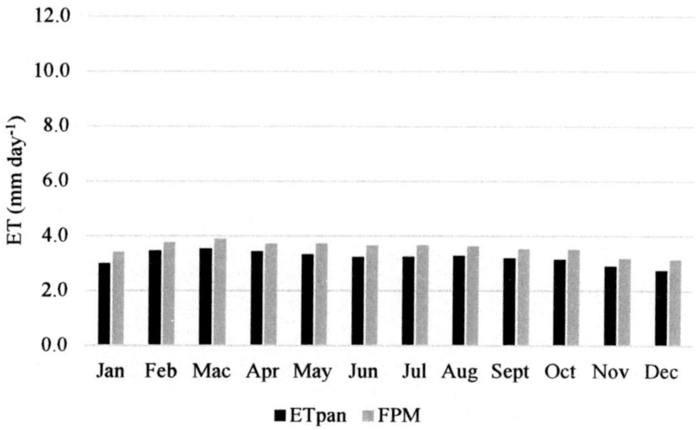

Fig. 5 Comparison of 20-year mean monthly FAO Penman–Monteith (FPM) and ET_{pan}

Table 6 Statistical performance of the FAO Penman–Monteith (FPM) versus ET_{pan}

Model	% error in total estimate	% error in max estimate	% error in min estimate	% error in mean estimate	RMSE
FAO Penman–Monteith	11.3	22.7	980.2	11.3	1.03

4 Conclusion

In the present study, the Conchrane-Orcutt 2014 and FPM models work well in estimating ET_0. Percentage of error in total, maximum, minimum, and mean value estimates along with the RMSE and the paired student's *t*-test based on the prediction accuracy can be adopted as the criteria to select the best performance in ET_0 estimation. Considering the small difference of RMSE values, the number of required meteorological variables, and the total predicted ET_0, the Conchrane-Orcutt 2014 model can be considered as the best model and is recommended for future research and practical hydrological application.

The estimation of ET_0 value still needs to be improved. The present study only provides limited information in applying point meteorological stations in ET_0 estimation. More meteorological stations and ET_0 estimation methods are required to better understand and estimate the spatial variability of the ET_0 values.

Acknowledgements The authors would like to acknowledge the financial support from the Ministry of Higher Education, Malaysia, under the Long-Term Research Grant (LRGS) No. 203/PKT/672004 entitled "Urban Water Cycle Processes, Management, and Societal Interactions: Crossing from Crisis to Sustainability" which is funded under a subproject entitled Urban Ecohydrology for Resilient Environment (UCOREN). The authors also wish to thank the Universitas Muhammadiyah Yogyakarta and the Universiti Malaysia Sabah for their support in the present study.

References

1. Al-Shibli, F., Ottom, M., Saoub, H., Al-Weshah, R.: Comparative analysis of potential evapotranspiration calculation methods with era-reanalysis climate models' projections in Western Asia, Jordan. Appl. Ecol. Environ. Res. **19**(6), 4849–4879 (2021)
2. Lu, J., Sun, G., McNulty, S.G., Amatya, D.M.: A comparison of six potential evapotranspiration methods for regional use in the Southeastern United States. J. Am. Water Resources Assoc. **41**(3), 621–633 (2005)
3. Tajuddin, R.: Water Crisis in Selangor (2014) [cited 2021 January 5]
4. Dou, X., Yang, Y.: Evapotranspiration estimation using four different machine learning approaches in different terrestrial ecosystems. Comput. Electron. Agric. **148**, 95–106 (2018)
5. Fisher, J.B., Whittaker, R.J., Malhi, Y.: ET come home: potential evapotranspiration in geographical ecology. Global Ecol. Biogeogr. **20**(1), 1–18 (2011)
6. Sun, G., Alstad, K., Chen, J., Chen, S., Ford, C.R., Lin, G., Liu, C., Lu, N., McNulty, S.G., Miao, H.: A general predictive model for estimating monthly ecosystem evapotranspiration. Ecohydrology **4**(2), 245–255 (2011)
7. Theng Hue, H., Ng, J.L., Huang, Y.F., Tan, Y.X.: Evaluation of temporal variability and stationarity of potential evapotranspiration in Peninsular Malaysia. Water Supply (2021)
8. Band, L.E., Mackay, D.S., Creed, I.F., Semkin, R., Jeffries, D.: Ecosystem processes at the watershed scale: sensitivity to potential climate change. Limnol. Oceanogr. **41**(5), 928–938 (1996)
9. Hay, L.E., McCabe, G.J.: Spatial variability in water-balance model performance in the conterminous United States. J. Am. Water Resources Assoc. **38**(3), 847–860 (2002)

10. Najim, M.M.M., Aminul Haque, M., Lee, T.: Estimation of evapotranspiration and climate change in the west coast of peninsular Malaysia. In: Malaysian Science and Technology Congress. (2003)
11. Henson, I.E., Yahya, Z., Noor, M.R.M., Harun, M.H., Mohammed, A.T.: Predicting soil water status, evapotranspiration, growth and yield of young oil palm in a seasonally dry region of Malaysia. J. Oil Palm Res. **19**(2), 398–415 (2007)
12. Yusop, Z., Hui, C.M., Garusu, G.J., Katimon, A.: Estimation of evapotranspiration in oil palm catchments by short-time period water-budget method. Malays. J. Civ. Eng. **20**(2) (2008)
13. Ali, M.H., Shui, L.T.: Potential evapotranspiration model for Muda irrigation project, Malaysia. Water Resources Manage. **23**(1), 57–69 (2009)
14. Lee, T., Najim, M., Aminul, M.: Estimating evapotranspiration of irrigated rice at the West Coast of the Peninsular of Malaysia. (2004)
15. Tukimat, N.N.A., Harun, S., Shahid, S.: Comparison of different methods in estimating potential evapotranspiration at Muda Irrigation Scheme of Malaysia. J. Agric. Rural Dev. Tropics Subtropics **113**(1), 77–85 (2012)
16. Chen, Y., Day, S.D., Wick, A.F., McGuire, K.J.: Influence of urban land development and subsequent soil rehabilitation on soil aggregates, carbon, and hydraulic conductivity. Sci. Total Environ. **494**, 329–336 (2014)
17. Dom, N.C., Ahmad, A.H., Adawiyah, R., Ismail, R.: *Spatial mapping of temporal risk characteristic of dengue cases in Subang Jaya.* In: 2010 International Conference on Science and Social Research (CSSR 2010). IEEE (2010)
18. Allen, R.G., Pereira, L.S., Raes, D., Smith, M.: Crop evapotranspiration- Guidelines for computing crop water requirements-FAO Irrigation and Drainage Paper 56. FAO, Rome D05109 (1998)
19. Ahmad, N.F.A., Askari, M., Harun, S., Fadhil, A.B., Demun, A.S.: Sensitivity analysis of a FAO Penman Monteith for potential evapotranspiration to climate change. Jurnal Teknologi **79**(7) (2017)
20. Thirukumaran, S.: An alternate imputation technique of a mean method for missing values and comparative study with neighbor methods. Int. J. Appl. Eng. Res. **10**(4), 9969–9982 (2015)
21. Tukey, J.W.: Exploratory Data Analysis, pp. 43–44. Addison-Wesley (1977)
22. Tabari, H., Grismer, M.E., Trajkovic, S.: Comparative analysis of 31 reference evapotranspiration methods under humid conditions. Irrig. Sci. **31**(2), 107–117 (2013)
23. Trajkovic, S., Kolakovic, S.: Evaluation of reference evapotranspiration equations under humid conditions. Water Resour. Manage. **23**(14), 3057–3067 (2009)
24. Bormann, H.: Sensitivity analysis of 18 different potential evapotranspiration models to observed climatic change at German climate stations. Climatic Change **104**(3), 729–753 (2011)
25. Xu, C., Singh, V.: Evaluation and generalization of radiation-based methods for calculating evaporation. Hydrol. Process. **14**(2), 339–349 (2000)
26. Xystrakis, F., Matzarakis, A.: Evaluation of 13 empirical reference potential evapotranspiration equations on the Island of Crete in Southern Greece. J. Irrig. Drain. Eng. **137**(4), 211–222 (2011)
27. Oudin, L., Hervieu, F., Michel, C., Perrin, C., Andréassian, V., Anctil, F., Loumagne, C.: Which potential evapotranspiration input for a lumped rainfall–runoff model?: Part 2—towards a simple and efficient potential evapotranspiration model for rainfall–runoff modelling. J. hydrol. **303**(1–4), 290–306 (2005)
28. Abtew, W., Obeysekera, J., Irizarry-Ortiz, M., Lyons, D., Reardon, A.: Evapotranspiration estimation for south Florida. In: World Water & Environmental Resources Congress. (2003)
29. Irmak, S., Irmak, A., Allen, R., Jones, J.: Solar and net radiation-based equations to estimate reference evapotranspiration in humid climates. J. Irrig. Drain. Eng. **129**(5), 336–347 (2003)
30. Ozgur, K.: Comparison of different empirical methods for estimating daily reference evapotranspiration in mediterranean climate. J. Irrig. Drain. Eng. **140**(1), 04013002 (2014)
31. Bogawski, P., Bednorz, E.: Comparison and validation of selected evapotranspiration models for conditions in Poland (Central Europe). Water Resour. Manage. **28**(14), 5021–5038 (2014)
32. Green, I., Stephenson, D.: Criteria for comparison of single event models. Hydrol. Sci. J. **31**(3), 395–411 (1986)

Methodologies for Estimating the Hydraulic Efficiency of Non-surcharged Clean Grate Inlets, Clogged Grate Inlets and Continuous Transversal Grate Inlets

Ali Zaiter and Nuridah Sabtu

Abstract The urban drainage system of cities is composed of two major and minor subsystems linked by drain inlets to form the dual-drainage system. Therefore, the sewer inlet capacity has a significant influence on the dual-drainage system workability. However, most of the design models and studies focus on the subsurface discharge capacity and neglect the inlet capacity. A reliable drainage system design has to take into consideration the hydraulic efficiency of the lining elements (grate inlets) and the effect of clogging on their efficiency. In this study, the main methodologies for determining the hydraulic efficiency of grate inlets are presented. In a comparative study on the hydraulic efficiency of clean grate inlets, the general hydraulic efficiency formula of Eq. (1) and the formula of Eq. (2) adopted by British Standard and Design Manual for Roads and Bridges and the UPC methodology in Eq. (3) showed similar results with low variance of 2–7%. However, Eq. (3) of the UPC methodology is the most accurate as it considers the majority of the parameters involved in the hydraulic interaction at the grate inlet. Moreover, the UPC methodology of Eq. (3) is proved to be applicable for grate inlets under clogging conditions with high reliability. The approaches presented in this paper can be used in numerical models to provide a reliable accurate simulation of the hydraulic behaviour at the single and continuous grate inlets.

Keywords Resilient infrastructure · Hydraulic efficiency · Grate inlets · Clogged inlets · Transversal inlets

A. Zaiter · N. Sabtu (✉)
School of Civil Engineering, Universiti Sains Malaysia, 14300 Nibong Tebal, Malaysia
e-mail: nsabtu@usm.my

N. Sabtu (ed.), *Proceedings of AWAM International Conference on Civil Engineering 2022 - Volume 3*, Lecture Notes in Civil Engineering 386,
https://doi.org/10.1007/978-981-99-6026-2_53

1 Introduction

The interaction between the natural water cycle and human activities in developed urban areas can be sorted into two main categories. The first one is the extraction of water from nature for human life essentials, whereas the second category is the increase of the impervious areas and the reduction of the naturally vegetated areas [1]. These activities reduce the infiltration capacity of land surfaces; therefore, they increase surface runoff and flooding likelihood. These two categories of interaction between the natural water cycle and humans require efficient urban drainage systems in the infrastructure design of cities.

The severe consequences of urban flooding on individuals and societies raised concerns among hydraulic engineers on the necessity of designing proper drainage systems to cope with extreme rainfall events [2]. During storm events, the urban drainage systems of the city are responsible to collect and convey the runoff sewer from the road and critical areas to prevent urban flooding; therefore, minimizing the environmental and human life problems [1, 3–5]. Operating and maintaining the urban drainage system is essential to ascertain its proper function. An inadequate urban drainage system or any function failure can result in urban floods and cause damages, inconveniences, economic losses and health risks to individuals, infrastructure, buildings and the environment. The urban drainage system is comprised of two subsystems. The major system is the surface system of roads and gutters that convey the surface runoff during storm events to drain inlets. On the other hand, the system of underground sewer pipes represents the minor system. These two systems are linked by a drain inlet system comprised of all gullies, manholes, inlets, or grates installed between the two subsystems aiming to intercept the surface runoff to the underground sewer system. In the linkage system, there are various types of inlets such as grate inlets, kerb opening inlets, combination inlets and transverse inlets (Fig. 1). The two subsystems (major and minor) with their linkage constitute the dual-drainage system. Accurate determination of the inlet hydraulic efficiency is an essential part of the proper design of urban drainage systems to reduce sewer runoff during intense rain events and to understand the interaction between surface runoff and underground drainage [6, 7].

This paper aims to provide a short review on the main equations and approaches adopted in hydraulic engineering for determining the hydraulic efficiency of single and continuous grate inlets under clean and clogged conditions. The formulas

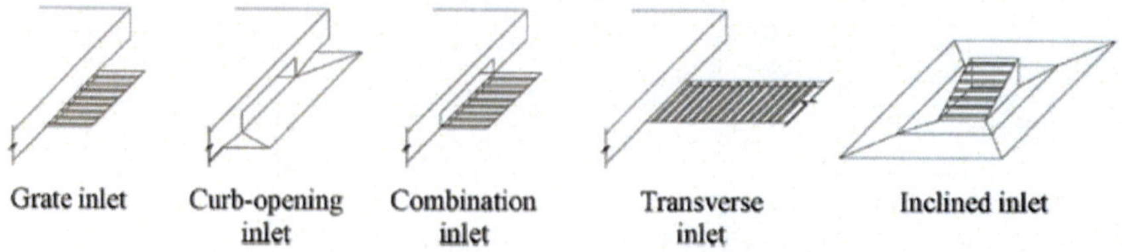

Fig. 1 Grate, kerb opening, combination, transverse and inclined inlets [8]

presented here can be used in numerical models to simulate the interception capacities of the different types of grate inlets.

2 Hydraulic Efficiency of Clean Grate Inlets

According to Comport and Thornton [9] and Guo and MacKenzie [10], the grate inlet hydraulic efficiency depends on its geometry as well as the external environment. The hydraulic behaviour of storm inlets can be determined by using several empirical equations, datasets or protocols developed through experimental facilities or by manufacturers. The inlet hydraulic efficiency (E) is defined as the ratio of the intercepted flow (Q_{i}) to the total flow approaching the inlet (Q_{a}) as shown in Eq. (1).

$$E = \frac{Q_{\mathrm{i}}}{Q_{\mathrm{a}}} \tag{1}$$

Spaliviero et al. [11] investigated the inlet efficiency of typical small grates in the UK and proposed a linear formula to count the efficiency of kerb inlets and gully grate inlets. However, this study was restricted to grates with a short length less than or equal to 50 cm and a low flow rate where the ratio of the flow to the water depth at the grate upstream was limited to 1.5 (suitable only for areas with low rainfall intensity). The reported results showed that the hydraulic efficiency of the kerb inlet and grate gully is a function of the water depth and approaching flow rate. The gully grate and kerb inlet hydraulic efficiency are determined as per Eqs. (2) and (3), respectively. These equations are adopted by BS EN 752:2008 [12] and the Design Manual for Roads and Bridges (DMRB) in the UK [13] approach for determining the hydraulic efficiency of grates and kerb inlets.

$$E = 100 - G\left(\frac{Q}{H}\right) \tag{2}$$

$$E = 100 - 36.1\left(\frac{Q}{H^{1.5}L_i}\right) \tag{3}$$

where E is the hydraulic efficiency (%), G is a geometric constant of the grate to be determined (refer BS EN 752), Q is the approaching flow rate (m^3/s), H is the water depth approaching the inlet (m), and L_i is the kerb opening length (m).

To cover the dimension and flow limitations of the study carried out by Spaliviero et al. [11], Gómez and Russo [14] carried an experimental investigation to assess the hydraulic efficiency of 11 different types of clean grate inlets under a large wide range of flow up to 0.2 m^3/s, cross slope between 0 and 4% and longitudinal slope between 0 and 10%. Results showed that for high longitudinal slopes, the inlet efficiency was reduced, whereas the inlet efficiency was enhanced for high transversal slopes. This agrees with the results reported by Dai et al. [15]. Based on the experimental results and validations, Gómez and Russo [14] proposed the UPC methodology for estimating the grate inlet efficiency to cover for high flows up to 0.2 m^3. The proposed equations can be applied to any similar non-tested inlet (width 33–63 cm and length 47.5–100 cm) and for each road geometry as shown in Eq. (4). The characteristic coefficient k (between zero and one) is determined from the road geometry whilst A and B are characteristic coefficients for the grate geometry determined from Eqs. (5) and (6), respectively.

$$E = \frac{Q_\mathrm{i}}{Q_\mathrm{a}} = kA\left(\frac{kQ_\mathrm{a}}{y}\right)^{-B} \tag{4}$$

$$A = \frac{1.988 \times A_\mathrm{g}^{0.403}}{\rho^{0.19} \times (n_\mathrm{t}+1)^{0.088} \times (n_\mathrm{l}+1)^{0.012} \times (n_\mathrm{d}+1)^{0.082}} \tag{5}$$

$$B = 1.346 \times \frac{L^{0.179}}{W^{0.394}} \tag{6}$$

where E is the dimensionless inlet hydraulic efficiency, Q_a is the flow approaching the inlet (m^3/s), Q_i is the flow intercepted by the inlet (m^3/s), y is the upstream flow depth (m), A_g is the smallest grate area containing all the holes (m^2), ρ is the ratio of the void area to A_g (%), n_t is the count of the grate transversal bars, n_l is the count of the grate longitudinal bars, n_d is the count of the grate diagonal bars, L is the grate length (cm), and W is the grate width (cm).

In the study of Wakif and Sabtu [7], the methodologies of Eqs. (1), (2) and (4) were validated with the experimental results of for clean grate inlet under different longitudinal slopes (flat, 1:100 and 1:50) and transversal slopes (flat, 1:40 and 1:80) where the flow ranged between 2 and 12 L/s. The results showed that the three equations showed similar values with the variance of 2–7% and Eq. (4) showed the most accurate efficiency followed by Eq. (2) and lastly Eq. (1).

Wu et al. [16] studied the interception efficiency of several gully grate inlets. The cases of gully without grating, grating with longitudinal bars, grating with longitudinal and transversal bars, and new grating with angled bars on a parallelogram cross-section were considered (Fig. 2). Results showed that at lower water depth and velocity, the grates showed almost similar efficiency of around 60%. On the other hand, the grate bars layout and cross-section effect on the inlet efficiency was distinguished when the inflow velocity increased as the kinetic energy dissipated by bars became more influential. The highest efficiency was recorded in the gully inlet without grating. This is attributed to the absence of bars obstructing the flow; therefore, lower dissipated kinetic energy. Even though gully with longitudinal bars showed a close efficiency to the gully with no grating, this kind of grates is not recommended for areas with bicycles access due to its inconvenience for cyclists, whereas it can be recommended for roads with no bicycles such as tunnels. The gully with angled bars on a parallelogram cross-section grate showed a lower efficiency than the grate with longitudinal bars and higher efficiency than grates with longitudinal and transversal bars. At high velocity, the overshoot of flow over the gully bars obstructing

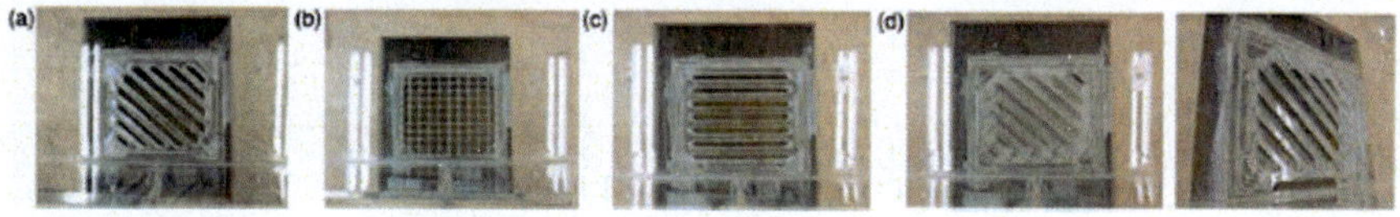

Fig. 2 **a** Existing grate, **b** grate with longitudinal bars, **c** grate with transversal bars and **d** new grate with angled bars on a parallelogram section [16]

the flow was more pronounced; therefore, a larger dissipated kinetic energy head and lower inlet efficiency were observed when longitudinal and transversal bars were installed in the gully grate. The aforementioned study proposed a relationship to compute the hydraulic interception efficiency of grated inlets as shown in Eq. (7).

$$E = a \times \left(\frac{2}{\sqrt{g}} \frac{Q}{H} \frac{h^{0.3}}{W L^{0.8}} \right) + b \quad (7)$$

where E is the gully grate efficiency, a and b are coefficients that depend on the void area ratio (ϕ) and the drag coefficient (C_d), Q is the approaching flow rate (m^3/s), g is the gravitational acceleration (m/s^2), H is the kerb water depth (m), h is the water depth at the gully (m), W is the width of the gully (m), and L is the length of the gully (m).

In a recent study, Kim et al. [17] conducted an experimental test on 1:2 scale models of three grate inlet sizes widely used in South Korea. The grate inlets of 40 $\times$ 50 cm^2, 40 $\times$ 100 cm^2 and 40 $\times$ 150 cm^2 sizes had the same grate width and varied grate length. The inlets with larger grate lengths exhibited higher interception efficiency compared to inlets with smaller grate lengths. This is attributed to the water splash-over phenomenon and the water film formed over the grate. For the small grate of 40 $\times$ 50 cm^2, the water film almost covered the whole grate and reduced its efficiency. However, for the larger grate sizes, the water did not cover more than half of the grate; therefore, they had larger interception efficiency. The gutter flow width was measured at 50 cm upstream of the installed grate. Results showed that when the road transversal slope increased, the surface runoff conveyed to the gutter increased and the gutter flow width was reduced. Consequently, the gutter flow and grate hydraulic efficiency were increased. Based on the experimental results, the intercepted flow of similar grates can be counted from Eq. (8).

$$Q_{in} = K' \times Q_m^{k_1} \times S_C^{k_2} \times S_L^{k_3} \quad (8)$$

where Q_{in} is the intercepted flow; Q_m is the total runoff amount; K', k_1, k_2, k_3 are empirical constants (refer to Table 1); S_C is the transversal slope; and S_L is the longitudinal slope.

Table 1 Empirical constants of Eq. (8)

Inlet size (cm^2)	K'	k_1	k_2	k_3
40 × 50	3.243	0.892	0.317	0.059
40 × 100	2.525	0.939	0.244	0.058
40 × 150	1.748	0.952	0.159	0.012

3 Hydraulic Efficiency of Clogged Grate Inlets

Most of the recorded studies on inlets hydraulic capacity tested the models under clean inlet condition and ignored the actual state on inlets where debris, dust, sand or tree leaves could accumulate at the grate inlet due to surface, environmental reasons or lack of proper maintenance [14, 18, 19]. In many recorded flooding cases, the subsurface drainage system was not surcharged and did not overflow. Instead, flooding was due to gully clogging with debris, dust, trash after first rain flush or tree leaves (Fig. 3). Consequently, localized surface water flooding was exhibited as the inlet maximum discharge capacity was restricted due to clogging or due to insufficient hydraulic capacity or distribution of the installed hydraulic inlets. The actual hydraulic efficiency of the inlet needs to be considered in the urban drainage system design. The results of tests under clean inlet conditions constitute the upper bound of the inlet efficiency. In fact, they are unrealistic as they overestimate the actual hydraulic efficiency of the clogged inlets. To cope with the efficiency reduction in clogged inlets, Guo [20] proposed the concept of a debris decay ratio known as clogging factor (c_0) between zero and one to count for the actual hydraulic efficiency of clogged inlets.

Fig. 3 Grate inlets clogged with debris and tree leaves [21]

The experimental study of Gómez et al. [21] on the hydraulic efficiency of partially clogged inlets in the urban basin of Riera Blanca (Barcelona) showed that the UPC methodology applied for clean inlets in the study of Gómez and Russo [14] is still applicable to the case of partially clogged inlets. The equation of the clogging factor c_0 was proposed as a function of the clean inlet hydraulic efficiency (E_{clean}) and the clogged inlet hydraulic efficiency (E_{clo}) as shown in Eqs. (9) and (10). The experimental results showed that for the same type of inlet under the same clogging pattern, the energy ratio $\Delta E/E$ and the clogging factor c_0 tend to a constant value as the flow to depth ratio (Q/y) increases. It can be said that under high flow rates, the hydraulic efficiency reduction was a function of the clogging pattern despite the flow conditions on the street.

$$E_{\text{clo}} = (1 - c_0)E_{\text{clean}} \tag{9}$$

$$\frac{\Delta E}{E} = \frac{E_{\text{clean}} - E_{\text{clo}}}{E_{\text{clean}}} = c_0 \tag{10}$$

There are various types and designs of inlets that differ remarkably between countries and cities based on their date of installation, shape, sizes and bar configuration. Numerous studies have been conducted to determine the surface parameters related to drain inlet efficiency such as type of inlet, road longitudinal slope and road transversal slope. For instance, the effects of road longitudinal slope, road transversal slope and grate clogging on the grate inlet hydraulic efficiency under various simulated storm water events were experimentally assessed in the research of Veerappan and Le [22]. Three grate inlets with transversal bars, longitudinal bars and transversal bars extended to the kerb opening were tested (Fig. 4). The clogging factor ranged between 50, 100 and over 100% (the case of clogging above the horizontal plane over kerb opening). Results showed that the grate efficiency increased for steeper transversal slopes and decreased for steeper longitudinal slopes. This is attributed to the larger runoff conveyed to the gutter for higher transversal slopes (interception efficiency increase) and larger runoff bypassing the grate inlet for higher longitudinal slopes (interception efficiency reduction). Moreover, the effect of the road transversal slope was more dominant on the grate hydraulic efficiency than the longitudinal road slope. Under clogging and non-clogging conditions, the hydraulic efficiency was reduced slightly when the rain intensity increased and the rain flow intensity impact was insignificant. The three types of inlets (with transversal bars, longitudinal bars and transversal bars extended to the kerb opening) showed a similar efficiency under non-clogging conditions. Under non-clogging conditions and clogging conditions up to 100%, the grate with longitudinal bars hydraulic efficiency was higher than the grate with transversal bars. However, when the clogging was extended beyond the horizontal plane to cover the kerb opening, the hydraulic efficiency of the grate with horizontal bars extended to the kerb opening was more than twice the grates with horizontal bars and longitudinal bars. Consequently, it is recommended to extend the grate horizontal bars to the kerb opening at grate inlets prone to clogging.

Fig. 4 Three grate inlets with longitudinal bars, transversal bars and transversal bars extended over kerb opening [22]

In a similar study, Gómez et al. [23] investigated a quantification methodology to evaluate the hydraulic efficiency of clogged inlets in Sant Marti urban catchment (Barcelona). The catchment inlets were visually inspected to investigate the inlet clogging patterns under wet and dry monsoons. In general, the catchment inlets were classified into seven major models covering 80% of the catchment inlets. The clogging condition of each model was classified based on the ratio of the clogged area (A_c) to the total area of holes for each inlet (A_h). Inlets with $A_c/A_h < 10\%$ were considered clean and inlets with $A_c/A_h \geq 80\%$ were considered out of order. The clogging patterns $C1$, $C2$ and $C3$ were referred to $10\% \leq A_c/A_h < 25\%$, $25\% \leq A_c/A_h < 50\%$ and $50\% \leq A_c/A_h < 80\%$, respectively. It is reported that the clogging patterns of inlets showed similar behaviour despite the dry or wet weather conditions. This can be due to the regular cleaning services of the catchment inlets. The UPC methodology developed by Gómez and Russo [14] to determine the hydraulic efficiency of clean inlets is still applicable to the case of clogged inlets by considering the effective hole areas of the clogged inlet. Moreover, the clogging coefficient increased as the flow rate increased; therefore, the design values of clogging coefficient c_0 should be determined at higher flow rates. However, the value of the clogging coefficient c_0 varied based on the inlet model (classified based on their geometry) and clogging pattern ($C1$, $C2$ or $C3$). The clogging coefficient c_0 of two selected inlet models showed that c_0 was 0.25–0.45, 0.4–0.5 and 0.67 for clogging patterns $C1$, $C2$ and $C3$, respectively. A constant value of clogging coefficient $c_0 = 0.5$ is usually addressed to take the effect of clogging on the inlet hydraulic efficiency into consideration [21]. The results of this study illustrate that adopting a fixed clogging coefficient c_0 of 0.5 can overestimate the effect of clogging on the inlet efficiency (case of clogging pattern $C1$) or underestimate its effect (case of clogging pattern $C3$). Therefore, it is recommended to use a clogging coefficient between 0.25 and 0.5 for normal clogging and a value close to 0.7 for the case of severe clogging.

4 Hydraulic Efficiency of Transversal Grate Inlets

Continuous traverse grates are usually used in areas where the isolated grate inlets are not effective as airports, parks and walkway areas where the transverse profile does not have a fixed or forced directional gradient that can direct the runoff flow to isolated grates as in the case of paved road lanes (Fig. 5). In this situation, the

Fig. 5 Typical continuous transverse grates [19]

continuous transverse grates are more efficient to collect the runoff flow. Russo, et al. [19] studied the hydraulic efficiency (E) of seven continuous transversal grates and proposed an equation for determining the hydraulic efficiency E per unit width of the grate. The experimental results showed that the hydraulic efficiency of the transversal grate decreases when the flow rate increases. Similarly, when the grate bars were in the form of fishbone, the hydraulic efficiency was lower than the grates with bars parallel to the water flow. This is attributed to the larger water splash in the fishbone grates; therefore, lower intercepted flow. The hydraulic efficiency E was expressed as a dimensionless function of flow depth, Froude number and geometrical characteristics of the grate as shown in Eq. (11).

$$E = \frac{Q_{\mathrm{i}}}{Q_{\mathrm{a}}} = \alpha \times F \times \left(\frac{y}{L}\right)^{0.812} + \beta \tag{11}$$

where α and β coefficients are calculated from the grate geometry (grate length in the flow direction, void area, number of diagonal, number of transversal bars and number of longitudinal bars), F is the Froude number, and y/L the normalized flow depth (y) with respect to the length of the grate opening in the flow direction (L).

5 Conclusion

This paper presents the most relevant methodologies for determining the hydraulic efficiency of clean grate inlets, clogged grate inlets and continuous transversal grates. The findings of this research can be summarized as follows:

- When the transversal slope increases, the conveyed water from the street to the grate increases; therefore, the grate hydraulic efficiency is enhanced.
- When the longitudinal slope increases, the flow velocity and water splash-over phenomenon over the grate increases; therefore, the grate hydraulic efficiency is reduced.
- The three Eqs. (1), (2) and (4) for clean inlet hydraulic efficiency showed almost similar values with small differences.

- Even though Eq. (2) developed by DMRB considers the grate coefficient parameter, the UPC methodology in Eq. (4) is recommended and considered the most accurate as it takes the number of longitudinal bars, transversal bars, diagonal bars, grate length, grate width and void areas into consideration for the grate inlet hydraulic efficiency.
- The same UPC methodology of Eq. (4) can be applied to the case of clogged inlets after determining the real void non-clogged area of the grate.
- For clogged inlets, a debris decay ratio known as the clogging factor (c_0) between zero and one is applied to determine the clogged inlet efficiency. A clogging factor of 0.25–0.5 can be adopted for normal clogging, whereas a 0.7 clogging factor can be used for severe cases.
- A similar approach based on the number and shape of grate bars, grate dimensions and the normalized depth (y) to grate length (L) is adopted to count for continuous transversal grate efficiency as shown in Eq. (11).
- This paper shows the reliability of using different approaches in numerical model simulations for determining the grate inlet hydraulic efficiency. The presented equations are function of the upstream flow depth as well as the grate and road geometries.

Acknowledgements The authors wish to express gratitude to the Ministry of Higher Education Malaysia for the financial support under Fundamental Research Grant Scheme with Project Code: FRGS/1/2021/WAB02/USM/03/2 for this study.

References

1. Butler, D., Davies, J.W.: Urban Drainage, 3rd edn. Spon Press, Abingdon (2011)
2. Wakif, S., Sabtu, N.: The effect of depression for a terminal gully inlet. Int. J. Civ. Eng. Technol. **9**(7), 1920–1926 (2018)
3. Leitão, J.P., Almeida, M.d.C., Simões, N.E., Martins, A.: Methodology for qualitative urban flooding risk assessment. Water Sci. Technol. **68**(4), 829–838 (2013)
4. Zhou, Q.: A review of sustainable urban drainage systems considering the climate change and urbanization impacts. Water **6**(4), 976–992 (2014)
5. Wakif, S., Sabtu, N.: Hydraulic performance of vertically depressed and non-depressed grate. Urban Water J. **16**(8), 554–563 (2019)
6. Kemper, S., Schlenkhoff, A.: Experimental study on the hydraulic capacity of grate inlets with supercritical surface flow conditions. Water Sci. Technol. **79**(9), 1717–1726 (2019)
7. Wakif, S., Sabtu, N.: Comparison of different methodologies for determining the efficiency of gully inlets. In: AWAM International Conference on Civil Engineering, pp. 1275–1283. Springer, Penang (2019)
8. Cárdenas-Quintero, M., Carvajal-Serna, F.: Review of the hydraulic capacity of urban grate inlet: a global and Latin American perspective. Water Sci. Technol. **83**(11), 2575–2596 (2021)
9. Comport, B.C., Thornton, C.I.: Hydraulic efficiency of grate and curb inlets for urban storm drainage. J. Hydraul. Eng. **138**(10), 878–884 (2012)
10. Guo, J.C., MacKenzie, K.: Hydraulic efficiency of grate and curb-opening inlets under clogging effect. University of Colorado Denver, CDOT-2012-3. Colorado (2012)

11. Spaliviero, F., May, R., Escarameia, M.: Spacing of road gullies hydraulic performance of BS EN 124 gully gratings and kerb inlets. Highways Agency of the Department of Transport, SR 533. Wallingford (2000)
12. British standard BS EN 752.: Drain and Sewer Systems Outside Buildings (2008)
13. Standard UK.: Design manual for road and bridges (DMRB)—spacing of road gullies: part 3, vol 4, Section 2, CD 526 (1999)
14. Gómez, M., Russo, B.: Methodology to estimate hydraulic efficiency of drain inlets. In: Proceedings of the Institution of Civil Engineers-Water Management, pp. 81–90. Thomas Telford Ltd, London (2011)
15. Dai, S., Jin, S., Qian, C., Yang, N., Ma, Y., Liang, C.: Interception efficiency of grate inlets for sustainable urban drainage systems design under different road slopes and approaching discharges. Urban Water J. 1–12 (2021)
16. Wu, C., Yu, J.T., Wong, A.S., Li, C.W.: Hydraulic interception efficiency of gully gratings on steep roads. HKIE Trans. **22**(3), 192–198 (2015)
17. Kim, J.S., Kwak, C.J., Jo, J.B.: Enhanced method for estimation of flow intercepted by drainage grate inlets on roads. J. Environ. Manage. **279**(2021), 1–9 (2021)
18. Lee, S., Nakagawa, H., Kawaike, K., Zhang, H.: Study on inlet discharge coefficient through the different shapes of storm drains for urban inundation analysis. J. Jpn. Soc. Civ. Eng. Ser. B1 (Hydraul. Eng.) **68**(4), 31–36 (2012)
19. Russo, B., Gómez, M., Tellez, J.: Methodology to estimate the hydraulic efficiency of nontested continuous transverse grates. J. Irrig. Drain. Eng. **139**(10), 864–871 (2013)
20. Guo, J.C.: Design of street curb opening inlets using a decay-based clogging factor. J. Hydraul. Eng. **132**(11), 1237–1241 (2006)
21. Gómez, M., Rabasseda, G.H., Russo, B.: Experimental campaign to determine grated inlet clogging factors in an urban catchment of Barcelona. Urban Water J. **10**(1), 50–61 (2013)
22. Veerappan, R., Le, J.: Hydraulic efficiency of road drainage inlets for storm drainage system under clogging effect. In: 5th International Conference on Flood Risk Management and Response, pp. 271–281. WIT Press, San Servolo (2016)
23. Gómez, M., Parés, J., Russo, B., Martínez-Gomariz, E.: Methodology to quantify clogging coefficients for grated inlets. Application to SANT MARTI catchment (Barcelona). J. Flood Risk Manage. **12**(4), 1–14 (2019)